MW00814553

Lecture Notes in Electrical Engineering 169

For further volumes:
http://www.springer.com/series/7818

Guennadi A. Kouzaev

Applications of Advanced Electromagnetics

Components and Systems

Springer

Author
Prof. Guennadi A. Kouzaev
Department of Electronics and Telecommunications
Norwegian University of Science and Technology-NTNU
Trondheim
Norway

ISSN 1876-1100 e-ISSN 1876-1119
ISBN 978-3-642-30309-8 e-ISBN 978-3-642-30310-4
DOI 10.1007/978-3-642-30310-4
Springer Heidelberg New York Dordrecht London

Library of Congress Control Number: 2012938038

Printed on acid-free paper

Springer is part of Springer Science+Business Media (www.springer.com)

Preface

In this book, the applications of advanced electromagnetics to the integrated components and to some systems are reviewed.

Today's technology is reaching its ultimate stage; further hardware improvements are with the sophisticated theories, techniques, materials, and macro- and microphysical effects. Especially, it is pressing problem for the micro- and millimeter-wave and high-speed circuits, which are the place of complicated electromagnetic (EM) and quantum phenomena. It is highly desirable to review and explain the application principles of advanced electromagnetics to the practicing microwave specialists and students who wish to be acknowledged on the contemporary state of the developments scattered in multiple journals and conference proceedings. The touched area is very large, and the choice of the reviewed material is dictated by its importance and the research and teaching experience of the Author, although the elements of his subjectivism is being minimized.

In the first four Chapters on the waves, particles, microwave components, and computational methods, the basic EM theory is given required for further understanding of advanced electromagnetics. Additionally to the field and wave equations, boundary conditions, and the methods of boundary value problem (BVP) solutions, some aspects of classical and quantum dynamics of particles are considered, including the recently proposed EM Hertz-quantum equations. For practicing engineers, the method of equivalent lines is described and applied to the interconnects and their elementary discontinuities. Although, it is a powerful tool known for decades, the developments of advanced components require deeper understanding of the EM field physics. In connection with this, a strong attention is paid to the topological methods of the field analysis and solutions of BVPs. Topology deals with the skeletons of the field-force line maps. They, being less detailed than the full-field pictures, are still with the main features of the modeled phenomena, and they can be used as the elements of a field "language" for the EM analysis. The proposed topological approach is applied to the developments of fast EM models, and they are validated by measurements and known analytical and numerical data. The results of other authors in applications of topology in electromagnetics circuits, quantum physics are considered as well.

Chapters 5-8 of this book are dedicated to the EM and technological aspects of the integrated passive components of frequencies up to 100 GHz and beyond. These passives are the bottleneck of contemporary electronics being difficult to be improved, and only combinations of the advanced technologies and EM modeling allow for enhancing them. In the beginning, the known manufacturing technologies and packaging techniques are reviewed to introduce the needed technology alphabets. Several tens of silicon integrated components as the interconnects, via-holes, transitions, power dividers, directional couplers, and baluns developed for the increased frequencies are analyzed and compared, and this book's part is interesting for millimeter-wave designers as a mini handbook on these circuits.

It is supposed, that the future of electronics is with the further component size reduction and intensive use of quantum effects. It will require new tools for seamless simulation of these hybrid integrations by the same design system. For this purpose, a technique of representation of quantum-mechanical equations by circuit ones is considered in Chapter 9. This approach is applied to computation of linear and non-linear Schrödinger equations by a commercially available circuit simulator, and the validity of these calculations is carefully verified. One of the areas of applications of this computation technology is the modeling of the EM integrated circuits for trapping and handling of cold matter, and they are reviewed in this Chapter.

The EM waves can carry information not only by the wave's magnitude, phase and frequency, but, similarly to the optical range, the shape of waves can be modulated separately or combining the all mentioned parameters. A particular case of spatial modulation is considered in the Chapter 10, and it is with the digital signaling by topologically different EM field impulses propagating along the interconnects or in open space. A theory of these signals and some designed digital components are presented. A novel processor of predicate logic interesting in the artificial intelligence applications and in the noise-tolerant computing is described which is based on the spatial signaling. Similar approaches of other authors are analyzed as well in this Chapter.

Enriching the spatial spectrum of EM signals leads to the quasi-optical principles of their registration and processing, and this aspect is touched in the Chapter 11 considering the high-sensitive radiometric passive imagers. Initially, the principles of radiometry are explained, and the statistical characteristics of human-body thermal signals registered by a radiometer are analyzed. These signals are distorted by the digital and thermal noise of hardware, and further sensitivity increase is with the study of these distortions. It is realized using the methods of stochastic dynamics, and the traces of the deterministic noise are detected in radiometric signals as the attractors of their phase spaces. These attractors are detected, calculated and classed, which is important for further algorithmic enhancements of radiometers. The passive imagers composed of multiple radiometers are reviewed, and the results of imaging of human body by a developed and manufactured 16-channel millimeter-wave imager are considered.

Acknowledgements

The Author thanks all his colleagues who have worked and who are working together with him for their collaboration, support and understanding.

Especially he is grateful to his Co-authors Dr., Prof. M.J. Deen (McMaster University, Canada), to whom belongs, additionally, an idea to write a review on integrated passives, S.V. Kapranov (NTNU, Norway), Dr., Assist. Prof. A.N. Kostadinov (Technical University-Plovdiv, Bulgaria), Dr., Prof. G.S. Makeeva (Penza State Technical University, Russia), Dr., Prof. N.K. Nikolova (McMaster University, Canada), Dr. A. Rahal (Nanowave Corp., Canada), Dr. K.J. Sand (NTNU, Norway), G. Ying (NTNU, Norway).

Drs. I. Nefedov (Aalto University, Finland) and H. Wang (Georgia Institute of Technology, USA) carefully read several sections, contents and the book conception and made helpful comments; the Author acknowledges with sincere thanks their work.

The Author thanks the Faculty of Informatics, Mathematics and Electronics of the Norwegian University of Science and Technology (NTNU) and the European Research Consortium for Informatics and Mathematics (ERCIM) for several PhD and Postdoctoral Fellow research grants.

Some work on this book was performed during the Author's research stay at the European Laboratory for Non-linear Spectroscopy (University of Florence, Italy) and the Author thanks the Head of the Quantum Science and Technology Center Dr., Prof. Cataliotti and his colleagues for their patiency and understanding.

Author's thanks are to the Russian Ministry of Science and Technological Politics, the Russian Fund of Basic Research, and the Moscow State Institute of Electronics and Mathematics (Technical University) who supported the initial research on applications of ideas of topology in communications and computing.

Most of the educational material of this book is from the courses on electrodynamics, microwave techniques and passive integrated components, and microwave optics given by the Author at the Norwegian University of Science and Technology and at the Moscow State Institute of Electronics and Mathematics (1993-2000), and he is very thankful to these Universities for their support.

The Author would like to express his gratitude to those Readers who may wish to contact him on found misprints.

My endless thanks to my wife Nadia and my children Inessa and Alexei for their understandings.

Summer 2012 Guennadi A. Kouzaev

Contents

1 Basic Electromagnetics **1**
1.1 Scalars, Vectors, and Fields 1
1.2 Electric Field 4
1.3 Magnetic Field 4
1.4 Dynamic Theory of the EM Field 5
1.4.1 Time-harmonic EM Field Equations 6
1.4.2 Wave Equations 8
1.4.3 Plane TEM Wave 9
1.4.4 Ray Representation of TEM Wave in Open Space 12
1.5 Nonlinear Electromagnetism 13
1.5.1 Macro Electromagnetism of Nonlinear Medium 13
1.6 Dynamics of Charged Particles and Dipolar Molecules in EM Field 15
1.6.1 Charged Particle in EM Field 16
1.6.1.1 Classical Theory of Charged Particle Movement in EM Field 16
1.6.1.2 Quantum Theory for the Charged Particle Movement. Low-velocity Approximation 16
1.6.1.3 Maxwell-Schrödinger Equations for Spin-less Charge 17
1.6.1.4 Schrödinger-Hertz Equation 18
1.6.1.5 Pauli-Hertz Equation 21
1.6.1.6 Dirac-Hertz Equations 22
1.6.1.7 Klein-Gordon-Hertz Equation 23
1.6.2 Dipolar Molecules in EM Field 24
1.7 Boundary Value Problems of Electromagnetism 31
1.7.1 Boundary Conditions for the Electric and Magnetic Fields 31
1.7.2 Surface Impedance Boundary Conditions 32
1.7.3 Boundary Value Problems 34
1.7.4 Energy Conservation Law 35
1.7.5 Reflection of TEM Waves from the Boundary of Two Magnetodielectrics 37
1.7.5.1 Perpendicular Polarization of the Incident Wave 37
1.7.5.2 Parallel Polarization of the Incident Wave 40
1.7.5.3 Wave Reflection from Perfect Conductors. Guiding Effect 41
1.7.5.4 Wave Reflection from Non-perfect Conducting Surfaces 44

1.7.5.5 Full Transmission and the Brewster Angle 45
1.7.5.6 Total Reflection of the EM Waves from the Boundary of Two Magnetodielectrics 45
References 46

2 Theory of Waveguides 51
2.1 Analytical Treatment of Waveguides 51
2.1.1 Separation of the Variables Method 53
2.1.2 Field Series Expansion Method 56
2.1.3 Transverse Resonance Method and Analytical Treatment of Waveguides 57
2.1.4 Analytical Models of TEM Transmission Lines 61
2.1.4.1 Triplate Transmission Line 61
2.1.4.2 Coupled Strips Line 62
2.1.5 Modeling of Quasi-TEM Modes of Transmission Lines 64
2.1.5.1 Parallel-plate Model of Microstrip Line 64
2.1.5.2 Coupled Microstrips Line 69
2.1.5.3 Coplanar Waveguide Model 71
2.2 Numerical Methods Used in the Waveguide Theory 74
2.2.1 Finite Difference and Transmission Line Matrix Methods 74
2.2.2 Integral Equations for Waveguides and Their Numerical Treatment 76
2.2.3 Variational Functionals for Waveguides 86
References 88

3 Waveguide Discontinuities and Components 95
3.1 A More Detailed Theory of Regular Waveguides for the Discontinuity Treatment 95
3.2 Equivalent Transmission Line Approach and Microwave Networks 98
3.3 EM Computational Methods for Waveguide Discontinuities 101
3.3.1 Mode Matching Method to Calculate the Waveguide Junctions 101
3.3.2 Integral Equation Method for Waveguide Joints 103
3.3.3 Stationary Functionals and Equivalent Circuits of Discontinuities 108
3.3.4 Microstrip Discontinuities and Their Parallel-plate Models 110
3.3.4.1 Oliner's Approach to Strip and Microstrip Discontinuities 111
3.3.4.2 Parallel-plate Waveguides, Orthogonal Series Expansion Method, and Microstrip Discontinuities 114
3.3.5 Full-wave FDTD, TLM, FEM, and Integral Equation Treatment of Discontinuities 115
References 117

4 Geometro-topological Approaches to the EM Problems 119
4.1 Topological Approach to the Theory of Boundary Value Problems....... 121
4.1.1 Visualization and Topological Analysis of the EM Field Maps 122
4.1.2 Field Maps of Harmonic Fields and Their Topological Analysis 122
4.1.3 Maxwell Relationships of Topological Schemes of the Electric and Magnetic Harmonic Fields 128
4.1.4 Boundary Conditions and the Field-force Lines 132
4.1.5 Applications of the Topological Technique to the EM Boundary Value Problems 133
4.1.6 Applications of the Topological Approach for the Modeling of Microwave Components 151
4.1.6.1 Ridged Waveguide Model 152
4.1.6.2 Some Other Waveguides and Components Studied Using the Topological Approach 161
4.1.6.3 Modeling of Square-pad Via-holes and Patch Antennas Using the Topological Approach 161
4.2 Topological Analysis of the Time-dependent EM Field 172
4.3 Topological Theory of the EM Field with the Differential Forms 175
4.4 Topological Electromagnetism of Open-space Radiated Field 177
References 179

5 Technologies for Microwave and High-speed Electronics 187
5.1 RF Printed Circuit Boards 187
5.2 Microwave Hybrid ICs 189
5.3 Ceramic Thick-film Technologies 190
5.4 Microwave Three-dimensional Hybrid Integrated Circuits 192
5.5 Gallium Arsenide Monolithic Integrated Circuit Technology 197
5.6 Silicon Technologies for RF and Microwaves 198
5.7 Micromachining Technologies 203
5.8 RF Nanointegrations 205
5.9 Packaging of Microwave and Millimeter-wave Microchips 206
5.9.1 Packages for Millimeter-wave Integrations 206
5.9.2 Interconnecting of Chips 208
5.9.3 Motherboards and Materials 210
5.10 On the Evolution of Microwave and Millimeter-wave Electronics 213
References 214

6 Transmission Lines and Their EM Models for the Extended Frequency Bandwidth Applications 221
6.1 Microstrip Lines 221
6.1.1 Thin-film Microstrip Lines 221
6.1.2 Superstrated Thin-film Microstrip Line 229
6.1.3 Two-layer Substrate Microstrip Transmission Line 230

6.1.4 Si-microstrip Transmission Lines Shielded by Impedance Layers 235
6.1.5 Three-dimensional Modifications of Microstrip Line 237
6.1.5.1 Stacked Thin-film Microstrip Line 237
6.1.5.2 V-shaped Microstrip Line 238
6.1.5.3 Air-filled Strip and Buried Microstrip Transmission Lines for Monolithic Integrations 239
6.1.5.4 Fenced Microstrip and Strip Transmission Lines 240
6.1.5.5 Ridged Microstrip Transmission Line 242
6.1.5.6 Vertical Strip Transmission Lines 244
6.2 Multilayered CPW for Monolithic Applications 246
6.2.1 Quasistatic Model of Generalized CPW for Monolithic ICs 246
6.2.2 Elevated CPWs for Increased Frequencies 252
6.2.3 Coplanar Strip Line 255
6.2.4 Three-dimensional Modifications of CPW 257
6.2.4.1 "Overlay" CPWs 257
6.2.4.2 Three-dimensional CPWs 258
6.2.4.3 Microshield CPW 260
6.2.4.4 Membrane Supported CPW 262
6.2.4.5 Micromachined Trenched and Embedded CPWs 263
6.3 Micromachined Rectangular-coaxial and Rectangular Waveguides 265
6.4 Substrate Integrated Waveguides (SIW) 270
6.4.1 Review on the SIWs, Simulation Methods and Main Results 270
6.4.2 Modeling of Differential SIW 274
6.4.3 EM Model of Fenced Strip Line 279
6.5 Comparative Analysis of Integrated Transmission Lines 281
6.6 Interconnects for Optoelectronics 283
6.6.1 Optical Interconnects 283
6.6.2 Plasmon-polariton Interconnects and Components 284
6.6.2.1 Plasmon-polariton Interconnects 284
6.6.2.2 Components of Plasmon-polariton Integrations 288
6.6.3 On the EM Theory of Plasmon-polariton Interconnects and Components 289
6.7 Prospective Interconnects 290
References 292

7 Inter-component Transitions for Ultra-bandwidth Integrations 307
7.1 Planar Line Transitions 309
7.2 Quasi-planar Transitions 311
7.2.1 CPW-microstrip Transitions 311
7.2.2 Microstrip-CPS Transitions 313
7.2.3 CPW-CPS Transitions 314
7.3 Interlayer Transitions for 3-D Integrations 315
7.3.1 Silicon Via-holes and Transitions 315
7.3.2 EM Interlayer Transitions 318

7.4 Via-holes of Micro- and Millimeter-wave Motherboards.......................322
7.4.1 Grounding Via-holes.......................322
7.4.2 Cavity Model of Circular-pad Grounding Via-hole.......................324
7.4.3 Eccentric Circular-pad Grounding Via-hole.......................326
7.4.4 Modeling and Measurement Results for Grounding Via-holes....329
7.5 Through-substrate Via-holes and Their Modeling.......................332
References.......................335

8 Integrated Filters and Power Distribution Circuits.......................341
8.1 Integrated Filters.......................341
8.2 Power Dividers, Directional Couplers, and Baluns.......................346
8.2.1 Power Dividers and Directional Couplers on Discrete Integrated Components.......................346
8.2.2 Distributed Power Dividers and Directional Couplers.......................348
8.3 Integrated Baluns.......................352
References.......................353

9 Circuit Approach for Simulation of EM-quantum Components.............359
9.1 Circuit Models of Linear Schrödinger Equation.......................360
9.2 Circuit Models of Nonlinear Schrödinger Equations.......................361
9.3 Circuit Models of EM-quantum Equations.......................361
9.3.1 Circuit Model for Schrödinger-Hertz Equation.......................362
9.4 Verification of Circuit Model of Tunneling Problem.......................363
9.5 Solution of 2-D Nonlinear Schrödinger Equations by Circuit Simulators.......................368
9.5.1 Equivalent Circuit Model for Gross-Pitaevskii Equation.............369
9.5.1.1 Telegraph Equation for Two-dimensional Condensates Trapped in a Cylindrical Domain.............369
9.5.1.2 Domain Discretization and Boundary Conditions.........370
9.5.1.3 Equivalent Circuit for an Infinitesimal Domain............372
9.5.1.4 Cross-coupled Equivalent Circuit.......................374
9.5.1.5 Envelope Technique.......................378
9.5.1.6 Time-dependent Component Approach.......................379
9.5.1.7 Stability and Convergence of the Solution.......................381
9.5.2 Numerical Validation.......................381
9.5.2.1 Ground-state Condensate.......................382
9.5.2.2 Free Self-interacting Condensate.......................384
9.5.2.3 Comparison with the Experimental Data.......................386
9.5.3 Comparison of the Proposed Techniques.......................389
9.6 Developments and Design of Hardware for Cold Matter.......................391
9.6.1 Current State of Circuits for Trapping of Cold Matter.......................391
9.6.1.1 Laser Cooling of Neutral Atoms.......................392
9.6.1.2 Optical Trapping of Neutral Atoms.......................393

9.6.1.3 Magnetic and EM Traps 394
9.6.1.4 Transportation of Cold Matter 402
References 408

10 EM Topological Signaling and Computing 413
10.1 Introduction to Topological Signaling and Computing 413
10.2 Topology and EM Signaling 414
10.3 Topological Description of the Non-stationary EM Field 415
10.4 Topological Modulation of the Field 416
10.5 Studied Topologically Modulated Field Shapes 417
10.5.1 Quasi-TEM Signaling 417
10.5.2 Signaling by Non-separable Field Impulses 421
10.5.3 Noise Immunity 421
10.6 Passive Gating of Microwave Topologically Modulated Signals 422
10.7 Theoretical and Experimental Validations of the Proposed Signals and Their Processing 425
10.7.1 Passive Switches and Their Experimental and Theoretical Studies 425
10.7.2 Hardware for Measurements of Picosecond Circuits 436
10.7.3 Diode Equipped Switch 438
10.7.4 Pseudo-quantum Gates for Topologically Modulated Signals 443
10.7.5 Passive OR/AND Gate for Microwave Topologically Modulated Signals 448
10.8 Predicate Logic, Gates and Processor for Topologically Modulated Signals (TMS) 451
10.8.1 Predicate Logic Theory 451
10.8.2 Modeling of Predicate Logic by Unipolar Spatial Signals and Single-ended Gates 454
10.8.3 Predicate Logic Processor 458
10.9 Related Results on the Space-time Signaling and Computing 461
10.9.1 Open-space EM Spatial Signaling 461
10.9.2 Differential and Combined Differential/Common Mode Signaling 461
10.9.3 Differential or Dual-rail Logic 463
10.9.4 Ultrafast Gating Circuits Using Coupled Waveguides 465
10.9.5 Multimodal Data Transmission Using Hybrid Substrate Integrated Waveguides 466
10.9.6 Passive Frequency Multiplier on Coupled Microstrips Lines 468
10.9.7 Microwave Passive Logic for Phase-modulated Signals 469
10.9.8 Microwave Mode Selective Devices, Converters, and Multimodal Frequency Filters 471
References 484

11 EM Radiometry and Imaging......495
11.1 Principles of the EM Radiometry......495
11.2 Natural Microwave Radiation and Its Characterization......496
11.3 Calibration of Radiometers and Calculation of Physical Temperature......498
11.4 Sensitivity of Radiometers......499
11.5 Radiometric Studies of Thermal Human-body EM Radiation......500
11.5.1 Measurements and Statistical Analysis......500
11.5.2 Technique of Calculation of the Attractor Dimension......504
11.5.3 Calculation of Parameters of Multi-attractor Signals......514
11.5.4 Data Processing......518
11.5.5 Results and Discussion......521
11.6 Radiometric Imagers......524
References......528

Abbreviations

AC	alternating current
A/D	analog-to-digital
ADS	Advanced Design System
ALU	arithmetic logic unit
ASIP	application specific processor
BCB	benzo-cyclobutene
BiCMOS	bipolar CMOS
BVP	boundary value problem
CAD	computer aided design
CMOS	complementary symmetric metal oxide semiconductor
CPW	coplanar waveguide
COC	cyclic olefin copolymer
CPS	coplanar strips
DC	direct current
DRIE	deep-reactive ion etching
DSL	digital subscriber line
ECL	emitter coupled logic
EEG	electroencephalography
EDA	electronic design automation
EM	electromagnetic
FEM	finite element method
FDTD	finite difference time domain
FPGA	field programmable gate array
FSM	finite state machine
GaAs	gallium arsenide
GCPW	CPW with ground layer
GPP	general propose processor
HBT BiCMOS	heterojunction BiCMOS
HPPI	high performance parallel interface
HRSi	high-resistivity silicon
HTCC	high-temperature co-fired ceramic
IC	integrated circuit
IR	instruction register
KGHZ	Klein- Gordon -Hertz
LCP	liquid crystal polymer
LRSi	low-resistivity silicon
LTCC	low-temperature co-fired ceramic
LVDS	low-voltage differential signaling

MCM	multi chip module
MCM-C	MCM Ceramic
MCM-L	MCM Laminate
MESFET	metal semiconductor field effect transistor
MHEMT	metamorphic high electron mobility transistor
MHEPT	metamorphic hot electron phototransistor
MIM	metal-insulator-metal
MIMO	multi-input-multi-output
MS	microstrip
MSIEM	Moscow State Institute of Electronics and Mathematics
NTNU	Norwegian University of Science and Technology
PC	program counter
PCB	printed circuit board
PLA	programmable logic array
PLP	predicate logic processor
PLU	predicate logic unit
PTFE	polytetrafluoroethylene
RAM	random access memory
RF	radio frequency
RMS	root mean square
SIW	substrate integrated waveguide
S-matrix	scattering matrix
SNR	signal-to-noise ratio
SOI	silicon-on-insulator
SPP	single purpose processor
SQL	structural query language
Sub-mm	sub-millimeter
TE	transversal electric
TEM	transversal electromagnetic
TFML	thin film microstrip line
TLM	transmission line matrix
TM	transversal magnetic
TMS	topologically modulated signals
TU	technical university
TV	television
UV	ultraviolet
VSWR	voltage standing wave ratio
WDDL	wave dynamic differential logic
2-D	two-dimensional
3-D	three-dimensional
4-D	four-dimensional

1 Basic Electromagnetics

Abstract. This Chapter is on the basics of electromagnetism needed for further understanding of advanced electromagnetics and its applications. Taking into account a number of contributions in this field [1]-[25], our material is given in a concise manner to remind the Readers only the main electromagnetic (EM) equations. Among them are those given for static electricity, stationary magnetism, and the Maxwell and wave equations. The boundary conditions and boundary value problems are considered and the reflection of plane waves is studied as an example of these problems. Additionally to this material traditionally included into the books on electromagnetism, the motion of charged particles and dipoles is considered from the classical and semi-classical point of view, and new EM-quantum-mechanical equations based on the use of the Hertz vectors and the particle wave functions are introduced. References -75. Figures -13. Pages -49.

1.1 Scalars, Vectors, and Fields

In electromagnetics, different quantities describe the EM fields, and there are the scalars, vectors, tensors, etc. A quantity, which has no any direction in the space, is called a scalar. It is marked by an algebraic symbol *a*, *b*,.... A space, to each point of which a scalar is assigned, is scalar field. In electromagnetism, it is usually the field of the electric $\varphi_e(\mathbf{r})$ or magnetic $\varphi_m(\mathbf{r})$ potentials where $\mathbf{r}$ is the radius vector of a point of the field calculation.

In general, the EM field is described by vector quantities. Then a field has the magnitude and the direction in space. It can be represented by the arrowed field-force lines. The assigned to them numbers are the field magnitudes at the points where these directions are given. The field vectors are the functions of the spatial coordinate $\mathbf{r}$ and time t.

The electric field vectors are the intensity $\mathbf{E}(\mathbf{r},t)$ and the flux density $\mathbf{D}(\mathbf{r},t)$. The field symbols are bolded to show that these quantities are vectors. The magnetic field is described by its intensity $\mathbf{H}(\mathbf{r},t)$ and flux density $\mathbf{B}(\mathbf{r},t)$.

To operate these field vectors, the elementary knowledge of the vector algebra is required. The vectors can be represented using their coordinate form:

$$\mathbf{E} = E_x\mathbf{x}_0 + E_y\mathbf{y}_0 + E_z\mathbf{z}_0 \tag{1.1}$$

G.A. Kouzaev: Applications of Advanced Electromagnetics, LNEE 169, pp. 1–49.
springerlink.com

where E_x, E_y, E_z are the vector projections on the Cartesian coordinate system axes, and $\mathbf{x_0}$, $\mathbf{y_0}$, $\mathbf{z_0}$ are the unit vectors of them.

As mentioned, a vector is described by its magnitude $|\mathbf{E}|$ and its direction $\mathbf{e}$ which is the unit vector aligned according to **E**. They are calculated as

$$|\mathbf{E}| = \sqrt{(E_x)^2 + (E_y)^2 + (E_z)^2}, \tag{1.2}$$

$$\mathbf{e} = \frac{\mathbf{E}}{|\mathbf{E}|}. \tag{1.3}$$

Field vectors can be added to each other similarly to any vectors:

$$\mathbf{E}_3 = \mathbf{E}_1 + \mathbf{E}_2 = \left(E_{x_1} + E_{x_2}\right)\mathbf{x_0} + \left(E_{y_1} + E_{y_2}\right)\mathbf{y_0} + \left(E_{z_1} + E_{z_2}\right)\mathbf{z_0}. \tag{1.4}$$

The scalar multiplication of the field vectors is defined as

$$W = \mathbf{E}_1 \cdot \mathbf{E}_2 = \mathbf{E}_1\mathbf{E}_2 = \left(\mathbf{E}_1\mathbf{E}_2\right) = E_{x_1}E_{x_2} + E_{y_1}E_{y_2} + E_{z_1}E_{z_2} \tag{1.5}$$

where W is proportional to the field energy.

The vector multiplication is another important operation in electromagnetism:

$$\mathbf{P} = \left[\mathbf{E}\times\mathbf{H}\right] = \begin{vmatrix} \mathbf{x_0} & \mathbf{y_0} & \mathbf{z_0} \\ E_x & E_y & E_z \\ H_x & H_y & H_z \end{vmatrix} = \\ = \mathbf{x_0}\left(E_yH_z - E_zH_y\right) - \mathbf{y_0}\left(E_xH_z - E_zH_x\right) + \mathbf{z_0}\left(E_xH_y - E_yH_x\right). \tag{1.6}$$

Additionally, the field vectors can be multiplied by a scalar, and it influences the vector magnitude. Multiplication of a field vector by a scalar matrix changes its direction and its length.

The scalar and vector fields are represented by their images. For example, a scalar field is shown by its isopotential lines $\varphi_i(\mathbf{r}) = const$, and for their calculation the graphical tools of the Matlab or the MathCAD [26] can be used, for instance (Fig. 1.1).

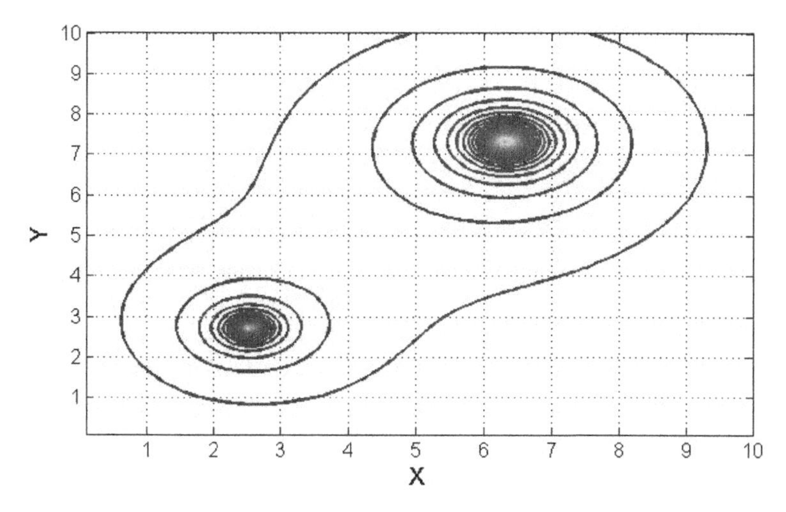

Fig. 1.1 Iso-potential representation of the scalar field of two charges (Matlab simulation)

The vector fields are visualized using the field-force line maps. A field-force line shows direction of the field at each space point. An example of this representation is Fig. 1.2 given for the electric field which corresponds to the above-shown iso-potential plot (Fig.1.1). More variants of graphical representation of fields are from publications on the EM field visualization.

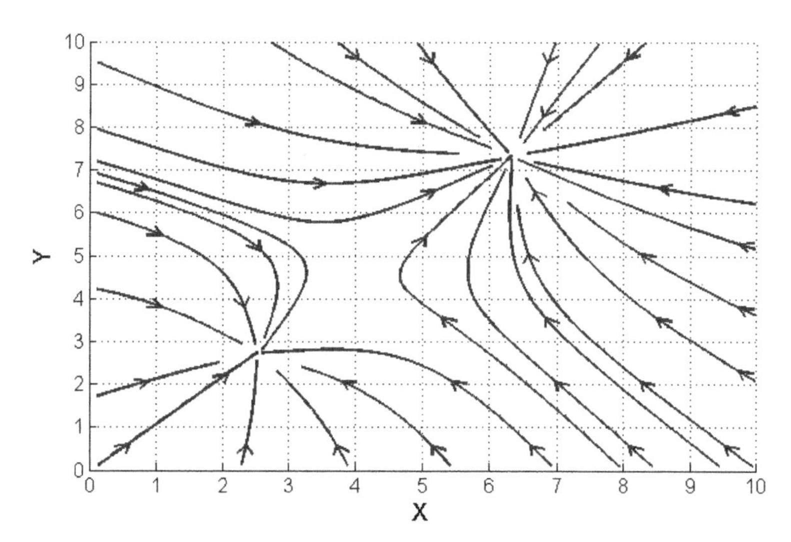

Fig. 1.2 Field-force line map of the electric field of two charges (Matlab simulation)

1.2 Electric Field

The source of static electric field is a charge usually marked as q and measured in *Coulomb*, $[\mathrm{C}]$. The alternating currents (AC) and magnetic fields generate the time-depending electric fields. The electric field intensity $\mathbf{E}, [\mathrm{V/m}]$ is measured by the force acting on the unit charge placed into this field. The space filling influences field, and the measure of it is the absolute permittivity of this medium ε. Vacuum has $\varepsilon = \varepsilon_0 \approx 8.8542 \cdot 10^{-12}$ F/m. A dielectric is described by the relative permittivity ε_r, and its absolute permittivity is $\varepsilon = \varepsilon_r \varepsilon_0$.

The electric flux density is $\mathbf{D} = \varepsilon \mathbf{E}, \left[\mathrm{C/m^2}\right]$ and it shows the medium influence on the electric phenomena. It can be derived from the equation (1.7) if the spatial charge density $\rho, \left[\mathrm{C/m^3}\right]$ is known:

$$\nabla \cdot \mathbf{D} = \rho \tag{1.7}$$

where $\nabla \equiv \frac{\partial}{\partial x}\mathbf{x}_0 + \frac{\partial}{\partial y}\mathbf{y}_0 + \frac{\partial}{\partial z}\mathbf{z}_0$.

Additionally, the scalar potential $\varphi_e, [\mathrm{V}]$ can be used for the electric field, and it is found from the Poisson equation:

$$\Delta \cdot \varphi_e = \rho \tag{1.8}$$

where $\Delta \equiv \nabla^2 \equiv \frac{\partial^2}{\partial x^2} + \frac{\partial^2}{\partial y^2} + \frac{\partial^2}{\partial z^2}$.

In the static, time-independent case $(q = const)$, the vectors $\mathbf{E}$ and $\mathbf{D}$ are derived from this potential φ_e and the charge density

$$\mathbf{E} = -\nabla \cdot \varphi_e,\ \mathbf{D} = \varepsilon_0 \varepsilon_r \mathbf{E}. \tag{1.9}$$

1.3 Magnetic Field

The source of magnetic field is the moving electricity: static and time-varying currents and time-depending electric fields. The magnetic field is described by several quantities. Among them are the magnetic field intensity $\mathbf{H}, [\mathrm{A/m}]$, magnetic field flux density $\mathbf{B}, [\mathrm{T}]$, magnetic vector potential $\mathbf{A}, [\mathrm{T \cdot m}]$, and the scalar magnetic potential φ_m.

The space filling influences magnetic field, and the measure of it is the absolute permeability μ of the medium. Vacuum has $\mu = \mu_0 = 4\pi \cdot 10^{-7}, [\mathrm{H/m}]$. Any medium is described by the relative permeability μ_r, and its absolute value is $\mu = \mu_r \mu_0$.

The vectors of the magnetic field **H** and **B** are coupled to each other through the permeability μ :

$$\mathbf{B} = \mu \mathbf{H}. \tag{1.10}$$

In the stationary field case, the field vectors $\mathbf{H}(\mathbf{r})$ and $\mathbf{B}(\mathbf{r})$ are calculated according to the densities of the time-independent source current $\mathbf{j}^{\text{source}}, \left[\mathrm{A/m^2}\right]$ and the conductivity current $\mathbf{j}_c, \left[\mathrm{A/m^2}\right]$:

$$\begin{aligned} &\nabla \times \mathbf{H} = \mathbf{j}^{(\text{source})} + \mathbf{j}_c, \ \ \mathbf{j}_c = \sigma \mathbf{E}, \\ &\nabla \cdot \mathbf{B} = 0, \qquad \mathbf{B} = \mu \mathbf{H} \end{aligned} \tag{1.11}$$

where $\sigma, [\mathrm{S/m}]$ is the medium conductivity.

In the spatial domains where the driving and conductivity currents are zero, the magnetic field is the magnetostatic one, and it satisfies the following equations:

$$\begin{aligned} &\nabla \times \mathbf{H} = 0, \\ &\nabla \cdot \mathbf{B} = 0, \qquad \mathbf{B} = \mu \mathbf{H}. \end{aligned} \tag{1.12}$$

They are convenient when the field of time-independent currents is calculated. The magnetic field can be visualized using the field-force line pictures similarly to the electric field.

1.4 Dynamic Theory of the EM Field

The time-varying electric and magnetic fields are not independent on each other. This coupling effect was found experimentally in the first half of the 19th Century, and it was described mathematically by J.C. Maxwell [1].

There are several forms of the Maxwell equations, and we are following the notations given by O. Heaviside [2]. One of them is written using the differential operators:

$$\begin{aligned} &\nabla \times \mathbf{H} = \mathbf{j}^{(\text{source})} + \mathbf{j}_c + \frac{\partial \mathbf{D}}{\partial t}, \ \ \mathbf{j}_c = \sigma \mathbf{E}, \\ &\nabla \times \mathbf{E} = -\frac{\partial \mathbf{B}}{\partial t}, \\ &\nabla \cdot \mathbf{D} = \rho, \qquad \mathbf{D} = \varepsilon \mathbf{E}, \\ &\nabla \cdot \mathbf{B} = 0, \qquad \mathbf{B} = \mu \mathbf{H}. \end{aligned} \tag{1.13}$$

It follows that the electric and magnetic fields are transformable to each other. The EM field is sourced by the charges and the electric currents. The medium influences the EM field, and the measure of it is the mentioned permittivity ε, permeability μ, and conductivity σ. In general case, all these medium parameters can depend on the spatial and time variables and be influenced by the field intensities.

Additionally, the EM field can be described by the scalar electric φ_e and the vector magnetic $\mathbf{A}$ potentials:

$$\mathbf{B} = \nabla \times \mathbf{A},$$
$$\mathbf{E} = -\nabla \cdot \varphi_e - \frac{\partial \mathbf{A}}{\partial t}. \tag{1.14}$$

The following differential equations are known for these quantities:

$$\nabla^2 \cdot \mathbf{A} - \mu\varepsilon \frac{\partial^2 \mathbf{A}}{\partial t^2} = -\mu \mathbf{j},$$
$$\nabla^2 \cdot \varphi_e - \mu\varepsilon \frac{\partial^2 \varphi_e}{\partial t^2} = -\frac{\rho}{\varepsilon}. \tag{1.15}$$

These potentials are not independent on each other, and they should satisfy the Lorentz gauge:

$$\nabla \cdot \mathbf{A} = -\mu\varepsilon \frac{\partial \varphi_e}{\partial t}. \tag{1.16}$$

They are the measurable quantities which are important for quantum-mechanical treatments and for calculation of radiated by antennas EM waves.

1.4.1 Time-harmonic EM Field Equations

The harmonic field varies with time according to the *sine* or *cosine* law. For example, the modulated signals are represented by sums of harmonic functions. For this type of the time dependence, a special mathematical apparatus is used called the *phasor notation*, which is similar to the mathematics used for the harmonic currents and voltages studied by the network theory. It requires the knowledge of complex numbers and operations with them.

The central idea is that the imaginary numbers exist, and the simplest one is the imaginary unit calculated as a square root $j = \sqrt{-1}$. Then an arbitrary complex number A is composed of real a' and imaginary a'' parts:

$$A = a' + ja''; \ \mathrm{Re}(A) = a'; \ \mathrm{Im}(A) = a''. \tag{1.17}$$

For complex numbers the sum and multiplication operations are introduced as

$$A + B = (a' + b') + j(a'' + b''); \tag{1.18}$$

$$A \cdot B = (a' + ja'')(b' + jb'') = a'b' + ja''b' + ja'b'' - a''b''. \tag{1.19}$$

Additional operation is the complex conjugation noted by the asterisk symbol (*):

$$A^* = (a' + ja'')^* = a' - ja'' = |A| e^{-j\phi} \tag{1.20}$$

where $|A|$ is the module of a complex number and $\phi = \arctan\left(\frac{a''}{a'}\right)$ is its argument. Then the module can be calculated as $|A| = \sqrt{AA^*}$. The operation of the complex conjugation is used to compute the energy and power of time-harmonic fields.

Complex or phasor notations can represent the time-harmonic functions:

$$\begin{Bmatrix} u_c(r,t) \\ u_s(r,t) \end{Bmatrix} = |U| \cdot \begin{Bmatrix} \cos(\omega t + \phi) \\ \sin(\omega t + \phi) \end{Bmatrix} = \begin{Bmatrix} \text{Re} \\ \text{Im} \end{Bmatrix} \{ |U| e^{j\phi} e^{j\omega t} \}. \tag{1.21}$$

In (1.21), the positive time-dependence is assumed as $e^{j\omega t}$. The term $U = |U| e^{j\phi}$ is the complex amplitude. The convenience of the complex notation is that instead of space-consuming integrals and derivatives the reduced notations are used

$$\begin{aligned} &\frac{\partial}{\partial t} \to j\omega, \\ &\int dt \to -j/\omega. \end{aligned} \tag{1.22}$$

All equations of electromagnetics are re-written for the time-harmonic fields, and, for example, the Maxwell equations are now regarding to the complex amplitudes of the fields and currents:

$$\begin{aligned} &\nabla \times \mathbf{H} = \mathbf{j}^{(\text{source})} + \sigma \mathbf{E} + j\omega\varepsilon_0\varepsilon_r \mathbf{E}, \\ &\nabla \times \mathbf{E} = -j\omega \mathbf{B}. \end{aligned} \tag{1.23}$$

Taking into account that the conductivity of a dielectric medium is with its loss, the complex permittivity $\tilde{\varepsilon}$ is introduced in this case:

$$\begin{aligned} &\nabla \times \mathbf{H} = \mathbf{j}^{(\text{source})} + \sigma \mathbf{E} + j\omega\varepsilon_0\varepsilon_r \mathbf{E}, \\ &\nabla \times \mathbf{H} = \mathbf{j}^{(\text{source})} + (\sigma + j\omega\varepsilon_0\varepsilon_r) \mathbf{E}, \\ &\nabla \times \mathbf{H} = \mathbf{j}^{(\text{source})} + j\omega\varepsilon_0 \left(\varepsilon_r - j\frac{\sigma}{\omega\varepsilon_0} \right) \mathbf{E}, \\ &\nabla \times \mathbf{H} = \mathbf{j}^{(\text{source})} + \tilde{\varepsilon} \mathbf{E}, \\ &\tilde{\varepsilon} = \varepsilon_0 \left(\varepsilon_r - j\frac{\sigma}{\omega\varepsilon_0} \right) = \varepsilon' - j\varepsilon''. \end{aligned} \tag{1.24}$$

The ratio of the imaginary and real parts of this permittivity is the tangent of the dielectric loss angle $\tan\delta$, which is the most important parameter to describe the lossy dielectrics:

$$\tan\delta = \frac{\varepsilon''}{\varepsilon'} = \frac{\sigma}{\omega\varepsilon_0\varepsilon_r}. \tag{1.25}$$

Usually, if $\tan\delta \ll 1$, then the material is a good dielectric. A poor dielectric has increased value of the dielectric loss tangent $(\tan\delta > 1)$. For metals, the dielectric permittivity is not applied, and the conductivity σ is the main factor allowing for calculation of thermal loss in them.

Magnetic medium has its own type of loss, and, similarly to the complex permittivity, the complex permeability $\tilde{\mu}$ is introduced as

$$\tilde{\mu} = \mu' - j\mu''. \tag{1.26}$$

Finally, the full system of the Maxwell equations for complex amplitudes of time-harmonic fields is re-written as

$$\begin{aligned} &\nabla\times\mathbf{H} = j\omega\tilde{\varepsilon}\mathbf{E} + \mathbf{j}^{(\text{source})},\\ &\nabla\times\mathbf{E} = -j\omega\mathbf{B},\\ &\nabla\cdot\mathbf{D} = \rho,\\ &\nabla\cdot\mathbf{B} = 0,\\ &\mathbf{B} = \tilde{\mu}\mathbf{H},\\ &\mathbf{D} = \tilde{\varepsilon}\mathbf{E}. \end{aligned} \tag{1.27}$$

1.4.2 Wave Equations

It follows from the Maxwell equations that the time-dependent electric and magnetic fields are transformed to each other, and this transformation is spreading away from the source. This process is called the wave radiation and propagation.

In the simplest time-harmonic case, the fields are transformed to each other with a period in the space, which is the wavelength λ:

$$\lambda = \frac{\omega\sqrt{\varepsilon_r\mu_r}}{c}$$

where c is the light velocity in vacuum. It seen that medium increases the wavelength λ and makes the wave slower because its velocity $v = \frac{c}{\sqrt{\varepsilon_r\mu_r}}$.

The space-time shape of waves depends on the excitation source, medium parameters, and the environment. The waves can be calculated by the Maxwell equations, but the full system of them (1.13) or (1.27) is not convenient for many practical needs. Very often, the wave second-order partial differential equations are used

$$\begin{aligned}
&\Delta\cdot\mathbf{H}-\frac{\varepsilon_r\mu_r}{c^2}\frac{\partial^2\mathbf{H}}{\partial t^2}=-\nabla\times\left(\mathbf{j}_c+\overset{\bullet}{\mathbf{j}}^{(\text{source})}\right),\\
&\Delta\cdot\mathbf{E}-\frac{\varepsilon_r\mu_r}{c^2}\frac{\partial^2\mathbf{E}}{\partial t^2}=\frac{1}{\varepsilon_0\varepsilon_r}\nabla\cdot\left(\rho+\rho^{(\text{source})}\right)+\mu_0\mu_r\frac{\partial}{\partial t}\left(\mathbf{j}_c+\overset{\bullet}{\mathbf{j}}^{(\text{source})}\right).
\end{aligned}\tag{1.28}$$

In the case of time-harmonic waves, the equations (1.28) are re-written as

$$\begin{aligned}
&\Delta\cdot\mathbf{H}+\omega^2\tilde{\varepsilon}\tilde{\mu}\mathbf{H}=-\nabla\times\overset{\bullet}{\mathbf{j}}^{(\text{source})},\\
&\Delta\cdot\mathbf{E}+\omega^2\tilde{\varepsilon}\tilde{\mu}\mathbf{E}=\frac{j}{\omega\tilde{\varepsilon}}\nabla\cdot\nabla\cdot\left(\overset{\bullet}{\mathbf{j}}^{(\text{source})}\right)+j\omega\tilde{\mu}\left(\overset{\bullet}{\mathbf{j}}^{(\text{source})}\right).
\end{aligned}\tag{1.29}$$

Additionally, they are called the non-homogeneous Helmholtz equations due to the driving current $\mathbf{j}^{(\text{source})}$ in their right parts. In the spatial domains where the source currents are zero, the homogeneous Helmholtz equations are applicable

$$\begin{aligned}
&\Delta\cdot\mathbf{H}+\tilde{k}^2\mathbf{H}=0,\\
&\Delta\cdot\mathbf{E}+\tilde{k}^2\mathbf{E}=0.
\end{aligned}\tag{1.30}$$

It is seen that the equations (1.30) depend on the medium parameters, and it can be estimated according to the complex wave constant $\tilde{k}=\omega\sqrt{\tilde{\varepsilon}\tilde{\mu}}$. Formally, the magnetic and electric fields follow independently the wave partial differential equations, but these vectors are coupled to each other through the Maxwell equations.

1.4.3 Plane TEM Wave

The simplest solution of (1.30) is the plane TEM (*transversal electromagnetic*) wave which propagates in open space with $\tilde{\mu}$ and $\tilde{\varepsilon}$. This wave has only two components of the electric E_x and magnetic H_y fields, and they do not depend on the coordinates (x, y) which are normal to the propagation direction $0\rightarrow z$. Their homogeneous Helmholtz equations written for complex amplitudes are

$$\begin{aligned}
&\frac{d^2E_x}{dz^2}+\tilde{k}_z^2E_x=0,\\
&\frac{d^2H_y}{dz^2}+\tilde{k}_z^2H_y=0
\end{aligned}\tag{1.31}$$

where $\tilde{k}_z = \tilde{k} = \omega\sqrt{\tilde{\varepsilon}\tilde{\mu}}$ is the complex longitudinal wave propagation constant that is analytically calculated in this simplest case. To calculate the space-dependent fields, only one equation from (1.31) can be solved taking into account the Maxwell equations:

$$\frac{d^2 E_x}{dz^2} + \tilde{k}_z^2 E_x = 0,$$
$$H_y = \frac{1}{\tilde{W}} E_x, \ \tilde{W} = \sqrt{\frac{\tilde{\mu}}{\tilde{\varepsilon}}} \tag{1.32}$$

where $\tilde{W}$ is the wave impedance. In this case, (1.32) is an ordinary differential equation, and its general solution gives two partial waves propagating from $z = \mp\infty$:

$$E_x = E_0^+ \exp\left(-j\tilde{k}_z z\right) + E_0^- \exp\left(+j\tilde{k}_z z\right) \tag{1.33}$$

where E_0^+ is the amplitude of the wave propagating along the positive direction of the z-axis (forward wave), and E_0^- is the amplitude of the wave coming from the opposite direction (backward wave).

A particular solution of (1.32) is derived by satisfying the *boundary condition at the infinity*. For example, we are assuming that the wave is not excited at the infinity $(z = \infty)$, and only one wave is propagating from $z = -\infty$ to $z = +\infty$. The wave amplitude E_0^+ is defined by the source, and it is not considered here, being arbitrary. The formulas to calculate this forward wave are

$$E_x = E_0^+ \exp\left(-j\tilde{k}_z z\right),$$
$$H_y = \frac{1}{\tilde{W}} E_0^+ \exp\left(-j\tilde{k}_z z\right). \tag{1.34}$$

The analysis of (1.34) shows that the fields depend exponentially on the z-coordinate. It means that the field is periodical along this coordinate if $\mathrm{Im}\left(\tilde{k}_z\right) = 0$. Its spatial period or wavelength is

$$\lambda = \frac{2\pi}{\mathrm{Re}\left(\tilde{k}_z\right)}. \tag{1.35}$$

The wave phase velocity v_p and its group velocity v_g are

$$v_p = \frac{\omega}{\mathrm{Re}\left(\tilde{k}_z\right)}, v_g = \frac{1}{\sqrt{\mathrm{Re}(\tilde{\varepsilon})\mathrm{Re}(\tilde{\mu})}}. \tag{1.36}$$

The EM field is a vector object the orientation and polarization of which is very important for many applications.

Let us consider the electric field as the reference component. If this component changes its direction only along the one axis ($\pm 0 \rightarrow x$, particularly) when the wave is propagating from $z = 0$ to $z = \infty$, then this wave is the linearly polarized one.

Two linearly polarized waves of different magnitudes and phase-shifted with respect to each other give another important case when the electric-field vector projection on the plane normal to the propagating direction will plot an ellipse on this plane. This wave is the elliptically polarized. A particular case is the circularly polarized one formed by two TEM waves of the equal magnitudes and having a phase difference. The effect of polarizations is used in microwave antennas, spatial filtering of waves, etc.

The EM waves carry a certain power taken from the source. This power flow has a direction in the space. In each point, the power spatial density, called as the Poynting vector $\mathbf{\Pi}$, is given

$$\mathbf{\Pi} = [\mathbf{E} \times \mathbf{H}], \ \mathbf{W}/\mathrm{m}^2. \tag{1.37}$$

In the case of the time-harmonic plane TEM, this wave propagating along the z-axis has the following averaged on the period Poynting vector $\tilde{\mathbf{\Pi}}$:

$$\hat{\mathbf{\Pi}} = \frac{1}{2}\mathrm{Re}\left[\mathbf{E} \times \mathbf{H}^*\right] = \frac{1}{2}\frac{\left(E_0^+\right)^2}{\left|\tilde{W}\right|}\mathbf{z}_0. \tag{1.38}$$

The averaged power $\hat{P}$ that passes through a surface S is $\hat{P} = \int_S \hat{\mathbf{\Pi}} d\mathbf{s}$.

In the case of lossless medium, the wave field span is not varied with the distance. In lossy medium, the energy of the wave is transformed into heat; the wave span is decreasing with the distance from the source exponentially:

$$\begin{aligned} E_x &\sim \exp(-jk_z' z)\exp(-k_z''|z|), \\ H_y &\sim \exp(-jk_z' z)\exp(-k_z''|z|) \end{aligned} \tag{1.39}$$

where the wave parameters valid for the lossy dielectrics are [12]

$$\begin{aligned} k_z' &\approx \omega\sqrt{\varepsilon'\mu'}, \ [\mathrm{rad/m}] \\ k_z'' &\approx \omega\frac{\sqrt{\varepsilon'\mu'}}{2}\tan\delta, \ [1/\mathrm{m}] \\ \tilde{W} &\approx \sqrt{\frac{\mu'}{\varepsilon'}}\left(1 + j\frac{\tan\delta}{2}\right), \ [\Omega]. \end{aligned} \tag{1.40}$$

In metals and semiconductors, the wave energy is spent fast for medium heating, and the wave is propagating only for a relative short distance from the surface where it is excited. Taking into account this concentration of microwave field in a narrow, surface layer of a metal, this phenomenon is called the *skin effect.*

The measure of it is the *skin depth* $\Delta^0 = \sqrt{2/\omega\mu\sigma}$. The wave propagated this distance decreases its magnitude span by e=2.71… times. Other parameters of wave propagating in a well-conducting medium are

$$\begin{aligned} k'_z &\approx k''_z \approx \sqrt{\frac{\omega\mu\sigma}{2}}, \\ \tilde{W} &\approx (1+j)\sqrt{\frac{\omega\mu}{2\sigma}}. \end{aligned} \tag{1.41}$$

It follows that due to the increased frequency ω and conductivity σ, the imaginary part of the propagation constant k''_z is rather large in metals, and the skin effect prevents penetration of microwaves into real conductors.

At the same time, concentration of microwave current in a narrow layer of a metal is the cause of increased conductor loss at high frequencies because the resistance of a metal sheet is proportional to the inverted square of the metal taken by current. To decrease the loss, a layer of silver or gold covers a poorly conducting metal where the main portion of the current is. A special attention should be paid to prevent the oxidation of the metal surface by covering it by gold, for instance. Additionally, the conductor surface is polished well to avoid the current scattering at micro-obstacles.

1.4.4 Ray Representation of TEM Wave in Open Space

In above, a plane wave propagating along the z-axis has been considered. Similarly, a wave propagating along an arbitrary direction $\boldsymbol{\zeta}$ can be treated. Supposing that the transversal to $\boldsymbol{\zeta}$-coordinate axis are the directions $\boldsymbol{\xi}_0$ and $\boldsymbol{\eta}_0$, the EM field associated with new-system of coordinates ξ, η, ζ is

$$\begin{aligned} \mathbf{E}(\xi,\eta,\zeta) &= \boldsymbol{\xi}_0 A e^{-jk\zeta}, \\ \mathbf{H}(\xi,\eta,\zeta) &= \boldsymbol{\eta}_0\, A/W\, e^{-jk\zeta} \end{aligned} \tag{1.42}$$

where A is the amplitude, W is the wave impedance, and $k = \omega\sqrt{\varepsilon\mu}$ is the wave constant. The argument of the exponent $-jk\zeta$ is the wave phase. The equation $\zeta(x, y, z) = const$ describes an iso-phase surface, and the vector $\boldsymbol{\zeta}(x, y, z)$ is normal to this surface. While the wave is propagating, this vector is moving along a curve in space. This oriented curve is called the *ray*. An arbitrary wave process can be shown by a system of rays similarly to the streamlines of the electric or magnetic fields.

The simplest ray picture is for the plane TEM wave. In this case, the phase and amplitude surfaces are the same, and they are the planes. Due to that, this wave is called the homogeneous TEM wave. The ray representations are popular in optics and quasioptics of microwave phenomena.

1.5 Nonlinear Electromagnetism[1]

Many applications of EM waves are with a linear medium which conductivity, permittivity, and permeability are field-independent. It is known that the motion of molecules in microwave field are, generally speaking nonlinear. The nonlinearity is especially well expressed in strong alternating field's, and it is known for some dielectrics, semiconductors, ferrites, and plasma.

Single molecule trajectories are described by nonlinear differential equations, and the multiple effects occur, including the multiplication and dropping of the rotation frequency regarding to the external driving field.

Another reason of nonlinearity is the quantum-mechanical properties of atoms and molecules. At certain frequencies, the energy of the EM field is comparable with the inter-level one of different quantum states, and the quantum absorption occurs. Here, the medium parameters depend on the energy of propagating waves, and the dispersion and nonlinear effects are strong at frequencies over several hundred Gigahertz. Another nonlinearity is with the ionization of medium and specific effects in plasma. The effects arising on the atomic and molecular levels are averaged, and they are taken into account by the field-dependent permittivity and permeability of medium. They, appearing in the Maxwell equations, make them nonlinear, and a special theory is needed to represent the EM field and solutions of the Maxwell and the Helmholtz equations for the developments of new microwave devices [27]. Below one of such theories is given.

1.5.1 Macro Electromagnetism of Nonlinear Medium

The mentioned EM phenomena are described by the Maxwell equation in their nonlinear form. In this case, the time–dependent equations are

$$\begin{aligned} \nabla\times\mathbf{H}(\mathbf{r},t) &= \frac{\partial\mathbf{D}(\mathbf{E}(\mathbf{r},t))}{\partial t}+\mathbf{J}(\mathbf{E}(\mathbf{r},t)), \\ \nabla\times\mathbf{E}(\mathbf{r},t) &= -\frac{\partial\mathbf{B}(\mathbf{H}(\mathbf{r},t))}{\partial t} \end{aligned} \tag{1.43}$$

where $\mathbf{E}(\mathbf{r},t)$ and $\mathbf{H}(\mathbf{r},t)$ are the electric and magnetic field intensity vectors, respectively, $\mathbf{D}(\mathbf{E}(\mathbf{r},t))$ is the electric-flux density vector depending on the electric field intensity vector, $\mathbf{B}(\mathbf{H}(\mathbf{r},t))$ is the magnetic induction vector depending on the magnetic field intensity vector, $\mathbf{J}(\mathbf{E}(\mathbf{r},t))$ is the electric current density vector depending on the electric field intensity vector. In general case of anisotropic nonlinear medium, $\mathbf{D}(\mathbf{E}(\mathbf{r},t))$, $\mathbf{B}(\mathbf{H}(\mathbf{r},t))$, and $\mathbf{J}(\mathbf{E}(\mathbf{r},t))$ are the vector functions of the vector arguments. They depend on scalar arguments if the field is considered in an isotropic nonlinear medium:

[1] Written by G.S. Makeeva and G.A. Kouzaev

$$\mathbf{D}(\mathbf{E}(\mathbf{r},t)) = \varepsilon_0 \varepsilon_r \left(|\mathbf{E}(\mathbf{r},t)| \right) \mathbf{E}(\mathbf{r},t),$$
$$\mathbf{B}(\mathbf{H}(\mathbf{r},t)) = \mu_0 \mu_r \left(|\mathbf{H}(\mathbf{r},t)| \right) \mathbf{H}(\mathbf{r},t), \tag{1.44}$$
$$\mathbf{J}(\mathbf{E}(\mathbf{r},t)) = \sigma\left(|\mathbf{E}(\mathbf{r},t)| \right) \mathbf{E}(\mathbf{r},t)$$

where $\varepsilon_r \left(|\mathbf{E}(\mathbf{r},t)| \right)$ and $\mu_r \left(|\mathbf{H}(\mathbf{r},t)| \right)$ are the nonlinear permittivity and permeability depending on the modulus of the electric and magnetic fields, respectively. Similarly, the nonlinear conductivity $\sigma\left(|\mathbf{E}(\mathbf{r},t)| \right)$ is introduced in (1.44).

Describing the medium by the phenomenological approach, at any space point, the considered fields and current density are expressed using polynomial expansions [28]-[30]:

$$\varepsilon_r \left(|\mathbf{E}(\mathbf{r},t)| \right) = \varepsilon_{r_1} + \varepsilon_{r_2} |\mathbf{E}(\mathbf{r},t)| + \varepsilon_{r_3} |\mathbf{E}(\mathbf{r},t)|^2 + \dots + \varepsilon_{r_n} |\mathbf{E}(\mathbf{r},t)|^{n-1}, \tag{1.45}$$

$$\mu_r (|\mathbf{H}(\mathbf{r},t)|) = \mu_{r_1} + \mu_{r_2} |\mathbf{H}(\mathbf{r},t)| + \mu_{r_3} |\mathbf{H}(\mathbf{r},t)|^2 + \dots + \mu_r |\mathbf{H}(\mathbf{r},t)|^{n-1}, \tag{1.46}$$

$$\sigma(|\mathbf{E}(\mathbf{r},t)|) = \sigma_1 + \sigma_2 |\mathbf{E}(\mathbf{r},t)| + \sigma_3 |\mathbf{E}(\mathbf{r},t)|^2 + \dots + \sigma_n |\mathbf{E}(\mathbf{r},t)|^{n-1} \tag{1.47}$$

where the coefficients of these expansions are determined experimentally or by the theoretical modelling of the medium.

Consider the Fourier representation of the time-dependent modulated field vectors generating a sum of time harmonics. Its spectrum consists of R discrete frequencies. Due to the medium nonlinearity, the response is the series of terms of combination frequencies:

$$\mathbf{E}(\mathbf{r},t) = \sum_{m=-\infty}^{\infty} \mathbf{E}(\mathbf{r},\omega_m) \exp(j\omega_m t),$$
$$\mathbf{H}(\mathbf{r},t) = \sum_{m=-\infty}^{\infty} \mathbf{H}(\mathbf{r},\omega_m) \exp(j\omega_m t),$$
$$|\mathbf{E}(\mathbf{r},t)| = \sum_{m=-\infty}^{\infty} E(\mathbf{r},\omega_m) \exp(j\omega_m t), \tag{1.48}$$
$$|\mathbf{H}(\mathbf{r},t)| = \sum_{m=-\infty}^{\infty} H(\mathbf{r},\omega_m) \exp(j\omega_m t)$$

where an index m is determined at the set of indices $\{m_1, m_2, \dots m_r\}$ and r is a number of a discrete frequency of the source signal. Substituting the time-dependent fields in (1.43) by their expressions (1.48), the nonlinear Maxwell equations are obtained written at the combination frequencies:

$$\begin{aligned}\nabla\times\mathbf{H}(\mathbf{r},\omega_m) &= j\omega_m\varepsilon_0\varepsilon_{r_1}(\omega_m)\mathbf{E}(\mathbf{r},\omega_m)+\mathbf{J}(\mathbf{r},\omega_m),\\ \nabla\times\mathbf{E}(\mathbf{r},\omega_m) &= -j\omega_m\mu_0\mu_{r_1}(\omega_m)\mathbf{H}(\mathbf{r},\omega_m)+\mathbf{Z}(\mathbf{r},\omega_m)\end{aligned} \tag{1.49}$$

where

$$\begin{aligned}\mathbf{J}(\mathbf{r},\omega_m) &= j\omega_m\varepsilon_0\Big(\sum_{k=-\infty}^{\infty}\sum_{n=-\infty}^{\infty}\tilde{\varepsilon}_{r_2}(\omega_m)\mathbf{E}(\mathbf{r},\omega_k)E(\mathbf{r},\omega_n)\gamma_{kn}+\\ &+\sum_{k=-\infty}^{\infty}\sum_{n=-\infty}^{\infty}\sum_{i=-\infty}^{\infty}\tilde{\varepsilon}_{r_3}(\omega_m)\mathbf{E}(\mathbf{r},\omega_k)E(\mathbf{r},\omega_n)E(\mathbf{r},\omega_i)\gamma_{kni}+\ldots+\\ &+\sum_{k=-\infty}^{\infty}\sum_{n=-\infty}^{\infty}\sum_{i=-\infty}^{\infty}\ldots\sum_{l=-\infty}^{\infty}\tilde{\varepsilon}_{r_n}(\omega_m)\mathbf{E}(\mathbf{r},\omega_m)E(\mathbf{r},\omega_n)E(\mathbf{r},\omega_i)\ldots E(\mathbf{r},\omega_l)\gamma_{kni\ldots l}\Big),\end{aligned} \tag{1.50}$$

$$\begin{aligned}\mathbf{Z}(\mathbf{r},\omega_m) &= j\omega_m\mu_0\Big(\sum_{k=-\infty}^{\infty}\sum_{n=-\infty}^{\infty}\mu_{r_2}(\omega_m)\mathbf{H}(\mathbf{r},\omega_k)H(\mathbf{r},\omega_n)\gamma_{kn}+\\ &+\sum_{k=-\infty}^{\infty}\sum_{n=-\infty}^{\infty}\sum_{i=-\infty}^{\infty}\mu_{r_3}(\omega_m)\mathbf{H}(\mathbf{r},\omega_k)H(\mathbf{r},\omega_n)H(\mathbf{r},\omega_j)\gamma_{knj}+\ldots+\\ &+\sum_{k=-\infty}^{\infty}\sum_{n=-\infty}^{\infty}\sum_{i=-\infty}^{\infty}\ldots\sum_{l=-\infty}^{\infty}\mu_{r_n}(\omega_m)\mathbf{H}(\mathbf{r},\omega_k)H(\mathbf{r},\omega_n)H(\mathbf{r},\omega_i)\ldots H(\mathbf{r},\omega_l)\gamma_{kni\ldots l}\Big)\end{aligned} \tag{1.51}$$

where

$$\tilde{\varepsilon}_{r_k}(\omega_m)=\varepsilon_{r_k}(\omega_m)-j\frac{\sigma_k(\omega_m)}{\varepsilon_0\omega_m},\quad k=1,2,\ldots,n\,,$$

$$\gamma_{kni\ldots l}=\begin{cases}0, \text{ if } & \omega_k+\omega_n+\omega_i+\ldots+\omega_l\neq\omega_m,\\ 1, \text{ if } & \omega_k+\omega_n+\omega_i+\ldots+\omega_l=\omega_m.\end{cases}$$

Thus, the considered formalism allows describing the nonlinear effects by calculation of each harmonics and composing the spectral contents of a signal propagating in a nonlinear medium. More detailed information on this approach and some applications are from [28]-[30].

1.6 Dynamics of Charged Particles and Dipolar Molecules in EM Field

The above-considered EM theory is based on macroscopic models of the medium. Nowadays, electronics tends to manipulate single electrons, atoms and molecules or their clusters [31],[32], and a theory of interaction of the EM field with these particles has to be seen in modern books on electromagnetics.

Unfortunately, classical theory of this sort is not always accurate, and the EM-quantum equations should be introduced to describe the physical effects on the nano level [3]. The presented here theory touches only some aspects of classical and semi-classical theories of the EM-particle interactions, including an approach to numerical simulations [33].

1.6.1 Charged Particle in EM Field

1.6.1.1 Classical Theory of Charged Particle Movement in EM Field

The simplest EM problem is the trajectory calculation of a moving charge in the external EM field. The trajectory of a moving charge q is derived from the Euler-Lagrange motion equation, and the relativistic Lagrangian L is [3],[34],[35]:

$$L = -mc^2\sqrt{1-|\mathbf{v}|^2/c^2} + q\mathbf{v}\mathbf{A}/c - q\Phi \tag{1.52}$$

where m is the particle mass, $\mathbf{v}$ is the particle velocity, respectively, and $\mathbf{A}$ and Φ are the external vector and scalar potentials, respectively. The moving accelerated charge q is the source of the time-dependent electric $\mathbf{E}$ and magnetic $\mathbf{H}$ fields:

$$\begin{aligned} &\mathbf{E}(\mathbf{r},t) = q\left[\frac{\mathbf{n}-\boldsymbol{\beta}}{\gamma^2(1-\boldsymbol{\beta}\mathbf{n})^3 R^2}\right] + \frac{q}{c}\left[\frac{\mathbf{n}\times(\mathbf{n}-\boldsymbol{\beta})\times\dot{\boldsymbol{\beta}}}{(1-\boldsymbol{\beta}\mathbf{n})^3 R}\right]\Bigg|_{t'=t-R/c}, \\ &\mathbf{H}(\mathbf{r},t) = [\mathbf{n}\times\mathbf{E}]\big|_{t'=t-R/c} \end{aligned} \tag{1.53}$$

where $\mathbf{r}_0$ and $\mathbf{r}$ are the coordinates of the particle and a watching point, respectively, $\mathbf{n} = (\mathbf{r}-\mathbf{r}_0)/R$, $R = |\mathbf{r}-\mathbf{r}_0|$, $\boldsymbol{\beta} = \mathbf{v}/c$, $\dot{\boldsymbol{\beta}} = \frac{\partial\boldsymbol{\beta}}{\partial t}$, and $\gamma = 1/\sqrt{1-\beta^2}$.

The field consists of two parts. They are the quasi-static time-dependent field $(\sim 1/R^2)$ and the "accelerated" one $(\sim 1/R)$. The radiated power and its spatial density depend on the charge kinetic energy and its acceleration. The charge's own field can influence the particle depending on its spatial shape. For instance, the spherically symmetric charges are not influenced by the electro-static self-force, but only by the radiated field. Unfortunately, classical theory of this motion has some limitations and contradictions [3], and the quantum-mechanical approach should be applied to this problem.

1.6.1.2 Quantum Theory for the Charged Particle Movement. Low-velocity Approximation

Quantum physics considers the particles as matter waves. For example, the low-energy quantum-mechanical effects are calculated by the Schrödinger equation [36] when the external potential Φ_0 is governing the particle probability density wave function Ψ:

$$j\hbar\frac{\partial\Psi}{\partial t}=-\frac{\hbar^2}{2m}\nabla^2\cdot\Psi+q\Phi_0\Psi \tag{1.54}$$

where $\hbar$ is the normalized Planck constant, and Ψ is normalized to the unit.

The moving charge produces the electric current and the quasi-static time-dependent electric and magnetic fields. Accelerated particles can radiate the EM waves, and, in the EM-quantum systems, the external and self-consistent fields influence these particles. It is described by the Maxwell-Schrödinger system of differential equations [37]-[42].

1.6.1.3 Maxwell-Schrödinger Equations for Spin-less Charge

Semi-classical approach of description of the EM-particle interaction consists of joint solution of Maxwell and Schrödinger equations. As the first-order approximation, the particle and the EM field are not quantified. A particular case [41] of this system is shown below:

$$\begin{aligned}
&j\hbar\frac{\partial\Psi}{\partial t}=-\frac{\hbar^2}{2m}\nabla^2\cdot\Psi-\frac{j\hbar q}{2m}(\nabla\cdot\mathbf{A})\Psi-\frac{j\hbar q}{m}\mathbf{A}(\nabla\cdot\Psi)+\frac{q^2}{2m}|\mathbf{A}|^2\Psi+q\varphi\Psi,\\
&\frac{\partial\mathbf{H}}{\partial t}=-\frac{1}{\mu_0}\nabla\times\mathbf{E},\\
&\frac{\partial\mathbf{E}}{\partial t}=\frac{1}{\varepsilon_0}\nabla\times\mathbf{H}-\frac{1}{\varepsilon_0}\mathbf{J}_e
\end{aligned} \tag{1.55}$$

where $\mathbf{A}$ is the self-consistent vector potential, φ is the sum of the external Φ_0 and self-consistent electric scalar potentials, and μ_0 and ε_0 are the vacuum permeability and permittivity, respectively. The electric current $\mathbf{J}_e$ of moving particles is calculated according to known formula [36]:

$$\mathbf{J}_e=q\left(\frac{\hbar}{2jm}\left(\Psi^*\nabla\cdot\Psi-\Psi\nabla\cdot\Psi^*\right)-\frac{q}{m}|\Psi|^2\mathbf{A}\right) \tag{1.56}$$

where the symbol * stands for complex conjugation. Taking into account that this current is proportional to the square of the wave function Ψ, the equations (1.55) are nonlinear, and they require complicated programming on their solutions.

This way of treating the EM-quantum-mechanical phenomena is chosen in [39]-[41]. For instance, the 1-D Maxwell-Schrödinger equations are solved by an iterative FDTD method [39]. A more detailed work on numerical simulations of EM and quantum-mechanical effects in carbon nanotube interconnects is performed in [40],[41]. The Maxwell equations are solved by the transmission line matrix method. The Schrödinger equation is splitted into the real and imaginary parts, and an FDTD approach is applied to approximate the Schrödinger differential operator. The simulations are verified by published measurements, and it is shown that the self-induced fields of a moving electron are needed for better correspondence of numerical and experimental data.

As seen, these approaches require complicated mathematical apparatus and careful control of stability and convergence of numerical simulations. Besides, the EM and quantum-mechanical equations (1.55) treat different quantities like the field vectors $\mathbf{E}$ and $\mathbf{H}$ and potentials φ and $\mathbf{A}$, and it requires additional intermediate computations.

Theoretical treatments are known [37],[38]-[42] for the equations regarding to the vector $\mathbf{A}$ and scalar φ potentials, their Lorentz gauge, and the probability density wave function Ψ:

$$
\begin{aligned}
& j\hbar\frac{\partial\Psi}{\partial t}=-\frac{\hbar^2}{2m}\nabla^2\cdot\Psi-\frac{j\hbar q}{2m}(\nabla\cdot\mathbf{A})\Psi-\frac{j\hbar q}{m}\mathbf{A}(\nabla\cdot\Psi)+\frac{q^2}{2m}|\mathbf{A}|^2\Psi+q\varphi\Psi \\
& \nabla^2\cdot\mathbf{A}-\frac{1}{c^2}\frac{\partial^2\mathbf{A}}{\partial t^2}=-\mu_0\mathbf{J}_e, \\
& \nabla^2\cdot\varphi-\frac{1}{c^2}\frac{\partial^2\varphi}{\partial t^2}=-\frac{\rho}{\varepsilon_0}, \\
& \frac{\partial\varphi}{\partial t}+\frac{1}{c^2}\nabla\cdot\mathbf{A}=0
\end{aligned}
\tag{1.57}
$$

where $\rho=q\Psi\Psi^*$ This considered system of nonlinear differential partial equations has a decreased number of unknown functions in comparison to (1.55), but it is still complicated for physical analysis.

1.6.1.4 Schrödinger-Hertz Equation

More flexibility is given by the use of the Hertz vectors to describe the EM part of the problem. There are two Hertz vectors, which are the electric $\mathbf{\Gamma}_e$ and magnetic $\mathbf{\Gamma}_m$ ones, and they "automatically" satisfy the Lorentz gauge [5],[6],[43]-[45]. Very often, only one or two components of these vectors can be chosen to describe the EM part of the problem in full.

Taking into account that $\varphi=-\nabla\cdot\mathbf{\Gamma}_e$ and $\mathbf{A}=\frac{1}{c^2}\frac{\partial\mathbf{\Gamma}_e}{\partial t}+\nabla\times\mathbf{\Gamma}_m$, the system (1.57) is re-written in its general form:

$$
\begin{aligned}
& j\hbar\frac{\partial\Psi}{\partial t}=-\frac{\hbar^2}{2m}\nabla^2\cdot\Psi-\frac{j\hbar q}{2m}\left(\nabla\cdot\left[\frac{1}{c^2}\frac{\partial\mathbf{\Gamma}_e}{\partial t}+(\nabla\times\mathbf{\Gamma}_m)\right]\right)\Psi- \\
& -\frac{j\hbar q}{m}\left[\frac{1}{c^2}\frac{\partial\mathbf{\Gamma}_e}{\partial t}+(\nabla\times\mathbf{\Gamma}_m)\right](\nabla\cdot\Psi)+ \\
& +\left[\frac{q^2}{2m}\left|\frac{1}{c^2}\frac{\partial\mathbf{\Gamma}_e}{\partial t}+(\nabla\times\mathbf{\Gamma}_m)\right|^2-q(\nabla\cdot\mathbf{\Gamma}_e)\right]\Psi, \\
& \nabla^2\cdot\mathbf{\Gamma}_e-\frac{1}{c^2}\frac{\partial^2\mathbf{\Gamma}_e}{\partial t^2}=-\frac{\mathbf{p}_e}{\varepsilon_0}, \\
& \nabla^2\cdot\mathbf{\Gamma}_m-\frac{1}{c^2}\frac{\partial^2\mathbf{\Gamma}_m}{\partial t^2}=-\mu_0\mathbf{p}_m
\end{aligned}
\tag{1.58}
$$

where $\mathbf{p}_e$ and $\mathbf{p}_m$ are the streams potentials calculated through the electric $\mathbf{J}_e$ and magnetic $\mathbf{J}_m$ currents:

$$\mathbf{p}_e = const + \int_0^t \mathbf{J}_e(t',\mathbf{r})dt',$$
$$\mathbf{p}_m = const + \int_0^t \mathbf{J}_m(t',\mathbf{r})dt'. \tag{1.59}$$

Taking into account that in open space

$$\mathbf{\Gamma}_e = \frac{1}{4\pi\varepsilon_0}\int_{V'} d\mathbf{r}' \frac{\mathbf{p}_e\left(t+\frac{|\mathbf{r}-\mathbf{r}'|}{c},\mathbf{r}'\right)}{(\mathbf{r}-\mathbf{r}')},$$
$$\mathbf{\Gamma}_m = \frac{\mu_0}{4\pi}\int_{V'} d\mathbf{r}' \frac{\mathbf{p}_m\left(t+\frac{|\mathbf{r}-\mathbf{r}'|}{c},\mathbf{r}'\right)}{(\mathbf{r}-\mathbf{r}')} \tag{1.60}$$

where V' is the volume occupied by magnetic and electric charges. The system (1.58) is transformed into a single integro-differential *Schrödinger-Hertz* equation:

$$j\hbar\frac{\partial\Psi}{\partial t} = -\frac{\hbar^2}{2m}\nabla^2\cdot\Psi - \frac{j\hbar q}{2m}\left(\nabla\cdot\left[\frac{1}{c^2}\frac{\partial\mathbf{\Gamma}_e}{\partial t}+\nabla\times\mathbf{\Gamma}_m\right]\right)\Psi -$$
$$-\frac{j\hbar q}{m}\left[\frac{1}{c^2}\frac{\partial\mathbf{\Gamma}_e}{\partial t}+\nabla\times\mathbf{\Gamma}_m\right](\nabla\cdot\Psi)+\frac{q^2}{2m}\left\|\left[\frac{1}{c^2}\frac{\partial\mathbf{\Gamma}_e}{\partial t}+\nabla\times\mathbf{\Gamma}_m\right]\right\|^2\Psi - q\nabla\cdot\mathbf{\Gamma}_e\Psi. \tag{1.61}$$

The use of the Hertz-vector conception allows for even more simplifications. For example, it was shown that the EM field in homogeneous, linear and isotropic medium could be described using only one electric Hertz vector [5],[43]-[45]. Additionally, the magnetic moments of currents of spin-less particles concentrated in the nanometric regions are negligible. Alternatively as it was shown in [45], the magnetic currents can be incorporated into the generalized current $\mathbf{J}$, and, again only one Hertz vector can be applied for solutions of many problems of electromagnetics and electrodynamics. Such a Hertz vector is denoted as $\mathbf{Z}$, and it satisfies the equation:

$$\nabla^2\cdot\mathbf{Z} - \frac{1}{c^2}\frac{\partial^2\mathbf{Z}}{\partial t^2} = -\frac{\mathbf{p}_e}{\varepsilon_0}. \tag{1.62}$$

The scalar φ and vector $\mathbf{A}$ potentials are expressed in this case:

$$\varphi = -\nabla\cdot\mathbf{Z},\ \mathbf{A} = \frac{1}{c^2}\frac{\partial\mathbf{Z}}{\partial t}. \tag{1.63}$$

Taking into account (1.59) and (1.60), the integral representation of the $\mathbf{Z}$ vector is

$$\mathbf{Z} = \frac{1}{4\pi\varepsilon_0} \int_{V'} d\mathbf{r}' \frac{\mathbf{p}_e\left(t + \frac{|\mathbf{r}-\mathbf{r}'|}{c}, \mathbf{r}'\right)}{(\mathbf{r}-\mathbf{r}')}. \tag{1.64}$$

Substituting the scalar φ and vector $\mathbf{A}$ potentials by their equivalent expressions (1.63), we rewrite (1.61) into a more compact differential-integral *Hertz* equation:

$$j\hbar\frac{\partial \Psi}{\partial t} = -\frac{\hbar^2}{2m}\nabla^2 \cdot \Psi - \frac{j\hbar q}{mc^2}\frac{\partial \mathbf{Z}}{\partial t}(\nabla \cdot \Psi) + \left\{\frac{q}{2mc^2}\left[q\left|\frac{\partial \mathbf{Z}}{\partial t}\right|^2 - j\hbar\left(\nabla \cdot \frac{\partial \mathbf{Z}}{\partial t}\right)\right] - q\nabla \cdot \mathbf{Z}\right\}\Psi$$

$$\tag{1.65}$$

The advantage of (1.61) and (1.65) regarding to the complicated systems (1.55) and (1.57) is obvious. For instance, it allows to estimate the influence of each equation term of the same origin and to develop a flexible approach to the solution of this equation depending on the energy of a moving particle and outer field. For instance, the first-order approximation of the Schrödinger-Hertz equation (1.66) is derived taking into account that $\hbar = 1.055 \cdot 10^{-34}\ \mathrm{J \cdot s}$, $m = m_e = 9.109 \cdot 10^{-31}\mathrm{kg}$, $q = e = -1.602 \cdot 10^{-19}\ \mathrm{C}$, and $c = 3 \cdot 10^8\ \mathrm{m/s}$, and treating properly the singular integral in (1.64):

$$j\hbar\frac{\partial \Psi}{\partial t} \approx -\frac{\hbar^2}{2m}\nabla^2 \cdot \Psi - q(\nabla \cdot \mathbf{Z})\Psi. \tag{1.66}$$

It is still a nonlinear one taking into account (1.56), (1.59), and (1.64), and it describes the low-energy spin-less charged particles. This equation is in good agreement with the classical and semi-classical studies stating that the influence of dynamic self-induced fields is strong enough only for the fast moving accelerated particles [3]. The zero-order approximation, when the self-consistent electric potential is zero gives the linear Schrödinger equation (1.54) governed by the external electric potential Φ_0. If the solution of this equation is known, then further treatment can be performed using a perturbation method, for instance.

The found currents and Hertz vectors are utilized further to derive the electric and magnetic fields [46]:

$$\mathbf{E} = \nabla \cdot (\nabla \cdot \mathbf{Z}) - \frac{1}{c^2}\frac{\partial^2 \mathbf{Z}}{\partial t^2},\ \mathbf{H} = \frac{1}{c^2}\frac{\partial}{\partial t}(\nabla \times \mathbf{Z}). \tag{1.67}$$

Additionally, these formulas can be used to get the boundary conditions for vector $\mathbf{Z}$.

1.6.1.5 Pauli-Hertz Equation

Electrons have intrinsic magnetic moment called spin $\mathbf{s}$, and the equations for such particles should take into account their orientation. Electrons aligned according to the magnetic field have positive spin $s = 1/2$. Other particles are in the spin-down state $s = -1/2$. Electrons can be in a mixed state called the qubit, which is not considered here. For instance, electronics employs intrinsic magnetic momentum of electrons in the devices based on the giant magnetoresistive effect. Recent promising findings are with the developments of spin-transistors.

In the case of spin particles, the Schrödinger equation is written in its spinor or Pauli form [36], and it is equipped by the potential equations:

$$\begin{aligned}
&j\hbar\frac{\partial\bar{\Psi}}{\partial t} = \left(\hat{H}_0 + \hat{H}_s\right)\bar{\Psi},\\
&\nabla^2\cdot\mathbf{A} - \frac{1}{c^2}\frac{\partial^2\mathbf{A}}{\partial t^2} = -\mu_0\mathbf{J}_e,\\
&\nabla^2\cdot\varphi - \frac{1}{c^2}\frac{\partial^2\varphi}{\partial t^2} = -\frac{\rho}{\varepsilon_0},\\
&\frac{\partial\varphi}{\partial t} + \frac{1}{c^2}\nabla\cdot\mathbf{A} = 0
\end{aligned} \tag{1.68}$$

where

$$\begin{aligned}
&\bar{\Psi} = \Psi_{\uparrow} + \Psi_{\downarrow},\\
&\hat{H}_0 = \frac{1}{2m_e}\begin{pmatrix} (\hat{\mathbf{p}} + e\mathbf{A})^2 & 0 \\ 0 & (\hat{\mathbf{p}} + e\mathbf{A})^2 \end{pmatrix} + e\begin{pmatrix} \varphi & 0 \\ 0 & \varphi \end{pmatrix},\\
&\hat{H}_s = \mu_B\left(\hat{\sigma}_x B_x + \hat{\sigma}_y B_y + \hat{\sigma}_z B_z\right), \mathbf{B} = \nabla\times\mathbf{A}, \hat{\mathbf{p}} = -j\hbar\nabla,\ \mu_B = e\hbar/2m,\\
&\hat{\sigma}_x = \begin{pmatrix} 0 & 1 \\ 1 & 0 \end{pmatrix},\ \hat{\sigma}_y = \begin{pmatrix} 0 & \text{-j} \\ \text{j} & 0 \end{pmatrix},\ \hat{\sigma}_z = \begin{pmatrix} 0 & 0 \\ 1 & \text{-1} \end{pmatrix}.
\end{aligned}$$

In (1.68), the vector potential $\mathbf{A}$ and associated magnetic field $\mathbf{B}$ are composed from the external and self-induced vectors, similarly as the electric potential φ.

Substituting the vectors $\mathbf{A}$, and $\mathbf{B}$, and potential φ by their expressions (1.63) and (1.67) in (1.68), the Pauli-Hertz equation written for electron in the external and self-induced magnetic and electric fields is derived. The Hertz vector requires the electric current taking into account the spinor nature of the probability density of spin particles:

$$\mathbf{J}'_e = e\left(\frac{\hbar}{2jm_e}\left(\bar{\Psi}^{\dagger}\nabla\cdot\bar{\Psi} - \bar{\Psi}\nabla\cdot\bar{\Psi}^{\dagger}\right) - \frac{e}{m_e}\bar{\Psi}^{\dagger}\bar{\Psi}\mathbf{A}\right) \tag{1.69}$$

where the symbol $\dagger$ is for Hermitian conjugation. Some corrections of this formula are available [47]. Then, the Hamiltonian is

$$\hat{H}_0+\hat{H}_s=\frac{1}{2m_e}\begin{pmatrix}\left(\hat{\mathbf{p}}+\frac{e}{c^2}\frac{\partial\mathbf{Z}}{\partial t}\right)^2 & 0\\ 0 & \left(\hat{\mathbf{p}}+\frac{e}{c^2}\frac{\partial\mathbf{Z}}{\partial t}\right)^2\end{pmatrix}-e\begin{pmatrix}\nabla\cdot\mathbf{Z} & 0\\ 0 & \nabla\cdot\mathbf{Z}\end{pmatrix}+$$
$$+\frac{\mu_B\mu_0}{c^2}\frac{\partial}{\partial t}\left\{\sigma_x[\nabla\times\mathbf{Z}]_x+\sigma_y[\nabla\times\mathbf{Z}]_y+\sigma_z[\nabla\times\mathbf{Z}]_z\right\}. \tag{1.70}$$

Comparing the coefficients of the members of this Hamiltonian and leaving only the largest ones, the first-order approximation of the *Pauli-Hertz* equation is derived

$$\begin{aligned} j\hbar\frac{\partial\Psi_\uparrow}{\partial t}&\approx-\frac{\hbar^2}{2m_e}\nabla^2\cdot\Psi_\uparrow-e(\nabla\cdot\mathbf{Z})\Psi_\uparrow,\\ j\hbar\frac{\partial\Psi_\downarrow}{\partial t}&\approx-\frac{\hbar^2}{2m_e}\nabla^2\cdot\Psi_\downarrow-e(\nabla\cdot\mathbf{Z})\Psi_\downarrow,\end{aligned} \tag{1.71}$$

or

$$j\hbar\frac{\partial\bar{\Psi}}{\partial t}\approx-\frac{\hbar^2}{2m_e}\nabla^2\cdot\bar{\Psi}-e\begin{pmatrix}\nabla\cdot\mathbf{Z} & 0\\ 0 & \nabla\cdot\mathbf{Z}\end{pmatrix}\bar{\Psi}. \tag{1.72}$$

Taking into account that the Hertz vector is a functional regarding to the Pauli quantum current $\mathbf{J}'_e$

$$\mathbf{Z}=F\left(\mathbf{J}'_e=e\left(\frac{\hbar}{2jm_e}\left(\bar{\Psi}^\dagger\nabla\cdot\bar{\Psi}-\bar{\Psi}\nabla\cdot\bar{\Psi}^\dagger\right)-\frac{e}{m_e}\bar{\Psi}^\dagger\bar{\Psi}\mathbf{A}\right)\right) \tag{1.73}$$

then the derived equations (1.71) for up- and down spinor components $\Psi_\uparrow$ and $\Psi_\downarrow$ are still coupled, and they are nonlinear if the EM self-influence is considered.

This performed analytical treatment allows reducing the complexity of the used equations for calculation of quantum-EM field phenomena according to the physics-based criteria. They are convenient for numerical or even semi-analytical treatment of quantum effects and nano-devices reducing the necessary computation resources.

1.6.1.6 Dirac-Hertz Equations

Similarly to the Schrödinger-Maxwell and Pauli-Maxwell equations, the Dirac-Maxwell one can be simplified using the Hertz vector. These equations describe relativistic particles in the external and self-induced EM fields [48],[49]. In contrast to the Pauli equation, the Dirac one is given for a pair of spin-having particle/anti-particle, i.e. the probability density wave function $\bar{\Psi}$ has four components. The Maxwell-Dirac equation is written using the vector and scalar potentials:

$$
\begin{aligned}
& j\hbar\frac{\partial\bar{\Psi}}{\partial t} = \sum_{k=1}^{3}\alpha^{k}\left(-j\hbar c\frac{\partial}{\partial x_k} - eA_k\right)\bar{\Psi} + \left(eA_0 + mc\beta\right)\bar{\Psi}, \\
& \nabla^2\cdot\mathbf{A} - \frac{1}{c^2}\frac{\partial^2\mathbf{A}}{\partial t^2} = -\mu_0\mathbf{J}_e, \\
& \nabla^2\cdot\varphi - \frac{1}{c^2}\frac{\partial^2\varphi}{\partial t^2} = -\frac{\rho}{\varepsilon_0}, \\
& \frac{\partial\varphi}{\partial t} + \frac{1}{c^2}\nabla\cdot\mathbf{A} = 0.
\end{aligned}
\tag{1.74}
$$

where A_k is a component of the 4-potential, $\mathbf{A} = \mathbf{x}_0\mathrm{A}_1 + \mathbf{y}_0\mathrm{A}_2 + \mathbf{z}_0\mathrm{A}_3$ and $\varphi=\mathrm{A}_0$ are composed of the external and self-consistent potentials, $\alpha_{k=1,2,3} = \begin{pmatrix} 0 & \sigma_k \\ \sigma_k & 0 \end{pmatrix}$, σ_k is the Pauli matrix, β is the Dirac matrix, $\rho = e\sum_{k=1}^{4}\left|\Psi_k\right|^2$, and $J_{e_k} = ec\bar{\Psi}^{\dagger}\alpha^k\bar{\Psi}$.

Supposing that the components of the 4-vector A_k is expressed through the Hertz vector $\mathbf{Z}$, the system (1.74) is transformed to one nonlinear differential-integral *Dirac-Hertz* equation:

$$
j\hbar\frac{\partial\bar{\Psi}}{\partial t} = \sum_{k=1}^{3}\alpha^{k}\left(-j\hbar c\frac{\partial}{\partial x_k} - eA_k(\mathbf{Z})\right)\bar{\Psi} + \left(eA_0(\mathbf{Z}) + mc\beta\right)\bar{\Psi}. \tag{1.75}
$$

1.6.1.7 Klein-Gordon-Hertz Equation

A particular case of the Maxwell-Dirac equation is the Klein-Gordon-Maxwell equation derived for relativistic spineless charges [50]:

$$
\begin{aligned}
& \left[\left(-j\nabla - e\mathbf{A}\right)^2 - \left(j\frac{\partial}{\partial t} - e\varphi\right)^2 + m_e^2\right]\Psi = 0, \\
& \nabla^2\cdot\mathbf{A} - \frac{1}{c^2}\frac{\partial^2\mathbf{A}}{\partial t^2} = -\mu_0\mathbf{J}_e, \\
& \nabla^2\cdot\varphi - \frac{1}{c^2}\frac{\partial^2\varphi}{\partial t^2} = -\frac{\rho}{\varepsilon_0}, \\
& \frac{\partial\varphi}{\partial t} + \frac{1}{c^2}\nabla\cdot\mathbf{A} = 0.
\end{aligned}
\tag{1.76}
$$

Taking into account that the current $\mathbf{J}_e$ and charge ρ are proportional to the squared particle probability density wave function, the system is nonlinear with respect to this function. Substituting the vector and scalar potentials by their expressions(1.63), the integral-differential *Klein-Gordon-Hertz* (KGHZ) equation is derived

$$\frac{\partial^2 \Psi}{\partial t^2} - \nabla^2 \cdot \Psi + \left\{ -e^2 \left[(\nabla \cdot \mathbf{Z})^2 - \frac{1}{c^4} \left(\frac{\partial \mathbf{Z}}{\partial t} \right)^2 \right] - 2je \left(\nabla \cdot \frac{\partial \mathbf{Z}}{\partial t} \right) + m_e^2 \right\} \Psi = 0. \quad (1.77)$$

Ignoring the terms with small coefficients, we derive an approximated nonlinear KGHZ equation:

$$\nabla^2 \cdot \Psi - \frac{\partial^2 \Psi}{\partial t^2} - \left\{ e^2 (\nabla \cdot \mathbf{Z})^2 + 2je \left(\nabla \cdot \frac{\partial \mathbf{Z}}{\partial t} \right) \right\} \Psi \approx 0. \quad (1.78)$$

Thus, the use of the Hertz vector formalism allows for obtaining very compact differential-integral EM-quantum equations interesting in the modeling and design of prospective quantum nano-electronics. Some simulation aspects of these equations are in the Chapter 9 of this book.

1.6.2 Dipolar Molecules in EM Field[2]

Studying the motion of molecules in the EM field is interesting in many areas. For instance, the obtained knowledge allows for better understanding of effects on the microlevel, and it is for the developments of EM-mechanical components for prospective molecular electronics. The space-time averaging of the studied effects allows for calculating the medium's macro-parameters as the complex permittivity [51]. This knowledge is useful to derive the microwave loss of dielectrics, and, in the future, to understand the molecular mechanisms of microwave heating and chemical reactions. Unfortunately, the basic theory developed by Debye, Cole, Davidson, and many other scientists is not enough to solve contemporary problems arisen in the theory of materials, especially, in the nano-engineered ones.

It is interesting that the theory of particles having the electric $(\mathbf{p}_e)$ and the magnetic $(\mathbf{p}_m)$ moments placed in general EM field has not been developed well to this moment, and discussion on it is still going on. For instance, an analytical formula for the force acting on an electric/magnetic dipole is given in [52], and it has been corrected only recently in [53]. According to the last paper, the force $\mathbf{F}$ on a moving with the velocity $\mathbf{v}$ dipole in EM field is

$$\mathbf{F} = \nabla \cdot (\mathbf{p}_e \cdot \mathbf{E}) + \nabla \cdot (\mathbf{p}_m \cdot \mathbf{B}) + \frac{1}{c} \frac{\partial}{\partial t} (\mathbf{p} \times \mathbf{B}). \quad (1.79)$$

This expression is derived under assumption that the dipole is of the negligible small size in comparison to relative change of the electric and magnetic fields along the dipolar axis. Some restrictions on calculation of motional trajectory using (1.79) can be found in [53]. In the mentioned paper, the relativistic formulas

[2] Written by S.V. Kapranov and G.A. Kouzaev

for the moments $(\mathbf{p}_e)$ and $(\mathbf{p}_m)$ are given to calculate the force in the frame of observation.

Very often, the molecular calculations are with the slowly moving electric dipoles in low-magnitude EM fields, and the force formula is simple

$$\mathbf{F} = \nabla \cdot (\mathbf{p}_e \cdot \mathbf{E}). \tag{1.80}$$

In inhomogeneous AC fields, a very interesting effect occurs called the pondermotive move of charges and dipoles. The trajectory equations obtained using (1.80) are nonlinear, and the dipolar motion is divided into two parts. One of them is the oscillatory movement according to the AC field, and the second one is with translation motion due to the spatial inhomogeneity of the field.

Many results are obtained for the polarizable atoms and molecules [54]-[56]. An equation of pomdermotive motion for them is given, for instance in [56]:

$$M\ddot{R}(t) = -\frac{1}{4m_e\omega^2}\nabla \cdot |\mathbf{E}_0|^2 \tag{1.81}$$

where M and m_e are the masses of atom and electron, correspondingly, given in atomic units, $\ddot{R}(t)$ is the second derivative of the center-of-mass position, ω is the cycling frequency of the electric field envelope $\mathbf{E_0}(R,t,\omega)$. In the cited paper, additionally to the theoretical analysis, the effect of ultrastrong acceleration was discovered for He and Ne atoms in strong short pulse laser fields, and such acceleration with the magnitude as high as 10^{14} was found in the radial direction of the laser light beam. It is supposed that the effects with the pondermotive force are promising for the advanced manipulations of atoms and molecules.

Our own results touch only some aspects of this problem, and they concern the motion of a single dipole molecule placed in the static (DC) and alternating (AC) electric fields. The theory was created using the analogies with the pendulum's one [57], and it includes the linear and nonlinear behavior of a molecule and its stochastization [58].

Even simple molecules have rather complicated spatial structures. The positive charges, associated with the atomic nuclei, are localized in space and glued together by clouds of electrons. The averaged centers of the positive and negative electricity take different coordinates, in general. Atoms have spherical symmetry of their electricity, but they are polarized in external electric field.

In the first-order approximation, the molecules can be represented by dipoles in which positive and negative charges are concentrated at a certain distance from each other (Fig. 1.3a). The electrical charge multiplied by this vector distance is called the dipole moment $\mathbf{p} = q\mathbf{l}_e$. Due to the certain mass of the molecule, it has a moment of inertia denoted here as I.

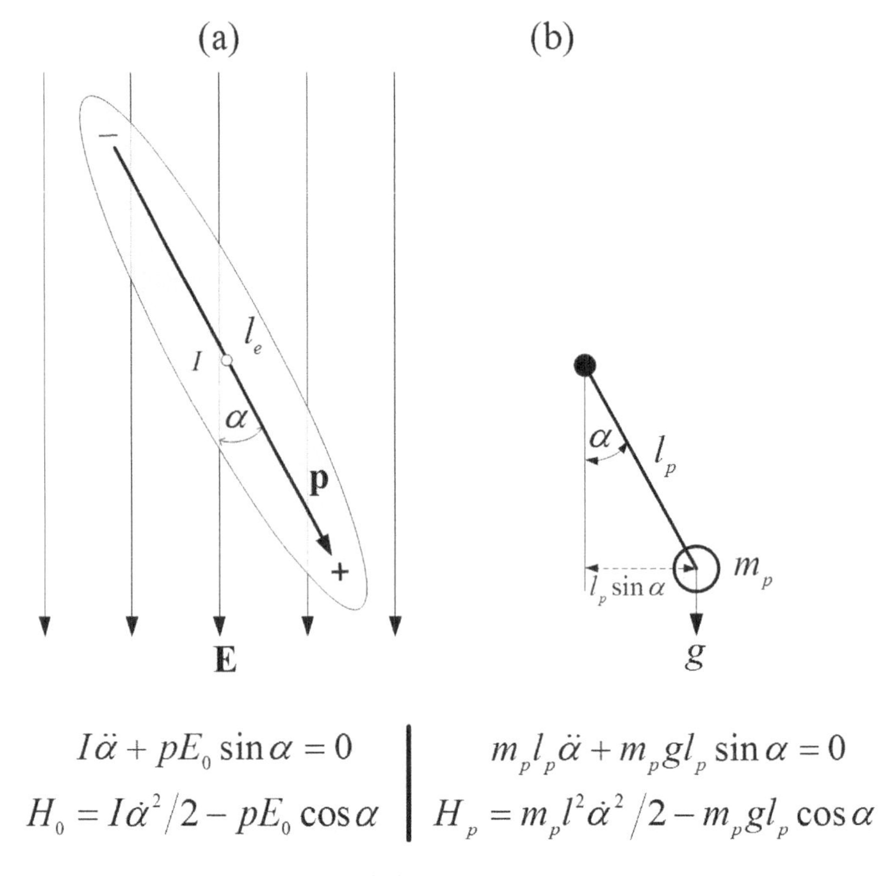

Fig. 1.3 Dipole molecule in the electric field (a) and gravitational pendulum (b)

The electric force applied to the dipole forges it to be turned along the field and around the central point of the dipole, and it occurs not instantaneously due to the inertia of the molecule. The additionally applied alternating electric field $\mathbf{E}(t)$ can cause much more complicated effects, and a theory of them is based on the analogy with the behavior of a pendulum in the gravity field.

The pendulum (Fig. 1.3b) has a fixed axis around which it rotates if a strong external alternating force is exerted. In the case of a small alternating force, the pendulum oscillates close to its lowest position. In general, the pendulum behavior is described by nonlinear differential equations.

For the pendular motion of the dipole, the static electric field plays the role of the gravitational one, and the charge q substitutes the mass of the pendulum bob m_p. The motion equations and the Hamiltonians for both cases are shown in the body of this figure; there is the complete analogy between the dipole and pendulum placed in the electric and gravity fields, correspondingly.

To derive the dipole motion equation, the Hamiltonian approach is used. The Hamiltonian of a moving system is a sum of its kinetic and potential energies. A dipole placed in the static electric field has the following Hamiltonian:

$$H_0 = I\dot{\alpha}^2/2 - pE_0 \cos\alpha \tag{1.82}$$

where I is the dipole's moment of inertia. For convenience, the normalized Hamiltonian is introduced as $\bar{H}_0 = H_0/I$. Following the theory of pendulum [59]-[61], the molecule rotating frequency ω is derived from the Hamiltonian, and it is a nonlinear equation with respect to this parameter:

$$\omega(\bar{H}_0) = \frac{\pi}{2}\omega_0 \begin{cases} 1/F(\pi/2;\kappa), \ \kappa < 1, \\ \kappa/F(\pi/2;1/\kappa), \ \kappa > 1 \end{cases} \tag{1.83}$$

where $\kappa^2 = (\omega_0^2 + \bar{H}_0)/2\omega_0^2$ and F is the elliptic integral of first kind.

When dipolar molecule oscillates close to an equilibrium point with the small amplitudes, its oscillations are linear, and the eigenfrequency is $\omega_0 = \sqrt{pE_0/I}$. When the oscillation amplitude is high, the non-linear motion equation derived from the Hamiltonian is

$$\ddot{\alpha} + \omega_0^2 \sin\alpha = 0. \tag{1.84}$$

Its analytical solution for the highest-amplitude oscillation gives [59]:

$$\begin{aligned} \dot{\alpha} &= \pm 2\omega_0 \cos(0.5\alpha), \\ \alpha &= 4\tan^{-1}\left(e^{\pm\omega_0 t}\right) - \pi. \end{aligned} \tag{1.85}$$

An example of the phase plane $\alpha(\dot{\alpha})$ is shown in Fig. 1.4.

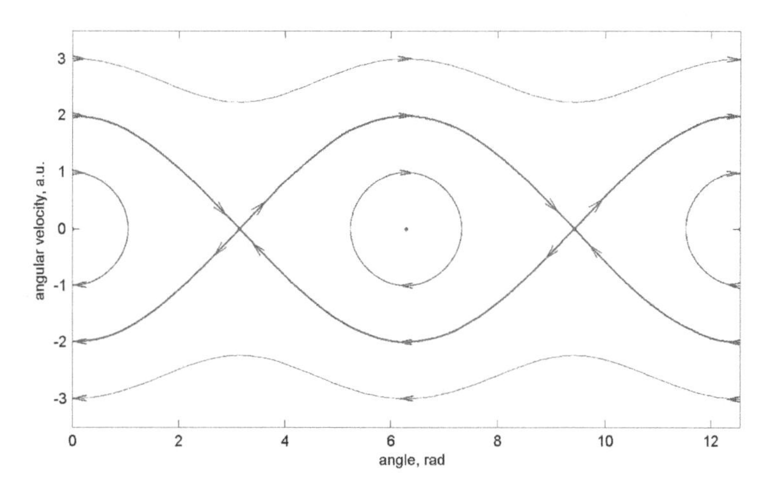

Fig. 1.4 Phase plane of oscillations of a dipole molecule

There is an equilibrium point in the center of the loop, and the molecule is not oscillating there at all, being oriented along the static field $\mathbf{E}_0$. In the areas inside the separatrix loops, which are shown by bold curves, the molecule is oscillating. The separatrices are crossed at the saddle points where the equilibrium is unstable and the dipole is oriented in the direction opposite to the electric field. At the areas outside the separatrices loops, this molecule is rotating around its inertia center. The movement of molecule along the separatrices is aperiodical.

As seen, the parameters of motion are regulated by the external static electric field. To analyze the large amplitude eigen-motion of molecules, the general equation (1.83) should be used.

One of the important effects, which is beyond the scope of the above simple analysis, is the chaotization of molecular movement close to the separatrices under even small outer perturbation by an AC electric field. It means that the molecular angular coordinate and angular velocity vary chaotically, and this effect is shown in Fig. 1.5 as a layer inside which the phase-space points can lie, instead of being on the separatrix.

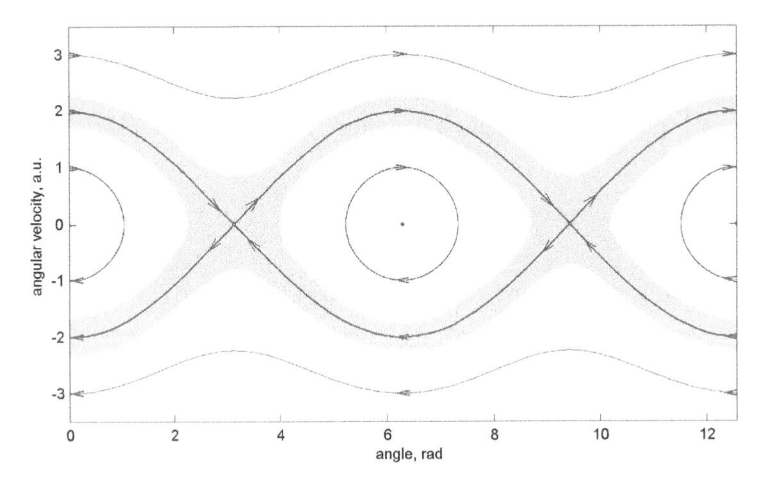

Fig. 1.5 Perturbed molecule's phase portrait. The molecule motion is chaotic in a layer which is close to the separatrix

The parameters of this layer can be varied by the external alternating field, and this is promising for the control of motion of single molecules and their clusters [32]. A theory describing the effect of the controllable chaos can be built on the base of analogy of the gravitational pendulum and the dipole in the electric field.

Following [59], the Hamiltonian for the perturbed problem is written as

$$H = H_0(\alpha, \dot{\alpha}) + \delta \cdot V(\alpha, t) \tag{1.86}$$

where $V(\alpha,t)=V\left(\alpha,t+\dfrac{2\pi}{\nu}\right)$ is the periodical perturbation part, and δ is the perturbation parameter. For the relative cycling frequency $\nu/\omega_0 \ll 1$, the maximum width Δ of the stochastisity layer is

$$\Delta \approx \delta \cdot \frac{\nu}{\omega_0} > \frac{|H-H_s|}{H_s} \tag{1.87}$$

where H_s is the Hamiltonian value calculated exactly on the separatrix of the phase space. The high enough frequency ν decreases the stochastic layer width Δ, and it exponentially approaches zero at $\nu \gg \omega_0$. The same effect occurs if the driving frequency ν tends to zero. In general, the stochasticity width Δ is a multi-extremum function regarding to ratio of ν and ω_0. A more detailed study is performed in [58] where the following interesting effects for a single dipolar molecule are shown. The maximum width $\Delta_{\max}$ of the stochastic layer near a perturbed pendulum separatrix is of the order of δ. Remarkably, this width varies with the perturbation frequency and it has a maximum at frequencies comparable with ω_0. The location of this maximum varies for different types of perturbation fields, and it can be tuned by changing the intensity of the static field applied. Nonlinear resonance gives rise to another interesting effect of bifurcation of the stochastic layer width for equal values of δ.

Example 1.1. Stochastic layer width calculation for a molecule in the TEM linearly polarized wave field [59],[60]. Assume a linearly-polarized EM wave propagating with zero losses in dielectric. Then $k_z''=0$, and the perturbation electric field $\mathbf{E}(z)$ is

$$\mathbf{E}(z)=\mathbf{x_0}e_0\sin(\nu t-k_z'z+\varphi) \tag{1.88}$$

where e_0 is the wave magnitude, and φ is the wave phase. Let a dipole be located at a fixed point with z=const. In this case:

$$\mathbf{E}=\mathbf{x_0}e_0\sin(\nu t+\varphi_0) \tag{1.89}$$

where $\varphi_0=-k_z'z+\varphi$.

When the static electric field $\mathbf{E_0}=\mathbf{x_0}E_0$ collinear to the alternating field $\mathbf{e}$ acts upon a dipole and makes angle α with its direction, the Hamiltonian of the dipole's angular motion is

$$\begin{aligned} H &= I\dot{\alpha}^2/2-\mathbf{p}\cdot(\mathbf{E_0}+\mathbf{E}) = I\dot{\alpha}^2/2 - pE_0\cos\alpha - \\ &- pe_0\cos\alpha\sin(\nu t+\varphi_0). \end{aligned} \tag{1.90}$$

If $\epsilon = e_0/E_0$, the normalized Hamiltonian is

$$\bar{H} = H/I = \dot{\alpha}^2/2 - \frac{pE_0}{I}\cos\alpha - \frac{pe_0}{I}\cos\alpha\sin(\nu t + \varphi_0) = \\ = \dot{\alpha}^2/2 - \omega_0^2\cos\alpha - \epsilon\omega_0^2\cos\alpha\sin(\nu t + \varphi_0). \tag{1.91}$$

Using the standard mapping procedure described by G.M. Zaslavsky [59],[60] and B.V. Chirikov [61], the dimensionless maximum energy width Δ of the stochastic layer formed by the linearly-polarized perturbation of the pendular dipole at the separatrix is derived [58] as

$$\frac{|H - H_s|}{H_s} \le \Delta = 2\epsilon\frac{\nu}{\omega_0}\left|\int_{-\infty}^{\infty}\frac{\sin[4\arctan\exp(\tau)]\sin(\nu\tau/\omega_0 + \varphi_0)}{\cosh\tau}d\tau\right| = \\ = \left|\frac{2\pi\epsilon(\nu/\omega_0)^3}{\sinh(\pi\nu/2\omega_0)}\sin\varphi_0\right|$$

where φ_0 is the perturbation phase with respect to the equilibrium point of the unperturbed dipole. The plot of the absolute amplitude value of the stochastic layer width is shown in Fig. 1.6 against the frequency ratio ν/ω_0 and given for $\epsilon = 1$. The maximum of the stochastic layer lies at $\nu/\omega_0 \approx 1.90$. It follows that the width is controllable by both frequencies and ratio of magnitudes of the alternating e_0 and static E_0 fields. More data on the controllable chaos near sepatrices of nonlinear oscillations of a dipolar molecule is in the cited work [58]. Although, the considered example is on an idealized case when the dipole is fixed on the plane at its "axis", and no thermal rotations and translational motion have been considered, it shows the difficulties to control such tiny objects like single molecules. More information on molecular theory, including quantum aspects of molecular vibrations and interaction with EM field, can be found for instance, in [62]-[64].

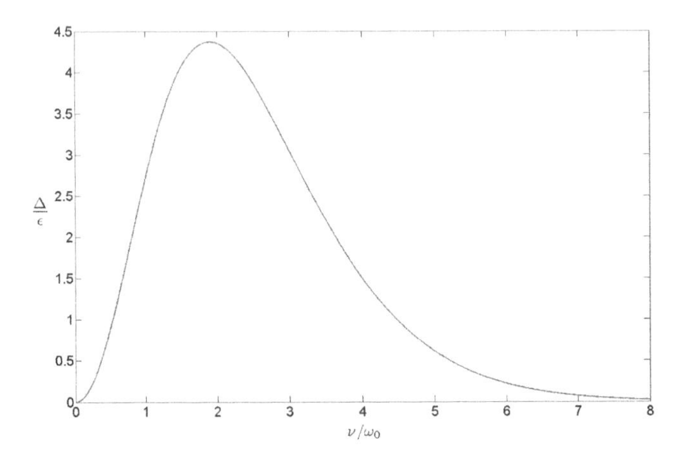

Fig. 1.6 Maximum stochastic layer width as a function of the perturbing and fundamental frequencies ratio ν/ω_0. The perturbation is caused by the linearly-polarized EM field

Although only a single dipole is considered here, such effects of chaotization can appear in a more realistic system of coupled molecular oscillators and this approach may allow for controlling transformation of the microwave energy into heat by proper tuning the DC and AC field intensities and frequency.

1.7 Boundary Value Problems of Electromagnetism

1.7.1 Boundary Conditions for the Electric and Magnetic Fields

The above-considered plane TEM wave (Section 1.4.3) propagates in open space. In practice, the EM waves are often guided by a stratified medium, channelized by guides, or confined by cavities. Due to the multiple reflections from the boundaries and diffraction of the elementary TEM waves, a very complicated spatio-time structure of the EM field is arising. To calculate it, the behavior of the field on the boundaries should be ascertained.

The fields on the interface of two domains can have abrupt behavior and the derivatives are singular there. Then the differential equations should be equipped by the *boundary conditions* for the field components. These conditions are obtained from the equations for the tangential and normal-to-the-boundary field components and the Maxwell equations in the integral form are used [3],[4].

Let us consider a plane boundary between two domains with the different material parameters ε_1, μ_1 and ε_2, μ_2 (Fig. 1.7).

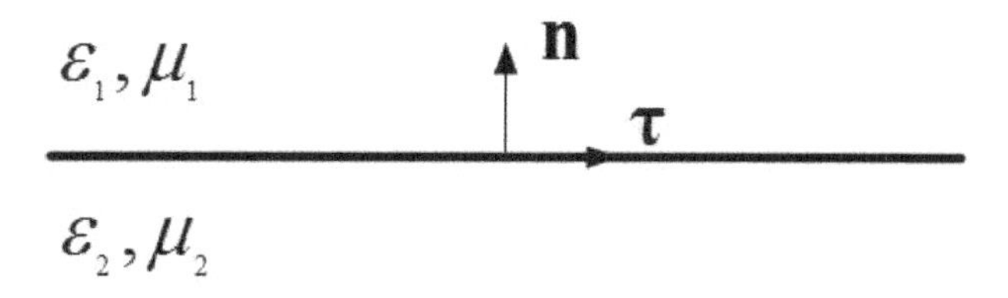

Fig. 1.7 Boundary between two domains

On this boundary, two unit vectors are defined. One of them is the normal-to-the boundary vector **n**; the second one is the tangential-to-the-boundary vector $\boldsymbol{\tau}$.

Using the Maxwell equations in the integral form, the following set of expressions for the tangential and normal-to-the-boundary electric field components are derived (see [3] for details)

$$\begin{aligned} \mathbf{E}_{1_\tau} - \mathbf{E}_{2_\tau} &= 0, \\ D_{1_n} - D_{2_n} &= \rho_s \end{aligned} \qquad (1.92)$$

where $\rho_s, \left[\mathrm{C/m^2}\right]$ is the surface charge density if the boundary is charged. From these equations, it follows that on the boundary of two different magnetodielectrics the tangential components of the electric field intensity are continuous. The normal-to-the-boundary components of the electrical flux density $\mathbf{D}_{1,2}$ are different from each other. The step between these components is caused by the surface charge. In general case, the normal components of the electric field intensity $\mathbf{E}_{1,2}$ are different from each other, too:

$$\frac{E_{1_n}}{E_{2_n}} = \frac{\varepsilon_2}{\varepsilon_1}. \tag{1.93}$$

The magnetic field components can show abrupt behavior on the boundary:

$$\begin{aligned} \mathbf{H}_{1_\tau} - \mathbf{H}_{2_\tau} &= \mathbf{j}_s, \\ B_{1_n} - B_{2_n} &= 0 \end{aligned} \tag{1.94}$$

where $\mathbf{j}_s, \left[\mathrm{A/m^2}\right]$ is the electric surface current density. If the boundary surface is free of currents, then the tangential magnetic field intensity components are continuous, and

$$\mathbf{H}_{1_\tau} - \mathbf{H}_{2_\tau} = 0. \tag{1.95}$$

In general case, the normal components of these vectors are different from each other:

$$\frac{H_{1_n}}{H_{2_n}} = \frac{\mu_2}{\mu_1}. \tag{1.96}$$

In electromagnetics, the ideal mediums are used frequently. One of them is the ideal conductor with $\sigma = \infty$. The boundary conditions on the surface between it and a dielectric are

$$\begin{aligned} \mathbf{E}_\tau &= 0,\ D_n = \rho_s, \\ \mathbf{H}_\tau &= \mathbf{j}_s,\ B_n = 0. \end{aligned} \tag{1.97}$$

From these equations, it follows that the electric field-force lines are normal to the surface of ideal conductor. The magnetic ones are tangential to it.

1.7.2 Surface Impedance Boundary Conditions

High conductivity of metals at microwave frequencies allows for using the approximate boundary condition introduced by M.A. Leontovich at the end of the 30s of the last Century [65]. Due to increased conductivity of metals, inside them, the normal-to-the-surface derivatives of the EM field components are larger than those in the tangential directions. Then, as it is found from the Maxwell equations, the tangential components of the electric $\mathbf{E}_\tau^{(\mathrm{m})}$ and magnetic $\mathbf{H}_\tau^{(\mathrm{m})}$ fields inside the metal are coupled to each at the surface as

$$\mathbf{E}_\tau^{(\mathrm{m})} + \eta_\mathrm{m}\left[\mathbf{H}_\tau^{(\mathrm{m})} \times \mathbf{n}\right] = 0 \tag{1.98}$$

where $\eta_m = \sqrt{\mu_m/\varepsilon_m}$ is the surface impedance of a metal, μ_m and ε_m are its permeability and permittivity, respectively, and $\mathbf{n}$ is the unit vector which is normal to the metal surface. Taking into account that the tangential components of the fields in the air and in the conductor are equal to each other, and then, in the air, the similar boundary condition should be fulfilled at the boundary:

$$\mathbf{E}_\tau^{(\mathrm{air})} + \eta_\mathrm{m}\left[\mathbf{H}_\tau^{(\mathrm{air})} \times \mathbf{n}\right] = 0. \tag{1.99}$$

It allows for considering only the fields out of the metal, and it makes the EM problems easier to be solved. Very often, the perturbation method is applied when the fields are obtained initially using the ideal boundary conditions (1.97) on the surface of metal. The found fields are substituted into stationary functionals which take into account the real conductivity through the Leontovich boundary condition, and the loss of waves or oscillating modes can be found. The boundary condition (1.99) was introduced for TEM waves, initially, but it can be used for the loss calculation of more complicated waves under certain limitations.

For instance, in [66], the applicability of the Leontovich boundary condition is considered for the cylinder and spherical lossy bodies. It is shown that this condition is valid if

$$\mathrm{Im}(N) \geq \frac{2.3}{k_0 a} \tag{1.100}$$

where N is the refractive index, a is the radius of a cylinder or sphere, and $k_0 = \omega/c$.

The Leontovich boundary condition is generalized for the surfaces, which can be described by an impedance of an arbitrary nature. Among them are, for instance, the surfaces of periodically perturbed metals and dielectrics [67],[68] and thin dielectric sheets [69]. The improved accuracy Leontovich's formulas are obtained in [70] for the boundary of air and a high-permittivity dielectric. In general, the medium can be anisotropic, and on the boundaries of such high-permeability or high-permittivity magnetodielectrics, the anisotropic Leontovich boundary condition is written, and it used for calculation of the ferrite thin-film components [71],[72].

In [72], it is shown that the Leontovich's boundary condition can be generalized further. In general, the tangential fields $\mathbf{E}_\tau$ and $\mathbf{H}_\tau$ are connected to each other through an integro-differential operator $\hat{Z}$:

$$\mathbf{E}_\tau = \hat{Z}\mathbf{H}_\tau. \tag{1.101}$$

This equation can be transformed into a matrix one if the fields are represented as the series of eigenfunctions.

Interesting surface boundary conditions are introduced by M.I. Kontorovich in 1939 [73] for the grid structures of different geometry and parameters, which are used now for calculation of grid antennas, photonic bandgap structures, fenced waveguides, etc. Additional information on the surface boundary conditions can be found in [74], for instance.

1.7.3 Boundary Value Problems

The EM phenomena are described by partial differential equations. They have general solutions which are independent on the boundaries and initial conditions. The solutions in the finite size domains are obtained from the boundary value problems, which are the differential equations, domain geometry, and the restraints given on the domain boundary.

An example of a boundary value problem is shown in Fig. 1.8 where the EM field is calculated inside a metalized cavity V of a known geometry filled by a magnetodielectric $(\tilde{\mu}, \tilde{\varepsilon})$ and excited by a source current with its volume density $\mathbf{J}$.

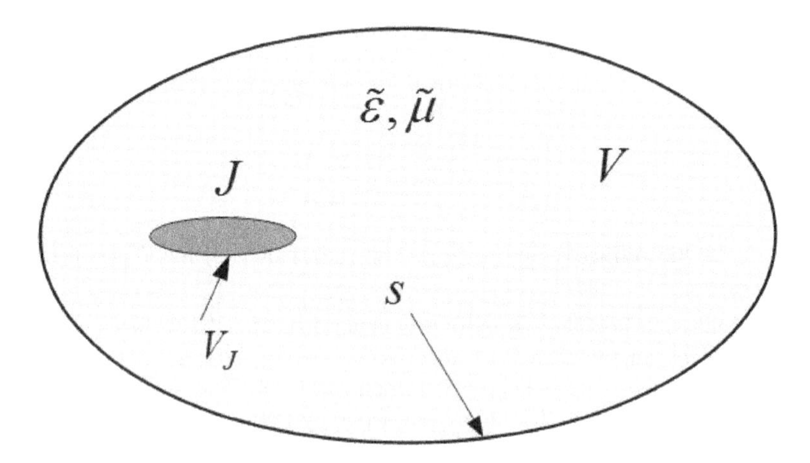

Fig. 1.8 Geometry for a boundary value problem

Assuming ideal cavity shielding, the boundary condition can be written for the surface electric field as $\mathbf{E}_\tau(\mathbf{r} \in S) = 0$. Besides, the cavity field should satisfy a certain boundary condition on the surface of the current volume V_J. Defining the Helmholtz equation for the electric field, the boundary value problem is written as

$$\begin{aligned}
&\Delta \cdot \mathbf{E}(\mathbf{r}) + \tilde{k}^2 \mathbf{E}(\mathbf{r}) = 0, \ |\mathbf{r}| \notin V_J, \\
&\Delta \cdot \mathbf{E}(\mathbf{r}) + \tilde{k}^2 \mathbf{E}(\mathbf{r}) = \frac{1}{\tilde{\varepsilon}} \nabla \cdot \rho(\mathbf{r}) + j\omega\tilde{\mu} J(\mathbf{r}), |\mathbf{r}| \in V_J, \\
&\mathbf{E}_\tau(\mathbf{r}) = 0, \ |\mathbf{r}| \in S, \ \mathbf{E}_\tau^V(\mathbf{r}) - \mathbf{E}_\tau^{V_J}(\mathbf{r}) = 0, \ |\mathbf{r}| \in S_J.
\end{aligned} \tag{1.102}$$

In some cases, the problem is solved analytically, and solution for the electric field depends on the geometry of the volume, its material filling, and the boundary condition. The magnetic field is derived using the Maxwell equations and the obtained electric field. Similarly to the electric boundary value problem, it can be formulated for the magnetic field.

Unfortunately, most domain geometries do not allow for analytical solutions, and the numerical methods are used. The unknown fields are derived approximately, and the criterion of accuracy of the solutions is needed. Very often, it is the *energy conservation law* written for the calculated field.

1.7.4 Energy Conservation Law

The electric and magnetic fields have certain energy of the potential and the dynamic types. For example, the static electric field has only potential energy, and the field's work on moving a charge along a closed trajectory is zero. The electric field induced by the time-varying magnetic field can have non-potential character, and the charge increases its velocity moving even on a closed trajectory.

At each point of space where the fields are defined, the energy densities of the electric w_e and magnetic w_m fields are introduced

$$w_e = \frac{\mathbf{ED}}{2}, \left[\mathrm{J/m^3}\right], \; w_m = \frac{\mathbf{HB}}{2}, \left[\mathrm{J/m^3}\right]. \tag{1.103}$$

Energy stored in a certain volume V is derived by integration of density functions (1.103).

The EM field is generated by a source current concentrated in a certain part of the space. At each point of the source area, the power density can be written as

$$p^{(\mathrm{source})} = \mathbf{J}^{(\mathrm{source})}\mathbf{E}, \left[\mathrm{J/m^3 s}\right]. \tag{1.104}$$

The energy conservation law answers the question on which phenomena the source power is spent. For the arbitrary time-varying fields, it is written for each moment of time. The time-harmonic fields are described by the energy-conservation law defined for the averaged-on-the-period quantities of the energy and power.

Consider an arbitrary domain of space limited by a surface s. The power of the source is spent on pumping the energy into the EM field and on radiation to the outer space through this boundary s. Some amount of energy can be spent on heating of the medium due to the Ohmic loss:

$$\oint_s [\mathbf{E}\times\mathbf{H}]d\mathbf{s} = -\int_V \left(\mathbf{H}\frac{\partial \mathbf{B}}{\partial t} + \mathbf{E}\frac{\partial \mathbf{D}}{\partial t}\right) dv - \int_V \sigma \mathbf{E}^2 dv - \int_V \mathbf{J}^{(\mathrm{source})}\mathbf{E} dv \tag{1.105}$$

where the integral on left of the expression is the power radiated through the boundary s. The first integral in the right part of it is the power pumped into the energy of the EM field at each moment of time. A part of the source power $P^{(\text{source})} = \int_V \mathbf{J}^{(\text{source})}\mathbf{E}dv < 0$ is spent on the heating due the ohmic loss, for example, and it is taken into account by the second term $\int_V \sigma\mathbf{E}^2 dv$ of the right part of (1.105).

At each point of the considered domain, the differential energy conservation law is valid

$$\nabla\cdot[\mathbf{E}\times\mathbf{H}] = -\left(\mathbf{H}\frac{\partial\mathbf{B}}{\partial t} + \mathbf{E}\frac{\partial\mathbf{D}}{\partial t}\right) - \sigma\mathbf{E}^2 - \mathbf{J}^{(\text{source})}\mathbf{E}. \tag{1.106}$$

The quantity $\mathbf{\Pi} = [\mathbf{E}\times\mathbf{H}]$ is the Poynting vector and it defines the radiation density at a certain point of space.

For time-harmonic fields, the averaged-on-the-period quantities are introduced

$$\begin{aligned}
&\langle\mathbf{E}^2\rangle = \frac{1}{2}\mathbf{E}\mathbf{E}^*,\ \langle\mathbf{H}^2\rangle = \frac{1}{2}\mathbf{H}\mathbf{H}^*,\\
&\langle\mathbf{J}\mathbf{E}\rangle = \frac{1}{2}\mathrm{Re}(\mathbf{J}\mathbf{E}^*) = \frac{1}{2}\mathrm{Re}(\mathbf{J}^*\mathbf{E}),\\
&\langle[\mathbf{E}\times\mathbf{H}]\rangle = \frac{1}{2}\mathrm{Re}([\mathbf{E}\times\mathbf{H}^*]) = \frac{1}{2}\mathrm{Re}([\mathbf{E}^*\times\mathbf{H}]).
\end{aligned} \tag{1.107}$$

In this case, the energy conservation law is written regarding to the real and imaginary parts of the power. The first of them describes the energy conservation law for the irreversible processes, i.e. how the source energy is spent for the loss and radiation:

$$\mathrm{Re}\oint_s \mathbf{\Pi}d\mathbf{s} = -\frac{\omega}{2}\int_V(\varepsilon''\mathbf{E}^*\mathbf{E} + \mu''\mathbf{H}\mathbf{H}^*)dv - \mathrm{Re}\,P^{(\text{source})}. \tag{1.108}$$

Beside the time-irreversible processes, the periodical phenomena occur in the EM systems. Roughly speaking, any of them has certain capacitance and inductance, and the energy can be pumped to these reactivities or taken back to the source. For this process, the energy conservation law must be fulfilled for the imaginary parts of the energy and power:

$$\mathrm{Im}\oint_s \mathbf{\Pi}d\mathbf{s} = \frac{\omega}{2}\int_V(\varepsilon'\mathbf{E}^*\mathbf{E} - \mu'\mathbf{H}\mathbf{H}^*)dv - \mathrm{Im}\,P^{(\text{source})}. \tag{1.109}$$

The differential forms of the energy conservation law for both parts of the complex power are

$$\nabla\cdot(\mathrm{Re}\,\mathbf{\Pi}) = -\frac{\omega}{2}\left(\varepsilon''\mathbf{E}^{*}\mathbf{E} + \mu''\mathbf{H}\mathbf{H}^{*}\right) - \mathrm{Re}\,p^{(\mathrm{source})},$$
$$\nabla\cdot(\mathrm{Im}\,\mathbf{\Pi}) = \frac{\omega}{2}\left(\varepsilon'\mathbf{E}^{*}\mathbf{E} - \mu'\mathbf{H}\mathbf{H}^{*}\right) - \mathrm{Im}\,p^{(\mathrm{source})}. \tag{1.110}$$

Any analytical or numerical models and experimental results must satisfy the energy conservation law with maximum accuracy.

1.7.5 Reflection of TEM Waves from the Boundary of Two Magnetodielectrics

The above-considered plane TEM waves propagate in space without any boundary. Such waves are called the free or regular ones. Very often, the EM waves are reflected from different objects or/and penetrate them. The reflected and transmitted fields can have complicated space-time structures, and they depend on the medium and the boundary.

Initially, to understand this phenomenon, the reflection and transmission effects are studied for the plane TEM waves and isotropic dielectric medium, only. Considering the boundary is infinitely wide, the reflection is described on the base of the ray theory (Section 1.4.4), and some elements of it is given below.

We assume that in both domains, the Helmholtz equations have been solved, and we know the fields and the propagation constants of the incident, reflected, and transmitted TEM waves, excepting the amplitudes of the last ones, which should be calculated, i.e. the general solutions of the Helmholtz equation are known in both domains.

The incident wave can be of an arbitrary polarization, but only two of them are considered here. The first one has the electric field vector oriented normally to the incident plane, and the wave is of the perpendicular polarization (Fig. 1.9). Another case is the wave which electric field is parallel to the incident plane (Fig. 1.10). Consider both of them following [4],[6],[12],[75].

1.7.5.1 Perpendicular Polarization of the Incident Wave

This case is shown in Fig. 1.9 where the electric field of the incident wave is normally oriented towards the picture plane.

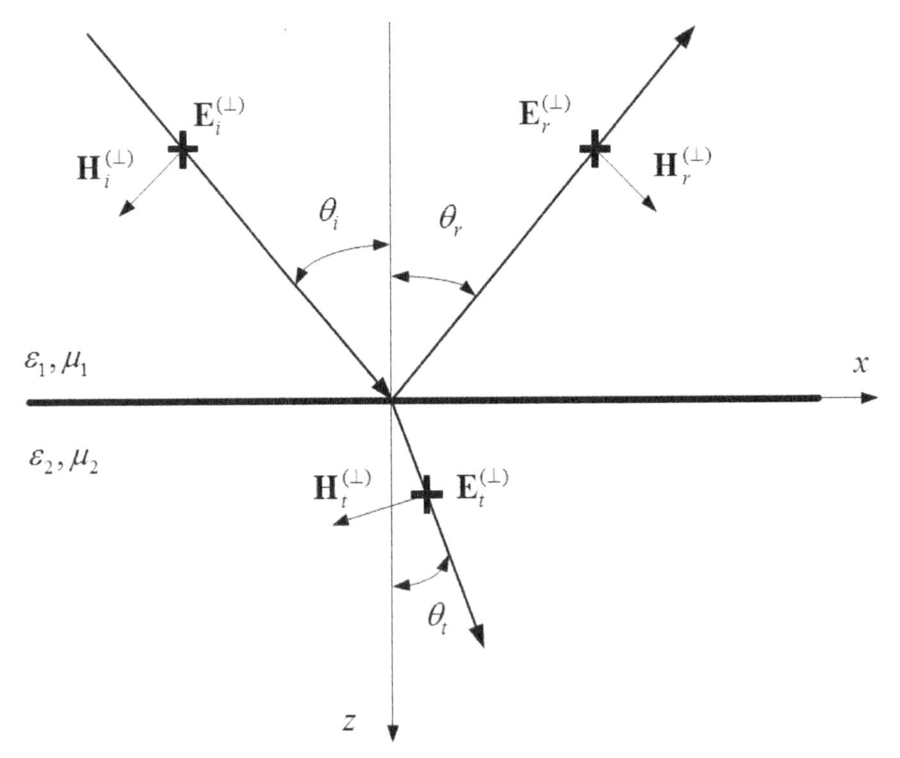

Fig. 1.9 Reflection of a perpendicularly-polarized TEM wave from the boundary of two dielectrics

The incident electric $\mathbf{E}_i^{(\perp)}$ and magnetic $\mathbf{H}_i^{(\perp)}$ fields are

$$\mathbf{E}_i^{(\perp)} = E_i^{(\perp)} \mathbf{y}_0 e^{-jk_1(z\cos\theta_i + x\sin\theta_i)}, \tag{1.111}$$

$$\mathbf{H}_i^{(\perp)} = \frac{E_i^{(\perp)}}{W_1}(-\mathbf{x}_0 \cos\theta_i + \mathbf{z}_0 \sin\theta_i) e^{-jk_1(z\cos\theta_i + x\sin\theta_i)} \tag{1.112}$$

where $E_i^{(\perp)}$ is the incident wave amplitude, $k_1 = \omega\sqrt{\varepsilon_1\mu_1}$, $W_1 = \sqrt{\frac{\mu_1}{\varepsilon_1}}$, and θ_i is the incident angle.

The incident wave is reflected only partly from the surface, and the fields $\mathbf{E}_r^{(\perp)}$, and $\mathbf{H}_r^{(\perp)}$ of the reflected wave which has the same polarization with the incident one are

$$\mathbf{E}_r^{(\perp)} = \Gamma_\perp E_i^{(\perp)} \mathbf{y}_0 e^{-jk_1(-z\cos\theta_r + x\sin\theta_r)}, \tag{1.113}$$

$$\mathbf{H}_r^{(\perp)} = \frac{\Gamma_\perp E_i^{(\perp)}}{W_1}(\mathbf{x}_0 \cos\theta_r + \mathbf{z}_0 \sin\theta_r) e^{-jk_1(-z\cos\theta_r + x\sin\theta_r)} \tag{1.114}$$

where $\Gamma_{\perp}$ is the reflection coefficient, and θ_r is the reflection angle (Fig. 1.9).

A part of the power of the incident wave penetrates the surface, and it is transmitted into the second dielectric with the same polarization as the incident one. The field components $\mathbf{E}_t^{(\perp)}$ and $\mathbf{H}_t^{(\perp)}$ of this wave are

$$\mathbf{E}_t^{(\perp)} = T_{\perp} E_i^{(\perp)} \mathbf{y}_0 e^{-jk_2(z\cos\theta_t + x\sin\theta_t)}, \tag{1.115}$$

$$\mathbf{H}_t^{(\perp)} = \frac{T_{\perp} E_i^{(\perp)}}{W_2} \left(-\mathbf{x}_0 \cos\theta_t + \mathbf{z}_0 \sin\theta_t\right) e^{-jk_2(z\cos\theta_t + x\sin\theta_t)} \tag{1.116}$$

where $T_{\perp}$ is the transmission coefficient, $k_2 = \omega\sqrt{\varepsilon_2\mu_2}$, $W_2 = \sqrt{\frac{\mu_2}{\varepsilon_2}}$, and θ_t is the transmission angle.

All waves are coupled to each other on the boundary, and the tangential components of the electric and magnetic fields should be matched at $z = 0$:

$$\begin{aligned} E_{i_y}^{(\perp)}(z=0) + E_{r_y}^{(\perp)}(z=0) &= E_{t_y}^{(\perp)}(z=0), \\ H_{i_x}^{(\perp)}(z=0) + H_{r_x}^{(\perp)}(z=0) &= H_{t_x}^{(\perp)}(z=0). \end{aligned} \tag{1.117}$$

It gives the equations to obtain the unknown coefficients $\Gamma_{\perp}$ and $T_{\perp}$:

$$\begin{aligned} &E_i^{(\perp)} e^{-jk_1(x\sin\theta_i)} + \Gamma_{\perp} E_i^{(\perp)} e^{-jk_1(x\sin\theta_r)} = T_{\perp} E_i^{(\perp)} e^{-jk_2(x\sin\theta_t)}, \\ &\frac{E_i^{(\perp)}}{W_1}\left(-\cos\theta_i\right) e^{-jk_1(x\sin\theta_i)} + \frac{\Gamma_{\perp} E_i^{(\perp)}}{W_1}\left(\cos\theta_r\right) e^{-jk_1(x\sin\theta_r)} = \\ &= \frac{T_{\perp} E_i^{(\perp)}}{W_2}\left(-\cos\theta_t\right) e^{-jk_2(x\sin\theta_t)}. \end{aligned} \tag{1.118}$$

The boundary conditions (1.118) are satisfied if the phase coefficients are matched to each other:

$$e^{-jk_1(x\sin\theta_i)} = e^{-jk_1(x\sin\theta_r)} = e^{-jk_2(x\sin\theta_t)}. \tag{1.119}$$

It is valid if

$$\begin{aligned} \theta_i &= \theta_r, \\ \sin\theta_t &= \frac{k_1}{k_2}\sin\theta_i. \end{aligned} \tag{1.120}$$

This relationship (1.120) is the Snell's law defined for the wave angles.

Taking into account (1.120), the boundary conditions (1.118) give the expressions (1.121) for the reflection and transmission coefficients as the result of the solution of the non-homogeneous system of linear algebraic equations:

$$\Gamma_{\perp} = \frac{W_2 \cos\theta_i - W_1 \cos\theta_t}{W_2 \cos\theta_i + W_1 \cos\theta_t}, \ T_{\perp} = \frac{2W_2 \cos\theta_i}{W_2 \cos\theta_i + W_1 \cos\theta_t}. \tag{1.121}$$

The formulas (1.120) and (1.121) allow to study all phenomena and calculate the field in both mediums. The considered procedure of the solution of this boundary value problem is typical, and it is called the *field matching method*.

1.7.5.2 Parallel Polarization of the Incident Wave

The ray picture is shown in Fig. 1.10 where the electric fields of all waves are parallel to the picture plane.

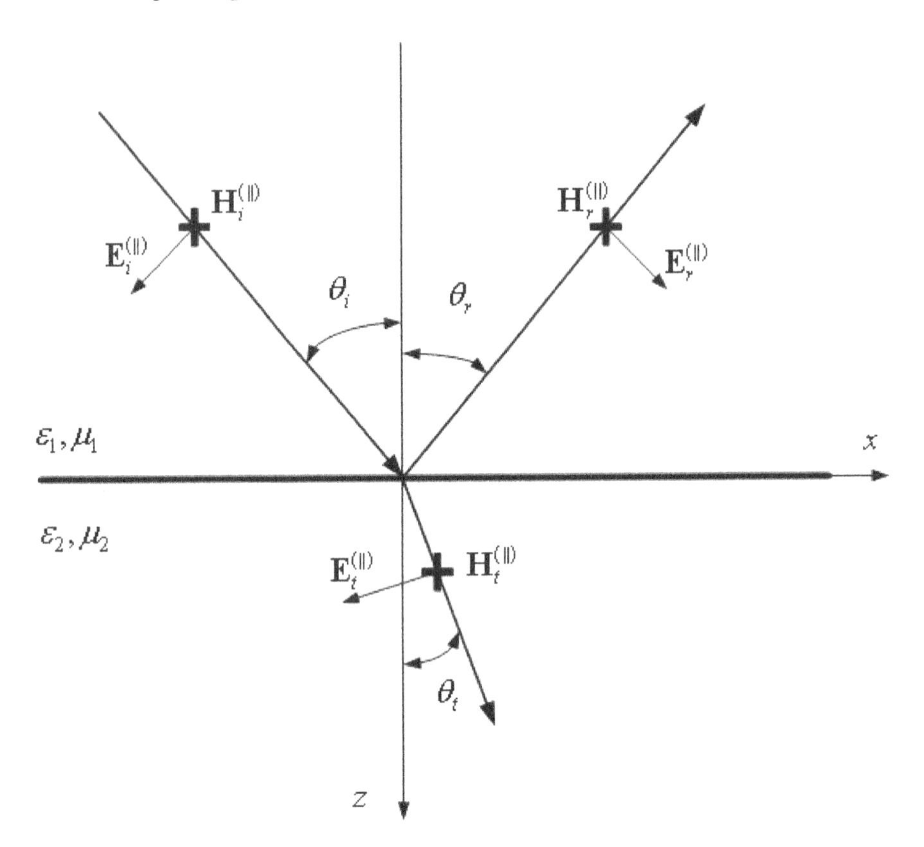

Fig. 1.10 Reflection of a parallelly-polarized TEM wave from the boundary of two dielectrics

In this case, the incident $\left(\mathbf{E}_i^{(\|)}, \mathbf{H}_i^{(\|)}\right)$, reflected $\left(\mathbf{E}_r^{(\|)}, \mathbf{H}_r^{(\|)}\right)$, and transmitted $\left(\mathbf{E}_t^{(\|)}, \mathbf{H}_t^{(\|)}\right)$ fields are

$$\mathbf{E}_i^{(\|)} = E_i^{(\|)}\left(\mathbf{x}_0 \cos\theta_i - \mathbf{z}_0 \sin\theta_i\right) e^{-jk_1(z\cos\theta_i + x\sin\theta_i)},$$
$$\mathbf{H}_i^{(\|)} = \frac{E_i^{(\|)}}{\dot{W}_1} \mathbf{y}_0 e^{-jk_1(z\cos\theta_i + x\sin\theta_i)}, \tag{1.122}$$

$$\mathbf{E}_r^{(\|)} = \Gamma_\| E_i^{(\|)}\left(\mathbf{x}_0 \cos\theta_r + \mathbf{z}_0 \sin\theta_r\right) e^{-jk_1(-z\cos\theta_r + x\sin\theta_r)}.$$
$$\mathbf{H}_r^{(\|)} = -\frac{\Gamma_\| E_i^{(\|)}}{W_1} \mathbf{y}_0 e^{-jk_1(-z\cos\theta_r + x\sin\theta_r)}, \tag{1.123}$$

$$\mathbf{E}_t^{(\|)} = T_\| E_i^{(\|)}\left(\mathbf{x}_0 \cos\theta_t - \mathbf{z}_0 \sin\theta_t\right) e^{-jk_2(z\cos\theta_t + x\sin\theta_t)},$$
$$\mathbf{H}_t^{(\|)} = \frac{T_\| E_i^{(\|)}}{W_2} \mathbf{y}_0 e^{-jk_2(z\cos\theta_t + x\sin\theta_t)} \tag{1.124}$$

where $\Gamma_\|$ and $T_\|$ are the reflection and transmission coefficients, respectively. The field matching at the boundary $z = 0$, similarly to the previous case, gives the following relations for the angles and coefficients under the treatment:

$$\theta_i = \theta_r,$$
$$\sin\theta_t = \frac{k_1}{k_2}\sin\theta_i \tag{1.125}$$

and

$$\Gamma_\| = \frac{W_2\cos\theta_t - W_1\cos\theta_i}{W_2\cos\theta_t + W_1\cos\theta_i}, \quad T_\| = \frac{2W_2\cos\theta_i}{W_2\cos\theta_t + W_1\cos\theta_i}. \tag{1.126}$$

Comparing (1.126) and (1.121) we find that these equations are similar but not identical due to the different polarizations of waves.

1.7.5.3 Wave Reflection from Perfect Conductors. Guiding Effect

The perfect conductors reflect completely the waves of any polarization, and the reflection and transmission coefficients are independent on the incident angle:

$$\Gamma_\perp = -1;\ T_\perp = 0;$$
$$\Gamma_\| = -1;\ T_\| = 0. \tag{1.127}$$

It means that the reflected wave has the opposite phase regarding to the incident one, and the interference of these waves gives the standing wave in the direction that is normal to the conductor surface.

The interfered field is composed of the incident and reflected waves and it is moving along the x-direction (Fig. 1.11) if the incident angle is different from zero. In the case of parallel polarization, it is called the *transverse magnetic* (TM) wave because it has only the electric field component $E_x^{\|}$ along the corresponding propagation direction $0 \to x$.

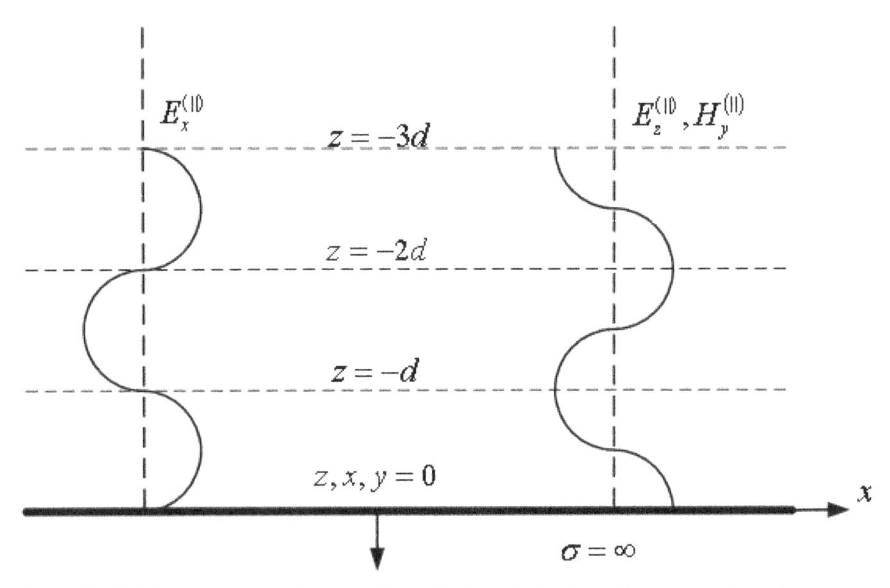

Fig. 1.11 Field components of a sum of parallelly-polarized incident and reflected waves

The normally-polarized wave reflection from the ideal metal surface gives similar field picture for other components (Fig. 1.12)

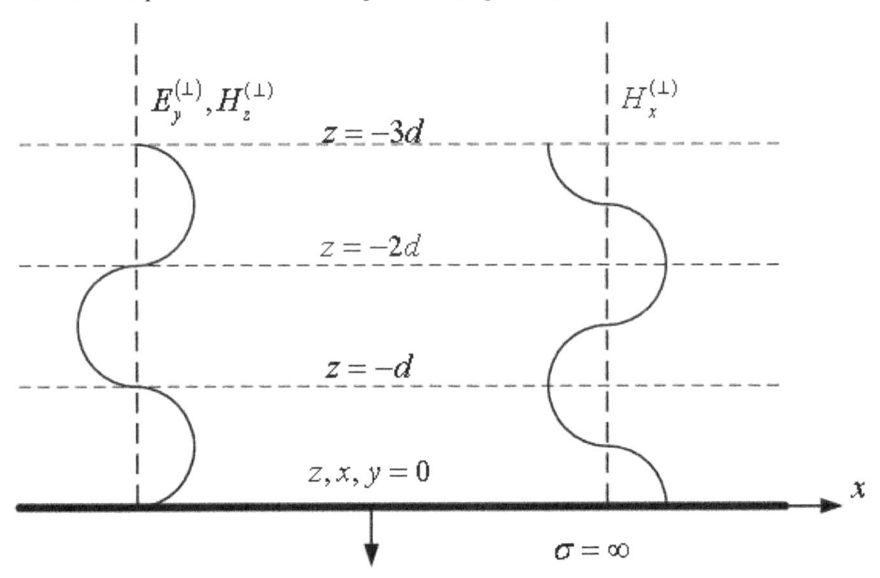

Fig. 1.12 Field components of a sum of normally-polarized incident and reflected waves

Due to the different orientation of the field compared to the case of parallel polarization, the interfered wave has the longitudinal magnetic field component $H_x^{(\perp)}$, and it is called the *transverse electric* (TE) wave. Similarly to the previous case, it moves along the x-axis if the incident angle is different from zero.

It is seen that the nodes of the electric field $(\mathbf{E}=0)$ have fixed coordinates $z_n = (n\pi/k_1 \sin\theta_i) = -nd$; $n = 0,1,2,...,\infty$ (Figs. 1.11 and 1.12). At each *n-th* distance, a new ideally conducting surface can be placed, and Fig. 1.13 shows it for the field of parallel polarization. The partial TEM waves are reflected from both surfaces *z=0* and $z = z_n$, and they form a field considered as a joined process or mode. It propagates between these two parallel plates along the *x*-axis. This structure is called the *parallel-plate waveguide*.

Any mode is described by its modal propagation constant γ_x:

$$\gamma_x = \sqrt{k_1^2 - (\pi n/d)^2}. \qquad (1.128)$$

It allows for calculating the modal velocity along the *x*-axis, its wavelength, and the modal impedance, which is different from the wave one $W = \sqrt{\mu/\varepsilon}$, in general. The discrete quantity *n* means the number of field variations along the *z-axis* (Fig. 1.13), and it is called the *wave number.*

The lowest or the fundamental mode with *n=0* is formed by a single TEM wave normally incident to both plates but propagating along the *x*-axis due to its oriented excitation (Fig. 1.13a). Its propagation constant is equivalent to the one for a TEM wave in the open space $\gamma_x = k = \omega\sqrt{\varepsilon\mu}$.

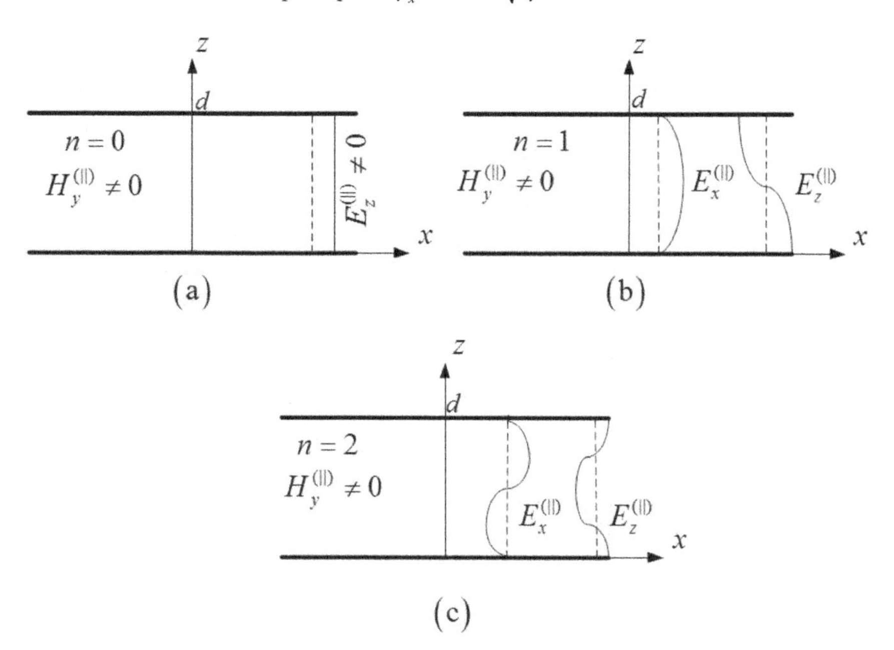

Fig. 1.13 Modal fields of a parallel-plate waveguide given for the first three transverse-magnetic modes

The number of modes, which can be excited in this waveguide, is infinite, i.e. $n = 0,1,2,...,\infty$. The higher-order modes with $n = 1,2,...,\infty$ start propagating from the cut-off frequencies $\omega_c^{(n)} = \frac{c}{\sqrt{\varepsilon_r \mu_r}} \frac{\pi n}{d}$. Over these frequencies, their propagation constants are real, and the modal longitudinal dependence is typical for the propagating modes: $\mathbf{E}, \mathbf{H} \sim e^{\pm j\gamma_x x}$ where the positive sign is for the waves propagating along the x-axis. Below these cut-off frequencies, the propagation constants are imaginary: $\mathbf{E}, \mathbf{H} \sim e^{\mp|\gamma_x|x}$, and the fields decrease fast with the distance. Such waves or modes are called the evanescent ones.

Similar analysis can be performed in the case of perpendicularly-polarized waves guided by two ideal conducting plates.

1.7.5.4 Wave Reflection from Non-perfect Conducting Surfaces

Real metals are not perfect conductors, and their conductivity varies within $10^6 - 10^7$ S/m. It means that the microwave field penetrates the metal surface, at the difference to ideal conducting medium.

In a lossy metal, the TEM wave constant is complex, and $\mathrm{Re}\left(\tilde{k}_2\right) \approx \sqrt{\omega\mu_2\sigma_2/2} \gg k_1$. Due to that the transmission angle $\theta_t \approx 0$, and the penetrated field has, practically, only the electric and magnetic field components which are tangential to the conductor surface. The magnitude of the field decreases exponentially from the surface, as it was shown in the Section 1.4.3, and the measure of it is the skin depth $\Delta^0 = \sqrt{2/\omega\mu_2\sigma_2}$. Practically, the field is negligible at the distance $10\Delta_0$ from the conductor surface.

In this considered case, a metal can be substituted by an equivalent impedance surface, and only the fields in the dielectrics are considered. This influence of the non-perfect conductivity is taken into account by the Leontovitch boundary condition [65] (See Section 1.7.2) written for the tangential components of the electric $\mathbf{E}_{i_\tau}$ and magnetic $\mathbf{H}_{i_\tau}$ fields of the incident wave:

$$\mathbf{E}_{i_\tau} = -Z_s \left[\mathbf{H}_{i_\tau} \times \mathbf{n}_0\right] \tag{1.129}$$

where $Z_s = (1+j)/(\sigma_2\Delta_0)$ is the complex surface impedance, and $\mathbf{n}_0$ is the normal vector to the conductor surface.

The idea of the equivalent surface impedance is very effective in electromagnetism. It allows not to calculate the fields inside non-perfect conductor where it is difficult to do. Besides, it opens a way to use the perturbation approach when the fields in the ideal case $(\sigma = \infty)$ initially are calculated, and they are substituted into some simplified formulas to calculate the loss in conductors. In this case, we need to know only the fields of the incident and reflected waves due to the use of Leontovitch boundary condition:

$$P = \frac{\operatorname{Re}(Z_s)}{2} \int_S \mathbf{H}_\tau^2 ds \tag{1.130}$$

where $\mathbf{H}_\tau$ is the sum of the magnetic fields of the incident and reflected waves tangential to the conductor surface. These components are calculated for the reflection from the ideal conductor. Today practically all analytical formulas, available for loss calculations, are originated from this approach. Only one limitation is that the current distribution along the conductor surface should be smooth, and the conductor thickness is of several skin depths, although more generalization can be found [74].

1.7.5.5 Full Transmission and the Brewster Angle

An analysis of the reflection coefficient formula given for the parallel-polarized waves (1.126) shows a case when all power of the incident wave is transmitted into the second medium: $\Gamma_{\|} = \frac{W_2 \cos\theta_t - W_1 \cos\theta_i}{W_2 \cos\theta_t + W_1 \cos\theta_i} = 0$. The numerator in this expression is zero if the incident angle θ_i satisfies the following condition:

$$\sin\theta_i^{(B)} = \sqrt{\frac{1}{1+\varepsilon_1/\varepsilon_2}}. \tag{1.131}$$

It follows that $\theta_i^{(B)} = \sin^{-1}\left(\sqrt{\frac{1}{1+\varepsilon_1/\varepsilon_2}}\right)$. The rays incident on the medium at this Brewster angle $\theta_i^{(B)}$ *are transmitted completely to the second region*, and the medium domains are matched to each other in this case. The effect is used to design low-reflection materials and EM shields.

1.7.5.6 Total Reflection of the EM Waves from the Boundary of Two Magnetodielectrics

A very interesting and useful effect occurs when the wave is incident from a medium of an increased permittivity $\varepsilon_r^{(1)} > \varepsilon_r^{(2)}$ or/and increased $\mu_r^{(1)} > \mu_r^{(2)}$. The incident angle $\theta_i = \theta_i^{(c)}$ results $\theta_t = 90^\circ$, and the transmitted ray is directed along the surface. Additionally, the modules of the reflection coefficients in both polarization cases are equal to the unit, and it corresponds to the *full wave reflection*. The critical incident angle $\theta_i = \theta_i^{(c)}$ is calculated as

$$\theta_i^{(c)} = \sin^{-1}\sqrt{\left(\frac{\varepsilon_2}{\varepsilon_1}\right)}. \tag{1.132}$$

Over this critical angle $\theta_i^{(c)}$, when $\sin\theta_t > 1$, the transmission angle is imaginary, being still applicable for calculation of the reflection and transmission coefficients. Another "paradox" is that $|R| = 1$, but $|T| \neq 0$ in this case, and it is formally in contradiction with the energy conservation law. Fortunately, the coefficients R and T are written regarding to the fields, only, and the power treatment on the boundary shows complete correctness of these formulas [75].

References

[1] Maxwell, J.C.: A Treatise on Electricity and Magnetism. Dover (1954)
[2] Heaviside, O.: Electromagnetic Waves. Taylor & Francis (1889), `http://www.archive.org`
[3] Jackson, J.D.: Classical Electrodynamics. John Wiley & Sons (1999)
[4] Balanis, C.A.: Advanced Engineering Electromagnetics. John Wiley (1989)
[5] Smythe, W.R.: Static and Dynamic Electricity. McGraw-Hill (1968)
[6] Stewart, J.V.: Intermediate Electromagnetic Theory. World Scientific (2001)
[7] Shekunoff, S.A.: Electromagnetic Field. Blaisdell Publ. Comp. (1963)
[8] Marcuvitz, N.: Waveguide Handbook. Inst. of Eng. and Techn. Publ. (1986)
[9] Mashkovzev, B.M., Zibisov, K.N., Emelin, B.F.: Theory of Waveguides. Nauka, Moscow (1966) (in Russian)
[10] Katsenelenbaum, B.Z.: High-frequency Electrodynamics. Wiley-vch Verlag Gmbh (2006)
[11] Vaganov, R.B., Katsenelenbaum, B.Z.: Foundation of the Diffraction Theory. Nauka, Moscow (1982) (in Russian)
[12] Nikolskyi, V.V., Nikolskaya, T.I.: Electrodynamics and Wave Propagation. Nauka, Moscow (1987) (in Russian)
[13] Collin, R.E.: Foundation of Microwave Engineering. John Wiley & Sons (2001)
[14] Felsen, L.B., Marcuvitz, N.: Radiation and Scattering of Waves. Prentice-Hall (1973)
[15] Taflove, A.T., Hagness, S.C.: Computational Electrodynamics. Artech House (2005)
[16] Sadiku, M.N.O.: Numerical Techniques in Electromagnetics. CRC Press (2001)
[17] Lewin, L.: Theory of Waveguides. Newnes-Buttertworths, London (1975)
[18] Yee, K.S.: Numerical solution of initial boundary value problems involving Maxwell's equations in isotropic medium. IEEE Trans., Antennas Propag. 14, 302–307 (1966)
[19] Mittra, R. (ed.): Computer Techniques for Electromagnetics. Pergamon Press (1973)
[20] Garg, R.: Analytical and Computational Methods in Electromagnetics. Artech House (2009)
[21] Nikolskii, V.V.: Variational Methods for Inner Boundary Value Problems of Electrodynamics. Nauka Publ., Moscow (1967) (in Russian)
[22] Harrington, R.F.: Field Computation by Moment Methods. IEEE Press (1993)
[23] Hoffmann, R.K.: Handbook of Microwave Integrated Circuits. Artech House (1987)
[24] Kompa, G.: Practical Microstrip Design and Applications. Artech House (2005)
[25] Edwards, T.C., Steer, M.B.: Foundation of Interconnect and Microstrip Design. J. Wiley & Sons, Ltd. (2000)
[26] Whites, K.W.: Visual Electromagnetics for MathCAD: Electronic Supplement for Introduction to Electromagnetics of K.W. Whites. McGraw-Hill (1998)

[27] Marcelli, R., Nikitov, S.A. (eds.): Nonlinear Microwave Signal Processing: Towards a New Range of Devices. NATO ASI Series, 3. High Technology, vol. 20. Kluwer Academic Publishing, Dordrecht (1996)
[28] Golovanov, Makeeva, G.S.: Simulations of Electromagnetic Wave Interactions with Nano-grids in Microwave and Terahertz Frequencies. Nauka (in print, 2012) (in Russian)
[29] Makeeva, G.S., Golovanov, O.A., Pardavi-Horvath, M., Kouzaev, G.A.: A method of autonomous blocks partially filled by nonlinear gyromagnetic medium for nanoelectromagnetic applications. In: Proc. 8th Int. Conf. Appl. of El. Eng., Houston, USA, April 30-May 2, pp. 204–207 (2009)
[30] Makeeva, G.S., Golovanov, O.A., Pardavi-Horvath, M., Kouzaev, G.A.: Decomposition approach to nonlinear diffraction problems of nanoelectromagnetics and nanophotonics using autonomous blocks with Floquet channels. In: Proc. 7th Int. Conf. Appl. El. Eng., AEE 2008, pp. 31–35 (2008)
[31] Deleonibus, S. (ed.): Electronic Device Architectures for the Nano-CMOS Era. World Sci (2008)
[32] Tour, J.M.: Molecular Electronics. World Sci. (2003)
[33] Kouzaev, G.A.: Hertz vectors and the electromagnetic-quantum equations. Modern Phys. Lett. B 24, 2117–2129 (2010)
[34] Levine, H., Moniz, E.J., Sharp, D.H.: Motion of extended charges in classical electrodynamics. Am. J. Phys. 45, 75–78 (1977)
[35] Rohlich, F.: The dynamics of a charged particle and the electron. Am. J. Phys. 65, 1051–1056 (1997)
[36] Kroemer, H.: Quantum Mechanics. Prentice-Hall (1994)
[37] Benci, V., Fortunato, D.: An eigenvalue problem for the Schrödinger-Maxwell equations. Topological Methods in Nonlinear Analysis 11, 283–293 (1998)
[38] Ginbre, J., Velo, G.: Long range scattering for the Maxwell-Schrödinger system with large magnetic field data and small Schrödinger data, vol. 42, pp. 421-459. Publ. RIMS, Kyoto Univ. (2006)
[39] Yang, J., Sui, W.: Solving Maxwell-Schrödinger equations for analyses of nano-scale devices. In: Proc. 37th Eur. Microw. Conf., pp. 154–157 (2007)
[40] Pierantoni, L., Mencarelli, D., Rozzi, T.: A new 3-D transmission line matrix scheme for the combined Schrödinger-Maxwell problem in the electronic/electromagnetic characterization of nanodevices. IEEE Trans., Microw. Theory Tech. 56, 654–662 (2008)
[41] Pieratoni, L., Mencarelli, D., Rozzi, T.: Boundary immitance operators for the Schrödinger-Maxwell problem of carrier dynamics in nanodevices. IEEE Trans., Microw. Theory Tech. 57, 1147–1155 (2009)
[42] Mastorakis, N.E.: Solution of the Schrödinger-Maxwell equations via finite elements and genetic algorithms with Nelder-Mead. WSEAS Trans. Math. 8, 169–176 (2009)
[43] Attaf, M.T.: Error analysis and Hertz vector approach for an electromagnetic interaction between a line current and a conducting plate. Int. J. Numer. 16, 249–260 (2003)
[44] Gough, W.: An alternative approach to the Hertz vector. PIER 12, 205–217 (1996)
[45] Sein, J.J.: Solutions to time-harmonic Maxwell equations with a Hertz vector. Am. J. Phys. 57, 834–839 (1989)
[46] Born, M., Wolf, E.: Principles of Optics. Electromagnetic Theory of Propagation, Interference and Diffraction of Light, 7th edn. Cambridge University Press (2000)
[47] Nowakowski, M.: The quantum mechanical current of the Pauli equation. Am. J. Phys. 67, 916–919 (1999)

[48] Esteban, M.J., Georgiev, V., Séré, E.: Stationary solutions of the Maxwell-Dirac and the Klein-Gordon-Dirac equations. Calc. Var. 4, 265–281 (1996)
[49] Bao, W., Li, X.-G.: An efficient and stable numerical method for the Maxwell-Dirac system. J. Comput. Phys. 199, 663–687 (2004)
[50] Balasubramanian, K.: Relativistic Effects in Chemistry, Part A. John Wiley & Sons, Inc. (1997)
[51] Kapranov, S.A., Kouzaev, G.A.: Relaxation mechanism of microwave heating of near-critical polar gases. Int. J. Thermal Sciences 49(12), 2319–2330 (2010)
[52] Vekstein, G.E.: On the electromagnetic force on a moving dipole. Eur. J. Phys. 18, 113–117 (1997)
[53] Kholometskii, I.L., Missevitch, O.V., Yarman, T.: Electromagnetic force on a moving dipole. Eur. J. Phys. 32, 873–881 (2011)
[54] Gao, Z.: Nonlinear pondermotive force by low frequency waves and nonresonant current drive. Physics of Plasmas 13, 112307-1–112307-6 (2006)
[55] Dodin, I.Y., Fisch, N.J.: Particle manipulation with nonadiabatic pondermotive forces. Physics of Plasmas 14, 055901-1–055901-6 (2007)
[56] Eichmann, U., Nubbemeyer, T., Rottke, H., et al.: Acceleration of neutral atoms in strong strong short-pulse laser field. Nature 461, 1261–1264 (2009)
[57] Block, P.A., Bohac, E.A., Miller, R.E.: Spectroscopy of pendular states: The use of molecular complexes in achieving orientation. Phys. Rev. Lett. 68, 1303–1306 (1992)
[58] Kapranov, S.V., Kouzaev, G.A.: Stochasticity in nonlinear pendulum motion of dipoles in electric field. In: Recent Advances in Systems Engineering and Applied Mathematics, pp. 107–111 (2008)
[59] Zaslavsky, G.M.: Physics of Chaos in Hamilton Systems. Imperial College Press (1998)
[60] Zaslavsky, G.M., Sagdeev, R.Z., Usikov, D.A., et al.: Weak Chaos and Quasi-Regular Patterns. Cambridge University Press (1991)
[61] Chirikov, B.V.: Nonlinear Resonance. Novosibirsk State University Publ. (1977) (in Russian)
[62] Fradkov, I.: Cybernetical Physics. Springer (2006) (in Russian)
[63] Barkai, E., Brown, F., Orrit, M., Yang, H. (eds.): Theory and Evaluation of Single Molecule Signals. World Scientific (2008)
[64] Wu, G.: Nonlinearity and Chaos in Molecular Vibrations. Elsevier (2005)
[65] Leontovich, M.A.: Investigation on Radiowave Propagation, Part II. Academy of Sci., Moscow (1948)
[66] Wang, D.-S.: Limits and validity of the impedance boundary condition on penetrable surfaces. IEEE Trans., Antennas Propag. 35, 453–457 (1987)
[67] Dybdal, R.B., Peters, L., Peake, W.H.: Rectangular waveguides with impedance walls. IEEE Trans., Microw. Theory Tech. 19, 2–9 (1971)
[68] Senior, T.B.A.: Approximate boundary condition. IEEE Trans., Antennas Propag. 29, 826–829 (1981)
[69] Karlsson, I.: Approximate boundary conditions for thin structures. IEEE Trans., Antennas Propag. 57, 144–148 (2009)
[70] Alshits, V.I., Lyubimov, V.N.: Generalization of the Leontovich approximation for electromagnetic fields on a dielectric-metal interface. Physics-Uspekhi 52, 815–820 (2009)

[71] Kurushin, E.P., Nefedov, E.I., Fialkovskyi, A.T.: Diffraction of Waves on Anisotropic Structures. Nauka, Moscow (1975) (in Russian)
[72] Kurushin, E.P., Nefedov, E.I.: Electrodynamics of Anisotropic Waveguiding Structures. Nauka, Moscow (1983) (in Russian)
[73] Kontorovich, M.I., Astrakhan, M.I., Akimov, V.P., et al.: Electrodynamics of Grid Structures. Radio i Svayaz, Moscow (1987) (in Russian)
[74] Yuferev, S.V.: Surface Impedance Boundary Conditions. CRC Press (2010)
[75] Demarest, K.R.: Engineering Electromagnetics. Prentice-Hall Int. (1997)

2 Theory of Waveguides

Abstract. The analytical and numerical models and methods of waveguides and integrated transmission lines are reviewed in this Chapter. Among them are the separation of the variables method and the transverse resonance one. Engineering formulas obtained by the conformal technique for most used integrated transmission lines are given and the accuracy of them are considered. The strong numerical EM methods are represented here by the finite difference time domain techniques, transmission line matrix method, finite element method, and the integral equation models of transmission lines. There are 101 references given for the Readers who wish to obtain more knowledge on the EM theory of waveguides and transmission lines. 17 figures are included into the text of 43 pages to explain the waveguides and integrated transmission lines.

2.1 Analytical Treatment of Waveguides

Not only parallel conductor plates are guiding the modes, as it was shown in the previous Chapter, but all regular structures separated by a boundary from the open space can support the traveling waves, and some of them, which are the hollow circular and rectangular waveguides, are shown in (Fig. 2.1).

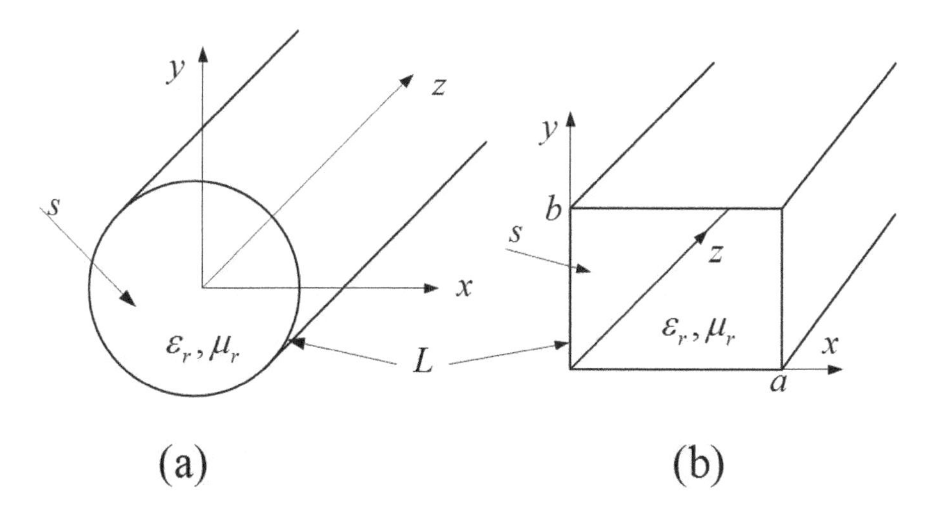

Fig. 2.1 Hollow circular (a) and rectangular (b) waveguides

G.A. Kouzaev: Applications of Advanced Electromagnetics, LNEE 169, pp. 51–93.
springerlink.com

Additionally, a dielectric-air interface guides the waves due to the effect of full reflection, and the dielectric sheet waveguides are known. This sheet can carry a conductor strip and a ground layer, and it is a microstrip line which guides the modes. The field of non-radiating modes concentrates close to the strip and inside this dielectric substrate.

All these effects are described by the Maxwell or wave differential equations together with the boundary conditions on the inter-dielectric and dielectric-conductor interfaces. In some cases, the differential equations are transformed into the integral ones or stationary functionals, and different analytical and numerical techniques can be applied to solve them and to calculate the modes. The guiding effect and computation techniques are the subject of the waveguide theory which some introductory elements are considered here.

The modes have periodical dependence of their fields along the longitudinal direction $0 \to z$, and this spatial period is the wavelength $\Lambda = 2\pi/k_z$ expressed through the modal propagation constant k_z. It is obtained by solution of a boundary value problem formulated for the considered waveguide.

Taking into account the traveling nature of waves, the modal fields are written as

$$\begin{aligned} \mathbf{E} &= \mathbf{E}(x, y)e^{-jk_z z}, \\ \mathbf{H} &= \mathbf{H}(x, y)e^{-jk_z z} \end{aligned} \tag{2.1}$$

where $\mathbf{E}(x, y)$ and $\mathbf{H}(x, y)$ are the transversally-dependent functions which are found by solutions of the 2-D Helmholtz equations additionally equipped by the conditions on the boundary L of the waveguide cross-section s (Fig. 2.1):

$$\begin{aligned} &\nabla^2 \cdot \mathbf{E}(x, y) + \chi^2 \mathbf{E}(x, y) = 0, \\ &\mathbf{E}_\tau(x, y) = 0,\ x, y \in L, \\ &\text{or} \\ &\nabla^2 \cdot \mathbf{H}(x, y) + \chi^2 \mathbf{H}(x, y) = 0, \\ &(\nabla \times \mathbf{H}(x, y))_\tau = 0,\ x, y \in L \end{aligned} \tag{2.2}$$

where $\chi^2 = k_0^2 \varepsilon_r \mu_r - k_z^2$.

It is found that the number of eigenmodes in a waveguide is infinite. In some transmission lines, which have the multi-connected cross-sections, several N_0 modes propagate since zero frequency, and they are, mostly, of the TEM or quasi-TEM types in multi-conductor lines. N_0 is calculated from the number of conductors N_c including the grounded one: $N_0 = N_c - 1$. Other higher-order modes of these lines propagate starting at certain frequencies, which are called the cut-off ones. Usually, one of these TEM or quasi-TEM modes is considered as the main wave, and it has the simplest spatial design of the modal field.

Many waveguides do not support the TEM or quasi-TEM modes at all, and a mode with the lowest cut-off frequency is considered as the fundamental one. The quantitative information on modes and their parameters are obtained by analytical or numerical treatments of the boundary value problems (2.2).

The EM modes of waveguides are solved by different methods. The most valuable of them are those which allow for the analytical obtaining of the exact or approximate formulas for the propagation constants. Several such methods are known, or a few of them are considered below.

2.1.1 Separation of the Variables Method

The availability of analytical solutions of boundary value problems depends on the geometry of the waveguide cross-sections and conditions defined on their boundaries. One of the popular methods is the separation of the variables technique described in many textbooks [1],[2]. Only a few waveguide cross-sections allow for this analytical method to be applied, but the structures that are more complicated can be substituted by the equivalent separable waveguides treated by this method.

The *method of separation of the variables* consists of representation of the field under the search as a product of functions. Each of them depends only on one variable. For instance, in the Cartesian coordinate system, the longitudinal field components are represented as

$$\begin{aligned} H_z(x,y,z) &= X^{(\mathrm{TE})}(x)Y^{(\mathrm{TE})}(y)e^{-jk_z z}, \\ E_z(x,y,z) &= X^{(\mathrm{TM})}(x)Y^{(\mathrm{TM})}(y)e^{-jk_z z}. \end{aligned} \tag{2.3}$$

The transversal field components are obtained using the Maxwell equations:

$$\begin{aligned} E_x &= -j\frac{k_z}{\chi^2}\frac{\partial E_z}{\partial x} - j\frac{\omega\mu_0\mu_r}{\chi^2}\frac{\partial H_z}{\partial y}, \\ E_y &= -j\frac{k_z}{\chi^2}\frac{\partial E_z}{\partial y} + j\frac{\omega\mu_0\mu_r}{\chi^2}\frac{\partial H_z}{\partial x}, \\ H_x &= j\frac{\omega\varepsilon_0\varepsilon_r}{\chi^2}\frac{\partial E_z}{\partial y} - j\frac{k_z}{\chi^2}\frac{\partial H_z}{\partial x}, \\ H_y &= -j\frac{\omega\varepsilon_0\varepsilon_r}{\chi^2}\frac{\partial E_z}{\partial x} - j\frac{k_z}{\chi^2}\frac{\partial H_z}{\partial y}. \end{aligned} \tag{2.4}$$

The waveguides, homogeneously filled by a dielectric, allow for considering the modes with only longitudinal magnetic field, and they are the transversal electric (TE_z) ones. The others in them, with only longitudinal electric field component, are the transversal magnetic or TM_z modes.

The TE_z modes are derived solving the Neumann boundary value problem if the waveguide channel is shielded by perfect conductor:

$$\nabla^2 \cdot H_z(x,y) + \chi^2 H_z(x,y) = 0,$$
$$\frac{\partial H_z(x,y)}{\partial v} = 0,\ x, y \in L. \tag{2.5}$$

The *Dirichlet* boundary condition is used to derive the TM_z modes of the same waveguide:

$$\nabla^2 \cdot E_z(x,y) + \chi^2 E_z(x,y) = 0,$$
$$E_z = 0,\ x, y \in L. \tag{2.6}$$

The variables in (2.5) and (2.6) are separated if the cross-section s is one of the following shapes: rectangular, circular, and elliptical one. More shapes for this method are from [2].

Consider a rectangular waveguide Fig. 2.1b in the Cartesian system of coordinates. An ordinary differential equation for each product from (2.3) can be written as

$$\frac{\partial^2 X^{(\mathrm{TE,TM})}}{\partial x^2} = \left(k_x^{(\mathrm{TE,TM})}\right)^2 X^{(\mathrm{TE,TM})},$$
$$\frac{\partial^2 Y^{(\mathrm{TE,TM})}}{\partial y^2} = \left(k_y^{(\mathrm{TE,TM})}\right)^2 Y^{(\mathrm{TE,TM})}, \tag{2.7}$$
$$\left(k_x^{(\mathrm{TE,TM})}\right)^2 + \left(k_y^{(\mathrm{TE,TM})}\right)^2 = \chi^2_{\mathrm{TE,TM}}.$$

General solutions of (2.7) are

$$X^{(\mathrm{TE,TM})} = A^{(\mathrm{TE,TM})} e^{jk_x^{(\mathrm{TE,TM})}x} + B^{(\mathrm{TE,TM})} e^{-jk_x^{(\mathrm{TE,TM})}x},$$
$$Y^{(\mathrm{TE,TM})} = C^{(\mathrm{TE,TM})} e^{jk_y^{(\mathrm{TE,TM})}y} + D^{(\mathrm{TE,TM})} e^{-jk_y^{(\mathrm{TE,TM})}y} \tag{2.8}$$

where A, B, C, and D are the unknown coefficients. The wavenumbers k_x, k_y, and k_z are obtained analytically using the boundary conditions on the waveguide wall, and they are

$$k_z^{\mathrm{TE,TM}} = \sqrt{k_0^2 \varepsilon_r \mu_r - \left(k_x^{(\mathrm{TE,TM})}\right)^2 - \left(k_y^{(\mathrm{TE,TM})}\right)^2},$$
$$k_x^{(\mathrm{TE,TM})} = \frac{m\pi}{a}, m = 1, 2, 3, \ldots \infty,$$
$$k_y^{(\mathrm{TE})} = \frac{n\pi}{a}, n = 0, 1, 2, \ldots \infty, \tag{2.9}$$
$$k_y^{(\mathrm{TM})} = \frac{n\pi}{a}, n = 1, 2, 3, \ldots \infty.$$

Satisfying the boundary conditions on the waveguide wall, the functions $X^{(\mathrm{TE,TM})}$ and $Y^{(\mathrm{TE,TM})}$ are found as the periodic ones, and the number of them of both sorts is infinite. It means that the waveguide supports infinite number of modes which are independent on each other. The modal field expressions for the TE waves are

$$
\begin{aligned}
H_z^{(mn)} &= H_0^{(mn)} \cos(k_x x)\cos(k_y y)e^{-jk_z z},\\
H_x^{(mn)} &= H_0^{(mn)} j\frac{k_z k_x}{\chi^2}\sin(k_x x)\cos(k_y y)e^{-jk_z z},\\
H_y^{(mn)} &= H_0^{(mn)} j\frac{k_z k_y}{\chi^2}\cos(k_x x)\sin(k_y y)e^{-jk_z z},\\
E_x^{(mn)} &= jW_{mn}^{(\mathrm{TE})} H_0^{(mn)} \frac{k_z k_y}{\chi^2}\cos(k_x x)\sin(k_y y)e^{-jk_z z},\\
E_y^{(mn)} &= -jW_{mn}^{(\mathrm{TE})} H_0^{(mn)} \frac{k_z k_x}{\chi^2}\sin(k_x x)\cos(k_y y)e^{-jk_z z}.
\end{aligned}
\tag{2.10}
$$

where $H_0^{(mn)}$ is the unknown modal amplitudes, and $W_{mn}^{(\mathrm{TE})} = \omega\mu_0\mu_r / k_z$ is the TE modal wave impedance.

Similarly, the field components of the TM modes are derived using the found expression for the longitudinal electric field and the Maxwell equations:

$$
\begin{aligned}
E_z^{(mn)} &= E_0^{(mn)} \sin(k_x x)\sin(k_y y)e^{-jk_z z},\\
E_x^{(mn)} &= -j\frac{k_x k_z}{\chi^2} E_0^{(mn)} \cos(k_x x)\sin(k_y y)e^{-jk_z z},\\
E_y^{(mn)} &= -j\frac{k_y k_z}{\chi^2} E_0^{(mn)} \sin(k_x x)\cos(k_y y)e^{-jk_z z},\\
H_x^{(mn)} &= j\frac{E_0^{(mn)}}{W_{mn}^{(\mathrm{TM})}}\frac{k_z k_y}{\chi^2}\sin(k_x x)\cos(k_y y)e^{-jk_z z},\\
H_y^{(mn)} &= -j\frac{E_0^{(mn)}}{W_{mn}^{(\mathrm{TM})}}\frac{k_z k_x}{\chi^2}\cos(k_x x)\sin(k_y y)e^{-jk_z z}
\end{aligned}
\tag{2.11}
$$

where $E_0^{(mn)}$ is the modal unknown amplitude, and $W_{mn}^{(\mathrm{TM})} = k_z / \omega\varepsilon_0\varepsilon_r$ is the TM modal wave impedance.

It is seen only the waveguide geometry, dielectric filling, and the frequency define the modal parameters. Similarly to the rectangular waveguides, other cross-section shapes of the separable type are solved.

In more complicated waveguides, the modal field can be represented as a series of these elementary modes with the following satisfaction of the boundary conditions. It allows for obtaining the equations for the eigenmodal parameters of these waveguides. This universal method is applicable for the waveguides of arbitrary shapes, and it is considered below.

2.1.2 Field Series Expansion Method

In most cases, the variables of the boundary value problems are not separable due to the cross-section shape or boundary conditions, and the *field series expansion method* is used. This approach originates from the works of Rayleigh on the diffraction of waves. Some additional information on this method can be found from [1],[3]-[5].

Consider a domain S on which boundary part l_1 a nonhomogeneous Dirichlet condition is defined

$$u(\mathbf{r} \in L) = \begin{cases} U_0(\mathbf{r}), \mathbf{r} \in l_1 \\ 0, \ \mathbf{r} \in l_2 \end{cases} \tag{2.12}$$

where $U_0(\mathbf{r} \in l_1)$ is a known function and $L = l_1 + l_2$.

In this case, a solution of the boundary value problem can be represented by an infinite series of the linearly independent functions $u_n(\mathbf{r})$ which of them is a general solution of the corresponding wave equation:

$$u(\mathbf{r}) = \sum_{n=0}^{\infty} a_n u_n(\mathbf{r}). \tag{2.13}$$

To derive the unknown coefficients a_n, the field on the boundary is matched to the boundary condition by the *minimization of the residue method* [4], for instance,

$$\int_L \left[\sum_{n=0}^{\infty} a_n u_n(\mathbf{r} \in l_1) - U_0(\mathbf{r} \in l_1) \right] u_m(\mathbf{r} \in l_1) dl, \ m = 0,1,2,...\infty. \tag{2.14}$$

To obtain the finite system of linear algebraic equations regarding to the unknown coefficients a_n, (2.14) is truncated limiting the number of basis (n) and weighting (m) functions. Then a modal eigenvalue is obtained solving the truncated system (2.14) by a pertinent algorithm.

The method is applicable to the eigenvalue problems. For instance, the line cross-section can consist of several sub-domains in which an expansion similar to (2.13) can be written. Usually, the sub-domain expansion functions are chosen to satisfy the boundary conditions on the most parts of the sub-domain boundaries. For other boundary parts, the conditions are satisfied similarly to (2.14), and a system of homogeneous linear algebraic equations is derived. The determinant of this system equalized to zero gives an eigenvalue equation to calculate the modal propagation constants.

2.1.3 Transverse Resonance Method and Analytical Treatment of Waveguides

Another approach to the analytical derivation of the eigenvalue equations is the *transverse resonance method* [6],[7]. Although the analytical treatment by this technique is not applicable to all geometries of transmission lines, the method, since its invention, has been evolved to a full-wave algorithm distinguished by its flexibility and accuracy [8].

Initially this method was applied to the multilayered dielectric waveguides, a particular case of which is shown in (Fig. 2.2).

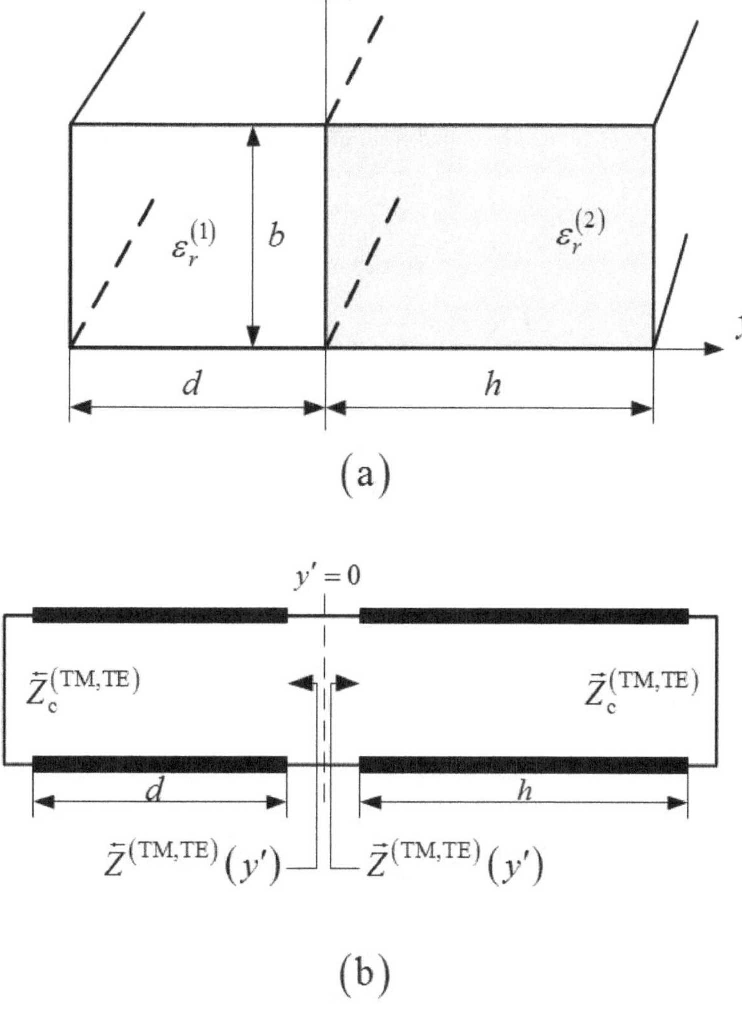

Fig. 2.2 A two-layer rectangular waveguide (a) and its equivalent circuit (b) for the transverse resonance method

For these waveguides the known classification of the modes on the TM_z and TE_z types is not correct. New two separate sets of modes are introduced for them according to the field components which are transversal to the propagation direction $0 \rightarrow z$. The modes having only the electric field component parallel to the axis $0 \rightarrow y$ and normal to the dielectric layers are called the TM_y ones $(E_y, E_x, E_z; H_x, H_z)$. Other modes have only the magnetic field component along the y-axis $(E_x, E_z; H_y, H_x, H_z)$, and they are of the TE_y type. To these y-components of the electric and magnetic fields, the method of the separation of variables is applied

$$E_y(x,y,z) = V(y)E_y(x,z),$$
$$H_y(x,y,z) = I(y)H_y(x,z) \tag{2.15}$$

where $V(y)$ and $I(y)$ will be found from the transverse resonance condition later. Additionally,

$$E_y(x,z) = \frac{1}{j\omega\varepsilon_0\varepsilon_r}\nabla_{(x,z)} \cdot \left(\mathbf{H}_{(x,z)} \times \mathbf{y}_0\right),$$
$$H_y(x,z) = \frac{1}{j\omega\mu_0\mu_r}\nabla_{(x,z)} \cdot \left(\mathbf{y}_0 \times \mathbf{E}_{(x,z)}\right) \tag{2.16}$$

where $\nabla_{(x,z)} = \mathbf{x}_0 \frac{\partial}{\partial x} + \mathbf{z}_0 \frac{\partial}{\partial z}$, $\mathbf{H}_{(x,z)} = \mathbf{x}_0 H_x + \mathbf{z}_0 H_z$, and $\mathbf{E}_{(x,z)} = \mathbf{x}_0 E_x + \mathbf{z}_0 E_z$. These components satisfy the following boundary conditions on the perfect conductor surface:

$$\mathbf{v} \times \mathbf{E}_{(x,z)} = 0,$$
$$\nabla_{(x,z)} \cdot \left(\mathbf{H}_{(x,z)} \times \mathbf{y}_0\right) = 0 \tag{2.17}$$

where $\mathbf{v}$ is the unit vector normal to the perfect conductor surface.

For the TM_y modes:

$$\mathbf{E}^{(TM)}_{(x,z)}(x,y,z) = V^{(TM)}(y)\mathbf{e}^{(TM)}(x,z),$$
$$\mathbf{H}^{(TM)}_{(x,z)}(x,y,z) = I^{(TM)}(y)\mathbf{h}^{(TM)}(x,z). \tag{2.18}$$

For the TE_y modes:

$$\mathbf{E}^{(TE)}_{(x,z)}(x,y,z) = V^{(TE)}(y)\mathbf{e}^{(TE)}(x,z),$$
$$\mathbf{H}^{(TE)}_{(x,z)}(x,y,z) = I^{(TE)}(y)\mathbf{h}^{(TE)}(x,z). \tag{2.19}$$

In (2.18), the vector functions $\mathbf{e}^{(\mathrm{TM})}(x,z)$ and $\mathbf{h}^{(\mathrm{TM})}(x,z)$ are found from the following equations:

$$\begin{aligned} &\nabla^2_{(x,z)} \cdot \mathbf{e}^{(\mathrm{TM})} + \left(k^{(\mathrm{TM})}\right)^2 \mathbf{e}^{(\mathrm{TM})} = 0, \\ &\mathbf{h}^{(\mathrm{TM})} = \mathbf{y}_0 \times \mathbf{e}^{(\mathrm{TM})}. \end{aligned} \tag{2.20}$$

The vector functions $\mathbf{h}^{(\mathrm{TE})}(x,z)$ and $\mathbf{e}^{(\mathrm{TE})}$ are derived from:

$$\begin{aligned} &\nabla^2_{(x,z)} \cdot \mathbf{h}^{(\mathrm{TE})} + \left(k^{(\mathrm{TE})}\right)^2 \mathbf{h}^{(\mathrm{TE})} = 0, \\ &\mathbf{e}^{(\mathrm{TE})} = \mathbf{y}_0 \times \mathbf{h}^{(\mathrm{TE})} \end{aligned} \tag{2.21}$$

where all functions $\mathbf{e}(x,y)$ and $\mathbf{h}(x,y)$ are normalized on the cross-section of waveguide:

$$\begin{aligned} &\int_S \left(\mathbf{e}^{(\mathrm{TM})} \cdot \mathbf{e}^{(\mathrm{TM})*}\right) ds = \int_S \left(\mathbf{e}^{(\mathrm{TE})} \cdot \mathbf{e}^{(\mathrm{TE})*}\right) ds = \\ &= \int_S \left(\mathbf{h}^{(\mathrm{TM})} \cdot \mathbf{h}^{(\mathrm{TM})*}\right) ds = \int_S \left(\mathbf{h}^{(\mathrm{TE})} \cdot \mathbf{h}^{(\mathrm{TE})*}\right) ds = 1, \\ &\int_S \left(\mathbf{e}^{(\mathrm{TM})} \cdot \mathbf{e}^{(\mathrm{TE})*}\right) ds = \int_S \left(\mathbf{e}^{(\mathrm{TE})} \cdot \mathbf{e}^{(\mathrm{TM})*}\right) ds = \\ &= \int_S \left(\mathbf{h}^{(\mathrm{TM})} \cdot \mathbf{h}^{(\mathrm{TE})*}\right) ds = \int_S \left(\mathbf{h}^{(\mathrm{TE})} \cdot \mathbf{h}^{(\mathrm{TM})*}\right) ds = 0. \end{aligned} \tag{2.22}$$

It is shown [6] that the modal amplitudes $V^{(\mathrm{TM,TE})}$ and $I^{(\mathrm{TM,TE})}$ are derived from the telegraph equations:

$$\begin{aligned} \frac{dV^{(\mathrm{TM,TE})}}{dy} &= jk_y^{(\mathrm{TM,TE})} Z_c^{(\mathrm{TM,TE})} I^{(\mathrm{TM,TE})}, \\ -\frac{dI^{(\mathrm{TM,TE})}}{dy} &= jk_y^{(\mathrm{TM,TE})} Y_c^{(\mathrm{TM,TE})} V^{(\mathrm{TM,TE})} \end{aligned} \tag{2.23}$$

where $Z_c^{(\mathrm{TM})} = \dfrac{1}{Y_c^{(\mathrm{TM})}} = \dfrac{k_y^{(\mathrm{TM})}}{\omega\varepsilon_0\varepsilon_r}$, $Z_c^{(\mathrm{TE})} = \dfrac{1}{Y_c^{(\mathrm{TE})}} = \dfrac{\omega\mu_0\mu_r}{k_y^{(\mathrm{TE})}}$, and $k_y^{(\mathrm{TM,TE})} = \sqrt{k_0^2\varepsilon_r\mu_r - \left(k_x^{(\mathrm{TM,TE})}\right)^2 - \left(k_z^{(\mathrm{TM,TE})}\right)^2}$.

Very often, the wave constant k_x is known because the layers are uniform along the x-axis (Fig. 2.2a). Then the equivalent lines can be introduced along the y-axis (Fig. 2.2b).

At any arbitrary point y' of this transversal line, the left $\overleftarrow{Z}^{(\mathrm{TM,TE})}$ and the right $\overrightarrow{Z}^{(\mathrm{TM,TE})}$ input impedances are written, and they are calculated by the known formulas if the characteristic impedances $Z_{\mathrm{c}}^{(\mathrm{TM,TE})}$ and boundary conditions at the ends of the transversal line are known.

The propagation condition along the z-axis requires the resonance along the transversal y-axis, and it gives the eigenvalue equation for $k_z^{(\mathrm{TM,TE})}$ written at an arbitrary point y':

$$\overleftarrow{Z}^{(\mathrm{TM,TE})}(y') + \overrightarrow{Z}^{(\mathrm{TM,TE})}(y') = 0. \tag{2.24}$$

Taking into account the analytical form of these impedances, this equation can be rather simple, and it is solved for some geometry even analytically.

An example is given in Fig. 2.2 where a 2-layer rectangular waveguide is shown. The left and right impedances for the TM_y and TE_y modes calculated at $y' = 0$ are

$$\begin{aligned} \overleftarrow{Z}^{(\mathrm{TM,TE})} &= j\overleftarrow{Z}_{\mathrm{c}}^{(\mathrm{TM,TE})} \tan\left(\overleftarrow{k}_y^{(\mathrm{TM,TE})} d\right), \\ \overrightarrow{Z}^{(\mathrm{TM,TE})} &= j\overrightarrow{Z}_{\mathrm{c}}^{(\mathrm{TM,TE})} \tan\left(\overrightarrow{k}_y^{(\mathrm{TM,TE})} h\right) \end{aligned} \tag{2.25}$$

where $\overleftarrow{k}_y^{(\mathrm{TM,TE})} = \sqrt{k_0^2 \varepsilon_r^{(1)} \mu_r^{(1)} - \left(k_x^{(\mathrm{TM,TE})}\right)^2 - \left(k_z^{(\mathrm{TM,TE})}\right)^2}$,

$\overrightarrow{k}_y^{(\mathrm{TM,TE})} = \sqrt{k_0^2 \varepsilon_r^{(2)} \mu_r^{(2)} - \left(k_x^{(\mathrm{TM,TE})}\right)^2 - \left(k_z^{(\mathrm{TM,TE})}\right)^2}$, $\overleftarrow{Z}_{\mathrm{c}}^{(\mathrm{TM})} = \dfrac{\overleftarrow{k}_y^{(\mathrm{TM})}}{\omega\varepsilon_0\varepsilon_r^{(1)}}$, $\overleftarrow{Z}_{\mathrm{c}}^{(\mathrm{TE})} = \dfrac{\omega\mu_0\mu_r^{(1)}}{\overleftarrow{k}_y^{(\mathrm{TE})}}$,

$\overrightarrow{Z}_{\mathrm{c}}^{(\mathrm{TM})} = \dfrac{\overrightarrow{k}_y^{(\mathrm{TM})}}{\omega\varepsilon_0\varepsilon_r^{(2)}}$, and $\overrightarrow{Z}_{\mathrm{c}}^{(\mathrm{TE})} = \dfrac{\omega\mu_0\mu_r^{(2)}}{\overrightarrow{k}_y^{(\mathrm{TE})}}$. Then the resonance equations for these modes written at $y' = 0$ are

$$\overleftarrow{Z}^{(\mathrm{TM,TE})} + \overrightarrow{Z}^{(\mathrm{TM,TE})} = j\overleftarrow{Z}_{\mathrm{c}}^{(\mathrm{TM,TE})} \tan\left(\overleftarrow{k}_y^{(\mathrm{TM,TE})} d\right) + j\overrightarrow{Z}_{\mathrm{c}}^{(\mathrm{TM,TE})} \tan\left(\overrightarrow{k}_y^{(\mathrm{TM,TE})} h\right) = 0 . \tag{2.26}$$

In general, these equations are transcendental, and they can be solved regarding to $k_z^{(\mathrm{TM,TE})}$ using numerical algorithms for the root searching.

Different waveguides and traveling wave antennas can be treated by this technique. For instance, the dielectric layers are separated additionally by thin conductor membranes or impedance walls, and the method allows for including the equivalent lumped components to model these waveguide elements if they do not transform the TM and TE modes to each other [7].

The method is applicable in the general case when the TM and TE modes are mixed due to the geometry of the conducting enclosures and dielectric layers, and the approach is transformed into one, which is called now the *transverse resonance matching method* [8].

2.1.4 Analytical Models of TEM Transmission Lines

2.1.4.1 Triplate Transmission Line

The multi-connected cross-section of a triplate line is shown in Fig. 2.3. It consists of the central strip embedded into a shielded dielectric sheet, and it supports propagation of the TEM mode which is fundamental regarding the higher-order TE- and TM ones.

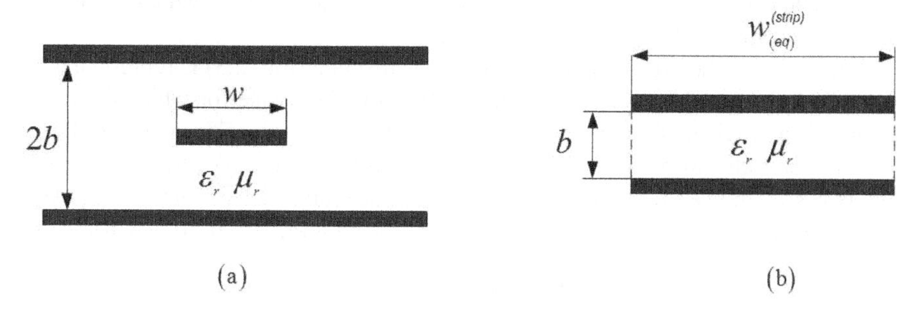

Fig. 2.3 Triplate strip line (a) and its parallel-plate model (b) with $w_{\mathrm{eq}}^{(\mathrm{strip})} = 60\pi b/\sqrt{\varepsilon_r}\, Z_{\mathrm{c}}$

The propagation constant of the TEM mode is $k_z = k_0\sqrt{\varepsilon_r \mu_r}$. The normalized value of this propagation constant k_z/k_0 does not depend on frequency, and such waves and modes are very interesting because they support ultra-bandwidth signaling.

In the TEM case, the Helmholtz equations (2.2) are transformed to the Laplace ones because $\chi^2 = 0$:

$$\begin{aligned} \nabla^2 \cdot \mathbf{E}(x, y) &= 0, \\ \nabla^2 \cdot \mathbf{H}(x, y) &= 0. \end{aligned} \tag{2.27}$$

Instead of solving the vector equations (2.27), the electric φ_e and magnetic φ_m potentials and corresponding to them scalar Laplace equations are introduced:

$$\begin{aligned} \nabla^2 \cdot \varphi_e &= 0, \\ \nabla^2 \cdot \varphi_m &= 0. \end{aligned} \tag{2.28}$$

The electric and magnetic fields are obtained as

$$\begin{aligned} \mathbf{E} &= -\nabla \cdot \varphi_e, \\ \mathbf{H} &= -\nabla \cdot \varphi_m. \end{aligned} \tag{2.29}$$

As known, the Laplace equation is for the static electricity phenomena. Here, the harmonically time-dependent electric and magnetic fields are described by the same equations being coupled to each other through the Maxwell ones. It explains the popularity of the quasi-static models for some transmission lines where the TEM modes are fundamental. They propagate along the transmission lines composed of several conductors placed inside a homogeneous shielded dielectric. Such modes can be calculated by the *conformal mapping approach* [9]-[11]. By this method, the cross-section of the modeled line is considered on a 2-D complex plane. Then using several steps, it is transformed to a new one, which allows for analytical solution of the Laplace equation.

For instance, the cross-section of a triplate strip line (Fig. 2.3) is represented by a parallel plate of the equivalent width on the complex plane for which the capacitance is analytical. The backward transformation to the real Cartesian coordinate system gives the strip capacitance C that takes into account the stray fields from the edges of the central conductor. Taking into account that for the TEM modes $k_z = \omega\sqrt{\varepsilon_r \mu_r} = \omega\sqrt{LC}$, the inductance L per unit length can be found from this expression, and the characteristic impedance is calculated as $Z_c = \sqrt{L/C}$. A simple formula is known for this parameter [12],[13]:

$$Z_c = \frac{30\pi}{\sqrt{\varepsilon_r}} \frac{K(k)}{K(k')}, \ [\Omega] \tag{2.30}$$

where $K(k)$ is the elliptic integral, $k = \operatorname{sech}(\pi w/4b)$, and $k' = \tanh(\pi w/4b)$.

The single and coupled microstrip lines, coplanar waveguides, and some other transmission lines are modeled by conformal mapping approach as well if the hybrid nature of the modal fields caused by the non-uniform dielectric filling is ignored. As a rule, the quasistatic formulas can be improved further taking into account the effects which cause the frequency dependence of the line parameters.

2.1.4.2 Coupled Strips Line

The coupled strips line consists of two conducting strips placed inside a shielded dielectric slab (Fig. 2.4). It supports propagation of two TEM modes – the odd (o) and even (e) ones classified according to the symmetry plane of the cross-section (Fig. 2.5).

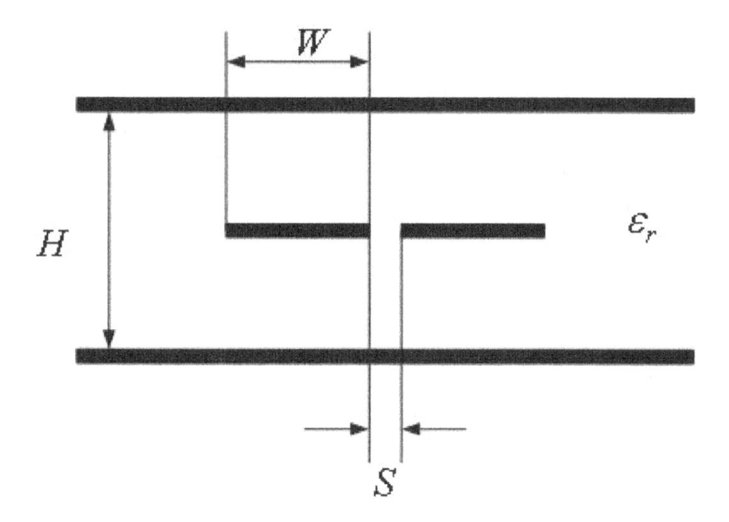

Fig. 2.4 Coupled strips transmission line

The higher-order modes of the TE and TM types propagate since their cut-off frequencies which depend on the geometry and dielectric filling of this line. The strips of the line, placed close to each other, are coupled, and the excitation of one of the strips leads to the excitation of both TEM modes, which modal field pictures are shown in Fig. 2.5. This line is used in a multiple microwave circuits, including filters, directional couplers, differential interconnect, and the lines for the mixed signaling by the even and the odd mode impulses.

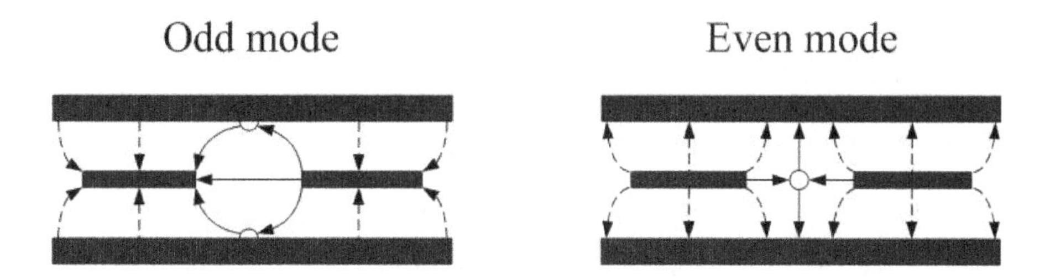

Fig. 2.5 Electric field maps of the odd and even modes of coupled strips line

Both modes, being of the TEM type, have the same propagation constants $k_z^{(o)} = k_z^{(e)} = k_0\sqrt{\varepsilon_r \mu_r}$ and different characteristic impedances $Z_c^{(o)}$ and $Z_c^{(e)}$. These modes are calculated by the conformal method or engineering formulas. Taking into account the importance of this line for today's microwave and high-speed electronics, the formulas for the modal characteristic impedances $Z_c^{(o)}$ and $Z_c^{(e)}$ from [13] are given below:

$$Z_c^{(o)} = \frac{30\pi}{\sqrt{\varepsilon_r}} \frac{K(k_o')}{K(k_o)}, [\Omega]; \; Z_c^{(e)} = \frac{30\pi}{\sqrt{\varepsilon_r}} \frac{K(k_e')}{K(k_e)}, [\Omega] \tag{2.31}$$

where $k_e = \tanh\left(\frac{\pi}{2}\frac{W}{H}\right)\tanh\left(\frac{\pi}{2}\frac{W+S}{H}\right)$, $k_o = \tanh\left(\frac{\pi}{2}\frac{W}{H}\right)\coth\left(\frac{\pi}{2}\frac{W+S}{H}\right)$, and $k'_{e,o} = \sqrt{1-k_{e,o}^2}$.

Additionally to the planarly coupled lines, more modifications are known. Among them are the broadside coupled strips lines of the horizontal (Fig. 2.6a) and vertical (Fig. 2.6b) designs with analytical formulas given for characteristic impedances [14]-[19].

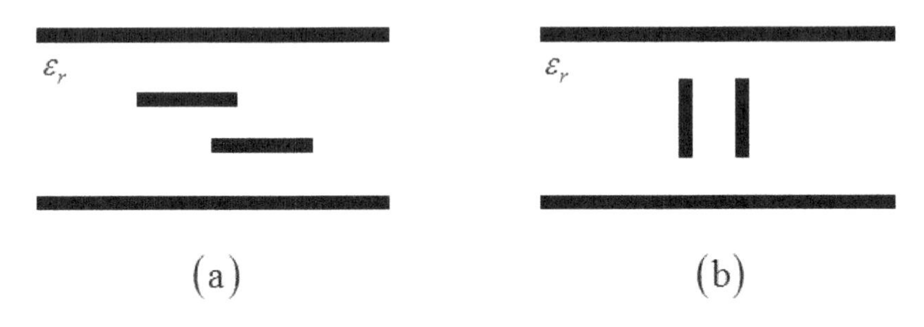

Fig. 2.6 Broadside coupled strip lines of the horizontal (a) and vertical (b) designs

2.1.5 Modeling of Quasi-TEM Modes of Transmission Lines

2.1.5.1 Parallel-plate Model of Microstrip Line

The above-considered multi-conductor lines support propagation of the TEM modes due to their homogeneous dielectric filling. Some effects, which are more complicated, take place in microstrip (MS) lines which conductors placed over a dielectric substrate. It leads to the hybridization of the modes, and they have all six field components. At low frequencies, the longitudinal field components are small enough, and the microstrip fundamental mode of the quasi-TEM type can be calculated using a modified technique applied earlier to the strip transmission lines.

Initially, the air-filled microstrip (Fig. 2.7a) is mapped onto a parallel plate waveguide (Fig. 2.7b), and the equivalent width $w_{eq}^{(air)}$ of it and the characteristic impedance Z_c formulas derived by Wheeler in [10],[11] are:

$$w_{eq}^{(air)} = w + \frac{2h}{\pi}\left\{\ln\left[2\pi e\left(\frac{w}{2h}+0.92\right)\right]\right\}, \tag{2.32}$$

$$Z_c = \eta \frac{h}{w_{eq}^{(air)}}, \; [\Omega] \tag{2.33}$$

where $\eta = \sqrt{\mu_0/\varepsilon_0}$.

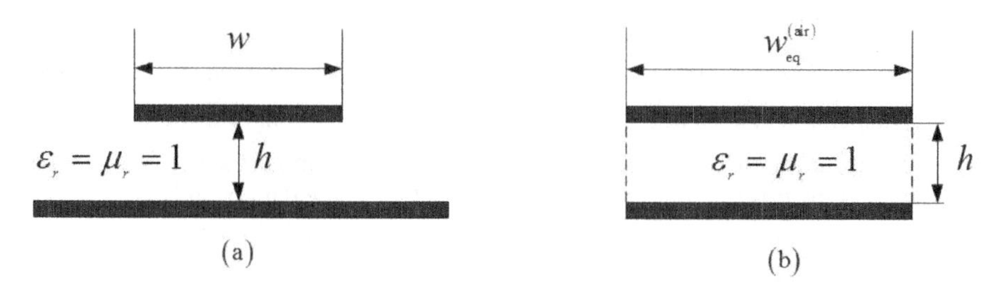

Fig. 2.7 Air-filled microstrip (a) and its parallel-plate model (b)

These formulas can be modified for the microstrip transmission line (Fig. 2.8). As has been mentioned, the fundamental mode of it is of the hybrid type, and it relates to the HE_0 one, i.e. its field has all six components of the EM field which is not alternating below microstrip conductor. At low frequencies, the longitudinal components can be neglected, and, formally, the quasistatic analysis can be applied to calculate the parameters of microstrip.

In comparison to strip line, the EM field of a microstrip has different concentrations in the air- and dielectric-filled regions, and it should be taken into account. In this case, the modal propagation constant will be proportional to a square root of the equivalent permittivity $k_z = k_0\sqrt{\varepsilon_{eq}}$, and this parameter takes into account the non-uniform distribution of the electric field caused by the layered dielectric and the edges of the strip conductor.

There are several engineering formulas used for calculation of the equivalent permittivity [10],[11],[20]-[23], and one of them, derived by Hammerstad and Jensen [22], is used throughout this book:

$$\varepsilon_{eq} = \frac{\varepsilon_r + 1}{2} + \frac{\varepsilon_r - 1}{2}\left(1 + \frac{10}{u}\right)^{-ab} \tag{2.34}$$

where

$$a = 1 + \frac{1}{49}\ln\left[\frac{u^4 + (u/\ 52)^2}{u^4 + 0.432}\right] + \frac{1}{18.7}\ln\left[1 + \left(\frac{u}{18.1}\right)^3\right],$$

$$b = 0.564\left(\frac{\varepsilon_r - 0.9}{\varepsilon_r + 3}\right), \; u = \frac{w}{h}.$$

The characteristic impedance Z_c of the microstrip transmission line is

$$Z_c = \frac{\eta}{2\pi\sqrt{\varepsilon_{eq}}} \ln\left[F_1/u + \sqrt{1+4/u^2} \right], \ [\Omega] \tag{2.35}$$

where $F_1 = 6 + (2\pi - 6)\exp\left[-(30.666/u)^{0.7528}\right]$.

The error in ε_{eq} of this formula is about 0.2% for low frequencies, where the modal dispersion (frequency dependence of ε_{eq}) is not strong. At these frequencies, the accuracy of the formula for the characteristic impedance (2.35) is high enough, and the error is no more than 1%, according to the data from [22].

The equivalent width w_{eq} of the parallel-plate waveguide given for microstrip (Fig. 2.8) is calculated from the idea about the equivalency of the transmitted power through the cross-sections of the microstrip line and parallel-plate waveguide, and, in the first approximation, it does not depend on frequency [24]:

$$w_{eq}(f=0) = \frac{\eta h}{\sqrt{\varepsilon_{eq}} Z_c}. \tag{2.36}$$

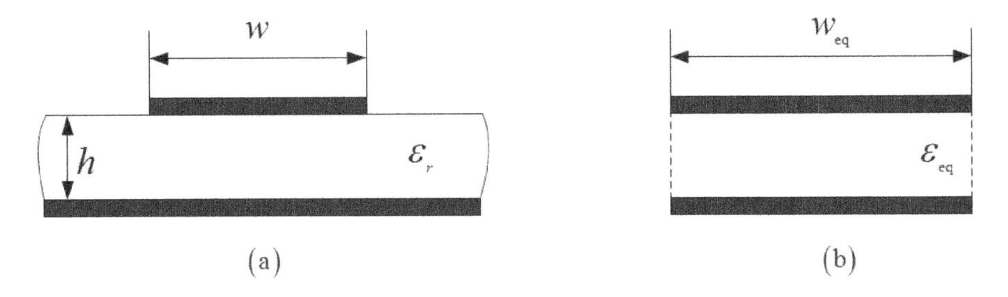

Fig. 2.8 Microstrip line (a) and its parallel-plate model (b)

At increased frequencies, the hybrid nature of the microstrip fundamental mode of the HE_0 type appears stronger, and the quasi-statically derived parameters have poor accuracy. These quasistatic formulas can be corrected using the frequency-dependent perturbations.

Several models have been developed which take into account the EM-effect caused dispersion of the fundamental microstrip mode [25]-[34]. One of such formulas for the frequency-dependent effective permittivity $\sqrt{\varepsilon_{eff}}$ is given by Yamashita and coauthors [31]:

$$\sqrt{\varepsilon_{eff}(f)} = \frac{\sqrt{\varepsilon_r} - \sqrt{\varepsilon_{eq}}}{1 + 4F^{-1.5}} + \sqrt{\varepsilon_{eq}} \tag{2.37}$$

where $F = \frac{4h\sqrt{\varepsilon_r - 1}}{\lambda_0}\left[0.5 + \left\{1 + 2\log_{10}\left(1 + \frac{w}{h}\right)\right\}^2\right]$ and $\lambda_0 = \frac{2\pi c}{\omega}$.

The applicable range of this formula defined by the authors of the cited paper is $2 < \varepsilon_r < 16,\ 0.06 < w/h < 16$, and $0.1\ \text{GHz} < f < 100\ \text{GHz}$. At high frequencies, the characteristic impedance Z_c is frequency-dependent, and it can be calculated according to the parallel-plate model:

$$Z_c(f) = \frac{\eta}{\sqrt{\varepsilon_{\text{eff}}(f)}} \frac{h}{w_{\text{eq}}(f)},\ [\Omega] \tag{2.38}$$

where the frequency-dependent effective width $w_{\text{eq}}(f)$ is

$$w_{\text{eq}}(f) = w + \frac{w_{\text{eq}}(f=0) - w}{1 + (f/f_c)^2} \tag{2.39}$$

where $f_c = c / \left(2w_{\text{eq}}(f=0)\sqrt{\varepsilon_{\text{eq}}}\right)$.

Similarly to any waveguide, the microstrip line supports the higher-order modes [35]. They are of the hybrid nature, and the first of them can be denoted as the modes having variations of the field along the strip width only: EH_1, EH_2,... (Fig. 2.9).

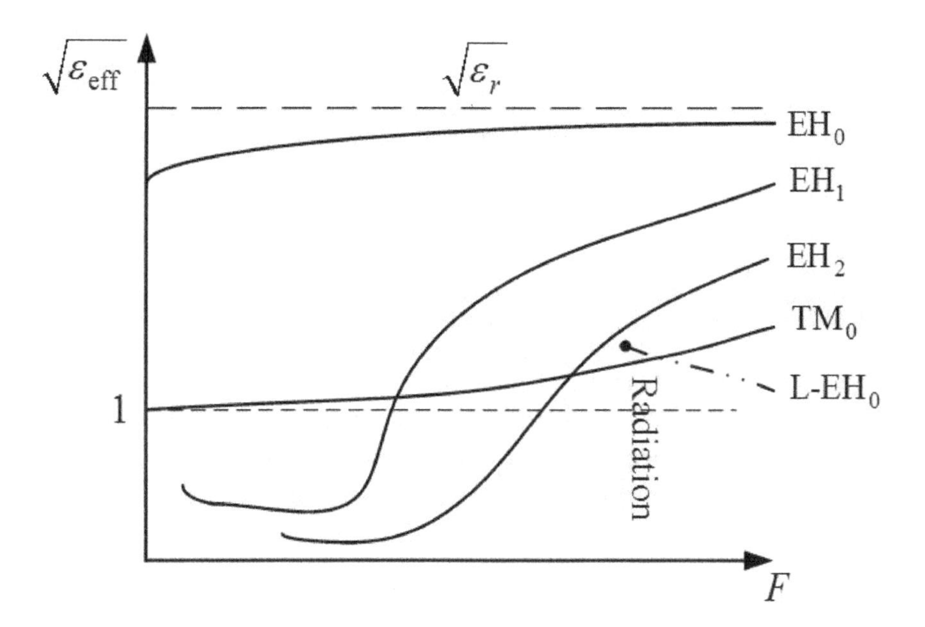

Fig. 2.9 Dispersion curves of the fundamental EH_0 and several first higher-order modes of a microstrip. Adapted from [35]-[39]

The parallel-plate approach allows for modeling of these strip modes by corresponding eigenfunctions of the equivalent magnetic-wall waveguide, and the experimentally verified simulations of discontinuities [24],[32],[33] show the validity of this technique, excepting the cases where the radiation from microstrip elements is very strong. There, the full-wave theory of microstrip line corrects this shielded model and explains the radiation physics of strip discontinuities.

At increased frequencies, other modes having more complicated fields can propagate along this line. It follows from the EM theory that open waveguides, additionally to the discrete spectrum eigenmodes, can support the waves of continuous spectrum, i.e. the radiation field propagating away from the regular open lines [34]. Additionally, the higher-order strip modes can have complex propagation constants due to radiation.

Fig. 2.9 shows a spectrum of a microstrip drawn after [35]-[39]. The main mode of this line is of the quasi-TEM type (EH_0), and it does not radiate. The higher-order strip modes tend to their cut-off frequencies with the frequency decrease. Up to the surface-mode TM_0 propagation region (solid curve), they have real propagation constants. Below this curve, their constants are complex, and the power tends away from the strip along the substrate, including along the transversal direction. Below the dashed straight line $\left(\sqrt{\varepsilon_{\mathrm{eff}}}=1\right)$, where the modal phase constant is less than k_0, these modes radiate into the space and along the surface of the substrate. The relative portion of two these radiations depends on the proximity of the modal normalized wave constant to the zero value. Additionally, at increased frequencies, the leaky mode of the fundamental type (L-EH_0) starts to propagate [37]. It has a complex modal constant, and it radiates into the substrate surface TM_0 mode.

This theory explains well the radiation of microstrip antennas. The incident main mode of a microstrip patch is diffracted at the edge, and a part of its power is transformed into the higher-order modes which have spatial and surface leakage. The last one taken by another patch is the cause of surface wave parasitic coupling distorting the parameters of multi-element antennas, for instance.

The higher-order modes can be found numerically [35],[36]-[42], analytically [28],[43], or using engineering formulas [44]. Some works are on measurements of the higher-order modes [41],[42]. For instance, a simple formula for complex propagation constant for the first higher-order mode is published in [44] where the complex effective width $\tilde{W}_e$ is introduced to model the radiation of the first higher-order mode. The propagation constant of this mode is calculated similarly to the main one, but the effective width $\tilde{W}_e$ is complex due to the radiation:

$$k_z=\sqrt{k_0^2\varepsilon_{\mathrm{eq}}-\left(\pi/\tilde{W}_e\right)^2} \tag{2.40}$$

where $\tilde{W}_e=W+2(\Delta W-jb)$ and $b=h/\left(2\varepsilon_{\mathrm{eq}}\right)$. A new formula for the width extension ΔW is introduced by the authors of the cited paper:

$$\Delta W = h\left[1+0.8874\left(\frac{1}{\varepsilon_r+0.375}\right)\left(\frac{W/h-11}{W/h+20.26}\right)\right]. \quad (2.41)$$

The results derived by these engineering formulas show the average relative error for the phase and loss constants in the limits of 2-10% depending on geometry and substrate permittivity of microstrip $(W/h \geq 5.5,\ \varepsilon_r \leq 12)$ and driving frequency. An example of the calculations of the real and imaginary parts of the propagation constant of this mode is shown in Fig. 2.10.

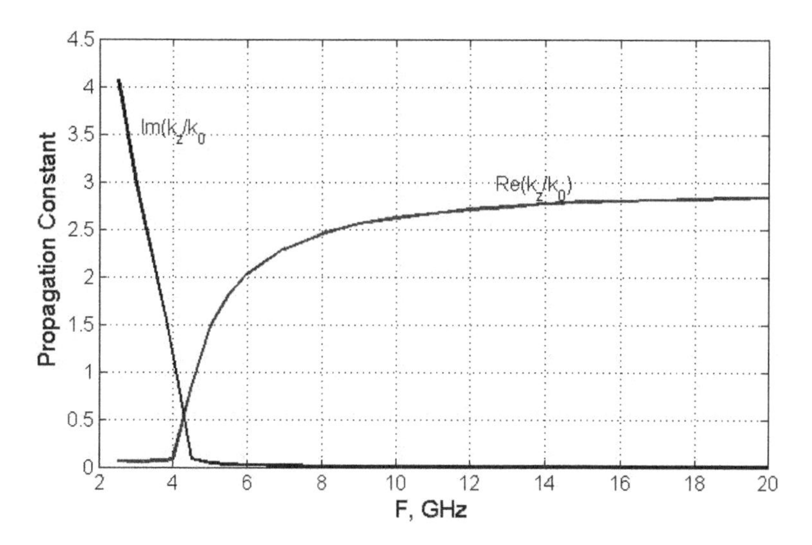

Fig. 2.10 Frequency dependence of the real and imaginary parts of the modal propagation constant of the EH_1 microstrip mode. Parameters of microstrip are: $h = 1$ mm, $W = 10h$, and $\varepsilon_r = 9.9$

It is seen that below the cut-off frequency $(f_c \approx 4$ GHz$)$, the phase constant is not zero due to the radiation which is increasing with the frequency decrease.

Calculations of the parameters of higher-order modes are important for proper modeling of microstrip discontinuities, antennas, and for a recently proposed oversized microstrip line serving as a differential waveguide [42].

2.1.5.2 Coupled Microstrips Line

One of the most popular basic elements of microwave and high-speed electronics is the coupled microstrips line shown in Fig. 2.11. It consists of two conducting strips placed close to each other for prescribed capacitive coupling.

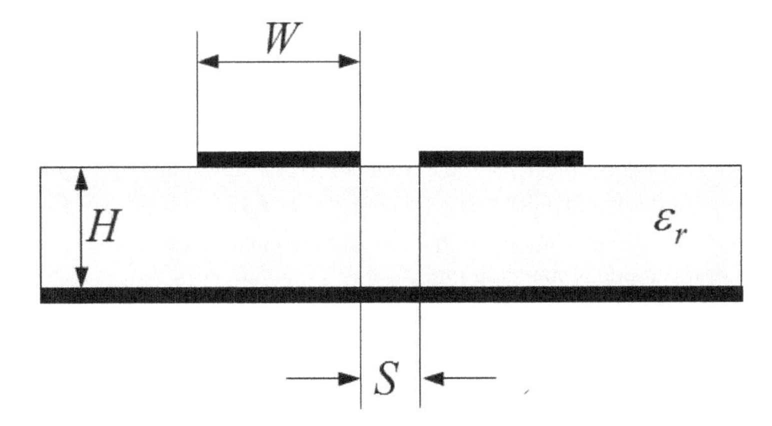

Fig. 2.11 Coupled microstrips line

This line supports two quasi-TEM modes, the odd and even, which have no cut-off frequencies. The propagating higher-order modes can be excited at increased frequencies similarly to other waveguides. The above-mentioned quasi-TEM modes propagate with the different phase velocities, and this difference can be around 10-15% depending on the geometry and substrate permittivity.

In quasi-static approximations, the modal equivalent circuits are composed of capacitors, which values are calculated analytically using the variational technique or conformal mapping approach [45]-[49]. In this case, the modal parameters are frequency-independent, of course. One set of formulas of the modal characteristic impedances and effective permittivities validated by full-wave simulations and measurements are given here from [49]:

$$Z_{\mathrm{c}}^{(e)} = \frac{1}{c\sqrt{C_{ae}C_{e}}}, \; [\Omega] \tag{2.42}$$

$$Z_{\mathrm{c}}^{(o)} = \frac{1}{c\sqrt{C_{ao}C_{o}}}, \; [\Omega] \tag{2.43}$$

$$\varepsilon_{\mathrm{eq}}^{(e)} = \frac{C_{e}}{C_{ae}}, \tag{2.44}$$

$$\varepsilon_{\mathrm{eq}}^{(o)} = \frac{C_{o}}{C_{ao}} \tag{2.45}$$

where $C_{e,o}$ are the modal capacitances of the even and odd modes calculated for a dielectric filled line cross-section, and $C_{ae,ao}$ are the corresponding capacitances of air-filled line. They are calculated by the following formulas which are the same for the dielectric (ε_r) and air-filled $(\varepsilon_r = 1)$ lines:

$$C_e = 2\varepsilon_0 \left[\varepsilon_r \frac{K(k_1)}{K(k_1')} + \frac{K(k_2)}{K(k_2')} \right], \tag{2.46}$$

$$C_o = 2\varepsilon_0 \left[\varepsilon_r \frac{K(k_3)}{K(k_3')} + \frac{K(k_4)}{K(k_4')} \right] \tag{2.47}$$

where

$$\begin{aligned} k_1 &= \tanh\left(\frac{\pi W}{4H}\right) \tanh\left(\frac{\pi(W+S)}{4H}\right), \\ k_2 &= \tanh\left(\frac{\pi W}{4(H+\pi W)}\right) \tanh\left(\frac{\pi(W+S)}{4(H+\pi W)}\right), \end{aligned} \tag{2.48}$$

$$k_3 = \tanh\left(\frac{\pi W}{4H}\right) \coth\left(\frac{\pi(W+S)}{4H}\right), \tag{2.49}$$

$$k_4 = \frac{W}{W+S}, \text{ and } k'_{1,2,3,4} = \sqrt{1-k_{1,2,3,4}^2}. \tag{2.50}$$

At low frequencies, this model is appropriate, but at increased frequencies, these modes are dispersive, i.e. their parameters, as the effective permittivities and characteristic impedances, are frequency-dependent. Further improvements of accuracy in a wide frequency band can be reached taking into account this dispersion. For instance, in [50], the equivalent circuits of modes are introduced consisting of capacitors and inductors, and they model the dispersion effect. In [51], it is reached by introducing the frequency-dependent partial capacitances of the strips and their effective widths. These semi-empirical formulas show good accuracy up to 10-15 GHz in comparison with the full-wave simulations.

In [46],[52], the empirically found formulas for the frequency-dependent modal effective permittivities are given. They are [46]

$$\varepsilon_{\text{eff}}^{(e,o)}(f) = \varepsilon_r - \frac{\varepsilon_r - \varepsilon_{\text{eq}}^{(e,o)}}{1 + G^{(e,o)} \cdot \left(f / f_p^{(e,0)}\right)^2} \tag{2.51}$$

where $G^{(e)} = 0.6 + 0.0045 Z_c^{(e)}$, $G^{(o)} = 0.6 + 0.018 Z_c^{(o)}$, $f_p^{(e)} = 7.83 Z_c^{(e)} / H$, and $f_p^{(o)} = 31.32 Z_c^{(o)} / H$. In the same work, the empirical expressions are given for frequency-dependent characteristic impedances.

2.1.5.3 Coplanar Waveguide Model

The coplanar waveguide (CPW) is a transmission line whose all conductors are placed on the top surface of a dielectric substrate [53]. Some of these transmission lines are shown in Fig. 2.12.

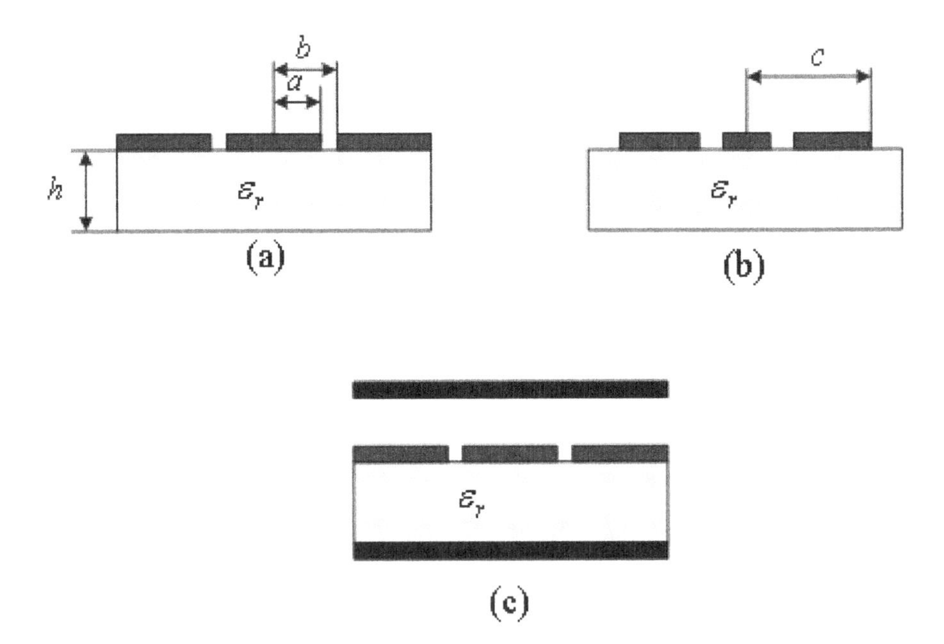

Fig. 2.12 Coplanar waveguides. (a)- CPW with infinite wide ground planes; (b)- CPW with finite lateral ground planes; (c)- Shielded CPW

Coplanar transmission lines have some technological advantages comparing to the microstrips because of their planar design. In most cases, CPWs have the same characteristics as the microstrips or even better. Coplanar waveguides can be connected to other transmission lines using the planar transitions, via-holes, or vertical EM connectors. An increased interest in the CPW starts with the monolithic ICs where this line demonstrates reduced loss of some geometries at microwave and millimeter-wave frequencies.

Initially, the CPW was invented for the hybrid ICs, and it was assigned as a line with the elliptical polarization of the quasi-TEM modal magnetic field that could be convenient for the ferrite devices. It was supposed that this line could be placed on a surface of a low-loss dielectric (Fig. 2.12), and several designs were studied and modeled by the conformal mapping method due to the quasi-TEM nature of the fundamental mode of this line.

Many formulas for calculation of the effective permittivity ε_{eq} and characteristic impedance Z_c are available now [13],[14],[54]-[59]. Additionally, this line is modeled by different numerical methods [60]-[64].

The CPWs of classical design, shown in Fig. 2.12, are calculated with formulas obtained using a conformal mapping approach [54]. There are two quasi-TEM modes, which have different modal field distributions. The fundamental or even mode has the magnetic plane of symmetry at the center of signal conductor, and this plane oriented along the longitudinal axis which is normal to picture plane. This mode is supported by geometrical symmetry of the line cross-section, and the potentials of ground planes are equalized using the air bridges, for instance. The second mode, which has zero cut-off frequency, can be effectively excited at the discontinuities, and careful potential symmetrization is needed to suppress this parasitic transformation of the even mode to the odd one.

Consider the quasistatic formulas [54] for the even mode of the CPW shown in Fig. 2.12b. The effective dielectric permittivity $\varepsilon_{\mathrm{eq}}$ of this line is calculated as a ratio of the capacitance C_{e} of the line filled with the dielectric ε_r to the capacitance C_0 of the air-filled line:

$$\varepsilon_{\mathrm{eq}} = \frac{C_{\mathrm{e}}}{C_0}. \tag{2.52}$$

The capacitances C_0 and C_{e} need the elliptical functions or their approximations. For a CPW shown in Fig. 2.12b, the formulas for calculation of the effective permittivity take into account the limited width of the ground conductors, and they are

$$\varepsilon_{\mathrm{eq}} = \frac{C_{\mathrm{e}}}{C_0} = 1 + q(\varepsilon_r - 1) \tag{2.53}$$

where

$$q = \frac{1}{2}\frac{K(k_6)}{K(k_6')}\frac{K(k_5')}{K(k_5)}. \tag{2.54}$$

The arguments k_5 and k_6 of the elliptic integrals K are

$$k_{5,6}' = \sqrt{1 - (k_{5,6})^2},\ k_5 = \frac{a}{b}\sqrt{\frac{1 - b^2/c^2}{1 - a^2/c^2}},$$
$$k_6 = \frac{\sinh(\pi a/2h)}{\sinh(\pi b/2h)}\sqrt{\frac{1 - \sinh^2(\pi b/2h)/\sinh^2(\pi c/2h)}{1 - \sinh^2(\pi a/2h)/\sinh^2(\pi c/2h)}}. \tag{2.55}$$

The characteristic impedance Z_{c} of the CPW is

$$Z_{\mathrm{c}} = \frac{1}{c\sqrt{\varepsilon_{\mathrm{eq}}}C_{\mathrm{e}}^2} = \frac{30\pi}{\sqrt{\varepsilon_{\mathrm{eq}}}}\frac{K(k_5')}{K(k_5)},\ [\Omega]. \tag{2.56}$$

The influence of the conductor-backing and upper shielding (Fig. 2.12c) is studied in [57]. More information about such a CPW and its modifications, components, and full-wave simulations is from [32],[59], for instance.

The above-given formulas are derived under the assumption of the quasi-TEM character of the main mode. Meanwhile, the fundamental modal field is composed from many partial waves, including the modes of the substrate. At increased frequencies, some energy portion is pumped into these partial modes, and it leads to the frequency dependence of the fundamental mode parameters and the modal radiation. These effects are modeled by full-wave simulations [13],[32]-[59],[60]-[64], and, for some cases, the approximate formulas are given to calculate the frequencies where strong radiation starts and for the radiation loss constants [13].

The conductor and dielectric losses are studied in many contributions, including [13]. A large amount of information on the CPW and components and their full-wave simulations and measurements is in [63],[64].

More than four decades of exploiting the CPW show that this line is distinguished by its reduced dispersion and radiation loss, the extended frequency band, low fabrication cost, easier grounding of components placed over the CPW, weak dependence of the CPW parameters on the substrate thickness and permittivity, and the availability of analytical formulas for the CPW parameters [14],[32],[59].

The disadvantage of this line is its increased space, excitation of the high-order modes due to the inaccuracy of the manufacturing and asymmetry of the ground planes. Additionally, at very increased frequencies, the main mode is radiating, and it leads to parasitic coupling of integrated components.

2.2 Numerical Methods Used in the Waveguide Theory

A large number of EM problems are solved only numerically due to their domain geometry and boundary conditions which do not allow for the separation of the variables. Very often, the above-considered *field expansion method* is with the time-consuming calculations of sub-domain functions and more effective numerical methods are applied in electromagnetics.

2.2.1 Finite Difference and Transmission Line Matrix Methods

In this case, one of the applicable methods of solution of differential equations is the *finite difference method* which allows for deriving the algebraic equations directly from the differential ones [4],[65],[66]. The method is used for both linear and nonlinear ordinary and partial differential equations. Although it is simple to be programmed, some problems should be solved with the increased stability of calculations and quality of models of the boundary shapes and conditions on them.

The method is applied directly to the shielded waveguides and components. The external problems of electromagnetism are difficult to be calculated due to increased needed dimension of the resulting system of algebraic equations. To avoid these problems, a calculated domain is surrounded by, for instance, an absorbing boundary imitating the open space. It allows for the modeling the field phenomena in open space using a domain of a reduced size.

The modeled space area and its boundary are divided into a grid of $i, j-$ nodes, and, at each node, a differential operator is substituted by a finite difference expression. For instance, the finite differences of the 2nd order written at the $i, j-th$ node are

$$\frac{\partial^2 u(x,y)}{\partial x^2} \approx \frac{u_{i+1,j} - 2u_{i,j} + u_{i-1,j}}{(\Delta x)^2},$$
$$\frac{\partial^2 u(x,y)}{\partial y^2} \approx \frac{u_{i,j+1} - 2u_{i,j} + u_{i,j-1}}{(\Delta y)^2} \tag{2.57}$$

where $u_{i,j} = u(x_i, y_j)$.

Substituting the differential operators in (2.2) by their finite difference formulas and taking into account the given boundary condition, a system of linear algebraic equations is derived regarding to the field values at the $i, j-th$ nodes. The eigenvalue equation is the determinant of this homogeneous system. Its roots are the modal propagation constants.

The time-depending wave effects are calculated by discretization of the time-differential operator, and the second time derivative expressed through the values of $u(x,y)$ in the neighboring nodes is

$$\frac{\partial^2 u}{\partial t^2} \approx \frac{u_{i,j+1} - 2u_{i,j} + u_{i,j-1}}{(\Delta t)^2}. \tag{2.58}$$

The frequency properties of the modeled waveguides and components are found by the Fourier transformations of the calculated time-depending data.

In many cases, the EM problems are with the hybrid fields consisting of all six components. Today, they can be solved with the Maxwell equations in their finite difference formulation. The origin of this technique relates to the work of Yee published in 1966 [67], and it is called as the *finite difference time domain* technique (FDTD) after the works of Taflove [66]. Today many commercially available software tools are known to simulate the Maxwell, elliptic, parabolic, and hyperbolic type of partial differential equations. Continuous progressing of computers allows for solving the electrically large EM problems.

There is another approach representing the modeled area and the EM field by a mesh of transmission lines and named as the *transmission-line-matrix (TLM) method*. These short mesh lines $(l \ll \lambda)$ are modeled by the elementary networks of lumped components, and the wave phenomenon is modeled as the electric pulse propagation along this 3-D network.

An initial idea on the representation of the Maxwell equations by the equivalent networks was published by G. Kron [68], and a more detailed work in this area belongs to P. Johns and R. Beurle [69]. Since that time, the method has been evolved from the research-level codes to the well-developed commercially available software tools. Today this approach is used not only for modeling of the EM phenomena, but also for simulating of quantum-mechanical effects (see Chpt. 9).

A comparison of the FDTD and TLM methods is considered in [4],[70] where it is shown that the FDTD and some cases of the TLM are equivalent to each other, and they can be derived from each other. Although the FDTD requires less memory and it is faster, the TLM has advantage to treat the EM problems of complicated boundaries and boundary conditions.

2.2.2 Integral Equations for Waveguides and Their Numerical Treatment

Additionally to the differential equations, the integral ones are used in electromagnetics [1],[71]. Two main types of integral equations are common for electromagnetism:

$$\int J(x')g(x,x')dx' = f(x) \tag{2.59}$$

and

$$\lambda \int J(x')g(x,x')dx' = J(x) + f(x). \tag{2.60}$$

In (2.59) and (2.60), $J(x)$ is the unknown function, $f(x)$ is the known or driving function which is zero for the eigenvalue problems, and λ is the parameter. The other function $g(x,x')$ is the integral equation kernel that can be infinite at $x = x'$. Then this integral equation is the singular one, and a special technique is used to solve it [72],[73].

The integral equation of the first kind (2.59) is more typical for the EM problems. The integral equation of the second kind (2.60) can be derived using different ways, including the Green function approach, or it is obtained from (2.59) by a special technique called the regularization [74]-[76], and it is more preferable for the use due to increased stability of numerical solutions.

The attractive feature of the integral equations is that they decrease the dimension of EM problems. For instance, a 2-D problem (Fig. 2.1) is transformed into a 1-D integral equation defined only on the boundary.

To explain the integral equation method, consider initially a 2-D domain S where a scalar Helmholtz equation regarding to the unknown function $u(\mathbf{r})$ is defined

$$\nabla^2 \cdot u(\mathbf{r}) + \chi^2 u(\mathbf{r}) = f(\mathbf{r}). \tag{2.61}$$

The integral equation is derived using the Green's function $G(\mathbf{r},\mathbf{r}')$ which is obtained from the following equation:

$$\nabla^2 \cdot G(\mathbf{r},\mathbf{r}') + \chi^2 G(\mathbf{r},\mathbf{r}') = \delta(\mathbf{r}-\mathbf{r}') \tag{2.62}$$

where $\delta(\mathbf{r}-\mathbf{r}')$ is the Dirac function. At the difference to (2.61), the boundary condition on L for this Green's function in (2.62) can be chosen arbitrary, and it allows for obtaining analytical solutions of (2.62) if the domain geometry is separable, for instance.

Using the second Green's formula [1], the field at any arbitrary point of the domain is calculated if $u(\mathbf{r}' \in L)$ is known

$$u(\mathbf{r}) = \int_S f(\mathbf{r}')G(\mathbf{r},\mathbf{r}')ds' + \oint_L \left[u(\mathbf{r}') \frac{\partial G(\mathbf{r},\mathbf{r}')}{\partial \nu} - G(\mathbf{r},\mathbf{r}') \frac{\partial u(\mathbf{r}')}{\partial \nu} \right] dl' \tag{2.63}$$

where ν is the normal direction to the boundary *L*. As a rule, $u(\mathbf{r} \in L)$ is the electric field tangential to the boundary, and $\frac{\partial u(\mathbf{r} \in L)}{\partial \nu}$ corresponds to the current on the conducting part of *L*, respectively. Suppose $\mathbf{r} \in L$ and $u(\mathbf{r} \in L) = 0$, the simplest integral equation of the first kind is derived

$$\oint_L \left[G(\mathbf{r},\mathbf{r}') \frac{\partial u(\mathbf{r}')}{\partial \nu} \right] dl' = \int_S f(\mathbf{r}')G(\mathbf{r},\mathbf{r}')ds', \mathbf{r} \in L. \tag{2.64}$$

In the case of the eigenvalue problem, it gives

$$\oint_L \left[G(\mathbf{r},\mathbf{r}') \frac{\partial u(\mathbf{r}')}{\partial \nu} \right] = 0, \ \mathbf{r} \in L. \tag{2.65}$$

As seen, (2.65) is the integral equation of the 1st kind regarding to the surface current on the conducting contour *L*. Very often, it is transformed into an integral equation of the 2nd kind to avoid the numerical complications due to the poor stability of the solutions of (2.65). Similarly to this algorithm, more complicated multi-domain cross-sections are treated by the integral equations derived by satisfaction of the boundary conditions on the inter-domain boundaries.

In the case of a vector boundary value problem, the scalar Green's function is transformed into the tensor one [3],[77]-[79]. In shielded sub-domains, the Green's functions are expressed through the sub-domain eigenfunctions, which can be found analytically, and the techniques to compose the Green's functions are published in many books and papers [4]. One of the problems arisen using the Green's functions is the singularity of kernels and caused by it poor convergence of eigenfunction expansions. Then the integral equations of this sort are to be regularized.

As a rule, the integral equations are hard to be solved analytically, and a number of different numerical methods are applied, including the widely known *method of moments* [4],[71],[80],[81]. According to this method, the unknown function $\upsilon(\mathbf{r}' \in L) = \frac{\partial u(\mathbf{r}' \in L)}{\partial \nu}$ in (2.65), for instance, is represented by a finite sum of basis functions $\upsilon_n(\mathbf{r} \in L)$, which amplitudes a_n are unknown, and these functions are linearly independent on each other:

$$\upsilon(\mathbf{r}' \in L) = \sum_{n=1}^{N} a_n \upsilon_n(\mathbf{r}' \in L). \tag{2.66}$$

The basis functions υ_n can be defined on the whole L or exist on the non-overlapped segments of this boundary being the piecewise functions. Additionally to the set of basis functions, the weighting functions w_m should be defined. In a particular case $w_m = \upsilon_m$, the algorithm is in its Galerkin formulation named after B.G. Galerkin [81]. Then the integral equation (2.65) is transformed into a system of linear algebraic equations regarding to the unknown coefficients a_n by calculation of moments of the integral equation to each weighting function w_m if the unknown function is represented by (2.66):

$$\sum_{n=1}^{N} a_n \oint_L \oint_L \left[G(\mathbf{r},\mathbf{r}') \upsilon_n (\mathbf{r}' \in L) \right] dl' w_m (\mathbf{r} \in L) dl = 0, \quad n=1,2,3,...N;\ m=1,2,3,...N. \tag{2.67}$$

The determinant of this homogeneous system gives an equation for calculation of modal propagation constants.

The convergence of this numerical algorithm depends on the chosen functions υ_n and w_m and on the geometry of the waveguide cross-section. Additionally to the mentioned Galerkin algorithm, a number of different approaches of algebraization of the integral equations are known. Among them are the sub-domain, collocation, least-square methods, *etc.* [4],[71],[78],[80]-[88].

Example 2.1. Integral Equation System for a Shielded Microstrip Line. An example of a shielded microstrip line is shown in Fig. 2.13. Particularly, the strip can be placed symmetrically in the box, and only a half of the structure is considered for treatment. In this case, a magnetic wall is placed at the center $(x=0)$ of the waveguide box where the tangential to this plane magnetic field components are zero.

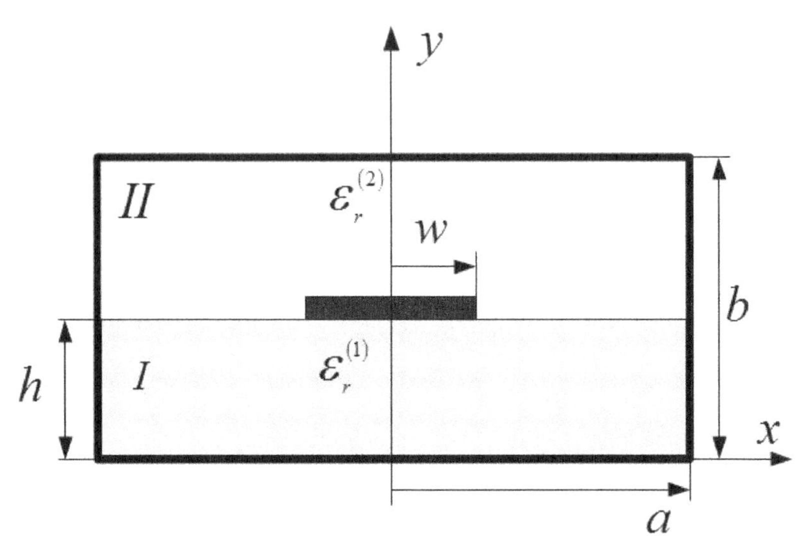

Fig. 2.13 Shielded microstrip line

The air-dielectric interface is the boundary between two domains I and II where the fields are represented as the domain modal series expansions. Satisfaction of the boundary conditions at $(y=h)$ for the electric and magnetic fields allows for obtaining the modal propagation constants and modal fields. Below, the integral equation method is considered in details, and it gives an algorithm how to treat more complicated transmission lines of planar and multilayered designs.

In general case, the microstrip modes have all six components of the EM field. Similarly to rectangular waveguide, the longitudinal electric and magnetic field components should be initially derived. They are general solutions of the Helmholtz equations defined in each sub-domain. The boundary condition on the air-dielectric interface cannot be satisfied by single domain functions, and the longitudinal fields are represented by their expansions. The longitudinal electric field components in both sub-domains are

$$
\begin{aligned}
E_z^{(1)} &= \sqrt{\frac{2}{a}}\sum_{m=0}^{\infty}\left[A_m \sin\left(k_{y,m}^{(1)} y\right)\cos\left(k_{x,m}x\right)\right]\exp(-jk_z z),\\
E_z^{(2)} &= \sqrt{\frac{2}{a}}\sum_{m=0}^{\infty}\left[B_m \sin\left(k_{y,m}^{(2)}(b-y)\right)\cos\left(k_{x,m}x\right)\right]\exp(-jk_z z).
\end{aligned}
\tag{2.68}
$$

where A_m and B_m are unknown amplitudes, and k_z is the unknown modal propagation constant, $k_{y,m}^{(1,2)} = \sqrt{k_0^2 \varepsilon_r^{(1,2)} \mu_r^{(1,2)} - \left(k_{x,m}\right)^2}$, and $k_{x,m} = \dfrac{(2m+1)\pi}{2a}$. These field components satisfy the boundary conditions $E_z^{(1,2)}(x=\pm a, y)=0$, $E_z^{(1)}(x, y=0)=0$, and $E_z^{(2)}(x, y=b)=0$. At the magnetic wall $E_z^{(1,2)}(x=0) \sim (-1)^{2m+1}$.

Similarly, the longitudinal components of the magnetic fields are written in both sub-domains:

$$
\begin{aligned}
H_z^{(1)} &= \sqrt{\frac{2}{a}}\sum_{m=0}^{\infty}\left[C_m \cos\left(k_{y,m}^{(1)} y\right)\sin\left(k_{x,m}x\right)\right]\exp(-jk_z z),\\
H_z^{(2)} &= \sqrt{\frac{2}{a}}\sum_{m=0}^{\infty}\left[D_m \cos\left(k_{y,m}^{(2)}(b-y)\right)\sin\left(k_{x,m}x\right)\right]\exp(-jk_z z).
\end{aligned}
\tag{2.69}
$$

Here, the amplitudes C_m and D_m are unknown, and the components satisfy the corresponding boundary conditions everywhere excepting this air-dielectric interface where the conducting strip is placed $(y=h)$.

The transversal field components are found using (2.4), (2.68), and (2.69):

$$
\begin{aligned}
E_x^{(1)} &= \sqrt{\frac{2}{a}}\, j\sum_{m=0}^{\infty}\left[\frac{k_z k_{x,m}}{\chi_1^2}A_m + \frac{\omega\mu_0\mu_r^{(1)}k_{y,m}^{(1)}}{\chi_1^2}C_m\right]\sin\left(k_{y,m}^{(1)} y\right)\sin\left(k_{x,m}x\right)e^{-jk_z z},\\
E_x^{(2)} &= \sqrt{\frac{2}{a}}\, j\sum_{m=0}^{\infty}\left[\frac{k_z k_{x,m}}{\chi_2^2}B_m - \frac{\omega\mu_0\mu_r^{(2)}k_{y,m}^{(2)}}{\chi_2^2}D_m\right]\sin\left(k_{y,m}^{(2)}(b-y)\right)\sin\left(k_{x,m}x\right)e^{-jk_z z},
\end{aligned}
\tag{2.70}
$$

$$
\begin{aligned}
H_x^{(1)} &= \sqrt{\frac{2}{a}}\, j\sum_{m=0}^{\infty}\left[\frac{\omega\varepsilon_0\varepsilon_r^{(1)}k_{y,m}^{(1)}}{\chi_1^2}A_m-\frac{k_z k_{x,m}}{\chi_1^2}C_m\right]\cos\left(k_{y,m}^{(1)}y\right)\cos\left(k_{x,m}x\right)e^{-jk_z z},\\
H_x^{(2)} &= -\sqrt{\frac{2}{a}}\, j\sum_{m=0}^{\infty}\left[\frac{\omega\varepsilon_0\varepsilon_r^{(2)}k_{y,m}^{(2)}}{\chi_2^2}B_m+\frac{k_z k_{x,m}}{\chi_2^2}D_m\right]\cos\left(k_{y,m}^{(2)}(b-y)\right)\cos\left(k_{x,m}x\right)e^{-jk_z z}.
\end{aligned}
\tag{2.71}
$$

To derive the integral equations, the unknown amplitudes A_m, B_m, C_m, and D_m are expressed through the electric field components on the slot between the conductor edge $(x=w)$ and the shield $(x=a)$ using the orthogonality of domain eigenfunctions on $x=(0,a)$. Initially, the electric series field expansions are matched to the unknown fields $e_z(x)$ and $e_x(x)$ on the slot:

$$
\begin{aligned}
E_z^{(1)} &= \sqrt{\frac{2}{a}}\sum_{m=0}^{\infty}\left[A_m\sin\left(k_{y,m}^{(1)}h\right)\cos\left(k_{x,m}x\right)\right]=e_z(x),\\
E_z^{(2)} &= \sqrt{\frac{2}{a}}\sum_{m=0}^{\infty}\left[B_m\sin\left(k_{y,m}^{(2)}(b-h)\right)\cos\left(k_{x,m}x\right)\right]=e_z(x),\\
E_x^{(1)} &= \sqrt{\frac{2}{a}}\, j\sum_{m=0}^{\infty}\left[\frac{k_z k_{x,m}}{\chi_1^2}A_m+\frac{\omega\mu_0\mu_r^{(1)}k_{y,m}^{(1)}}{\chi_1^2}C_m\right]\sin\left(k_{y,m}^{(1)}h\right)\sin\left(k_{x,m}x\right)=e_x(x),\\
E_x^{(2)} &= \sqrt{\frac{2}{a}}\, j\sum_{m=0}^{\infty}\left[\frac{k_z k_{x,m}}{\chi_2^2}B_m-\frac{\omega\mu_0\mu_r^{(2)}k_{y,m}^{(2)}}{\chi_2^2}D_m\right]\sin\left(k_{y,m}^{(2)}(b-h)\right)\sin\left(k_{x,m}x\right)=e_x(x)\\
&x\in(0,a).
\end{aligned}
\tag{2.72}
$$

In these expressions and below, the exponent $e^{-jk_z z}$ is omitted.

Taking into account the orthogonality of the sub-domain eigenfunctions on $x\in(0,a)$, the expressions for unknown amplitudes are obtained

$$
\begin{aligned}
A_m &= \frac{\mathrm{E}_{z,m}}{\sin\left(k_{y,m}^{(1)}h\right)},\quad B_m=\frac{\mathrm{E}_{z,m}}{\sin\left(k_{y,m}^{(2)}(b-h)\right)},\\
C_m &= -\frac{j\chi_1^2\mathrm{E}_{x,m}+k_z k_{x,m}\mathrm{E}_{z,m}}{\omega\mu_0\mu_r^{(1)}k_{y,m}^{(1)}\sin\left(k_{y,m}^{(1)}h\right)},\quad D_m=\frac{j\chi_2^2\mathrm{E}_{x,m}+k_z k_{x,m}\mathrm{E}_{z,m}}{\omega\mu_0\mu_r^{(2)}k_{y,m}^{(2)}\sin\left(k_{y,m}^{(2)}(b-h)\right)}
\end{aligned}
\tag{2.73}
$$

where

$$
\begin{aligned}
\mathrm{E}_{x,m} &= \sqrt{\frac{2}{a}}\int_w^a e_x(x')\sin\left(k_{x,m}x'\right)dx',\\
\mathrm{E}_{z,m} &= \sqrt{\frac{2}{a}}\int_w^a e_z(x')\cos\left(k_{x,m}x'\right)dx'.
\end{aligned}
\tag{2.74}
$$

Finally, to derive the integral equations, the boundary conditions for the magnetic field components on the inter-domain interface are used

$$H_z^{(2)} - H_z^{(1)} = j_x(x), \\ H_x^{(2)} - H_x^{(1)} = -j_z(x), \tag{2.75}$$

and the coefficients (2.73) are substituted into (2.69), (2.71) and (2.75). After some elementary treatments, the integral equations regarding to the unknown electric field (e_x, e_y) on the slot between the strip edge $(x = w)$ and the shield wall $(x = a)$ are obtained

$$\frac{2}{a}\sum_{m=0}^{\infty}\left[g_{11}^{(m)}\int_w^a e_x(x')\sin(k_{x,m}x') + g_{12}^{(m)}\int_w^a e_z(x')\cos(k_{x,m}x')\right]\sin(k_{x,m}x)dx' = j_x(x), \\ \frac{2}{a}\sum_{m=0}^{\infty}\left[g_{21}^{(m)}\int_w^a e_x(x')\sin(k_{x,m}x') + g_{22}^{(m)}\int_w^a e_z(x')\cos(k_{x,m}x')\right]\cos(k_{x,m}x)dx' = -j_z(x) \tag{2.76}$$

where $g_{11}^{(m)} = \frac{j\chi_2^2}{\omega\mu_0\mu_r^{(2)}k_{y,m}^{(2)}}\cot\left(k_{y,m}^{(2)}(b-h)\right) + \frac{j\chi_1^2}{\omega\mu_0\mu_r^{(1)}k_{y,m}^{(1)}}\cot\left(k_{y,m}^{(1)}h\right)$,

$$g_{12}^{(m)} = g_{21}^{(m)} = \frac{k_z k_{x,m}}{\omega\mu_0\mu_r^{(2)}k_{y,m}^{(2)}}\cot\left(k_{y,m}^{(2)}(b-h)\right) + \frac{k_z k_{x,m}}{\omega\mu_0\mu_r^{(1)}k_{y,m}^{(1)}}\cot\left(k_{y,m}^{(1)}h\right),$$

$$g_{22}^{(m)} = -j\left(\frac{(k_z k_{x,m})^2}{\chi_2^2\omega\mu_0\mu_r^{(2)}k_{y,m}^{(2)}} + \frac{\omega\varepsilon_0\varepsilon_r^{(2)}k_{y,m}^{(2)}}{\chi_2^2}\right)\cot\left(k_{y,m}^{(2)}(b-h)\right) - \\ -j\left(\frac{(k_z k_{x,m})^2}{\chi_1^2\omega\mu_0\mu_r^{(1)}k_{y,m}^{(1)}} + \frac{\omega\varepsilon_0\varepsilon_r^{(1)}k_{y,m}^{(1)}}{\chi_1^2}\right)\cot\left(k_{y,m}^{(1)}h\right).$$

These integral relations transform the electric slot field into the currents on the strip, and due to that they belong to the admittance type of EM integral equations [78],[89].

In some cases, the integral equations should be written regarding to the currents on conductors, and these equations can be derived in a similar way. The unknown coefficients A_m, B_m, C_m, and D_m are expressed through the strip currents $j_x(x)$ and $j_z(z)$, and the integral equations of the impedance type are obtained. They are preferable if the width of conductors exceed the one of the slots.

Numerical simulations of the integral equations can be performed with the help of the Galerkin method, and a various basis functions are in use. For instance, the basis functions take into account the singularity of fields at the conductor edges, and the algorithms are free from the relative convergence phenomenon [86]. The discretely defined functions provide flexibility in the choice of discrete polynomials. In this case, a pertinent mesh that can take into account the singular behavior of the fields close to the conductor or/and dielectric edges can be used to improve the convergence of the algorithm. A pertinent basis function sets can be composed of the trigonometrical or linear and constant pulse functions [71],[87],[89]-[93]. The last set of functions for the considered example is

$$e_x(x)=\sum_{k=1}^{K}a_k u_k(x),\qquad u_k(x)=\begin{cases}1, & x\in(x_{k-1},x_k)\\ 0, & x\notin(x_{k-1},x_k)\end{cases},\tag{2.77}$$

$$e_z(x)=\sum_{k=1}^{K-1}b_k \upsilon_k(x),\qquad \upsilon_k(x)=\begin{cases}(x-x_{k-1})/\Delta x_k, & x\in(x_{k-1},x_k)\\ (x_{k+1}-x)/\Delta x_{k+1}, & x\in(x_k,x_{k+1})\\ 0, & x\notin(x_{k-1},x_{k+1})\end{cases}\tag{2.78}$$

where a_k and b_k are the unknown coefficients, u_k and υ_k are the basis functions, $\Delta x_k=x_k-x_{k-1}$, and $\Delta x_{k+1}=x_{k+1}-x_k$. It is seen that the quasistatic relation $e_x(x)\sim\dfrac{\partial e_z(x)}{\partial x}$ between these two electric field components is supported that improves the accuracy and convergence of the algorithm. This technique is common now to solve the integral equations for more complicated transmission lines (see, for instance, [90]-[93]) and it allows to calculate the modal propagation constants and modal fields.

Example 2.2. Integral Equations for Antipodal Slot Transmission Line. The cross-section of this transmission line is shown in Fig. 2.14. It consists of a dielectric layer, which both sides are covered by the conducting sheets shorted to the walls of the shield. The attractive feature of this line is that the overlapped configuration (Fig. 2.14a) provides the modal propagation constant and the characteristic impedance close to the ones available from a microstrip line. The non-overlapped geometry (Fig. 2.14b) gives increased characteristic impedance and decreased propagation constant, which are typical for a slot line.

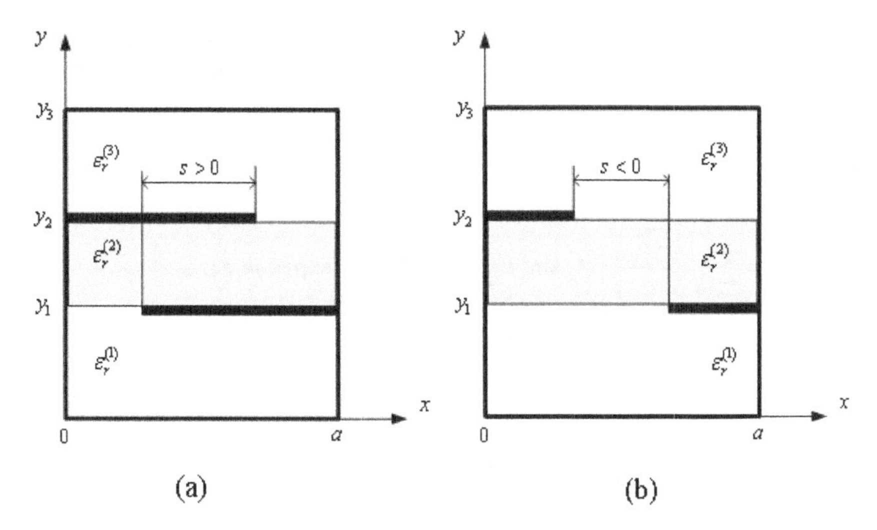

Fig. 2.14 Shielded antipodal slot line

One time this line was popular for millimeter-wave applications and for inter-layer transitions in microwave hybrid 3-D integrations. It was modeled electro-magnetically in several papers [90]-[94]. General geometry requires solving a system of four integral equations regarding to the vector slot electric fields, although the decreasing the dimension twice is possible for the symmetric configurations [90]. The coupled multilayered lines composed of these slots (Fig. 2.15) are described by eight integral equations [92].

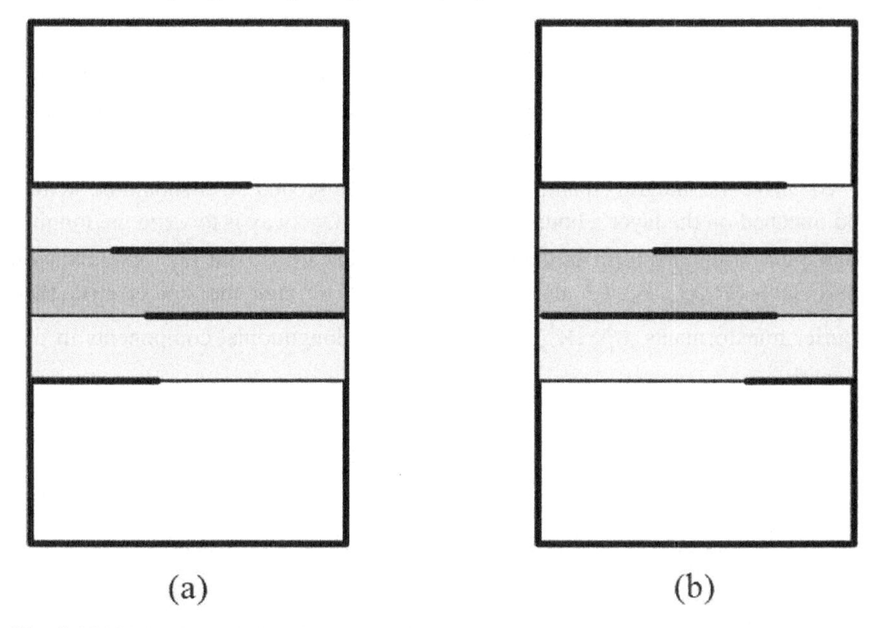

Fig. 2.15 Shielded coupled antipodal slot lines

The integral equations for antipodal slot transmission line are derived similarly to the above-considered algorithm for shielded microstrip line. Taking into account the multilayered design (Fig. 2.14), the fields of each mode are of the hybrid nature, and they have all six components. Initially, the longitudinal components of the EM field are written for each layer, and they are the following for the layers 1 and 3:

$$\begin{aligned} E_z^{(1)} &= \sum_{m=1}^{\infty} \sqrt{\frac{2-\delta_{0,m}}{a}} \left[A_m \sin\left(k_{y_m}^{(1)} y\right) \sin\left(k_{x,m} x\right) \right] \exp(-jk_z z), \\ H_z^{(1)} &= \sum_{m=0}^{\infty} \sqrt{\frac{2-\delta_{0,m}}{a}} \left[C_m \cos\left(k_{y,m}^{(1)} y\right) \cos\left(k_{x,m} x\right) \right] \exp(-jk_z z), \\ E_z^{(3)} &= \sum_{m=1}^{\infty} \sqrt{\frac{2-\delta_{0,m}}{a}} \left[B_m \sin\left(k_{y,m}^{(3)} (y_3 - y)\right) \sin\left(k_{x,m} x\right) \right] \exp(-jk_z z), \\ H_z^{(3)} &= \sum_{m=0}^{\infty} \sqrt{\frac{2-\delta_{0,m}}{a}} \left[D_m \cos\left(k_{y,m}^{(3)} (y_3 - y)\right) \cos\left(k_{x,m} x\right) \right] \exp(-jk_z z) \end{aligned} \tag{2.79}$$

where A_m, B_m, C_m, D_m are the unknown coefficients, k_z is the unknown longitudinal propagation constant, $k_{x,m} = m\pi/a$, $k_{y,m}^{(1,3)} = \sqrt{k_0^2 \varepsilon_r^{(1,3)} \mu_r^{(1,3)} - k_{x,m}^2 - k_z^2}$, and $\delta_{0,m} = \begin{cases} 1, m = 0 \\ 0, m \neq 0 \end{cases}$.

Other components of the electric and magnetic fields are derived using the Maxwell equations, and all these components, including the longitudinal ones, satisfy the boundary conditions on the ideal conductor shield at $x = 0, a$ and $y = 0, y_3$. Similarly to microstrip line, the unknown coefficients A_m, B_m, C_m, D_m are expressed through the Fourier transformants of the slot electric fields $\mathrm{E}_{x,m}^{(1,2)}$, $\mathrm{E}_{z,m}^{(1,2)}$ for $y = y_1$ and $\mathrm{E}_{x,m}^{(2,3)}$, $\mathrm{E}_{z,m}^{(2,3)}$ for $y = y_2$.

To derive the integral equations, the fields in the second layer should be written and matched on the layer's boundaries $y = y_1, y_2$. One way is to write the longitudinal field components in the central layer which are depent on four unknown coefficients P_m, Q_m, R_m, S_m and to express them through the slot electric field Fourier transformants $\mathrm{E}_{x,m}^{(1,2)}, \mathrm{E}_{z,m}^{(1,2)}, \mathrm{E}_{x,m}^{(2,3)}, \mathrm{E}_{z,m}^{(2,3)}$. The longitudinal components in this case are

$$\begin{aligned} E_z^{(2)} &= \sum_{m=1}^{\infty} \sqrt{\frac{2-\delta_{0,m}}{a}} \left[P_m \sin\left(k_{y_m}^{(2)} y\right) + Q_m \cos\left(k_{y,m}^{(2)} y\right) \right] \sin\left(k_{x,m} x\right) e^{-jk_z z}, \\ H_z^{(2)} &= \sum_{m=0}^{\infty} \sqrt{\frac{2-\delta_{0,m}}{a}} \left[R_m \sin\left(k_{y,m}^{(2)} y\right) + S_m \cos\left(k_{y,m}^{(2)} y\right) \right] \cos\left(k_{x,m} x\right) e^{-jk_z z} \end{aligned} \tag{2.80}$$

with $k_{y,m}^{(2)} = \sqrt{k_0^2 \varepsilon_r^{(2)} \mu_r^{(2)} - k_{x,m}^2 - k_z^2}$.

Other components of the EM field in this layer are derived using the Maxwell equations, similarly to the above-considered microstrip line case. Matching the magnetic fields on the slots, a system of four integral equations regarding to the unknown slot electric fields $e_x^{(1)}, e_z^{(1)}, e_x^{(2)}, e_z^{(2)}$ is derived

$$\begin{aligned}
H_z^{(2)}(x,y_1)-H_z^{(1)}(x,y_1)&=\sum_{m=0}^{\infty}\int_0^{w_1}\begin{bmatrix} e_x^{(1)}(x')\,g_{11}^{(m)}(x,x')+e_z^{(1)}(x')\,g_{12}^{(m)}(x,x')+ \\ +e_x^{(2)}(x')\,g_{13}^{(m)}(x,x')+e_z^{(2)}(x')\,g_{14}^{(m)}(x,x') \end{bmatrix}dx'=j_x^{(1)}(x),\\
H_x^{(2)}(x,y_1)-H_x^{(1)}(x,y_1)&=\sum_{m=1}^{\infty}\int_0^{w_1}\begin{bmatrix} e_x^{(1)}(x')\,g_{21}^{(m)}(x,x')+e_z^{(1)}(x')\,g_{22}^{(m)}(x,x')+ \\ +e_x^{(2)}(x')\,g_{23}^{(m)}(x,x')+e_z^{(2)}(x')\,g_{24}^{(m)}(x,x') \end{bmatrix}dx'=-j_z^{(1)}(x),\\
H_z^{(3)}(x,y_2)-H_z^{(2)}(x,y_2)&=\sum_{m=0}^{\infty}\int_{w_2}^{a}\begin{bmatrix} e_x^{(1)}(x')\,g_{31}^{(m)}(x,x')+e_z^{(1)}(x')\,g_{32}^{(m)}(x,x')+ \\ +e_x^{(2)}(x')\,g_{33}^{(m)}(x,x')+e_z^{(2)}(x')\,g_{34}^{(m)}(x,x') \end{bmatrix}dx'=j_x^{(2)}(x),\\
H_x^{(3)}(x,y_2)-H_x^{(2)}(x,y_2)&=\sum_{m=1}^{\infty}\int_{w_2}^{a}\begin{bmatrix} e_x^{(1)}(x')\,g_{41}^{(m)}(x,x')+e_z^{(1)}(x')\,g_{42}^{(m)}(x,x')+ \\ +e_x^{(2)}(x')\,g_{43}^{(m)}(x,x')+e_z^{(2)}(x')\,g_{44}^{(m)}(x,x') \end{bmatrix}dx'=-j_z^{(2)}(x)
\end{aligned} \tag{2.81}$$

where $g_{ij}^{(m)}(x,x')$ are known components of the tensor Green function of this system of integral equations. They can be solved by the Galerkin method, and similarly to microstrip line and [91]-[93], their unknown slot electric fields are approximated by discrete (pulse) functions (2.77) and (2.78). Some results are in the cited papers, and a qualitative dependence of the normalized propagation constant of the fundamental mode versus the line geometry is shown in Fig. 2.16.

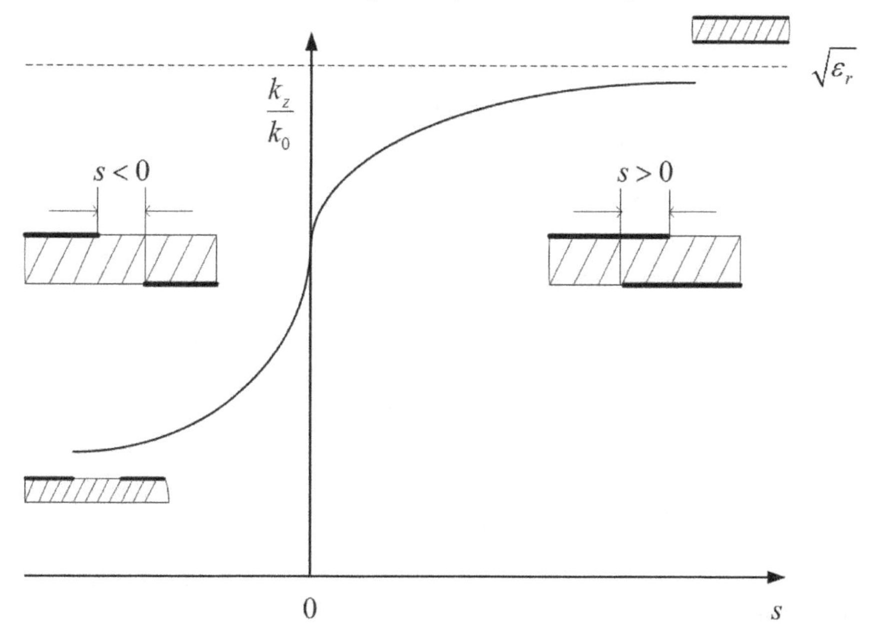

Fig. 2.16 Normalized propagation constant k_z/k_0 of the fundamental mode of an antipodal slot line versus its geometry

It is seen that the line with the overlapped conductors (in right) provides the propagation constant tending to the one of a parallel plate waveguide due to the field concentration between conductors. A non-overlapped configuration (in left) is tending to the slot line configuration and, finally, to a 3-layer rectangular waveguide [90]-[93].

Fig. 2.17 shows an electric field map of the fundamental mode in a shielded antipodal slot line calculated by the solution of the integral equations (2.81) and computations of the field-force lines according to the field components and found modal propagation constant.

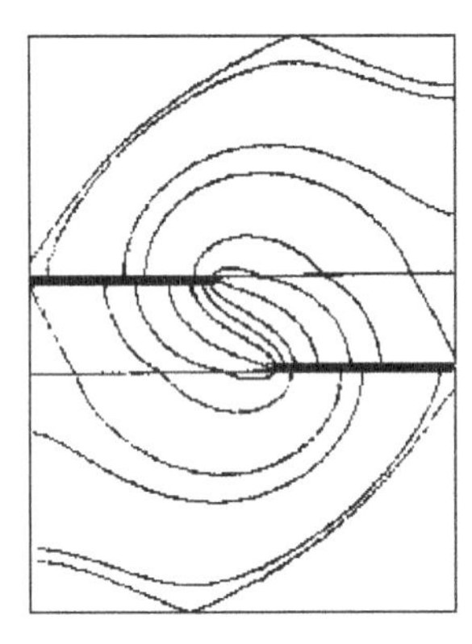

Fig. 2.17 Electric field-force line map calculated for the fundamental mode of a shielded antipodal slot line (calculated by M.I. Utkin)

Additionally to the fundamental mode, the higher-order modes can propagate, and their cut-off frequencies depend on the geometry of conductors, substrate permittivity, and the size of the shield [91]-[93]. Some microwave components, developed using the antipodal slot line, are considered in [95]-[98].

2.2.3 Variational Functionals for Waveguides

The main idea of the variational analysis is that the value under the search is represented as an integral on a function $u(\mathbf{r})$. This integral is able to reach an extremum when the function is the exact solution of considered boundary value problem $u(\mathbf{r}) = u_0(\mathbf{r})$. Representing the function as a series of basis functions and varying the expansion's coefficients, an extremum is sought as a point where all derivatives of the integral are zero. In some cases, a careful analysis of the problem and a found simple initial approximation allows to get a good solution from the first attempt. Unfortunately, not all problems in electromagnetism can be represented by stationary functionals which have extrema. The ones having them are called the quadratic functionals, and the task of variational analysis is to find their extrema. The details of the applications of variational techniques can be found in many books and papers [4],[77],[78],[83],[85],[87] and the idea is considered below shortly.

The boundary value problems, associated with a 2-D Helmholtz equation, can be put into a correspondence a functional $I(u)$ on the waveguide cross-section S where $\chi^2 = k_0^2\varepsilon_r\mu_r - k_z^2$ and $f(\mathbf{r})$ is known excitation function [4]:

$$I(u) = \frac{1}{2}\iint_S \left[\left| \nabla \cdot u(\mathbf{r}) \right|^2 - \chi^2 u^2(\mathbf{r}) + 2u(\mathbf{r}) f(\mathbf{r}) \right] ds. \tag{2.82}$$

For the time-dependent wave equation, a stationary functional is given in [4] where the integration is performed over a certain time span $(0, t_0)$ and space volume V:

$$I(u) = \int_0^{t_0}\int_V \left[\left| \nabla \cdot u(\mathbf{r},t) \right|^2 - \left[\frac{\sqrt{\varepsilon_r}}{c} \frac{\partial^2 u(\mathbf{r},t)}{\partial t^2} \right]^2 \right] dv dt. \tag{2.83}$$

Many other functionals, including those which are $I(u_0(\mathbf{r})) = \chi^2$, are available from the books on the *finite element method (FEM)* or on the variational methods for waveguides and resonators [1],[5],[87].

For instance, the transversal wave number χ can be found from the stationary functional [77]:

$$\chi^2 = \min \frac{\int_S u(\mathbf{r}) \nabla^2 \cdot u(\mathbf{r}) ds}{\iint_S \left[u(\mathbf{r}) \right]^2 ds}. \tag{2.84}$$

For numerical finding of the minimal functional value, the Ritz method [82] is used. The unknown function is approximated by a finite series, which every member $u_n(\mathbf{r})$ can satisfy the boundary conditions, and the expansion coefficients α_n are unknown

$$u(\mathbf{r}) = \sum_{n=0}^{N} \alpha_n u_n(\mathbf{r}). \tag{2.85}$$

The minimum of the functional $\chi^2 = \min(I(u))$ is reached for any α_n when

$$\frac{\partial I}{\partial \alpha_n} = 0. \tag{2.86}$$

It is a homogeneous system of linear equations regarding to α_n, and its determinant gives the eigenvalue equation or even an analytical expression for the propagation constant.

A particular case of the variational analysis is the above-mentioned FEM when the basis functions are defined on 2-D or 3-D segments, and they are the sub-domain-defined polynomials [88].

The stationary functionals can be written regarding to the electric or magnetic currents on the strips or slots cut in metallization layers, correspondingly, and it can increase the computational effectiveness. An interested reader can be referred to the papers and books [4],[82],[83],[85], [99]. An example of the use of a stationary functional obtained using the reaction theorem [3],[100] can be found in [89] where a shielded slot line is calculated.

An attractive feature of the variational analysis is that the stationary functionals allow for increased accuracy of calculation of their values even the initial approximation of the unknown function $u^{(0)}(\mathbf{r})$ is very far from the exact. Additionally, the lower and upper bounds of the eigenvalues can be obtained, and the accuracy of approximate solutions can be estimated a priory. An analysis of the effectiveness of the variational and the method of moments applied to the equivalent integral equations can be found, for instance, in [101].

The proper choice of $u^{(0)}(\mathbf{r})$ from the solutions of the corresponding or simplified boundary value problems is able to provide acceptable accuracy of analytical formulas for stationary values using even these first-step approximations. Although this search may be considered close to the art than a technique, some approaches, based on the qualitative and topologo-geometrical analysis of the boundary value problems and their solutions, provide the increased accuracy of these first-shot formulas for calculation of the eigenmodal parameters. A review on applications of the topological methods in electromagnetism and computational electromagnetics is given in Chpt. 4.

References

[1] Morse, P.M., Feshbach, H.: Methods of Theoretical Physics. McGraw Hill Co. (1953)
[2] Miller, W.: Symmetry and Separation of Variables. Addison-Wesley (1977)
[3] Balanis, C.A.: Advanced Engineering Electromagnetics. John Wiley (1989)
[4] Sadiku, M.N.O.: Numerical Techniques in Electromagnetics. CRC Press (2001)
[5] Garg, R.: Analytical and Computational Methods in Electromagnetics. Artech House (2009)
[6] Felsen, L.B., Marcuvitz, N.: Radiation and Scattering of Waves. Prentice-Hall (1973)
[7] Walter, C.H.: Traveling Wave Antennas. Dover Publ, N.Y (1970, 1965)
[8] Bates, B.D., Staines, G.W.: Transverse Resonance Analysis Technique for Microwave and Millimeter-wave Circuits, DSTO-RR-0027, Dept. of Defence, USA (1995)
[9] Gibbs, W.J.: Conformal Transformations in Electrical Engineering. Chapman & Hall, London (1958)

[10] Wheeler, H.A.: Transmission-line properties of parallel wide strips by a conformal mapping approximation. IEEE Trans., Microw. Theory Tech. 12, 280–289 (1964)
[11] Wheeler, H.A.: Transmission-line properties of parallel strips separated by a dielectric sheet. IEEE Trans., Microw. Theory Tech. 13, 172–185 (1965)
[12] Cohn, S.B.: Characteristic impedance of the shielded strip transmission line. IRE Trans., Microw. Theory Tech. 2, 52–57 (1954)
[13] Gupta, K.C., Garg, R., Bahl, I.J.: Microstrip Lines and Slot Lines. Artech House Inc., Dedham (1979)
[14] Wadell, B.C.: Transmission Line Design Handbook. Artech House (1991)
[15] Mathaei, G., Young, L., Johnes, E.M.T.: Microwave Filters, Impedance-Matching Networks, and Coupling Structures. Artech House (1980)
[16] Gvozdev, V.I., Kouzaev, G.A., Nefedov, E.I.: Filters on multilayered microwave integrated circuits for antennas applications. In: Proc. Conf. Design and Computation of Strip Transmission Line Antennas, Sverdlovsk, Russia, pp. 72–76 (1982) (in Russian)
[17] May, J.W., Rebeiz, G.M.: A 40-50-GHz SiGe 1:8 differential power divider using shielded broadside-coupled striplines. IEEE Trans., Microw. Theory Tech. 56, 1575–1581 (2008)
[18] Chirala, M.K., Nguen, C.: Multilayer design techniques for extremely miniaturized CMOS microwave and millimeter-wave distributed passive circuits. IEEE Trans., Microw. Theory Tech. 54, 4218–4224 (2006)
[19] Chirala, M.K., Guan, X., Nguyen, C.: Integrated multilayered on-chip inductors for compact CMOS RFICs and their use in a miniature distributed low-noise-amplifier design for ultra-wideband applications. IEEE Trans., Microw. Theory Tech. 56, 1783–1789 (2008)
[20] Oliner, A.A.: Equivalent circuits for discontinuities in balanced strip transmission line. IRE Trans. 3, 134–143 (1955)
[21] Menzel, W., Wolff, I.: A method for calculating the frequency-dependent properties of microstrip discontinuities. IEEE Trans., Microw. Theory Tech. 25, 107–112 (1977)
[22] Hammerstad, E., Jensen, O.: Accurate models for microstrip computer-aided design. In: 1980 IEEE MTT-S Int. Microw. Symp. Dig., pp. 407–409 (1980)
[23] Asbesh, C.B., Garg, R.: Conformal mapping analysis of microstrip with finite strip thickness. In: Proc. APSYM 2006, Dept. of Electronics, CUSAT, India, December 14-16, pp. 27–30 (2006)
[24] Wolff, I., Kompa, G., Mehran, R.: Calculation method for microstrip discontinuities and T-junctions. El. Lett. 8, 177–179 (1972)
[25] Getsinger, W.: Microstrip dispersion model. IEEE Trans., Microw. Theory Tech. 21, 34–39 (1973)
[26] Carlin, H.J.: A simplified circuit model for microstrip. IEEE Trans., Microw. Theory Tech. 21, 589–591 (1973)
[27] Kobayashi, M.: A dispersion formula satisfying recent requirements in microstrip lines. IEEE Trans., Microw. Theory Tech. 36, 1246–1250 (1988)
[28] Nefeodov, E.I., Fialkovskyi, A.T.: Strip Transmission Lines. Nauka, Moscow (1980) (in Russian)
[29] Pramanick, P., Bharatia, P.: A new microstrip dispersion model. IEEE Trans., Microw. Theory Tech. 32, 1379–1384 (1984)

[30] Verma, A.K., Kumar, R.: New empirical unified dispersion model for shielded-, suspended-, and composite-substrate microstrip line for microwave and mm-wave applications. IEEE Trans., Microw. Theory Tech. 46, 1187–1192 (1998)
[31] Yamashita, E., Atsuki, K., Hirahata, T.: Microstrip dispersion in a wide-frequency band. IEEE Trans., Microw. Theory Tech. 29, 610–611 (1981)
[32] Hoffmann, R.K.: Handbook of Microwave Integrated Circuits. Artech House (1987)
[33] Kompa, G.: Practical Microstrip Design and Applications. Artech House (2005)
[34] Schevchenko, V.V.: Continuous Transitions in Open Waveguides (Electromagnetics). Golem Press (1972)
[35] Oliner, A.A., Lee, K.S.: The nature of the leakage from higher modes on microstrip line. In: 1986 IEEE MTT-S Dig., pp. 57–60 (1986)
[36] Michalski, K.A., Zheng, D.: Rigorous analysis of open microstrip lines of arbitrary cross-section in bound and leaky regimes. In: 1989 IEEE MTT-S Dig., pp. 787–790 (1989)
[37] Ngihiem, D., Williams, J.T., Jackson, D.R., Oliner, A.A.: Existence of a leaky dominant mode on microstrip line with an isotropic substrate: Theory and measurement. In: 1993 IEEE MTT-S Dig., pp. 1291–1294 (1993)
[38] Bagby, J.S., Lee, C.-H., Nyquist, D.P., Yuan, Y.: Identification of propagation regimes on integrated microstrip transmission lines. IEEE Trans., Microw. Theory Tech. 41, 1881–1894 (1993)
[39] Liu, J., Jackson, D.R., Liu, P., et al.: The propagation wavenumber for microstrip line in the first higher-order mode. In: ICMMT 2010 Proc., pp. 965–968 (2010)
[40] van de Capelle, A.R., Luypaert, P.J.: Fundamental- and higher-order modes in open microstrip lines. El. Lett. 9(15), 345–346 (1973)
[41] Chen, S.-D., Tzuang, C.K.C.: Characteristic impedance and propagation of the first higher order microstrip mode in frequency and time domain. IEEE Trans., Microw. Theory Tech. 50, 1370–1379 (2002)
[42] Chiu, L.: Oversized microstrip line as differential guided-wave structure. El. Lett. 46(2), 144–145 (2010)
[43] Nefeodov, E.I.: Technical Electrodynamics. Academia Publ., Moscow (2008) (in Russian)
[44] Liu, J., Long, Y.: Formulas for complex propagation constant of first higher mode of microstrip line. El. Lett. 44, 261–262 (2008)
[45] Rizzoli, V.: A unified variational solution to microstrip array problems. IEEE Trans., Microw. Theory Tech. 23, 223–234 (1975)
[46] Garg, R., Bhal, I.J.: Characteristics of coupled microstriplines. IEEE Trans., Microw. Theory Tech. 27, 700–705 (1979)
[47] Bedair, S.S., Sobhy, M.I.: Accurate formulas for computer-aided design of shielded microstrip lines. In: IEE Proc., vol. 127, pt. H, pp. 305–308 (December 1980)
[48] Wan, C.: Analytically and accurately determined quasi-static parameters of coupled lines. IEEE Trans., Microw. Theory Tech. 44, 75–80 (1996)
[49] Abbosh, A.M.: Analytical closed-form solutions for different configurations of parallel-coupled microstrip lines. IET Microw. Antennas Propag. 3, 137–147 (2009)
[50] Carlin, H.J., Civalleri, P.P.: A coupled-line model for dispersion in parallel-coupled microstrips. IEEE Trans., Microw. Theory Tech. 23, 444–446 (1975)
[51] Tripathi, V.K.: A dispersion model for coupled microstrips. IEEE Trans., Microw. Theory Tech. 34, 66–71 (1986)

[52] Getsinger, W.J.: Dispersion of parallel-coupled microstrip. IEEE Trans., Microw. Theory Tech. 21, 144–145 (1973)
[53] Wen, C.P.: Coplanar waveguide: A surface strip transmission line suitable for nonreciprocal gyromagnetic device applications. IEEE Trans., Microw. Theory Tech. 17, 1087–1090 (1969)
[54] Ghione, G., Naldi, C.U.: Coplanar waveguides for MMIC applications: Effect of upper shielding, conductor backing, finite extent ground planes, and line-to-line coupling. IEEE Trans., Microw. Theory Tech. 35, 260–267 (1987)
[55] Riazat, M., Majidi-Ahy, R., Feng, I.-J.: Propagation modes and dispersion characteristics of coplanar waveguides. IEEE Trans., Microw. Theory Tech. 38, 245–251 (1990)
[56] Ghione, G., Goano, M.: A closed-form CAD-oriented model for the high-frequency conductor attenuation of symmetrical coupled coplanar waveguides. IEEE Trans., Microw. Theory Tech. 45, 1065–1070 (1997)
[57] Ghione, G., Goano, M.: The influence of ground plane width on the ohmic losses of coplanar waveguides with finite lateral ground planes. IEEE Trans., Microw. Theory Tech. 45, 1640–1642 (1997)
[58] Gorur, A., Karpuz, C.: Analytical formulas for conductor-backed asymmetric CPW with one lateral ground plane. Microw. Opt. Lett. 22, 123–126 (1999)
[59] Edwards, T.C., Steer, M.B.: Foundation of Interconnect and Microstrip Design. J. Wiley & Sons, Ltd. (2000)
[60] Jackson, R.W.: Coplanar waveguide vs. microstrip for millimeter wave integrated circuits. In: 1986 MTT-S Dig., pp. 699–702 (1986)
[61] Chang, C.-N., Wong, Y.-C., Chen, C.H.: Full-wave analysis of coplanar waveguides by variational conformal mapping technique. IEEE Trans., Microw. Theory Tech. 38, 1339–1344 (1990)
[62] Ke, J.-Y., Chen, C.H.: Dispersion and attenuation characteristics of coplanar waveguides with finite metallization thickness and conductivity. IEEE Trans., Microw. Theory Tech. 43, 1128–1135 (1995)
[63] Simons, R.N.: Coplanar Waveguide Circuits, Components & Systems. J. Wiley & Sons (2001)
[64] Wolff, I.: Coplanar Microwave Integrated Circuits. Wiley Interscience (2006)
[65] Uzunoglu, N.K., Nikita, K.S., Kaklamani, D.I. (eds.): Applied Computational Electromagnetics. Springer (1999)
[66] Taflove, A.T., Hagness, S.C.: Computational Electrodynamics. Artech House (2005)
[67] Yee, K.S.: Numerical solution of initial boundary value problems involving Maxwell's equations in isotropic medium. IEEE Trans., Antennas Propag. 14, 302–307 (1966)
[68] Kron, G.: Equivalent circuit of the field equations of Maxwell. In: Proc. IRE, vol. 32, pp. 289–299 (May 1944)
[69] Johns, P., Beurle, R.: Numerical solution of 2-dimensional scattering problems using a transmission-line matrix. Proc. IEEE 118, 1203–1208 (1971)
[70] Johns, P.B.: On the relationships between TLM and finite-difference methods for Maxwell equations. IEEE Trans., Microw. Theory Tech. 35, 60–61 (1987)
[71] Harrington, R.F.: Field Computation by Moment Methods. IEEE Press (1993)
[72] Muskhelishvili, N.I.: Singular Integral Equations. Wolters-Noordhoff (1972)
[73] Gakhov, F.D.: Boundary Value Problems. Pergamon Press (1966)

[74] Lavrentiev, M.M.: Ill-posed Problems of Mathematical Physics and Analysis. Am. Math. Soc. (1986)
[75] Tikhonov, A.N., Arsenin, V.Y.: Solutions of Ill Posed Problems. V.H. Winston and Sons (1977)
[76] Leonov, A.S.: On quasi-optimal choice of the regularization parameter in the Lavrentiev's method. Siberian Math. J. 34(4), 117–126 (1993) (in Russian)
[77] Vaganov, R.B., Katsenelenbaum, B.Z.: Foundation of the Diffraction Theory. Nauka, Moscow (1982) (in Russian)
[78] Nikolskyi, V.V., Nikolskaya, T.I.: Electrodynamics and Wave Propagation. Nauka, Moscow (1987) (in Russian)
[79] Tai, C.T.: Dyadic Green's Functions in Electromagnetic Theory. Intex Educational Publ., Scranton (1971)
[80] Kantorovich, L.V., Krylov, V.I.: Approximate Methods of Higher Analysis. John Wiley (1964) (Translated from Russian)
[81] Krylov, A.N., et al.: Academician B. G. Galerkin. On the seventieth Anniversary of his birth. Vestnik Akademii Nauk SSSR 4, 91–94 (1941)
[82] Zhang, W.-X.: Engineering Electromagnetism: Functional Method. Ellis Horwood (1991)
[83] Marcuvitz, N.: Waveguide Handbook. Inst. of Eng. and Techn. Publ. (1986)
[84] Mashkovzev, B.M., Zibisov, K.N., Emelin, B.F.: Theory of Waveguides. Nauka, Moscow (1966) (in Russian)
[85] Lewin, L.: Theory of Waveguides. Newnes-Buttertworths, London (1975)
[86] Mittra, R. (ed.): Computer Techniques for Electromagnetics. Pergamon Press (1973)
[87] Nikolskii, V.V.: Variational Methods for Inner Boundary Value Problems of Electrodynamics. Nauka Publ., Moscow (1967) (in Russian)
[88] Silvester, P.P., Ferrari, R.L.: Finite Elements for Electrical Engineers. Cambridge University Press, Cambridge (1983)
[89] Kouzaev, G.A., Kurushin, E.P., Neganov, V.A.: Numerical computations of a slot-transmission line. Izv. Vysshikh Utchebnykh Zavedeniy Radiofizika (Radiophysics) 23, 1041–1042 (1981) (in Russian)
[90] Hofmann, H., Meinel, H., Adelseck, B.: New integrated components mm-wave components using finlines. In: 1978 IEEE MTT-S Microw. Symp. Dig., pp. 21–23 (1978)
[91] Kouzaev, G.A.: Balanced slotted line. In: Gvozdev, V.I., Nefedov, E.I. (eds.) Microwave Three-Dimensional Integrated Circuits, pp. 45–50. Nauka Publ, Moscow (1985) (Invited Chapter,in Russian)
[92] Gvozdev, V.I., Kouzaev, G.A., Nefedov, E.I.: Balanced slotted line. Theory and experiment. Radio Eng. Electron. Physics (Radiotekhnika i Elektronika) 30, 1050–1057 (1985)
[93] Gvozdev, V.I., Kouzaev, G.A., Nefedov, E.I., Utkin, M.I.: Electrodynamical calculation of microwave volume integrated circuit components based on a balanced slotted line. J. Commun. Techn. Electronics (Radiotekhnika i Electronika) 33, 39–43 (1989)
[94] Lerer, A.M., Mikhalevskyi, V.S., Zvetkovskaya, S.M.: Ribbed transmission line. Izv. Vuzov, Radioeklektronika 10(10), 46–50 (1981)
[95] Kurushin, E.P., Kouzaev, G.A., Neganov, V.A., et al.: Microwave circulator. USSR Invention Certificate No 1080689 dated on, June 14 (1982)
[96] Gvozdev, V.I., Golovinskaja, S.Y., Kouzaev, G.A., et al.: "Circulator," USSR Invention Certificate, No 1712989 dated on, May 31 (1990)

[97] Gazarov, V.M., Gvozdev, V.I., Kouzaev, G.A., et al.: Oscillator for microwave 3D-ICs. USSR Invention Certificate No 1830555 dated on, June 12 (1990)
[98] Gvozdev, V.I., Gluschenko, A.G., Kouzaev, G.A., et al.: Amplifier. USSR Invention Certificate No 1775845 dated on, June 21 (1991)
[99] Schwinger, J., Saxon, D.S.: Discontinuities in Waveguides. Gordon and Breach Sci. Publ. (1968)
[100] Monteath, G.D.: Applications of the Electromagnetic Reciprocity Principle. Pergamon Press (1973)
[101] Richmond, J.H.: On the variational aspects of the moment method. IEEE Trans., Antennas Propag. 39, 473–479 (1991)

3 Waveguide Discontinuities and Components

Abstract. The considered in this Chapter method of treatment of waveguide discontinuities is with the theory of diffraction of modes at the obstacles in waveguides and transmission lines, which gives clear understanding of this effect. The modes in waveguides are associated with the equivalent transmission lines, and the one-modal approximation is used widely in microwave techniques. The multimodal representation of the diffracted fields requires more complicated matching of them at the discontinuities, and the integral and stationary functional methods are used to obtain the equivalent circuit models of the obstacles. This idea, related to the founders of the waveguide theory, is additionally required for interpretations of the results obtained by those numerical methods which calculate the field in the whole discontinuity domain without using the modal expansion method. References -35. Figures -7. Pages -24.

3.1 A More Detailed Theory of Regular Waveguides for the Discontinuity Treatment

Very complicated effects appear when a waveguide mode encounters an obstacle or deformation of the waveguide shape. Similarly to the propagation of modes in uniform waveguides, the Maxwell or wave equations describe this effect and it can be solved by the above-considered methods modified for the 3-D or 4-D simulations.

Some of them provide clear understanding of the EM phenomena caused by diffraction of waves. An introduction to these methods requires more detailed explanation of the modal theory of waveguides and its relation to the formalism of the equivalent transmission lines and lumped circuits used to explain and model the waveguide discontinuities and components. Below, some of the most important details are explained based on the contributions of Collin [1], Felsen & Marcuvitz [2], Marcuvitz [3], Schwinger [4], Mashkovzev, et al. [5], Nikolskyi & Nikolskaya [6], Heaviside [7], Shelkunoff [8], et al.

Initially, a theory of discontinuities was established for the shielded transmission lines. The modal spectrum of these waveguides is discrete, i.e. it is infinite, but countable. At discontinuities, these modes are coupled to each other, and the diffraction field consists of all modes of the waveguide spectrum. Neglecting this effect or an asymptotic way of taking into account leads to the unimodal models of the discontinuities, and the method of equivalent lines and circuits is used in its simplest variant.

G.A. Kouzaev: Applications of Advanced Electromagnetics, LNEE 169, pp. 95–118.
springerlink.com

Consider again a shielded transmission line (Fig. 2.1). The field components $\mathbf{H}_t$ and $\mathbf{E}_t$, which are transversal to the propagation direction $0 \to z$, are expressed as

$$jk_0\sqrt{\frac{\mu_0\mu_r}{\varepsilon_0\varepsilon_r}}E_z = \nabla_{xy}\cdot(\mathbf{H}_t\times\mathbf{z}_0),$$
$$jk_0\sqrt{\frac{\mu_0\mu_r}{\varepsilon_0\varepsilon_r}}H_z = \nabla_{xy}\cdot(\mathbf{z}_0\times\mathbf{E}_t). \tag{3.1}$$

In general, when the geometry of a line is varied with the longitudinal coordinate, the waveguide field consists of infinite sum of coupled modes. They can have all six field components, and it is more convenient to derive the expressions for the transversal fields $\mathbf{E}_t$ and $\mathbf{H}_t$ instead of the longitudinal ones E_z and H_z:

$$\mathbf{E}_t = \sum_i V_i'(z)\mathbf{e}_i'(x,y) + V_i''(z)\mathbf{e}_i''(x,y),$$
$$\mathbf{H}_t = \sum_i I_i'(z)\mathbf{h}_i'(x,y) + I_i''(z)\mathbf{h}_i''(x,y). \tag{3.2}$$

The modal electric $(\mathbf{e}_i',\mathbf{e}_i'')$ and magnetic $(\mathbf{h}_i',\mathbf{h}_i'')$ functions are derived from

$$\mathbf{e}_i' = -\nabla_{xy}\cdot\Phi_i,$$
$$\mathbf{h}_i' = \mathbf{z}_0\times\mathbf{e}_i', \tag{3.3}$$

and

$$\mathbf{e}_i'' = \mathbf{z}_0\times\nabla_{xy}\cdot\Psi_i,$$
$$\mathbf{h}_i'' = \mathbf{z}_0\times\mathbf{e}_i''. \tag{3.4}$$

The scalar functions Φ and Ψ are the solutions of the following boundary value problems:

$$\nabla_{xy}^2\cdot\Phi_i + (\chi')^2\Phi_i = 0,$$
$$\Phi_i(x,y) = 0,\ (x,y)\in L \tag{3.5}$$

and

$$\nabla_{xy}^2\cdot\Psi_i + (\chi'')^2\Psi_i = 0,$$
$$\frac{\partial\Psi_i(x,y)}{\partial\nu} = 0,\ (x,y)\in L. \tag{3.6}$$

The case $\chi^2 = k_0^2\varepsilon_r\mu_r - k_z^2 = 0$ corresponds to the multiply connected cross-sections supporting the TEM or quasi-TEM modes. Then to the above-considered solutions of (3.5) or (3.6), the functions obtained from the Laplace equations $\nabla^2\cdot\Phi = 0$ or $\nabla^2\cdot\Psi = 0$ should be added to the modal field expansion (3.2).

It is interesting to note that the modal electric $(\mathbf{e}_i', \mathbf{e}_i'')$ and magnetic $(\mathbf{h}_i', \mathbf{h}_i'')$ functions are orthogonal to each other, and it manifests their independency, in spite of the corresponding modes can be coupled due to the non-uniformity of waveguide in the longitudinal direction. The orthogonality condition is defined on a waveguide cross-section:

$$\int_S \mathbf{e}_i' \cdot \mathbf{e}_j' ds = \int_S \mathbf{e}_i'' \cdot \mathbf{e}_j'' ds = \delta_{ij},$$
$$\int_S \mathbf{e}_i' \cdot \mathbf{e}_j'' ds = 0, \tag{3.7}$$
$$\delta_{ij} = \begin{cases} 1, & i = j \\ 0 & i \neq j \end{cases}.$$

The unknown modal amplitudes V_i and I_i are expressed using the introduced field orthogonal functions:

$$V_i' = \int_S \mathbf{E}_t \cdot \mathbf{e}_i'' ds,\ V_i'' = \int_S \mathbf{E}_t \cdot \mathbf{e}_i'' ds,$$
$$I_i' = \int_S \mathbf{H}_t \cdot \mathbf{h}' ds,\ I_i'' = \int_S \mathbf{H}_t \cdot \mathbf{h}'' ds. \tag{3.8}$$

Now the longitudinal components of the EM field are

$$E_z = -\frac{j}{k_0}\sqrt{\frac{\varepsilon_0\varepsilon_r}{\mu_0\mu_r}}\sum_i I_i'(z)\left(\chi_i'\right)^2 \Phi_i,$$
$$H_z = -\frac{j}{k_0}\sqrt{\frac{\mu_0\mu_r}{\varepsilon_0\varepsilon_r}}\sum_i V_i''(z)\left(\chi_i''\right)^2 \Psi_i. \tag{3.9}$$

As seen from the above-shown equations, the voltages V_i and currents I_i are still unknown, and they should be found substituting the fields $\mathbf{E}$ and $\mathbf{H}$ into the Maxwell equations and the boundary conditions.

In general, if a waveguide is non-uniform, the Maxwell equations allow for an infinite system of coupled differential equations regarding to these voltages and currents. It means that the modes are coupled to each other due to the waveguide longitudinal non-uniformity. Anyway, some of such waveguides are solved analytically, and among them are the radial waveguides [3], for instance.

In spite of a number of successful applications of non-uniform waveguides and microstrip transmission lines, the most known components are composed of regular transmission lines and their abrupt discontinuities. The EM theory supposes the full-wave treatment of discontinuities, i.e. the excited at the waveguide obstacles higher-order modes should be taken into account. At low frequencies, their influence can be considered as negligible, and a unimodal theory of obstacles is used. This theory is known as the equivalent transmission line approach, and its origin relates to the works of O. Heaviside [7], S. Shelkunoff [8], et al.

3.2 Equivalent Transmission Line Approach and Microwave Networks

In a uniform waveguide, all its modes are independent. They are described by the uncoupled pairs of equations (3.10) valid for both TM $\left(V_i', I_i'\right)$ and TE $\left(V_i'', I_i''\right)$ modes:

$$\frac{dV_i}{dz} = -jk_z Z_i I_i, \qquad \frac{dI_i}{dz} = -jk_z Y_i V_i \tag{3.10}$$

where $Z_i' = \frac{1}{Y_i'} = \frac{k_z'}{\omega\varepsilon_0\varepsilon_r}$, and $Z_i'' = \frac{1}{Y_i''} = \frac{\omega\mu_0\mu_r}{k_z''}$.

The equations (3.10) can be re-written into a set of differential ones of the second order, and they are the *telegraph equations* which are the same for the TM, TE, and TEM modes

$$\frac{d^2V_i}{dz^2} + \left(k_z^{(i)}\right)^2 V_i = 0, \qquad \frac{d^2I_i}{dz^2} + \left(k_z^{(i)}\right)^2 I_i = 0. \tag{3.11}$$

If the ratio $Z_c^{(i)} = V_i/I_i$ of the modal voltages and currents is known, then only one equation from (3.11) can be used.

An equivalent two-conductor transmission line corresponds to a waveguide mode if they both support the same averaged modal power (Fig. 3.1):

$$\frac{1}{2}\mathrm{Re}\int_S \left(\mathbf{E}_t^{(i)} \times \mathbf{H}_t^{(i)*}\right) ds = \frac{1}{2}\mathrm{Re}\left(V_i I_i^*\right). \tag{3.12}$$

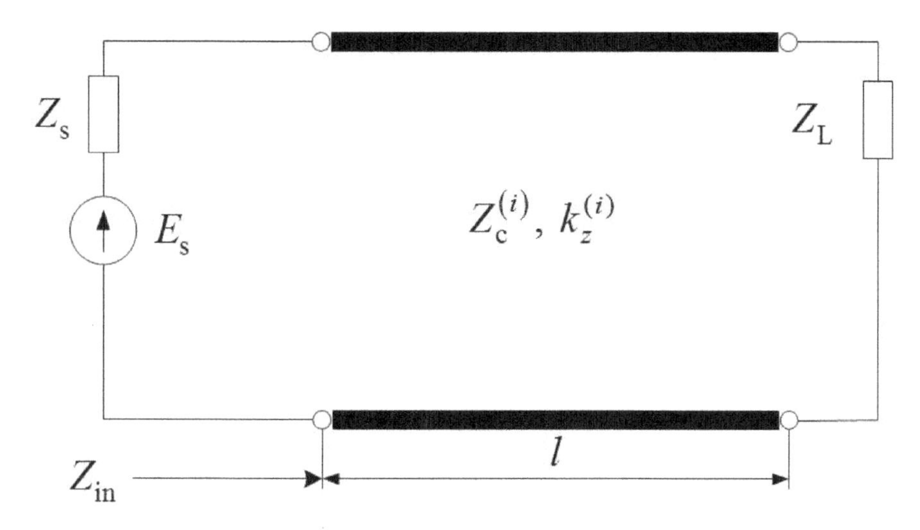

Fig. 3.1 Equivalent transmission line for the $i-th$ mode of a waveguide

It means that the modeled and modeling lines should have the same propagation constants and characteristic impedances.

Supposing $k_z^{(i)} = const$ regarding to the z-variable, (3.11) are solved analytically in a simple way, and the solutions of these equations are the voltages and currents of the forward and backward waves of unknown amplitudes. They are defined applying the boundary conditions at the source or load terminals for obtained general solutions.

A transmission line is matched if the load Z_L and the characteristic impedance are equal to each other. The reflection happens from the load if $Z_c^{(i)} \neq Z_L$. The loaded transmission line of the length l is described by the input impedance Z_{in} and the reflection coefficient Γ measured at the input port as the ratio of the reflected $\left(V^{(-)}\right)$ to the incident $\left(V^{(+)}\right)$ voltages (Fig. 3.1):

$$Z_{in} = Z_c \frac{Z_L \cos k_z l + jZ_c \sin k_z l}{Z_c \cos k_z l + jZ_L \sin k_z l}, \tag{3.13}$$

$$\Gamma = \frac{V^{(-)}}{V^{(+)}} = \frac{Z_L - Z_c}{Z_L + Z_c} e^{-2jk_z l}. \tag{3.14}$$

Very often, another parameter is used instead of the reflection coefficient, and it is the voltage standing wave ratio (VSWR):

$$\text{VSWR} = \frac{1+|\Gamma|}{1-|\Gamma|}. \tag{3.15}$$

For the multiport circuits, the S-matrix is used, and it is composed of the reflection S_{kk} and transmission S_{kl} coefficients normalized to the corresponding port's characteristic impedance or to the reference 50-Ω load. They are defined as the ratios of the amplitudes of waves appeared at the ports of a network. For instance, a 2-port network (Fig. 3.2), that can be composed of transmission lines and discrete components, is described by a 2-dimensional *S*-matrix:

$$\begin{pmatrix} V_1^{(-)} \\ V_2^{(-)} \end{pmatrix} = \begin{pmatrix} S_{11} & S_{12} \\ S_{21} & S_{22} \end{pmatrix} \cdot \begin{pmatrix} V_1^{(+)} \\ V_2^{(+)} \end{pmatrix} \tag{3.16}$$

where $V_{1,2}^{(-)}$ and $V_{1,2}^{(+)}$ are the voltages of reflected and incident waves, correspondingly.

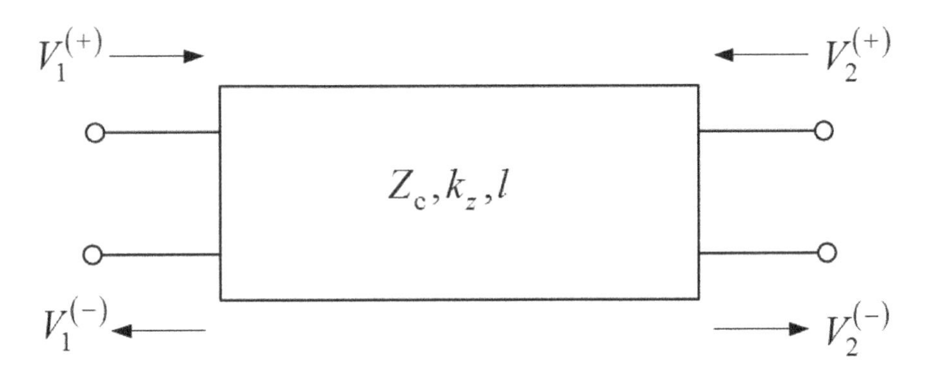

Fig. 3.2 Two-port network corresponding to Fig. 3.1

The elements of the *S*-matrix are defined

$$\begin{aligned} S_{11} &= \frac{V_1^{(-)}}{V_1^{(+)}}, \; V_2^{(+)} = 0; \text{ Port II is matched,} \\ S_{22} &= \frac{V_2^{(-)}}{V_2^{(+)}}, \; V_1^{(+)} = 0; \text{ Port I is matched,} \\ S_{12} &= \frac{V_1^{(-)}}{V_2^{(+)}}, \; V_1^{(+)} = 0; \text{ Port I is matched,} \\ S_{21} &= \frac{V_2^{(-)}}{V_1^{(+)}}, \; V_2^{(+)} = 0; \text{ Port II is matched.} \end{aligned} \tag{3.17}$$

Similarly, the *S*-matrix can be introduced for any multiport network.

The *S*-matrices of many elementary components are known from the analytical treatments or measurements, and they are included into the component libraries of most commercially available design tools. More complicated circuits composed of these elementary elements are described by calculation of a united *S*-matrix, and the algorithms of such calculations are considered, for instance in [9]-[10].

Additionally to the scattering matrix formalism, the microwave networks are described by Z-, Y- and ABCD-matrices that represent the relationships between the currents and voltages at the ports of circuits. All matrices are transformable to each other, and the transition formulas can be found, for instance, in [9].

Unfortunately, the component models based on the unimodal transmission lines are of low accuracy, and more advanced theories are used to study the effects of multimodal diffraction. Some algorithms are based on the representation of the field near a discontinuity by an infinite sum of the diffracted or scattered modes, and they provide the clearest representation of modal diffraction. Others are based on the direct numerical solutions of the Maxwell or Helmholtz equations and the following treatments to obtain the scattering matrices of discontinuities.

3.3 EM Computational Methods for Waveguide Discontinuities

3.3.1 Mode Matching Method to Calculate the Waveguide Junctions

The mode matching method is a powerful tool to calculate the waveguide discontinuities that has been proofed for many waveguides and components [1],[5],[6],[11]-[13].

Consider a junction of two uniform waveguides of an arbitrary design (Fig. 3.3). In particular, they can be the integrated lines or hollow waveguides. From the above-considered theory (Chpt. 2) it follows that the uniform lines support infinite number of independent modes, including the propagating and the evanescent ones. Some waveguides support the modes having complex propagation constants in the absence of any loss [14].

The independency of modes is expressed by the orthogonality relation on a waveguide cross-section s (Fig. 3.3):

$$\int_s \left(\mathbf{E}_n \times \mathbf{H}_k^*\right) \mathbf{ds} = \delta_{nk} = \begin{cases} 1, & n = k \\ 0, & n \neq k \end{cases} \tag{3.18}$$

where the indexes n, k are the modal numbers, and $\mathbf{ds}$ is the normal-to-s unit vector. At the discontinuity, an incident wave excites infinite number of scattered forward and backward modes, which are coupled to each other at the discontinuity plane $s^{(I,II)}$, and the boundary conditions written at this plane are

$$\begin{aligned}
&\mathbf{E}^{(I)}(x,y,z=0) = \mathbf{E}_{m=1}^{(I,+)}(x,y,z=0) + \sum_{m=1}^{\infty} a_m \mathbf{E}_m^{(I,-)}(x,y,z=0),\\
&\mathbf{H}^{(I)}(x,y,z=0) = \mathbf{H}_{m=1}^{(I,+)}(x,y,z=0) + \sum_{m=1}^{\infty} a_m \mathbf{H}_m^{(I,-)}(x,y,z=0),\\
&\mathbf{E}^{(II)}(x,y,z=0) = \sum_{n=1}^{\infty} b_n \mathbf{E}_n^{(II,+)}(x,y,z=0),\\
&\mathbf{H}^{(II)}(x,y,z=0) = \sum_{n=1}^{\infty} b_n \mathbf{H}_n^{(II,+)}(x,y,z=0)
\end{aligned} \tag{3.19}$$

where $\mathbf{E}^{(I)},\mathbf{H}^{(I)}$ and $\mathbf{E}^{(II)},\mathbf{H}^{(II)}$ are the fields in the left and right waveguides, respectively (Fig. 3.3), $\mathbf{E}_{m=1}^{(I,+)},\mathbf{H}_{m=1}^{(I,+)}$ are the fields of the incident main mode propagating along the z-axis in the first waveguide, $\mathbf{E}_m^{(I,-)},\mathbf{H}_m^{(I,-)}$ are the fields of a backward scattered mode in the first waveguide, $\mathbf{E}_n^{(II,+)},\mathbf{H}_n^{(II,+)}$ are the fields of a forward scattered mode in the second waveguide, and a_m,b_n are the unknown modal amplitudes.

At the aperture $s^{(I,II)}$ of the connection of these two waveguides, the tangential to this surface fields are equal to each other:

$$\begin{aligned}&\mathbf{E}_\tau^{(I)}(x,y,z=0)-\mathbf{E}_\tau^{(II)}(x,y,z=0)=0,\\&\mathbf{H}_\tau^{(I)}(x,y,z=0)-\mathbf{H}_\tau^{(II)}(x,y,z=0)=0,\end{aligned}\tag{3.20}$$

on the other hand:

$$\begin{aligned}&\mathbf{E}_{\tau_{m=1}}^{(I,+)}+\sum_{m=1}^{\infty}a_m\mathbf{E}_{\tau_m}^{(I,-)}-\sum_{n=1}^{\infty}b_n\mathbf{E}_{\tau_n}^{(II,+)}=0,\\&\mathbf{H}_{\tau_{m=1}}^{(I,+)}+\sum_{m=1}^{\infty}a_m\mathbf{H}_{\tau_m}^{(I,-)}-\sum_{n=1}^{\infty}b_n\mathbf{H}_{\tau_n}^{(II,+)}=0.\end{aligned}\tag{3.21}$$

The unknown excitation modal coefficients a_m and b_n are found using the Galerkin method supposing that the modal tangential fields are the vector basis functions, and calculating the projections of (3.21) on each of modes gives

$$\begin{aligned}&\int_{s^{(I)}}\left[\left(\mathbf{E}_{\tau_{m=1}}^{(I,+)}+\sum_{m=1}^{\infty}a_m\mathbf{E}_{\tau_m}^{(I,-)}-\sum_{n=1}^{\infty}b_n\mathbf{E}_{\tau_n}^{(II,+)}\right)\times\mathbf{H}_{\tau_m}^{(I,-)*}\right]\mathbf{ds}=0,\\&m=1,2,3,..\infty;\ n=1,2,...,\infty,\\&\int_{s^{(II)}}\left[\mathbf{E}_{\tau_n}^{(II,+)*}\times\left(\mathbf{H}_{\tau_{m=1}}^{(I,+)}+\sum_{m=1}^{\infty}a_m\mathbf{H}_{\tau_m}^{(I,-)}-\sum_{n=1}^{\infty}b_n\mathbf{H}_{\tau_n}^{(II,+)}\right)\right]\mathbf{ds}=0,\\&m=1,2,3,..\infty;\ n=1,2,...\infty.\end{aligned}\tag{3.22}$$

This infinite system of linear algebraic equations regarding to unknown modal amplitudes a_m,b_n is reduced according to the convergence criterion to the overall size $N+M$, and a pertinent algorithm solves it. The mentioned coefficients are the elements of a multi-modal S-matrix if the modal fields are properly normalized according to (3.18): $a_m=S_{m,1}^{(I,I)}$ and $b_n=S_{n,1}^{(II,I)}$. The full multi-modal S-matrix of the $(N+M)$-size is derived considering diffraction of all modes from this $(N+M)$ modal set [5].

Very often, only the reflection $S_{m=1}^{(I,I)}$ and transmission $S_{m=1}^{(II,I)}$ coefficients of the fundamental mode are needed but the multimodal expansion allows for reaching more accurate values of these coefficients in comparison with the monomodal calculations. Besides, taking into account the higher-order modes essentially increases accuracy of calculation of the arguments of the complex elements of the scattering matrix, which is very important for many microwave applications. An additional warning with this and similar methods is the relative convergence of the algorithms when the increase of taken modes leads to the results, which are different from the exact ones [12]. To avoid this effect, the algorithm should be regularized. For instance, one way is to take into account the singularities of the fields at the edges of conductors and at the aperture $s^{(I,II)}$.

3.3.2 Integral Equation Method for Waveguide Joints

The algorithms based on the integral equations for the aperture fields are more flexible. For instance, these fields can be represented by a variety of the base and weighting functions. Besides, the integral equations of the first kind can be transformed into the ones of the second kind providing more stable results and improved convergence of calculations [15].

Consider the electric fields at the connection plane $s^{(I,II)}$ of a joint of two waveguides (Fig. 3.3).

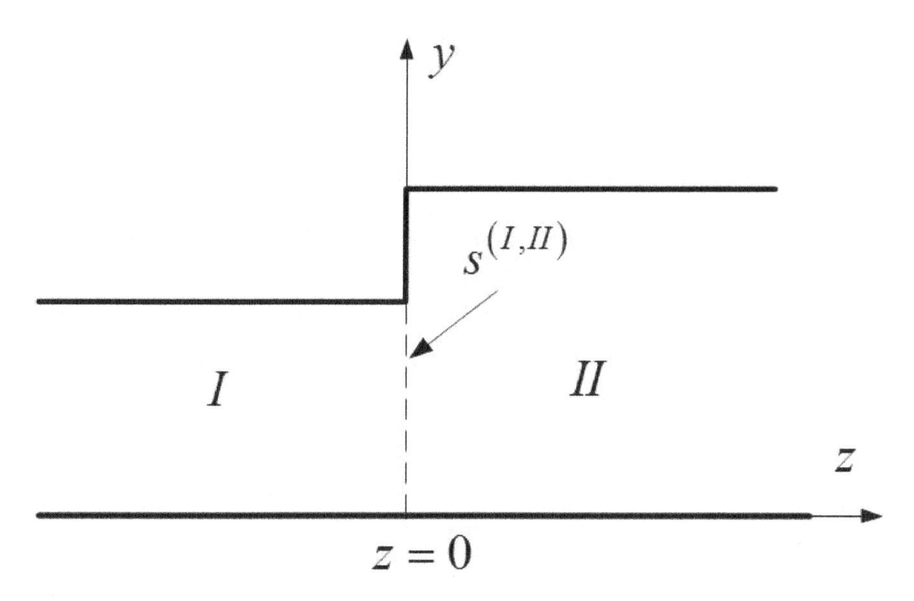

Fig. 3.3 Joint of two waveguides of different geometry

The electric fields from the left and right sides are equal to the tangential field $\mathbf{E}_\tau(x,y)$ at the aperture $s^{(I,II)}$:

$$\mathbf{E}_{\tau_{m=1}}^{(I,+)}(x,y,z=0)+\sum_{m=1}^{\infty}a_m\mathbf{E}_{\tau_m}^{(I,-)}(x,y,z=0)=\mathbf{E}_\tau(x,y),$$
$$\sum_{n=1}^{\infty}b_n\mathbf{E}_{\tau_n}^{(II,+)}(x,y,z=0)=\mathbf{E}_\tau(x,y). \tag{3.23}$$

Taking into account that the modal fields are orthogonal to each other (3.7) and $\mathbf{E}_{\tau_m}^{(I,-)}=\mathbf{E}_{\tau_m}^{(I,+)}$ at $z=0$, the unknown coefficients a_m and b_n for normalized modes are

$$a_{m=1}=\int_{s^{(I,II)}}\mathbf{E}_\tau(x',y')\mathbf{E}_{\tau_{m=1}}^{(I,+)}(x',y')dx'dy'-1,$$
$$a_m=\int_{s^{(I,II)}}\mathbf{E}_\tau(x',y')\mathbf{E}_{\tau_m}^{(I,-)}(x',y')dx'dy',\ m=2,3,\ldots,\infty, \tag{3.24}$$
$$b_n=\int_{s^{(I,II)}}\mathbf{E}_\tau(x',y')\mathbf{E}_{\tau_n}^{(II,+)}(x',y')dx'dy',\ n=1,2,\ldots,\infty.$$

Additionally, the magnetic modal field can be derived using the Maxwell equations, the electric field, and the modal wave impedance: $\left(\mathbf{H}_{\tau_k}\times\mathbf{z}_0\right)=Y_k\mathbf{E}_{\tau_k}$.

The boundary condition for the magnetic field is written at the aperture $s^{(I,II)}$ taking into account that $\mathbf{H}_{\tau_k}^{(-)}=-\mathbf{H}_{\tau_k}^{(+)}$ at $z=0$:

$$\left[\left(\mathbf{H}_{\tau_{m=1}}^{(I,+)}-\sum_{m=1}^{\infty}a_m\mathbf{H}_{\tau_m}^{(I,+)}-\sum_{n=1}^{\infty}b_n\mathbf{H}_{\tau_n}^{(II,+)}\right)\times\mathbf{z}_0\right]=$$
$$=Y_1^{(I)}\mathbf{E}_{\tau_{m=1}}^{(I,+)}-\sum_{m=1}^{\infty}a_mY_m^{(I)}\mathbf{E}_{\tau_m}^{(I,+)}-\sum_{n=1}^{\infty}b_nY_n^{(II)}\mathbf{E}_{\tau_n}^{(II,+)}=0. \tag{3.25}$$

By substituting a_m, b_n from (3.24) to (3.25), this boundary condition is re-written as a non-homogeneous integral equation of the first kind:

$$2Y_1^{(I)}\mathbf{E}_{\tau_{m=1}}^{(I,+)}(x,y)=Y_1^{(I)}\left(\int_{s^{(I,II)}}\mathbf{E}_\tau(x',y')\mathbf{E}_{\tau_{m=1}}^{(I,+)}(x',y')dx'dy'\right)\mathbf{E}_{\tau_{m=1}}^{(I,+)}(x,y)+$$
$$+\sum_{m=2}^{\infty}Y_m^{(I)}\left(\int_{s^{(I,II)}}\mathbf{E}_\tau(x',y')\mathbf{E}_{\tau_m}^{(I,+)}(x',y')dx'dy'\right)\mathbf{E}_{\tau_m}^{(I,-)}(x,y)+ \tag{3.26}$$
$$+\sum_{n=1}^{\infty}Y_n^{(II)}\left(\int_{s^{(I,II)}}\mathbf{E}_\tau(x',y')\mathbf{E}_{\tau_n}^{(II,+)}(x',y')dx'dy'\right)\mathbf{E}_{\tau_n}^{(II,+)}(x,y)=0.$$

Now the unknown aperture field $\mathbf{E}_\tau(x',y')$ can be represented by an expansion of the basis functions $\mathbf{e}_k(x',y')$:

$$\mathbf{E}_\tau(x', y') = \sum_{k=1}^{K} \alpha_k \mathbf{e}_k(x', y'). \quad (3.27)$$

Applying to (3.26) the Galerkin method, a system of linear algebraic equations regarding to α_k is obtained. This system can be solved by any pertinent numerical algorithm, and the modal amplitudes a_m, b_n are calculated.

The advantage of the integral equation approach to the discontinuity problem is that there is large choice in the approximation of the aperture electric field. In particular, the basis functions are from the 2-D Helmholtz equation solutions; they can be the polynomial ones and be defined on the aperture discrete sub-domains. The basis functions can take into account the edge field behavior using the singular functions, or a tighter mesh close to the edge in the case of the discretely defined base and weighting functions can be used. In addition, different methods of analytical solutions of integral equation (3.26) are used to derive the approximate values of the equivalent reactivity of the joint and the reflection and transmission coefficients. Of course, other discontinuities can be considered and treated by the integral equation method [5],[12],[13],[16]-[19].

Example 3.1. Diffraction at the Joint of Three Parallel-Plate Waveguides (Fig. 3.4).

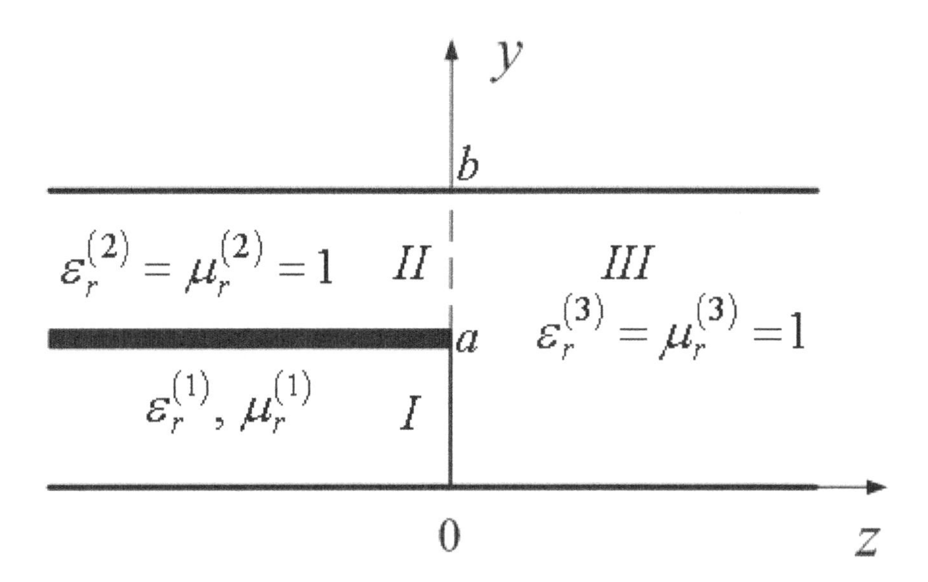

Fig. 3.4 Joint of three parallel-plate waveguides

It consists of a parallel-plate waveguide (*I*) filled by a dielectric connected to another air-filled parallel-plate waveguide (*III*). The incident main mode of the first waveguide is diffracted at the joint, and the scattered modes of the same polarization are propagating back in the same waveguide. Besides, the modes are excited in waveguides *II* and *III*. This problem can be solved by the integral equation method. The electric and magnetic field components in each waveguide are derived from the following expressions:

$$E_y = \frac{j}{\omega\varepsilon_0\varepsilon_r}\frac{\partial^2 F}{\partial z^2}, \quad H_x = -\frac{\partial F}{\partial z}, \quad F = \varphi(y)e^{\pm jk_z z}. \tag{3.28}$$

In the first waveguide, the field consists of the incident main mode *(l=0)* and reflected ones:

$$\begin{aligned} E_y^{(I)}(y,z) &= -\frac{j\left(k_{z,0}^{(1)}\right)^2}{\omega\varepsilon_0\varepsilon_r^{(1)}}\varphi_0^{(1)}(y)e^{-jk_{z,0}^{(1)}z} - \\ &\quad -\frac{j}{\omega\varepsilon_0\varepsilon_r^{(1)}}\sum_{l=0}^{\infty}\left(k_{z,l}^{(1)}\right)^2 A_l\varphi_l^{(1)}(y)e^{jk_{z,l}^{(1)}z}, \\ H_x^{(I)}(y,z) &= jk_{z,0}^{(1)}\varphi_0^{(1)}e^{-jk_{z,0}^{(1)}z} - j\sum_{l=0}^{\infty}k_{z,l}^{(1)}A_l\varphi_l^{(1)}e^{jk_{z,l}^{(1)}z} \end{aligned} \tag{3.29}$$

where $\varphi_l^{(1)}(y) = \sqrt{\frac{2-\delta_{0,l}}{a}}\cos\left(\frac{l\pi y}{a}\right)$, $\delta_{0l} = \begin{cases}1, l=0\\ 0, l\neq 0\end{cases}$, $k_{z,l}^{(1)} = \sqrt{k_0^2\varepsilon_r^{(1)}\mu_r^{(1)} - \left(\frac{l\pi}{a}\right)^2}$.

In the second waveguide, the modes propagate against the z-axis:

$$\begin{aligned} E_y^{(II)}(y,z) &= \frac{-j}{\omega\varepsilon_0\varepsilon_r^{(2)}}\sum_{m=0}^{\infty}\left(k_{z_m}^{(2)}\right)^2 B_m\varphi_m^{(2)}(y)e^{jk_{z_m}^{(2)}z}, \\ H_x^{(II)}(y,z) &= -j\sum_{m=0}^{\infty}k_{z,m}^{(2)}B_m\varphi_m^{(2)}e^{jk_{z,m}^{(2)}z} \end{aligned} \tag{3.30}$$

where

$$\varphi_m^{(2)}(y) = \sqrt{\frac{2-\delta_{0,m}}{b-a}}\cos\left(\frac{m\pi y}{b-a}\right), \ \delta_{0m} = \begin{cases}1, m=0\\ 0, m\neq 0\end{cases}, \ k_{z,m}^{(2)} = \sqrt{k_0^2\varepsilon_r^{(2)}\mu_r^{(2)} - \left(\frac{m\pi}{b-a}\right)^2}.$$

The excited modes in the third waveguide are written as

$$\begin{aligned} E_y^{(III)}(y,z) &= \frac{-j}{\omega\varepsilon_0\varepsilon_r^{(3)}}\sum_{n=0}^{\infty}\left(k_{z,n}^{(3)}\right)^2 C_n\varphi_n^{(3)}(y)e^{-jk_{z,n}^{(3)}z}, \\ H_x^{(III)}(y,z) &= j\sum_{n=0}^{\infty}k_{z,n}^{(3)}C_n\varphi_n^{(3)}e^{-jk_{z,n}^{(3)}z}. \end{aligned} \tag{3.31}$$

where $\varphi_n^{(3)}(y)=\sqrt{\frac{2-\delta_{0,n}}{b}}\cos\left(\frac{n\pi y}{b}\right)$, $\delta_{0n}=\begin{cases}1, n=0\\0, n\neq 0\end{cases}$, $k_{z,n}^{(3)}=\sqrt{k_0^2\varepsilon_r^{(3)}\mu_r^{(3)}-\left(\frac{n\pi}{b}\right)^2}$.

The coefficients A_n, B_l, C_m in (3.29)-(3.31) are the unknown amplitudes of the scattered modes. They are expressed through the electric field at the apertures similarly as (3.24)

$$
\begin{aligned}
A_{l=0} &= \frac{j\omega\varepsilon_0\varepsilon_r^{(1)}}{\left(k_{z,0}^{(1)}\right)^2}\int_0^a e_y^{(1)}(y')\varphi_{l=0}^{(1)}(y')dy'-1,\\
A_{l>0} &= \frac{j\omega\varepsilon_0\varepsilon_r^{(1)}}{\left(k_{z,l}^{(1)}\right)^2}\int_0^a e_y^{(1)}(y')\varphi_l^{(1)}(y')dy',\\
B_m &= \frac{j\omega\varepsilon_0\varepsilon_r^{(2)}}{\left(k_{z,m}^{(2)}\right)^2}\int_a^b e_y^{(2)}(y')\varphi_m^{(2)}(y')dy',\\
C_n &= \frac{j\omega\varepsilon_0\varepsilon_r^{(3)}}{\left(k_{z,n}^{(3)}\right)^2}\int_0^b e_y^{(3)}(y')\varphi_n^{(3)}(y')dy'.
\end{aligned}
\tag{3.32}
$$

The integral equation is obtained by substituting of these coefficients to the magnetic field expressions (3.29)-(3.31) and satisfying the boundary condition at the joint plane $z=0$:

$$
\begin{aligned}
&H_x^{(III)}(y,z=0)=H_x^{(I)}(y,z=0), y\in(0,a)\\
&e_y^{(3)}(y)=e_y^{(1)}(y), y\in(0,a),\\
&H_x^{(III)}(y,z=0)=H_x^{(II)}(y,z=0),\ y\in(a,b),\\
&e_y^{(3)}(y)=e_y^{(2)}(y), y\in(a,b).
\end{aligned}
\tag{3.33}
$$

Then

$$
\begin{aligned}
H_x^{(I)}-H_x^{(III)} &= 2j\varphi_0^{(1)}k_{z,0}^{(1)}+\omega\varepsilon_0\varepsilon_r^{(1)}\sum_{l=0}^{\infty}\frac{\varphi_l^{(1)}(y)}{k_{z,l}^{(1)}}\int_0^a e_y^{(3)}(y')\varphi_l^{(1)}(y')dy'+\\
&+\omega\varepsilon_0\varepsilon_r^{(3)}\sum_{n=0}^{\infty}\frac{\varphi_n^{(3)}(y)}{k_{z,n}^{(3)}}\int_0^a e_y^{(3)}(y')\varphi_n^{(3)}(y')dy'=0,\ y\in(0,a);\\
H_x^{(II)}-H_x^{(III)} &= \omega\varepsilon_0\varepsilon_r^{(2)}\sum_{m=0}^{\infty}\frac{\varphi_m^{(2)}(y)}{k_{z,m}^{(2)}}\int_a^b e_y^{(3)}(y')\varphi_m^{(2)}(y')dy'+\\
&+\omega\varepsilon_0\varepsilon_r^{(3)}\sum_{n=0}^{\infty}\frac{\varphi_n^{(3)}(y)}{k_{z,n}^{(3)}}\int_a^b e_y^{(3)}(y')\varphi_n^{(3)}(y')dy'=0,\ y\in(a,b).
\end{aligned}
\tag{3.34}
$$

Similar equation is solved by the Galerkin method in [20]. The unknown electric field $e_y^{(3)}\left(y' \in 0, b\right)$ is represented by a sum of pulse functions:

$$e_y^{(3)}\left(y'\right) \approx \sum_{k=1}^{K} \alpha_k e_k \tag{3.35}$$

where

$$e_k = \begin{cases} 1, y' \in \left(y'_{k-1}, y'_k\right) \\ 0, y' \notin \left(y'_{k-1}, y'_k\right) \end{cases}. \tag{3.36}$$

The series expansions in (3.34) are truncated, and the convergence of the algorithm is studied regarding to the stabilization of the reflection coefficient of the main incident mode of the waveguide *I* depending on the truncation numbers K, L, M, N of the corresponding series (3.35) and (3.34):

$$R = A_{l=0} = \frac{j\omega\varepsilon_0\varepsilon_r^{(1)}}{\left(k_{z,0}^{(1)}\right)^2} \int_0^a e_y^{(3)}\left(y'\right)\varphi_{l=0}^{(1)}\left(y'\right)dy' - 1. \tag{3.37}$$

It was found that the convergence occurs if $L = M = N \geq 100$ and $K \geq 8$.

3.3.3 *Stationary Functionals and Equivalent Circuits of Discontinuities*

Additionally to the integral equation technique applied to the obstacle problems, the variational method is used often. It demonstrates the same flexibility, and provides the estimations of low and upper bonds of the stationary functionals, and, due to that, the accuracy of numerical models can be predicted.

In the case of waveguide obstacles, a stationary functional can be chosen to be proportional to the normalized reactivity of the considered obstacle [1],[3],[5]. The method describes mathematically that the incident-propagating mode excites the diffracted EM field represented by the infinite expansions of the backward and forward scattered modes. The non-propagating modes decrease fast within the obstacle, and they form its reactive field described by a lumped equivalent circuit. A review on the contemporary state of this technique and on other numerical methods is in [18].

To calculate the reactivity of an obstacle, a stationary functional should be composed. One of these techniques is using the Rayleigh quotient to obtain the stationary functionals for the reactivity of obstacles, eigenvalues of waveguides, and eigenfrequencies of cavities. This quotient *R* is formulated for an abstract operator *A* and the unknown function u as

$$R = \frac{\left(u^* \cdot Au\right)}{\left(u^* \cdot u\right)}. \tag{3.38}$$

One of such functionals is considered in [3], and it is obtained for a joint of two waveguides through a capacitive window (Fig. 3.5). The incident wave is the TE_{10} mode, and other excited modes of the higher order are non-propagating.

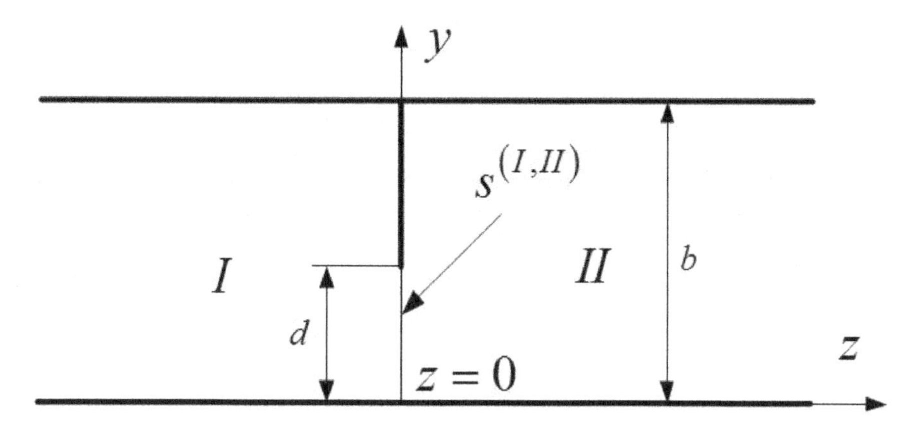

Fig. 3.5 Joint of two waveguides through a capacitive window

This stationary expression for the window susceptance B is

$$j\frac{B}{2}=\sum_{n=1}^{\infty}Y_n\left(\frac{\int_d E(y')h_n(y')dy'}{\int_d E(y')h(y')dy'}\right)^2 \tag{3.39}$$

where $E(y')$ is the unknown window electric field, $Y_n=\omega\varepsilon_0\varepsilon_r\Big/\sqrt{k_0^2\varepsilon_r\mu_r-\left(\frac{n\pi}{b}\right)^2}$, $h(y')$ is the magnetic field of the incident main mode, and $h_n(y')$ is the modal magnetic field. The studies of (3.39) show its stationary character, i.e. the first-order approximation of $E(y')$ gives the second-order accuracy in calculation of B.

Different methods can be applied for minimization of this functional. One can try to use the Ritz method [18], and simple constant or polynomial trial functions for $E(y')$ may be effective. Additionally, the number of the modal functions can be truncated according to the convergence of the derived stationary value. In some cases, the modal expansions can be calculated asymptotically, and the analytical or semi-analytical expressions are obtained [3],[16]. For example, if the window is a narrow one, then a simple approximation $E(y')=const$ gives

$$\frac{B}{Y_0} = \frac{8b}{\lambda_0} \sum_{n=1}^{\infty} \frac{1}{\sqrt{n^2 - \left(\frac{2b}{\lambda_0}\right)^2}} \left(\frac{\sin\left(\frac{n\pi d}{b}\right)}{\frac{n\pi d}{b}} \right)^2 . \tag{3.40}$$

For some geometries of waveguide and capacitive window, even several first members of (3.40) are enough to derive the susceptance value with a good accuracy. Electrically narrow windows cause poor convergence of (3.40), and the method of asymptotical improvements of the convergence is used. For instance, it gives the well-known first-order approximation for the capacitive window [3]:

$$\frac{B}{Y_0} \approx \frac{8b}{\lambda_0} \ln\left(\frac{2b}{\pi d}\right). \tag{3.41}$$

To this time, many waveguide obstacles have been treated using the integral equation method and the variational technique, and the equivalent circuits and formulas for calculating of their components can be found in the waveguide handbooks [3]. They are in the component libraries of many commercially available software tools. As a rule, they are used to describe the unimodal devices when only the main mode is propagating. The accuracy of formulas depends on the number of taken evanescent modes and on the quality of approximation of the field on the obstacle. An additional limitation is for the obstacles placed close to each other when the interaction of evanescent modes is strong, and it should be taken into account [18],[21]. Another attractive feature of this approach is that some integrated lines can be treated similarly if they allow to be put into a correspondence with the rectangular waveguides. The method was firstly proposed by A.A. Oliner [22], and it is considered shortly below.

3.3.4 Microstrip Discontinuities and Their Parallel-plate Models

Taking into account the significance of the integrated lines, an introductory review on the methods of calculation of their discontinuities is given here. Most methods developed and employed for hollow waveguides can be used for the integrated transmission lines, but their design, open to space, hinders the applications of many analytical methods. Anyway, some of them are based on the transformation of the strip and microstrip lines to the equivalent parallel-plate and magnetic wall waveguides and on the following applications of earlier developed techniques of obtaining of equivalent circuits in rectangular waveguides. Since their first applications known from the last Century, the methods have been matured well, and they are used by most commercially available microwave design tools now.

3.3.4.1 Oliner's Approach to Strip and Microstrip Discontinuities

The Oliner's findings, published in 1955, defined the research directions of printed lines for many decades [22]. It was proposed to use the equivalency between the parallel-plate and the triplate waveguides for treatment of strip line discontinuities. A triplate line is substituted by a parallel-plate waveguide of the halved height and which width is limited by perfect magnetic walls placed at the distance w_{eq} from each other (Fig. 2.3). This effective width takes into account the fringing fields of the strip edges and it is calculated as

$$w_{\text{eq}} = 60\pi b / \sqrt{\varepsilon_r}\, Z_{\text{c}}. \tag{3.42}$$

The strip line components are modeled as joints of these waveguides, for which calculation the approximate formulas are obtained similarly as for the joints of rectangular waveguides.

Fig. 3.6 shows a joint of two strip lines with different characteristic impedances. The reactivity of this obstacle is inductive, and its normalized value X/Z_{c_1} is [22]

$$\frac{X}{Z_{c_1}} = \frac{2w_{\text{eq}_1}}{\lambda_0} \ln\left(\csc\left(\frac{\pi w_{\text{eq}_2}}{2w_{\text{eq}_1}}\right)\right) \tag{3.43}$$

where the characteristic impedances $Z_{c_{1,2}}$ are calculated according (2.30).

A simple study shows the validity of this formula in the asymptotical cases when $w_{\text{eq}_1} \to w_{\text{eq}_2}$ and/or $\lambda_0 \to \infty$. The comparisons to the measurements at low microwave frequencies are shown by Oliner in the above-cited paper. A more complete library of the discontinuities and their models is available from [22],[23].

As seen, similarly to the waveguide discontinuity models, the Oliner's ones are compatible with the equivalent network representation of microwave devices, but these models provide better accuracy due to taking into account the parasitic reactivities of components. Being semi-empirical, they have limitations in accuracy, and the measurements and full-wave analysis are to verify and improve these equivalent circuits.

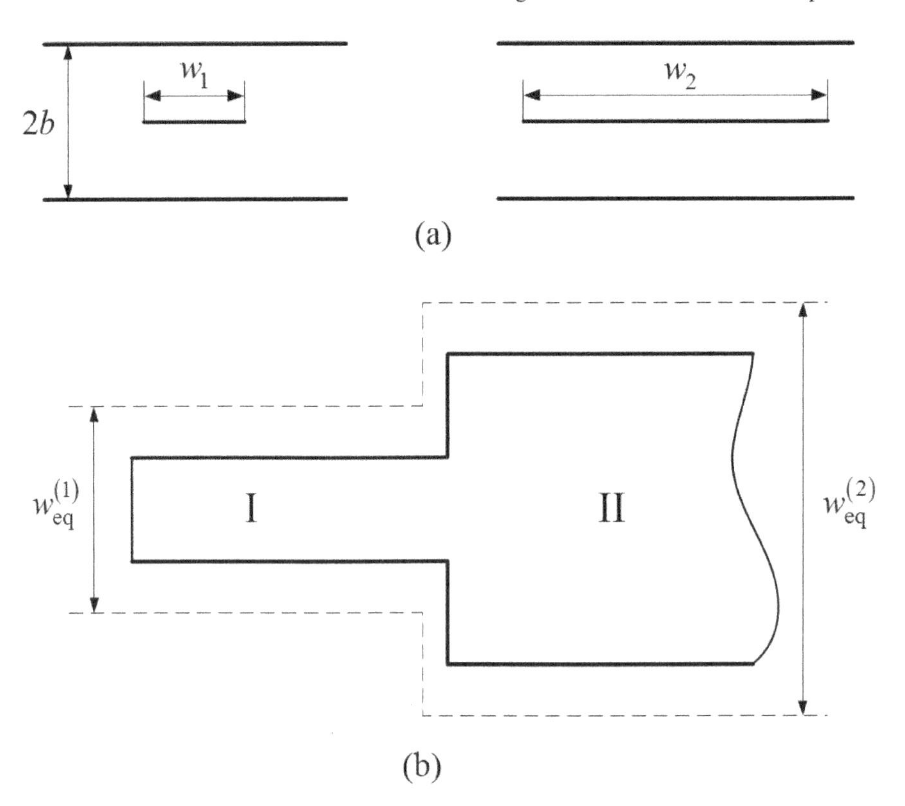

Fig. 3.6 Joint of two triplate lines of different characteristic impedances. (a)- Line cross-sections; (b)- Joint of triplates

The Oliner's approach is applicable to the microstrip line discontinuities. In this case, the magnetic-wall parallel-plate waveguide is filled by the effective frequency-dependent medium which permittivity ε_{eq} can be calculated according to (2.34), for instance. One of the first works in this area was on the modeling of T-junctions and microstrip directional couplers where the Oliner's formulas were modified according to the idea of the effective dielectric filling [24]. Later, it was found that the accuracy of equivalent circuits could be improved further if their elements were calculated by numerical means instead of using modified parallel-plate waveguide formulas [25]-[28].

The results of some numerical simulations are fitted by engineering formulas in [9],[24]-[30], and they are used for practical simulations and design of microwave microstrip circuits. An example of these equivalent circuits is given here for a microstrip step (Fig. 3.7).

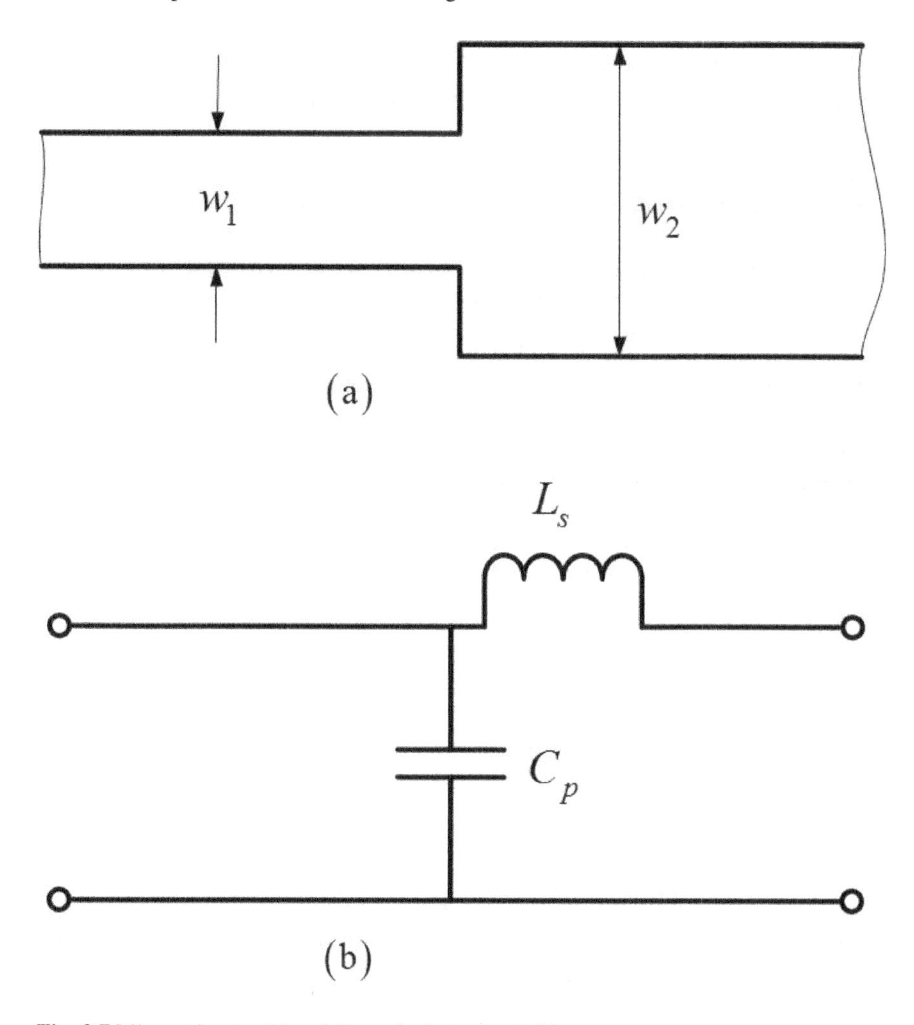

Fig. 3.7 Microstrip step (a) and its equivalent circuit (b)

In this circuit, the inductivity L_s takes into account the current crowding effect which is strong close to the step, and the capacitance C_p is for modeling of the stray electric field concentrated close to the conductor edges. The quasistatic formulas for L_s and C_p obtained by a curve-fitting technique in [23] are

$$\begin{aligned} L_s &= 0.987h\left(1-\frac{Z_{c_1}}{Z_{c_2}}\sqrt{\frac{\varepsilon_{\mathrm{eq}_1}}{\varepsilon_{\mathrm{eq}_2}}}\right)^2, \text{ [nH]} \\ C_p &= 1.37h\frac{\sqrt{\varepsilon_{\mathrm{eq}_1}}}{Z_{c_1}}\left(1-\frac{w_2}{w_1}\right)\frac{\varepsilon_{\mathrm{eq}_1}+0.3}{\varepsilon_{\mathrm{eq}_1}-0.258}\frac{w_1/h+0.264}{w_1/h+0.8}, \text{ [pF]}. \end{aligned} \tag{3.44}$$

where h is the substrate height, and the effective permittivities $\varepsilon_{eq_{1,2}}$ and the characteristic impedances $Z_{c_{1,2}}$ are calculated using (2.34) and (2.35), for instance. Today the models of this type can be improved further by full-wave simulations, measurements, and curve-fitting codes allowing for generating of approximate analytical multi-parametrical expressions of improved accuracy.

3.3.4.2 Parallel-plate Waveguides, Orthogonal Series Expansion Method, and Microstrip Discontinuities

Full wave simulation of microstrip line and components is a time-consuming task even now, and then the use of analytical models or formulas to enhance the modeling of microwave devices is an attracting goal.

The orthogonal series expansion method applied by Kompa to the modeling of microstrip discontinuities unifies the analytical theory of microstrip lines and numerical modeling of discontinuities [30]. To avoid the time-consuming computations of microstrip modes, the modeled components are represented as joints of parallel-plate magnetic-wall waveguides. Their modal fields and parameters are calculated analytically. For instance, the tangential to the joint plane and normal to the strip electric field $\mathbf{E}_{\tau_m}$ of the *m-th* mode is represented by a simple formula instead of the numerically derived microstrip modal field:

$$\mathbf{E}_{\tau_m} = \sqrt{\frac{\xi_m}{w_{\text{eff}}(f)h}} \cos\left[\frac{m\pi}{w_{\text{eff}}(f)}\left(x + \frac{w_{\text{eff}}(f)}{2}\right)\right]\mathbf{y}_0,$$
$$\xi_m = \begin{cases} 1, & m = 0 \\ 2, & m \neq 0 \end{cases}. \tag{3.45}$$

These analytical modal field expressions are used by the orthogonal series expansion method applied to the modeling of equivalent parallel-plate waveguides instead of microstrips. According to this method, the incident and scattered fields composed of multiple parallel-plate modes, are matched at the joint plane by the Galerkin method, and a system of linear algebraic equations is obtained which solution gives the unknown amplitudes of scattered modes.

To increase the accuracy of this algorithm further, an additional analytical technique is applied [30]. Microwave components are composed of strips of different widths. At the steps, their sharp edges essentially disturb the electric and magnetic fields. It causes slow convergence of the algorithms based on the orthogonal series expansion method. To enhance these algorithms it is proposed to take into account the edge field singularity. The portion of energy and equivalent capacitance related to the edge's stray fields are calculated quasistatically, and the effective step extension Δl_s is given for wider microstrip of this joint [30]:

$$\Delta l_s = \frac{1}{2}\left(\frac{\varepsilon_{\text{eff}_2}(f)}{\varepsilon_r} w_{\text{eff}_2}(f) - w_2\right). \tag{3.46}$$

This method of semi-analytical treatment of microstrip discontinuities has been tested by multiple measurements, and it shows good accuracy even at increased frequencies where the higher-order propagating modes disturb the characteristics of microwave devices [30].

3.3.5 *Full-wave FDTD, TLM, FEM, and Integral Equation Treatment of Discontinuities*

The general approach to simulation of discontinuities by the mentioned methods is based on the numerical treatment of the Maxwell equations or equivalent stationary functionals. As a rule, a discontinuity is surrounded by a surface on which some boundary conditions are set. For instance, the open space is simulated by an absorbing boundary condition. The components can be placed into a perfect conductor box. Instead of modeling of the adjoin waveguides, the corresponding boundary conditions are set on the walls where these waveguides join the modeling area.

Taking into account that a full-wave treatment is time-consuming process, any symmetry of the modeled area or obstacle should be used to decrease the simulation time. The symmetry planes, which are equivalent to the perfect electric $(\mathbf{E}_\tau = 0)$ or magnetic $(\mathbf{H}_\tau = 0)$ walls, allow to calculate a subdomain of reduced size.

From practical reasons, the size of spatial steps in the case of the FDTD should be essentially less than the wavelength and the minimal size of the modeled body's details. Especial attention should be paid to the modeling of real conductors inside of which the EM field decreases quickly. The typical spatial step should be less than the skin depth to provide appropriate accuracy of lossy components computations. Additionally to the above-considered qualitative recommendations, the spatial parameter $\delta = \Delta x = \Delta y = \Delta z$ should be coordinated to the time discretization step Δt to provide stability of calculations. One of the rules [31] is

$$\frac{u_{\max}\Delta t}{\delta} \le \frac{1}{\sqrt{n}} \tag{3.47}$$

where $u_{\max}$ is the maximum wave velocity within the model, and n is the number of the space dimension. Taking into account that the TLM and FDTD methods have some commons in their nature [32], the criterion (3.47) can be used to provide the needed accuracy and stability of the transmission line matrix simulations.

The FEM method of treatment of the vector 3-D or 4-D EM problems is distinguished by increased flexibility in the choice of a stationary functional, the basis polynomial functions, and the shapes of elements. One of the stationary functionals considered in [31] is given for the harmonically dependent electric field $\mathbf{E}$ excited by $\mathbf{E}_{\text{inc}}$ in a volume V with the boundary S, on which the Sommerfeld radiation condition [27] is defined

$$F(\mathbf{E}) = \frac{1}{2}\iiint\limits_V \left[\frac{1}{\mu_r}(\nabla\cdot\mathbf{E})\cdot(\nabla\cdot\mathbf{E}) - k_0^2\varepsilon_r(\mathbf{E}\cdot\mathbf{E})\right]dV + \\ + \iint\limits_S \left[\frac{jk_0}{2}(\nu\times\mathbf{E})\cdot(\nu\times\mathbf{E}) + (\mathbf{E}\cdot\mathbf{U})\right]dS \tag{3.48}$$

where $\mathbf{U} = \nu\times(\nabla\times\mathbf{E}_{\text{inc}}) + jk_0(\nu\times\mathbf{E}_{\text{inc}})$, and $\mathbf{v}$ is the normal vector to the boundary S.

To construct an FEM model, the volume V should be divided into N of 3-D elements of the tetrahedral shape, for instance. Then on each element a vector basis function $\mathbf{N}_i$ is defined even it does not satisfy the Helmholtz equation. The field in the volume V is represented by

$$\mathbf{E} = \sum_{1}^{N} E_i\mathbf{N}_i. \tag{3.49}$$

To derive a system of linear equations regarding to the unknown basis function amplitudes E_i, the representation of the field (3.49) is substituted into (3.48), and the Ritz method is applied

$$\frac{\partial F}{\partial E_i} = 0,\ i = 1, 2, ..., N. \tag{3.50}$$

This system can be solved by a pertinent algorithm, and the field inside the volume and on the boundary is derived. Magnetic field is obtained using the Maxwell equations and the calculated electric one.

The waveguide problems are with the boundaries where the conditions are established for the incident and scattered fields. One of the pertinent algorithms is a combination of the FEM and the boundary integral equations [19].

In general, the FEM based methods are very flexible, accurate, and they allow for simulating multiple physical effects coupled to each other. Several commercially available tools are known and used in practice. The details on the FEM in its different formulations are available from many books and guides [31]-[33] and they are not considered here.

One of the methods used to calculate the integrated microwave components is the 2-D integral equation technique used now by several commercially available tools. The method is adapted to the planar design of integrated circuits, which can consist of multiple dielectric and conductor layers. Some software tools additionally simulate the vertical via-holes of different geometrical shapes used to connect several layers of integrations. As known, the planar integral equations decrease the dimension of a 3-D EM problem. Instead of solving the Maxwell equations in a 3-D domain, the integral ones are written regarding to the surface electric currents on the conductors, for instance.

The integral equations are obtained from the vector wave equations with a help of the Green functions. They can satisfy all boundary conditions excepting the interfaces with the signal conductors which currents are the subject to obtain.

The techniques to find the Green functions are highlighted in many publications, and some of them are found in [17],[18],[34]. Very often, the components of Green functions are costly to calculate, and their low- or high-frequency asymptotics are preferable to use. Another idea is to neglect the dependence of the Green function components along an electrically thin substrate [35].

These integral equations can be solved by the Galerkin method in its 2-D modification. The algorithms are sensitive regarding to the proper choice of the basis functions. One recommendation is that the basis functions given for different components of the electric current should satisfy the relations obtained from the Maxwell equations.

The electric current components, which are parallel to the edges, have singularities of their distributions at these edges, and a good idea is to take into account these singularities in the sub-domains which are close to the conductor edges. There are several ways exist. One of them is to use a tighter mesh close to the edges. The second one is using the weighted singular functions. In addition, the different ways of "smart" meshing can decrease the time simulation by the integral equation method, but it requires increased level of knowledge of the EM physics and the numerical methods. Always, proper testing and comparisons of numerical results with the analytical and experimental ones are required.

References

[1] Collin, R.E.: Foundation of Microwave Engineering. John Wiley & Sons (2001)
[2] Felsen, L.B., Marcuvitz, N.: Radiation and Scattering of Waves. Prentice-Hall (1973)
[3] Marcuvitz, N.: Waveguide Handbook. Inst. of Eng. and Techn. Publ. (1986)
[4] Schwinger, J., Saxon, D.S.: Discontinuities in Waveguides. Gordon and Breach Sci. Publ. (1968)
[5] Mashkovzev, B.M., Zibisov, K.N., Emelin, B.F.: Theory of Waveguides. Nauka, Moscow (1966) (in Russian)
[6] Nikolskyi, V.V., Nikolskaya, T.I.: Electrodynamics and Wave Propagation. Nauka, Moscow (1987) (in Russian)
[7] Heaviside, O.: Electromagnetic Waves. Taylor & Francis (1889), http://www.archive.org
[8] Shelkunoff, S.A.: Electromagnetic Field. Blaisdell Publ. Comp. (1963)
[9] Gupta, K.C., Garg, R., Chadha, R.: Computer Aided Design of Microwave Circuits. Artech House (1981)
[10] Helszajn, J.: Passive and Active Microwave Circuits. J. Wiley and Sons (1978)
[11] Balanis, C.A.: Advanced Engineering Electromagnetics. John Wiley (1989)
[12] Mittra, R. (ed.): Computer Techniques for Electromagnetics. Pergamon Press (1973)
[13] Vaganov, R.B., Katsenelenbaum, B.Z.: Foundation of the Diffraction Theory. Nauka, Moscow (1982) (in Russian)
[14] Mrozovski, M.: Guided Electromagnetic Waves. Properties and Analysis. RSP Press Ltd. (1997)
[15] Neganov, V.A.: Physical Regularization of some Ill-posed Problems of Electromagnetism. Nauka Publ., Moscow (2007) (in Russian)
[16] Lewin, L.: Theory of Waveguides. Newnes-Buttertworths, London (1975)

[17] Tai, C.T.: Dyadic Green's Functions in Electromagnetic Theory. Intex Educational Publ., Scranton (1971)
[18] Zhang, W.-X.: Engineering Electromagnetism: Functional Method. Ellis Horwood (1991)
[19] Silvester, P.P., Ferrari, R.L.: Finite Elements for Electrical Engineers. Cambridge University Press, Cambridge (1983)
[20] Gvozdev, V.I., Kouzaev, G.A., Kulevatov, M.V.: Narrow band-pass microwave filter. Telecommun. Radio Eng. 49, 1–5 (1995)
[21] Kapilevich, B.Y.: Waveguide Dielectric Filters. Svayz Publ. Comp. (1980)
[22] Oliner, A.A.: Equivalent circuits for discontinuities in balanced strip transmission line. IRE Trans. 3, 134–143 (1955)
[23] Gupta, K.C., Garg, R., Bahl, I.J.: Microstrip Lines and Slot Lines. Artech House Inc., Dedham (1979)
[24] Leighton Jr., W.H., Milnes, A.G.: Junction reactance and dimensional tolerance effects on X-band 3-dB directional couplers. IEEE Trans., Microw. Theory Tech. 19, 818–824 (1971)
[25] Easter, B.: The equivalent circuit of some microstrip discontinuities. IEEE Trans., Microw. Theory Tech. 23, 655–660 (1975)
[26] Silvester, P., Benedek, P.: Microstrip discontinuity capacitances for right-angle bends, T junctions, and crossings. IEEE Trans., Microw. Theory Tech. 23, 341–346 (1973)
[27] Jin, J.: The Finite Element Method in Electromagnetics. J. Wiley & Sons, Inc. (1993)
[28] Nefedov, E.I., Fialkovsky, A.T.: Strip Transmission Lines. Nauka, Moscow (1980) (in Russian)
[29] Hoffmann, R.K.: Handbook of Microwave Integrated Circuits. Artech House (1987)
[30] Kompa, G.: Practical Microstrip Design and Applications. Artech House (2005)
[31] Sadiku, M.N.O.: Numerical Techniques in Electromagnetics. CRC Press (2001)
[32] Johns, P.B.: On the relationships between TLM and finite-difference methods for Maxwell equations. IEEE Trans., Microw. Theory Tech. 35, 60–61 (1987)
[33] Taflove, A.T., Hagness, S.C.: Computational Electrodynamics. Artech House (2005)
[34] Nikolskyi, V.V. (ed.): Computer Aided Design of Microwave Devices. Radio and Svayz Publ. (1982) (in Russian)
[35] Okoshi, T.: Planar Circuits for Microwaves and Light Waves. Springer, Berlin (1985)

4 Geometro-topological Approaches to the EM Problems

Abstract. In this Chapter, several geometro-topological theories of the EM field and boundary value problems are considered. The main attention is paid to the topological approach based on the representation of fields by their skeletons or topological charts of field-force line maps. These charts are associated with the main features of components, and these skeletons and their bifurcations schemes are applicable for the qualitative analysis of the EM effects in microwave elements. Additionally, an approach is proposed allowing simplified semi-analytical or numerical solutions of 2-D and 3-D boundary value problems, which is promising for the development of fast EM models. In the end of this Chapter, the results of other authors on the topological theory of the EM field and its visualization and computing are considered. References -111. Figures -35, Pages -67.

"Physics is but geometry in act."
H. Weyl

Since the beginning of the theory of electricity, the field quantities were associated with the geometrical elements of space [1]. For instance, following to M. Faraday, the vector fields are represented by their force lines, which are the motion trajectories of an elementary probe particle. The Maxwell equations were formulated using these geometric representations of the electric and magnetic fields, and computation of the field means the calculation of its vector geometry additionally to its magnitude spatial distribution.

In opposite to geometry, topology is not interested in exact geometrical properties of figures. Only the ones preserved under the bi-continuous transformations are the subject of topology. Very often, topology is called the fuzzy geometry. In physics, the topological features correspond to the most important properties of the studied objects, and topology can be used to create the qualitative models of physical effects.

Topology related solutions and problems are often arisen in physics, and the topological methods are rather powerful in quantum mechanics and in general field theory [2]-[8]. For instance, topology can be applied to the multi-domain Hamiltonian systems in which the inter-domain barriers are not removed by smooth transformations, and the particles stay in their domains forever.

G.A. Kouzaev: Applications of Advanced Electromagnetics, LNEE 169, pp. 119–185.
springerlink.com

Another topology-related problem is arisen in the general theory of fields where the topologically non-trivial fields are derived from the condition of relative minima of the action functionals. These fields are characterized by certain parameters called the topological invariants which are preserved if the by-continuous transformations are applied to these fields. More information on the use of topology in quantum mechanics and general field theory are from the above-cited publications.

The applications of topological methods in electromagnetism are less known. The lower interest of the researchers is explained by the preference to the well-developed representation of the EM problems by the equivalent circuits and their integral parameters as the modal propagation constants and impedances, scattering matrices of obstacles in waveguides, etc.

The recent needs in manipulations of nanoparticles, single molecules, and atoms turn the attention to the field calculation and control of their spatio-time distributions [9]-[12]. Here the particle motion in known fields is described by the dynamical systems which phase space can have rather complicated multi-cell structure. These cells are separated by separatrices, and interception of them by the particles is prohibited due to the energy reasons. Then the phase space of the system is of non-trivial topology, and its qualitative theory can be used following to A.A. Andronov and other authors [13]-[15],[16].

The field-force lines of the electric and magnetic fields are considered as the trajectories of elementary electric and magnetic charges, respectively, and the use of the field-force lines maps in electromagnetism is known from the Faraday's time. The equations of these trajectories are the autonomous dynamical systems. Time-dependent fields and their maps are described by the non-autonomous dynamical systems. Taking into account the complexity of the EM field maps, the only visualization of the field geometry, which is now common, is not enough, and the attention should be paid to the developments of the "field-oriented language" to explain the EM phenomena [16]-[19].

One of the approaches is with the use of topological means. Following to A.A. Andronov [13], the maps of the autonomous dynamical systems are described by their skeletons or topological charts. In electromagnetics, the field topology depends on the domain's geometry, boundary conditions, and driving frequency. Our first work in this direction was published in 1988 [20] where we proposed a topological method of the automated field analysis and the Maxwell-like relations between the topological charts of the electric and the magnetic fields. Later, an approach to the qualitative solution of boundary problems was proposed and applied to several transmission lines and components [21],[22].

For further developments of this approach, some techniques from general theory of the autonomous dynamical systems are interesting. Among them are the algorithms of automated phase portrait analysis and a method of topological description of time-dependent vector fields [23]-[26].

Another theory of topological origin (1989) relates to the contributions of A.F. Ranada who proposed the nonlinear Maxwell-like equations for the field-force lines of radiated fields [27], and this idea is considered in this book, too.

An interesting approach to the geometrization of electromagnetism is with the use of differential forms. In this theory, the EM quantities are associated with the spatial shapes of different topology, and the Maxwell equations establish the metric correspondence between these shapes. This theory, which origin relates to the 30s of the last Century [28], was considered by many authors [29]-[33], although its practical applications in the EM computations are still in the future.

The necessity of using topological ideas to explain the classical electromagnetism and quantum-electromagnetic effects is expressed by many authors. For example, T.W. Barret has recently analyzed some known problems in physics. He concludes that the local Maxwell theory does not explain them well, and the potentials $\mathbf{A}_{\mu}$ are to be used to describe them correctly [34]. These potentials are the local operators mapping the global spatiotemporal conditions onto the local EM field. Then the EM fundamental equations are not locally defined because, in opposite case, they are not able to describe in full all known physical effects. Probably, the reason of the necessity to use the topological field descriptions is originated from the quantum level, where, for instance, photons are not localizable in free space and calculation of their density is with the integrations over the whole space [3]. The above-mentioned ideas and theories are considered below with the different degree of attention. More information on them can be found from the above-cited books and papers.

4.1 Topological Approach to the Theory of Boundary Value Problems

The attention of the Author to the topological aspects of the boundary problems of electromagnetism was initiated by problems in the EM modeling of the microwave 3-D integrated circuits [35],[36]. It was found that many 3-D circuit components needed the EM means to be explained and designed. The topological description of the field and components was considered as a complimentary theory to the already existed to-that-moment and powerful EM analytical and numerical methods.

The first our paper on topological modeling of the boundary value problems was published in 1988 in a book of proceedings of a scientific conference hold in contemporary Georgia, a former Republic of the USSR [17]. Theoretical aspects of this approach were considered further during several years when the Author worked in the space industry and at the Moscow State Institute of Electronics and Mathematics (Technical University) [21],[22],[37]-[49]. Later, the attention of the Author was switched to the signaling and computing with the topologically modulated signals and to the developments of logical circuitry for them. Additional work in electromagnetics was performed at the McMaster University (Canada) on the developments of EM models of via-holes and antennas using topological approach [50]. Resent results, obtained at the Norwegian University of Science and Technology–NTNU, are with the development, design, and modeling of processors, which idea originated from our earlier papers in topological computing (1991-1992). The Author thanks all colleagues from the above-mentioned institutions for their support of his research activity in this field.

4.1.1 Visualization and Topological Analysis of the EM Field Maps

The field-force line maps were initially introduced by M. Faraday and considered by him as the real objects of the ether, which vibrations could explain the EM phenomena. Later, these field-force lines, hydraulic and mechanistic analogies were used by J.C. Maxwell to compose his famous 20 equations of electromagnetism [1]. Now the field-force lines are viewed only as a convenient tool for the visualization of fields and they have lost their initial analytic power. Meanwhile, the complexity of the EM phenomena and new possibilities in visualization provided by modern computers and software tools allow for paying more increased attention to this matter to obtain new effective results and solutions.

4.1.2 Field Maps of Harmonic Fields and Their Topological Analysis

Consider the field maps of microwave harmonic fields. The equations for the field-force lines are

$$\frac{d\mathbf{r}_{e,h}}{dt_{e,h}} = \mathrm{Re}\left[\mathbf{E}(\mathbf{r}_e), \mathbf{H}(\mathbf{r}_h)\right] \tag{4.1}$$

where $\mathbf{r}_{e,h}$ is the radius-vector of a field-force line point of the real part of the electric $\mathbf{E}(\mathbf{r}_e)$ or magnetic $\mathbf{H}(\mathbf{r}_h)$ fields in their phasor notations, and $t_{e,h}$ is the parametric variable which is not connected to time in general. Computation of maps requires preliminary knowledge of these fields, and, due to that, for a long period, these maps have played only an illustrative role in electromagnetics.

Following to H. Poincare and A.A. Andronov, the qualitative analysis is the calculation of *the equilibrium points of these dynamical systems, their separatrices and composing of topological charts and their bifurcations caused by the variation of the boundary conditions and driving frequency*. The topological charts and their transformations relate to the parameters of the modeled components such as the scattering matrices, impedances, and propagation constants of the modeled waveguides.

Equilibriums of fields are the points where a field is equal to zero: $\mathbf{E}\left(\mathbf{r}_0^{(i)}\right) = 0$ or $\mathbf{H}\left(\mathbf{r}_0^{(j)}\right) = 0$. Their coordinates are found by a search algorithm if the field is known. An equilibrium point is described by its determinant given for the real fields:

$$D_e = \det\begin{pmatrix} \dfrac{\partial E_x\left(\mathbf{r}_0^{(i)}\right)}{\partial x} & \dfrac{\partial E_x\left(\mathbf{r}_0^{(i)}\right)}{\partial y} & \dfrac{\partial E_x\left(\mathbf{r}_0^{(i)}\right)}{\partial z} \\ \dfrac{\partial E_y\left(\mathbf{r}_0^{(i)}\right)}{\partial x} & \dfrac{\partial E_y\left(\mathbf{r}_0^{(i)}\right)}{\partial y} & \dfrac{\partial E_y\left(\mathbf{r}_0^{(i)}\right)}{\partial z} \\ \dfrac{\partial E_z\left(\mathbf{r}_0^{(i)}\right)}{\partial x} & \dfrac{\partial E_z\left(\mathbf{r}_0^{(i)}\right)}{\partial y} & \dfrac{\partial E_z\left(\mathbf{r}_0^{(i)}\right)}{\partial z} \end{pmatrix}. \tag{4.2}$$

If it is different from zero, the equilibrium point relates to the simple one according to the classification introduced by A.A. Andronov [13]. In this case, all roots $\lambda_e^{(1,2,3)}$ of the characteristic equation (4.3) are not equal to zero

$$\det\begin{pmatrix} \dfrac{\partial E_x\left(\mathbf{r}_0^{(i)}\right)}{\partial x} - \lambda_e & \dfrac{\partial E_x\left(\mathbf{r}_0^{(i)}\right)}{\partial y} & \dfrac{\partial E_x\left(\mathbf{r}_0^{(i)}\right)}{\partial z} \\ \dfrac{\partial E_y\left(\mathbf{r}_0^{(i)}\right)}{\partial x} & \dfrac{\partial E_y\left(\mathbf{r}_0^{(i)}\right)}{\partial y} - \lambda_e & \dfrac{\partial E_y\left(\mathbf{r}_0^{(i)}\right)}{\partial z} \\ \dfrac{\partial E_z\left(\mathbf{r}_0^{(i)}\right)}{\partial x} & \dfrac{\partial E_z\left(\mathbf{r}_0^{(i)}\right)}{\partial y} & \dfrac{\partial E_z\left(\mathbf{r}_0^{(i)}\right)}{\partial z} - \lambda_e \end{pmatrix} = 0. \tag{4.3}$$

The complex equilibrium point has, at minimum, one zero root of (4.3), and these equilibriums are unstable regarding to perturbations of the fields in (4.1). A similar equation can be written for a magnetic field equilibrium point:

$$\det\begin{pmatrix} \dfrac{\partial H_x\left(\mathbf{r}_0^{(j)}\right)}{\partial x} - \lambda_h & \dfrac{\partial H_x\left(\mathbf{r}_0^{(j)}\right)}{\partial y} & \dfrac{\partial H_x\left(\mathbf{r}_0^{(j)}\right)}{\partial z} \\ \dfrac{\partial H_y\left(\mathbf{r}_0^{(j)}\right)}{\partial x} & \dfrac{\partial H_y\left(\mathbf{r}_0^{(j)}\right)}{\partial y} - \lambda_h & \dfrac{\partial H_y\left(\mathbf{r}_0^{(j)}\right)}{\partial z} \\ \dfrac{\partial H_z\left(\mathbf{r}_0^{(j)}\right)}{\partial x} & \dfrac{\partial H_z\left(\mathbf{r}_0^{(j)}\right)}{\partial y} & \dfrac{\partial H_z\left(\mathbf{r}_0^{(j)}\right)}{\partial z} - \lambda_h \end{pmatrix} = 0. \tag{4.4}$$

Depending on the position of λ on a complex plane, there are nine simple equilibrium points for a 3-D dynamical system [51]. The plane systems have only four equilibrium points, and they are the node, focus, center, and saddle (Fig. 4.1).

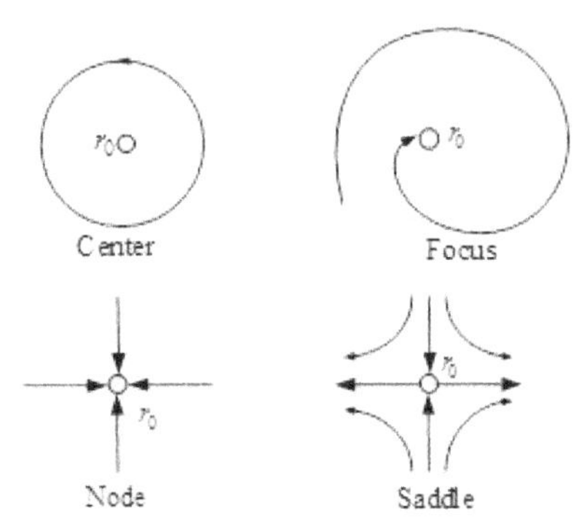

Fig. 4.1 Equilibrium points of a plane autonomous dynamical system. Separatrices are shown in bold

In the case of modal fields in waveguides, resonators, and other microwave components, the equilibrium points can be placed on the symmetry lines, which compose 1-D equilibrium manifolds (Fig. 4.2).

The equilibrium points can be structurally stable or "coarse" according to A.A. Andronov, and these ones are not destroyed by smooth perturbations of the field. Some equilibrium points, especially, complex ones, are not stable, and the effect of their transformation or disappearing is called the *local bifurcation*.

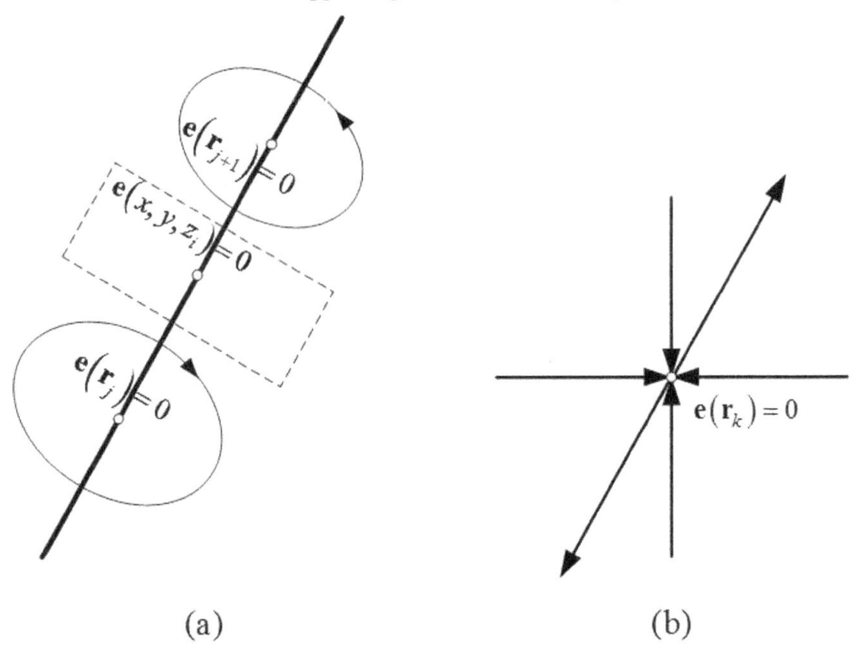

Fig. 4.2 Elements of 3-D field maps. (a)- Zero field axis surrounded by circular field-force lines and a plane of symmetry (dashed) where all field components are zero; (b)- 3-D saddle point

The equilibriums are connected to each other by special field-force lines called the *separatrices*, and they take the manifolds which separate the phase space into a number of cells where the field-force lines geometry is the same. The direction of a separatrix can be calculated analyzing the field close to an equilibrium point [13].

Although the 3-D dynamical systems are more complicated for qualitative analysis, the EM origin limits the number of types of equilibrium points. For instance, in waveguides, the most frequently encountered equilibrium positions are the centers and saddles placed along the lines or curves (Fig. 4.2a). Additional elements are the planes of symmetry where the fields are zero (Fig. 4.2a). Another element, commonly found in the fields maps of waveguide modes, is a 3-D saddle point (Fig. 4.2b). The full classification of elements of the phase space of the EM field has not been finished yet, and much more work and simulations are needed.

The analysis shows that the electric/magnetic field maps have certain spatial structures or skeletons, or following to A.A. Andronov, topological schemes. These schemes and their bifurcations can be calculated qualitatively according to the given fields without detailed solutions of (4.1), and it is the topological analysis of these equations and field maps. This analysis is now popular in many sciences were the vector fields are described by autonomous dynamical systems [23]-[26]. The process of composing of topological schemes can be programmed, and the algorithms of the design of components according to the given topological charts can be developed [40].

Consider the field maps and their topological schemes $T_{H,E}$ for rectangular waveguides for which the analytical expressions for field components are known. The differential equations to calculate the field-force lines of any mode are written in the reduced phase space of (4.1). For instance, the TE mode equations valid on a half of the spatial period along the longitudinal axis $0 \to z$ are derived taking into account (2.10):

$$\begin{aligned}
\frac{\partial y_h^{(\mathrm{TE})}}{\partial x} &= \frac{H_y}{H_x} = \frac{k_y}{k_x}\cot(k_x x)\tan(k_y y), \\
\frac{\partial z_h^{(\mathrm{TE})}}{\partial x} &= \frac{H_z}{H_x} = \frac{\chi^2}{k_x k_z}\cot(k_x x)\cot(k_z z), \\
\frac{\partial y_e^{(\mathrm{TE})}}{\partial x} &= \frac{E_y}{E_x} = -\frac{k_x}{k_y}\tan(k_x x)\cot(k_y y)
\end{aligned} \tag{4.5}$$

where $k_x = m\pi/a$, $k_y = n\pi/b$, $\chi^2 = k_x^2 + k_y^2$, $k_z = \sqrt{k_0^2 \varepsilon_r \mu_r - \chi^2}$, $m = 1,2,3,\ldots,\infty$, and $n = 1,2,3,\ldots,\infty$.

Analyzing the right parts of these equations, the equilibrium manifolds are obtained immediately. For instance, the equilibriums of electric field are at the points $(k_x x) = (k-1)\pi, k = 1,2,3,\ldots$ or $(k_y y) = \dfrac{(2l-1)}{2}, l = 1,2,3,\ldots$

The magnetic field has the equilibrium points of the coordinates $(k_x x) = \frac{(2i-1)}{2}\pi, i = 1,2,3,...$ and $(k_z z) = \frac{(2j-1)}{2}\pi, j = 1,2,3,...$.

Studying the characteristic equations at the equilibrium points, the type of them can be defined. Particularly, (4.5) are solved analytically due to their separated type, and the curves, to which the vectors of corresponding fields are tangential, are calculated according to the following formulas [17],[22]:

$$y_h^{(\mathrm{TE})}(x) = \frac{1}{k_y}\arcsin\left\{\frac{\left[\sin(k_x x)\cdot\sin(k_x x_0)\right]^{(k_y/k_x)^2}}{\sin(k_y y_0)}\right\},$$
$$z_h^{(\mathrm{TE})}(x) = \frac{1}{k_z}\arccos\left[\cos(k_z z_0)\cdot\left[\sin(k_x x)\cdot\sin(k_x x_0)\right]^{-\chi^2/k_x^2}\right], \tag{4.6}$$
$$y_e^{(\mathrm{TE})}(x) = \frac{1}{k_y}\arccos\left(\frac{\cos(k_y y_0)\cos(k_x x_0)}{\cos(k_x x)}\right)$$

where x_0, y_0, z_0 are the coordinates of the starting point of calculation of a field-force line.

The TM modes of a rectangular waveguide are studied similarly. The field-force line equations are derived using the field components (2.11):

$$\frac{\partial y_e^{(\mathrm{TM})}}{\partial x} = \frac{E_y}{E_x} = \frac{k_y}{k_x}\tan(k_x x)\cot(k_y y),$$
$$\frac{\partial z_e^{(\mathrm{TM})}}{\partial x} = \frac{E_z}{E_x} = \frac{\chi^2}{k_x k_z}\tan(k_x x)\cot(k_z z), \tag{4.7}$$
$$\frac{\partial y_h^{(\mathrm{TM})}}{\partial x} = \frac{H_y}{H_x} = -\frac{k_x}{k_y}\cot(k_x x)\tan(k_y y).$$

The magnetic field has the equilibrium lines in $(k_x x) = \frac{(2k-1)}{2}, k = 1,2,3,...$ and $(k_y y) = (l-1)\pi, l = 1,2,3,...$. The electric field has the equilibrium points at $(k_x x) = (j-1)\pi, j = 1,2,3,...$ and $(k_z z) = \frac{2i-1}{2}\pi, i = 1,2,3,...$.

The differential equations (4.7) are solved analytically

$$y_e^{(\mathrm{TM})}(x) = \frac{1}{k_y}\arccos\left(\cos(k_y y_0)\left(\frac{\cos(k_x x)}{\cos(k_x x_0)}\right)^{\left(\frac{k_y}{k_x}\right)^2}\right),$$

$$z_e^{(\mathrm{TM})}(x) = \frac{1}{k_z}\arccos\left(\cos(k_z z_0)\left(\frac{\cos(k_x x)}{\cos(k_x x_0)}\right)^{\frac{\chi^2}{k_x^2}}\right), \tag{4.8}$$

$$y_h^{(\mathrm{TM})}(x) = \frac{1}{k_y}\arcsin\left(\frac{\sin(k_x x_0)}{\sin(k_x x)\sin(k_y y_0)}\right).$$

Fig 4.3 shows, as an example, the topological charts of the magnetic (Fig. 4.3a) and the electric (Fig. 4.3b) fields of the TM_{11} mode in a rectangular waveguide.

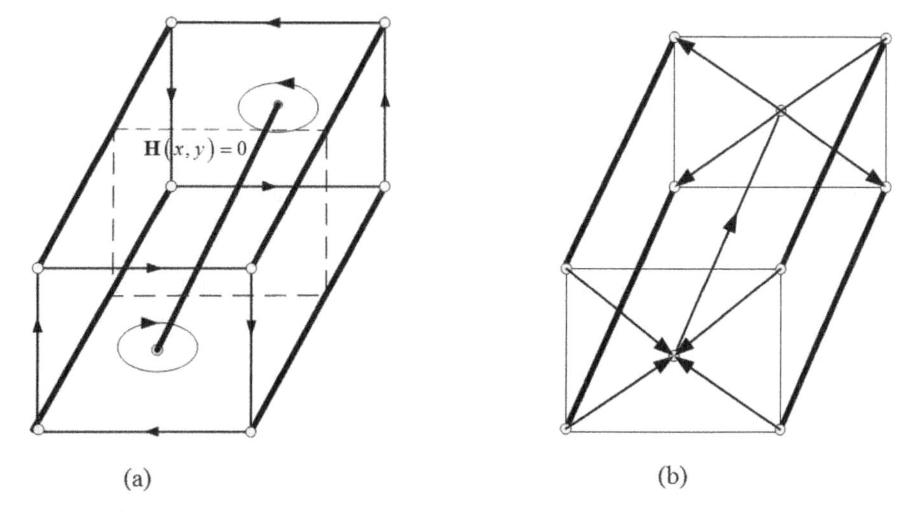

Fig. 4.3 Topological charts of the magnetic (a) and electric (b) fields of the TM_{11} mode

The magnetic field has a line of equilibrium at the center axis of the waveguide. At the corners, the lines of the saddle equilibrium are. Additionally, a plane of the zero field (shown by dashed lines) divides the volume. The separatrices are placed on the sides of rectangular waveguide. The electric field chart has more complicated structure. The 3-D saddle points are placed on each half of the wavelength along the central waveguide axis, and they are connected to each other by separatrices. At the corners, the saddle equilibrium lines are placed.

It is interesting to study the behavior of the modal topological schemes while the driving frequency tends to the cut-off one where $k_z = 0$. It is a bifurcation point of the field maps. At this frequency, the formulas for $z_e(x)$ and $z_h(x)$ give the infinite values, but the transverse structure of field-force lines is not varied according to (4.7) and (4.8). Before their cut-off frequencies, the modes are non-propagating, and their field-force lines are calculated using formulas (4.6) and (4.8):

$$z_h^{(\mathrm{TE})}(x) = \frac{1}{k_z''}\operatorname{arcosh}\left\{\frac{\left[\sin(k_x x)\sin(k_x x_0)\right]^{-\frac{\chi^2}{k_x^2}}}{\cosh(k_z'' z_0)}\right\},$$
$$z_e^{(\mathrm{TM})}(x) = \frac{1}{k_z''}\operatorname{arcosh}\left\{\frac{1}{\cosh(k_z'' z_0)}\left[\frac{\cos(k_x x)}{\cos(k_x x_0)}\right]^{\frac{\chi^2}{k_x^2}}\right\} \quad (4.9)$$

where $k_z'' = \sqrt{\chi^2 - k_0^2\varepsilon_r\mu_r}$.

4.1.3 Maxwell Relationships of Topological Schemes of the Electric and Magnetic Harmonic Fields

Another step in the topological analysis is the comparison of the maps of electric and magnetic fields to find the analytical relations of their patterns. It is well known that the field-force line pictures of the electric and magnetic fields are coupled to each other, but not always, the geometry of one field can be easily reconstructed according to the geometry of another field. Especially, it is difficult to perform it for the 3-D and 3-D time-dependent fields, and a special theory should be created. One of these attempts was done in 1988 [17],[22], and it is a technique on establishing the relationships of topological schemes of the harmonic electric $(\mathbf{E},\mathbf{D})$ and magnetic $(\mathbf{H},\mathbf{B})$ fields:

$$T_H = F(T_D),$$
$$T_E = \Phi(T_B). \quad (4.10)$$

For the isotropic and source-free mediums $T_D \to T_E$ and $T_B \to T_H$, i.e. these schemes are isomorphic to each other, (4.10) is re-written as

$$T_H = F(T_E),$$
$$T_E = F^{-1}(T_H). \quad (4.11)$$

Taking into account that the topological schemes are derived from the nonlinear equations for field-force lines, the operator F should be nonlinear.

This idea was applied to the comparison of the already calculated fields with the purpose to establish the general relationship of topological schemes. To find the above-mentioned relations, the fact of the orthogonality of the electric and magnetic fields in the domains, which are free of conducting and source currents, can be used:

$$(\mathbf{E}\cdot\mathbf{B})=0. \tag{4.12}$$

In the case of 2-D problems and TEM fields, this orthogonality allows easily reconstructing the field picture of one field according to another if the spatial distribution of the reconstructed one is not important, and the fields are considered out of the source region.

In more general 2-D and 3-D cases, a theory of (4.10) should be developed. As was mentioned, this theory is nonlinear, and it can be elaborated analyzing the field-force line maps and applying the techniques taken from the autonomous dynamic system treatments. Some ideas of this sort are considered below, although a complete theory of (4.10) has not been created yet.

Initial idea was to analyze the relations of the parameters of common equilibrium points where the harmonic fields $\mathbf{E}(r_j)=\mathbf{H}(r_j)=0$. These equilibrium points are stable and they are invariant regarding to the Lorentz group infinitesimal transformations. The field Lagrangian is zero and invariant at these points, too.

At the equilibrium points of this sort, the harmonic fields are potential:

$$\nabla\times\mathbf{H}=0,\mathbf{H}=-\nabla\cdot\varphi_{\mathrm{m}}, \tag{4.13}$$

$$\nabla\times\mathbf{E}=0,\mathbf{E}=-\nabla\cdot\varphi_{\mathrm{e}}. \tag{4.14}$$

In spite of the vector fields are zero at these points, the potentials φ_m and φ_e can be different from zero, and they are calculated from the photon density at this point [3]. These calculations are not local, and the establishments of any parameters of the equilibrium points are connected, anyway, to the whole domain or whole (topological) properties of the EM field.

Compare the characteristic equations of both fields at an EM equilibrium point:

$$\det\begin{pmatrix} \frac{\partial E_x(\mathbf{r}_0^{(k)})}{\partial x}-\lambda_e^{(k)} & \frac{\partial E_x(\mathbf{r}_0^{(k)})}{\partial y} & \frac{\partial E_x(\mathbf{r}_0^{(k)})}{\partial z} \\ \frac{\partial E_y(\mathbf{r}_0^{(k)})}{\partial x} & \frac{\partial E_y(\mathbf{r}_0^{(k)})}{\partial y}-\lambda_e^{(k)} & \frac{\partial E_y(\mathbf{r}_0^{(k)})}{\partial z} \\ \frac{\partial E_z(\mathbf{r}_0^{(k)})}{\partial x} & \frac{\partial E_z(\mathbf{r}_0^{(k)})}{\partial y} & \frac{\partial E_z(\mathbf{r}_0^{(k)})}{\partial z}-\lambda_e^{(k)} \end{pmatrix}=0, \tag{4.15}$$

$$\det\begin{pmatrix} \frac{\partial H_x(\mathbf{r}_0^{(k)})}{\partial x}-\lambda_h^{(k)} & \frac{\partial H_x(\mathbf{r}_0^{(k)})}{\partial y} & \frac{\partial H_x(\mathbf{r}_0^{(k)})}{\partial z} \\ \frac{\partial H_y(\mathbf{r}_0^{(k)})}{\partial x} & \frac{\partial H_y(\mathbf{r}_0^{(k)})}{\partial y}-\lambda_h^{(k)} & \frac{\partial H_y(\mathbf{r}_0^{(k)})}{\partial z} \\ \frac{\partial H_z(\mathbf{r}_0^{(k)})}{\partial x} & \frac{\partial H_z(\mathbf{r}_0^{(k)})}{\partial y} & \frac{\partial H_z(\mathbf{r}_0^{(k)})}{\partial z}-\lambda_h^{(k)} \end{pmatrix}=0. \quad (4.16)$$

For the fields, satisfying the Maxwell equations,

$$\begin{aligned} \nabla\times\mathbf{H} &= j\omega\varepsilon_0\varepsilon_r\mathbf{E}, \\ \nabla\times\mathbf{E} &= -j\omega\mu_0\mu_r\mathbf{H} \end{aligned} \quad (4.17)$$

it follows that both determinant matrices are symmetric, and this is the manifestation of the fact of the spatial symmetry of the considered EM field having such an equilibrium point.

These determinants (4.15) and (4.16) allow for the relations between the field values and the equilibrium's characteristic numbers. For instance, the use of the first Maxwell equation and (4.15) gives an expression for the "electric" characteristic number $\lambda_e^{(k)}$ as a function of the magnetic field second-order derivatives at an EM equilibrium point:

$$\lambda_e^{(k)} = F_{\text{eh}}\left(\frac{\partial^2 H_x}{\partial x\partial y}\left(\mathbf{r}^{(k)}\right),\ldots\right). \quad (4.18)$$

Similarly, the "magnetic" characteristic number $\lambda_h^{(k)}$ is expressed as

$$\lambda_h^{(k)} = F_{\text{he}}\left(\frac{\partial^2 E_x}{\partial x\partial y}\left(\mathbf{r}^{(k)}\right),\ldots\right). \quad (4.19)$$

These formulas for the characteristic numbers (4.15), (4.16), (4.18), and (4.19) can be used to find the parameters of the separated equilibrium points of the electric and magnetic fields, namely their real $\mathrm{Re}(\mathbf{E},\mathbf{H})$ or imaginary $\mathrm{Im}(\mathbf{E},\mathbf{H})$ parts.

To define the type of an equilibrium point, it is enough to calculate only the position of the characteristic numbers on the complex plane λ, i.e. a rough approximation of field can be used instead of accurate one. It allows for the expressions of the characteristic numbers of one field through the characteristic ones of another field using the linear approximations of fields close to an EM equilibrium point:

$$\lambda_{e,h}^{(k)} = \mathbf{\Phi}_{e,h}\left(\lambda_{h,e}^{(k)}\right). \quad (4.20)$$

Some treatments of this kind are shown in [22] where one field is linearized close to the equilibrium, and it is expressed through the characteristic numbers of the

Jacobian matrix at the equilibrium point. Then the derived field is expressed using the characteristic numbers. This field is substituted to the Maxwell expression for the second one, and again the determinant of the last is calculated at the common for both equilibrium points. Although such a treatment is rather complicated, the possibility to derive the non-linear relations for the characteristic numbers of the EM equilibrium points was proofed by this way.

The above-considered Maxwell relations of the elements of topological schemes are obtained using strong mathematical treatments. Some empirical relations can be found by comparisons of the schemes, and they need further study to confirm their generality. It was found, for instance, that some separatrices of a field took the same place with the lines of equilibriums of other field. In this case, the first field along this line can be calculated as a gradient of a potential. As an example, the separatrix of the electric field of the TM_{11} on the waveguide axis can be considered. Along this line, the equilibriums of the magnetic field are placed, and $E_z\left(x=a/2, y=b/2\right)=-\frac{\partial}{\partial z}\varphi(z)$.

It was found that some other separatrices of both fields are oriented normally to each other, and they are placed at the maxima of fields. Consider the first Maxwell equation, as an example,

$$\nabla\times\mathbf{H}=|\mathbf{H}|\cdot\nabla\times\mathbf{h}+\nabla\cdot|\mathbf{H}|\times\mathbf{h}=j\omega\varepsilon_0\varepsilon_r\mathbf{E} \tag{4.21}$$

where $\mathbf{h}$ is the unit vector oriented along $\mathbf{H}$. It follows that in the coordinate of the local maximum of magnetic field $|\mathbf{H}|$, is the maximum of $\mathbf{E}$. Additionally, a large value of $\nabla\cdot|\mathbf{H}|$ means that the field-force lines changes their directions fast in the area close to this local maximum. The separatrix of the electric field is oriented normally to the one of the magnetic field according to (4.12). There is a need in further study of the generality of the found relations of the fields and their separatrices of this sort.

Another empirical relation found analyzing the modal fields in rectangular waveguides is with the mutual orientations of the one-field vectors on the loops of the separatrices of another [22]. Consider a fragment of Fig. 4.3, which is the cross-section of the waveguide with the shown separatrices of the magnetic and electric fields including the semi-sepatrices connecting the 3-D equilibrium point with the mirrored ones (Fig. 4.4).

They have four cross-points where the electric and magnetic fields are not zero. Let us calculate the signs of scalar products of the vector fields and the orts of the Cartesian system of coordinates in each cross-section point of the electric and magnetic separatrices following the contour shown in Fig. 4.4 by a pointed line.

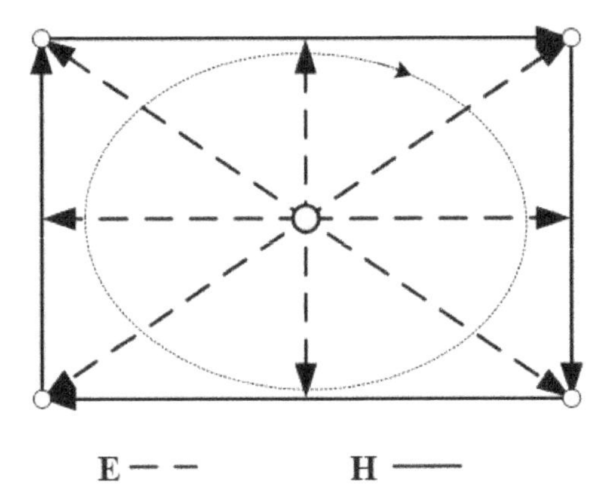

Fig. 4.4 Electric and magnetic separatrices and the calculation contour shown by pointed line

Due to the periodicity:

$$\sum_{1}^{8} \mathrm{sign}\left(\mathbf{E}_i \cdot \mathbf{n}_i\right) \mathrm{sign}\left(\mathbf{H}_i \cdot \boldsymbol{\tau}_i\right) = 0 \tag{4.22}$$

where $\mathbf{E}_i$ and $\mathbf{H}_i$ are the vectors at the $i-$th cross-point, $\mathbf{n}_i$ is the ort which normal to the boundary or magnetic field separatrix and oriented along the corresponding axis, and $\boldsymbol{\tau}_i$ is the ort of the Cartesian system of coordinates which is tangential (parallel or anti-parallel) to the magnetic field separatrix. Calculations of such sums for the TM-modes and similar for the TE modes show their equality to zero due to the periodicity of the fields and the Maxwell relations between them. This criterion was obtained empirically analyzing the waveguide modes, and its validity for arbitrary geometries and problems needs more research.

4.1.4 Boundary Conditions and the Field-force Lines

In many cases, a microwave transmission line or a resonator is composed of several sub-domains filled by different magnetodielectrics. The boundary conditions are established using Maxwell equations in their integral form, and they couple the field components of sub-domains at the boundaries (see Section 1.7.1). From these equations, written for the tangential and normal components of the fields, the geometrical boundary conditions for the field-force lines are obtained. They couple the angles $\alpha_{1,2}$ of the field-force lines at the boundary between two domains [22],[53] (Fig. 4.5).

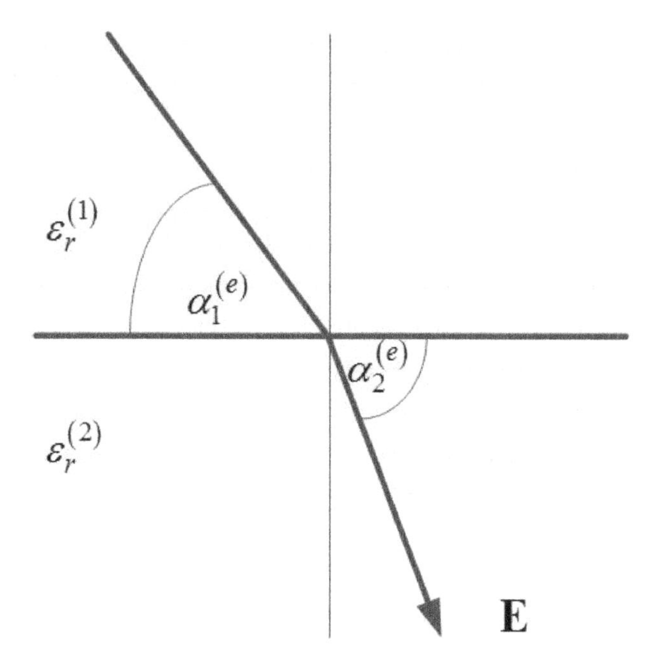

Fig. 4.5 Electric field-force line near a boundary of two dielectrics

For the electric field-force lines, it is:

$$\frac{\tan \alpha_1^{(e)}}{\tan \alpha_2^{(e)}} = \frac{\varepsilon_r^{(1)}}{\varepsilon_r^{(2)}}. \tag{4.23}$$

For the magnetic field-force lines the boundary conditions are written similarly if the mediums have different magnetic permeabilities:

$$\frac{\tan \alpha_1^{(h)}}{\tan \alpha_2^{(h)}} = \frac{\mu_r^{(1)}}{\mu_r^{(2)}}. \tag{4.24}$$

As known, the electric field-force lines are normal to ideal metal with $\alpha_1^{(e)} = \pi/2$, and the magnetic ones are parallel to its surface, then $\alpha_2^{(h)} = 0$.

4.1.5 Applications of the Topological Technique to the EM Boundary Value Problems

The above analysis was performed for the EM field for which the source and domain geometries were not specified [21],[22]. The main attention was paid to some relations of the characteristics of the equilibrium points. Anyway, it is underlined that the properties of equilibrium points are defined by global features of the EM field and the domain geometry.

For the developments of a practical technique on the qualitative or approximate calculations of the field and its topology according to the geometry and boundary conditions, a new topologo-geometrical object should be introduced which is the vector boundary graph. It can be defined for the boundary electric or magnetic field, and it is denoted as $\mathbf{\Gamma}_{\mathrm{e,m}}$, correspondingly. The geometry of $\mathbf{\Gamma}_{\mathrm{e,m}}$ is the same with the one of the domain. The electric boundary graph on which branches $\mathbf{E}_{\tau}=0$ is the unloaded graph, and it can be described by a scalar Γ_{e} graph. On the unloaded magnetic graph $\mathbf{\Gamma}_{\mathrm{m}}$ the normal component of the magnetic field $H_{\nu}=0$. A graph is normalized if its branches are multiplied by the wave number k_0. In this case, the notation $\tilde{\mathbf{\Gamma}}_{\mathrm{e,h}}$ is used. A graph is loaded if a field $\mathbf{e}_L$ or $\mathbf{h}_L$ is defined on an its branch, then $\mathbf{\Gamma}_{\mathrm{e,m}} \to \mathbf{\Gamma}_{\mathrm{e_L,m_L}}$.

An example of an electric field boundary graph is shown in Fig. 4.6 given for a rectangular resonator excited by a slot electric field $e_x(x,z)$.

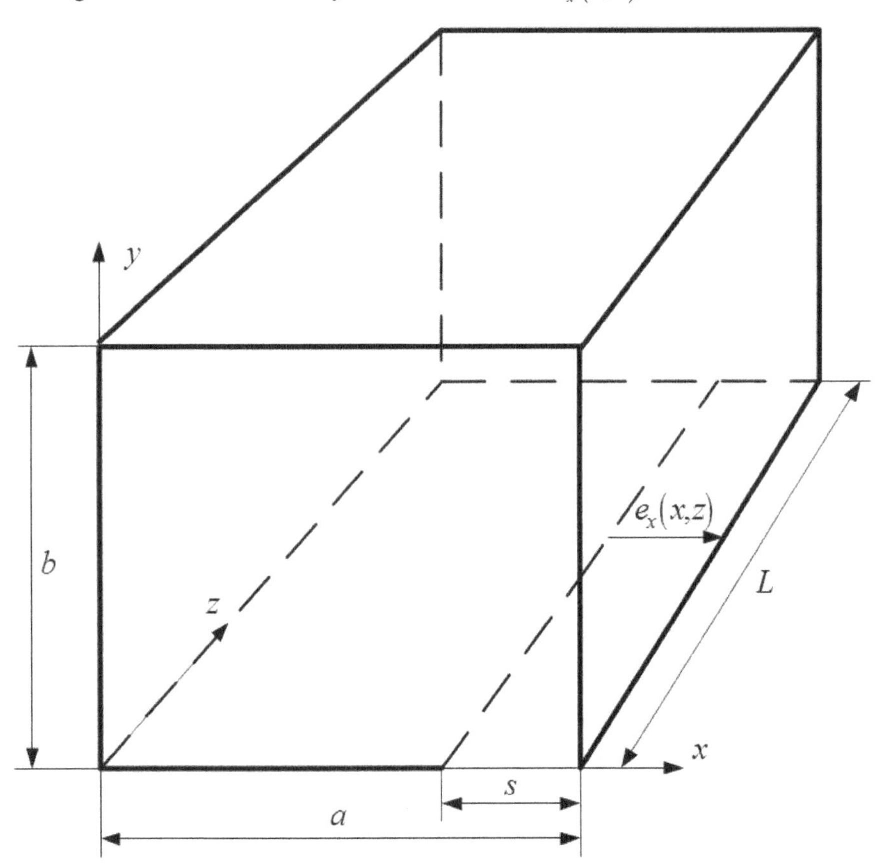

Fig. 4.6 Rectangular cavity resonator excited by slot electric field

The topological approach to the boundary value problems is the composition of the topological schemes of the electric and magnetic fields and their bifurcation diagrams according to the given normalized loaded boundary graph:

$$T_{E,H} = F\left(\mathbf{\Gamma}_{\mathrm{e_L}}\right) \tag{4.25}$$

or

$$T_{E,H} = \Phi\left(\mathbf{\Gamma}_{\mathrm{m_L}}\right). \tag{4.26}$$

Taking into account that the topological schemes can be defined qualitatively then (4.25) and (4.26) are considered as the qualitative solutions of the boundary value problems.

There are several ways how to solve them. One can start considering some direct ways to obtain the topological schemes and their bifurcations according to the given boundary graph and driving frequency. Another practical approach is used in [21],[22]. It is based on the representation of the excited field by a series of domain eigenmodes. Their amplitudes depend on the driving frequency and the exciting fields and currents.

As known, an excited field in a cavity is represented by a series of the cavity modes:

$$\begin{aligned} \mathbf{E}(\mathbf{r}) &= \sum_{\nu=1}^{\infty} a_\nu \mathbf{E}_\nu(\mathbf{r}), \\ \mathbf{H}(\mathbf{r}) &= \sum_{\nu=1}^{\infty} b_\nu \mathbf{H}_\nu(\mathbf{r}) \end{aligned} \tag{4.27}$$

where a_ν and b_ν are the modal amplitudes, $\mathbf{E}_\nu$ and $\mathbf{H}_\nu$ are the modal fields, and $\mathbf{r}$ is the spatial variable.

The modal amplitudes are estimated using the following formulas [52],[53]:

$$\begin{aligned} a_\nu &= \frac{|\tilde{\varepsilon}_r|\tilde{\mu}_r^*}{\tilde{\varepsilon}_r\tilde{\mu}_r} \frac{j\tilde{\omega}_\nu^* \int_S \left[\mathbf{\Gamma}_{e_L} \times \mathbf{H}_\nu^*(\mathbf{r})\right] d\mathbf{S}}{\left(\omega^2 - \tilde{\omega}_\nu^2\right)}, \\ b_\nu &= \frac{|\tilde{\mu}_r|\tilde{\varepsilon}_r^*}{\tilde{\mu}_r\tilde{\varepsilon}_r} \frac{j\omega \int_S \left[\mathbf{\Gamma}_{e_L} \times \mathbf{H}_\nu^*(\mathbf{r})\right] d\mathbf{S}}{\left(\omega^2 - \tilde{\omega}_\nu^2\right)} \end{aligned} \tag{4.28}$$

where ω and $\tilde{\omega}_\nu$ are the driving and complex eigenmodal frequencies, correspondingly.

The rotational system of the electric and magnetic eigenmodal fields written in the case of the H_z modes is [52],[53]

$$
\begin{aligned}
H_x &= B_x \sin(k_x x)\cos(k_y y)\cos(k_z z),\\
H_y &= B_y \cos(k_x x)\sin(k_y y)\cos(k_z z),\\
H_z &= B_z \cos(k_x x)\cos(k_y y)\sin(k_z z),\\
E_x &= A_x \cos(k_x x)\sin(k_y y)\sin(k_z z),\\
E_y &= A_y \sin(k_x x)\cos(k_y y)\sin(k_z z)
\end{aligned}
\tag{4.29}
$$

where $k_x = \dfrac{l\pi}{a}$, $k_y = \dfrac{k\pi}{b}$, $k_z = \dfrac{p\pi}{L}$. The electric and magnetic fields are ortho-normalized

$$
\begin{aligned}
&\int_V \mathbf{E}_\nu \mathbf{E}_\mu^* dV = \delta_{\mu\nu} / \varepsilon_0 |\mu_r|,\\
&\int_V \mathbf{H}_\nu \mathbf{H}_\mu^* dV = \delta_{\mu\nu} / \mu_0 |\mu_r|.
\end{aligned}
\tag{4.30}
$$

The modal coefficients in (4.29) are

$$
\begin{aligned}
B_x &= -jQk_x\delta_{m0}\delta_{n0}\delta_{p0}\sqrt{k_x^2+k_y^2}/\eta,\\
B_y &= -jQk_y\delta_{m0}\delta_{n0}\delta_{p0}\sqrt{k_x^2+k_y^2}/\eta,\\
B_z &= -jQk_z\delta_{m0}\delta_{n0}\delta_{p0}\sqrt{k_x^2+k_y^2}/\eta,\\
A_x &= -Qk_y\delta_{m0}\delta_{n0}\delta_{p0}\sqrt{\kappa^2},\\
A_y &= -Qk_x\delta_{m0}\delta_{n0}\delta_{p0}\sqrt{\kappa^2},\\
Q &= \frac{2\sqrt{2}}{\sqrt{\varepsilon_0|\varepsilon_r| abL\left(k_x^2+k_y^2\right)\kappa^2}}, \quad \kappa^2 = k_x^2+k_y^2+k_z^2,\\
\eta &= 120\pi\sqrt{\mu_r/\varepsilon_r},\\
\delta_{l0,k0,p0} &= \begin{cases} 1, & 1,k,p \neq 0 \\ 1/\sqrt{2}, & 1,k,p = 0 \end{cases}
\end{aligned}
\tag{4.31}
$$

where all numbers l, k, p should not be equal to zero at the same time.

Another way to represent the cavity field is the consideration of the modes, which are traveling from the exciting boundary $(y=0)$ and reflected from the opposite side $(y=b)$. In our considered cavity (Fig. 4.6), the fields of the TE type excited by the slot electric field $e_x(x,z)$ are

$$H_z(\mathbf{r}) = \sum_{m=0}^{\infty} \sqrt{\frac{2-\delta_{0m}}{aL}} D_m \cos\left[k_y^{(m)}(b-y)\right]\cos\left(k_x^{(m)}x\right)\sin\left(k_z z\right),$$

$$H_x(\mathbf{r}) = j\sum_{m=1}^{\infty} \sqrt{\frac{2-\delta_{0m}}{aL}} \frac{D_m k_z k_x^{(m)}}{\chi_m^2} \cos\left[k_y^{(m)}(b-y)\right]\sin\left(k_x^{(m)}x\right)\cos\left(k_z z\right),$$

$$H_y(\mathbf{r}) = -j\sum_{m=0}^{\infty} \sqrt{\frac{2-\delta_{0m}}{aL}} \frac{D_m k_z k_y^{(m)}}{\chi_m^2} \sin\left[k_y^{(m)}(b-y)\right]\cos\left(k_x^{(m)}x\right)\cos\left(k_z z\right), \quad (4.32)$$

$$E_x(\mathbf{r}) = -j\omega\mu_0\mu_r \sum_{m=0}^{\infty} \sqrt{\frac{2-\delta_{0m}}{aL}} \frac{D_m k_y^{(m)}}{\chi_m^2} \sin\left[k_y^{(m)}(b-y)\right]\cos\left(k_x^{(m)}x\right)\sin\left(k_z z\right),$$

$$E_y(\mathbf{r}) = -j\omega\mu_0\mu_r \sum_{m=1}^{\infty} \sqrt{\frac{2-\delta_{0m}}{aL}} \frac{D_m k_x^{(m)}}{\chi_m^2} \cos\left[k_y^{(m)}(b-y)\right]\sin\left(k_x^{(m)}x\right)\sin\left(k_z z\right)$$

where $k_x^{(m)} = \frac{m\pi}{a}$, $k_z = \frac{\pi}{L}$, $k_y^{(m)} = \sqrt{k_0^2\varepsilon_r\mu_r - \left(k_x^{(m)}\right)^2 - k_z^2}$, $\chi_m^2 = \left(k_x^{(m)}\right)^2 + \left(k_y^{(m)}\right)^2$,

and $\delta_{0m} = \begin{cases} 1, m = 0 \\ 0, m \neq 0 \end{cases}$.

The unknown coefficients D_m are derived by the field matching method according to the defined exciting electric field $e_x(x,z)$:

$$D_m = -\frac{\chi_m^2 \mathrm{E}_x^{(m)}}{j\omega\mu_0\mu_r k_y^{(m)} \sin\left(k_y^{(m)} b\right)},$$
$$\mathrm{E}_x^{(m)} = \sqrt{\frac{2-\delta_{0m}}{aL}} \int_S e(x',z')\cos\left(k_x^{(m)}x'\right)\sin\left(k_z z'\right)dx'dz'. \quad (4.33)$$

Both representations (4.29) and (4.32) have some advantages and disadvantages for the field calculations. For instance, (4.29) can be poor to model the fields close to the exciting boundary. The traveling modes in (4.32) are normalized along the $x-$axis and $z-$axis only.

Let us proof these statements calculating the field maps of a rectangular cavity excited by slots of different geometry. One of the studied geometries is shown in Fig. 4.6 where a prismatic resonator is excited by a driving electric field of the amplitude A:

$$e_x(x',z') = A\sqrt{\frac{1}{L}} e_x(x')\sin\left(\frac{\pi z'}{L}\right) \quad (4.34)$$

and defined on slot s which is regular in the longitudinal direction z.

To calculate the fields and to study different effects, the formulas (4.32) are used, in particular. The field calculations show that close to the boundary with the exciting electric field, the egenmodal expansions may suffer from severe Gibbs effect, and the field has some oscillations which amplitudes are increasing with the truncation number M [54]. It does not allow for concluding on the convergence of the field calculations and for performing even topological analysis of the excited field. To stabilize the calculations, each modal coefficient $D_{m\neq 0}$ is multiplied by a Lanczos factor of the third order [55]:

$$a_L = \left[\frac{M}{m\pi}\sin\left(\frac{m\pi}{M}\right)\right]^3, \tag{4.35}$$

which is zero if $m = M$.

Fig. 4.7 shows the electric field distribution on the excitation side of the cavity calculated by conventional (curve 1) and regularized (curve 2) summations.

A rectangular electric field pulse of the unit amplitude defined on a narrow slot excites the cavity field. It is seen that the Lanczos technique allows for avoiding the Gibbs effect, and the stabilized modal expansions can be used for topological and geometrical analysis of the field now.

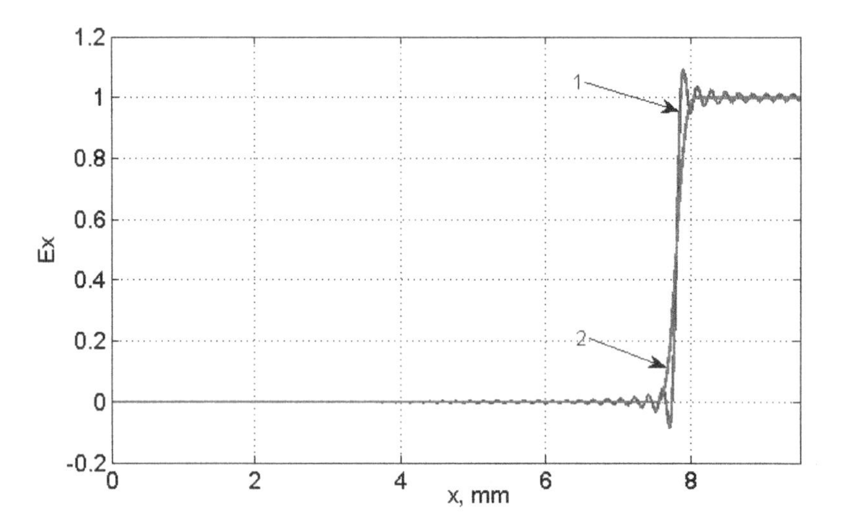

Fig. 4.7 Excited E_x electric field distribution calculated at $y=0$. Cavity geometry: $a\times b\times L = 9.5\times 9.5\times 16.76$ mm . Slot width $s=1.7$ mm , $F=10$ GHz , $\varepsilon_r=1$, $\mu_r=1$. 1- Non-stabilized summation; 2- Results obtained with the Lanczos stabilization factor

The field-force lines are calculated according to the differential equations which are the autonomous dynamical systems. The magnetic field-force line system of equations for the TE_z modes is

$$\frac{dy_h}{dx} = \frac{\mathrm{Re}(H_y(x,y))}{\mathrm{Re}(H_x(x,y))},$$
$$\frac{dz_h}{dx} = \frac{\mathrm{Re}(H_z(x,y))}{\mathrm{Re}(H_x(x,y))}. \tag{4.36}$$

The electric field phase map in this case is only 2-dimensional, and these field-force lines are derived using only one differential equation:

$$\frac{dy_e}{dx} = \frac{\mathrm{Re}(E_y(x,y))}{\mathrm{Re}(E_x(x,y))}. \tag{4.37}$$

Analyzing (4.32), (4.36), and (4.37), it is seen that these autonomous dynamical systems parametrically depend on the permittivity, permeability, boundary condition, and geometry of the calculated cavity resonator, and they are, in fact, the differential-integral equations taking into account (4.33).

Using a normalized to the wave number $\left(k = k_0\sqrt{\varepsilon_r\mu_r}\right)$ boundary graph $\tilde{\mathbf{\Gamma}}_{e_L}$, the equations (4.36) and (4.37) are re-written as one-parameter-dependent autonomous dynamical systems:

$$\frac{dy_h}{dx} = \frac{\mathrm{Re}\left(H_y\left(x,y,\tilde{\mathbf{\Gamma}}_{e_L}\right)\right)}{\mathrm{Re}\left(H_x\left(x,y,\tilde{\mathbf{\Gamma}}_{e_L}\right)\right)},$$
$$\frac{dz_h}{dx} = \frac{\mathrm{Re}\left(H_z\left(x,y,\tilde{\mathbf{\Gamma}}_{e_L}\right)\right)}{\mathrm{Re}\left(H_x\left(x,y,\tilde{\mathbf{\Gamma}}_{e_L}\right)\right)}, \tag{4.38}$$

and

$$\frac{dy_e}{dx} = \frac{\mathrm{Re}\left(E_y\left(x,y,\tilde{\mathbf{\Gamma}}_{e_L}\right)\right)}{\mathrm{Re}\left(E_x\left(x,y,\tilde{\mathbf{\Gamma}}_{e_L}\right)\right)}. \tag{4.39}$$

In this particular case, when only the TE_z field is considered, the qualitative analysis of (4.39) as an autonomous system on a plane is preferable. The qualitative theory of these systems is known which considers only the scalar bifurcation parameters. The use of the equations dependent on the distributed or non-local parameters $\tilde{\mathbf{\Gamma}}_{e_L}$ allows for decreasing the problem's co-dimension, which is the spatial dimension plus the number of bifurcation parameters of a dynamical system. It allows to hope on the developments of a more elegant bifurcation theory for the field maps generated by the EM boundary value problems. According to the best knowledge of the Author, this theory is on the initial stage only, including the below-given results originated from our preliminary ones dated by 1988.

Analyzing again (4.29) and (4.32), it is seen that, in both cases, a mode or even several modes can have increased amplitudes due to the frequency resonances, and $|a_\nu|,|b_\nu| \to \infty$, or $|D_m| \to \infty$ if the excited resonator is ideal, and the driving frequency ω is close to a resonant one ω_ν. Additionally, these amplitudes can be strong due to the spatial resonance when

$$\left| \iint_S \left[\mathbf{\Gamma}_{e_L} \times \mathbf{H}_\nu(\mathbf{r}) \right] d\mathbf{S} \right| = \max. \tag{4.40}$$

However, this effect is less expressive comparing to the frequency resonance.

In [21],[22], it is found that the field maps topology is formed by the modes which amplitudes are strong due to the above-mentioned resonances. Outside of the area of structural stability of these field maps, other higher-order modes with even negligible amplitudes can cause the bifurcations.

The structurally stable schemes are formed by a relatively small number of modes N_T which are close to the above-mentioned frequency and spatial resonances and having increased magnitudes. This number N_T is the dimension of the algebra of U_T defined as a set of the cavity modes $\{\mathbf{E}_\nu, \mathbf{H}_\nu\}$ with the defined operations of the addition of modes and their multiplication by a scalar a_ν.

As a rule, N_T is not so high due to structural stability of topological schemes which are out of the bifurcation points. A structurally stable topological scheme is not changed with the adding of additional series members δN_T:

$$T\left(U_T^{N_T}\right) \to T\left(U_T^{N_T+\delta N_T}\right). \tag{4.41}$$

Meanwhile, depending on the boundary graph and frequency, the field topology can have bifurcations keeping the same scheme topology dimension, i.e. $\dim(U_T) = \text{const}$. This sort of bifurcations is the *isomeric* one.

To proof the idea that the structurally stable or coarse topological schemes are formed by a few eigenmodes only, the above-considered resonator is simulated, and the convergence of the topology and geometry of field maps is studied for different exciting frequencies and excitation conditions. Each calculation of a field map is followed by comparative analysis of the resonant frequencies f_{lk1} of the rectangular waveguide cavity: $f_{lk1} = \dfrac{c}{2\pi\sqrt{\varepsilon_r \mu_r}} \sqrt{\left(\dfrac{l\pi}{a}\right)^2 + \left(\dfrac{k\pi}{b}\right)^2 + \left(\dfrac{\pi}{L}\right)^2}$ given in Table 4.1 and the magnitudes of the electric field components of each $m-th$ eigenfunction from (4.32):

$$d_x^{(m)} = \left| -j\omega\mu_0\mu_r \sqrt{\frac{2-\delta_{0m}}{aL}} D_m \frac{k_y^{(m)}}{\chi_m^2} \right|,$$
$$d_y^{(m)} = \left| -j\omega\mu_0\mu_r \sqrt{\frac{2-\delta_{0m}}{aL}} D_m \frac{k_x^{(m)}}{\chi_m^2} \right|. \tag{4.42}$$

The conclusion on the dimension of U_T is given according to the convergence of the topological schemes and the values of $d_{x,y}^{(m)}$.

Table 4.1 First four resonant frequencies of the studied resonator

Cavity Mode	TE_{101}	TE_{111}	TE_{201}	TE_{211}
f_{lk1}, GHz	16.4112	22.7735	31.8943	35.5887

Initially, the topology formation is studied at frequencies which are rather far from the resonant ones. It allows for discovering the influence of the spatial resonances on the field formation. Fig. 4.8 shows a field map obtained at $f = 2$ GHz using only two first modes of the field expansion (4.32). Additionally, the relative magnitudes $d_{x,y}^{(m)}/d_x^{(m=0)}$ are shown in the inset by the blue and white bars, correspondingly.

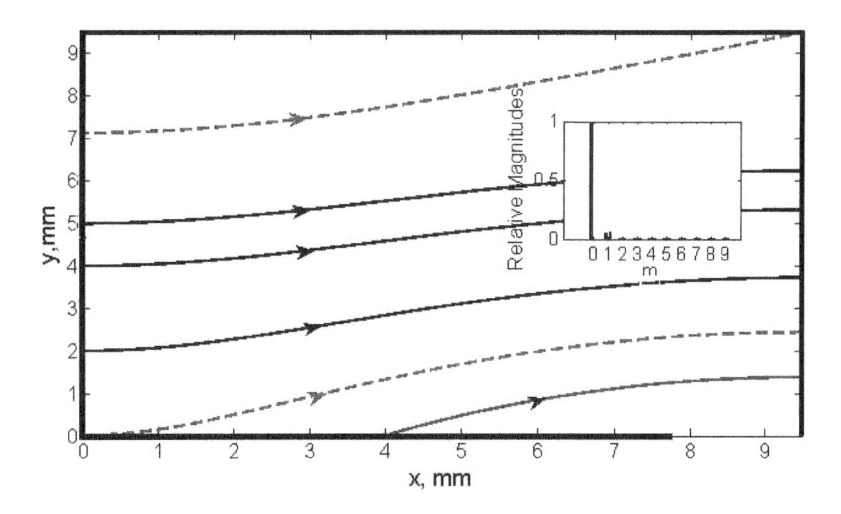

Fig. 4.8 Electric field map obtained using the first two modes $(m = 0,1)$. Cavity geometry: $a \times b \times L = 9.5 \times 9.5 \times 16.76$ mm. Slot width $s = 1.7$ mm, $F = 2$ GHz, $\varepsilon_r = 1$, $\mu_r = 1$

It is seen that the zero-number mode is of the dominating magnitude, and only two first modes form the topological chart of the excited field. Other modes do not influence the electric field topology in this case. For comparison, the field map derived by calculation using M=101 traveling modes is shown in Fig. 4.9.

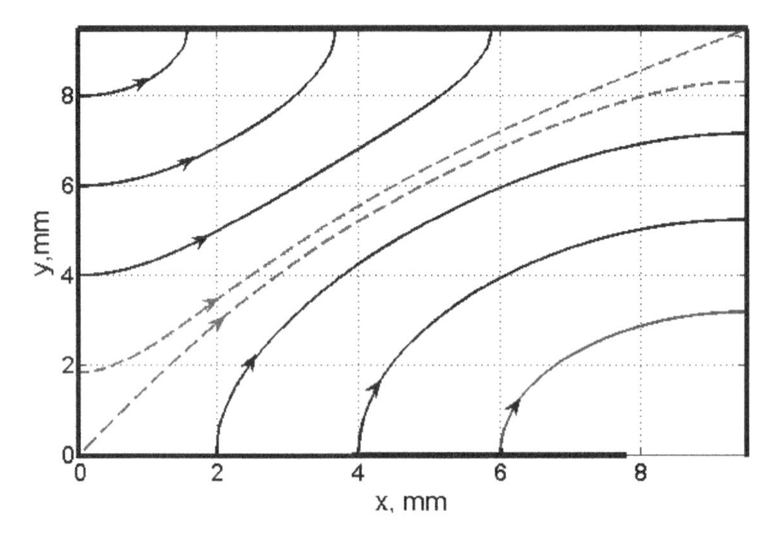

Fig. 4.9 Electric field map calculated using 101 modes. Cavity geometry: $a \times b \times L = 9.5 \times 9.5 \times 16.76$ mm. Slot width $s = 1.7$ mm, $F = 2$ GHz, $\varepsilon_r = 1$, $\mu_r = 1$

It follows from Figs 4.8 and 4.9 that the equilibriums of center type are at the left upper corners of plots. The saddle points are at other corners, and the separatrices (dashed curves) from them end at the same walls of the resonators. This analysis shows that both maps have similar topological charts: $T_e^{(M=2)} \rightarrow T_e^{(M=100)}$, and then the minimal dimension of the field skeleton is $N_T = \dim(U_T) = 2$.

At the same time, it is seen, that the 2-modal approximation (Fig. 4.8) is rather poor to provide accuracy of the boundary condition satisfaction at $y = 0$, and the field-force lines are not normal to this wall $(x < a - s)$ at the difference to the field map calculated with the increased number of traveling modes M=101 (Fig. 4.9).

Simulations show, that for this case and relatively low frequencies, the electric field topological scheme and its algebra U_T are invariant regarding to the geometry of the excitation slot variations if it is excited by a rectangular electric field pulse (Fig. 4.10).

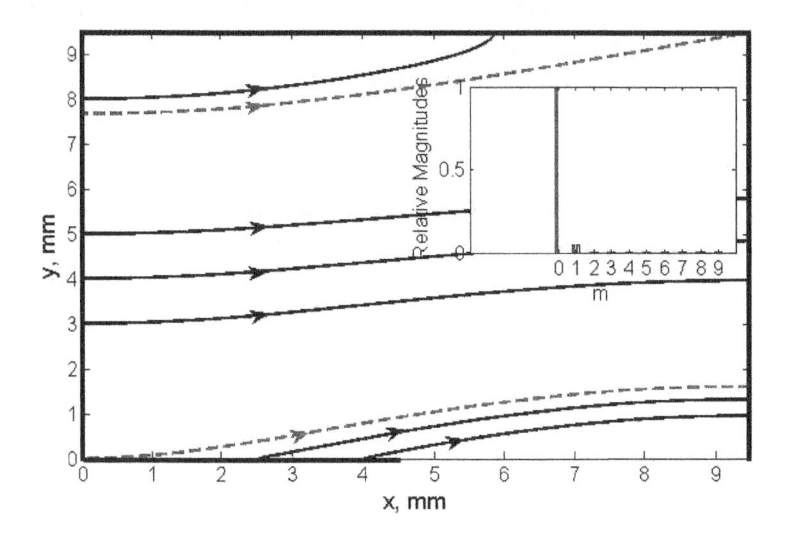

Fig. 4.10 Electric field map obtained using the first two modes $(m=0,1)$. Cavity geometry: $a \times b \times L = 9.5 \times 9.5 \times 16.76$ mm. Slot width $s = 5$ mm, $F = 2$ GHz, $\varepsilon_r = 1$, $\mu_r = 1$

The map calculated using 101 modes has the same field topology, and it is shown in Fig 4.11.

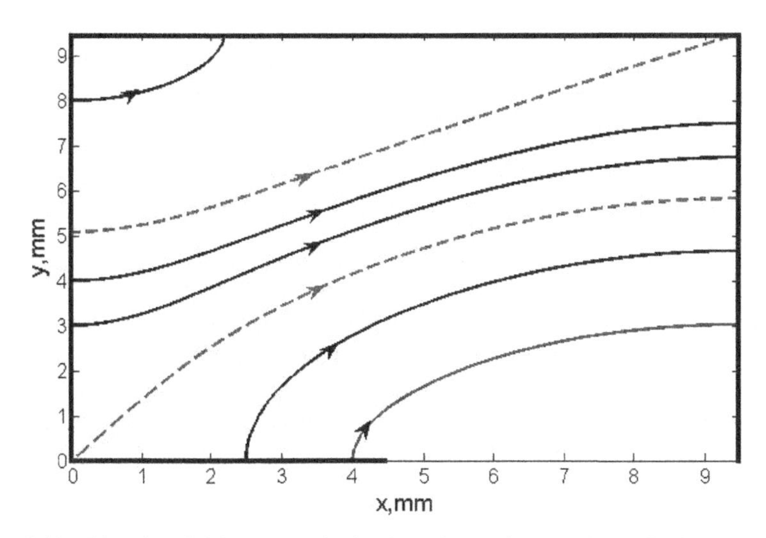

Fig. 4.11 Electric field map obtained using 101 modes. Cavity geometry: $a \times b \times L = 9.5 \times 9.5 \times 16.76$ mm. Slot width $s = 5$ mm, $F = 2$ GHz, $\varepsilon_r = 1$, $\mu_r = 1$

At the same time, the content of the algebra U_T and the topological schemes can be sensitive towards a variation of the size of the cavity resonator. Fig. 4.12 shows the electric field map calculated for a resonator which vertical size b is reduced down to 6.5 mm.

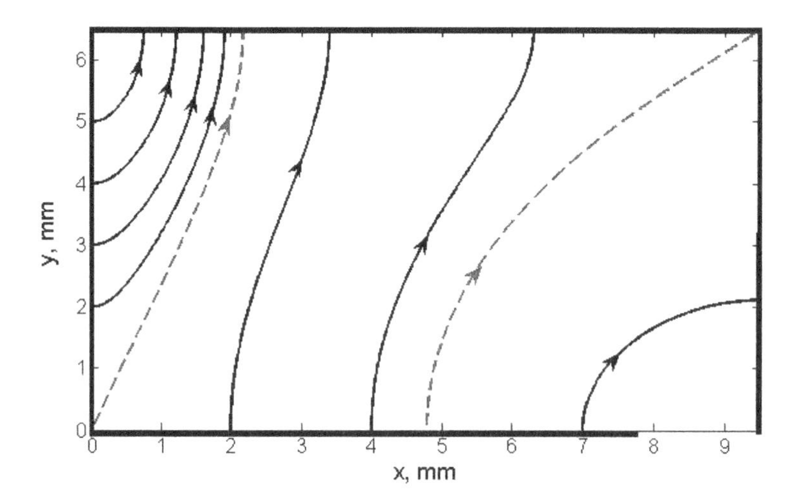

Fig. 4.12 Electric field map obtained using 101 modes. Cavity geometry: $a \times b \times L = 9.5 \times 6.5 \times 16.76$ mm . Slot width $s = 1.7$ mm , $F = 2$ GHz , $\varepsilon_r = 1$, $\mu_r = 1$

The electric field-force lines are shorted to the upper side of the cavity now and this global bifurcation occured because the separatrices, shown by dashed lines, have changed their attracting sides. Studying the field shown in Fig. 4.12, gives the minimal dimension of the algebra $\dim(U_T) = 3$.

Thus the topological schemes shown in Figs. 4.9 and 4.12 are not homeomorphic to each other although the number of equilibrium points has not been changed, i.e. $T_e(\mathbf{\Gamma}_{eL}) - / \rightarrow T_e(\mathbf{\Gamma}_{eL} + \delta\mathbf{\Gamma}_{e_L})$.

It is supposed that there the geometry(s) and driving frequency(s) exist when the separatrix connects two saddle points placed at the opposite corners. Fig. 4.13 shows one of the geometries where a single separatrix connects two saddle points. It is interesting to notice that this curve is the axis of curvilinear symmetry of this excited field map. This topological scheme is unstable, and it is destroyed by any small variation of geometry and frequency.

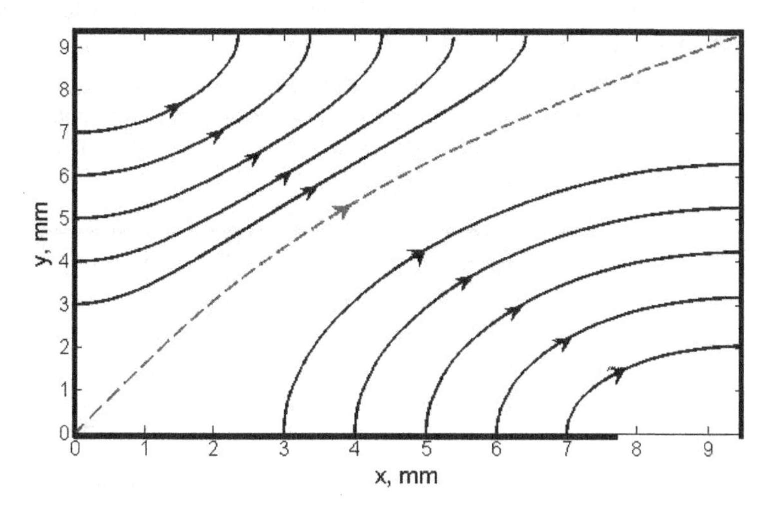

Fig. 4.13 Electric field map obtained using 101 modes. Cavity geometry: $a \times b \times L = 9.5 \times 9.331 \times 16.76$ mm. Slot width $s = 1.7$ mm, $F = 2$ GHz, $\varepsilon_r = 1$, $\mu_r = 1$

Additionally to the geometry, variation of driving frequency can cause a dramatic change of the field map and the spectral contents of topological schemes. Fig. 4.14 shows a converged field map for the case of a resonator excited by electric field of the frequency $F = 21$ GHz which is rather close to the resonant one of the TE_{111} mode.

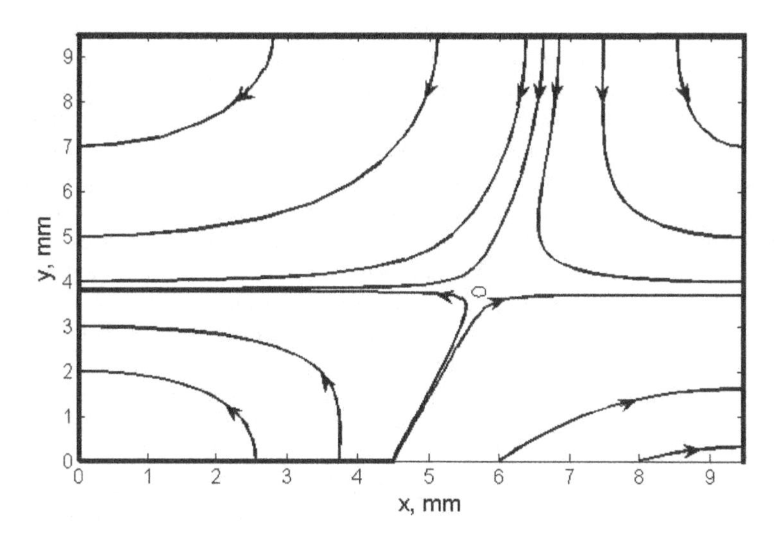

Fig. 4.14 Electric field map obtained using 101 modes. Cavity geometry: $a \times b \times L = 9.5 \times 9.5 \times 16.76$ mm. Slot width $s = 5$ mm, $F = 21$ GHz, $\varepsilon_r = 1$, $\mu_r = 1$

Similarly to the previous cases, the field topology is formed by a few modes, and $\dim(U_T)=5$ (Fig. 4.15).

Our study of the convergence of the field maps shows that, additionally to the modes with the largest magnitudes, some of them with essentially smaller ones can take part in the formation of topological schemes, especially, in the areas, which are rather far from the exciting boundary side.

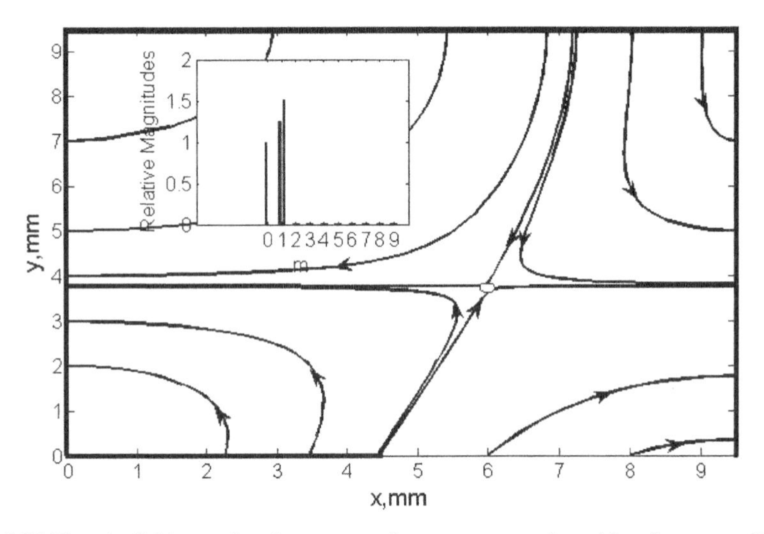

Fig. 4.15 Electric field map for the same cavity geometry and exciting frequency F as in Fig. 4.14 obtained using the first five modes. $N_T=\dim(U_T)=5$

It is explained by the non-monotonic dependence of the modal fields of traveling modes in the y-axis direction, and, in these areas, the higher-order modal fields can be even stronger of ones of lower indices. Then another criterion should be developed for the study of the convergence of the field maps. To this moment, a recommendation is that, additionally to the traveling modes of the largest magnitudes, several others should be added to provide proper convergence of the field geometry in the areas which are rather far from the exciting boundary parts. Especially, it is valid while frequency increasing and approaching to the higher-order mode resonances.

It is expected that the spectral contents of the field maps would be more complicated at increased frequencies due to denser spectrum of resonator. In practice, most resonators are used at frequencies which are in lower part of their spectrum, and the excited fields are formed by few modes, which can be defined using the topology stabilization criteria. Using it, simple equivalent circuits of resonators can be composed taking into account that the modes forming the field topology are of the most energy.

The found low dimensionality of topological schemes is confirmed by the modeling of more complicated resonator's boundary geometries. Fig. 4.16 shows a field map in a resonator excited by two slots. The total electric and magnetic fields are obtained using (4.32) for the fields excited separately by the vertical and horizontal slots and added to each other.

Due to the geometry of this excited resonator the field map has spatial symmetry regarding to the separatrix connecting two saddle equilibrium points placed at the corners of the resonator with the coordinates $x = y = 0$ and $x = a,\ y = b$. The minimal dimension of the topological scheme algebra is $N_T = \dim(U_T) = 1+1$, and only one mode (m=0) of each field is added to each other to derive the same topology field, as it seen from the comparison of Fig. 4.16 to Fig. 4.17.

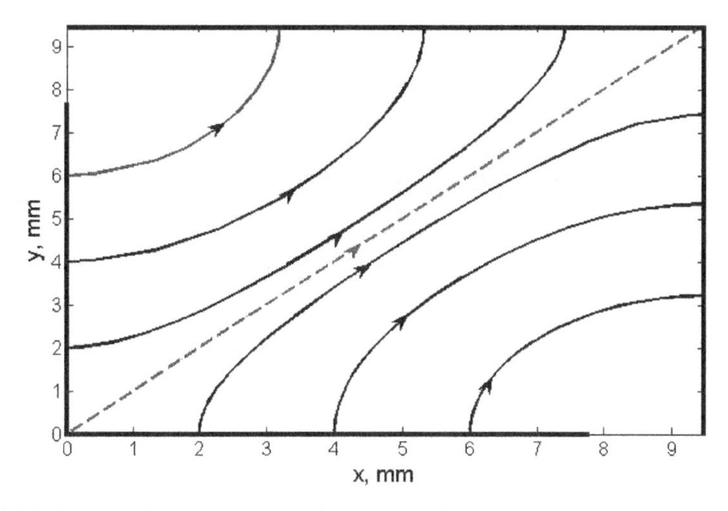

Fig. 4.16 Electric field map excited by two slots $(s_1 = s_2 = 1.7\ \text{mm})$ obtained using 2×101 modes. Cavity geometry: $a\times b\times L = 9.5\times9.5\times16.76\ \text{mm}$, $F = 2\ \text{GHz}$, $\varepsilon_r = 1$, $\mu_r = 1$

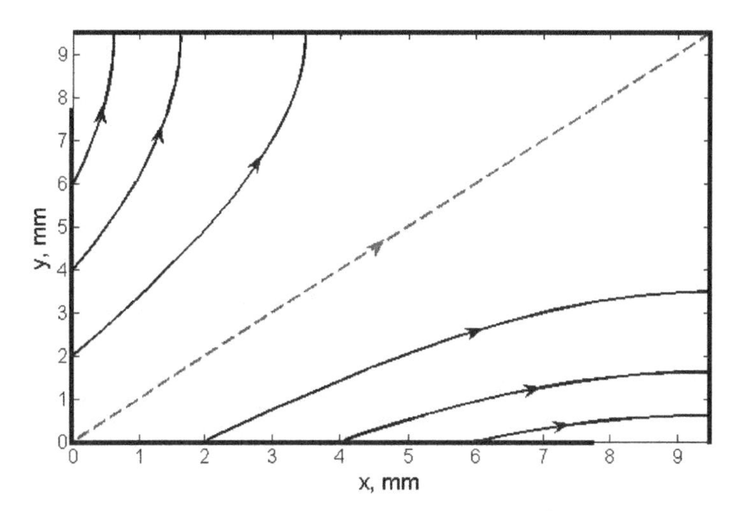

Fig. 4.17 Electric field map excited by two slots $(s_1 = s_2 = 1.7\ \text{mm})$ obtained using two first modes each of them excited by vertical and horizontal slots. Cavity geometry: $a\times b\times L = 9.5\times9.5\times16.76\ \text{mm}$, $F = 2\ \text{GHz}$, $\varepsilon_r = 1$, $\mu_r = 1$

A violation of the geometrical symmetry of the boundary graph $\mathbf{\Gamma}_{e_L}$ leads to the global bifurcation and splitting of the separatrix (Fig. 4.18 and 4.19).

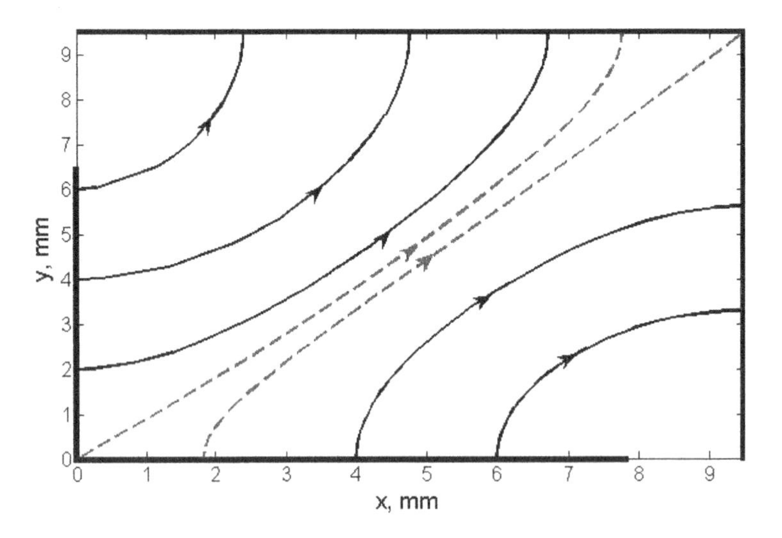

Fig. 4.18 Electric field map excited by two slots $(s_1 = 3 \text{ mm}, s_2 = 1.7 \text{ mm})$ obtained using 2×101 modes. Cavity geometry: $a \times b \times L = 9.5 \times 9.5 \times 16.76$ mm , $F = 2$ GHz , $\varepsilon_r = 1$, $\mu_r = 1$

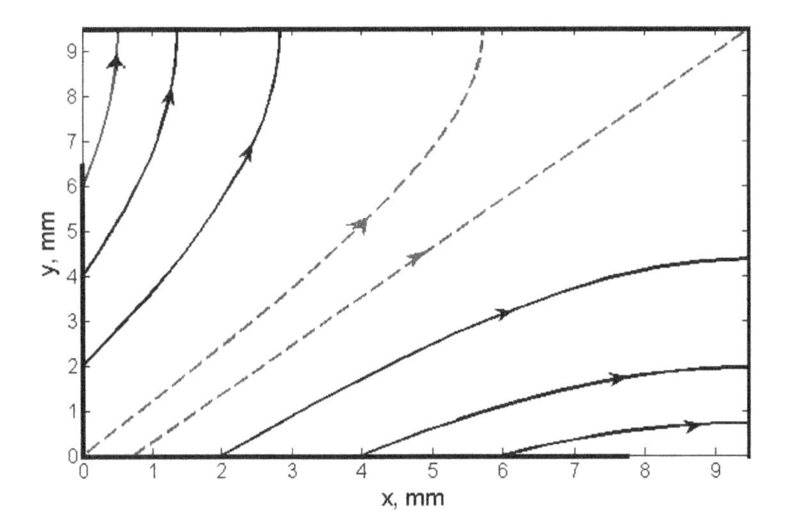

Fig. 4.19 Electric field map excited by two slots $(s_1 = 3 \text{ mm}, s_2 = 1.7 \text{ mm})$ obtained using the first two modes. Cavity geometry: $a \times b \times L = 9.5 \times 9.5 \times 16.76$ mm , $F = 2$ GHz , $\varepsilon_r = 1$, $\mu_r = 1$

Fig. 4.20 shows another bifurcated field map. This bifurcation is caused by transition from a square shape cross-section (Fig. 4.17) to a rectangular one. Minimal dimension of this map is again $N_T = 1+1$, and these modes provide a topological scheme which is homeomorphic to the one from Fig. 4.18.

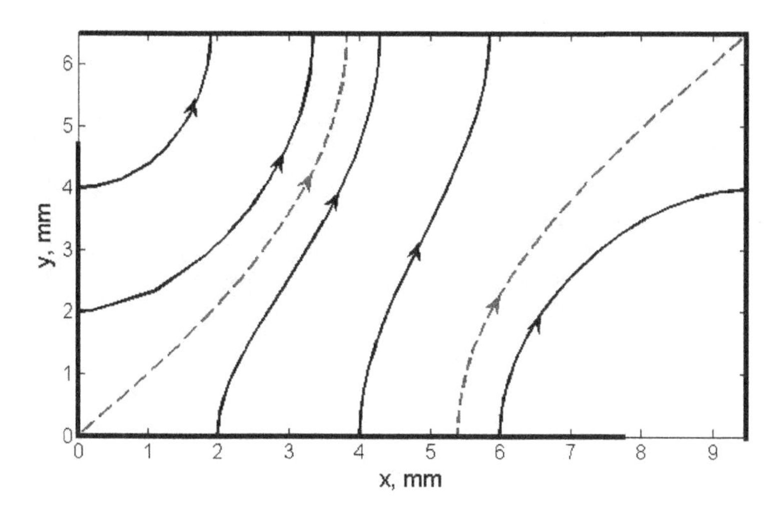

Fig. 4.20 Electric field map excited by two slots $(s_1 = s_2 = 1.7 \text{ mm})$ obtained using 2×101 modes. Cavity geometry: $a\times b\times L = 9.5\times6.5\times16.76$ mm, $F = 2$ GHz, $\varepsilon_r = 1$, $\mu_r = 1$

The same bifurcation occurs if the exciting electric field magnitudes on the slots are not equal to each other. Fig. 4.21 shows the electric field map in the case $e_y^{(1)}/e_x^{(2)} = 2$. It is seen that the obtained map has been bifurcated regarding to Fig. 4.16. It is homeomorphic to the one calculated for the rectangular cross-section resonator (Fig. 4.20). The minimal dimension of this map is again $N_T = 1+1$.

It is seen that these bifurcations occur, as was mentioned in [22], due to the change of the ratio of the magnitudes of two first excited modes in the resonator. Taking into account that the original (Fig. 4.17) and bifurcated maps are formed by the same modes ($N_T = \text{const}$), the bifurcations of this sort relate to the mentioned isomeric ones.

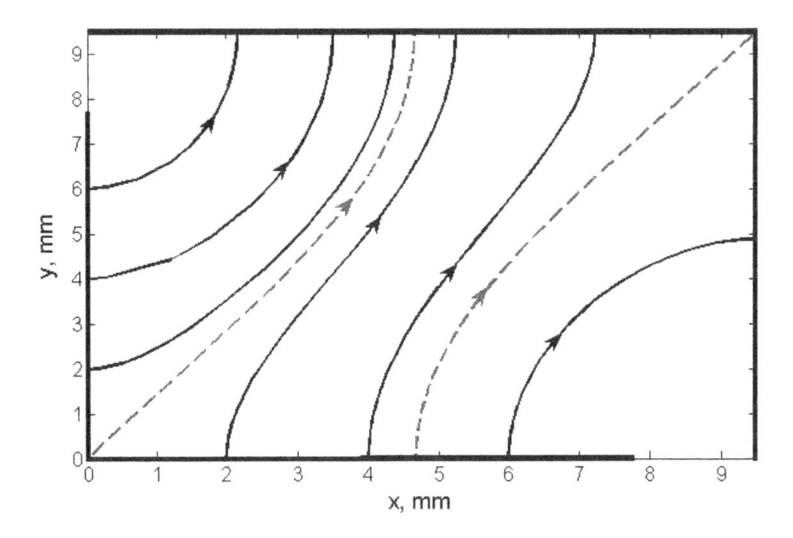

Fig. 4.21 Electric field map excited by two slots $(s_1 = s_2 = 1.7\ \text{mm})$ obtained using 2×101 modes and excited by the slot electric field of different magnitudes $\left(e_y^{(1)}/e_x^{(1)} = 2\right)$. Cavity geometry: $a\times b\times L = 9.5\times9.5\times16.76$ mm, $F = 2$ GHz, $\varepsilon_r = 1$, $\mu_r = 1$

Further simulations of the field maps allow for composition of a bifurcation diagram (Fig. 4.22). It is given for the frequency which is rather far from the resonant ones, and the influence of spatial properties of the boundary graph on the field topology formation is represented clearly. To understand the details, a two-slot symmetrical boundary graph is chosen as the initial one (Fig. 4.22, cell (1,4)). It is seen that this graph and the electric field have classical symmetry regarding to the main diagonal of the resonator cross-section. Perturbations of the boundary graph $\mathbf{\Gamma}_{e_L}$ lead to appearing of more complicated spatial symmetries of the field maps, including the curvilinear and topological ones [47].

As seen from Fig. 4.22, the other $i, j-th$ graphs (cells (1, 1-3), (1, 5-7), and (2, 3-5) are derived from the graph (1,4) by applying the transformation operators $\delta_{ij}^{(1,4)}$ to the initial graph $\mathbf{\Gamma}_{e_L}^{(1,4)}$. In general, $\Gamma_{e_L}^{(i,j)} = \delta_{i,j}^{(m,n)}\Gamma_{e_L}^{(m,n)}$. The graphs, numbered by (3,3) and (3,5), are obtained by the applications of two transformation operators to the initial graph $\mathbf{\Gamma}_{e_L}^{(1,4)}$: $\mathbf{\Gamma}_{e_L}^{(3,3)} = \delta_{1,1}^{(1,4)}\delta_{2,3}^{(1,4)}\mathbf{\Gamma}_{e_L}^{(1,4)}$ and $\mathbf{\Gamma}_{e_L}^{(3,5)} = \delta_{1,1}^{(1,4)}\delta_{4,5}^{(1,4)}\mathbf{\Gamma}_{e_L}^{(1,4)}$.

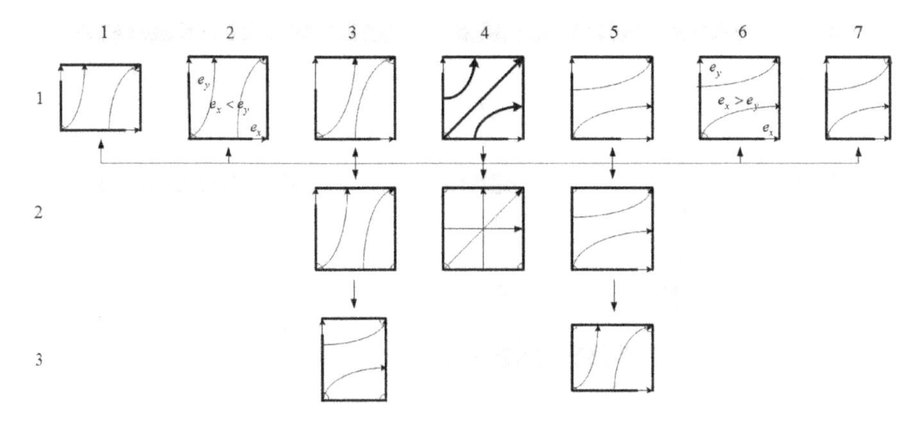

Fig. 4.22 Bifurcation diagram for the electric field excited in a rectangular cavity resonator

Selecting only those which $\delta_{i,j}^{(m,n)}\delta_{m,n}^{(i,j)} = 1$ and $\delta_{p,q,j}^{(k,l)}\left(\delta_{i,j}^{(m,n)}\delta_{m,n}^{(i,j)}\right) = \delta_{m,n}^{(i,j)}\left(\delta_{i,j}^{(m,n)}\delta_{p,q,j}^{(k,l)}\right)$, we can see that the topological schemes of the excited field from Fig. 4.22 are invariant regarding to the boundary graph transformation group Δ_Γ composed of the elements $\delta_{i,j}^{(m,n)}$ satisfying the group condition. The mentioned group Δ_Γ is composed of the continuous and discrete transformation elements, and searching for the conservation laws may give a way to direct qualitative solution of the boundary value problems formulated in the beginning of this Chapter and the earlier Author's works. One can try to create this theory using a combination of the qualitative theory of the autonomous systems, the Lie and the non-local group symmetry approaches [56],[57], and the discrete Noether theorem [58] applied for the above-considered field-force line equations. Although the qualitative analysis is performed here for a simple case, for which analytical solution is known, the idea can be used for computation of more geometrically complicated resonators and waveguides.

4.1.6 Applications of the Topological Approach for the Modeling of Microwave Components

In general, the resonators and waveguides can have rather complicated spatial shapes consisting of multiple sub-domains filled by air and dielectrics (Fig. 4.23). Very often, the initial study and modeling are in the obtaining of the eigenmodes of them and composition of equivalent circuits to model the excitation of these components. One of the approaches to the qualitative modeling of geometrically complicated waveguides and resonators was proposed in [41],[46],[48].

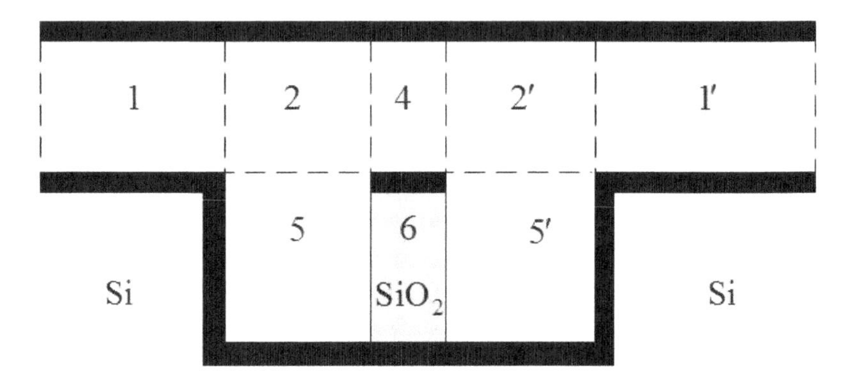

Fig. 4.23 Cross-section of a low-loss shielded microstrip line embedded into a silicon substrate. Central strip conductor is supported by a pedestal of silicon dioxide and isolated by conducting shield from the lossy silicon

Consider the cross-section of a waveguiding structure shown in Fig. 4.23. It consists of several sub-domains of known geometry and magneto-dielectric filling. A qualitative study starts with the defining of the electric fields on the coupling slots or electric currents on the strips. Very often, a constant distribution of them is a pertinent initial approximation for both quantities. The value of the modal propagation constant or resonant frequency can be chosen according to a rough approximation taking into account the structural stability of field maps.

The next step in the modeling of waveguide is the calculation of the modes forming the topological schemes of the electric and magnetic fields and study of bifurcations of them. Then the defined sub-domain modes are substituted into stationary functionals, for instance, to calculate a more accurate propagation constant or resonant frequency. It is supposed that the stationary functionals would use a decreased number of sub-domain modes due to the preliminary performed physical analysis, and it provides faster calculations of the mentioned parameters of waveguides or resonators regarding to the conventional computations. This idea was proofed by qualitative analyses of the modes of several waveguides for 3-D hybrid integrations [22],[36],[46],[48],[50]. These technique and results are considered below.

4.1.6.1 Ridged Waveguide Model

The cross-section of this waveguide is shown in Fig. 4.24. The infinitely thin ridge allows for regulating the cut-off frequency of the fundamental mode.

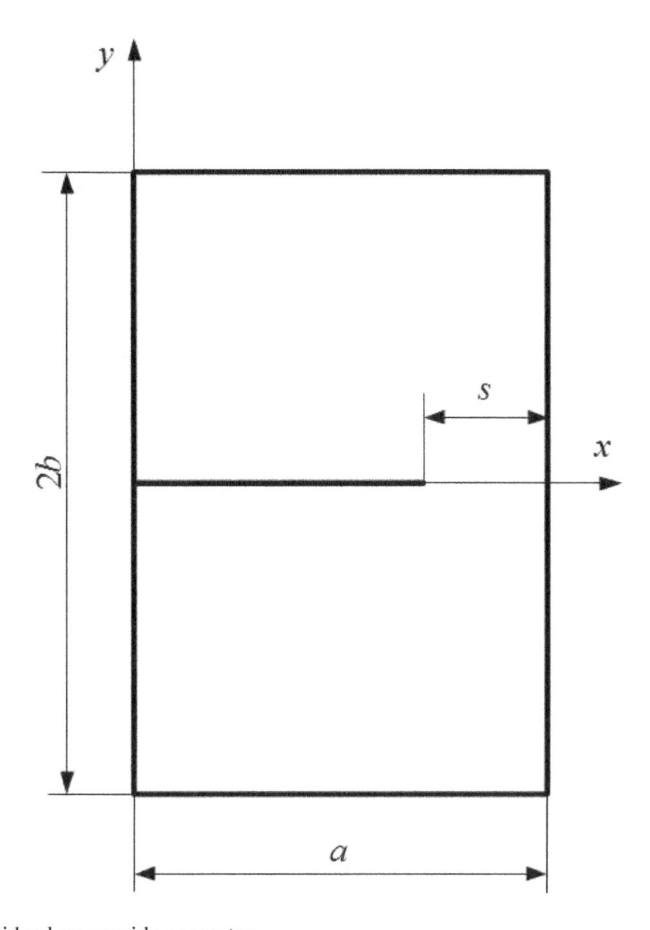

Fig. 4.24 Ridged waveguide geometry

This waveguide was studied in many papers, and one of them authored by Y. Utsumi [59] was chosen as a test one. In that paper, the waveguide with a ridge of a limited thickness is studied by a variational model. The cross-section of the ridged waveguide is divided into four sub-domains. Two of them are placed just under the ridge. The symmetry of cross-section allows for analyzing only a half of this waveguide that reduces the number of subdomain modes used in calculations. The propagation constant is derived using a stationary functional written for the transversal eigenvalue of this waveguide. The number of sub-domain modes varies within 20-200, and a good accuracy is reached in comparison with the experimental and theoretical data of other authors.

The goal of our study here is the development of a technique to obtain the stable, accurate, and reduced time-consuming models based on the topological analysis and application of the regularization methods.

At the difference to the above-mentioned paper, our ideas on the use of topological models are proofed using the integral equation method which is more sensitive towards the quality of the field approximations, and it can serve as the worst case to test the proposed calculation technique.

An integral equation regarding to the slot electric field $e_x(x)$ obtained by the modal expansion method is written for the infinitely thin ridge:

$$\frac{\chi^2}{j\omega\mu_0\mu_r}\sum_{m=0}^{\infty}\frac{2-\delta_{0m}}{a}\frac{\cot\left(k_y^{(m)}b\right)}{k_y^{(m)}}\cos\left(k_x^{(m)}x\right)\int_s e_x(x')\cos\left(k_x^{(m)}x'\right)dx' = j_x(x) \quad (4.43)$$

where k_z is the modal longitudinal propagation constant, $k_x^{(m)} = \frac{m\pi}{a}$, $\chi^2 = k_0^2\varepsilon_r\mu_r - k_z^2$, $k_y^{(m)} = \sqrt{k_0^2\varepsilon_r\mu_r - \left(k_x^{(m)}\right)^2 - k_z^2}$, and $j_x(x)$ is the electric current density on the ridge.

Before the use of any numerical method to solve the integral equation, the nature of possible sensitivity towards the approximations should be studied preliminary. The equ. (4.43) is an integral equation of the first kind, and it can be highly sensitive towards the quality of approximations of the integral kernel, the unknown fields, and the modal parameters under the search. Additionally, the integral kernel is singular, and the modal expansions can have poor convergence. The transversal electric field component defined on the slot has a singularity at the edge of the ridge, and the simple field approximations, which do not take into account this effect, can cause the relative convergence, and the obtained modal propagation constant can be of poor accuracy in some cases.

This integral equation is solved using the Galerkin method and the approximation of unknown electric field $e_x(x)$ at the slot by a series of discrete functions:

$$e_x(x) \approx \sum_{n=1}^{N} a_n u_n(x) \quad (4.44)$$

where

$$u_n(x) = \begin{cases} 1, & x \in (x_{n-1}, x_n) \\ 0, & x \notin (x_{n-1}, x_n) \end{cases}. \quad (4.45)$$

The number of subdomain eigenmodes M in (4.43) is chosen empirically, and usually $M \gg N$. Application of this method gives a homogeneous system of linear algebraic equations regarding to the unknown coefficients a_n, and the determinant of this system gives us the dispersion equation which roots are the modal propagation constants.

Several geometries and frequencies were studied to establish the convergence and the use of the stabilization of the field-map topology for the developments of simple and numerically effective models of guiding structures. The modal wavelength $\lambda_g = 2\pi/k_z$ is calculated at several frequencies and geometries of the slot.

Initial results are with the use of the simplest approximation of the slot electric field $e_x(x), x \in (a-s, a)$ by a constant (N=1 in (4.44)), and the test calculations are performed at F=10 and 14 GHz for a narrow slot s=1.7 mm. The difference between our numerical results and published Utsumi's data [59] is within 0.34-0.53% for these frequencies.

Fig. 4.25 shows two couples of convergence curves, which are the dependencies of the relative difference $\Delta = 2\left|\left(\lambda_g(M) - \lambda_g(M=101)\right)/\left(\lambda_g(M) + \lambda_g(M=101)\right)\right|, \%$ versus the number of subdomain modes M in (4.43). The first of them numbered by 1 and 2 are for a narrow slot geometry $(s = 1.7 \text{ mm})$. Others (3 and 4) are calculated for a wide slot s=4 mm.

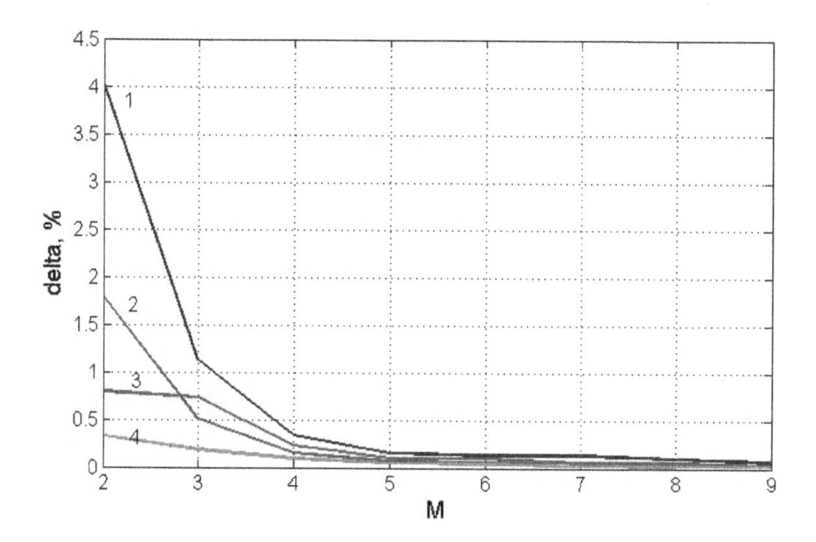

Fig. 4.25 Relative error in calculation of the main mode wavelength versus number of subdomain modes. 1- s=1.7 mm, F=10 GHz; 2- s=1.7 mm, F=14 GHz; 3- s=4 mm, F=10 GHz; 4- s=4 mm, F=14 GHz. Geometry of line: a=b=9.5 mm.

Although, the discrepancy between the initial approximation with M=2 and M=101 is no greater than several percent, the results are stabilized with M>2-4.

Interesting to notice that the electric field map topology stabilization occurs with M=2, i.e. only two modes m=0,1 of the field series with the improved convergence according to Lanczos technique allow for calculation of the map (Fig. 4.26) which is topologically equivalent to the converged field picture (Fig. 4.27, M=101).

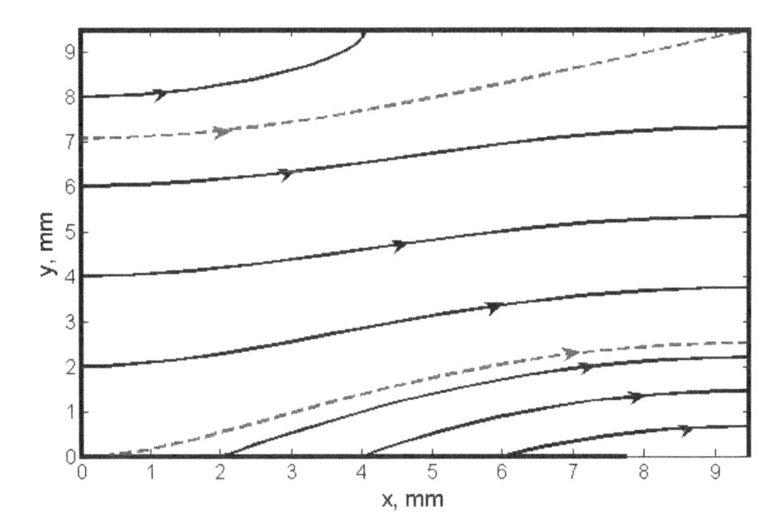

Fig. 4.26 Electric field map of the main mode calculated using only two first sub-domain modes $(m=0,1)$ in the upper part of the waveguide cross-section. Geometry of cross-section: $a \times b = 9.5 \times 9.5$ mm, s=1.7 mm, $\varepsilon_r = 1$, $\mu_r = 1$, $F = 10$ GHz.

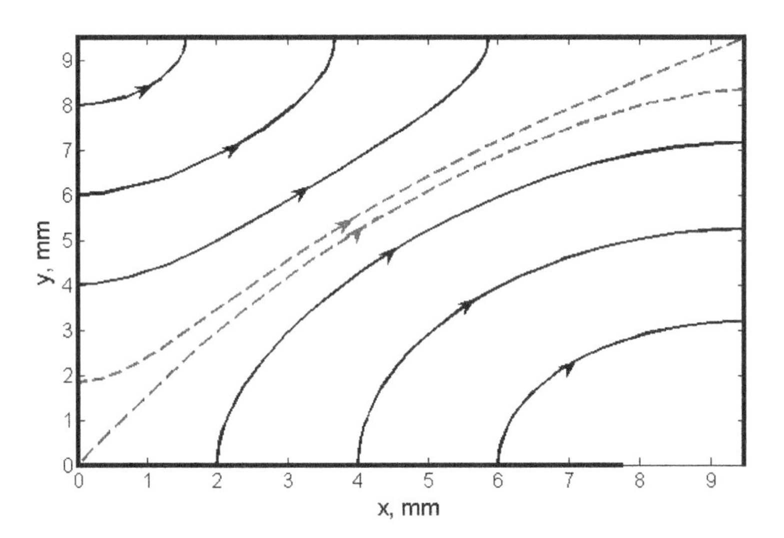

Fig. 4.27 Electric field map of the main mode calculated using 101 modes in the upper part of the waveguide cross-section. Geometry of cross-section: $a \times b = 9.5 \times 9.5$ mm, s=1.7 mm, $\varepsilon_r = 1$, $\mu_r = 1$, $F = 10$ GHz

As was mentioned, in the case of the cavity field calculations, the modal series being not corrected by the Lanczos factors suffer from severe Gibbs effect. Anyway, calculations show synchronization of the modal wavelength and field topology convergence if the field map is calculated at areas, which are far from the slot. Practical recommendation is to follow the convergence of the modal field map calculated with the Lanczos technique and to define a proper truncation number M_T. For reliability of the data and its accuracy, this number can be increased by additional 1-3 sub-domains modes to provide stabilization of topology at any area of the cross-section.

Qualitative studies of the multi-domain waveguides and their modes relate to the theory of multi-composed dynamical systems, and special conditions should be specified at the boundaries between the sub-domains of the studied waveguides. Among them are the geometrical orientations of the field-force lines at the boundaries.

A waveguide can consist of multiple conductor and dielectric components, which shapes are cornered, and the field components are singular at their edges which are the knots of the electric field-force lines. In electromagnetism, the Meixner condition limits the field behavior close to the edges [60]. According to it, the energy density close to the edge is an integrable value:

$$\int |\mathbf{E}|^2 dV < \infty, \quad \int |\mathbf{H}|^2 dV < \infty. \tag{4.46}$$

In addition, the flux of energy from an edge is zero. It leads to a certain type of the behavior of the fields and currents close to the edges. For example, in the integral equation (4.43), the probe function should be $e_x(x) \sim 1/\sqrt{x^2 - s^2}$.

In many cases, the simple slot field approximation $e_x(x) = \text{const}$ provides satisfying accuracy for calculation of modal propagation numbers. Unfortunately, on the field level, it leads to unphysical behavior of the field-force lines, as it is shown in Fig. 4.28 where, in spite of the symmetry of the geometry, the electric field lines cross the boundary instead being parallel to the slot plane. Besides, the ridge edge should play a role of a knot for the electric field-force lines. An additional study is required to control the violation of the energy conservation law in this case and its consequences. Thus, proper simulation of the field maps can serve a criterion for the waveguide and resonator modeling.

Better results are with the increase of the number of probe functions N. Usually, to obtain accurate results, the truncation numbers should be chosen as $M \gg N$, which makes the computations more time-consuming for the increased N. Unfortunately, the choice of N comparable with M or even less makes the calculations very unstable due to the nature of the mentioned integral equation of the 1st kind. For instance, Figs 4.29 and 4.30 illustrate this effect. The first picture shows stable calculations of the determinant $D(\lambda_g)$ of the system of linear algebraic equations obtained by the Galerkin method. A zero of this determinant is associated with the modal wavelength (given in millimeters below).

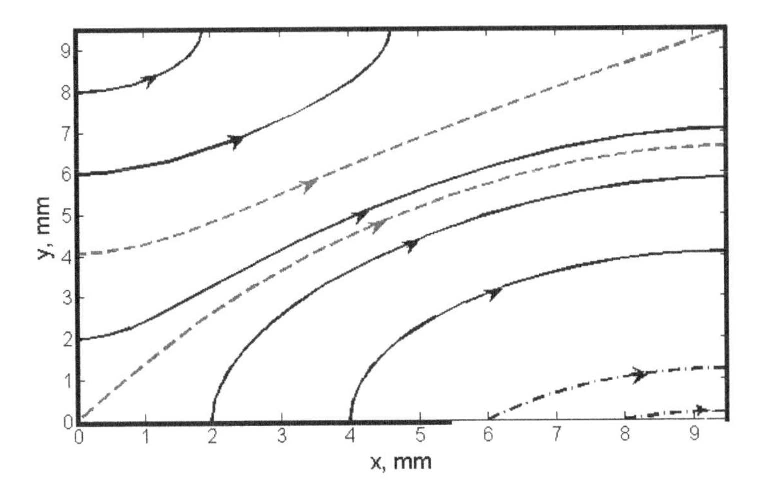

Fig. 4.28 Electric field map of the main mode calculated using only 101 modes in the upper part of the waveguide cross-section. Geometry of cross-section: $a \times b = 9.5 \times 9.5$ mm, s=4 mm, $\varepsilon_r = 1$, $\mu_r = 1$, $F = 10$ GHz

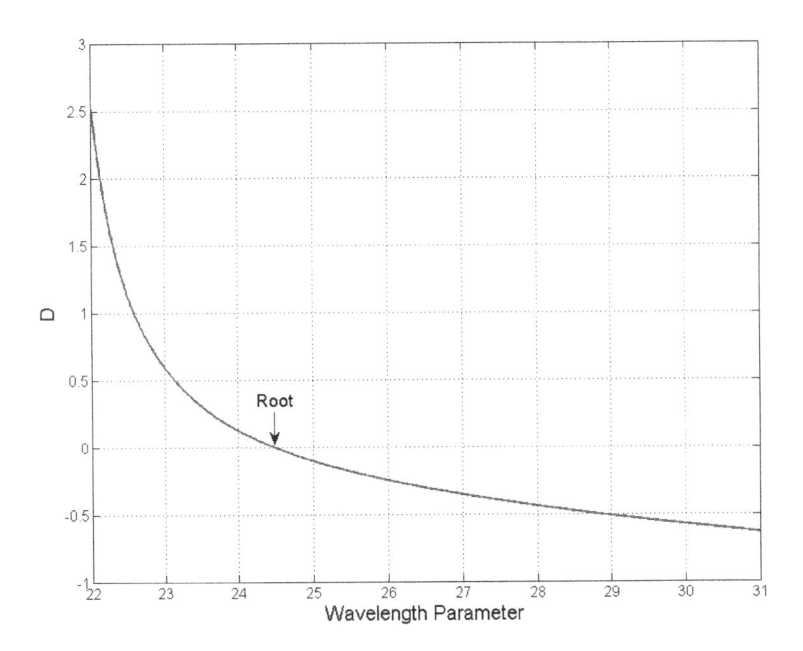

Fig. 4.29 Stable results of calculations of the determinant $D(\lambda_g)$ obtained for $M > 10$ and $N = 5$. Geometry of cross-section: $a \times b = 9.5 \times 9.5$ mm, $\varepsilon_r = 1$, $\mu_r = 1$, $F = 14$ GHz , s=4 mm

Taking into account the goal of our study is the fast models, the truncation number M is decreased, and since M=3 and $N=5$ the calculations start to be unstable (Fig. 4.30) which is seen from the chaotical behavior of the curve $D(\lambda_g)$.

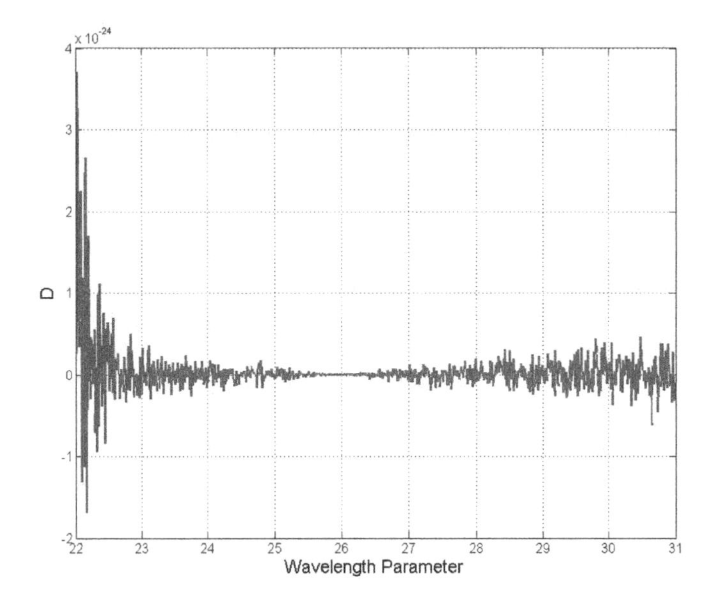

Fig. 4.30 Unstable calculations of the of the determinant $D(\lambda_g)$ obtained for $M=3$ and $N=5$. Geometry of cross-section: $a\times b=9.5\times9.5$ mm, $\varepsilon_r=1$, $\mu_r=1$, $F=14$ GHz , s=4 mm

From the first point of view, this numerical instability prevents using the ideas of stabilization of topology to obtain the low-time consuming models of wave-guides and transmission lines. A discussion of similar situation in the mode matching method occured many years ago. It is found that a simple increase of the number of taken modes in the 1st order problems leads to destabilization of calculations in many cases, and the nature of these effect, similarly to the Gibbs phenomenon, is numerical [61]. Earlier, Mittra and Lee proposed stabilization of calculations by taking into account the edge singularity [62]. Additionally, they introduced an empirical criterion for proper choosing of the number of modes in different waveguides, which joint is the subject of calculations.

In 1993, Yang and Omar showed that preliminary analysis of the multiple rectangular aperture irises in a rectangular waveguide and excluding of the non-interacting modes allows for saving in more 50% of computer storage and reduction, down to about 25% of processor time [63].

An interesting paper on the study of interacting waveguide discontinuities was published by Gessel and Ciric in 1994 [64] where they showed that taking into account an increased number of any-type modes can destabilize calculations, and all modes should be analyzed together with the geometry of discontinuity to exclude the non-interacting ones with the purpose to improve the convergence. Today it is concluded that the minimum number of modes is defined both by complicated interplay of the modeled geometry and by completely numerical effects, and proper problem formulation and adequate choosing of the method of the solution can decrease the complexity of calculations.

Additionally, different other regularization techniques can be applied to the integral equations of the first kind, including the Tikhonov's and Lavrentiev's methods [65],[66]. The application of the last one to the modeling of the ridged waveguide is considered in [22] and here in more details.

It was shown by M.M. Lavrentiev that an ill-posed problem could be substituted by an approximately equivalent one of the second kind:

$$Au(x) - f(x) = 0 \rightarrow Au(x) - f(x) - \alpha u(x) = 0 \tag{4.47}$$

where $u(x)$ is the unknown function, A is the operator, $f(x)$ is the known function, and α is the regularization parameter. The regularized equation is less sensitive, and it allows for using the operator composed of essentially decreased number of spatial harmonics in (4.43). One of the serious questions is finding in a proper manner the value of the regularization parameter α providing stabilization of calculations and acceptable accuracy.

Fig. 4.31 (compare Fig. 4.30) shows the results of calculation of the determinant $D(\lambda_g)$ for the regularized according to (4.47) integral equation with the empirically found parameter $\alpha = 0.35\,\chi^2 / j\omega\mu_0\mu_r$.

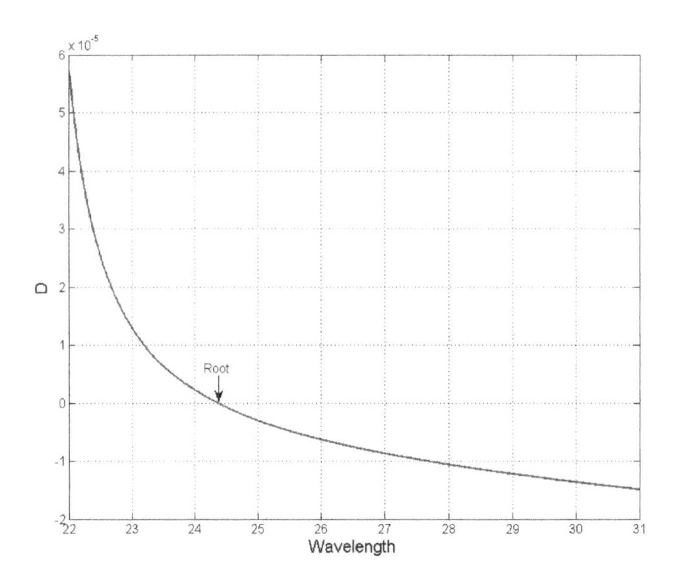

Fig. 4.31 Regularized dependence $D(\lambda_g)$ obtained for $M = 3$ and $N = 5$. Geometry of cross-section: $a \times b = 9.5 \times 9.5$ mm; Permittivity $\varepsilon_r = 1$ and permeability $\mu_r = 1$, $F = 14$ GHz , s=4 mm

It is seen that the Lavrentiev's regularization allows for calculating the modal propagation constants even for the unstable cases when $M < N$, as it was considered in [22]. Unfortunately, the empirically chosen regularization parameter α allows to calculate the line with good accuracy only for this given geometry or nearby, and a proofed technique to calculate this coefficient should be used instead of this empirical approach. One of them is with the a priori technique, and the regularization parameter α can be obtained analyzing the operator A and the errors in the unknown function $e(x)$ approximations [67].

A posteriori technique is with a functional, which is equivalent to the integral equation, although some doubts on the effectiveness of this approach can be found in [67]:

$$\int \left(Ae_x(x) - \alpha e_x(x) \right)^2 dx. \tag{4.48}$$

Minimization of (4.48) gives the needed parameter α, the unknown field coefficients a_n, and the modal propagation coefficient k_z.

For instance, an expression for α can be found analytically from (4.48), and it can be substituted into (4.47) to which the Galerikin method can be applied. A system of regularized equations is

$$\int \left(Ae_x(x) - \alpha e_x(x) \right) u_n(x) dx = 0. \tag{4.49}$$

This idea was proofed only in part for N=1 and M=2 in [22], and more study is required on this attractive technique to obtain fast models of components of microwave modules.

4.1.6.2 Some Other Waveguides and Components Studied Using the Topological Approach

Additionally to the ridged waveguide studied in the details to develop a topological technique for simulation of more complicated waveguiding structures, several transmission lines have been modeled using the above-mentioned ideas (Figs 5.4i-k, m, 5.7, 5.8 from Chapter 7).

All these lines were studied qualitatively when the electric fields on the slots modeled by constants. The initial values of propagation numbers were taken using ones from the lines of the geometries which are close to the studied waveguides. These obtained results on the field modeling were used for the developments of 3-D antennas, directional couplers, microwave sensors for the measurements of liquid dielectrics [68],[69], etc. Some results on applications of topological ideas to the developments of EM models of square-pad via-holes and shorted patch antennas confirmed by measurements are considered below in details.

4.1.6.3 Modeling of Square-pad Via-holes and Patch Antennas Using the Topological Approach[1]

In electromagnetism, the most complicated problem is the analytical calculation of 3-D components. The above-considered topological approach is able to provide

[1] Written together with M.J. Deen, N.K. Nikolova, and A. Rahal.

the approximate analytical or semi-analytical EM models of reduced complexity for some 3-D components, and it can be used to create effective models for EM software tools [50].

A couple of the examples of this kind are considered below, and the subject of the modeling is the square pad via-holes and shorted-patch microstrip antennas (Fig. 4.32).

4.1.6.3.1 Square-pad Via-holes

The analytical modeling of square-pad via-holes is a complicated task due to combination of cylindrical and square shapes (Fig. 4.32).

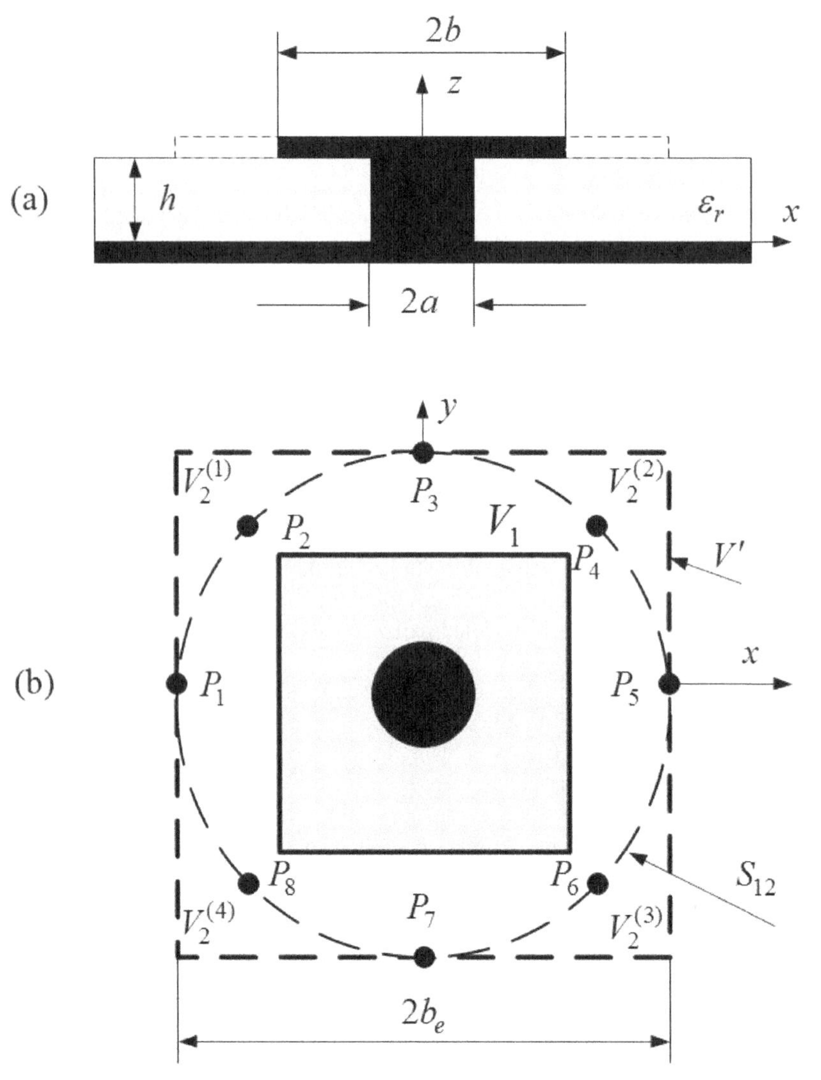

Fig. 4.32 Square-patch resonator [50]: (a)- Cross-section; (b)- Field matching scheme. Reprinted with permission of the IEEE, license # 2873190190705

Our model [50] is based on an approximation using the modes of a circular-pad via-hole and the modes of a square-patch planar resonator. Each modal field inside a square-pad via-hole is composed of the modal fields of the mentioned resonators if they have the same topology. Then the equivalent circuit parameters are calculated for these modes.

Initially, the modes of a circular-pad via-hole are calculated (see [50] and Section 7.4). They are different from each other by the angular (n), radial (m), and height $(k=0)$ numbers and the field distribution along the corresponding directions. To calculate the resonant frequencies, a via-hole cavity is surrounded by a magnetic wall of a certain effective radius b_e which value is defined by the fringing field and modal number. It is calculated according to the empirical formulas:

$$\begin{aligned} b^{(nm0)} &= b + 0.553h, \; n = 0, m = 1, \\ b^{(nm0)} &= b + 0.450h \; n \geq 1, m \geq 1. \end{aligned} \tag{4.50}$$

The segmentation of the grounded square-patch volume by two regular overlapping shapes, for which analytical solutions are available, is used. Solutions to problems involving arbitrary shapes have been constructed before by enclosing them in a regularly shaped housing, e.g. [70],[71]. There, an internal boundary is defined. It separates the regular shaped computational volume to two regions where the internal region contains the irregular shape. A numerical solution is found by mode matching or point matching at the internal boundary.

To obtain a system of fewer equations, an approach similar to the one proposed in [47]-[49] is used called as a topological method of the EM boundary problems. A physically based choice of matching subdomain eigenfunctions is made, which allows for the approximate analytical treatment of complicated boundary-value problems. This approach aims at finding the correlation between the map of the excited vector field and the excitation currents at the boundaries of the domain. It extends Poincare's qualitative theory of ordinary differential equations [13] to the vector partial differential equations. Since the approach operates with field topologies, it is named the topological electromagnetic theory of boundary-value problems.

Following this topological theory, the grounded square-patch effective volume is divided by two subdomains. The subdomain V_1 has a circular form. The rest is represented by the corner subdomains $V_2^{(i)}$ $i = 1, ..., 4$ (see Fig. 4.32). The housing volume V' is of a square shape and it does not contain the rod (Fig. 4.32b). In effect, it is a square microstrip resonator which is solved analytically for the electric field E'_z using the magnetic wall cavity representation [72]:

$$E'_z = \sum_{l,p} B_{lp0} \cos\left(\frac{l\pi x}{b_e}\right) \cos\left(\frac{p\pi y}{b_e}\right), \; (x, y \in V'). \tag{4.51}$$

Here, B_{lp0} are the unknown normalized modal amplitudes; and l and p are the numbers of field variation in the x- and y-directions, respectively. The third zero index shows that we consider only z-invariant modes.

The eigenfunctions of the circular subdomain V_1 are the modes of the grounded circular patch $E_{z_{nm0}}^{(I)}$ from [73] provided that the magnetic wall condition is valid on the internal effective boundary S_{12}. The field expansion for $E_z^{(I)}$ is written as

$$E_z^{(I)}(r,\phi)=\sum_{n,m} A_{nm0}E_{z_{nm0}}^{(I)}(r,\phi),\ (r,\phi)\in V_1 \tag{4.52}$$

where A_{nm0} is the unknown normalized modal amplitude; and n and m are the numbers of the angular and radial field variations, respectively.

Next, the field in the square-patch corner regions $V_2^{(i)}$ is considered. Using (4.51) it can be approximated by the modal representation of the field of the square microstrip housing:

$$E_z^{(II)}(r,\phi)\approx E_z'(x,y),\ (x,y)\text{ and }(r,\phi)\in V_2^{(i)},\ i=1,\ldots,4. \tag{4.53}$$

The eigenvalue equation is derived from the fields (4.52) and (4.53) as well as the boundary condition at S_{12} (see Fig. 4.32b):

$$E_z^{(I)}(r,\phi)-E_z^{(II)}(r,\phi)+M_\phi=0,\ (r,\phi)\in S_{12} \tag{4.54}$$

where M_ϕ is the fictitious magnetic current density component due to the step-wise change of the field at the boundary. Numerical solutions of (4.54) are possible using the method of moments. However, our goal is a semi-analytical approach, which avoids the solution of a large system of equations.

The equation in (4.54) is solved by point-matching properly chosen eigenfunctions in each subdomain. Consider the field given by the eigenfunctions with $v_I=(n,m,0)$ in the grounded circular-patch cavity and $v_{II}=(l,p,0)$ in the square-patch cavity:

$$\begin{aligned}E_{z_v}^{(I)}(r,\phi)&=A_{nm0}E_{z_{nm0}}^{(I)}(r,\phi),\\ E_{z_v}^{(II)}(r,\phi)&=B_{lp0}\cos\left(\frac{l\pi x}{b_e}\right)\cos\left(\frac{p\pi y}{b_e}\right).\end{aligned} \tag{4.55}$$

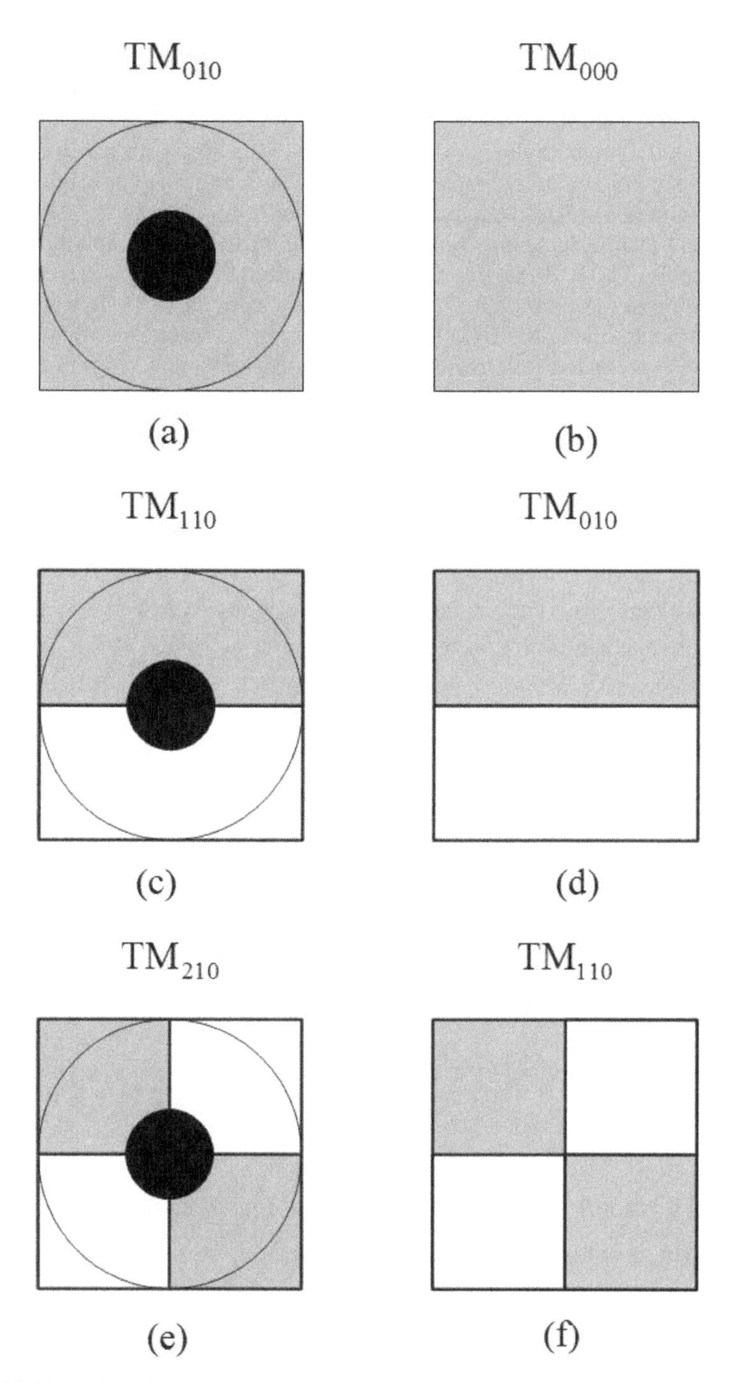

Fig. 4.33 Similarity of E_z modal field maps of square-pad and circular-pad via-holes (a, c, e) and a square patch microstrip resonator (b, d, f). Reprinted with permission of the IEEE, license #2873190190705 (see [50])

The mode numbers (n,m) of the grounded circular-patch cavity determine the modal field behavior of the grounded square-patch via-hole. The proper choice of matched eigenfunctions from each of the two subdomains is based on their similarity as illustrated by (Fig. 4.33a,c,e). There, the grey patterns correspond to positive values of the eigenfunctions and the white patterns correspond to their negative values. Similar modes of the grounded circular patch and the square patch have the same number of field variations along the angular and the radial variables (see Fig. 4.33b,d,f). For an efficient solution, the choice of eigenfunctions must ensure that the rotational symmetry of the modes of the square patch housing, which approximate the field in the corners, has to match the rotational symmetry of the grounded patch modes. Thus, once the numbers n and m of the cylindrical eigenfunction $E_{z_{nm0}}^{(I)}$ are chosen, the eigenfunction numbers of the square microstrip housing l and p can be determined. For example, for $n=0$ and $m=1$ (corresponding to the eigenfunction $E_{z_{nm0}}^{(I)}$), the respective values of l and p are $l=p=0$ (Fig. 4.33b); for $n=1$ and $m=1$ (TM_{110} mode), $l=0$ and $p=1$ (Fig. 4.30d); for $n=2$ and $m=1$ (TM_{210} mode), $l=1$ and $p=1$ (Fig. 4.33f). The higher-mode similarity of the circular and square domains can be found in [74].

The point-matching solution of (4.54) uses a set of points $P_j^{(n)}$ $(j=1,\dots,N_n)$ for each mode with an n-dependence on the angular variable. These points are chosen where the normalized field magnitude $|E_{z_v}^{(I)}(P_j^{(n)})|$ is equal to one, and the fictitious magnetic current density M_ϕ is set to zero. Thus, the voltage $U(P_j^{(n)})$ between the patch and the ground is

$$U(P_j^{(n)}) = U_{\text{edge}} = \int_0^h E_{z_v}^{(I)}\left(r_j,\phi_j\right)dz = \int_0^h E_{z_v}^{(II)}\left(r_j,\phi_j\right)dz. \tag{4.56}$$

To solve it for $U(P_j^{(n)})$, at least one point is needed. The coefficients A_{nm0} and B_{lp0} in (4.55) are found from $U(P_j^{(n)})$:

$$A_{nm0} = \frac{U(P_j^{(n)})}{hE_{z_{nmo}}^{(I)}\left(r_j,\phi_j\right)}, \quad B_{lp0} = \frac{U(P_j^{(n)})}{h\cos\left(\dfrac{l\pi x_j}{b_e}\right)\cos\left(\dfrac{p\pi y_j}{b_e}\right)}. \tag{4.57}$$

Finally, A_{nm0} and B_{lp0} are used to calculate the equivalent modal capacitance of the grounded square-patch resonator $C_{\text{sq}}^{(nm0)}$ [72]:

$$C_{\text{sq}}^{(nm0)} = 2W_{e_{\text{sq}}}^{(nm0)} / \left[U(P_j^{(n)})\right]^2 \tag{4.58}$$

where $W_{e_{\mathrm{sq}}}^{(nm0)}$ is the electric energy stored in the volume of the grounded square patch. The capacitance $C_{\mathrm{sq}}^{(nm0)}$ is written as a sum of two terms, which use the energy stored in the circular subdomain $W_{e_{\mathrm{cir}}}^{(nm0)}$ and the corner subdomains $W_{e_{\mathrm{corner}}}^{(lp0)}$:

$$C_{\mathrm{sq}}^{(nm0)} = \frac{2W_{e_{\mathrm{cir}}}^{(nm0)}}{\left(U(P_j^{(n)})\right)^2} + \frac{2W_{e_{\mathrm{corner}}}^{(lp0)}}{\left(U(P_j^{(n)})\right)^2} = C_{\mathrm{cir}}^{(nm0)} + \frac{2W_{e_{\mathrm{corner}}}^{(lp0)}}{\left(U(P_j^{(n)})\right)^2}. \tag{4.59}$$

The modal field distribution and, therefore, the energy $W_{e_{\mathrm{cir}}}^{(nm0)}$ are those corresponding to the magnetic-wall cavity solution in the circular region. The second term, associated with $W_{e_{\mathrm{corner}}}^{(lp0)}$, is computed using $E_{z_v}^{(II)}$ in (4.55) and B_{lp0} in (4.57) Note that the energy is proportional to $[U(P_j^{(j)})]^2$ because the modal coefficients are expressed according to (4.57). Thus, the modal voltage $U(P_j^{(n)})$ is only an auxiliary variable, which is set equal to one.

The modal resonant frequency ω_{nm0} is computed using the modal inductance $L_{\mathrm{sq}}^{(nm0)}$. The magnetic energy associated with the rod current is concentrated near the rod, and the portion in the corner volumes is negligible. Thus, the overall inductance of the grounded square patch $L_{\mathrm{sq}}^{(nm0)}$ is approximated with the grounded circular patch modal inductance $L_{\mathrm{cir}}^{(nm0)}$, i.e., $L_{\mathrm{sq}}^{(nm0)} \approx L_{\mathrm{cir}}^{(nm0)}$ [73]. With this approximation, the modal resonant frequency is calculated as

$$\omega_{nm0} = \left[\left(C_{\mathrm{cir}}^{(nm0)} + \frac{2W_{e_{\mathrm{corner}}}^{(lp0)}}{\left(U(P_j^{(n)})\right)^2}\right) L_{\mathrm{cir}}^{(nm0)}\right]^{-1/2}. \tag{4.60}$$

To validate our model, an experimental kit is manufactured (Fig. 4.34) where the square-pad via holes are excited through a capacitive gap.

Comparison of the results are shown in [50] for three geometries of the square-pad via holes and good corresdpondence of measurements and calculations is found with he error wich is no more than 2 %.

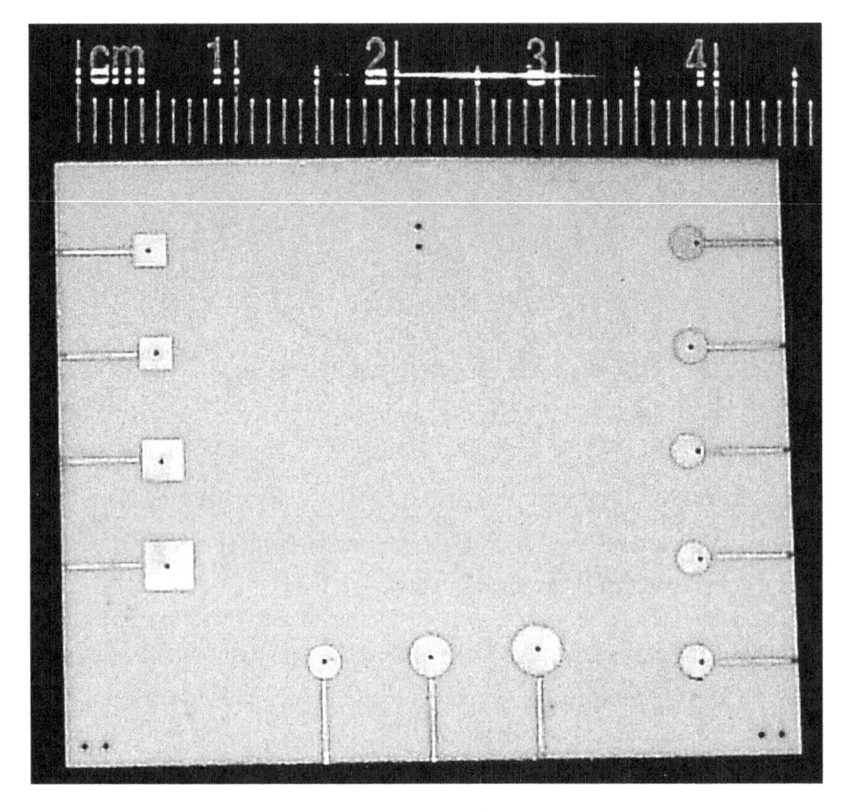

Fig. 4.34 A photograph of the experimental kit [50]. The square-pad via-holes (left) are fed through a capacitive gap except the uppermost sample, which is connected directly to the 50-Ohm microstrip line. The gap length is equal to the height of the alumina substrate. The parameters of the alumina substrate are: thickness 0.254 mm, relative permittivity 9.6. Reprinted with permission of the IEEE, license # 2873190848584

It is seen the negligible difference between the experimental and theoretical data. Additionally, our results (solid lines) were compared with the ones derived by the Agilent Momentum (squares) simulations, and these both results are shown in Fig. 4.35.

Using Momentum, the resonant frequencies are obtained by analysing the resonant curves of isolated via-holes excited by a microstrip capacitively coupled to the modeled component. It is seen a good correspondence between the simulations shown in Fig. 4.35.

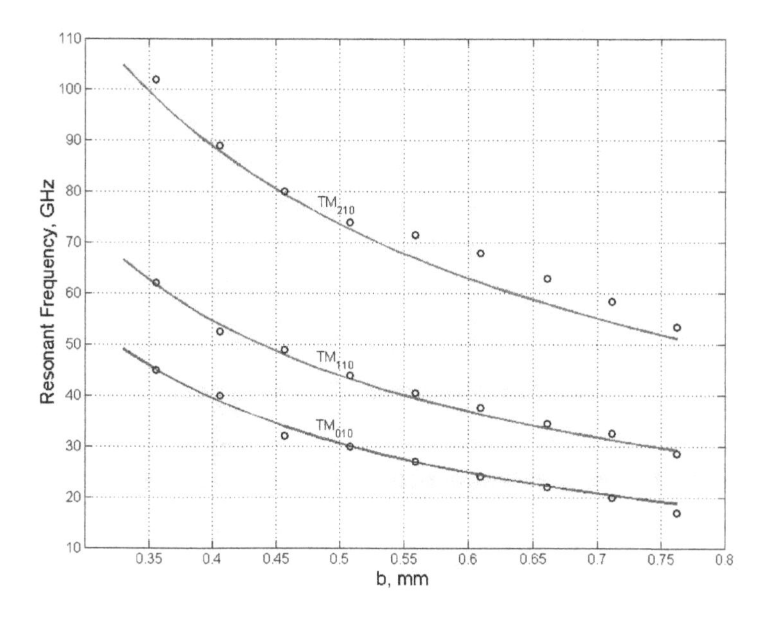

Fig. 4.35 Resonant frequencies of the TM_{010}, TM_{110} and TM_{210} modes versus the square-pad via-hole size *b* [50]: Circles – Momentum; Solid curves – our approach. Via-hole parameters: $a = 0.152$ mm, $h = 0.254$ mm, $\varepsilon_r = 9.9$. Reprinted with permission of the IEEE, license # 2873191334118

4.1.6.3.2 Shorted Square and Rectangular Patch Antennas

Shorted patch antennas are widely used in microwaves [75]-[82]. Shortening allows to decrease the antenna size due to the additional inductance provided by shortening rod inductance [75],[76]. The position of this rod allows regulating of the input impedance of antennas. The square patches are used for circularly polarised antennas fed by two input signals with 90^0 phase shift.

Rectangular shorted patches are more widely used than the square ones. The geometry of a shorted rectangular patch is shown in Fig. 4.36.

Our approach allows us to model such a component. First, an elliptical shorted patch cavity can be solved semi-analytically. The method of the solution is similar to [83] where a circular-elliptical waveguide is considered. In contrast to the solutions in [83], the eigenfunctions of our cavity satisfy the magnetic wall condition at the elliptical boundary whith the effective extension (4.50). This yields the eigenmodes expressed in terms of the Mathieu functions and allows for the computation of the modal capacitance and inductance. Then the elliptical modes are matched with the eigenfunctions of the rectangular housing, which has the same class of spatial symmetry. Subsequently, the modal resonant frequency can be computed similarly to (4.60).

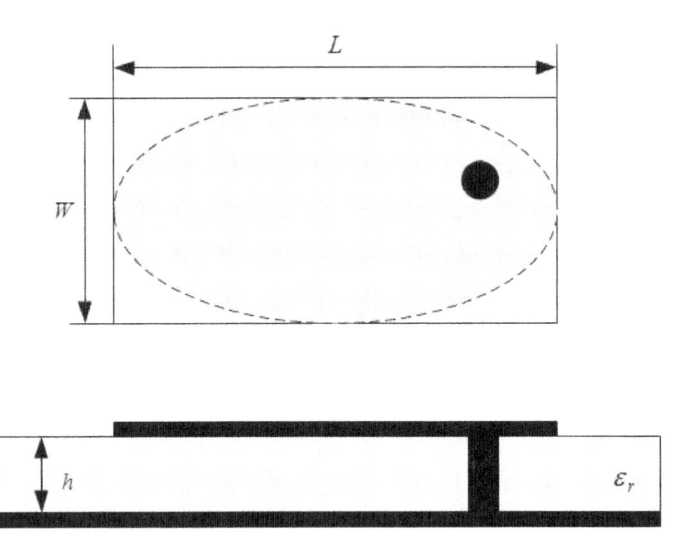

Fig. 4.36 Shorted rectangular patch and an elliptical subdomain (dashed) for semi-analytical computation of the eigenfunctions of the elliptical arbitrary shorted patch [50]. Reprinted with permission of the IEEE, license # 2873190848584

For the special case of a centrally shorted rectangular patch whose geometry is close to a square (~20%), a simplified approach was developed for the calculation of the fundamental mode resonant frequency, which is based on the perturbation theory.

Such a rectangular patch is regarded as a square one with slightly extended sides (Fig. 4.37). In addition to the corner capacitances found according to (4.59) an extension capacitance C_{ext} is calculated assuming a constant electric field at the extensions. The extension inductance is ignored. Then the fundamental mode resonant frequency is computed as

$$\omega_{010}^{(\text{rect})} = \left[\left(C_{\text{cir}}^{(010)} + 2C_{\text{ext}} + \frac{2W_{e_{\text{corner}}}^{(000)}}{\left(U(P_j^{(n)})\right)^2}\right) L_{\text{cir}}^{(010)}\right]^{-1/2}. \tag{4.61}$$

where

$$C_{\text{ext}} = \frac{1}{2}\frac{\varepsilon_0 \varepsilon_r}{h}(L-W)(W+1.106h). \tag{4.62}$$

Our model allows for the computation of the resonant frequencies of rectangular shorted patches if their shapes differ slightly (~20%) from a square. This approximation was verified by the analysis of the shorted patch antenna from [76]. Its geometry corresponds to that in Fig. 4.37. The shorting post with the radius a=0.5 mm is placed at the center of the patch whose parameters are: length L=16 mm, and width W=13 mm $(L/W = 1.3)$. The substrate height is h=0.8 mm and its permittivity is ε_r=2.68. Our results are in good agreement with the measurements provided in [76] and the data derived with IE3D and Momentum with the difference which is no more than 5%. More comparisons are from Fig. 4.38 where our simulations (solid curves) with the Momentum and IE3D are shown.

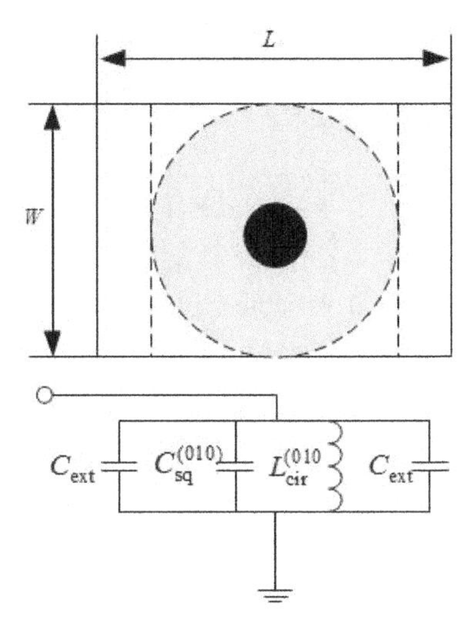

Fig. 4.37 Centrally shorted rectangular patch and the fundamental mode equivalent circuit [50]. Reprinted with permission of the IEEE, license # 2873190848584

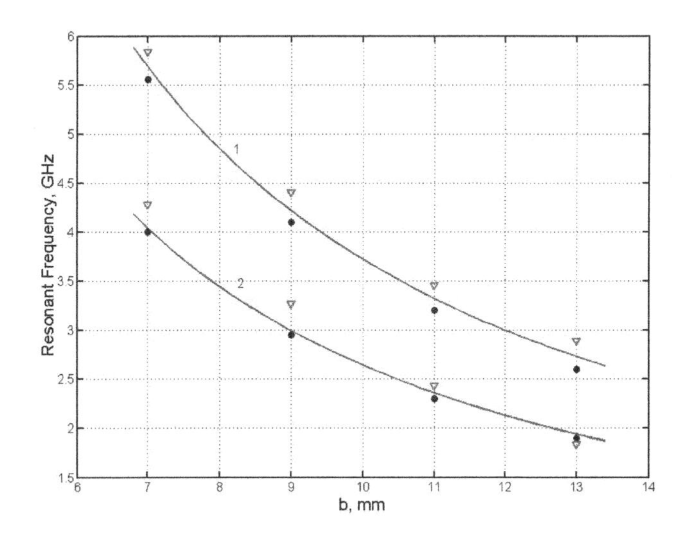

Fig. 4.38 Resonant frequency of the TM_{010} mode versus the side length 2*b* of a shorted square-patch resonator [50]: Solid curves – our model; Circles – Momentum; Triangles – IE3D; Curve 1 – $\varepsilon_r = 5$, $h = 0.631$ mm, $a = 0.5$ mm; Curve 2 – $\varepsilon_r = 9.9$, $h = 0.631$ mm, $a = 0.5$ mm. Reprinted with permission of the IEEE, license # 2873191334118

4.2 Topological Analysis of the Time-dependent EM Field

Previous analysis is with the time-harmonic electric and magnetic fields, and the considered field maps are given for any arbitrary moment of time. They are calculated using the real or imaginary parts of fields written using the phasor notation. Each force line is considered a trajectory of elementary electric or magnetic charge moving in the corresponding time-independent field and their equations are the autonomous dynamical systems:

$$\frac{d\mathbf{r}_{e,h}}{ds_{e,h}} = \mathrm{Re}\left[\mathbf{E}(\mathbf{r}_e), \mathbf{H}(\mathbf{r}_h)\right]. \tag{4.63}$$

where $s_{e,h}$ are the parametrical variables. Introducing the particle time t_e or t_h the equations (4.63) are rewritten as

$$\frac{d\mathbf{r}_{e,h}}{dt_{e,h}} = \mathrm{Re}\left[\frac{\mathbf{E}(\mathbf{r}_e)}{|\mathbf{E}(\mathbf{r}_e)|}\left(\frac{\partial s_e}{\partial t_e}\right), \frac{\mathbf{H}(\mathbf{r}_h)}{|\mathbf{H}(\mathbf{r}_h)|}\left(\frac{\partial s_h}{\partial t_h}\right)\right]. \tag{4.64}$$

In the case of static or time-harmonic fields, $ds_e/dt_e = const$ and $ds_h/dt_h = const$. These fields are studied by topological means, and the first results (1988) on topological analysis of EM fields and solutions of boundary value problems are from [20]-[22], for instance.

If the fields are transient, the phasor notations are not used, and then the equations (4.64) are the nonautonomous ones:

$$\frac{d\mathbf{r}_{e,h}}{dt_{e,h}} = \left[\frac{\mathbf{E}(\mathbf{r}_e, t_e)}{|\mathbf{E}(\mathbf{r}_e, t_e)|}\left(\frac{\partial s_e}{\partial t_e}\right), \frac{\mathbf{H}(\mathbf{r}_h, t_h)}{|\mathbf{H}(\mathbf{r}_h, t_h)|}\left(\frac{\partial s_h}{\partial t_h}\right)\right]. \tag{4.65}$$

These equations can be transformed into an autonomous system by the price of the system dimension, as it is considered in [13],[84]:

$$\begin{aligned} \frac{d\mathbf{r}_{e,h}}{d\tau_{e,h}} &= \mathrm{Re}\left[\frac{\mathbf{E}(\mathbf{r}_e, \tau_e)}{|\mathbf{E}(\mathbf{r}_e, \tau_e)|}\left(\frac{\partial s_e}{\partial \tau_e}\right), \frac{\mathbf{H}(\mathbf{r}_h, \tau_h)}{|\mathbf{H}(\mathbf{r}_h, \tau_h)|}\left(\frac{\partial s_h}{\partial \tau_h}\right)\right], \\ \frac{d\tau_{e,h}}{dt_{e,h}} &= 1. \end{aligned} \tag{4.66}$$

where τ is a new quasi-spatial variable. Unfortunately, according to the best knowledge of the Author, this system or a similar one, given for the time-dependent EM fields, has not been studied, although some ideas were proposed on the modulation of the 4-D phase of this system [84]. Some results on the visualization of radiated time-dependent fields can be found in papers on antennas, for instance [16],[18],[19].

To develop a topological theory of transient EM fields, the results of research performed in the area of general steady and time-dependent vector fields and their visualizations by topological means can be used [23]-[26],[85]-[98]. Taking into account the importance of this approach for the creation of a topological theory of the EM initial-boundary problems, a short review of the above-cited papers is given below.

The techniques, developed for study of steady vector fields, are important in this case, and the plane fields are the most studied. As was mentioned, they are described by autonomous 2-D dynamical systems, and since the Andronov's time [13], the visualization tools have been developed, including the software on the automatic composition of topological schemes of vector fields [23]-[26]. It is mentioned on the essential reducing of the analysis time of vector fields described using topological means.

Topological structures of 3-D vector fields have been less studied even now. Some interesting results are from [14],[15],[26],[87],[88], but, again, the qualitative theory of them touches the equilibrium points, separatrices and topological charts as was mentioned earlier in this Chapter. Very often, in this case, the separatrices are represented by surfaces, which hide each other, and the attractiveness of the topological schemes for visual analysis of topologically rich fields starts to be poor. In this case, the vector surfaces can be represented by several separatrix curves called the saddle connectors [87],[88]. Additionally, it allows decreasing the required memory required for representations of multiple surfaces of separatrices.

Non-stationary vector field visualization and the field topological analysis is still a challenging problem. The ideas can be borrowed from the general theory of differential equations and hydrodynamics where the stream, path, and streak lines are used to describe a flow or motion of an individual particle. Following to [88], for instance, consider the mathematical apparatus used to simulate the vector fields and to calculate the components of the field's topological schemes or skeletons.

The streamlines are described by a dynamic system at any fixed moment of time t_0:

$$\frac{d\mathbf{r}(\tau)}{d\tau} = \mathbf{v}(\mathbf{r}(\tau), t_0). \tag{4.67}$$

where $\mathbf{v}$ is the time-dependent vector field and τ is the particle self-time. To find the time dependence of the field maps, they should be calculated for a time series and the dependence of topological schemes on the time parameter t_0 can be studied as the parametrical bifurcations of the field maps.

Additionally, the non-steady flows can be described by the path lines, which are the trajectories of a massless particle in a vector field, and their equation is:

$$\frac{d\mathbf{r}}{dt} = \mathbf{v}(\mathbf{r}(t), t). \tag{4.68}$$

It is transformed to an autonomous system of equations of increased dimension and some methods of qualitative study can be applied

$$\begin{aligned} \frac{d\mathbf{r}}{dt} &= \mathbf{v}\left(\mathbf{r}(t), t\right) \\ \frac{dt}{dt} &= 1. \end{aligned} \tag{4.69}$$

The equations (4.68) and (4.69) correspond to the EM ones (4.65) and (4.66), respectively. The equations (4.65) and (4.69) have no zeroes of equilibrium points, and the known methods of qualitative study of dynamical systems are not applied in this case.

The equation (4.68) is transformed to a 4-D autonomous one for which the conventional topological study is applicable. The derived vector quantity is called the feature flow field $\mathbf{f}(x, y, z, t)$, and its streamline equation is [23]

$$\frac{d}{dt}\begin{pmatrix} x \\ y \\ z \\ t \end{pmatrix} = \mathbf{f}(x, y, z, t)\begin{pmatrix} \det(\mathbf{v}_y, \mathbf{v}_z, \mathbf{v}_t) \\ \det(\mathbf{v}_z, \mathbf{v}_t, \mathbf{v}_x) \\ \det(\mathbf{v}_t, \mathbf{v}_x, \mathbf{v}_y) \\ \det(\mathbf{v}_y, \mathbf{v}_z, \mathbf{v}_t) \end{pmatrix} \tag{4.70}$$

where $\mathbf{v}_{x,y,z,t}$ are the corresponding derivatives of vector $\mathbf{v}$. The feature flow field allows tracking the equilibrium points of the time-dependent field $\mathbf{v}$, but the separatrices and their time-evolutions cannot be calculated by this technique. For the electric and magnetic fields, the similar feature flow equations can be composed if they are differentiable.

The topological analysis of the considered equations is to calculate the equilibrium points, separatrices, and other elements and to compose the topological schemes and their bifurcations depending on the given parameters. The next important is the visualization of these 3-D or 4-D objects, and the developments of the analyzing techniques for this topologically organized data. For instance, the field maps can be sliced in time, and the equilibrium points of each slice are connected to each other. These tracking technique gives clear representation of the time-related bifurcation if the dynamical systems (4.65) and (4.68) are studied [94]. Similar techniques on tracking topology elements of scalar potential maps are used in [96].

Unfortunately, in the non-stationary electromagnetism, no any systematic topological study has been developed, according to the best knowledge of the Author, and such a research is in the near-future plans of the Author's group. Additionally to the topological study of non-stationary fields, an interest is in the development of a topological theory of the initial-boundary problems. Topology should not only help in analyses, but in the creation of more effective analytical and numerical methods of field computations, similarly as it has been done partly for the problems of the harmonic EM fields in this book.

4.3 Topological Theory of the EM Field with the Differential Forms

Previous treatment of boundary value problems was based on the representation of fields by vectors and the Maxwell equations in their metric form. The spatial characteristics, which are independent on the geometry, are derived analyzing the field metric-dependent field-force line maps. As was mentioned, this approach was considered as a complementary tool to the known analytical and numerical EM methods.

Another approach dealing with the topological representations of the field quantities and geometrical relations of them was developed by applications of the external differential forms, which theory was proposed by Grassmann in the 19th Century. It was applied to electromagnetism by G. A. Deshamps [29], D. Baldomir and P. Hammond [30], I.V. Lindell [31], P.W. Gross and P.R. Kotigua [32], F. H. Hehl, F.H. Hehl and Yu. Obukhov [33], W.L. Engl [99], K.F. Warnik and co-authors [100], R.M. Kiehn [101], D.H. Delphenich [102], et al.

The idea consists in representation of the EM quantities by the differential forms, which are the generalization of vectors. Consider the Maxwell equations in their integral form:

$$\begin{aligned}
&\oint_L \mathbf{H}\cdot d\mathbf{l} = \frac{d}{dt}\int_S \mathbf{D}\cdot \mathbf{ds} + \int_S \mathbf{J}\cdot \mathbf{ds},\\
&\oint_L \mathbf{E}\cdot d\mathbf{l} = -\frac{d}{dt}\int_S \mathbf{B}\cdot \mathbf{ds},\\
&\oint_\sigma \mathbf{D}\cdot \mathbf{d\sigma} = \int_V \rho dv,\\
&\oint_\sigma \mathbf{B}\mathbf{d\sigma} = 0
\end{aligned} \tag{4.71}$$

where S is a surface inside a closed curve L, and V is a volume inside a closed surface σ. Formally, these integral relations do not depend on the coordinate system, and they can be re-written at each point of space. For this, one can consider $A = \mathbf{E}d\mathbf{l}$ and re-write this expression through the components which are parallel to the ones of the vector $d\mathbf{l}$ at the given point:

$$\mathrm{E} = e_1 dx + e_2 dy + e_3 dz. \tag{4.72}$$

Then $e_{1,2,3}$ do not depend on the global system of coordinates, and they can be considered as of the topological origin. The product of integration E depends on the value of $\mathbf{E}$ at each point, and it can be shown by a surface in 3-D space. In the case of static electric field, E corresponds to an equipotential surface described by its tangential vectors. It means that a field-force line is substituted by an equivalent infinitesimally small surface with its tangential vectors.

An arbitrary expression $\mathrm{E} = e_1 dx + e_2 dy + e_3 dz$ is a 1-form, and $e_{1,2,3}$ are its components. Similarly, a 1-form for magnetic field intensity is introduced: $\mathrm{H} = h_1 dx + h_2 dy + h_3 dz$.

The vectors $\mathbf{B}$ and $\mathbf{D}$ are represented by 2-forms because they are integrated over the oriented infinitesimal surfaces $\mathbf{ds}$ or $\mathbf{d\sigma}$:

$$\begin{aligned} \mathrm{B} &= b_1 dx \wedge dz + b_2 dz \wedge dy + b_3 dx \wedge dy, \\ \mathrm{D} &= d_1 dx \wedge dz + d_2 dz \wedge dy + d_3 dx \wedge dy. \end{aligned} \tag{4.73}$$

The used wedge products allow to reflect that the surfaces $\mathbf{ds}$ or $\mathbf{d\sigma}$ are oriented, and the product of a vector and an oriented surface has a certain sign which depends on the mutual orientation of this vector and surface. Due that, the wedge or exterior product is defined as an antisymmetric value which sign depends on the order of composing components. Then the 2-forms, similarly to 1-forms, have a certain direction in space. The 2-form for current $\mathbf{J}$ is defined as:

$$\mathrm{J} = j_1 dx \wedge dz + j_2 dz \wedge dy + j_3 dx \wedge dy, \tag{4.74}$$

Additionally, the full charge Q is represented by its 3-form:

$$\mathrm{Q} = \rho dxdydz. \tag{4.75}$$

To all considered quantities, certain geometrical shapes can be put into correspondence. For instance, the field intensity forms correspond to the surfaces, field flux quantities are visualized by tubes, and 3-forms of charges and currents corresponds to boxes or equivalent spheres. It means that the EM field has certain global topology independently on its origin. The Maxwell equations should define the correspondence of these topological quantities. There are two ideas. First of them is to establish these relations by a metric-independent way, and the derived relations are the topological Maxwell equations. Some ideas on it were proposed by Delphenich in [102].

Another way to write the Maxwell equations is to use an external derivation operator d :

$$\mathrm{d} \equiv \frac{\partial}{\partial x} dx + \frac{\partial}{\partial y} dy + \frac{\partial}{\partial z} dz. \tag{4.76}$$

This operator maps a n-form to a $(n+1)$-form.

Finally, the Maxwell equations in their topological form are:

$$\begin{aligned} &\mathrm{dH} = \frac{\partial}{\partial t}\mathrm{D} + \mathrm{J}, \\ &\mathrm{dE} = -\frac{\partial}{\partial t}\mathrm{B} \\ &\mathrm{dD} = \mathrm{Q},\ \mathrm{D} = \varepsilon_0 \circ \mathrm{E}, \\ &\mathrm{dB} = 0,\quad \mathrm{B} = \mu_0 \circ \mathrm{H} \end{aligned} \tag{4.77}$$

where the symbol $\circ$ is for the Hodge operator establishing the isomorphism between the 1-forms and 2-forms.

Additionally to the Maxwell equations, the wave ones can be written using the differential forms [31]. The boundary conditions for differential forms are considered in many books and papers, and the boundary value problems formulated for the Maxwell equations and the wave ones [31],[33],[100]. Unfortunately, the analytical power of this theory on calculation of parameters of EM components has not been demonstrated in full, yet.

The Maxwell equations, written for the medium with the memory, are described by fractional differential forms in [103]. The variational functionals written regarding to the differential forms and some computational aspects are considered in [32]. The use of differential forms for creation of highly versatile computational EM codes is considered in [32],[104],[105]. In [106], it is communicated on the created code FEMSTER- an object class library of discrete differential forms. The used in this software apparatus allows to write the EM equations in an elegant way and to use the same formulation for the time-dependent and the static Maxwell equations. Differential geometry allows using in a simple way the finite element of arbitrary order and it reduces the required memory usage and simulation time. For instance, the differential forms allow in a simple way to describe the propagation of EM waves in the metamaterials with continuous transformation of the EM metric. These materials are prospective for the EM cloak blueprints, and one application and theory are described in [107] where the reflection-less waveguide bends are simulated using the differential form formalism and an FTDT method.

Concluding this short review on the applications of differential forms in electromagnetism, it is necessary to state that the potential of this theory is rather far from to be used in full, and more research is required in spite of several decades have been passed since the first publications on this and related topics.

4.4 Topological Electromagnetism of Open-space Radiated Field

Another theory, origin of which related to 1989, is the topological electromagnetism of A. F. Ranada who put into the correspondence to the magnetic and electric field-force lines the level curves of a pair of complex scalar fields ϕ and θ, respectively [108]. The EM equations regarding to these complex fields ϕ and θ are nonlinear, and the standard Maxwell equations are the linearization of them by change of variables.

In this theory, the EM field, namely, the dual Faraday tensors $F_{\mu\nu}$ and $G_{\mu\nu}$ are expressed through these complex scalar functions as

$$F_{\mu\nu} = \frac{\sqrt{a}}{2\pi j} \frac{\partial_\mu \phi^* \partial_\nu \phi - \partial_\nu \phi^* \partial_\mu \phi}{\left(1+\phi^* \phi\right)^2}, \tag{4.78}$$

$$G_{\mu\nu} = \frac{\sqrt{a}}{2\pi j} \frac{\partial_\mu \theta^* \partial_\nu \theta - \partial_\nu \theta^* \partial_\mu \theta}{\left(1+\theta^* \theta\right)^2} \tag{4.79}$$

where a is a constant to adjust the dimensions for the EM field. The duality of these tensors are expressed as

$$G_{\mu\nu} = \frac{1}{2} \varepsilon_{\mu\nu\alpha\beta} F^{\alpha\beta}, \ F_{\mu\nu} = -\frac{1}{2} \varepsilon_{\mu\nu\alpha\beta} G^{\alpha\beta} \tag{4.80}$$

where $\varepsilon_{\mu\nu\alpha\beta}$ is the Levi-Civita symbol.

It was shown that if these tensors satisfy two first Maxwell equations

$$\varepsilon^{\alpha\beta\gamma\delta} \partial_\beta F_{\gamma\delta} = 0, \ \varepsilon^{\alpha\beta\gamma\delta} \partial_\beta G_{\gamma\delta} = 0 \tag{4.81}$$

and two second ones:

$$\partial_\alpha F^{\alpha\beta} = 0, \ \partial_\alpha G^{\alpha\beta} = 0, \tag{4.82}$$

and the fields are orthogonal to each other $(\mathbf{E} \cdot \mathbf{B}) = 0$, then the complex functions ϕ and θ are dual:

$$\circ\phi^* \mathrm{d}\sigma = -\theta^* \mathrm{d}\sigma \tag{4.83}$$

where $\mathrm{d}\sigma$ the are 2-form on the complex surface. It is seen that the relations between ϕ and θ is the non-metric one, and some interesting consequences can be found from this fact.

To find the complex scalar functions ϕ and θ, an action integral should be written expressed through the tensors $F_{\mu\nu}$ and $G_{\mu\nu}$:

$$I = -\frac{1}{4} \iint \left[\mathrm{F}(\phi) \circ {}^*\mathrm{F}(\phi) + {}^*\mathrm{F}(\theta) \circ \mathrm{F}(\theta) \right] \tag{4.84}$$

where $\mathrm{F}(\phi) = -\sqrt{a}\phi^* \mathrm{d}\sigma$, ${}^*\mathrm{F}(\phi) = \sqrt{a}\theta^* \mathrm{d}\sigma$, $\mathrm{F}(\theta) = -\sqrt{a}\theta^* \mathrm{d}\sigma$, and ${}^*\mathrm{F}(\theta) = \sqrt{a}\phi^* \mathrm{d}\sigma$. Its first variation gives, in general, the nonlinear equations regarding ϕ and θ. It was found that (4.84) has multiple solutions discretely different from each other, and this difference is in a spatial shapes of the family of the field-force lines. A particular solution of (4.84) is given below [109]:

$$\mathbf{B}_{n=1}(\mathbf{r},t)=\frac{\sqrt{a}}{2\pi j\left(1+\phi\phi^*\right)^2}\nabla\phi\times\nabla\phi^*,$$
$$\mathbf{E}_{n=1}(\mathbf{r},t)=\frac{\sqrt{a}}{2\pi j\left(1+\theta\theta^*\right)^2}\nabla\theta^*\times\nabla\theta,$$
$$\phi_{n=1}=\frac{(AX-TZ)+j(AY+T(A-1))}{(AZ+TX)+j(A(A-1)-TY)},$$
$$\theta_{n=1}=\frac{(AY-T(A-1))+j(AZ+TX)}{(AX-TZ)+j(A(A-1)-TY)} \quad (4.85)$$

where X,Y,Z,T are the normalized coordinates, $A=\left(R^2-T^2+1\right)/2$, and $R=\sqrt{X^2+Y^2+Z^2}$. These solutions can be labeled by a topological constant n, and it is a linking number of the electric and magnetic field-force lines. It was found that the field-force lines of the radiated field compose the EM knots, and the time evolution does not change the linking number, i.e. a 3-D package is moving preserving the spatial topology of its field-force lines.

This curious feature of the radiated fields can be used to the confining and movement of matter. For instance, in [110], it is studied the confining plasma by knots of the electric field-force line and the plasma stream lines. Different aspects of the linked and knotted beams of light and their evolutions in the space-time are studied in [111] where it is noticed on the future of these EM field configurations for trapping of atomic and colloidal particle matter.

Recently, a paper of F. Tamburiny and his colleagues has been published where two-channel communication by the field impulses of different topologies was realized through open space [112].

It follows that the ideas of topology, after more than 30 years of the research started to be realized in practice: from the solutions of the boundary problems of electromagnetism to telecommunications and computing, and this topic will be considered in the details in Chpt. 10.

References

[1] Maxwell, J.C.: A Treatise on Electricity and Magnetism. Dover (1954)

[2] Schwarz, A.S.: Quantum Field Theory and Topology. Springer (1993) (Translation from Russian 1989)

[3] Afanasiev, G.N.: Topological Effects in Quantum Mechanics. Kluwer Academic Publ. (1999)

[4] Nash, C., Sen, S.: Topology and Geometry for Physicists. Academic Press (1983)

[5] Nakahara, M.: Geometry, Topology and Physics. IOP Publ. (2003)

[6] Eshrig, H.: Topology and Geometry for Physics. Lecture Notes, IWF Dresden (2008)

[7] Fraser, G. (ed.): The New Physics for the 21st Century. Cambridge University Press (2006)
[8] Akchurin, I.A.: Philosophy Problems of Particle Physics. Nauka Publ., Moscow (1994) (in Russian)
[9] Durach, M., Rusina, A., Stockman, M.I., Nelson, K.: Toward full spatiotemporal control on the nanoscale. Nanolett. 7(10), 3145–3149 (2007)
[10] Halas, N.J.: Connecting the dots: Reventing optics for nanoscale dimensions. PNAS 106(10), 3643–3644
[11] Huang, F.M., Zheludev, N.I.: Super-resolution without evanescent waves. Nanolett. 9(3), 1249–1254
[12] Gerke, T.D., Piestun, R.: Aperiodic volume optics. Nanoletters, Advance Online Publication, February 7 (2010)
[13] Andronov, A.A., Leontovich, E.A., Gordon, I.I., et al.: Qualitative Theory of Second Order Dynamical Systems. Halsted Press (1973)
[14] Bautin, N.N., Leontovitch, E.A.: Methods and Techniques of Qualitative Study of Dynamical Systems on Plane. Nauka Publ. (1991) (in Russian)
[15] Arrowsmith, D.K., Place, C.M.: Differential Equations, Maps and Chaotic Behaviour. Chapman and Hall (1992)
[16] Cole, R.W., Miller, E.K., Chakrabarti, S., et al.: Learning about fields and waves using visual electromagnetics. IEEE Trans., Educ. 33, 81–94 (1990)
[17] Gvozdev, V.I., Kouzaev, G.A.: A field approach to design of SHF 3D-ICs- a review. Zarubezhnaya Radioelektronika (Foreign Radio Electronics) (7), 29–35 (1990) (in Russian)
[18] Miller, E.K., Deadrick, F.A.: Visualizing near-field energy flow and radiation. IEEE Antennas Propag. Mag. 42, 46–54 (2000)
[19] Miller, E.K.: Electromagnetics without equations. IEEE Potentials, 17–20 (April/ (May 2001)
[20] Gvozdev, V.I., Kouzaev, G.A.: Field approach for CAD of microwave 3-D ICs. In: Proc. Conf. Microwave Three-Dimensional Integrated Circuits, Tbilisy, USSR, pp. 67–73 (1988) (in Russian)
[21] Kouzaev, G.A.: Mathematical fundamentals of topological electrodynamics and the three-dimensional microwave integrated circuits' simulation. In: Electrodynamics and Techniques of Microwaves and EHF, MIEM, pp. 37–44 (1991) (in Russian)
[22] Gvozdev, V.I., Kouzaev, G.A.: Physics and field topology of 3-D microwave circuits. Soviet Microelectronics 21, 1–17 (1992)
[23] Helman, J., Hesselink, L.: Representation and display of vector field topology in fluid data sets. IEEE Comp. 22, 27–36 (1989)
[24] Lee, W., Kuipers, B.J.: A qualitative method to construct phase portraits. In: Proc. 11th National Conf. Artificial Intelligence, AAAI 1993 (1993)
[25] Nishida, T.: Grammatical description of behaviours of ordinary differential equations in two-dimensional phase space. Artificial Intelligence 91, 3–32 (1997)
[26] Weinkauf, T., Theisel, H., Hege, H.-C., et al.: Topological construction and visualization of higher-order 3D vector fields. In: Proc. Eurographics Symp., pp. 469–478 (2004)
[27] Ranada, A.F.: A topological theory of the electromagnetic field. Lett. Math. Phys. 18, 97–106 (1989)
[28] Kähler, E.: Bemerkungen über die Maxwellschen gleichungen. Abh. Math. Seminar Univ. Hamburg 12(1), 1–28 (1938)

[29] Deshamps, G.A.: Electromagnetics and differential forms. IEEE Proc. 69, 676–696 (1981)

[30] Baldomir, B., Hammond, P.: Geometry of Electromagnetic Systems. Oxford Sci. Publ. (1995)

[31] Lindell, I.V.: Differential Forms in Electromagnetics. IEEE Press/Wiley Interscience (2004)

[32] Gross, P.W., Kotigua, P.R.: Electromagnetic Theory and Computations: A Topological Approach. Cambridge University Press (2004)

[33] Hehl, F.W., Hehl, F.W., Obukhov, Y.N.: Foundation of Classical Electrodynamics. Birchhauser Publ. (2003)

[34] Barrett, T.W.: Topological Foundation of Electromagnetism. World Scientific (2008)

[35] Gvozdev, V.I., Nefedov, E.I.: Three-Dimensional Microwave Integrated Circuits. Nauka Publ., Moscow (1985) (in Russian)

[36] Gvozdev, V.I., Kouzaev, G.A., Nefedov, E.I., et al.: Physical principles of the modeling of three-dimensional microwave and extremely high-frequency integrated circuits. Soviet Physics-Uspekhi 35, 212–230 (1992)

[37] Kouzaev, G.A.: Basics of topological electromagnetics and its applications, Plenary lecture. In: Proc. 8th Int. Conf. Applications of El. Eng., Houston, USA, April 30-May 2, pp. 195–198 (2009)

[38] Gvozdev, V.I., Kouzaev, G.A.: Topological schemes of the electromagnetic field and design applications. In: Proc. Conf. Theory and Math. Modeling of Microwave 3D ICs, Alma-Ata, USSR, pp. 29–41 (1989) (in Russian)

[39] Kouzaev, G.A.: Symmetry of electromagnetic fields in microwave three-dimensional integrated circuits. In: Proc. Conf. Math. Modeling and CAD of Microwave Three-Dimensional Integrated Circuits, Tula, USSR, pp. 60–79 (1990) (in Russian)

[40] Gvozdev, V.I., Kouzaev, G.A.: CAD for three-dimensional integrated circuits based on the topological approach. In: Bunkin, B.V., Gridin, V.N. (eds.) Intelligent Integrated CAD Systems for Radio-Electronic Devices and LSI, pp. 105–111. Nauka Publ., Moscow (1990) (in Russian)

[41] Kouzaev, G.A.: Topological approach for eigenmodal problems. In: Proc. Conf. Mathematical Simulation and CAD of Microwave and Millimeter-Wave Integrated Circuits, Volgograd, USSR, pp. 25–26 (1991) (in Russian)

[42] Gvozdev, V.I., Kouzaev, G.A.: Topology and physics of electromagnetic fields in three-dimensional components of microwave integrated circuits. In: Proc. Conf. Math. Simulation and CAD of Microwave and Millimeter-Wave Integrated Circuits, Volgograd, USSR, pp. 27–29 (1991) (in Russian)

[43] Kouzaev, G.A., Utkin, M.I.: Topological analysis of the dominant mode field of an antipodal slot transmission line. In: Proc. Conf. Progress in Analog and Digital Integrated Circuits Based on Three-Dimensional Components, Tula, USSR, pp. 113–114 (1991) (in Russian)

[44] Kouzaev, G.A.: Convergence of solutions in topological electrodynamics. In: Proc. Conf. Techniques, Theory, Math. Modeling and CAD for Ultra-High-Speed Three-Dimensional Microwave Integrated Circuits, Moscow, vol. 2, pp. 238–241 (1992) (in Russian)

[45] Gvozdev, V.I., Kouzaev, G.A.: The similarity principle for simulation of three-dimensional microwave integrated circuits. In: Proc. Conf. Techniques, Theory, Mathematical Modeling, and CAD for Ultra-High-Speed Three-Dimensional Microwave Integrated Circuits, Moscow, vol. 1, pp. 127–134 (1992) (in Russian)
[46] Gvozdev, V.I., Kouzaev, G.A., Tikhonov, A.N.: New transmission lines and electrodynamical models for three-dimensional microwave integrated circuits. Sov. Physics-Doklady 35, 675–677 (1990)
[47] Gvozdev, V.I., Kouzaev, G.A., et al.: Topological models of the natural modes in coupled corner transmission lines. J. Commun. Techn. Electron (Radiotekhnika i Elektronika) 37, 48–54 (1992)
[48] Gvozdev, V.I., Kouzaev, G.A., et al.: Directional couplers based on transmission lines with corners. J. Commun. Techn. Electron (Radiotekhnika i Elektronika) 37, 37–40 (1992)
[49] Gvozdev, V.I., Kouzaev, G.A., et al.: Element base and functional components of three-dimensional integrated circuits of the microwave range. In: Modeling and Design of Devices and Systems of Micro- and Nanoelectronics, pp. 5–10. MIET Publ., Moscow (1994) (in Russian)
[50] Kouzaev, G.A., Deen, M.J., Nikolova, N.K., Rahal, A.: Cavity models of planar components grounded by via-holes and their experimental verification. IEEE Trans., Microwave Theory Tech. 54, 1033–1042 (2006)
[51] Arnold, V.I.: Geometrical Methods in the Theory of Ordinary Differential Equations. Springer (1988)
[52] Nikolskii, V.V.: Variational Methods for Inner Boundary Value Problems of Electrodynamics. Nauka Publ., Moscow (1967) (in Russian)
[53] Nikolskyi, V.V., Nikolskaya, T.I.: Electrodynamics and Wave Propagation. Nauka, Moscow (1987) (in Russian)
[54] Omar, A.S., Jensen, E., Lutgert, S.: On the modal expansion of resonator field in the source region. IEEE Trans., Microw. Theory Tech. 40, 1730–1732 (1992)
[55] Jerry, A.J.: The Gibbs Phenomenon in Fourier Analysis, Splines and Wavelet Approximations. Kluver Academic Publ. (1998)
[56] Bluman, G.W., Anco, S.C.: Symmetry and Integration Methods for Differential Equations. Springer (2002)
[57] Bocharov, A.V., Verbovezyi, A.M., Vongradov, A.M., et al.: Symmetry and the Conservation Laws of the Equations of Mathematical Physics. Factorial-Press (2005) (in Russian)
[58] Mansfield, E.L.: Noether theorem for smooth, difference and finite element systems. In: Foundations of Computational Mathematics, Santander 2005, pp. 230–254. Cambridge University Press (2006)
[59] Utsumi, Y.: Variational analysis of ridged waveguide modes. IEEE Trans., Microw. Theory Tech. 33, 111–120 (1985)
[60] Meixner, J.: The behavior of electromagnetic fields at edges. IEEE Trans., Antennas Propag. 20, 442–446 (1972)
[61] Leroy, M.: On the convergence of numerical results in modal analysis. IEEE Trans., Antennas Propag. 31, 655–659 (1983)
[62] Mittra, R. (ed.): Computer Techniques for Electromagnetics. Pergamon Press (1973)
[63] Yang, R., Omar, A.S.: Investigation of multiple rectangular aperture irises in rectangular waveguide using TE^{x}_{mn} -modes. IEEE Trans., Microw. Theory Tech. 41, 1369–1374 (1993)

[64] Gessel, G.A., Ciric, I.R.: Multiple waveguide discontinuity modeling with restricted mode interaction. IEEE Trans., Microw. Theory Tech. 42, 351–353 (1994)
[65] Lavrentiev, M.M.: Ill-posed Problems of Mathematical Physics and Analysis. Am. Math. Soc. (1986)
[66] Tikhonov, A.N., Arsenin, V.Y.: Solutions of Ill Posed Problems. V.H. Winston and Sons (1977)
[67] Leonov, A.S.: On quasi-optimal choice of the regularization parameter in the Lavrentiev's method. Siberian Math. J. 34(4), 117–126 (1993) (in Russian)
[68] Gvozdev, V.I., Kouzaev, G.A., Linev, A.A., et al.: Sensor for measurements of the permittivity of a medium in closed systems. Measurement Techniques 39, 81–83 (1996)
[69] Gvozdev, V.I., Kouzaev, G.A., Linev, A.A., et al.: Sensor for measurement of physical parameters. RF Patent, No 2057325 dated on, February 26 (1993)
[70] Conciauro, G., Guglielmi, M., Sorrentino, R.: Advanced Modal Analysis. J. Willey, New York (1999)
[71] Wang, C.Y.: Frequencies of a truncated circular waveguide - method of internal matching. IEEE Trans., Microw. Theory Tech. 48, 1763–1765 (2000)
[72] Okoshi, T.: Planar Circuits for Microwaves and Light Waves. Springer, Berlin (1985)
[73] Kouzaev, G.A., Nikolova, N.K., Deen, M.J.: Circular-pad via model based on cavity field analysis. IEEE Microw. Wireless Lett. 13, 481–483 (2003)
[74] Larsen, T.: On the relation between modes in rectangular, elliptical and parabolic waveguides and a mode classification systems. IEEE Trans., Microw. Theory Tech. 20, 379–384 (1972)
[75] Waterhouse, R.B., Targonski, S.D.: Performance of microstrip patches incorporating a single shorting post. In: Antennas Propag. Soc. Int. Symp., AP-S Dig., July 21-26, vol. 1, pp. 29–32 (1996)
[76] Yan, Z.: A microstrip patch resonator with a via connecting ground. Microw. Opt. Tech. Lett. 32, 9–11 (2002)
[77] Vaughan, R.G.: Two-port higher mode circular microstrip antennas. IEEE Trans. Antennas Propag. 36, 309–321 (1988)
[78] Clavijo, S., Dias, R.E., McKinzie, W.E.: Design methodology for Sievenpiper high-impedance surfaces: an artificial magnetic conductor for positive gain electrically small antennas. IEEE Trans. Antennas Propag. 51, 2678–2690 (2003)
[79] Lin, S.-C., Wang, C.-H., Chen, C.H.: "Novel patch-via-spiral resonators for the development of miniaturized bandpass filters with transmission zeros. IEEE Trans., Microw. Theory Tech. 55, 137–146 (2007)
[80] Razalli, M.S., Ismail, A., Mahd, M.A., et al.: Compact configuration ultra-wideband microwave filter using quarter-wave lengthy short-circuited stub. In: Proc. Asia-Pacific Microw. Conf. (2007)
[81] Mahajan, M., Khah, S.K., Chakarvarty, T., et al.: Computation of resonant frequency of annular microstrip antenna loaded with multiple shorting posts. IET Microw. Antennas Propag. 2, 1–5 (2008)
[82] Kamgaing, T., Ramahi, O.M.: Multiband electromagnetic-bandgap structures for applications in small form-factor multichip module packages. IEEE Trans., Microw. Theory Tech. 56, 2293–2300 (2008)
[83] Roumeliotis, J.A., Savaidis, S.P.: Cuttoff frequencies of eccentric circular-elliptic waveguides. IEEE Trans., Microw. Theory Tech. 42, 2128–2138 (1994)

[84] Kouzaev, G.A.: Communications by vector manifolds. In: Mastorakis, M., Mladenov, V., Kontargry, V.T. (eds.) Proc. Eur. Computing Conf. LNEE, vol. 1, 27, ch. 6, pp. 617–624. Springer (2009) (invited paper)
[85] Globus, A., Levit, C., Lasinski, T.: A tool for visualizing the topology of three-dimensional vector fields. In: Proc. IEEE Visualization 1991, pp. 33–40 (1991)
[86] Weinkauf, T., Theisel, H., Hege, H.-C., et al.: Topological constructions and visualization of higher order 3D vector fields. Computer Graphics Forum 23(3), 469–478 (2004)
[87] Theisel, H., Rössl, C., Weinkauf, T.: Topological representation of vector fields. In: De Floriani, L., Spagnuolo, M. (eds.) Shape Analysis and Structuring, pp. 215–240. Springer (2008)
[88] Theisel, H., Weinkauf, T., Hege, H.-C., et al.: On the applicability of topological methods for complex flow data. In: Topology Based Methods in Visualization, pp. 105–120. Springer (2007)
[89] Aigner, W., Miksch, S., Müller, W., et al.: Visual methods for analyzing time-oriented data. IEEE Trans., Vis. Comp. Graphics 14, 47–60 (2008)
[90] Marchesin, S., Chen, C.-K., Ho, C., et al.: View-dependent streamlines for 3D vector fields. IEEE Trans., Visualization and Comp. Graph. 16, 1578–1586 (2010)
[91] Theisel, H., Seidel, H.-P.: Feature flow fields. In: Proc. Joint EUROGRAPHICS-IEEE TCVG Symp. Visualization, pp. 141–148 (2003)
[92] Tricoche, X., Scheuermann, G., Hagen, H.: Topology-based visualization of time-dependent 2D vector fields. In: Proc. Vis. Symp., Data Visualization 2001, pp. 117–126 (2001)
[93] Peng, Z., Laramee, R.S.: Higher dimensional field visualization: a survey. In: EG UK Theory and Practice of Computer Graphics, The European Eurographics Association (2009)
[94] Theisel, H., Wenkauf, T., Hege, H.C., et al.: Stream line and path line oriented topology for 2D time-dependent vector fields. In: Proc. IEEE Visualization 2004, Austin, USA, pp. 321–328 (October 2004)
[95] Smiths, A.J., Lin, T.T. (eds.): Flow Visualization: Techniques and Examples. World Scientific (2011)
[96] Sohn, B.-S., Bajaj, C.: Time-varying contour topology. IEEE Trans., Visualization and Comp. Graph. 12, 14–25 (2006)
[97] Wegenkittl, R., Loeffelmann, H., Groeller, E.: Visualizing the behavior of higher dimensional systems. In: Proc. IEEE Conf. Visualization 1997, pp. 119–125 (1997)
[98] Datta-Gupta, A., King, M.J.: Streamline Simulation: Theory and Practice, Society of Petroleum Engineers (2007)
[99] Engl, W.L.: Topology and geometry of the electromagnetic field. Radio Science 19, 1131–1138 (1984)
[100] Warnick, K.F., Selfridge, R.H., Arnold, D.V.: Teaching electromagnetic field theory using differential forms. IEEE Trans., Microw. Theory Tech. 40, 53–68 (1997)
[101] Kiehn, R.M.: Topological evolution of classical electromagnetic field and the photon. In: Dvoeglazov, V. (ed.) Photon and Poincaré Group, pp. 246–262. Nova Sci. Publ. Inc. (1999)
[102] Delphenich, D.H.: On the axioms of topological electromagnetism. Annalen der Physik 14, 347–377 (2005)
[103] Baleanu, D., Golmankhaneh, A.K., Baleanu, M.C.: Fractional electromagnetic equations using fractional forms. Int. J. Theor. Phys. 48, 3114–3123 (2009)

[104] Nicolet, A., Remacle, J.-F., Meis, B., et al.: Transformation methods in computational electromagnetism. J. Appl. Phys. 75(10), 6036–6038 (1994)
[105] Texeira, F.L., Chew, W.C.: Lattice theory from topological viewpoint. J. Math. Phys. 40, 169–187 (1999)
[106] Castilo, P., Rieben, R., White, D.: FEMSTER: An object oriented class library of discrete differential forms. Lawrence Livermore National Laboratory Report. UCRL-JC-150238-ABC, The Internet
[107] Donderici, B., Texeira, F.L.: Metamaterial blueprints for reflectionless waveguide bends. IEEE Microw. Wireless Comp. Lett. 18, 233–235 (2008)
[108] Ranada, A.F.: Topological electromagnetism with hidden nonlinearity. In: Evans, M.W. (ed.) Modern Nonlinear Optics, Part 3. Advances in Chem. Physics, 2nd edn., vol. 119, pp. 197–252. J. Wiley & Sons, Inc. (2001)
[109] Donoso, J.M., Ranada, A.F., Trueba, J.L.: On electromagnetic models of ball lightning with topological structure (2003), arXiv:physics/0312122v1
[110] Irvine, W.T.M., Bouweester, D.: Linked and knotted beams of light. Nature Physics 4, 716–720 (2008)
[111] Arrayas, M., Trueba, J.L.: Motion of charged particles in an electromagnetic knot (January 27, 2010), arXiv:1001.4985v1 (math-ph)
[112] Tamburini, F., Mari, E., Sponselli, A., et al.: Encoding many channels on the same frequency through radio vorcity: first experimental test. New J. Phys. 14, 1–17 (2012)

5 Technologies for Microwave and High-speed Electronics

Abstract. This Chapter is an introductory review on the contemporary technologies of manufacturing of microwave and millimeter-wave integrated circuits and their packaging. Among them are the RF and microwave laminated printed circuit boards, microwave hybrid integrated circuits, and monolithic integrations. Some attention is paid to 3-D integrations realized by different technologies. This review allows for better understanding the typical geometries of integrated circuits which is important for EM engineers and designers. References -109. Figures -19. Pages -34.

Although this book is mainly on the applications of theoretical methods to the simulation and design of components and systems, the today's EM engineers need knowledge on contemporary possibilities of technology of manufacturing of RF, microwave and millimeter-wave integrations. This Chapter provides only the elementary basic information on them. For further reading, some books relating to these technologies can be interesting [1]-[22]. The history of microwave technology is considered in [17]-[20]. The technologies and components of nano-CMOS and future nano-IC electronics are highlighted in [22]-[25], for instance.

5.1 RF Printed Circuit Boards

The printed circuit board (PCB) technology, which had been developed before the World War II, was the first step towards the integrated electronics, and it was for manufacturing of military hardware of improved characteristics. Soon, the technology started to be very popular in civilian radio and TV hardware because it allowed for automatization of manufacturing process with the essential decrease of the handwork amount.

In the 50s, the developments of compact on-board radars stimulated the work on the design of microwave PCBs, and the studies of printed strip and microstrip lines were performed [11]. It was found the criticality of loss of the used PCB materials and poor accuracy of the used at that moment technologies.

Today a commercially attractive RF PCB material is the FR-4 [3],[5],[10] with $\varepsilon_r \approx 4.3-4.8$ and $\tan\delta \approx 0.017-0.02$. A PCB layer consists of a woven glass fiber material, which is impingent into an organic resin. The surfaces of a dried

G.A. Kouzaev: Applications of Advanced Electromagnetics, LNEE 169, pp. 187–220.
springerlink.com

FR-4 sheet are covered by glue and two copper layers of the 10-50-μm thickness. This metalized laminate or core layer is then pressed and dried in a vacuum chamber. It allows avoiding the air bubbles between the dielectric surface and the copper layer. Then copper can be patterned using the photolithography according to the designer needs.

Additional sheet of dielectric is called clad, and it is embedded between two core layers. An exploded view of a PCB stack is shown in Fig. 5.1 where some layers are used for artworks, and a couple of them are for grounding and power distribution.

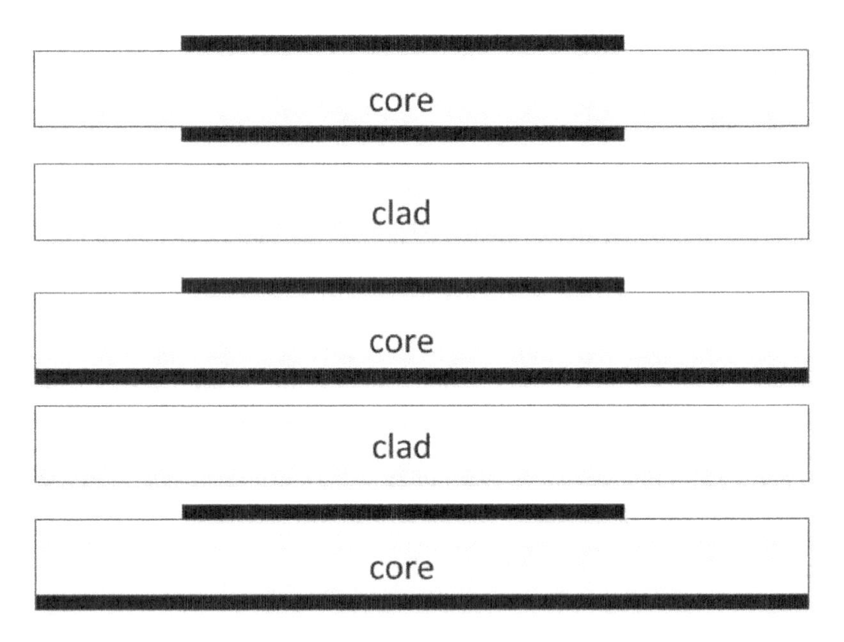

Fig. 5.1 Exploded view of a PCB stack

The via-holes, providing vertical interconnectivity, are manufactured using the mechanical, laser, or chemical etching [21]. Copper covers the walls of via-holes or fills them in full. The outer copper layers can be golden to prevent the copper's oxidation and caused by this degradation of the interconnect parameters. Several patterned layers are glued to each other, pressed, and fired in a vacuum chamber. Then these PCBs are populated by discrete components and microchips. Today this technology is highly automatized starting with the computer-aided design of PCBs and the following generation of codes for manufacturing of boards and populations of them by discrete components. The cost of PCBs is low, and this commercially attractive technology is used for many needs, including the wireless telecommunications. It is interesting to notice that the FR4 boards can be micromachined, and the limped components, microcircuits, optical cables are imbedded inside the PCB cores, which makes the boards more miniature and stacked in the vertical direction [26].

In microwaves, the main problem is with the increased loss of the FR-4 material and poor accuracy of interconnects. This technology has been adapted to high-frequency needs using better microwave dielectrics, which are considered below, in the Section 5.9.4. For instance, such a laminated material as the liquid crystal polymer (LCP) allows for working in frequencies over 100 GHz if the accuracy of used patterning technology is pertinent enough [27].

5.2 Microwave Hybrid ICs

The microwave hybrid ICs were started to be developed in the 50s of the last Century using the achievements of the low-frequency PCB technology [11]. These circuits integrate the distributed components which size is comparable with the wavelength, and the discrete or chip elements as the capacitors, resistors, inductors, semiconductors, and/or microchips, and this is the reason of the name of this technology.

The substrate material is a low-loss dielectric which permittivity is highly controllable and stable in time and regarding to the temperature variations. The laminated microwave materials and ceramics can be used in microwave hybrid integrations. The manufacturing of the microwave ICs based on the laminated materials is similar with the PCB process with only the increased requirements towards the geometry accuracy.

One of the technologies used with the ceramics, and, particularly, with the Alumina (Al_2O_3) substrates, is the thin-film technology [12]. The cross section of a microstrip line manufactured using this technology is shown in Fig. 5.2.

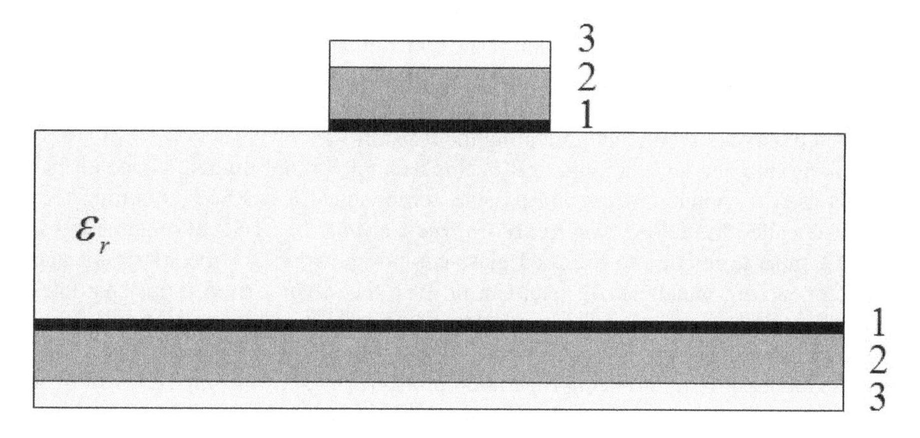

Fig. 5.2 Cross-section of a thin-film microstrip. 1- Ni/Cr layer; 2- Copper; 3- Gold

Initially, a substrate of Alumina consisting of 96-99.5% Al_2O_4 ($\varepsilon_r \approx 9.6-10.1$, $\tan\delta \approx 0.00015$) is manufactured. At millimeter-waves, other low-loss materials can be used with this technology, and one of them is the fused quartz.

Both surfaces of this sheet are well polished to avoid the unwanted roughness. A thin (100-200 nm) layer (1) of Cr/Ni is deposited in a vacuum chamber through a photomask on the surfaces of this substrate (Fig. 5.2). This layer has the increased adhesion to the ceramic. Then a copper layer (3-7 μm) is electroplated on these Cr/Ni spots (2), and they are covered by gold (3), which thickness is about of 1-2 μm that prevents the oxidation of copper. The surface of gold is polished to avoid the additional electron scattering increasing the conductor loss at microwaves. Usually, this technology has limitations on the minimal size of conductors and on the space between them (several tens of microns). Additionally to conductors, this technology allows for thin-film resistors deposited in vacuum chamber.

The manufactured plates are visually and electrically controlled, and they are populated by discrete components and microchips. The manufactured ICs are installed into a metallic box or its shielded section to avoid the EM cross talks.

5.3 Ceramic Thick-film Technologies

The above-described thin-film technology is rather expensive due to using the vacuum equipment and the electroplating. Thick film technology is with the screen-printing of conducting, resistive, and isolation materials in a paste form onto a ceramic substrate [3],[13]. The printed pasta is fired composing a conductor strip, and the firing temperature depends on the used materials. For instance, the conductors made of Mo, MoMn, and W particles need the firing temperature about of 1600°-1800° C. The low-firing temperature pastes are composed of Au, Ag, and Cu, and they are fired at 850°C. The thickness of fired conductors is about of 10-12 μm with the tendency to be reduced further. One of the disadvantages of this technology is with the multiple printing steps and firings. They are avoided in the *High Temperature Cofired Ceramic (HTCC)* technology when the conductor layers and the multi-layered ceramic plates are fired together at 1600°-1800° C making a 3-D module with the embedded interconnects, capacitors, inductors, and distributed passive components, excepting the resistors.

Taking into account the high cost of this technology and limited choice of materials used for interconnects and passive components, it has been modified to a new one called the *Low Temperature Cofired Ceramic* or *LTCC*. It requires lower firing temperatures due to the used glass-ceramic powders for the substrates and well-conducting metals as Ag, Au, Cu, or even the silver-coated copper particles for conductors.

A typical technology flow described here after the DuPont™ GreenTape™ one (www.dupont.com/mcm) is in preparation of a "green" substrate tape consisting of ceramic/glass particles and an organic binder. This tape is blanked into a multiple sheets marked by registration pins. According to the design file, the plates are mechanically punched to make the via-holes connecting the components placed on different layers. After optical inspection, these holes are filled by a conducting paste, and the substrate is dried.

The next technological step is with the printing of conductors through the screens manufactured according to a design file. A screen is pressed to the sheet surface, and a conducting paste, consisting of metal particles and a semi-liquid organic binder, fills the empty holes of the screen and covers this "green" ceramic sheet in these spots. After optical inspection, the sheets are collated, pressed, and co-fired using the programmable furnaces with the maximum temperature of 850°C. The cross-section of an LTCC module is shown in Fig. 5.3.

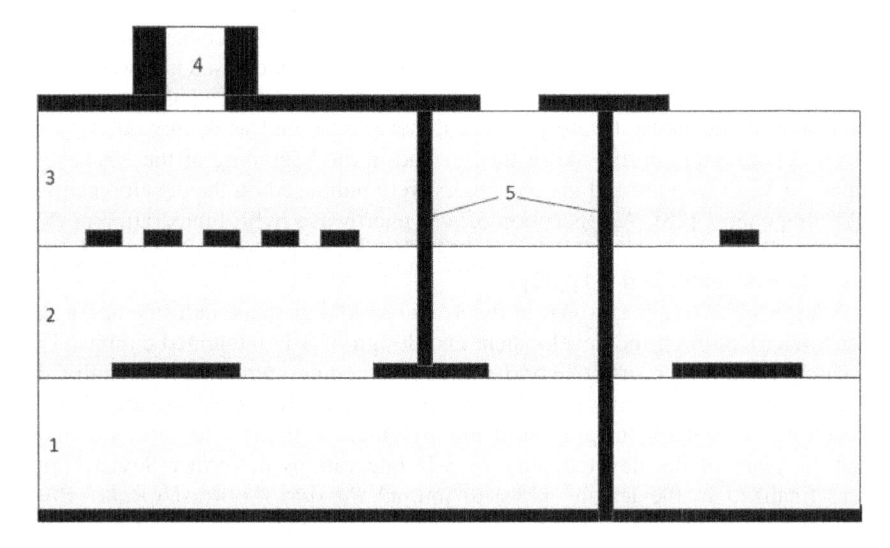

Fig. 5.3 Cross-section of an LTCC module (Conductors are shown in black). 1-3- Ceramic layers (can be of different permittivity); 4- Chip element; 5- Via-holes

Some additional components can be post-printed and post-fired on the surface of a co-fired module. Geometry of the components, which are placed on the outer surfaces of the module, can be trimmed by laser. All modules are inspected, electrically tested, and then they are populated by additional lumped passive and active components, including the integrated microcircuits.

Today many microwave and millimeter-wave LTCC modules are known with the competitive characteristics regarding to the thin-film ones [15]. A recent announcement is that a DuPont technology allows for manufacturing of the conductors and substrates workable in frequencies up to 100 GHz and beyond (DuPont™ GreenTape™ 9K7 Ceramic System), i.e. this technology starts to be competitive with the GaAs, LCP, and BCB (benzocyclobutene) based ones.

In addition to the acceptable electric performances, the co-fired modules have appropriate thermal properties, including the stable thermal coefficients of the resonant frequencies and the matched thermal-mechanical characteristics of the LTCC modules and the GaAs and Si chip components. These modules are distinguished by increased reliability in harsh environment, including towards vibrations and mechanical shocks.

5.4 Microwave Three-dimensional Hybrid Integrated Circuits

The initial microwave printed circuits were of the planar design in spite of their origin from the multilayer PCBs [11]. The microwave substrates are placed into separated by metallic walls cells to avoid the EM cross talks and to shield them from the environment. This design allows for the essential decrease of the mass and size of microwave modules compared to the waveguide based hardware.

While the hybrids matured, the limitations of this planar integration became clear, and the necessity to use the third dimension started to be obvious to engineers. For instance, in the end of the 70s and in the beginning of the 80s several papers of V.I. Gvozdev and his co-authors were published on the developments of 3-D components [28]. A conception of 3-D microwave hybrid integration or "volumetrical" integration is published in [29],[30]. Soon, a couple of monographs in this field were published [31],[32].

A large work was performed in the academia and airspace industry of the former Soviet Union on the development and design of 3-D integrated circuits. This research allowed for many hybrid integrated modules with the essentially decreased mass, volume, and improved microwave parameters. Especially, this approach was useful for integration of multi-element systems. The results of more than 10 years of the developments of 3-D integrations in former Soviet Union were finalized in the leading physical journal *Russian Physics-Uspekhi* (1992) [33], and they were awarded by several prizes from the USSR and Russia governments.

According to [33],[34], *a three-dimensional microwave IC is a multilevel device with planar or 3-D-components placed in the module's volume and connected to each other electrically, electromagnetically, or quantum-mechanically.*

The first microwave hybrid 3-D modules integrated the distributed, planar thin-film, and lumped components into a multi-level module, and they were coupled to each other by the interconnects, the via-holes, and the EM 3-D transitions. The active lumped devices were connected to the ports placed on the outer planes of these 3-D modules. The mass and size reducing by 10-100 times was achieved due to the stacking of plates to a more compact module, which requires less heavy shield for the whole integrated module. Improving of electric characteristics was achieved due to the use of the optimal design of each planar or 3-D component. Besides, the 3-D design allowed for practically complete shielding of the integrated components. The strong broadside coupling of vertically integrated lines could help in the realization of wide bandwidth performance of components. The speed-action of digital microwave circuits was available using the shorted vertical interconnects and the EM transitions of improved characteristics [32].

Besides very promising results and many developed components and subsystems, some problems were left unsolved at those times. Among them were the technological problems with the stacking of laminates [35] and the EM simulation issues. The electromagnetism of 3-D microwave and millimeter-wave components is more complicated regarding to the planer ones, and the full-wave models of them are required for simulation and design.

A large work on the creations of EM models and methods for planar and 3-D microwave integrations was performed by E.I. Nefedov [36], V.A. Neganov [37], V.V. Nikolskyi [38]-[40], G.S. Makeeva [41], V.V Schestopalov [42], and many other researchers and engineers at those times in the former Soviet Union.

Both, the analytical and numerical methods were paid increased attention. For instance, it was found that better understanding was reachable if the EM field dynamics was studied in the details, and an approach on topological modeling of the field was proposed in [33]. It allows for operating the less-detailed EM field skeletons or topological schemes associated with the main EM features of the studied components, and this approach is promising for the development of fast component models. Later, this idea of operating of field skeletons was implemented for microwave signaling and digital computing.

Many scientists and researchers from different areas of electronics took part or supported the developments of microwave 3-D integrated circuits. Additionally to the above-mentioned V.I. Gvozdev and E.I. Nefedov, who were the initiators of the research, they are Eu.V. Armenskyi, G.M. Aristarkhov, D.V. Bykov, G.A. Kouzaev, A.A. Lebed, V.V. Litvinenko, I.V. Nazarov, E.D. Pozhidaev, V.S. Saenko, Yu.V. Schestopalov, V.V. Schevchenko, Yu.N. Schirokov, V.V. Sedlezkyi, V.A. Solntzev, V.V. Tchernyi, A.N. Tikhonov, V.S. Zhdanov, et al.

As an example, some of the passive microwave components studied by the Author with his colleagues are shown in Figs. 5.4-5.8 [43]-[65].

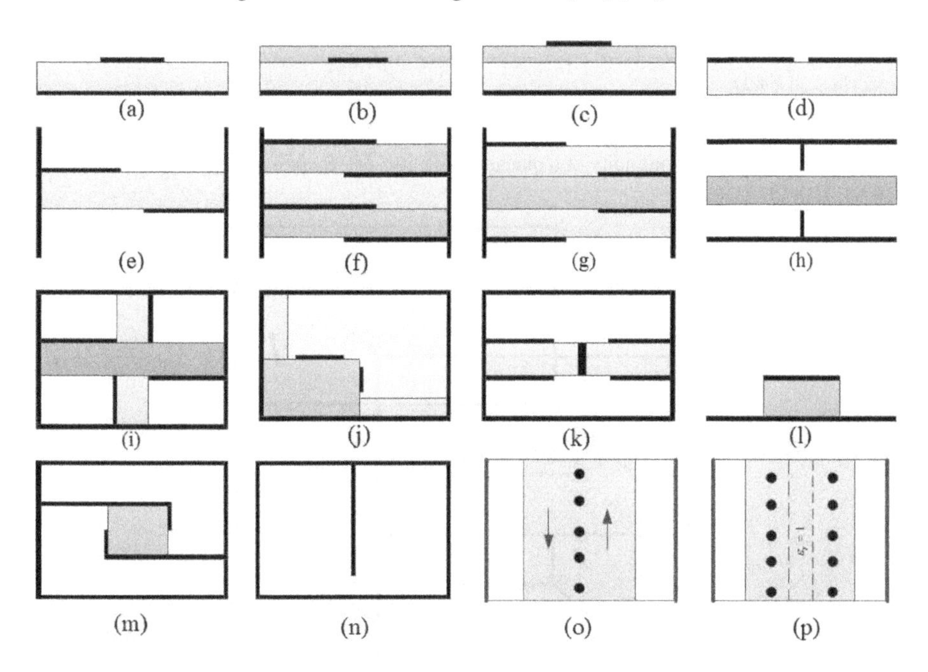

Fig. 5.4 Some transmission lines for hybrid planar and 3-D integrations studied by the Author of this book. (a-n)- Line cross-sections. (o, p)- Horizontal view of modeled fenced waveguides

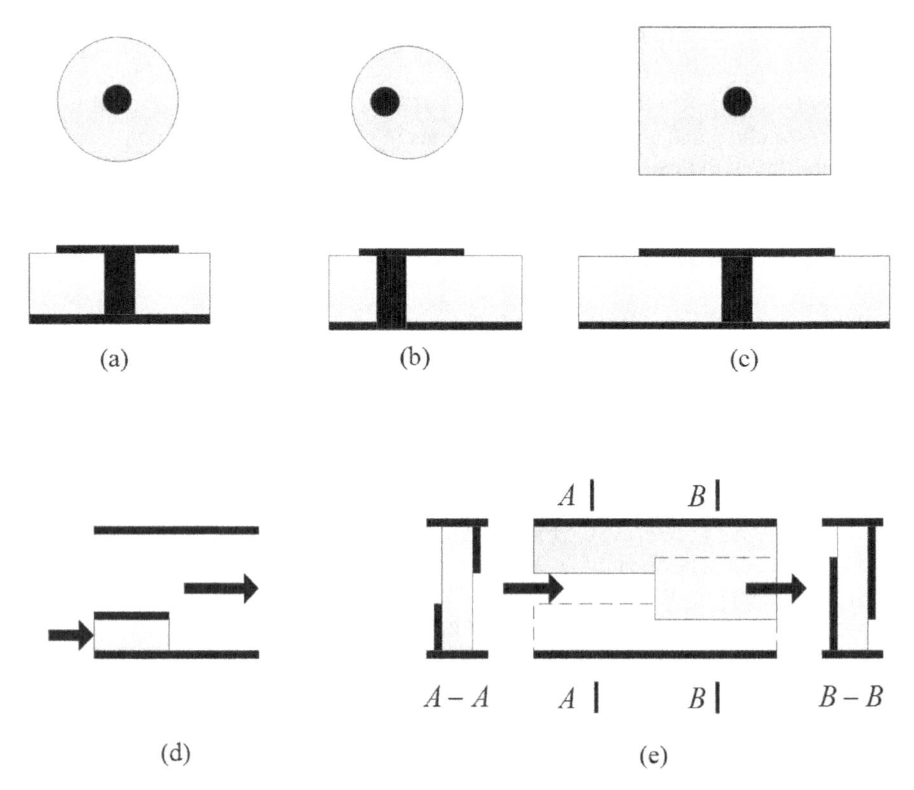

Fig. 5.5 Elementary components for hybrid 3-D integrations. (a) –(c) -Via-holes and shorted patch antennas; (d)- Discontinuity of a planar waveguide; (e)- Step of a balanced slot line

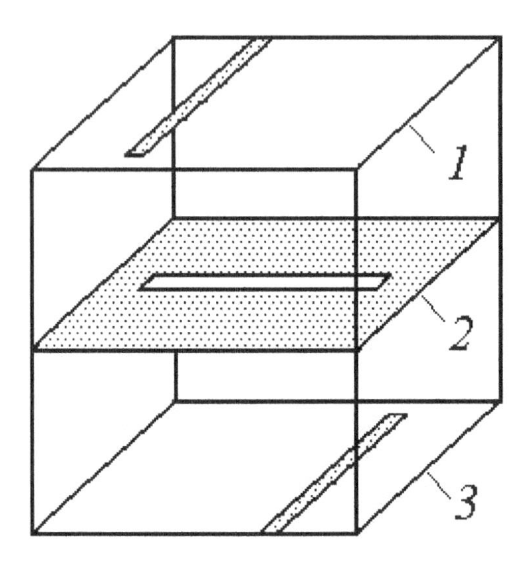

Fig. 5.6 Microstrip-slot line filter fragment. Adapted from [65]

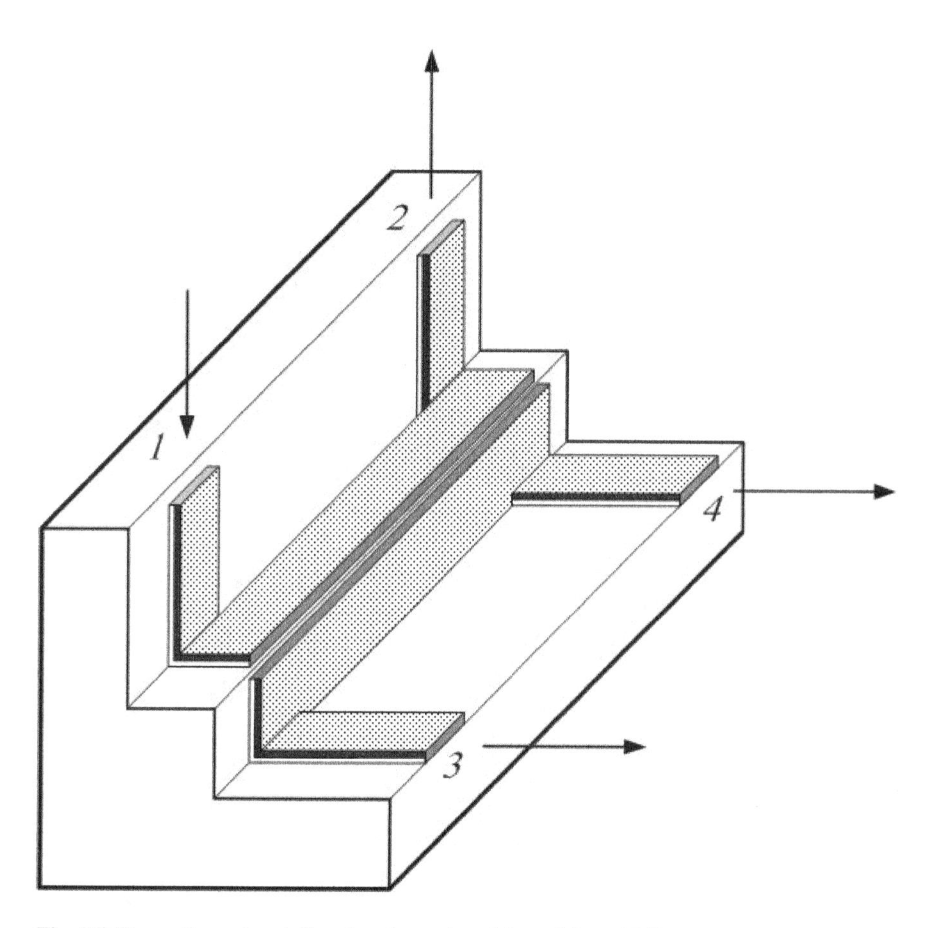

Fig. 5.7 Three-dimensional directional coupler. Adapted from [54]

Fig. 5.8 Three-dimensional power divider. Adapted from [50],[51]

Unfortunately, this promising research in Russia was terminated in the beginning of the 90s due to the collapse of the electronic industry of this country. Besides, several other objective reasons were at those times. Among them were the undeveloped 3-D technology, poor local industry prospective and the strong competition with the GaAs microcircuits.

The 3-D integration received a new breath with the invention of the multi-layer LTCC technology allowing for up to 6-60 conductor layers connected, as in the first 3-D hybrid integrations, by via-holes and EM transitions (Section 5.4). The integrated passive components are placed, similarly to the initial "volumetric" integrations, on different layers. As it was mentioned, the LTCC technology allows for mass-production of low-cost micro- and millimeter-wave modules for the civilian and military applications at frequencies up to 100 GHz.

Not so many years had been passed from the start of the mass-production of the first GaAs monolithic integrations, as the increased complexity of RF systems and high-cost of this material stimulated the developments of 3-D monolithic integrations in the middle of the 90s of the last Century. This architecture is now common and available for the commercial and research purposes from most foundries. Today's engineers and scientists are working on stacking of multiple semiconductor layers to improve the speed-action of digital processors and analogue circuits. Many scientists suppose that the future nano-integrations will be of the 3-D design [23]-[25].

5.5 Gallium Arsenide Monolithic Integrated Circuit Technology

The history of the GaAs integration technology is described in [17]. The initial works with this semiconductor material related to the 60s when gallium arsenide was used as a substrate semi-isolating material.

Many works were performed on the creation of GaAs discrete diodes and transistors that allowed for the millimeter-wave range of frequencies. In 1976, the first integrated GaAs amplifier was published by R.S. Pengelly and J.A. Turner. A GaAs power amplifier was developed by V. Sokolov et al. from Texas Instruments in 1979. Further designs were with a DARPA program on radar modules. Significant activity in this field was in Europe [20] and Japan.

Initially, these GaAs integrations were met with the increased skepticism of the engineering community at that time because the high cost of material and the low yield of production. Some concerns with the GaAs integrated circuits are up to now, and many efforts aims at the modification of GaAs technology for the needs of increased frequencies. For instance, the initial MESFET GaAs transistors, having low cut-off frequencies, were substituted by the pseudomorphic high electron mobility (HEMT) and the metamorphic (MHEPT) transistors working in frequencies over 100 GHz. The initial uniplanar architecture has been transformed into a multilayered design (Fig. 5.9).

A GaAs substrate is covered by multiple layers of an organic dielectric, for instance, polyimide, and the interconnect layers and planar passive components are placed between them [66]-[68]. The components are connected by via-holes and other vertically oriented elements. Some ideas on three-dimensional monolithic integrations and components on GaAs can be found in [69]. Today several other dielectrics are used in millimeter-wave GaAs integrations. For instance, the BCB substrates have acceptable properties up to 200-300 GHz. Some dielectrics, including the BCB, allow for micromachining, and the GaAs circuits can be equipped by the 3-D components placed over a surface of the substrate.

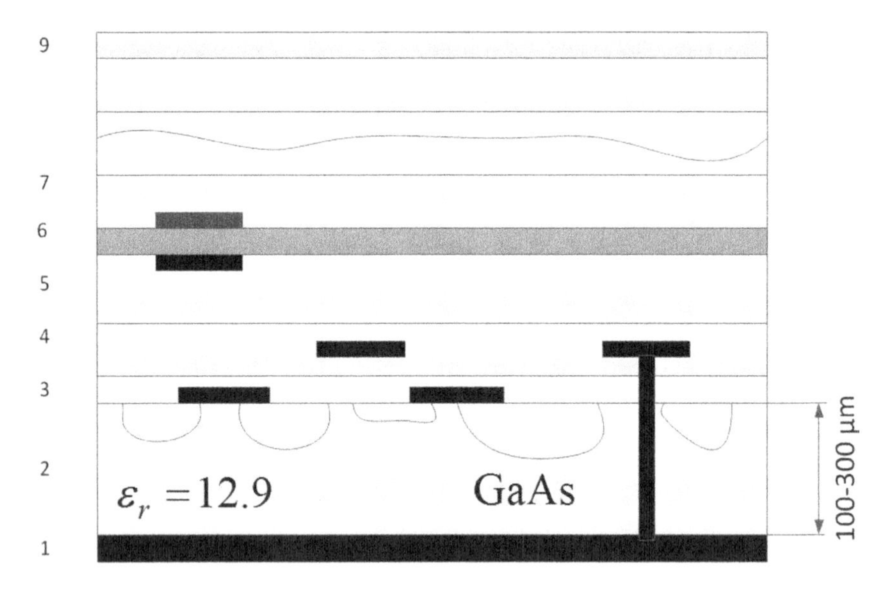

Fig. 5.9 GaAs microchip stack. 1- Ground layer (gold); 2- GaAs substrate; 3- Passivation layer (TiN); 4,5,7- Polyimide; 6- Increased permittivity dielectric (Si_3N_4); 9- Passivation layer (polyimide). Conductors (gold) are shown in black

5.6 Silicon Technologies for RF and Microwaves

Historically, the silicon technology was developed for the low-frequency and low-rate digital applications, and it has been matured well since the 60s. The initial limitations were with the upper frequencies of the bipolar transistors, and now they work in frequencies up to 20 GHz. A new "era" started with the invention of the CMOS and SiGe transistors, which upper frequency is in the terahertz region at this moment.

At the difference to transistors and diodes, the passive components are more conservative regarding to the improvements. Unfortunately, not all problems associated with them can be compensated by active components in full. For instance, the lossy inductors and interconnects define the poor signal-to-noise ratio (SNR) of the oscillators and amplifiers, and the best way is to enhance the above-mentioned passives. Unfortunately, this way requires strong knowledge of typical EM effects in the silicon-manufactured circuits and used materials at increased frequencies. Additionally, the passives' performance is affected by the chip architectonics. It is recognized, for example, that the today's problems in electronics are with the interconnects and the EM noise rather than with the transistors. Then these questions should be paid increased attention in literature.

A typical cross-section or a stack of a silicon chip is shown in Fig. 5.10. It consists of the doped grounded silicon substrate of the relative permittivity about of 11.7 and of the height of 100-300 μm, depending on the technology. The resistivity of this doped silicon varies within 5-50 Ω-cm , and the passive components placed close to its surface are very lossy. The low-loss silicon has its resistance greater than 1 KΩ-cm, and it is applicable in the 30-100-GHz band. The ion-implanted silicon shows resistance about of 1 MΩ-cm, and it is for the applications even beyond of 100 GHz.

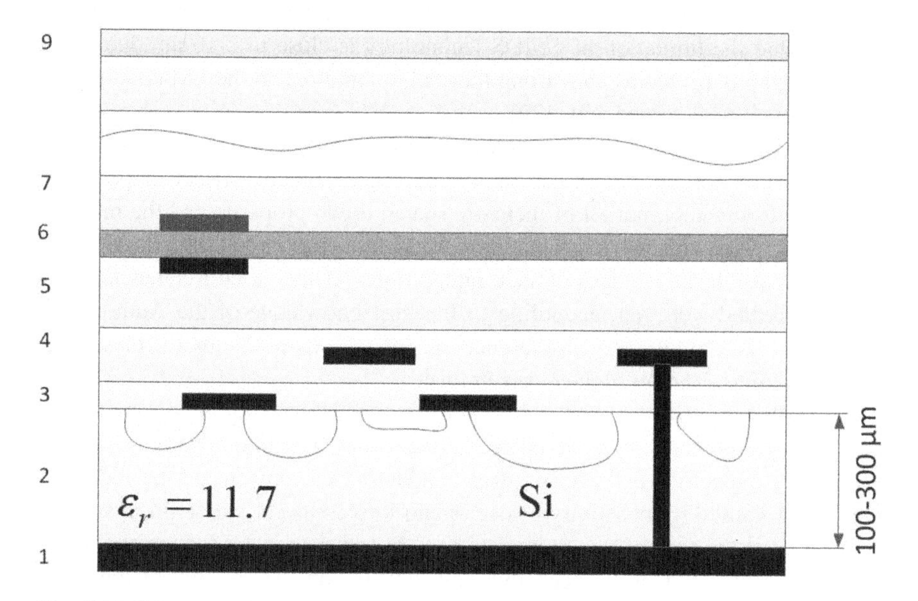

Fig. 5.10 Silicon chip stack. 1- Ground conductor (Al or Cu); 2- Silicon; 3- Passivation layer (SiO_2); 4,5,7- Dielectric layers; 6- MIM layer (Si_3N_4); 9- Passivation layer (SiO_2). Signal conductors (in black) are made of Al or Cu

The transistors are implemented into the upper layer of silicon. They are connected to each other by aluminum, gold, or copper wires that can be placed directly on the silicon surface. Additionally, it is a place for the implanted or mesa resistors, diodes, and varactors.

The substrate surface is covered by a passivation layer made of silicon dioxide SiO_2 $(\varepsilon_r = 4.5)$ or titanium nitride TiN $(\varepsilon_r = 6.5)$ to isolate other components from the lossy silicon. This passivated silicon can be completely isolated from the upper levels of the chip by a grounded metal layer of several micron height made of aluminum or copper.

Beside the silicon dioxide layers, the chips can have implemented several dielectric layers of increased permittivity made, for instance, of Si_3N_4 ($\varepsilon_r = 7.4$) for metal-insulator-metal (MIM) capacitors. To increase the capacitance, these layers can be made of $TiNbO_5$ films ($\varepsilon_r = 160-300$) of height 5-15 nm [70] announced recently by the International Center for Materials Nanoarchitectonics, Japan.

Depending on the technology and needs, a silicon chip can consist of 40 layers of dielectrics and conductors. For RF electronics, the most common technologies used nowadays are the 180-, 130-, 90-nm ones [71]. Today's silicon technology is in its nano-era, in-fact, when the 28-nm node has been reached recently, and intensive research is on the developments of 7-15-nm integration (Int. Sematech). It is supposed that the limitation of CMOS technology is close to 7-10 nm due to the photolithography problems and strong thermal, quantum, and the EM coupling effects influencing all components [22].

Scaling of micro- and millimeter-wave components down to the nanometric size (7-100 nm) formally decreases the typical high-frequency effects appeared in CMOS. Unfortunately, not all of them are scaled down properly, and the most serious effects are with the lossy interconnects, inductors, and parasitic EM and thermal noise. The advantages of such integrations for increased frequencies have not been studied well, yet, according to the best knowledge of the Author. It is predictable that the loss and interference are the serious problems for these integrations in micro- and millimeter-wave range.

The architecture of contemporary post-micron CMOS integration, practically, is identical to the above-considered GaAs one. Some attention is paid to the low-permittivity dielectrics and copper strip conductors to compensate the increased delay times caused by resistivity of deca-nano-interconnects. Some promising results are with the further developments of 3-D transistors. Unfortunately, many other results obtained in the academia and industry on the creation of this post-micron CMOS technology and the design tools are beyond the scope of this contribution. More information on the mentioned technology and prospective solutions for future nano-integrations can be found in [22]-[25].

The Si ICs' interconnects are of the strip, microstrip, and coplanar designs. Their wires are made of aluminum, copper, or gold. Their thickness is of several microns, and their performance is highly depended on the design, employed dielectrics, and used conductors. Additionally, the technological inaccuracy about of 10% makes hardly predicted behavior of the parameters of high-frequency components.

The major factor limiting the passive components performance at high frequencies is the semiconductor and dielectric loss. The non-perfect thin-film conductors are the cause for another trouble. The loss associated with them has rather complicated frequency behavior. Being explained in a short way, it is caused by the peculiarities of transformation of the EM energy into the chaotic movement of carriers and atoms of a metal. This movement is frequency dependent because the density of electron distribution is non-homogeneous due to the Lorentz effect. With the frequency, electrons are shifted towards the conductor surface. At high frequencies, the carriers are concentrated in a thin layer that is called the skin layer, and the conductor resistance is increasing with the frequency due to the growing concentration of carriers in this thin sheet. The depth of it is calculated analytically only in the case of the semi-infinitely thick metal slabs. In the wires of limited cross-section and in the tight environment, the current distribution is nonhomogeneous along the cross-section, and this factor essentially influences the accuracy of loss calculations. Due to that, the numerical simulations, which use a coarse discretization of the conductor cross-section, may have rather poor accuracy. Here the analytical models of transmission lines based on the accurate physical analysis are very important.

The microchips are the high-density integrations. At frequencies of millimeter-wave range, it leads to increased coupling of interconnects and severe cross-talks. Close to 100 GHz, the EM coupling is caused by excitation of dielectric modes and spatial radiation. All above-mentioned effects stimulate the analytical and numerical studies of passive components and whole integrations for smart design of ICs in frequencies of 30-100 GHz.

It is well recognized that the interconnects and passives are the bottleneck of contemporary electronics. To improve their bandwidth, different ideas and techniques are used. The most popular of them is the reconfiguration of the cross-section of a transmission line to optimize its characteristics using the low-cost techniques. For example, the three-dimensional design of silicon integrated circuits (Fig. 5.10) allows for a large variety of transmission lines. Some of them are shown in Fig. 5.11. It is the multilayer microstrip lines which conductors are separated from lossy silicon by a dioxide layer (Fig. 5.11a-c), the coplanar waveguide (Fig. 5.11d), the coplanar strip line (Fig. 5.11e), and the thin-film microstrip line (Fig. 5.11f).

The basic of them is a microstrip line on Si-SiO_2 substrate (Fig. 5.11a). Although, this line is known by increased loss even at low microwave frequencies, a short review of its waveguiding properties can explain the problems arisen in silicon integration. The signal conductor is placed over a two-layer substrate. One of these layers is a low resistivity $(5-20\ \Omega\text{-cm})$ silicon, and the second one is an isolating SiO_2 sheet. Additionally, this line can be covered by several dielectric layers, as seen in Fig. 5.10 and 5.11.

From the EM point of view, the fundamental mode of this line is a hybrid one, and it has all six components of the EM field. The properties of this mode depend on the geometry and parameters of the two-layer substrate and upper dielectric layers.

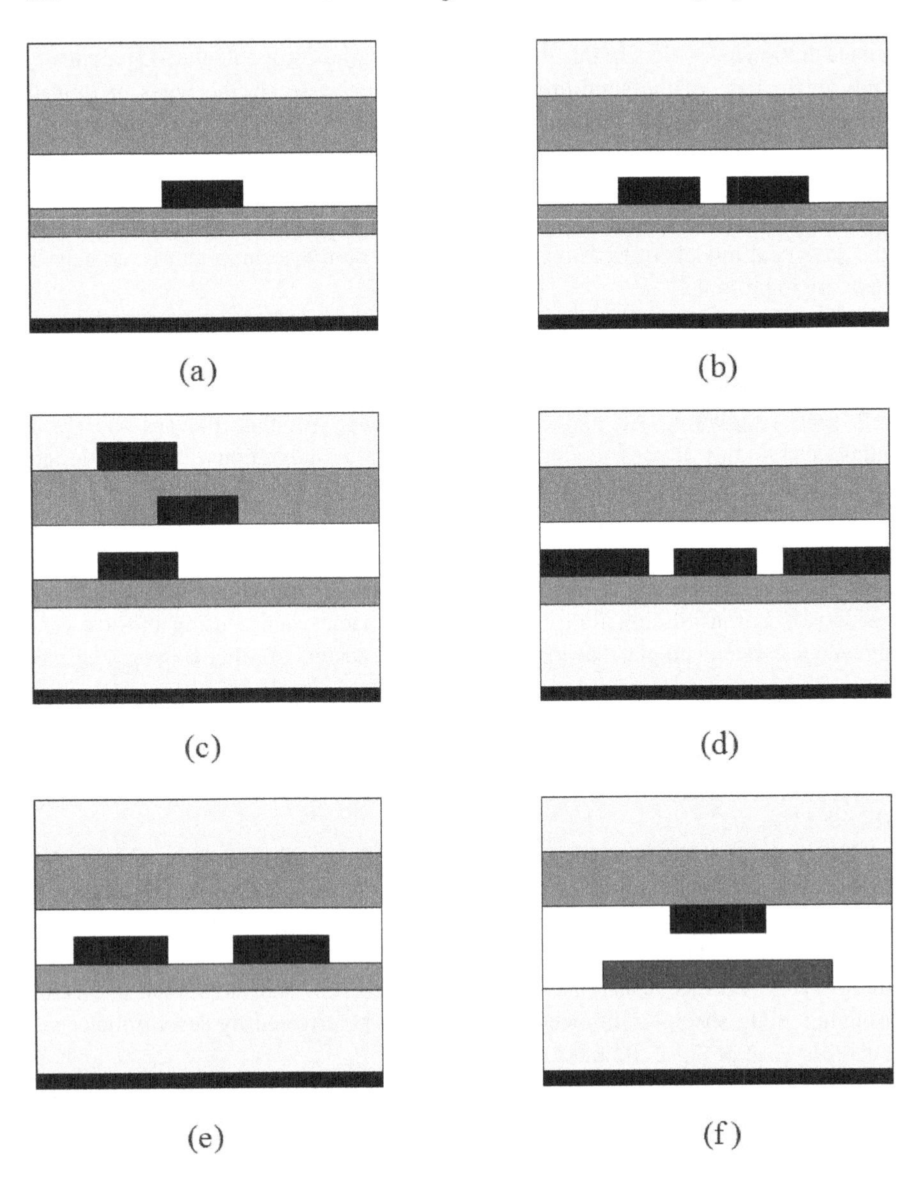

Fig. 5.11 Transmission lines used in silicon integrated circuits. (a)- Multilayer microstrip; (b)- Planar coupled multilayer microstrips; (c)- Broadside coupled microstrips; (d)- Coplanar waveguide (CPW), (e)- Coplanar strip lines (CPS); (f)- Thin-film microstrip line (TFML)

The first studies of this line are performed in [72],[73] where its parallel-plate model is considered allowing for an analytical theory. It was found that, depending on the frequency, geometry, and conductivity of the silicon substrate, this microstrip fundamental mode had its three forms.

At low frequencies and for high-resistivity substrates, the longitudinal components of the field are not strong, and the mode is *the quasi-TEM* one. The idealized analytical models can describe it, and a perturbation approach to calculate the substrate and conductor losses is used.

At high frequencies, the modal field concentrates close to the separation surface between the low permittivity dioxide and the silicon, and it is *the surface or slow-wave form of the fundamental mode.*

If the resistivity of the silicon is low, an additional effect happens when the field decreases towards the grounded silicon side according to the skin effect. The longitudinal electric field is strong in this case, and the quasi-TEM models are not correct any more. Thus, this mode has increased frequency dependence of its parameters, and it is *the skin-effect* mode.

Very often, the line is described by three different models, and, in a mistake, it is stated the existence of three different modes of this line instead of the three limiting forms of the fundamental mode.

At the difference to classical microstrip, a fully analytical model of this line has not been created, according to the best Author's knowledge. Some numerical simulations of the Si-SiO_2 microstrip are performed up to 100 GHz in [74]. The theoretical and experimental studies of this line show its increased loss and its limited applicability for increased frequencies. Further improvements on this silicon-based microstrip line are with the high-resistivity substrates, with the isolation of the line from lossy substrate by a grounded metal sheet, and with the micromachining of silicon. At high frequencies, the coplanar waveguides of different configurations and coplanar strip line are an alternative solution instead of the microstrips.

5.7 Micromachining Technologies

Today many dreams of engineers worked for the developments of the first 3-D microwave integrations can be realized using the numerous technologies based on the 3-D etching and forming the 3-D structures in or on the substrates [75]-[77]. The interest in these sophisticated microtechnologies is with the manufacturing of IC components with the essentially improved parameters.

The micromachining technology, namely, its particular case of bulk processing, was born during the developments of silicon IC. It is with the formation of holes and trenches in bulk silicon by selective moving of the material using wet or dry etching. In the first case, the liquid etchants, like potassium hydroxide (KOH), ethylene diamine, etc. are used. Depending on the materials, the etching process is dependent on the crystal directions, and the vertical-wall holes or trenches in silicon are manufactured using combination of etches or even etching processes. The isotropic materials can be etched with the same dissolving speed rate at any directions, and the vertical structures are etched well. In silicon integrations, the etching is used, for instance, for via-hole manufacturing. In millimeter-wave electronics, the bulk etching is interesting in preparation of the trenches and cavities for the low-loss transmission lines and resonators isolated from the silicon [78],[79]. Additionally, the etching is one of the technological steps in the preparation of micromechanical components [77].

Additionally to wet etching, the dry one is used. There are several technologies of this type with the bombarding of silicon or other material by ions, plasma, or chemically active gases. Depending on material, it allows for the anisotropic or isotropic etching of most materials used in microelectronics.

During last time, the ultraviolet (UV) laser etching is used, and it is applicable, practically, for many materials of microelectronics [80]. This high-energy light with the wavelength of 157-248 nm disrupts the strong couplings of molecules or atoms, and the particles of materials taking energy of UV light are moved away from an illuminated spot with the minimal heating of the surrounding area. The ablation rate is regulated by the exposition time to the pulsed UV radiation. This technology allows for etching of materials with the minimal size of the etched spots about of several microns. The photoetching is used not only in microtechnology, but the PCBs can be drilled or trenched by UV light [21].

Additionally to the bulk micromachining associated with the etching of solid materials, another type of micromachining technologies is with forming of 3-D structures on the surface of substrates. Very often both bulk and surface micromachining are used together to create the 3-D integrated components of the needed geometry and electric characteristics.

One of the technologies is with the silicon dioxide selective depositing. A metal, with the following removal of dioxide, covers the layers, and the air 3-D structures supported by remains of dioxide are manufactured. It allows for obtaining increased quality factors of inductors, transformers, and transmission lines. Additionally, there are materials that can be used as the structural and sacrificial ones [77].

Structures that are more complicated are manufactured using the LIGA-like technologies, which initial variant was developed in the Research Karlsruhe Center [81]. It uses the X-ray lithography, galvanoforming, and molding to produce the microcomponents made of plastics, metals, and ceramics.

Another interesting technology is the microstereolithgraphy based on the use of UV laser light for selective polymerization of polymers. Some of them have low loss at millimeter-wave frequencies, and they are used as microsubstrates for transmission lines. Metals are deposited on polymers by electroplating technology, and different 3-D structures are manufactured. Metalized microslabs can be glued to each other, and the microcavities and microwaveguides are constructed on the substrate surface.

Contemporary 3-D technologies allow for micromachining of metals. For example, 3-D components are manufactured by the electroplating technology EFAB [82]-[85]. This technology consists of using a couple of metals of different reaction on an acid. One of them (Ni) is constructing, and another (Cu) is the sacrificial material. Combining the electroplating and removing the sacrificial material, different 3-D microstructures are manufactured, including for the RF and microwave applications. A couple examples are shown below. One of them is the 3-D inductors (solenoids) placed on a dielectric substrate (Fig. 5.12).

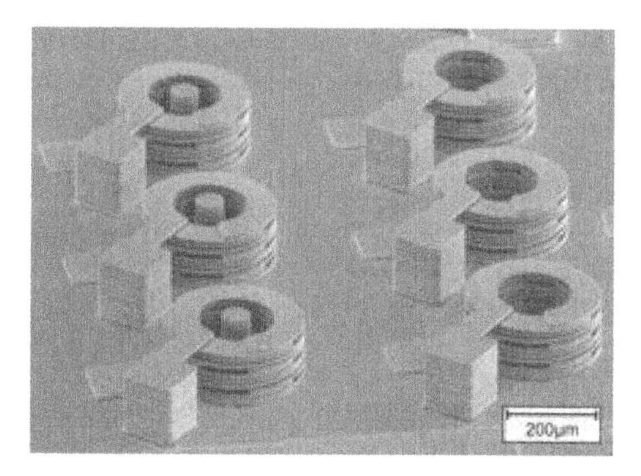

Fig. 5.12 Three-dimensional inductors. Reprinted with permission of Microfabrica Inc. (www.microfabrica.com)

A rectangular coaxial line is shown in Fig. 5.13, and the detailed characteristics of it and some components for the frequency range up to 100 GHz are published in [83].

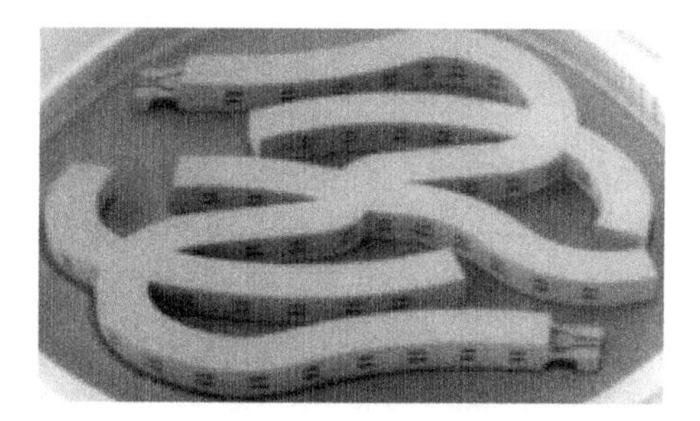

Fig. 5.13 Rectangular coaxial waveguide manufactured by electroplating technology. Reprinted with permission of Microfabrica Inc. (www.microfabrica.com)

5.8 RF Nanointegrations

Similarly to the post-micron ICs, the prospective nanointegrations are with new active devices like the graphene transistors, which will be able working up to the infrared light, with the well-conducting carbon tubes, graphene strips, and plasmonic waveguides as the interconnects. This technological node dealing with the feature size behind 10 nm is a milestone for contemporary science and technology [24],[25]. Now, it is, mostly, the place of ideas, theories, and only some experimental samples. A radical view on the integration is with the molecular electronics [24]. Taking into account the thermal movements of molecular components, they are predicted to be fixed in a material matrix to provide their workability at the room temperature and over.

The apologists of the 3-dimensionality show that the most effective structure of complex systems should be the spatial one which provides the maximum of connectivity and decreased time-delay of signals [24],[33]. There are different ways how to reach the ultimate 3-D on nanolevel. Recent studies on graphene interconnects and single-electron transistors placed over the well-known silicon substrates show that the multi-layer technology is still strong enough to create the 3-D integrations composed of planar atomic or monomolecular layers vertically connected by carbon nanotubes or molecular chains [24],[86],[87]. Another technology called the nanoimprint one has been matured strong enough to manufacture the multilayer and machined nano-integrations [24].

Besides the technological and architectural problems, there are many other challenges with the thermal noise, and EM and quantum interference. A special attention should be paid to the theory of nano-integrations and to the creation of new design tools.

5.9 Packaging of Microwave and Millimeter-wave Microchips

Packaging is the technique to assemble and to connect the IC components and/or assembled chips into a radio electronic system [5],[8],[88],[89]. Among the problems involved into the packaging is the architecture of chips and coupling of their components, package issue, and the connection of multiple chips into a millimeter-wave front-end.

Two approaches are on the way now. The first of them is the System-on-a-Chip technique. It is supposed to integrate into a single chip a whole system consisting of a millimeter-wave front-end and the digital hardware, including the microprocessors. Here, the whole-silicon approach and 3-D integration are very advantageous.

The second idea is the System-in-a-Package when the multiple chips are connected to each other being placed on a common plate, and it is called the Multi Chip Module (MCM). Additionally, the MCMs are different on the used board material and technology. Both approaches need solving numerous problems complicated by increased frequencies and wide bandwidth. Below only some of them are considered in short.

5.9.1 Packages for Millimeter-wave Integrations

A single chip can be encapsulated into a hermetic package [90]. This package should provide signal and power distribution and the heat evacuation. Additionally, it protects the chip from the humidity, the aggressive environment, and the EM and particle radiation. The chip housing should be manufactured using specially chosen materials. Among them are the plastics, glass-frit, ceramic, and silicon [90]-[91]. The first two of them demonstrate degradation of its characteristics with the frequency, and the ceramic or silicon packages are preferable.

The packages should provide communications with outer world minimally distorting the millimeter-wave signals. Very often, the size of a package is comparable with the wavelength, and it can demonstrate the resonance-like behavior distorting the chip performance. Additional source of signal distortions is the transitions to the board, and not all known solutions as the wire bonding are applicable at millimeter waves [92],[93].

A simplified draft of such a silicon housing is shown in Fig. 5.14 according to [93]. It consists of a Si micromachined cap and silicon-dioxide substrate. The chip is communicating through a one-mode CPW at frequencies up to 110 GHz. A gold ring placed on the bottom of silicon cap and connected to the CPW ground conductors provides the monomodal regime. It is shown that this package gives 0.05-0.26 dB insertion loss per transition in 0-110 GHz with a return loss better than -20 dB.

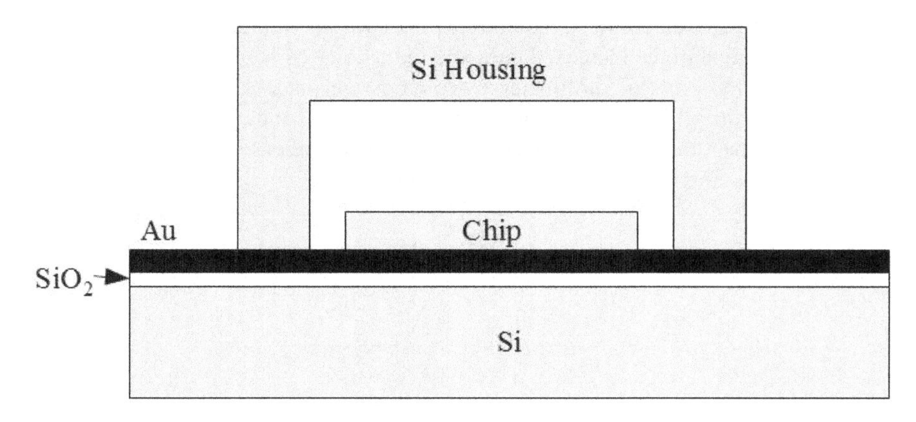

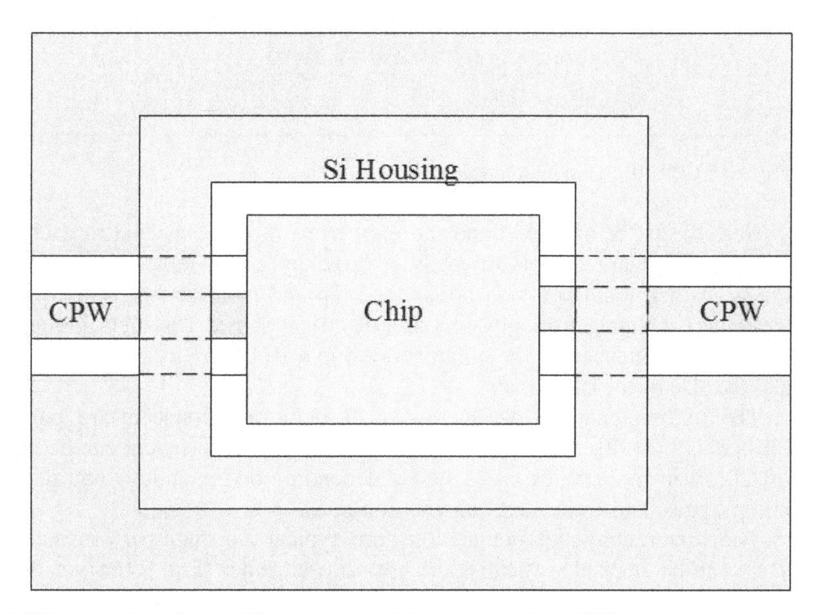

Fig. 5.14 Silicon package for a millimeter-wave chip. Adapted from [93]

The analysis of the packaging issues shows that the design of millimeter-wave packages is a rather complicated task, and the EM software should be involved for simulations.

Another problem arising in packaging is the interconnects from a chip to the motherboard or chip carrier. The chips can be connected directly to each other, and the chip-chip transitions are used.

5.9.2 Interconnecting of Chips

In millimeter-wave electronics, the conventional wire bonding to connect a chip is not applicable due to the increased parasitic inductivity of wires and their coupling [94]. Mostly, the ports of millimeter-wave chips are made of the CPW, CPS, or inverted microstrip line, and one of the best solutions is the use of flip-chip technology to connect the chips to motherboard or to each other [88],[95]-[99]. An example of a flip-chip interconnect is shown in Fig. 5.15.

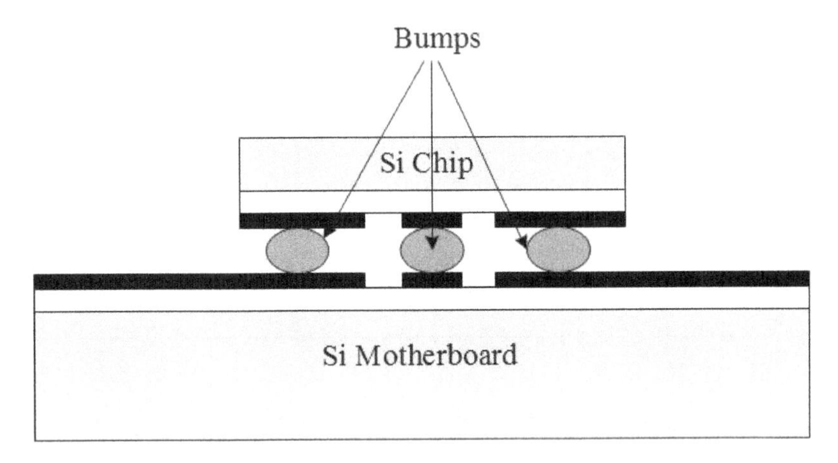

Fig. 5.15 CPW flip-chip interconnects

Here the CPW of a chip and the motherboard are connected to each other by bumps, which are the plastic balls covered by a low-temperature InSn solder. Firing or thermocompression bonds the chip and motherboard. This flip-chip interconnect is small in its physical and electrical sense. The high-frequency parameters are influenced by the bump height (up to 100 μm), its diameter (20-80 μm), and the size of the CPW pads.

The millimeter-wave parameters of flip-chip-interconnects are published in [88],[90],[95]-[98], and it follows that the reflection coefficient can be kept about of -20 dB in frequencies up 110 GHz depending on technology and geometry of bumps, pads, and connected transmission lines.

The direct chip-chip connections are typical for high-performance MCMs. They can be stacked vertically, directly to each other (Fig. 5.16a), or through an intermediate dielectric substrate (Fig. 5.16b) by via-holes. The bumps are placed over the dielectric substrate and under it [99].

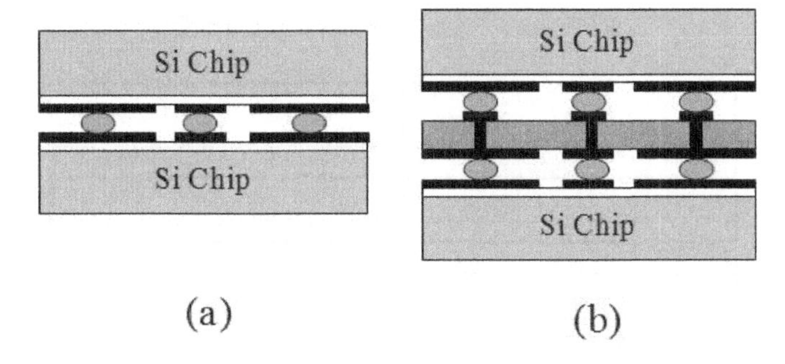

Fig. 5.16 Vertically stacked silicon chips. (a)- Directly connected chips; (b)- Chips connected through a dielectric intermediate substrate

Fig. 5.17 shows the bumps placed on the backside of a silicon substrate.

Fig. 5.17 Bumps on the backside of a silicon substrate. Printed with the permission of Allvia Inc. (www.allvia.com)

At the difference to the conventional planar integrated chips, the 3-D stacking provides reduced time-delay of signals, and this type of integration is preferable for the high-speed and ultra-wide band systems. Recently, a new planar ultra-bandwidth integration of chips has been proposed [100]. The semiconductor chips are integrated into a quilt consisting of a number of analogue, digital, and optical microchips (Fig. 5.18). They are placed close to each other on a holder-substrate, and an air-gap between them is no greater than 50 μm.

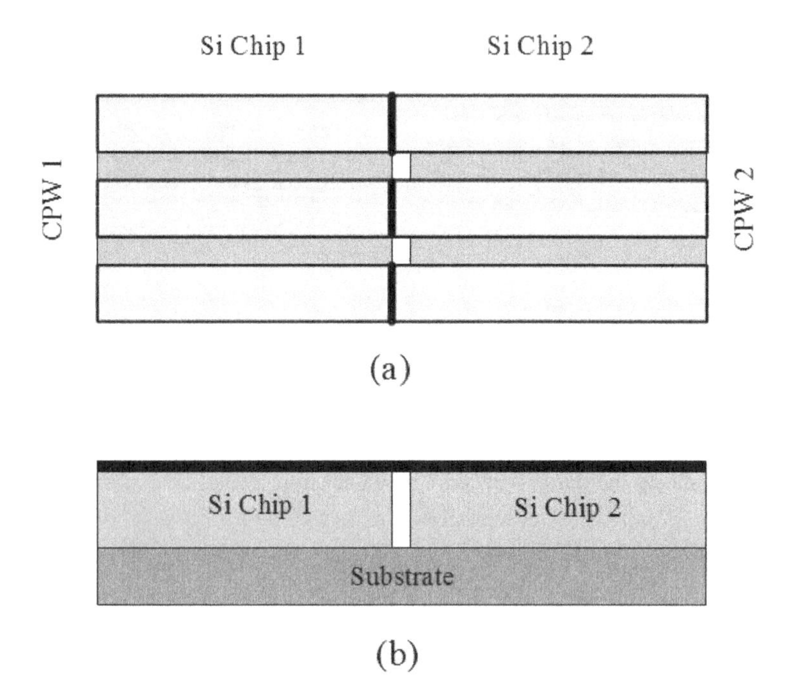

Fig. 5.18 Quilt packaging of chips. (a)- Horizontal view of a package; (b)- Vertical view of the packaged chips. Adapted from [100]

The interconnects of a microstrip or CPW design are on the surface of the integrated chips, and they are connected to each other by welding providing the galvanic contacts of conductors. It was found that this type of inter-chip connecting could provide the ultra-wide bandwidth communications up to 250 GHz, and the measurements results were given up to 40 GHz. The frequency limitations are due to the air-gap between the connected chips, and they can be diminished by proper matching of the Si- and air-filled CPWs (Fig. 5.18).

5.9.3 Motherbords and Materials

Today the chips, lumped components, and active elements can be integrated by several technologies. The first of them is the printed circuit board (PCB) technology. In this case, a motherboard is made of a low-cost glass-organic material covered by copper metallization. The interconnects are etched lithographically, and the via-holes are made by mechanical or laser drilling, or even punching [25]. Then each metalized layer is glued to each other, pressed in vacuum, and fired. Unfortunately, the manufactured stacks are distinguished by poor control of the dielectric permittivity, its anisotropy, and increased dielectric loss, which is not acceptable for millimeter-wave applications. As a rule, the accuracy of this lithography is not enough for the millimeter-wave applications.

The most common material for low frequencies is the FR4, and it is based on the woven glass and epoxy. Its permittivity is between 3.8-6.3 depending on the ratio of the used glass and epoxy. Some PCBs, based on enhanced FR4 materials and assigned for mass-production, show stable electrical, mechanical, and thermal characteristics in frequencies up to 10 GHz [101].

In micro- and millimeter-waves with their increased requirements to the accuracy of interconnects, the PCB technology has been transformed into the multichip one. The motherboards are made of low-loss materials of well-controlled dielectric parameters. It is preferred to avoid the integrating a large number of passive and active lumped components additionally to the integrated chips. The interconnects should be able to transfer a higher amount of data than it is typical for low-performance PCBs.

As the materials of the motherboards it can be used the *laminated* high-frequency PTFE filled substrates, and the MCM is marked as the MCM-L in this case. The *ceramic* boards (MCM-C), fired at low or high temperatures, can be used for millimeter-wave applications, and the contemporary upper frequencies of them are greater than 100 GHz.

If a thick-film technology used for MCM-C provides insufficient accuracy of interconnects, the monolithic technologies can be used (MCM-D). In this case, a motherboard of silicon or other semiconductor or polyimide is covered by a thick layer of a metal using a thin-film technology. On this ground, the multiple dielectric layers are deposited with the interconnects and some passives. The surface micromachining can be applied to manufacture the components of the improved parameters. As dielectrics, the silicon dioxide and different organic layers can be used. Among the last of them are the polyimide, BCB, and PTFE based materials, etc., which show the acceptable loss parameters even at very increased frequencies.

In the case of MCM-L, the low-loss laminated materials can be used, and, be based, for instance, on the PTFE (polytetrafluoroethylene) as the epoxy that decreases the dielectric loss. Additionally, the non-woven materials, which consist of the PTFE epoxy and randomly oriented glass microfibers, can provide the decreased laminate anisotropy. The ceramic PCBs are made of alumina particles and PTFE, and they have increased permittivity.

The results of the measurements of some such materials in frequencies of 29-39 GHz are published in [91], for instance. The best material from the studied ones is the Sheldahl Comclad XFx10mil with its lowest $\tan\delta$=0.00024-0.00055 and $\varepsilon_r = 2.24 - 2.26$. These parameters are measured in the vertical direction regarding to the surface of dielectric plates.

In many cases, the thermal stability of dielectrics is very important, and their characterization is on demand. In the above-cited paper, the thermal drift of the dielectric parameters of several materials is given in the range of -50°C - +70°C. In [102], a Duroid material is studied at temperatures of 20°-200° and frequencies of 30-70 GHz, and the remarkably stable thermal properties are registered.

A very interesting laminated material is the LCP [27],[103],[104], which has stable electric parameters $(\varepsilon_r = 3.1, \tan\delta = 0.002 - 0.0045)$ in frequencies up to 110 GHz. These low-cost laminates are manufactured using conventional RF PCB technologies, and they are compatible with copper, silicon, and GaAs. Conventional PCB technology allows for the 30-55-μm distance between conductors, and more advanced ones are to manufacture the high-precision components and integrations. The laminates are stacked, and the three-dimensional motherboards are available. Many passive components have been already designed using the LCP for the micro- and millimeter-waves, including the packages for these frequencies [27].

Measurements of parameters of these dielectrics in a wide frequency band are not a trivial task, and several techniques are used, including the resonance methods and high-precession transmission line measurements. For instance, in [105], the permittivity of silica $(\varepsilon_r \sim 3.8)$ and BCB $(\varepsilon_r \sim 2.7)$ is measured in frequencies up to 40 GHz using the microstrip lines. It is shown that the error of measurements is in the limits of 5% for both dielectrics. Other results of measurements of a BCB material are in [106] where a conductor-backed CPW is used. It is shown that the studied material has permittivity 2.65 in the entire frequency band of 11-65 GHz, and its loss tangent is varied within 0.001-0.009.

Many materials are measured in the frequency range of 20-100 GHz in [107] using the short waveguide, long waveguide, and transmission line methods. Among them are fused silica, PTFE, AF45 glass, and several organic materials. It is shown that at frequencies over 60 GHz, the dielectric tangent loss of the most of these packaging materials is no less than 0.01. Only fused silica shows $\tan\delta < 0.01$ for frequencies up to 90 GHz. It follows that the millimeter-wave packaging requires an intensive research on the developments of new materials with decreased loss, predictable permittivity, and acceptable thermal parameters. Recently, new dielectrics presented by Rogers and Park Nelko have been characterized in [108]. One of them, RO3003 $(\varepsilon_r \sim 3.01 - 3.15)$, shows $\tan\delta < 0.003$ in frequencies up to 67 GHz.

The materials define the parameters of interconnects, and the additional information on dielectrics can be found from the known manufactures of dielectric materials as DuPont, Arlon LLC, Rogers, or from the books on microwave materials, packaging, integrations, and components [1],[3],[4],[7],[8],[25],[109].

The most applicable interconnects for motherboards are the strip and microstrip lines, CPW, substrate integrated waveguides, and dielectric waveguides. The choice of the interconnects is dictated by the used technology, bandwidth, and cost. Different interconnects should be compared on the provided performance. For example, CPW has reduced conductor loss compared to the microstrip lines only for certain geometrical parameters. A large amount of information on the transmission lines and components are published in the papers and books on micro- and millimeter-wave hybrid ICs (see References).

5.10 On the Evolution of Microwave and Millimeter-wave Electronics

The overall review of the evolution of microwave and millimeter wave electronics shows that it can be illustrated by a multi-branch curve (Fig. 5.19) [34]. Each of its branches is considered, as a trajectory for a certain type of technology, and the measure of this evolution is the achieved complexity of a module per the volume unit. Such characteristics are qualitatively described by the complexity coefficient C:

$$C = \frac{N_e N_{ic} N_c \alpha_q}{\frac{V}{\lambda^3} \frac{M}{m}} \tag{5.1}$$

where

- C shows the potential ability of a certain technology to provide a solution with the acceptable quality
- N_e is the number of functional elements in a module
- N_{ic} is the number of interconnections
- N_c is the number of connectors
- V is the module volume
- λ is the wave length on the central or clock frequency
- M is the module mass
- m is the average mass of a component
- α_q is the quality coefficient varied from one technology to another.

According to this formula, the most advanced technology realizes the miniaturized electronic modules with the highest number of components, interconnections, input/outputs, etc. per unit volume.

The ultimate stage of the development of 3-D integrations is with molecular, nanotechnology, and hybrid CMOS-Carbon Nanotube technologies. They, being hypothetical when the Fig. 5.19 was drown [34], have been manufactured recently [86],[87].

Increasing the frequencies and density of 3-D integrations stimulates to employ new ways for the signal coding which are immune to the EM noise. For example, these circuits allow to write digital information into the discrete spatial structure (topology) of the field impulses and to double the noise immunity of signals.

The high-density silicon integrations of the 30-100 GHz range are a place of complicated EM effects, and an increased attention should be paid to analytical and numerical modeling of these circuits. Unfortunately, the interest in the EM research of transmission lines and passive components has been essentially decreased and new efforts in this direction are required.

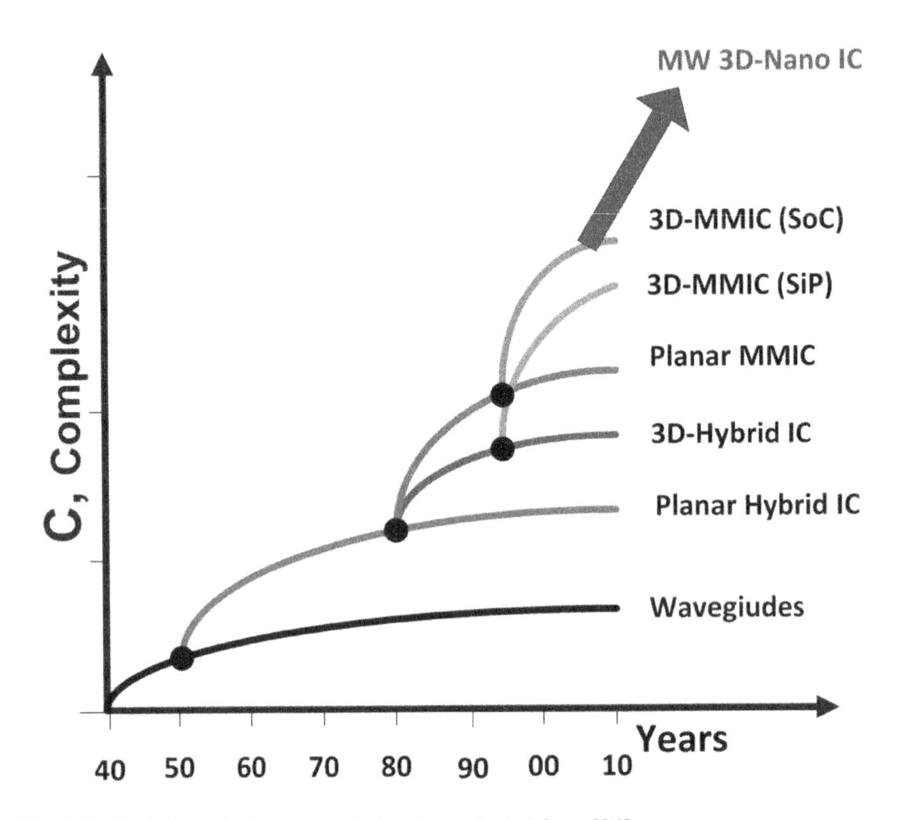

Fig. 5.19 Evolution of microwave electronics. Adapted from [34]

References

[1] Robertson, I.D., Lucyszyn, S. (eds.): RFIC and MMIC Design and Technology. IEE Press (2001)

[2] Liu, D., Pfeiffer, U., Grzyb, J., Gaucher, B. (eds.): Advanced Millimeter-wave Technologies: Antennas, Packaging and Circuits. Wiley (2009)

[3] Harper, C.A.: Electronic Packaging and Interconnection Handbook. McGraw-Hill (2005)

[4] Johansson, C.: High Frequency Electronic Packaging and Components: Characterization, Simulation, Materials and Processing, Linkøping University (2007)

[5] Kuang, K., Kim, F., Gahill, S.S. (eds.): RF and Microwave Microelectronics Packaging. Springer (2009)

[6] Edwards, T.C., Steer, M.B.: Foundations of Interconnects and Microstrip Design. Wiley (2000)

[7] Ulrich, R.K., Schrader, L.W. (eds.): Integrated Passive Component Technology. IEEE Press (2003)

[8] Tummala, R.R. (ed.): Fundamentals of Microsystems Packaging. McGraw-Hill, New York (2001)

[9] Cruickshank, D.: Microwave Materials for Wireless Applications. Artech House (2011)
[10] Gaynor, M.P.: System-in-Package RF Design and Applications. Artech House (2007)
[11] Barret, R.M.: Microwave printed circuits-the early years. IEEE Trans., Microw. Theory Tech. 32, 983–990 (1984)
[12] Elshabuni-Riad, A.A.R., Barlow III, F.D.: Thin Film Technology Handbook. McGraw-Hill (1997)
[13] Pitt, K.E.G. (ed.): Handbook of Thick Film Technology. Electrochemical Publications, Port Eirin (2005)
[14] Volman, V.I. (ed.): Handbook on the Calculation and Design of Microwave Strip Circuits. Radio i Svayz Publ. Comp, Moscow (1982) (in Russian)
[15] Imanaka, Y.: Multilayered Low Temperature Cofired Ceramics (LTCC) Technology. Springer (2010)
[16] Deen, M.J. (ed): Silicon Based Millimeter-wave Technology. Submitted to Elsevier (2012)
[17] Mcquiddy Jr., D.N., Wassel, J.W., LaGrange, J.B., et al.: Monolithic microwave integrated circuits: An historical perspective. IEEE Trans., Microw. Theory Tech. 32, 997–1008 (1984)
[18] Niehenke, E.C., Pucel, R.A., Bahl, I.J.: Microwave and millimeter-wave integrated circuits. IEEE Trans., Microw. Theory Tech. 50, 846–857 (2002)
[19] Sobol, H., Tomiyasu, K.: Milestones of microwaves. IEEE Trans., Microw. Theory Tech. 50, 594–611 (2002)
[20] Sorrentino, R., Oxley, T., Salmer, G., et al.: Microwaves in Europe. IEEE Trans., Microw. Theory Tech. 50, 1056–1071 (2002)
[21] Lau, J.H., Lee, S.W.R.: Microvias for Low Cost, High Density Interconnects. McGraw-Hill (2001)
[22] Wong, B.P., Mittal, A., Cao, Y., Starr, G.: Nano-CMOS Circuits and Physical Design. Wiley Interscience (2005)
[23] Bakir, M.S., Meindi, J.D. (eds.): Integrated Interconnect Technologies for 3D Nanoelectronic Systems. Artech House (2009)
[24] Crawley, D., Nikolić, K., Forshaw, M. (eds.): 3D Nanoelectronic Computer Architects and Implementation. IOP Publ (2005)
[25] Ryzhii, M., Ryzhii, V. (eds.): Physics and Modeling of Tera- and Nanodevices. Imperial College Press (2008)
[26] Palm, P., Tuominen, R., Kivikero, A.: Integrated Module Board (IMB); an advanced manufacturing technology for embedding active components inside organic substrate. In: Proc. 54th IEEE Conf. Electronic Components and Technology, vol. 2, pp. 1227–1231 (2004)
[27] Thompson, D., Tantot, O., Jallageas, H., et al.: Characterization of liquid crystal polymer (LCP) material and transmission lines on LCP substrates from 30-110 GHz. IEEE Trans., Microw. Theory Tech. 52, 1343–1352 (2004)
[28] Gvozdev, V.I.: Use of the unbalanced slotted line in SHF microcircuits. Radioeng. Electron Physics (Radiotekhnika i Elektronika) 27(11), 42–47 (1981)
[29] Gvozdev, V.I., Nefedov, E.I.: Some possibilities of three-dimensional integrated UHF structures. Sov. Phys. Dokl. 27, 959–960 (1982)
[30] Gvozdev, V.I., Gulayev, Y.V., Nefedov, E.I.: Possible use of the principles of three-dimensional integrated microwave circuits in the design of ultra-fast digital computers. Sov. Phys. Dokl. 31, 760–761 (1986)

[31] Gvozdev, V.I., Nefedov, E.I.: Three-Dimensional Microwave Integrated Circuits. Nauka Publ., Moscow (1985) (in Russian)
[32] Gvozdev, V.I., Nefedov, E.I.: Volumetrical Microwave Integrated Circuits–Element Base of Analog and Digital Radioelectronics. Nauka Publ., Moscow (1987) (in Russian)
[33] Gvozdev, V.I., Kouzaev, G.A., Nefedov, E.I., et al.: Physical principles of the modeling of three-dimensional microwave and extremely high-frequency integrated circuits. Soviet Physics-Uspekhi 35, 212–230 (1992)
[34] Kouzaev, G.A., Deen, M.J.: 3D-integrated circuits. Part. 1. High-frequency 3D-hybrid ICs. An analytical review. Microelectronics Res. Lab. Report, McMaster University, Canada (2002)
[35] Bykov, D.V., Vorob'evsky, E.M., Gvozdev, V.I., Kouzaev, G.A., et al.: Technology for three-dimensional microwave integrated circuits. Zarubezhnaya Radioelektronika (Foreign Radio Electronics) (11), 49–65 (1992) (in Russian)
[36] Nefedov, E.I., Fialkovskyi, A.T.: Strip Transmission Lines: Electromagnetic Basics of the CAD of Microwave Integrated Circuits. Nauka (1980) (in Russian)
[37] Neganov, V.A.: Physical Regularization of Ill-posed Problems of Electrodynamics. Science-Press (2008) (in Russian)
[38] Nikolskyi, V.V. (ed.): Computer Aided Design of Microwave Devices. Nauka (1980) (in Russian)
[39] Nikolskyi, V.V., Nikolskaya, T.A.: Electrodynamics and Radio Wave Propagation. Nauka (1986) (in Russian)
[40] Nikolskyi, V.V., Nikolskaya, T.A.: Decomposition Approach to Electrodynamics Problems. Nauka (1983) (in Russian)
[41] Golovanov, Makeeva, G.S.: Simulations of Electromagnetic Wave Interactions with Nano-grids in Microwave and Terahertz Frequencies. Nauka (2011) (in print) (in Russian)
[42] Shestopalov, Y., Shestopalov, V.: Spectral Theory and Excitation of Open Structures. Peter Peregrinus Ltd. (1996)
[43] Kouzaev, G.A., Deen, M.J., Nikolova, N.K.: A parallel-plate waveguide model of lossy microstrip transmission line. IEEE Microw. Wireless Comp. Lett. 15, 27–29 (2005)
[44] Kouzaev, G.A., Deen, M.J., Nikolova, N.K., Rahal, A.: An approximate parallel-plate waveguide model of a lossy multilayered microstrip line. Microw. Opt. Tech. Lett. 45, 23–26 (2005)
[45] Kouzaev, G.A., Kurushin, E.P., Neganov, V.A.: Numerical calculations of a slot-transmission line. In: Izv. Vysshikh Utchebnykh Zavedeniy Radiofizika (Radiophysics), vol. 23, pp. 1041–1042 (1981) (in Russian)
[46] Kouzaev, G.A.: Quasistatic model of ribbed nonsymmetrical slotted line. Radio Eng. Electron. Phys. 28, 137–138 (1983)
[47] Kouzaev, G.A.: Balanced slotted line. In: Gvozdev, V.I., Nefedov, E.I. (eds.) Microwave Three-Dimensional Integrated Circuits, pp. 45–50. Nauka Publ., Moscow (1985) (Invited Chapter, in Russian)
[48] Gvozdev, V.I., Kouzaev, G.A., Nefedov, E.I.: Balanced slotted line. Theory and experiment. Radio Eng. Electron. Physics (Radiotekhnika i Elektronika) 30, 1050–1057 (1985)
[49] Gvozdev, V.I., Kouzaev, G.A., Nefedov, E.I., Utkin, M.I.: Electrodynamical calculation of microwave volume integrated circuit components based on a balanced slotted line. J. Commun. Techn. Electronics (Radiotekhnika i Electronika) 33, 39–43 (1989)

[50] Gvozdev, V.I., Kouzaev, G.A., et al.: A fin-slot line: theory, experiment, and structures. J. Commun. Technology and Electronics (Radiotekhnika i Elektronika) 35, 81–85 (1990)
[51] Gvozdev, V.I., Kouzaev, G.A., Pozhydaev, E.D., et al.: Asymmetrical slot transmission line. USSR Invention Certificate No 1730692 dated on, August 22 (1989)
[52] Gvozdev, V.I., Kouzaev, G.A., et al.: Topological models of the natural modes in coupled corner transmission lines. J. Commun. Techn. Electron (Radiotekhnika i Elektronika) 37, 48–54 (1992)
[53] Gvozdev, V.I., Kouzaev, G.A., et al.: Directional couplers based on transmission lines with corners. J. Commun. Techn. Electron (Radiotekhnika i Elektronika) 37, 37–40 (1992)
[54] Gvozdev, V.I., Kolosov, S.A., Kouzaev, G.A., et al.: Directional coupler. USSR Invention Certificate No 1786561 dated on, February 14 (1990)
[55] Kolosov, S.A., Kouzaev, G.A., Skulakov, P.I., et al.: Slot transmission line. USSR Invention Certificate No 1683100 dated on, May 17 (1989)
[56] Gvozdev, V.I., Kouzaev, G.A., Kulevatov, M.V.: Narrow band-pass microwave filter. Telecommun. Radio Eng. 49, 1–5 (1995)
[57] Gvozdev, V.I., Kouzaev, G.A., Tikhonov, A.N.: New transmission lines and electrodynamical models for three-dimensional microwave integrated circuits. Sov. Physics-Doklady 35, 675–677 (1990)
[58] Gvozdev, V.I., Kouzaev, G.A., Nefedov, E.I., et al.: Slot transmission line. USSR Invention Certificate No 1626281 dated on, March 31 (1989)
[59] Gvozdev, V.I., Kouzaev, G.A.: Physics and field topology of 3-D microwave circuits. Soviet Microelectronics 21, 1–17 (1992)
[60] Kouzaev, G.A.: Electromagnetic model of differential substrate integrated waveguide. In: Proc. Eur. Computing Conf., Paris, France, April 27-29, pp. 282–284 (2011)
[61] Kouzaev, G.A.: Oscillations of cylindrical ferrite resonators covered by a thick semiconductor layer. In: Electromagnetic Fundamentals for the CAD of Microwave Integrated Circuits, pp. 56–64. USSR Acad. Sci, Moscow (1981) (in Russian)
[62] Kouzaev, G.A., Nikolova, N.K., Deen, M.J.: Circular-pad via model based on cavity field analysis. IEEE Microw. Wireless Lett. 13, 481–483 (2003)
[63] Kouzaev, G.A., Deen, M.J., Nikolova, N.K., Rahal, A.: Influence of eccentricity on the frequency limitations of circular-pad via-holes. IEEE Microw. Wireless Lett. 14, 265–267 (2004)
[64] Kouzaev, G.A., Deen, M.J., Nikolova, N.K., Rahal, A.: Cavity models of planar components grounded by via-holes and their experimental verification. IEEE Trans., Microwave Theory Tech. 54, 1033–1042 (2006)
[65] Gvozdev, V.I., Kouzaev, G.A., Nefedov, E.I., Fomina, L.M.: Band-pass filter. USSR Invention Certificate, No 1185440 dated on, October 1 (1982)
[66] Tokumitsu, T., Nishikawa, K., Kamogawa, K., et al.: Three-dimensional MMIC technology for multifunction integration and its possible application to masterslice MMIC. In: Proc. 1996 IEEE Microwave and Millimeter-Wave Monolithic Circ. Symp. Dig., pp. 85–88 (June 1996)
[67] Onodera, K., Hirano, M., Tokimutsu, M., et al.: Folded U-shaped microwave technology for ultra-compact three-dimensional MMIC's. IEEE Trans., Microw. Theory Tech. 44, 2347–2353 (1996)
[68] Toyoda, F., Nishikawa, K., Tokumitsu, T., et al.: Three-dimensional Masterslice MMIC on Si substrate. IEEE Trans., Microw. Theory Tech. 45, 2524–2530 (1997)

[69] Gvozdev, V.I., Kouzaev, G.A., Podkovyrin, S.I.: Microwave volume-metrical monolithic integrated circuits. In: Proc. Trans. Black Sea Region Symp. Appl. Electromagnetism, Metsovo, Epirus-Hellas, Athens, Greece, April 17-19, p. 173 (1996)
[70] Happich, J.: Nanometer-thin film enables highest permittivity capacitors. In: EETimes Europe, January 4 (2012)
[71] Heydari, B., Bohsali, M., Adabi, E., et al.: Milimeter-wave devices and circuit blocks up to 104 GHz in 90 nm CMOS. IEEE J. Solid-State Circ. 42, 2893–2903 (2007)
[72] Guckel, H., Brennan, P.A., Paloscz, I.: A parallel-plate waveguide approach to micro-miniaturized, planar transmission lines for integrated circuits. IEEE Trans., Microw. Theory Tech. 15, 468–476 (1967)
[73] Hasegawa, H., Furukawa, M., Yanai, H.: Properties of microstrip line on Si-SiO_2system. IEEE Trans., Microw. Theory Tech. 19, 869–881 (1971)
[74] Demeester, T., De Zutter, D.: Quasi-TM-transmission line parameters of coupled lossy lines based on the Dirichlet to Neumann boundary operator. IEEE Trans., Microw. Theory Tech. 56, 1649–1659 (2008)
[75] Al-Sarawi, S.F., Abbot, D., Franzon, P.D.: A review of 3-D packaging technology. IEEE Trans., Comp. Pack., Manuf. Techn.-Part B 21, 2–14 (1998)
[76] 3-D Integration technologies. In: Maurelli, D., Belot, D., Camparado, G. (eds.) Proc. IEEE, vol. 97, pp. 1–193 (2009)
[77] Varadan, V.K., Jiang, X., Varadan, V.V.: Microstereolithography and other Fabrication Techniques for 3D MEMS. Wiley (2001)
[78] Herrick, K.J., Yook, J.-G., Katehi, L.P.B.: Microtechnology in the development of three-dimensional circuits. IEEE Trans., Microw. Theory Tech. 46, 1832–1844 (1998)
[79] Margomenos, A., Herrik, K.J., Herman, M.I., et al.: Isolation in three-dimensional integrated circuits. IEEE Trans., Microw. Theory Tech. 51, 25–32 (2009)
[80] Sercel, J.P.: Ultraviolet laser-based MOEMS and MEMS micromachining. An alternative to wet processing. In: Advanced Packaging, pp. 29–31 (April 2004)
[81] Guckel, H.: High-aspect-ratio micromachining via deep X-ray lithography. Proc. IEEE 86, 1586–1593 (1998)
[82] Cohen, A., et al.: EFAB: rapid, low-cost desktop micromachining of high-aspect ratio true 3-D MEMS. In: Proc. IEEE MEMS, pp. 244–251 (1999)
[83] Reid, J.R., Marsh, E.D., Webster, R.T.: Micromachined rectangular-coaxial transmission lines. IEEE Trans., Microw. Theory Tech. 54, 3433–3442 (2006)
[84] Yoon, J.-B., Kim, B.-I., Choi, Y.-S., et al.: 3-D construction of monolithic passive components for RF and microwave ICs using thick-metal surface micromachining technology. IEEE Trans., Microw. Theory Tech. 51, 279–288 (2003)
[85] Wang, Y., Ke, M., Lancaster, M.J., et al.: Micromachined millimeter-wave rectangular-coaxial branch-line coupler with enhanced bandwidth. IEEE Trans., Microw. Theory Tech. 57, 1655–1660 (2000)
[86] Akinwande, D., Yasuda, S., Paul, B., et al.: Monolithic integration of CMOS VLSI and carbon nanotubes for hybrid nanotechnology applications. IEEE Trans., Nanotechn. 7, 636–639 (2008)
[87] Lu, J.-Q.: 3-D hyperintegration and packaging technologies for micro-nanosystems. Proc. IEEE 97, 18–30 (2009)
[88] Heinrich, W.: The flip-chip approach for millimeter-wave packaging. In: IEEE Microw. Mag., pp. 36–45 (September 2005)

[89] Jentzsch, A., Heinrich, W.: Theory and measurements of flip-chip interconnects for frequencies up to 100 GHz. IEEE Trans., Microw. Theory Tech. 49, 871–878 (2001)

[90] Herman, M.I., Lee, K.A., Kolawa, E.A., et al.: Novel techniques for millimeter-wave packages. IEEE Trans., Microw. Theory Tech. 43, 1516–1523 (1995)

[91] Egorov, V.N., Maslov, V.L., Nefyodov, Y.A., et al.: Dielectric constant, loss tangent and surface resistance of PCB materials at *K*-band frequencies. IEEE Trans., Microw. Theory Tech. 53, 627–635 (2005)

[92] Henderson, R.M., Katehi, L.P.B.: Silicon based micromachined packages for high-frequency applications. IEEE Trans., Microw. Theory Tech. 47, 1563–1569 (1999)

[93] Min, B.-W., Rebeiz, G.M.: A low-loss silicon-on-silicon DC-110-GHz resonance-free package. IEEE Trans., Microw. Theory Tech. 54, 710–716 (2006)

[94] Lim, J., Kim, G., Hwang, S.: Suppression of microwave resonances in wirebond transitions between conductor-backed coplanar waveguides. IEEE Microw. Wireless Comp. Lett. 18, 31–33 (2008)

[95] Masuda, S., Takahashi, T., Joshin, K.: An over-110-GHz InP HEMT flip-chip distributed baseband amplifier with inverted microstrip line structure for optical transmission system. IEEE J. Solid-State Cir. 38, 1479–1484 (2003)

[96] Song, Y.K., Lee, C.C.: Millimeter-wave coplanar strip (CPS) line flip chip packaging on PCBs. In: Proc. 2005 El. Comp. Techn. Conf., pp. 1807–1813 (2005)

[97] Chu, K.-M., Choi, J.-H., Lee, J.-S., et al.: Optoelectronic and microwave transmission characteristics of indium solder bumps for low-temperature flip-chip applications. IEEE Trans., Adv. Pack. 29, 409–414 (2006)

[98] Kangasvieri, T., Komulainen, M., Jantunen, H., et al.: Low-loss and wideband package transitions for microwave and millimeter-wave MCMs. IEEE Trans., Adv. Pack. 31, 170–181 (2006)

[99] Lee, B.-W., Tsai, J.-Y., Jin, H., et al.: New 3-D chip stacking architectures by wire-on-bump and bump-on-flex. IEEE Trans., Adv. Pack. 31, 367–376 (2008)

[100] Bernstein, G.H., Liu, Q., Yan, M., et al.: Quilt packaging: high-density, high-speed interchip communications. IEEE Trans., Adv. Pack. 30, 731–740 (2007)

[101] Cauwe, M., De Baets, J.: Broadband material parameter characterization for practical high-speed interconnects on printed circuit board. IEEE Trans., Adv. Pack. 31, 649–656 (2008)

[102] Morcillo, C.D., Bhattacharya, S.K., Horn, A., et al.: Thermal stability of the dielectric properties of the low-loss, organic material RT/Duroid 6002 from 30 GHz to 70 GHz. In: Proc. 2010 El. Comp. Techn. Conf., pp. 1830–1833 (2010)

[103] Vyas, R., Rida, A., Bhattacharya, S.K., et al.: Liquid crystal polymer (LCP): The ultimate solution for low cost RF flexible electronics and antennas. In: Proc. IEEE APS Int. Symp., pp. 1729–1732 (2007)

[104] McGrath, M.P., Aihara, K., Chen, M.J., et al.: Liquid crystal polymer for RF and millimeter-wave multi-layer hermetic packages and modules. In: Kuang, K., et al. (eds.) RF and Microwave Microelectronics Packaging, ch. 5, pp. 91–113. Springer Science + Business Media, LLC (2010)

[105] Janezic, M.D., Williams, D.F., Blaschke, V., et al.: Permittivity characterization of low-*k* thin films from transmission-line method. IEEE Trans., Microw. Theory Tech. 51, 132–136 (2003)

[106] Costanzo, S., Venneri, I., Di Massa, G., et al.: Benzocyclobutene as substrate material for planar millimeter-wave structures: dielectric characterization and application. J. Infrared Milli Terahz Waves 31, 66–77 (2010)

[107] Zwick, T., Chandrasekhar, A., Baks, C.W., et al.: Determination of the complex permittivity of packaging materials at millimeter-wave frequencies. IEEE Trans., Microw. Theory Tech. 54, 1001–1010 (2006)
[108] Lopez, A.L.V., Bhattacharya, S.K., Morcillo, C.A.D.: Novel low loss thin film materials for wireless 60 GHz application. In: Proc. 2010 El. Comp. Techn. Conf., pp. 1990–1995 (2010)
[109] Cruickshank, D.: Microwave Materials for Wireless Applications. Artech House (2011)

6 Transmission Lines and Their EM Models for the Extended Frequency Bandwidth Applications

Abstract. In this Chapter, an analytical review on interated transmission lines for millimeter-wave applications is given. Several tens of lines are analyzed and their characteristics are compared. The results of comparison are in a table showing the transmission lines' loss, used materials, technologies, etc. Interconnects based on new effects are considered in this Chapter as well. References -252. Figures -47. Pages -85.

6.1 Microstrip Lines

Microstrip lines are the ones of the oldest interconnects for microwave and digital signaling known since the 40s of the last Century [1]. Many contributions are on physics, technology and EM modeling of these lines [2]-[7]. The presented here results are only on some theoretical aspects which are important for the extended frequency bandwidth applications.

6.1.1 Thin-film Microstrip Lines[1]

The line appears when the increased loss caused by the semiconductor substrate started to be troublesome at high frequencies. To avoid it, the interconnects are isolated from the silicon substrate by the ground plane of several micron height, and this thickness is comparable with the height of the dielectric layer between the ground plane and signal conductor (Fig. 5.11f). Very often, the line is called as the TFML. This line demonstrates rather complicated frequency behavior towards their hybrid-technology counterparts. It is caused that its conductor thickness is comparable with the skin depth at low frequencies. At these frequencies, the field of the dominant mode has the increased longitudinal component of the electric field inside the lossy conductors, and the mode is of the quasi-TM type one. Here this line is similar to a lossy-dielectric sandwich waveguide. It leads to a non-monotonic behavior of the phase constant and increased loss of the main mode. Below, this effect is demonstrated by detailed modeling of TFML.

Our approximate analytical model of this transmission line valid up to 100 GHz is developed in [8]. The studied geometry is shown in Fig. 6.1. It is a microstrip

[1] Sections 6.1.1-6.1.3 written together with M.J. Deen, N.K. Nikolova, and A. Rahal.

G.A. Kouzaev: Applications of Advanced Electromagnetics, LNEE 169, pp. 221–305.
springerlink.com

line which conductors have limited heights and conductivity. Their heights t_1 (assigned for the ground layer) and t_2 (assigned for the signal conductor) are comparable with the thickness of the dielectric h which is often the silicon dioxide.

Ignoring the dispersion and loss effects, the line can be described by a quasi-static model (2.34) which supposes that the fundamental mode of this line is a quasi-TEM mode having the negligible longitudinal electric and magnetic field components. The modal or line parameters are calculated by a quasistatic approach when the propagation parameters are derived from the solution of the Laplace equation. At high frequencies, the microstrip mode has dispersion, and the quasistatic formulas should be improved taking into account the frequency dependence of modal parameters (2.37).

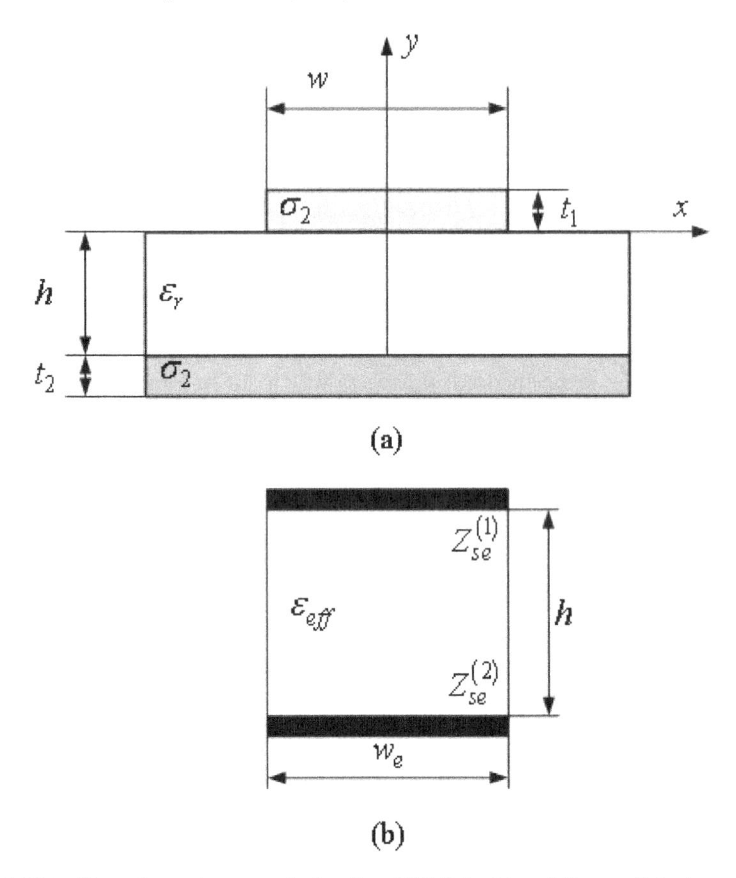

Fig. 6.1 Thin-film microstrip transmission line (TFML) –(a) and its parallel-plate model –(b). Adapted from [8]

The imperfect conductors and substrates cause the loss in microstrip transmission lines. Computation of modal loss in a wide frequency band is one of the most complicated tasks. An adequate modeling can be produced by the EM simulation when conductors are considered as lossy medium, and the penetrated field is simulated by a fine mesh inside of thin conductor strips [9]-[12]. Unfortunately, EM simulations are time-consuming, and a number of empirical and semi-empirical models were developed. Among them are the models based on the Wheeler's incremental inductance rule [13]-[15], conformal mapping approach [16], and models based on curve-fitting techniques applied to electromagnetically derived data [9],[17]. A few papers are for precise measurements of loss [18]-[21].

The models based on the incremental inductance rule are working well for thick conductors when their thickness is higher than the skin-depth. At low frequencies, the EM field penetrates deeply the thin strips, and these models are not correct. In this case, the modal field has the longitudinal E_z component [9]-[11]. The mode can still be considered to be of the quasi-TEM type, with strong E_z component. Thus, the mode is defined more exactly as a quasi-TM one [22],[23]. At higher frequencies, the field is pushed outside the conductors, and the skin-effect accounts for the loss of the line.

Our wide frequency band model uses an equivalent parallel-plate waveguide representing microstrip line (Fig. 6.1). Unlike in [22], our model depends on an effective width w_e and an effective dielectric constant ε_{eff}, which are crucial for the accurate computation of the complex propagation constant.

The two lossy plates in Fig. 6.1b are characterized by the complex surface impedances $Z_{\text{se}}^{(1)}$ and $Z_{\text{se}}^{(2)}$. The plates cover a dielectric slab of the height h and permittivity ε_{eff}, which is calculated as (2.37). The fundamental mode is a TM mode, and the E_z field component is derived from the Helmholtz equation as

$$E_z = \left(A \sin k_y y + B \cos k_y y\right) e^{-j\tilde{k}_z z} \tag{6.1}$$

where A and B are unknown coefficients and $k_y = \sqrt{k_0^2 \varepsilon_{\text{eff}} - \tilde{k}_z^2}$. At the difference to the ideal microstrip line, the propagation constant now is complex $k_z = \tilde{k}_z$. The transversal magnetic field is derived from Maxwell's equations as

$$H_x = jY_{\text{W}} \left(A \cos k_y y - B \sin k_y y\right) e^{-j\tilde{k}_z z}. \tag{6.2}$$

Here, $Y_{\text{W}} = \omega \varepsilon_{\text{eff}} \varepsilon_0 / k_y$.

To calculate the unknown propagation constant, the approximate Leontovitch boundary condition is used

$$E_z(z=0) = Z_{\text{se}}^{(1)} H_x(z=0), \;\; E_z(z=h) = -Z_{\text{se}}^{(2)} H_x(z=h). \tag{6.3}$$

It is valid for small surface impedances and yields two linear equations for the unknown coefficients A and B. The system of these equations has a unique solution if its determinant is zero. This gives the eigenvalue equation for the propagation constant $\tilde{k}_z$:

$$\left(\tan k_y h - jZ_{\text{se}}^{(2)} Y_{\text{W}}\right) - jZ_{\text{se}}^{(1)} Y_W \left(1 - jZ_{\text{se}}^{(2)} Y_{\text{W}} \tan k_y h\right) = 0. \tag{6.4}$$

We notice in (6.4) that the absolute value of the wave number k_y is small because $\text{Re}(\tilde{k}_z^2) \approx k_0^2 \varepsilon_{\text{eff}}$ and $\text{Im}(\tilde{k}_z^2 / k_0^2) < 1$ for practical microstrips. Besides, thin substrates yield $|k_y h| \ll 1$ and $\tan(k_y h) \approx k_y h$ [22]-[24]. Substituting this linear approximation of the tangent function in (6.4), we get an approximate expression for the propagation constant $\tilde{k}_z$:

$$\tilde{k}_z^2 \approx k_0^2 \varepsilon_{\text{eff}}(f) - \left(Z_{\text{se}}^{(1)} + Z_{\text{se}}^{(2)}\right) \frac{j\omega\varepsilon_0\varepsilon_{\text{eff}}(f)}{h} + Z_{\text{se}}^{(1)} Z_{\text{se}}^{(2)} \left(\omega\varepsilon_0\varepsilon_{\text{eff}}(f)\right)^2. \tag{6.5}$$

Beside the conductor loss, the microstrip lines suffer from substrate loss. It is comparable with the conductor loss only at high frequencies when the modal field is close to the TEM type. To compute the respective attenuation per unit length α_d, the quasi-TEM formula originated from [25],[26] is used

$$\alpha_d = 27.3 \cdot \frac{\varepsilon_r}{\varepsilon_r - 1} \cdot \frac{\varepsilon_{\text{eff}}(f) - 1}{\sqrt{\varepsilon_{\text{eff}}(f)}} \cdot \frac{\tan\delta}{\lambda_0[\text{mm}]}, \quad [\text{dB/mm}]. \tag{6.6}$$

In (6.6), $\tan\delta$ is the dielectric loss tangent, $\varepsilon_{\text{eff}}(f)$ is calculated according to (2.37). Finally, the total loss α_t of a microstrip transmission line is

$$\alpha_t = -10\log_{10}\left[\exp(2\,\text{Im}(\tilde{k}_z) \cdot l[\text{mm}])\right] + \alpha_d, \ [\text{ dB/mm}] \tag{6.7}$$

where l is the length of a transmission line.

Using (6.2), and (6.3), we derive the unknown coefficients A and B; we compute the field of the fundamental mode and its complex characteristic impedance $\tilde{Z}_c$ using the linear approximation of trigonometric functions:

$$\tilde{Z}_c(f) = Z_{\text{TM}} \frac{h}{w_e(f)} \cdot \frac{1}{1 - jZ_{\text{se}}^{(1)} \omega\varepsilon_0\varepsilon_{\text{eff}}(f) h}, \ [\Omega] \tag{6.8}$$

where $Z_{\text{TM}} = \tilde{k}_z / \omega\varepsilon_0\varepsilon_{\text{eff}}(f)$, and $w_e(f)$ is calculated according to (2.39).

The accuracy of the derived formulas depends on the width of the microstrip. For wide ones with $w/h > 3.5$, the used parallel-plate model shows acceptable accuracy, and the surface impedances of the signal and ground conductors are derived from the infinite sheet model [22]:

$$Z_{\text{se}}^{(i)} = -jZ_{0\text{M}}^{(i)} \cot k_i t_i \tag{6.9}$$

where $i = 1$ for the ground conductor, $i = 2$ for the upper plate, $\sigma_{1,2}$ are the specific conductivities of the conductor plates, $t_{1,2}$ are their thicknesses,

$k_i = \sqrt{-j\omega\mu^{(i)}\sigma_i}$, $Z_{0M}^{(i)} = (1+j)\sqrt{\omega\mu^{(i)}/\sigma_i}$, and $\mu^{(i)}$ is the absolute conductor's permeability.

Our wide-microstrip model has the following range of applicability: width of the microstrip is $w/h \geq 3.5$; conductor height is $0.03 \leq t/h \leq 0.472$; dielectric permittivity is $2.7 \leq \varepsilon_r \leq 12.9$. Its accuracy is verified with theoretical and experimental results [9],[10],[27],[28] for frequencies up to 100 GHz, see Figs. 6.2-6.6.

Fig. 6.2 shows the phase and attenuation constants of a microstrip line which parameters are: $w/h = 4.71$, $\varepsilon_r = 2.7$, $\tan\delta = 0.015$, $h = 1.7$ μm, and $\sigma_{1,2} = 2.5\times10^7$ S/m, $t_{1,2} = 0.8$ μm, and $\mu^{(1,2)} = \mu_0$.

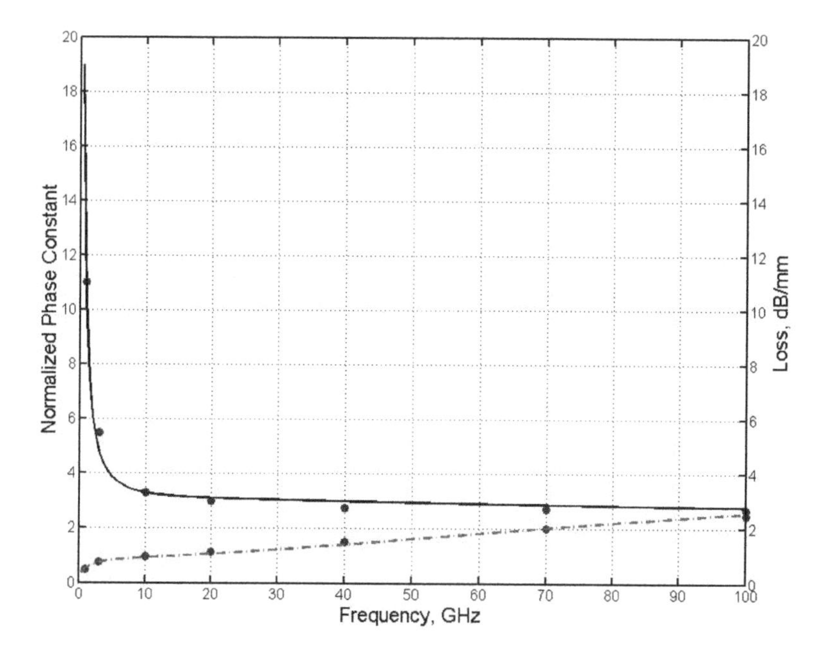

Fig. 6.2 Normalized phase constant, $\mathrm{Re}(\tilde{k}_z/k_0)$ (solid curve) and total loss α_t (dash-dotted curve) of a wide microstrip line for monolithic IC applications [8]. Circles – full-wave data [9]. Reprinted with permission of IEEE. License # 2873061358030

At low frequencies, the loss, mostly, is due to the modal field inside the conductors. The longitudinal electric field is not negligible and the modal phase velocity is very low [9]. The phase constant $\mathrm{Re}(\tilde{k}_z/k_0)$ decreases fast with frequency. The modal characteristic impedance $\tilde{Z}_c$ is complex and strongly frequency-dependent (Fig. 6.3).

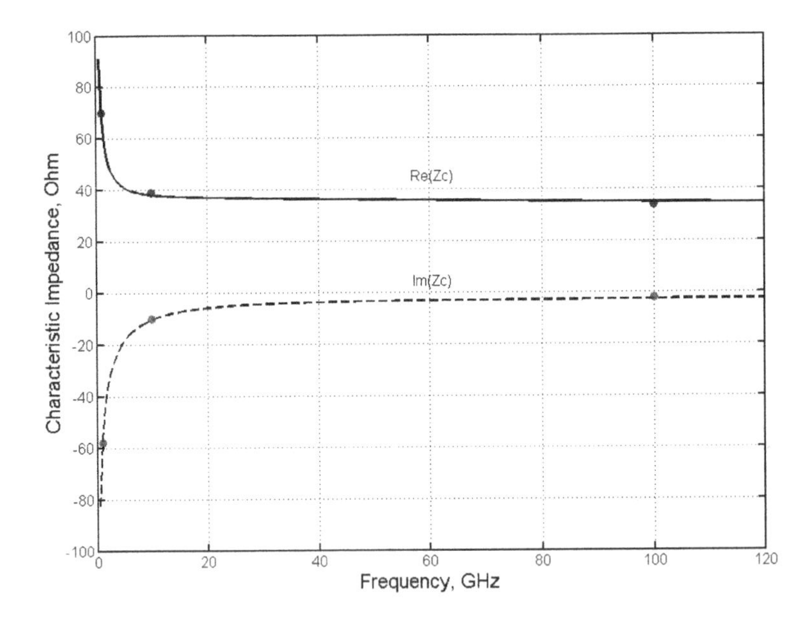

Fig. 6.3 Complex characteristic impedance of a wide lossy microstrip line for monolithic ICs [8]: Curves – calculations using (6.8); Circles – full-wave data [9]. Reprinted with permission of the IEEE. License # 2873061358030

At higher frequencies, the loss increases linearly mostly due to the dielectric. The characteristic impedance $\tilde{Z}_c$ is almost real and frequency independent. The difference between our results and the full-wave data from [9] does not exceed several percent up to 100 GHz.

In addition, we carried out comparisons with the full-wave data from [10] at frequencies up to 100 GHz, and a good agreement to within several percent has been observed for the complex propagation constant and characteristic impedance. Figs. 6.4 and 6.5 show the phase and attenuation constants of a microstrip line, which parameters are: $w/h = 4.724$, $\varepsilon_r = 9.6$, $\tan\delta = 0$, $h = 0.635$ mm, $\sigma_1 = \infty$, $\sigma_2 = 100$ kS/m, and $t_2 = 0.3$ mm.

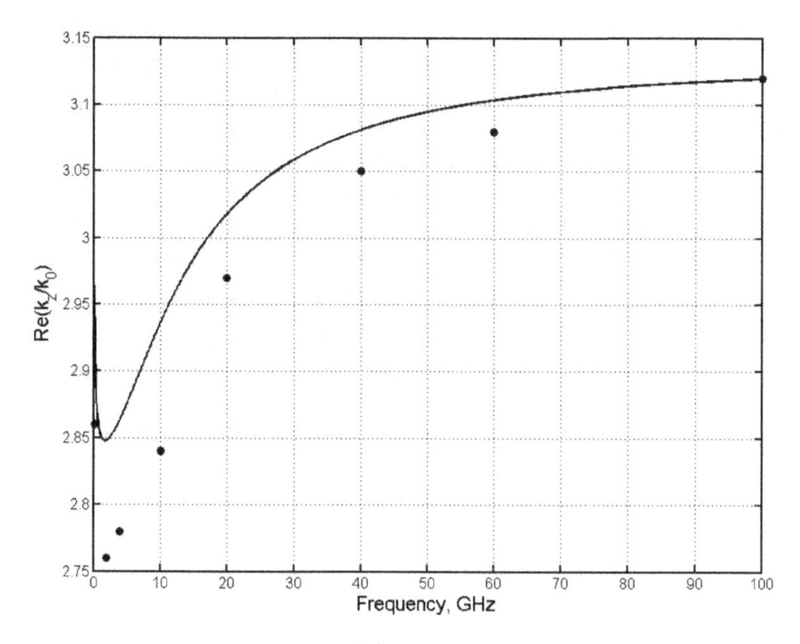

Fig. 6.4 Normalized phase constant $\mathrm{Re}(\tilde{k}_z/k_0)$ of a wide microstrip line for monolithic IC applications. Solid curve- our results; Circles- full-wave data [10]

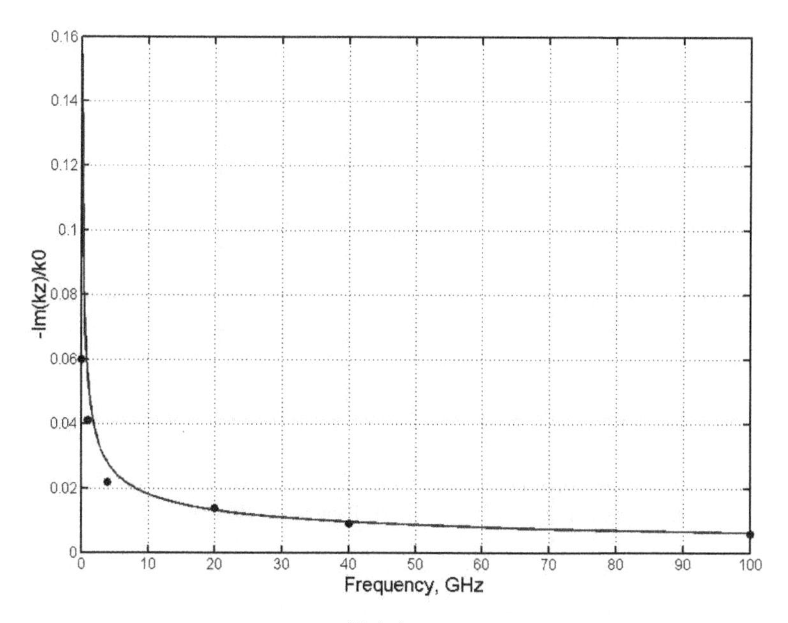

Fig. 6.5 Normalized imaginary part $\mathrm{Im}\left(\tilde{k}_z/k_0\right)$ of the complex propagation constant (solid line) of a wide microstrip line for monolithic IC applications. Circles – full-wave data [10]

The phase constant is influenced by the loss of conductors at low frequencies where the longitudinal electric field component is rather strong (Fig. 6.4). The field is concentrated inside the conductor and the modal phase constant is high. The modal effective permittivity can be over ε_r (see Figs. 6.2 and 6.4) that is in the contradiction with the theory considering the strips as ideal conductors. The maximal difference between full-wave and our data does not exceed 3.5%.

Recently, some experimental data was published for a thin-film Ti (200 nm, $\sigma_1 = 2.84\times10^6$ S/m)/Au (1 μm, $\sigma_2 = 2.94\times10^7$ S/m) microstrip realized on low-loss cyclic-olefin copolymer $(\varepsilon_r = 2.33,\ \tan\delta = 0.0005,\ w/h = 3.1)$ at frequencies up to 220 GHz where the maximum loss was registered about of 1.75 dB/mm at 220 GHz [29]. The measurements were compared with the data obtained with formulas from [9].

To compare our theory with the measurements, the surface impedance formulas for the signal Z_s and ground Z_g conductors are modified to take into account their layered cross-sections:

$$Z_s = Z_{s_1}^{(\infty)} \frac{Z_{2_s}\cos(\chi_{y_s}^{(1)} t_{1_s}) + jZ_{s_1}^{(\infty)}\sin(\chi_{y_s}^{(1)} t_{1_s})}{Z_{s_1}^{(\infty)}\cos(\chi_{y_s}^{(1)} t_{1_s}) + jZ_{2_s}\sin(\chi_{y_s}^{(1)} t_{1_s})} \qquad (6.10)$$

where $Z_{2s} = -jZ_{s_2}^{(\infty)}\cot(\chi_{y_s}^{(2)} t_{2_s})$, $Z_{s_{1,2}}^{(\infty)} = (1+j)\sqrt{\omega\mu_s^{(1,2)}/\sigma_{s_{1,2}}}$, $\chi_{y_s}^{(1,2)} = (1-j)/\Delta_{s_{1,2}}^0$, and

$\Delta_{s_{1,2}}^0 = \sqrt{2/\omega\mu_s^{(1,2)}\sigma_s^{(1,2)}}$.

$$Z_g = Z_{g_1}^{(\infty)} \frac{Z_{2_g}\cos(\chi_{y_g}^{(1)} t_{1_g}) + jZ_{g_1}^{(\infty)}\sin(\chi_{y_g}^{(1)} t_{1_g})}{Z_{g_1}^{(\infty)}\cos(\chi_{y_g}^{(1)} t_{1_g}) + jZ_{2_g}\sin(\chi_{y_g}^{(1)} t_{1_g})} \qquad (6.11)$$

where

$Z_{2g} = -jZ_{g_2}^{(\infty)}\cot(\chi_{y_g}^{(2)} t_{2_g})$, $Z_{g_{1,2}}^{(\infty)} = (1+j)\sqrt{\omega\mu_g^{(1,2)}/\sigma_{g_{1,2}}}$, $\chi_{y_g}^{(1,2)} = (1-j)/\Delta_{g_{1,2}}^0$,

$\Delta_{g_{1,2}}^0 = \sqrt{2/\omega\mu_g^{(1,2)}\sigma_g^{(1,2)}}$.

The complex propagation coefficient formula (6.5) is used again with Z_s and Z_g:

$$\tilde{k}_z^2 \approx k_0^2\varepsilon_{\text{eff}}(f) - (Z_s + Z_g)\frac{j\omega\varepsilon_0\varepsilon_{\text{eff}}(f)}{h} + Z_s Z_g\left(\omega\varepsilon_0\varepsilon_{\text{eff}}(f)\right)^2. \qquad (6.12)$$

Calculations of loss according to (6.12) and (6.7), experimental and theoretical data from [29] are shown together in Fig. 6.6. It is seen, that both theoretical models give comparable accuracy towards the measurements [29], but they have increased difference at frequencies over 150 GHz. It might be that the reason of this error is in the real properties of the titanium/dielectric interface and in the microgranularity of metals which have not been taken by the used simple models of conductors.

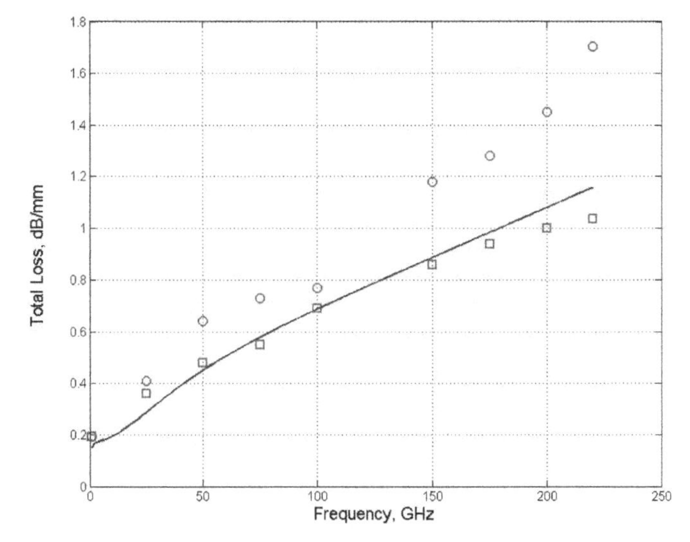

Fig. 6.6 Comparison of our data (solid curve), measurements [29] (circles), and calculations obtained by the authors of [29] using formulas from [9] (squares)

It is expected that the accuracy of (6.5) and (6.12) would be worse for narrow microstrips due to strong influence of the current crowding effect disturbing the conductor surface impedance and the applicable width range should be established for the mentioned formulas comparing the full-wave simulations and our formula's data.

6.1.2 Superstrated Thin-film Microstrip Line

In superstrated microstrip lines, the upper conductor is covered by a dielectric layer (Fig. 6.7a). Such a design is used in antennas, filters, and high-power couplers [15],[30]-[37]. In monolithic ICs, the signal conductors are buried inside a layered dielectric made of silicon, silicon oxide, or a passivation layer having different dielectric permittivities and loss tangents.

At high frequencies, it leads to complicated loss mechanism, and the layered structure should be taken into account by any software. An effective means to simulate the multilayered microstrips is a full-wave approach based on the mode matching techniques or the FDTD analysis. Circuit oriented software tools need analytical expressions valid in a wide frequency band and developed for an extended range of geometrical and physical parameters.

An analytical model of a lossy superstrated microstrip line is elaborated similarly to the previous one [8]. At the first, to this line an equivalent parallel plate waveguide model filled by the effective permittivity $\varepsilon_{\text{eff}}^{(s)}$ is put into the correspondence (Fig. 6.7b) [15],[31],[36]. At the second, the plates are substituted by the equivalent impedance walls of the surface impedances $Z_{\text{se}}^{(1,2)}$. The dielectric loss is calculated according to [25],[26]. It is supposed that the accuracy of the model depends on the formulas for the effective permittivity and dielectric loss, and again, the workability of them is expected up to 100 GHz.

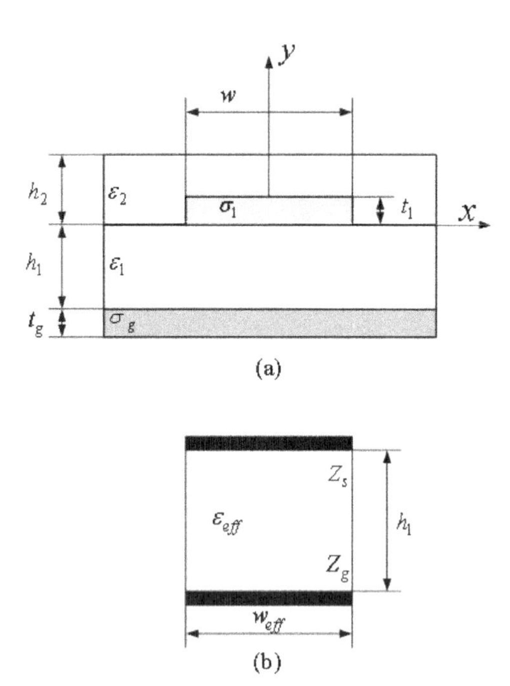

Fig. 6.7 Superstrated microstrip transmission line. (a)- Cross-section; (b)- Parallel-plate model

6.1.3 Two-layer Substrate Microstrip Transmission Line

The widely used silicon substrates have low resistivity about of 5-20 Ω-cm , and it is the cause of increased loss of microstrip lines and coplanar waveguides placed over these substrates. To avoid it, the silicon substrate is covered by a dioxide layer (Fig. 6.8a), and it decreases the loss. For instance, some measurements are given in where the substrate is covered by 1.5-μm dioxide film, and it allows for the loss within 28-67 dB/cm for the aluminum microstrip conductor of the height 4 μm [38],[39].

Further improvement of microstrips is with increasing the substrate resistivity. For instance, silicon can be bombarded by protons or ions, and semi-isolating high-resistivity $\left(10^6\ \Omega\text{-cm}\right)$ regions are formed in it. Microstrips placed over these spots have loss less than 4 dB/cm at 50-GHz frequency [38],[39]. Taking into account the semi-insulating substrates, the lines can be analyzed by techniques developed for the multi-layer dielectric microstrips. They have been studied by numerical methods in [27]-[30]. A few papers are on the measurement of loss in such transmission lines [18],[38],[39]. Analytical quasistatic models are published in [15] and [31] where the effective dielectric constant and the characteristic impedance are obtained by the conformal mapping approach. In [15], the loss is computed by the use of the incremental inductance rule that is applicable only to the thick-conductor lines.

The studied by us microstrip [40] consists of a two-layer high-resistivity substrate, a sandwiched conducting strip, and a ground layer (Fig. 6.8a). The sandwiched conductors are composed of metal layers of different specific conductivities σ_1 and σ_2. At low frequencies, the heights of the layers are comparable to the skin-depth.

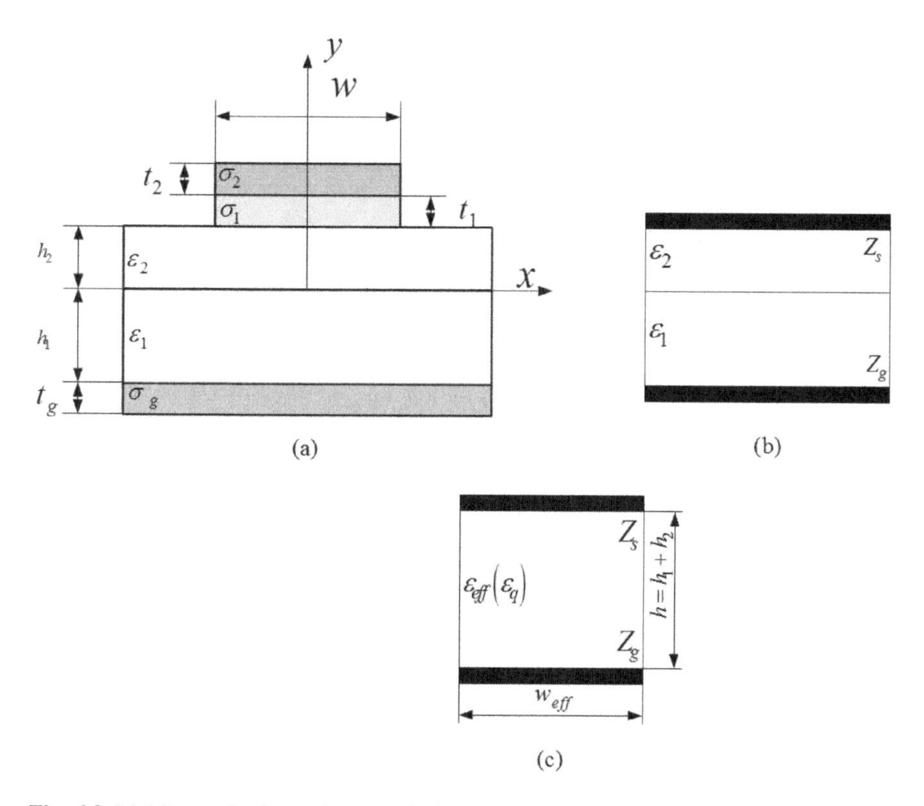

Fig. 6.8. Multilayered microstrip transmission line [40]. (a)- Cross-section; (b)- Multilayer parallel-plate equivalent model; (c)- Monolayer equivalent model. This material is reprinted with permission of John Wiley & Sons, Inc.

In the proposed model, a multilayered microstrip is represented by a parallel-plate multilayered waveguide (Fig. 6.8b). The multilayered conductors are substituted by the walls with the assigned to them equivalent surface impedances Z_s and Z_g. The fundamental mode field is described by the TM_y field, and the transverse resonant method is used for derivation of the eigenvalue equation regarding to the complex longitudinal propagation constant $\tilde{k}_z$. The resonant condition written at $y = 0$ is

$$Z_{\downarrow} + Z_{\uparrow} = 0 \tag{6.13}$$

where

$$Z_{\uparrow} = Z_{c_2} \frac{Z_s \cos(k_y^{(2)} h_2) + jZ_{c_2} \sin(k_y^{(2)} h_2)}{Z_{c_2} \cos(k_y^{(2)} h_2) + jZ_s \sin(k_y^{(2)} h_2)}, \tag{6.14}$$

$$Z_{\downarrow} = Z_{c_1} \frac{Z_g \cos(k_y^{(1)} h_1) + jZ_{c_1} \sin(k_y^{(1)} h_1)}{Z_{c_1} \cos(k_y^{(1)} h_1) + jZ_g \sin(k_y^{(1)} h_1)}, \tag{6.15}$$

$$Z_{c_1} = k_y^{(1)} / \omega \varepsilon_r^{(1)} \varepsilon_0, \ Z_{c_2} = k_y^{(2)} / \omega \varepsilon_r^{(2)} \varepsilon_0. \tag{6.16}$$

In (6.13) to (6.16), $k_y^{(1,2)} = \sqrt{k_0^2 \varepsilon_r^{(1,2)} - \tilde{k}_z^2}$, $\varepsilon_r^{(1,2)}$ are the relative permittivities of the dielectric layers, and Z_s, Z_g are the surface impedances of the upper and lower conductors, respectively.

The expression (6.13) is a transcendental equation, which can be solved regarding to $\tilde{k}_z$ only approximately. For this purpose, we assume that the specific conductivities of the thin layers of metals are high, the surface impedances Z_s and Z_g are low, and they do not depend on the propagation constant $\tilde{k}_z$. We suppose that the dielectric layers are electrically thin: $k_0 h_1 < 1$ and $k_0 h_2 < 1$. In this case, the trigonometric functions in (6.14) and (6.15) are substituted with $\cos(k_y^{(1)} h_1) \approx \cos(k_y^{(2)} h_2) \approx 1$, and $\sin(k_y^{(1)} h_1) \approx k_y^{(1)} h_1$, $\sin(k_y^{(2)} h_2) \approx k_y^{(2)} h_2$,[22]-[24]. It allows for transforming (6.13) into the following equation:

$$\begin{aligned} &Z_{c_1} Z_{c_2} \left(Z_s + Z_g + jZ_{c_1} k_y^{(1)} h_1 - jZ_{c_2} k_y^{(2)} h_2 \right) \\ &+ jZ_g Z_s \left(Z_{c_1} k_y^{(2)} h_1 + Z_{c_2} k_y^{(1)} h_2 \right) \approx 0. \end{aligned} \tag{6.17}$$

Taking into account $Z_g Z_s \approx 0$, and substituting Z_{c_1} and Z_{c_2} with (6.16) we get the first-order approximation for the propagation constant:

$$\tilde{k}_z^2 \approx k_0^2 \varepsilon_e - \left(Z_s + Z_g \right) j \omega \varepsilon_0 \varepsilon_e / h \tag{6.18}$$

where

$$\varepsilon_e = \varepsilon_r^{(1)} \varepsilon_r^{(2)} h \Big/ \left(h_1 \varepsilon_r^{(2)} + h_2 \varepsilon_r^{(2)} \right). \tag{6.19}$$

Here, $h = h_1 + h_2$. The expression (6.18) is similar to a formula (6.5) derived for an impedance-wall parallel-plate waveguide of the height h and filled by a medium of the permittivity ε_e. Unfortunately, this formula [35] does not take into account the fringing field, and it can have low accuracy.

To improve (6.18) the parallel-plate multilayered waveguide (Fig. 6.8b) is substituted by an equivalent monolayer line filled by the equivalent medium of ε_{eq}, which takes into account the fringing fields. Then, this line is represented by a parallel-plate waveguide (Fig. 6.8c) filled by the effective frequency dependent permittivity TE_{20}. In this case, (6.18) becomes

$$\tilde{k}_z^2 \approx k_0^2 \varepsilon_{eff}\left(\varepsilon_{eq}, w, f\right) - \left(Z_s + Z_g\right) \frac{j\omega\varepsilon_0 \varepsilon_{eff}\left(\varepsilon_{eq}, w, f\right)}{h} \tag{6.20}$$

where $\varepsilon_{eff}\left(\varepsilon_{eq}, w, f\right)$ is computed according to (2.37).

The conductor loss computation by (6.20) requires the equivalent impedances of the upper (strip) and lower (ground) plates. The upper conductor consists of two layers of different specific conductivities σ_1 and σ_2. The effective surface impedance Z_s is approximated by the expression (6.21) that takes into account the multilayer structure of the strip and the current crowding effect influencing narrow conductors:

$$Z_s = \begin{cases} Z_w = Z_{s_1}^{(\infty)} \dfrac{Z_2 \cos(\chi_y^{(1)} t_1) + jZ_{s_1}^{(\infty)} \sin(\chi_y^{(1)} t_1)}{Z_{s_1}^{(\infty)} \cos(\chi_y^{(1)} t_1) + jZ_2 \sin(\chi_y^{(1)} t_1)}; & \dfrac{w}{h} \geq 3.5 \\ Z_w \left[0.7\, w_{eq}(f) / w_{eq_t}\right]; & 0.1 \leq w/h \leq 3.5 \end{cases} \tag{6.21}$$

where $Z_{s_{1,2}}^{(\infty)} = (1+j)\sqrt{\omega\mu^{(1,2)}/\sigma_{1,2}}$, $\chi_y^{(1,2)} = (1-j)/\Delta_{1,2}^0$, $\Delta_{1,2}^0 = \sqrt{2/\omega\mu^{(1,2)}\sigma_{1,2}}$, and $Z_2 = -jZ_{s_2}^{(\infty)} \cot(\chi_y^{(2)} t_2)$. The extended due to the fringing effect width w_{eq_t} can be calculated according to [41].

Here, $\mu^{(1,2)}$ are the absolute permeabilities of microstrip conducting layers, and $w_{eq}(f)$ is the effective frequency-dependent width computed according to (2.39). The ground conductor of limited thickness t_g is substituted by a surface of the effective impedance Z_g that is computed according to

$$Z_g = -jZ_g^{(\infty)} \cot(\chi_y^{(g)} t_g) \tag{6.22}$$

where $Z_g^{(\infty)} = (1+j)\sqrt{\omega\mu^{(g)}/\sigma_g}$, $\chi_y^{(g)} = (1-j)\Big/\sqrt{\omega\mu^{(g)}\sigma_g/2}$, $\mu^{(g)}$ and σ_g are the absolute permeability and the specific conductivity of the ground conductor, respectively.

Beside the conductor loss, the microstrip lines suffer from the substrate loss α_d. This loss is comparable with the conductor loss only at high frequencies where the modal field is close to the TEM type. To calculate this loss, the perturbation formula (6.6) can be used, but it requires calculating the equivalent loss tangent of this multilayered substrate. It is performed using the Schneider's formula [25],[26]:

$$\left(\tan\delta\right)_{\text{eq}} = \frac{1}{\varepsilon_{\text{eq}}}\sum_{n=1}^{2}\varepsilon_n \frac{\partial\varepsilon_{\text{eq}}}{\partial\varepsilon_n}\tan\delta_n \tag{6.23}$$

Finally, the total loss is calculated according to (6.7).

The ability of the developed equivalent monolayer model to represent the multilayered microstrip lines was verified as the first step. For this purpose, our results for $\varepsilon_{\text{eff}}\left(\varepsilon_{\text{eq}}, w, f\right)$ were compared with the quasistatic data derived in [37].

At low frequencies, the difference is within 0.9-6 % for $0.1 < w/h < 10$ in spite the use of a simple formula for ε_{eq} and $\left(\tan\delta\right)_{\text{eq}}$. More accuracy is possible to achieve using the advanced formulas for the equivalent quasistatic dielectric permittivity from [15] and [31].

The calculations of loss α_t (Fig. 6.9) are verified with experimental and theoretical results from [18],[27],[28] given for a two-layer dielectric microstrip. The signal strip conductor of this transmission line consists of two metal layers. The geometrical and physical parameters are $\varepsilon_r^{(1)} = 12.9$, $\tan\delta_1 = 10^{-4}$, $h_1 = 100$ μm, $\varepsilon_r^{(2)} = 6.5$, $\tan\delta_2 = 0$, $h_2 = 2$ μm, $t_1 = 0.4$ μm, $t_2 = 3$ μm, $t_g = 12$ μm, $\sigma_1 = 2.1\times10^6$ S/m, and $\sigma_2 = \sigma_g = 4.1\times10^7$ S/m.

Fig. 6.9 shows a comparison between our calculation (lines) with experimental data (upward and downward triangles) from [18].

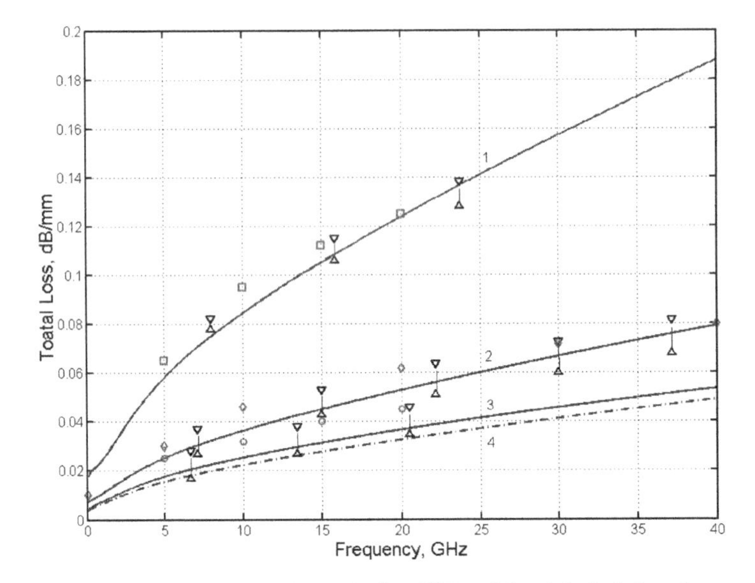

Fig. 6.9. Total loss α_t of a narrow microstrip line [40]: solid and dashed-dotted curves – our model; Downward triangles – upper level of measured loss [18]; Upward triangles – lower level of the measured loss [18]. Squares – full-wave data [27]; Diamonds - full-wave data [28]; Circles – full-wave data [27]. Microstrip line parameters: (1) $w/h = 0.1$; (2) $w/h = 0.7$; (3),(4) $w/h = 3.5$. This material is reprinted with permission of John Wiley & Sons, Inc.

As well, the full-wave results of [28] and [27] are presented by diamonds and circles, correspondingly. For comparison, a wide microstrip line $(w/h = 3.5)$ was computed using two formulas for the equivalent surface impedance. The dash-dotted line represents the results derived by the impedance formula valid for wide microstrips. The solid line shows the loss dependence calculated with the empirically corrected formula that takes into account the current crowding effect. Both formulas give comparable results with measurements.

The narrow microstrips $(w/h < 3.5)$ were computed using the surface impedance formula multiplied by a correction factor from (6.21). The comparison with measurements and theoretical data shows a good agreement.

The accuracy of the phase computation was studied by comparison of our results with full-wave data derived in [28] for the multilayered microstrip line. The line has the same geometrical and physical parameters with the line considered above. It has been found that for the given data the difference does not exceed several per cent (Fig. 6.10).

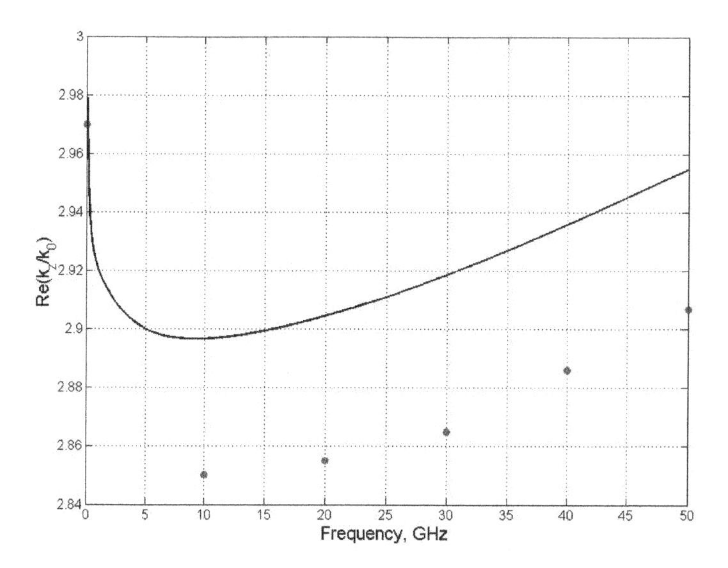

Fig. 6.10 Normalized phase constant versus frequency [40]. Solid line- our model; Circles- full-wave data [28]. Microstrip line normalized width: $w/h = 0.7$. This material is reprinted with permission of John Wiley & Sons, Inc.

6.1.4 Si-microstrip Transmission Lines Shielded by Impedance Layers

As it has been already mentioned above, the microstrips placed over the silicon or silicon oxide/silicon substrates suffer from increased loss. One of the ideas of further modification of silicon-based microstrip lines is with the shielding of lossy silicon by a doped semiconductor layer with increased conductivity or by patterned metal layers. For instance, in [42], the silicon is covered by a n-type layer doped to 8 Ω/sq. and connected to the ground (Fig. 6.11). This layer is patterned to prevent the longitudinal current. Besides, it plays a role of capacitive shield towards lossy silicon and decreases the induced current in the substrate, which causes the loss. Microstrip of the conductor thickness 3 μm and width 25 μm is placed over a dioxide layer of the thickness 4 μm.

It is shown by measurements that this shielding allows to decrease the loss down to 1.4 dB/cm versus 8.3 dB/cm of an unshielded microstrip at 6 GHz. Besides, this line can provide increased characteristic impedance in comparison to the TFML due to the decreased capacitance to the ground.

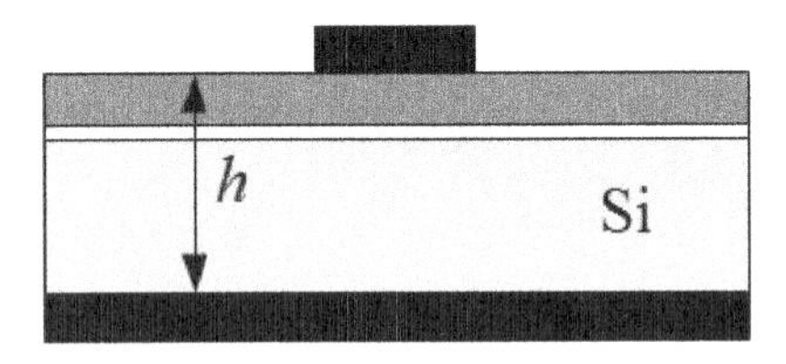

Fig. 6.11 Doped Si-SiO_2 microstrip transmission line. The n-type layer is shown in white

A modification of n-doped Si-SiO_2 microstrip line is described in [43] for millimeter-wave frequency range. Lossy silicon is separated from the dioxide layer by a thin (0.001-1 μm) well-conducting Nepi-layer (in white color) which decreases penetration of the EM field deeply inside lossy silicon. Additionally, the signal conductor has decreased capacitance to the ground due to this shielding. It allows using this transmission line for the millimeter-wave spiral inductors of increased quality factors (up 20 at 30 GHz) and of the self-resonance frequencies up to 64 GHz.

A full theory of this transmission line has not been created yet, but the quasi-TEM fundamental mode can be calculated similarly to the models developed for mono- and multilayered microstrips (see above). For this purpose, the n-doped line (Fig. 6.11) is transformed into an equivalent monolayer microstrip using the approach proposed by J. Svačina [31]. In this case, the equivalent quasi-static permittivity $\varepsilon_{\mathrm{eq}}$ is

$$\varepsilon_{\mathrm{eq}} = \frac{\left(\sum_{1}^{M} q_i\right)^2}{\sum_{1}^{M} \frac{q_i}{\varepsilon_r^{(i)}}} + \frac{\left(\sum_{j=m+1}^{N} q_j\right)^2}{\sum_{j=M+1}^{N} \frac{q_j}{\varepsilon_r^{(j)}}} \tag{6.24}$$

where $M = 3$ and $N = 4$, and q_i are available from [31]. To calculate the loss caused by the substrate layers and conductors, the equivalent monolayer microstrip transmission line can be transformed to an equivalent parallel-plate waveguide.

Then the phase constant and the conductor and dielectric loss constants are calculated according to (6.5), (6.6), and [25],[26], correspondingly. Accuracy of calculations depends on the theory [31] as well.

At frequencies, where the skin effect in semiconductor layers is strong and the *n*-doped layer influences the fundamental mode, the equivalent impedance model of well-conducting layers is useful. Further data on this line can be obtained by EM numerical simulations.

Instead of *n*-doped silicon layer, a patterned conductor shield can be used placed between the semiconductor and dioxide layers. The grounded strips are orthogonal to the signal conductor, and it prevents the penetration of the field to silicon to reduce the loss. An analytical model based on the Agilent Momentum simulation data of this line is given in [32] where a significant decrease of the frequency dependence of the line's shunt capacitance and inductance is shown.

6.1.5 Three-dimensional Modifications of Microstrip Line

High loss of conventional microstrip lines in millimeter-wave range stimulates further research on their modifications. Some lines on multilayered substrates or/and superstrates have been considered above. The 3-D technologies allow modifying the microstrip cross-section to decrease the loss and parasitic coupling. Some results are with increasing the slow-factors of 3-D transmission lines [33],[34], which is interesting in realization of distributed circuits by monolithic technologies.

6.1.5.1 Stacked Thin-film Microstrip Line

The above-considered results on simulation of thin-film microstrip lines show a strong role of conductor loss at low- and high-frequencies. Today's tendency of integrated technology is towards the use of nanometric components and interconnects, and, soon, the fundamental limits of CMOS technology would be reached. The interconnects and passive components, being responsible for the loss, dispersion, and parasitic resonances, seriously undermine the attempts to improve the parameters of millimeter- and sub millimeter-wave integrations. Some problems can be avoided by further decrease of the size of capacitors and inductors, but, at the same time, it causes increased conductor loss due to smaller and smaller interconnect cross-sections. For instance, a contemporary 45-nm digital silicon technology allows for the thinnest signal conductor (Al) about of 1 μm only, while a 180-nm technology is with the 3-μm top metal (Al) thickness. Other metal layers are thinner, but they can be manufactured using copper, which only partly improves the situation with the conductor loss.

To increase the cross-section of conductors, in many papers, the stacking of ground layer and signal strips are proposed. In [44], a stacked ground layer microstrip is studied which is manufactured using a 45-nm digital CMOS technology. An Al signal conductor is of 1-μm height, and eleven Cu layers are connected to each other by via-holes to make a stacked ground. The measurements show the loss of microstrip below 15 dB/cm at 110 GHz. Unfortunately, more detailed information on the geometry of this measured line is not available. This loss data is in some accordance with [45], where the stacked microstrips were manufactured using the STMicroelectronics CMOS 32-nm technology. The signal strip consists of two AL and Cu layers. Ground plane is of six perforated Cu-layers. The measurements show the loss constant close to 16 dB/cm for a 70-Ω microstrip line at 110 GHz.

6.1.5.2 V-shaped Microstrip Line

One of these ideas to decrease the loss is the microshielding of the microstrip cross-section from the substrate [46]-[49]. According to one of these technologies, initially, a V-shaped cavity is formed in silicon by etching. This cavity, covered by a metal, is filled by a low-loss dielectric, and the signal conductor of this microstrip is placed over it, being isolated from the lossy silicon (Fig. 6.12). It is seen that this transmission line is a transient one to a coplanar waveguide, and it has all its features as the compatibility with the planar semiconductor components.

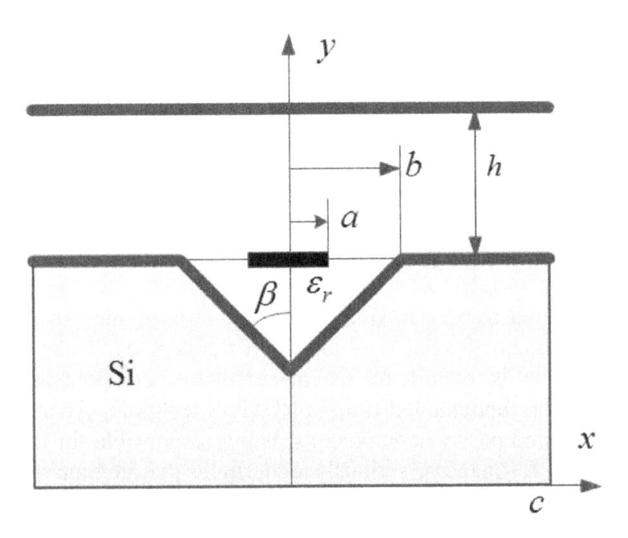

Fig. 6.12 V-shaped microstrip transmission line. Adapted from [48].

This V-shaped line has been studied using a conformal mapping approach taking into account the quasi-static nature of its fundamental mode [47]-[49]. The quasistatic effective permittivity ε_{eq} of this line is calculated as a ratio of the capacitances derived with a two stage conformal transformation:

$$\varepsilon_{\text{eq}} = \frac{C_v(\varepsilon_r)}{C_v(\varepsilon_r = 1)} \tag{6.25}$$

where $C_v = C_1 + C_2$, $C_1 = 2\varepsilon_0 K(k_1)/K(k_1')$, $C_2 = 2\varepsilon_r K(k_2)/K(k_2')$, and K is the elliptical integral. Its argument k_1 is computed with $k_1 = \tanh(\pi a/2h)/\tanh(\pi b/2h)$. Calculation of the second argument k_2 is more complicated, and the following integral is involved

$$\frac{a}{b} = \int_0^{k_2} f(t)dt \Big/ \int_0^1 f(t)dt \tag{6.26}$$

where

$$f(t) = (t^2 - 1)^{-(2\beta+\pi)/2\pi}. \tag{6.27}$$

The modified arguments are calculated as $k_{1,2}' = \sqrt{1-k_{1,2}^2}$. The characteristic impedance Z_c is

$$Z_c = \frac{60\pi}{\sqrt{\varepsilon_{eq}}}\left[\frac{K(k_1)}{K(k_1')} + \frac{K(k_2)}{K(k_2')}\right]^{-\frac{1}{2}}, \ [\Omega]. \tag{6.28}$$

It is shown that Z_c increases with the angle β. The influence of the slot between the strip and edge of the grounded cavity is essential only if this slot is a narrow one. The simulations performed in [47]-[49] show that the characteristic impedance of this line can vary within 20-150 Ω.

6.1.5.3 Air-filled Strip and Buried Microstrip Transmission Lines for Monolithic Integrations

Although the design of these lines is traditional and known for decades, they can be used in monolithic integrated circuits being manufactured by different 3-D technologies. The air-filling allows for decreasing the dielectric loss and for isolating the signal conductors from lossy silicon substrate. One of the first attempts to manufacture these lines and to explore their loss parameters is published in [50] where a thick dioxide layer is covered by a copper ground layer and then by the porous silicon dioxide. Using the techniques of photolithography, electroplating, and wet dissolving of porous silicon, the air-filled strip (Fig. 6.13a) and the microstrip (Fig. 6.13b) lines are manufactured.

The signal conductors made of copper are supported by metallic pedestals placed on the silicon dioxide substrate through the holes in the ground copper layer. In the case of strip line, a dissolver removes the porous silicon dioxide through the technological windows in the upper shield to make the cross-section free of this dielectric. A 50-Ω strip line (w= 26 μm, t=12 μm, $2b$=36 μm) is measured in frequencies up 40 GHz, and the loss less than 3 dB/cm is found. The loss of buried microstrip line (Fig. 6.13b) is less than 1.2 dB/cm in the frequency range of 1-40 GHz (w=30 μm, t=12 μm, h=25 μm).

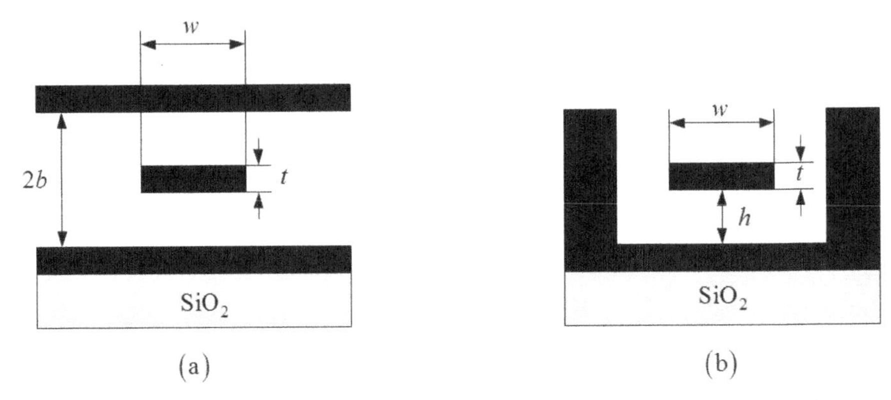

Fig. 6.13 Air-filled strip (a) and buried microstrip (b) transmission lines available for monolithic integrations. Adapted from [50]

To calculate these lines, the known formulas can be used taking into account the comparability of the thickness of central conductors to the height of the air-gap between them and the ground layers. In the case of buried microstrip (Fig. 6.13b), the influence of sidewalls should be taken into account additionally.

6.1.5.4 Fenced Microstrip and Strip Transmission Lines

The increasing density of hybrid and monolithic ICs leads to a high level of cross talks of components and transmission lines. Contemporary microwave modules should have reduced coupling between the components. One of the techniques is the isolation of transmission lines from each other and from the lossy silicon.

Another idea is the use of via-hole fencing [51]-[57]. Via-hole fence reduces coupling through the shortening of the electrical field to the ground (Fig. 6.14).

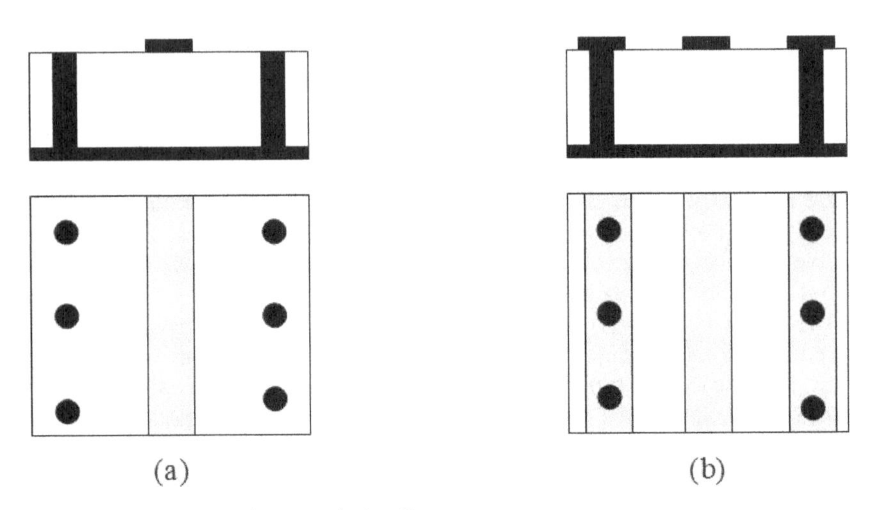

Fig. 6.14 Fenced microstrip transmission lines

In [51], two variants of fenced microstrip line are studied. The fence of the first of them (Fig. 6.14a) consists of via-holes shorted to the ground plane. It is shown that this fence is not effective enough to shield the modal electric field of microstrip line. In some cases, the fence is excited, and parasitic radiation increases the coupling of microstrip lines separated from each other by this fence. To decrease this effect, the upper ends of the via-holes are connected to each other by a conducting strip (Fig. 6.14b). An additional idea to increase the isolation is making a two-row fence. Usually, it gives additional several dB of isolation [52]. Next technique is placing signal strips on different levels of inside of neighboring fenced channels [54].

In some cases, the via-hole fence influences the microstrip, and the best distance between the strip and the fence is about three heights of the substrate. To avoid serious radiation, the distance between the via-holes should be less than a half of the modal wavelength.

Fencing is used for isolation not only of microstrips but for shielding of strip lines where the parasitic coupling is supported with a parallel-plate mode. In [55], two modifications of this transmission line are studied (Fig. 6.15).

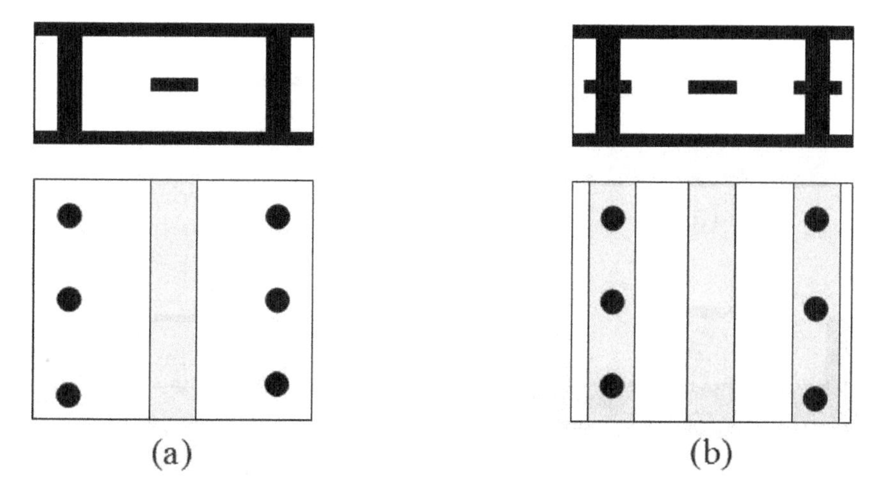

Fig. 6.15 Fenced strip transmission lines. Adapted from [55].

The first of them uses a fence consisting of a single row of via-holes (Fig. 6.15a). Numerical simulations show that the near-end coupling is reduced in this case. At the same time, the far-end coupling can be increased through this via-hole fence. To suppress this parasitic effect, the strips should be placed at a certain distance from the fence. Another approach, studied by the authors of [55], is the connecting of via-holes by conducting strips placed at the same level as the signal one (Fig. 6.15b). It decreases coupling down to -30 dB according to the experimental and theoretical data from [55].

The above-described techniques are applied to a silicon integrated power-divider network in 40-50-GHz band [56]. The isolation of the network arms exceeds 25 dB at 45 GHz.

The low-loss coupled fenced microstrips placed inside a silicon dioxide layer are studied in [57] up to 50 GHz. It is shown that due to fencing by via-holes and isolation of the strips from the lossy silicon by ground layer, the loss close to 10 dB/cm has been reached, and some analytical formulas verified by measurements are given for calculation of the common and differential modes of the coupled microstrip lines.

6.1.5.5 Ridged Microstrip Transmission Line

The ridged or truncated microstrip was considered in [58] and studied initially by a numerical quasi-static method. This transmission line consists of a grounded dielectric slab fully or partially covered by a conducting strip (Fig. 6.16a). It was found that the effective permittivity was decreased by 30% versus a conventional microstrip if the edges of microstrip were placed at a distance less than a half of the strip width: $W_t < 0.5W$.

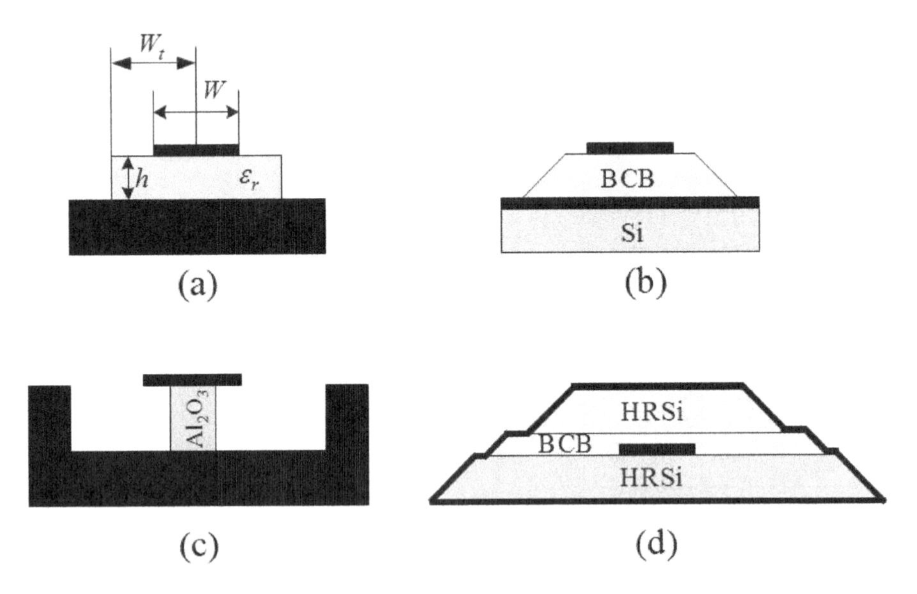

Fig. 6.16 Ridged microstrip transmission line (a) and its monolithic thin-film variant (b-d)

These ridged microstrips are interesting in many microwave and high-speed applications including the narrow-bandwidth filters, multi-element antennas with decreased parasitic coupling between the ridged microstrip patches, *etc*. These lines, being even placed close to each other, do not support the surface waves, which are typical for the microstrip patches placed on common substrate, and the ridged patches are highly isolated from each other even at millimeter-wave frequencies. This effect, the modal phase constants, and the characteristic impedances of a single and coupled ridged waveguides are calculated using the integral equation-mode matching method up to 105 GHz in [59]. Simple analytical parameters of this line were derived by J. Svacina in 1993.

One of these applications is considered in [60] where the miniature filters on ridged resonators and microstrips are investigated. Initially, the diffraction of the main mode at the edge of the ridged resonator is studied to define the resonant frequencies and the radiation from the edges. Then several ridged square-patch resonators ($\varepsilon_r = 100$) are coupled to each other and fed by a ridged waveguide ($\varepsilon_r = 10$). The filters have been tuned experimentally, and they show the elliptical type of the frequency response with the bandwidth of 5% at 4.85 GHz.

In [61], some quasistatic analytical formulas are derived using an asymptotic technique for a theory of ridged microstrip lines. To compute the effective permittivity ε_{eq} and the characteristic impedance Z_c of this line, a formula for the line capacitance C is calculated initially:

$$C = C_\infty - A_0/W_t^3 + A_1/W_t^5 \tag{6.29}$$

where $A_0 = \dfrac{2(\varepsilon_r - 1)^2 q \cdot h^2}{3\pi^2 \varepsilon_r^2} W$, $A_1 = (0.5W)^5 \left[C_{a,t} - C_\infty + A_0 \cdot (0.5W)^{-3} \right]$,

$$C_{a,t} = \frac{2\pi\varepsilon_0}{\ln\left(\dfrac{8h}{W} + 1\right)} + \frac{\varepsilon_0 W}{h}\left(\varepsilon_r - \frac{\pi}{4}\right).$$

In addition, q should be a unit charge [61]. The effective permittivity is calculated as a ratio of the capacitance C of the line filled by a dielectric to the capacitance C_0 of the air-filled line: $\varepsilon_{eq} = C/C_0$ where $C_0 = \dfrac{\varepsilon_0 W_{eq}}{h}$ with $W_{eq} = W + \dfrac{2h}{\pi}\{\ln[2\pi e \cdot (W/2h + 0.92)]\}$. The characteristic impedance of the ridged microstrip line is $Z_c = Z_0/\sqrt{\varepsilon_{eq}}$ where $Z_0 = 1/cC_0$.

Additionally, C needs the capacitance of the non-ridged microstrip $C_\infty = \sqrt{\varepsilon_{eq}^{(\mathrm{mstrip})}}/cZ_c^{(\mathrm{mstrip})}$ where $\varepsilon_{eq}^{(\mathrm{mstrip})}$ and $Z_c^{(\mathrm{mstrip})}$ are calculated using (2.34) and (2.35).

The maximal error estimated by the authors of [61] in comparison to the data from [58] is -6% for the characteristic impedance Z_c and -12% for the effective permittivity ε_{eq} given for $\varepsilon_r = 24$ and $W_t = h$.

In [62], a microstrip with the truncated dielectric and ground plane is studied numerically, and this study is useful for monolithic applications where the ground plane sheets are of a limited width comparable with the signal conductor size.

A monolithic modification of the ridged microstrip (Fig. 6.16b) workable up to 220 GHz and manufactured by a silicon technology is studied in [63],[64]. The signal strip conductor made of a gold layer of the thickness 3 μm is placed on the top of a BCB dielectric of the trapezoidal shape and of the height $h = 20$ μm ($\varepsilon_r = 2.65$, $\tan\delta = 0.002$). This substrate is isolated from a low-resistivity silicon by a thick (10 μm) gold layer of ground. In a wide frequency band up to 220 GHz, a 50-Ω transmission line shows the loss which is less than 6 dB/cm. Using this line, a 180-GHz branch-line directional coupler is designed and measured with good agreement with the Agilent ADS simulations and experimental data [64].

Several variants of the ridged microstrip manufactured using the selective anodizing processes are considered in [65]. The cross-section of a modified line is shown in Fig. 6.16c. It is placed inside an air-filled microcavity etched in a thick aluminum sheet. The ridge is of the porous alumina having low permittivity and loss. The air-filled cavity can be covered by a shield. It is found that the 50-Ω microstrips at frequency 40 GHz have low loss about 2-2.6 dB/cm (unshielded variant) and about 0.43-0.76 dB/cm for a microstrip covered by a metallic shield.

The microstrip line shown in Fig 6.16d can be considered as a shielded variant of ridged line [66]. The signal conductor is placed inside a sandwich of high-resistivity silicon (HRSi) and BCB layers. The shield is grounded by via-holes (are not shown in Fig. 6.16d). It is stated that it prevents the excitation of higher-order modes, and the parameters of this line are measured in frequencies up 20 GHz with the found loss up to 1.45 dB/cm.

6.1.5.6 Vertical Strip Transmission Lines

The vertical strip transmission lines are known for decades [67]. Their strips are placed normally to the conductor plate to decrease the capacitance to the ground (Fig. 6.17a). At the same time, the inductivity of this line is about equal to the one of the planar strip transmission line (Fig. 6.17b).

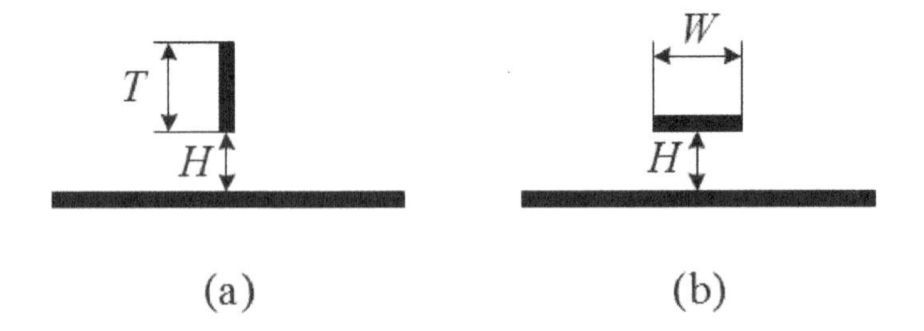

Fig. 6.17 Vertical (a) and horizontal (b) strip lines

This feature is used in the broadband directional couplers and in the 3-D filters [67],[68]. One of such filters is described in [69] where the vertical strip resonators placed on several layers of an IC. Additionally to the strong coupling between the resonators due to their broadside orientation, the parasitic capacitance to the ground is reduced by vertical orientation of the resonator's conductors.

The simplest geometry of a vertical line is calculated analytically with the conformal mapping approach [70]. Compare the capacitances of the infinitely thin strips of vertical (Fig. 6.17a) and horizontal (Fig. 6.17b) orientations. For the vertically oriented strip, the capacitance is

$$C_{\mathrm{VP}} = \frac{\varepsilon_r \varepsilon_0}{2} M(k) \tag{6.30}$$

where ε_r is the relative permittivity of the dielectric in which the strip is embedded. The function $M(k)$ is given by

$$\begin{cases} \dfrac{2\pi}{\ln\left[2\dfrac{1+\sqrt[4]{1-k^2}}{1-\sqrt[4]{1-k^2}}\right]} & 0 \le k \le \dfrac{1}{\sqrt{2}} \\ \dfrac{2}{\pi}\ln\left[2\dfrac{1+\sqrt{k}}{1-\sqrt{k}}\right] & \dfrac{1}{\sqrt{2}} \le k \le 1 \end{cases} \tag{6.31}$$

where

$$k = \sqrt{1 - \frac{1}{(1+T/H)^2}}. \tag{6.32}$$

The capacitance C_{HP} of the horizontal strip (Fig. 6.17b) is

$$C_{\mathrm{HP}} \cong \varepsilon_0 \varepsilon_r \left[\frac{W}{H} + \frac{4\ln(2)}{\pi}\right], \; W \gg H \tag{6.33}$$

Comparing C_{VP} and C_{HP} it is found that $C_{\mathrm{VP}}/C_{\mathrm{HP}} \approx 0.5$, i.e. the vertical orientation of the strip decreases twice the capacitance of the line.

Another design of the vertically oriented strip line is considered in [71] where its strip is placed orthogonally to a supporting dielectric slab made of a low permittivity dielectric (Fig. 6.18). It has decreased parasitic capacitance to the ground and increased phase velocity regarding to the air-filled analog. This transmission line can be used in hybrid ICs for high-speed interconnections where the time delay is an important issue.

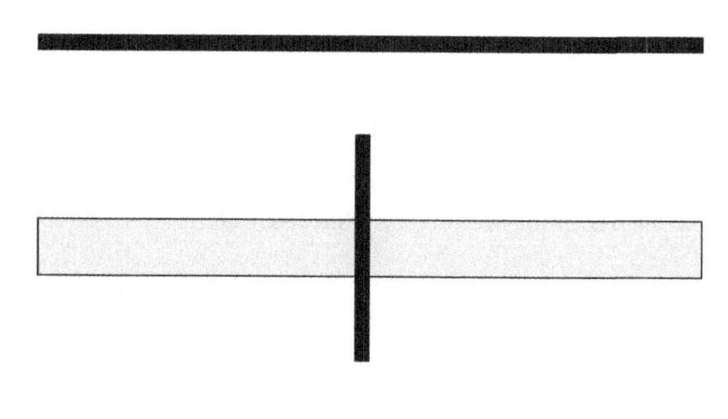

Fig. 6.18 Vertical strip line for high-speed applications

In practice, even more complicated lines with vertically mounted strips are used to reach the optimal coupling between the elements of 3-D modules or to isolate them from the environment (see Section 5.4 and [72]-[74]).

A vertically oriented strip is embedded into a non-irradiative dielectric (NRD) waveguide, which, usually, consists of a dielectric slab placed between two conducting plates. Unfortunately, this waveguide can suffer from parasitic excitation of two modes whose propagation constants are close to each other. In [75], it was found that this effect could be eliminated by placing a vertical strip at the center of the waveguide (Fig. 6.19).

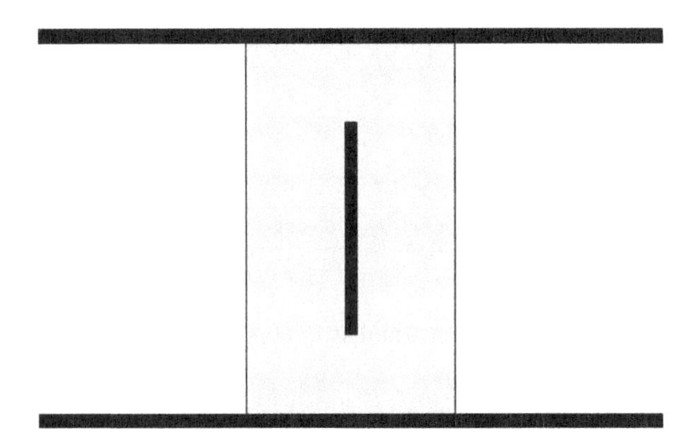

Fig. 6.19 NRD guide with a vertically imbedded conducting strip

The authors of [75] show that, in the case of a narrow strip, the quasi- LSE_{01} even mode is excited mostly. For wide microstrips, the excitation of the quasi-LSM_{11} odd mode is more typical.

The vertically oriented microstrips can be manufactured by monolithic technologies. One of the designs is considered in [76]-[78] where the vertical conductors are placed over a dielectric substrate and used for couplers and inductors. In directional couplers, the extension of the frequency band is realized by broadside coupling of vertical microstrips. The developed inductors have the reduced real-estate due to this vertical design and the decreased loss thanks to the increased cross sections of conductors. Another application of this line and this published technology is shielding of microstrip interconnects by vertical conductors shorted to the ground plane. It is shown that this technique improves isolation of two parallel strips up to 10 dB in the frequency band 0-25 GHz.

6.2 Multilayered CPW for Monolithic Applications

6.2.1 Quasistatic Model of Generalized CPW for Monolithic ICs

In monolithic applications, the CPW conductors are placed into a multilayered substrate (Fig. 6.20). Unfortunately, this line has been studied less than the classical CPW, and the numerical simulations, analytical, and experimental results can be found, for example, in [79]-[100].

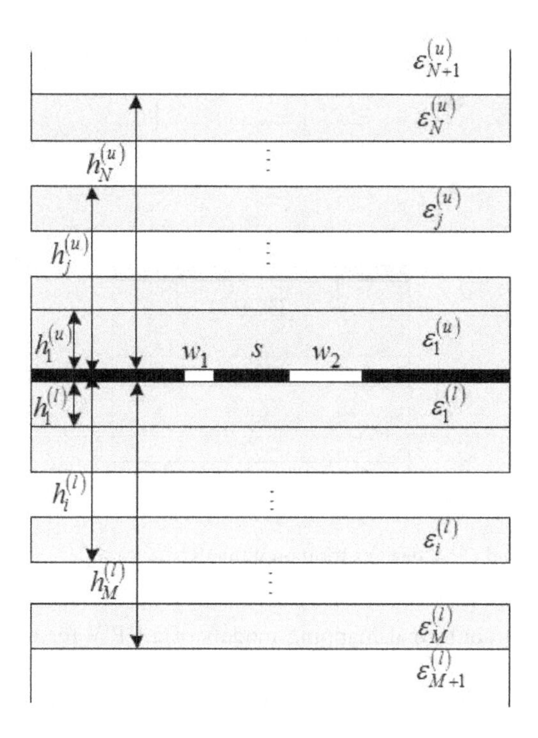

Fig. 6.20 CPW in a multilayered dielectric. Adapted from [79]

Some practically used multilayered geometries of CPW for monolithic applications are shown in Fig. 6.21.

The analytical modeling of CPW is based on the assumption that the CPW's fundamental mode is of the quasi-TEM type, and the conformal mapping can be applied to model this line. Then the loss of CPW can be calculated by a perturbation approach supposing that the silicon conductivity does not influence essentially the modal field and the propagation constant.

The accuracy of this approach is questionable, especially, at increased frequencies due to hybridization of the main mode field and for the lines where the modal field penetrates deeply the low-resistivity semiconductor substrates. In these cases, the large longitudinal electric field might be arisen, and the mode cannot be any more of the quasi-TEM type.

Quasistatic approach does not answer the question on the radiation of the fundamental mode that is strong at increased frequencies and for the thick-substrate CPWs. Practically, all analytical models of the quasi-TEM approximation should be carefully tested by measurements and full-wave simulations.

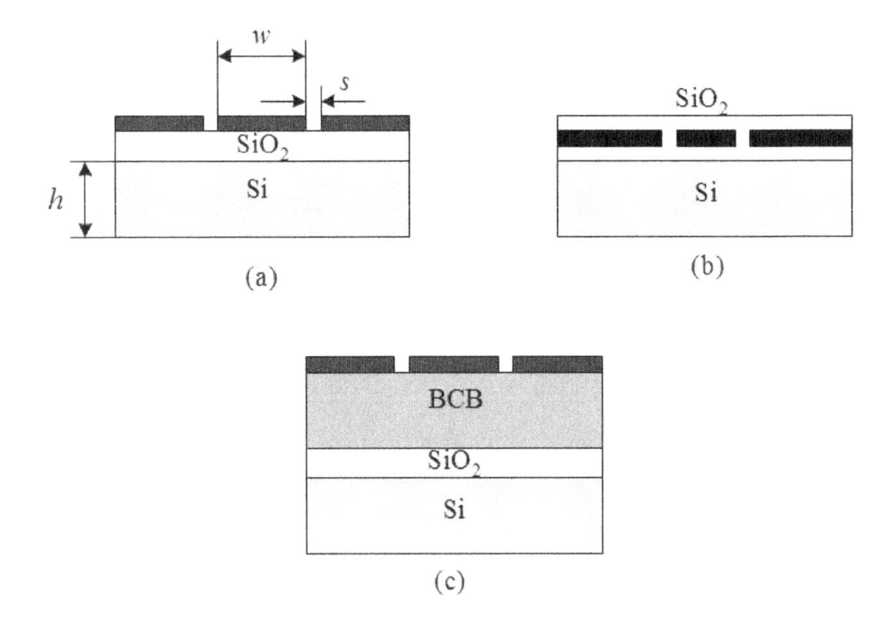

Fig. 6.21 Multilayered CPW designs for monolithic ICs

One of the first conformal mapping models of a CPW for monolithic applications was published by Svačina [79]. It was modeled the generalized structure shown in Fig. 6.20 where the CPW's slots $w_1 \neq w_2$. The quasistatic permittivity ε_{eq} of this line is

$$\varepsilon_{\text{eq}} = \sum_{i=1}^{M+1} \varepsilon_{r_i}^{(l)} q_i + \sum_{j=1}^{N+1} \varepsilon_{r_j}^{(u)} q_j \tag{6.34}$$

where $q_i = \dfrac{K'(k)}{2K(k)}\left[\dfrac{K(k_i)}{K'(k_i)} - \dfrac{K(k_{i-1})}{K'(k_{i-1})}\right]$, $k_i^2 = \dfrac{2\left(k_{i_1}+k_{i_2}\right)}{\left(1+k_{i_1}\right)\left(1+k_{i_2}\right)}$,

$$k_{i_1} = \frac{\sinh\left(\dfrac{\pi s}{4h_i}\right)}{\sinh\left(\dfrac{\pi(s+2w_1)}{4h_i}\right)}, \quad k_{i_2} = \frac{\sinh\left(\dfrac{\pi s}{4h_i}\right)}{\sinh\left(\dfrac{\pi(s+2w_2)}{4h_i}\right)}, \quad \text{and } k = k_{M+1} = k_{N+1}.$$

Here K and $K'\left(\sqrt{1-k_i^2}\right)$ are the elliptical integrals. The expression (6.34) tends to be identical with the formulas given by S.S. Bedair and I. Wolff for the symmetrical three-layer CPW [100].

The knowledge of the effective permittivity ε_{eq} allows for calculating the conductor and substrate loss constants α_c and α_s, correspondingly, under the assumption that the lossy conductors and substrate and the multilayered design do not transform the quasi-TEM mode to a hybrid one, and the frequency is below the coupling frequencies of the substrate partial modes. For instance, the conductor loss α_c for a three-layer CPW (Fig. 6.21a) can be calculated with the following formula for conductors which height t is larger than three skin-depths [85]:

$$\alpha_c = \frac{8.68 R_s \sqrt{\varepsilon_{\text{eq}}}}{240\pi K(k) K(k')(1-k^2)} \left[\Phi(w) + \Phi(d)\right], \ [\text{dB/cm}] \tag{6.35}$$

where d=2s+w, $k = w/d$, $k' = \sqrt{1-k^2}$, $R_S = \sqrt{\pi\rho/\varepsilon_0 c\lambda_0}$ is the surface resistance per square, and ρ is the specific resistivity of the conductor. The function Φ is defined as $x\Phi(x) = \pi + \ln\left(4\pi \frac{x}{t}\frac{1-k}{1+k}\right)$.

The substrate loss α_s is caused mostly by free charges in the silicon:

$$\alpha_s = 8.68 \frac{30\pi}{\sqrt{\varepsilon_{\text{eq}}}\rho_{\text{Si}}}, \ [\text{dB/length unit}] \tag{6.36}$$

where ρ_{Si} is the specific resistivity of this semiconductor. For calculation of these parameters, the equivalent permittivity is needed [85]

$$\varepsilon_{\text{eq}} = 1 + \frac{\varepsilon_r - 1}{2} \frac{K(k_1)/K(k_1')}{K(k)/K(k')} \tag{6.37}$$

where $k = w/d$, $k_1 = \sinh\left(\frac{\pi w}{4h}\right) \Big/ \sinh\left(\frac{\pi d}{4h}\right)$, $k' = \sqrt{1-k^2}$, $k_1' = \sqrt{1-k_1^2}$, and h is the silicon layer height. The thin dioxide layer (200 nm) is neglected in calculations of the loss and the effective permittivity.

The characteristic impedance Z_c of a 3-layer CPW can be calculated using a simplified formula from [85]:

$$Z_c = \frac{30\pi}{\sqrt{\varepsilon_{\text{eq}}}} \frac{K(k')}{K(k)}, \ [\Omega]. \tag{6.38}$$

Additionally to the substrate, the loss is caused by the charges moving on the interface between the silicon and the silicon dioxide, and it depends on the bias voltage applied to the silicon substrate. Unfortunately, the analytical formulas for calculation of this loss are unknown, but experiments show an essential additional loss due to this interface mechanism [85]. In that paper, several CPWs are measured. One set of the results are given for the silicon of $\rho > 1000$ Ω-cm covered by a thin (200 nm) dioxide layer, and the measurements are for the 50-Ω lines of different conductor (Al) widths and the applied-to-the-substrate bias voltages V. The loss minimum 4.8 dB/cm at 40 GHz is reached with V=-3 V and the central conductor width w=45 μm of the height 1 μm. The same waveguide of the lower specific silicon resistivity $\rho > 50\ \Omega\text{-cm}$ shows loss about 6.7 dB/cm at 40 GHz measured for a bias voltage -1 V.

Interesting data for a CPW and a thin-film microstrip is published in [105] where a commercial SiGe HBT BiCMOS technology is used for manufacturing these lines and for comparison of their characteristics. The conductors of these 50-Ω transmission lines are made of aluminum $\sigma = 2.3\times10^{7}\ \text{S/m}$, and they are placed on the top level of integration. The conductor layers are separated by silicon dioxide, which loss tangent is estimated as 0.02. Used slotted shield does not separate the CPW conductors from the lossy silicon, and the loss of a 50-Ω line is about 80 dB/cm at 110 GHz. The thin-film microstrip shows better performance being isolated from silicon by a solid part of ground layer, and the line's loss is close to 11.5 dB/cm. Further improving of CPW characteristics is with covering the semiconductor by a thick polyimide or BCB layer (Fig. 6.21c) [82]-[86], which have less loss than the silicon dioxide.

In [82], the loss of 50-Ω CPWs is about 5.5-12 dB/cm at 40 GHz, which is measured for silicon of $\rho < 20\ \Omega\text{-cm}$. The polyimide $(\tan\delta \approx 10^{-3})$ layer's thickness is varied within 20.15-8.83 μm, and the height of gold conductors is 1.5 μm.

The BCB material $(\tan\delta \approx 10^{-4})$, which has better parameters compared to polyimide, allows for loss about 3 dB/cm at 30 GHz for the 20-Ω-cm silicon substrate covered by the 20-μm BCB with the CPW $(Z_c \approx 50\ \Omega)$ of the copper conductors of the 6-μm height [86].

Further improvements of CPWs can be reached using the high-resistivity $(\rho \geq 1\ \text{k}\Omega\text{-cm})$ and the ion-implanted $(\rho \geq 1\ \text{M}\Omega\text{-cm})$ silicon substrates [38],[39]. For instance, the last technology allows for manufacturing the Si/SiO_2 CPWs with a low loss about 6 dB/cm at 110 GHz. The measured items are placed on a thin dioxide silicon layer $(1.5\ \mu\text{m})$ covering the ion-implanted semiconductor. The 50-Ω CPW's strips are made of 4-μm aluminum. The same line placed on the non-implanted Si/SiO_2 substrate $(\rho_{\text{Si}} = 10\ \Omega\text{-cm})$ shows loss about 50 dB/cm at 110 GHz.

Similarly to microstrip line, the parameters of Si/SiO_2 CPW can be improved using the impedance layers between the silicon and the silicon dioxide. For instance, in [101], a patterned shield consisting of strips orthogonal to the line direction and placed over a GaAs substrate allows for increasing the slow factor of this line and for decreasing the loss. This idea is found applications in silicon technology [102]-[106]. For instance, in [106], it is achieved the effective permittivity equal to 48 for hundreds of MHz to 40 GHz with essential improving of the quality factor regarding to classical Si/SiO_2 CPW.

Additionally, substitution of Al or Cu conductors by silver ones can improve the Q-factors of lines. A silver CPW is studied in [107], which is manufactured on a quartz substrate. The results are compared with the one, which conductors are of copper. It is found the decreasing of distributed resistance of the silver CPW regarding to the copper line on 2-3 Ω/cm at frequencies up to 20 GHz.

Another effect that should be taken into account in the applications over 100 GHz is the main mode dispersion and radiation [108]-[111]. This CPW's mode is a quasi-TEM one, i.e. its transversal components of the fields averaged on the space are stronger than the longitudinal ones: $\langle|\mathbf{E}_t|\rangle \gg \langle|E_z|\rangle$ and $\langle|\mathbf{H}_t|\rangle \gg \langle|H_z|\rangle$. In a CPW, which substrate and conductors are perfect, the fundamental mode is only slightly dispersive due to that, and the above-considered quasi-TEM models can be used up to 40-100 GHz depending on the geometry and the substrate dielectrics.

The CPW dominant mode field is composed of a spectrum of partial waves. Among them are the substrate modes and the waves of continuous spectrum if the cross-section of this line is open to space. In [110] and [111], it has been shown that the longitudinal field components are negligible if the distance d between the most distant conductors is less than the empirically found geometrical parameter $d_{\max}$:

$$d_{\max} = \frac{15}{f\sqrt{\varepsilon_{rs}}}\left[1+3.24\cdot 10^{8}\left(\frac{\sigma}{f\varepsilon_{rs}}\right)^{2}\right]^{-\frac{1}{4}} \tag{6.39}$$

where f is the frequency in GHz, ε_{rs} is the relative permittivity of a substrate, and σ is the substrate conductivity given for a millimeter. If this criterion is violated, the quasi-TEM model does not take into account the longitudinal component of the electric modal field in the lossy layer and the Joule loss caused by this component. Thus computations of the loss and phase constants with this quasi-TEM model are not accurate in this case.

One of the components of the CPW's field is the TM_0 partial wave of completely grounded dielectric slab of the height h. This partial wave spreads in the transversal and longitudinal directions. At certain critical frequency, the phase constant β of the fundamental CPW mode is equal to the modal constant β_{TM_0} of the TM_0 partial wave. The leaky partial wave TM_0 starts to play a more intensive role in the modal CPW field formation, and this mode becomes complex, frequency-dependent, and radiating [108]-[111].

With this critical frequency, a CPW can have strong parasitic coupling to other IC components through this field leakage. Due to that a digital signal which spectrum is wider than the CPW's frequency band limited by the considered critical frequency, has additional distortions [98]. Some ideas to decrease them are considered in [110].

Another cause that transforms the quasi-TEM mode into a hybrid one is the layered structure of the CPW cross-section and the loss in its conductors and in the dielectric layers. Analytical formulas for calculations of critical coupling frequencies can be found for a one-layer [112] and a two-layer conductor-backed CPW [113]. Consider this last case which is important for monolithic applications. The CPW geometry is shown in Fig. 6.22.

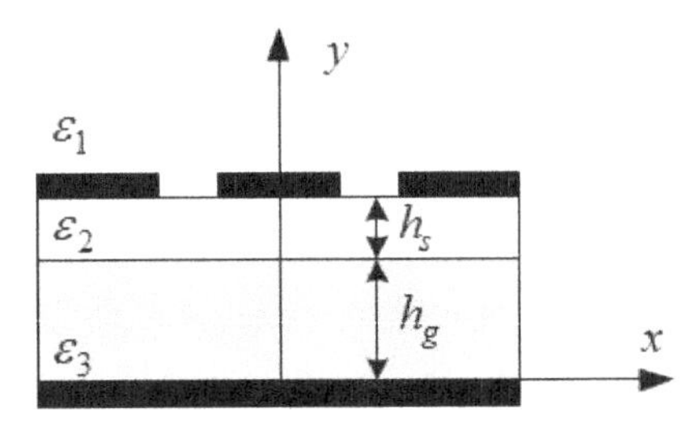

Fig. 6.22 Conductor-backed two-layer CPW where $\varepsilon_{1,2,3}$ are the relative dielectric permittivities

The substrate consists of two dielectric layers. One of them is covered by a sheet of ideal conductor which is the ground. The CPW conductors are placed on another side of the substrate. It is found that the CPW quasi-static mode is strongly coupled to the TM_y modes. The characteristic equation for the critical frequency $f_{c(\text{TM})}$ is [113]

$$f_{c(\text{TM})} = \frac{2c\tan^{-1}\left[\dfrac{A\cdot B-1+\sqrt{(A\cdot B)^2+A^2+B^2+1}}{A+B}\right]}{\pi h_s\sqrt{2(\varepsilon_2-\varepsilon_1)}} \tag{6.40}$$

where $B=(\varepsilon_2/\varepsilon_3)\tanh(\alpha_3 h_g)\sqrt{(\varepsilon_1+\varepsilon_2-2\varepsilon_3)/(\varepsilon_2-\varepsilon_1)}$,
$\alpha_3=(\pi f_{c(\text{TM})}/c)\sqrt{2(\varepsilon_1+\varepsilon_2-2\varepsilon_3)}$, and $A=\varepsilon_2/\varepsilon_1$.

This equation is simplified if $h_g \gg h_s$. In this case, $B=\dfrac{\varepsilon_2}{\varepsilon_3}\sqrt{\dfrac{\varepsilon_1+\varepsilon_2-2\varepsilon_3}{\varepsilon_2-\varepsilon_1}}$, and the critical frequency is calculated analytically. Similar formulas are derived in the cited paper for coupling of the coplanar mode with the TE_y mode. This theory was confirmed by measurements up to 110 GHz where the strong radiation started close to the mentioned critical frequencies [112],[113]. More measurements of radiation performed for GaAs CPWs in frequencies 140-220 GHz are in [114].

6.2.2 Elevated CPWs for Increased Frequencies

The above-considered CPWs are not alone to solve the problem of high loss and limited range of the characteristic impedance variation [115]-[122]. A couple of designs are shown in Fig. 6.23. One of them is a CPW which central conductor is elevated by conducting posts placed over the substrate at the distance about several tens of microns (Fig. 6.23a). The central conductor of the second line (Fig. 6.23b) is on the surface of a low-loss dielectric bar (in grey color).

In both cases, the substrate consists of a semiconductor layer (in grey color, Si, GaAs, or InP) and a passivation layer placed over it (SiO_2 or polyimide, in white color).

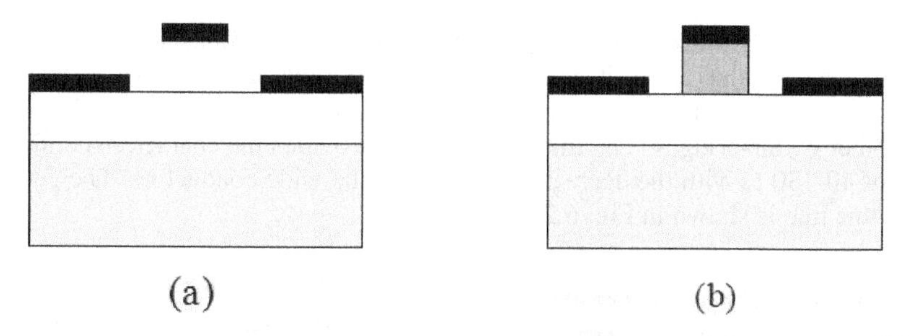

Fig. 6.23 Elevated signal-strip coplanar waveguides. (a)- Central conductor is supported by metallic pedestals; (b)- Central conductor is placed on a low-loss dielectric slab

These transmission lines are studied by numerical simulations and by measurements in [115]-[119] up to 300 GHz, and they show decreased loss by two - five times regarding to conventional CPW for GaAs and silicon substrates. It is noticed that the effective permittivity can be reduced four times, which is interesting in high-speed applications. The characteristic impedance varies within 25-110 Ω.

In [120], a resonator of this elevated type for 80 GHz manufactured by a standard CMOS technology is considered. In that paper, the lossy silicon is isolated from the elevated CPW by a patterned conducting plate placed below the dioxide layer. This CPW has the quality factor Q varying within 6.5-11 at 80 GHz depending on the geometry. A traveling-wave amplifier for 73.5 GHz is described as well that uses the above-mentioned line.

Another modification of the elevated CPW is in [121] where the conductors are placed on a thick (6.35-20.15 μm) polyimide layer deposited on the 1 Ω-cm silicon wafer. Afterwards, the polyimide layer is dry-etched, and the CPW conductors are supported by remains of the polyimide under the conductors made of Ti and Au. The line is simulated, measured, and it demonstrates the 2.5-decreased loss in comparison to the CPW of the non-etched design (0.8-3.5 dB/cm loss for frequencies 1-40 GHz). It is caused by weaker concentration of the electric field in the silicon substrate due to the thick, separated from each other polyimide pedestals. Some analytical formulas for a CPW with the elevated central conductor are in [122].

All conductors of CPW can be lifted (Fig. 6.24a), and one of the designs with the measurement results is published in [50]. The ground copper layer is placed over a silicon dioxide substrate. The conductors of the thickness 12 μm are elevated at the height 12 μm and supported by metallic pedestals, which are placed on the substrate through the holes in the ground layer. The width of the central conductor is 35 μm with the gap 13 μm, and the line provides the 50-Ω characteristic impedance with the loss below 0.7 dB/cm in frequencies up to 40 GHz. The line parameters can be calculated using formulas from [123] for the conductor-backed CPW with some corrections which take into account the thickness of conductors.

A new dielectric - cyclic-olefin copolymer (COC) having low loss ($\tan\delta = 0.0005$) up to 220 GHz is used in [29] for a conductor-backed CPW (Fig. 6.24b). The 50-Ω CPW, which composite conductor made of Ti/Au, has the loss of 5-13.5 dB/cm at 220 GHz depending on the line geometry.

Another design of the elevated coplanar waveguide is published, for instance, in [124]. All conductors of it, including the ground planes, are elevated over the substrate, and they are supported by the posts. The ground planes are shorted to each other by air-bridges. The transmission line provides the characteristic impedance of 40-130 Ω with the decreased loss due to the wide conductors. The geometry of this line is shown in Fig. 6.24c.

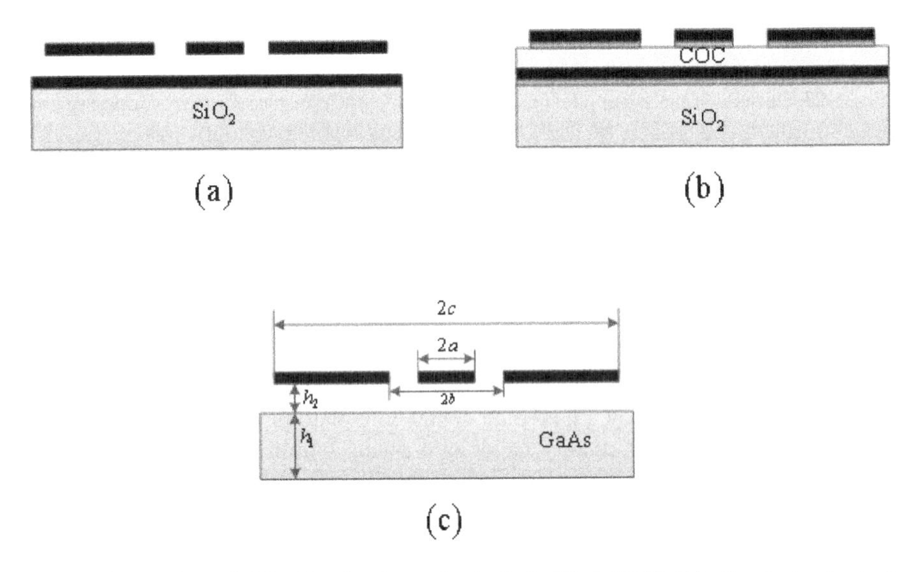

Fig. 6.24 Low-loss CPW. (a)- Conductor backed air-filled CPW; (b)- Conductor backed COC-filled CPW; (c)- Elevated 3-layer CPW

Quasistatic analytical formulas for CPW from Fig. 6.24c compared with full-wave simulations and measurements are published in [124]. The effective permittivity ε_{eq} is calculated as a ratio of the total capacitance C_{tot} to the air-filled line total capacitance C_{totair}:

$$\varepsilon_{\text{eq}} = \frac{C_{\text{tot}}}{C_{\text{totair}}}. \tag{6.41}$$

The characteristic impedance Z_{c} is

$$Z_{\text{c}} = \frac{1}{c\sqrt{C_{\text{tot}}C_{\text{totair}}}}, \; [\Omega] \tag{6.42}$$

where the capacitances C_{tot} and C_{totair} are calculated as

$$C_{\text{tot}} = C_{\text{air}} + C_{\text{pp}} + C_2(1-\varepsilon_r) + C_3(\varepsilon_r - 1), \tag{6.43}$$

$$C_{\text{totair}} = C_{\text{air}} + C_{\text{pp}}, \tag{6.44}$$

$$C_{pp} = 2\varepsilon_0 \frac{t}{b-a}, \tag{6.45}$$

$$C_{air} = 4\varepsilon_0 \frac{K(k_1)}{K(k_1')}, \; k_1 = \frac{a}{b}\sqrt{\frac{1-\left(\frac{b}{c}\right)^2}{1-\left(\frac{a}{c}\right)^2}}, \tag{6.46}$$

$$C_2 = 2\varepsilon_0 \frac{K(k_2)}{K(k_2')}, \; k_2 = \frac{\sinh(x\cdot a)}{\sinh(x\cdot b)}\sqrt{\frac{1-\frac{\sinh^2(x\cdot b)}{\sinh^2(x\cdot c)}}{1-\frac{\sinh^2(x\cdot a)}{\sinh^2(x\cdot c)}}} \tag{6.47}$$

where

$$x = \frac{\sqrt[\pi]{\left(\frac{\varepsilon_r^2}{23}h_2 + 0.176t + 0.5a\right)\frac{1-\frac{a}{2b}}{\varepsilon_r - \frac{a}{2b}}}}{2h_2\sqrt{2b}}, \tag{6.48}$$

$$C_3 = 2\varepsilon_0 \frac{K(k_3)}{K(k_3')} \tag{6.49}$$

where t is the conductor height, ε_r is the substrate permittivity, $k_3 = k_2(x = x_3)$, $x_3 = \pi/2(h_1 + h_2)$, and $k'_{1,2,3} = \sqrt{1-k_{1,2,3}^2}$.

6.2.3 Coplanar Strip Line

Another important uniplanar waveguide is the coplanar strip line (CPS) introduced by Kneppo and Gotzman [125]. The cross-sections of a couple applicable lines are shown in Fig. 6.25. A line consists of a dielectric or a composite dielectric-semiconductor substrate and two strips. One of them is grounded, and the waveguide is considered as a "2/3 CPW". Additionally, the line can be used for differential signaling in which case the strips are excited by the opposite voltages. Similarly to CPW, it can be distinguished by reduced loss and weak modal dispersion.

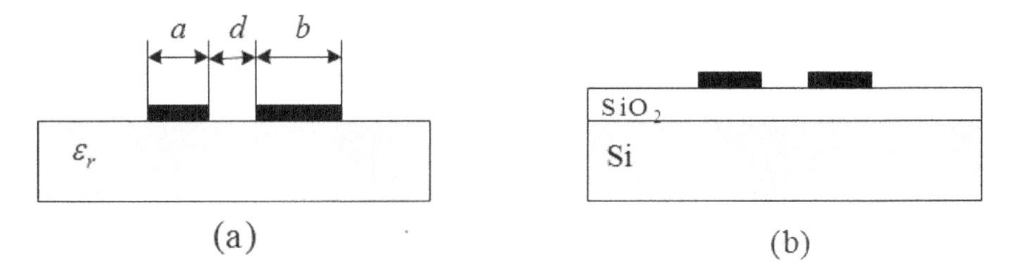

Fig. 6.25 (a)- Dielectric substrate CPS; (b)- Si-SiO_2 substrate CPS

To estimate the relative quasistatic effective permittivity ε_{eq} and the characteristic impedance Z_{c} of the CPS shown in Fig. 6.25a, the following formulas [125] can be used

$$\varepsilon_{\text{eq}} = \frac{\left(\sqrt{\varepsilon_r}+1\right)^2}{4}, \tag{6.50}$$

$$Z_{\text{c}} = \frac{120\pi}{1+\sqrt{\varepsilon_r}} \frac{K(k)}{K'(k)}, \ [\Omega] \tag{6.51}$$

where $K(k)$ and $K' = K\left(\sqrt{1-k^2}\right)$ are the complete elliptical integral of the first kind and its supplement, respectively. The argument k is

$$k^2 = \frac{1+a/d+b/d}{(1+b/d)(1+a/d)} \tag{6.52}$$

where a and b are the widths of strips, and d is the distance between them. Unfortunately, these simple formulas do not take into account the height h of the substrate, and a set of new formulas is published in [126],[127]. According to them, the effective permittivity ε_{eq} and the characteristic impedance Z_{c} are calculated as

$$\varepsilon_{\text{eq}} = 1 + \frac{\varepsilon_r - 1}{2} \frac{K(k')K(k_1)}{K(k)K(k_1')}, \tag{6.53}$$

$$Z_{\text{c}} = \frac{120\pi}{\sqrt{\varepsilon_{\text{eq}}}} \frac{K(k)}{K(k')}, \ [\Omega]. \tag{6.54}$$

If $a = b$, then $k = d/(d+2a)$, $k_1 = \sinh\left(\frac{\pi d}{4h}\right) \Big/ \sinh\left(\frac{\pi(d+2a)}{4h}\right)$, $k' = \sqrt{1-k^2}$,

and $k_1' = \sqrt{1-k_1^2}$.

These formulas are verified by measurements of transients induced by laser light pulses of 150-fs duration in CPS (300/2500 Å Ti/Au) placed over a thin (0.5 μm) silicon layer, which is on a thick (430 μm) sapphire. The results are available in the frequencies up to 800 GHz [127]. Theory shows an error only up to 6% in frequencies below 100 GHz. It increases with the frequency, being rather small up to 500 GHz. These formulas can be improved taking into account the dispersion of the CPS mode, as it is performed in [128] where some semi-empirical formulas are introduced to calculate the frequency-dependent effective permittivity and the loss constant.

Additionally, the loss in a CPS placed over a two-layer substrate (SiO_2 / Si, Fig. 6.25b) is measured in [129]. In this case, a 0.2-μm silicon dioxide layer is over a high-resistivity ($\rho > 2000$ Ω-cm) semiconductor. The strips are of a multilayer design, and they are composed of thin layers of Cr (20 nm) and Au (400 nm). The measurements show loss up to 5 Neper/cm at frequencies up to 500 GHz.

The loss of CPS can be decreased twice by proper design of this line, as it is demonstrated in [130], by simulations and measurements in frequencies up to 40 GHz. For instance, the narrow-strip lines show less concentration of the field in silicon, and they can demonstrate the above-mentioned effect of the loss decreasing. Using this technique, the loss was decreased down to 8-9 dB/cm for a 15-μm SU-8 isolation layer ($\varepsilon_r = 3.5$, $\tan\delta = 0.043$) placed over a low-resistivity silicon ($\rho < 10$ Ω-cm). Its gold conductors are of the thickness 1 μm. The proper geometry for the given characteristic impedance and the effective permittivity can be found according to the CPS synthesis formulas from [131].

Some CPS circuit components are considered in [132]-[135] for 0.1-110 GHz. Several ideas on high-frequency designs based on distributed CPS components are from [136],[137].

6.2.4 Three-dimensional Modifications of CPW

6.2.4.1 "Overlay" CPWs

The need to decrease the loss in CPW in a wide frequency band has stimulated many attempts to modify the well-known design of CPW. In [138], two new structures are proposed which have decreased loss at frequencies up 100 GHz. The first transmission line is an inverted "overlay" CPW (Fig. 6.26a), and the second one is with the partially elevated signal conductor (Fig. 6.26b).

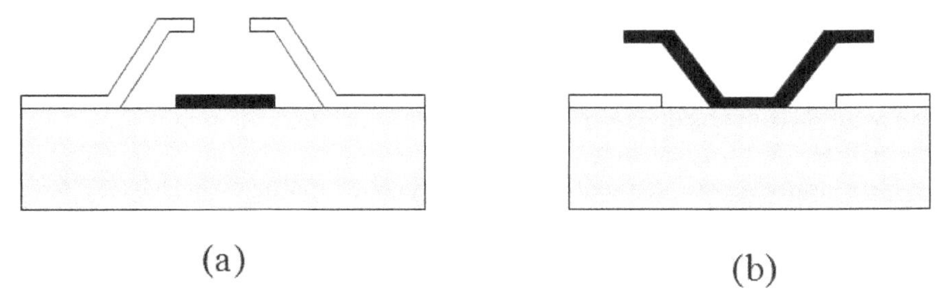

Fig. 6.26 Overlay coplanar waveguides. Adapted from [138].

The signal conductor (in black) of the first line is in a shell formed with the elevated ground conductors (in white) and the substrate. These two ground conductors are connected to each other by air-bridges. The quasi-TEM modal field is, mostly, in the air gap between the signal and ground conductors that reduces the substrate loss.

In the second design (Fig. 6.26b), the most part of the modal energy concentrates in two gaps formed by the signal conductor wings and the ground conductors, i.e. outside the substrate.

In [138], the conductors were placed over a 520-μm quartz substrate. The elevation of conductors is 15 μm. According to [138], the characteristic impedance Z_c of this line can vary in a wide 25-80-Ω range. Another advantage of these lines is their technology that is simpler in comparison to the micromachined membrane supported CPW (see Section 6.2.4.4). As well, this line has increased mechanical reliability compared to the membrane supported CPWs. Although these micron-sized lines are manufactured over a quartz substrate, the design is prospective for Si-based integrations.

The measurements of these two lines show that the first of them (Fig. 6.26a) provides the loss which is less than 2 dB/cm up to 110 GHz regarding to the conventional CPW (3 dB/cm) of the same impedance $Z_c = 36\ \Omega$. The high-impedance elevated CPW (Fig. 6.26a) with $Z_c > 50\ \Omega$ has the loss which is comparable with one of conventional CPW, and it is less than 1 dB/cm at 50 GHz. Note that all these figures are given with the loss of two transitions.

6.2.4.2 Three-dimensional CPWs

The transmission lines with vertically oriented conductors were developed to obtain the parameters varying in a wide range [67]. Later, they have been found some applications in 3-D hybrid integrations for filters and directional couplers [69],[94],[95].

An example of this sort of transmission lines is shown in Fig. 6.27 [139]. The line consists of two pairs of grounded planar conductors. The central signal conductor is a vertically oriented strip. Due to this design, the field is confined inside the cross-section, and the central conductor connects the components which are placed on both sides of this substrate. The fundamental mode of this line is of the quasi-TEM type, and its electric field is calculated in [140].

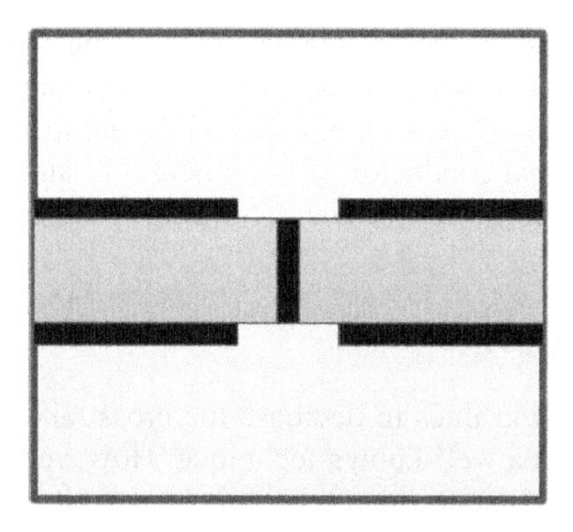

Fig. 6.27 Two-side shielded coplanar waveguide. Adapted from [139]

Additionally, some research of 3-D CPWs is performed in [141],[142], for instance, where the authors designed the CPWs which central conductors have modified cross-sections to reach the extended range of characteristic impedances, decreased loss, and to provide the access to the functional surface of the GaAs substrate.

The lines from [142] are shown in Fig. 6.28, and they, together with the classical counterparts, are studied in the cited paper by measurements and simulations at frequencies up to 20 GHz. The gold conductors are formed inside the polyimide layers (5-μm height) placed over a semi-isolating GaAs substrate. The central 3-D conductor of the transmission line, shown in Fig. 6.28a, has no access to the functional level of the substrate. Another line (Fig. 6.28b) is designed to contact the components formed in the semiconductor and on its surface.

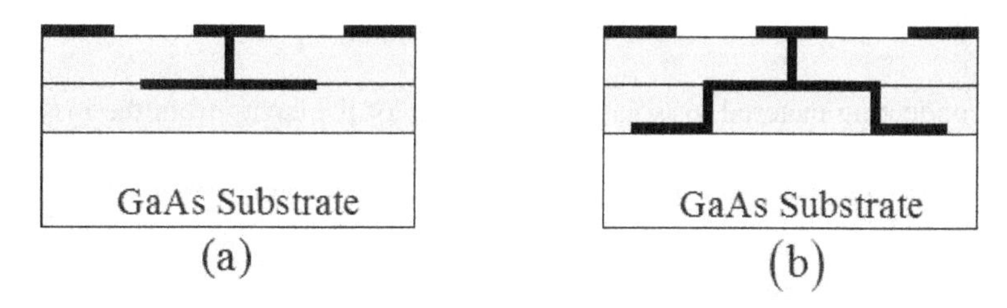

Fig. 6.28 Three-dimensional coplanar waveguides. Adapted from [142]

The studied transmission lines have reduced loss compared to the CPW placed directly on the semiconductor surface due to the increased cross-sections of the central conductors and isolation of them from the lossy substrate. The maximum loss of these lines at frequencies up to 15 GHz is within 1-9 dB/cm (Fig. 6.28a, $Z_c < 50\ \Omega$) and 0.5-4 dB/cm (Fig. 6.28b, $Z_c < 50\ \Omega$). It means that, in the three-layer line (Fig. 6.28b), the current is more effectively distributed along the conductor cross-section, and it reduces the loss. It is noticed that the flexibility of the geometry allows for the characteristic impedance about 20-80 Ω with decreased loss.

Further conductor loss reducing is with the stacking the ground or/and signal layers to increase the cross-sections of CPW's conductors. For instance, a semi-coaxial CPW manufactured by a 45-nm CMOS digital technology is described in [44]. The CPWs Al-signal conductor of 1-μm height is surrounded by the vertically stacked Cu ground forming a niche. The messaged loss of this line is about 10 dB/cm at 110 GHz.

6.2.4.3 Microshield CPW

Shielding the transmission lines to decrease the cross talk or isolating these lines from the environment is a well-known technique. However, this approach has taken a renaissance thanks to the micromachining technology and the needs to decrease the loss in monolithic Si-based integrated circuits [46]-[49].

The considered here microshield coplanar waveguide is formed on the surface of a filled by a low-loss dielectric cavity that is etched in the semiconductor substrate (Fig. 6.29).

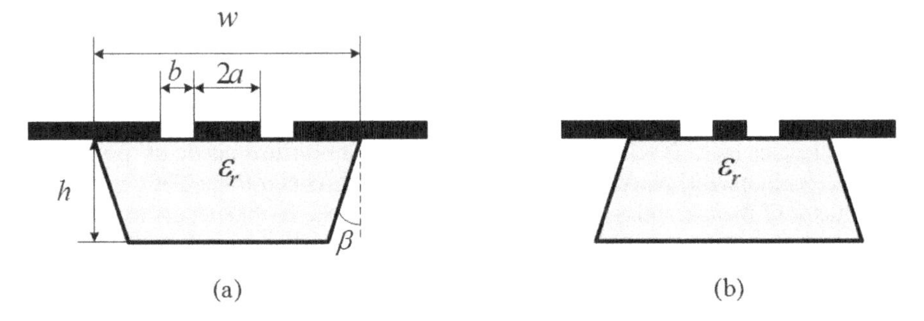

Fig. 6.29 Microshield CPW of a trapezoidal cross-section. Adapted from [47]

A conducting material to isolate the dielectric of the cavity from the lossy substrate covers the walls of this cavity. In the ideal case, this channel is of the rectangular shape. Due to the anisotropy of the wet- or ion-etching, the walls of the cavity have the negative (Fig. 6.29a, $\beta > 0$) or the positive slopes (Fig. 6.29b, $\beta < 0$).

This trapezoidal cross-section line is modeled in [47] using a conformal-mapping approach. The quasistatic effective permittivity $\varepsilon_{\mathrm{eq}}$ of the line is computed as a ratio of the filled-line capacitance $C_{\mathrm{T}}(\varepsilon_r)$ and the air-filled-line capacitance $C_{\mathrm{T}}(\varepsilon_r = 1)$:

$$\varepsilon_{\mathrm{eq}} = \frac{C_{\mathrm{T}}(\varepsilon_r)}{C_{\mathrm{T}}(\varepsilon_r = 1)} \tag{6.55}$$

where

$$C_{\mathrm{T}}(\varepsilon_r) = 2\varepsilon_0 \frac{K(k_1)}{K(k_1')} + 2\varepsilon_0\varepsilon_r \frac{K(k_2)}{K(k_2')}. \tag{6.56}$$

Here $K(k)$ is the elliptical integral of the first kind, $k_1 = a/b$, $k_2 = t_a/t_b$, $k_i' = \sqrt{1-k_i^2}$, and i=1,2.

The parameters t_a, t_b, and t_c are calculated through the integral expressions, which are

$$\frac{2h}{w}\Delta = \cos\beta \int_0^{\operatorname{arc\,cosh}(1/t_c)} \frac{d\theta}{\sinh^{1-2p}\theta\left(1-t_c^2\cosh^2\theta\right)^p}, \tag{6.57}$$

$$\frac{2a}{w}\Delta = \int_0^{\arcsin(t_a/t_c)} \frac{d\theta}{\cos^{1-2p}\theta\left(1-t_c^2\sin^2\theta\right)^p}, \tag{6.58}$$

$$\frac{2b}{w}\Delta = \int_0^{\arcsin(t_b/t_c)} \frac{d\theta}{\cos^{1-2p}\theta\left(1-t_c^2\sin^2\theta\right)^p}, \tag{6.59}$$

$$\Delta = \int_0^{\pi/2} \frac{d\theta}{\cos^{1-2p}\theta\left(1-t_c^2\sin^2\theta\right)^p} \tag{6.60}$$

where $p = \frac{1}{2} - \frac{\beta}{\pi}$. The characteristic impedance Z_c is calculated with

$$Z_c = c^{-1}\left[C_T(\varepsilon_r)C_T(\varepsilon_r = 1)\right]^{-\frac{1}{2}},\ [\Omega]. \tag{6.61}$$

These formulas (6.55)-(6.61) allow for computing the characteristic impedance and the effective dielectric permittivity in a wide region of the line geometrical parameters. For studied in [47] geometries, the characteristic impedance varies within 65-120 Ω.

In [48], a V-shaped microshield CPW is modeled with the conformal approach, and some analytical formulas are given for characteristic impedance calculations (Fig. 6.30).

Fig. 6.30 V-shaped coplanar waveguide

The cavity area in this V-shaped line is etched in lossy silicon, covered by a metal, and filled with a microwave dielectric. The signal and grounding plates are placed above the opening of this cavity. Like in the previous case, this line has reduced loss due to isolation of the cavity and the strip from the semiconductor. According to the simulations from [48], the studied line provides a wide range of impedances within 30-130 Ω.

6.2.4.4 Membrane Supported CPW

The further modification of the CPW is with the membrane-supported conductors [46],[143]-[151]. In this case, a thin dielectric membrane is placed over an empty cavity formed in a lossy substrate by an etching technology. A thin layer of a metal covers the walls of the cavity. The membrane supports the central signal strip and two grounding plates connected to the cavity walls. All of the studied configurations of this line have reduced dielectric loss, negligible dispersion, and the one-modal transmission of the power in a wide frequency band up to 170-450 GHz. One of the possible geometries of this line is shown in Fig. 6.31a.

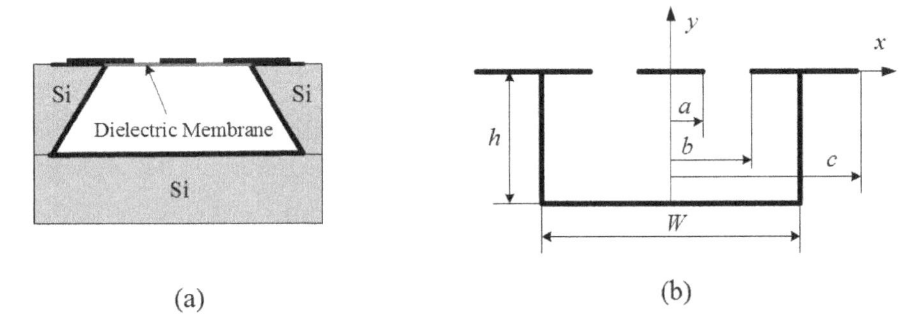

Fig. 6.31 Membrane supported CPW -(a); Simplified model -(b)

The fundamental mode of this line is of the quasi-TEM type, and its quasistatic model [143] was developed according to Fig. 6.31b. Additionally, a quasistatic theory of this line is available from [48],[144].

According to [143] , the characteristic impedance Z_c is

$$Z_c = \frac{1}{c(C_a + C_w)}, \ [\Omega] \tag{6.62}$$

where

$$C_a = 2\varepsilon_0 \frac{K(k)}{K(k')}, \tag{6.63}$$

$$C_w = 2\varepsilon_0 \frac{K(\varsigma)}{K(\varsigma')}. \tag{6.64}$$

In (6.63), the arguments of the elliptical integrals K are $k = \frac{a}{b}\sqrt{\frac{1-b^2/c^2}{1-a^2/c^2}}$ and $k' = \sqrt{1-k^2}$.

Formula (6.64) requires calculation of the arguments ς and ς', which are

$$\varsigma = \frac{sn(a/\beta)}{sn(b/\beta)}, \tag{6.65}$$

$$\varsigma' = \sqrt{1-\varsigma^2}. \tag{6.66}$$

In (6.65), $\beta = \frac{W}{2K(\gamma)}$ and $\gamma = \left[\frac{e^{\pi W/2h} - 2}{e^{\pi W/2h} + 2}\right]^2$, $sn(x)$ is the Jacobian elliptic function.

Initial measurements of the membrane supported coplanar waveguide and multiple components on its base are published in [145] for frequencies up to 40-60 GHz. The loss of the measured lines is within 0.2-0.7 dB/cm in 7-40 GHz.

Now the measurements of this line are known up to 170 GHz, and they demonstrate extremely low loss around 1 dB/cm due to isolation of the line from the lossy silicon and the concentration of the field in the air and low-loss dielectric membrane [145]-[149].

Some components using this line are considered in [145]-[152]. Among them are the short ends, open- and short-end stubs, right-angle bends, transitions, filters up to 150 GHz, and a frequency tripler for 50-450 GHz.

The known disadvantages of these waveguides and components are with the low mechanical reliability of them, increased modal wavelength, and their high cost.

6.2.4.5 Micromachined Trenched and Embedded CPWs

The three-dimensional shaping of substrates allows for reducing the field coupling to the semiconductor substrate and for improving the characteristics of transmission lines placed over semiconductors [153]-[156]. One of the first designs of this kind is from [153] where a trenched by plasma substrate CPW is proposed, modeled, and measured up to 30 GHz (Fig. 6.32a). In this case, an essential part of the electric field is out of silicon, and loss reducing is achieved.

It is shown by modeling and by measurements that the trenching (h=9 μm) and proper biasing of the p-type high-resistivity substrate ($\rho = 10$ kΩ-cm) allows for some loss reducing which is 1.6 dB/cm at 30 GHz.

A trenched CPW (Fig. 6.32b), manufactured by wet etching techniques, is studied at frequencies up to 300 GHz [154]. The trenches are of the depth h=10 μm, the gold conductors are of the height t=10 μm, and the signal conductor width is 9 μm. It is shown that the trenched 50-Ω CPW placed directly on high-resistivity silicon $(\mathrm{HRSi}, \rho = 1$ KΩ-cm$)$ provides the loss which is less than 35 dB/cm at 300 GHz.

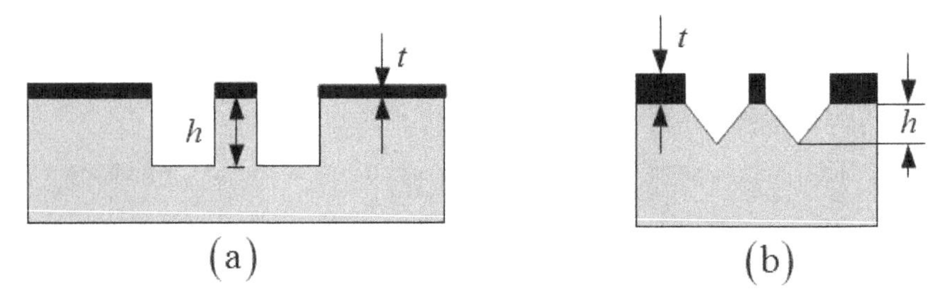

Fig. 6.32 Trenched high-resistivity silicon CPWs. Adapted from [153], [154]

Further improvements are with the isolation of conductors from the silicon by dioxide layers.

Another design of trenched CPW assigned to improve its parameters is with an embedded CPW geometry [156]. A high-resistivity silicon substrate of the thickness 525 μm is trenched with a micromachining technology, and the central copper conductor and copper ground planes of the thickness 6 μm are partially bended (Fig. 6.33a). The height of the trenches in silicon is varied from 10 to 240 μm. This design allows for less concentration of the field in silicon, and it decreases the substrate-associated loss. Additionally, proper choosing of the geometry can increase the usable impedance range.

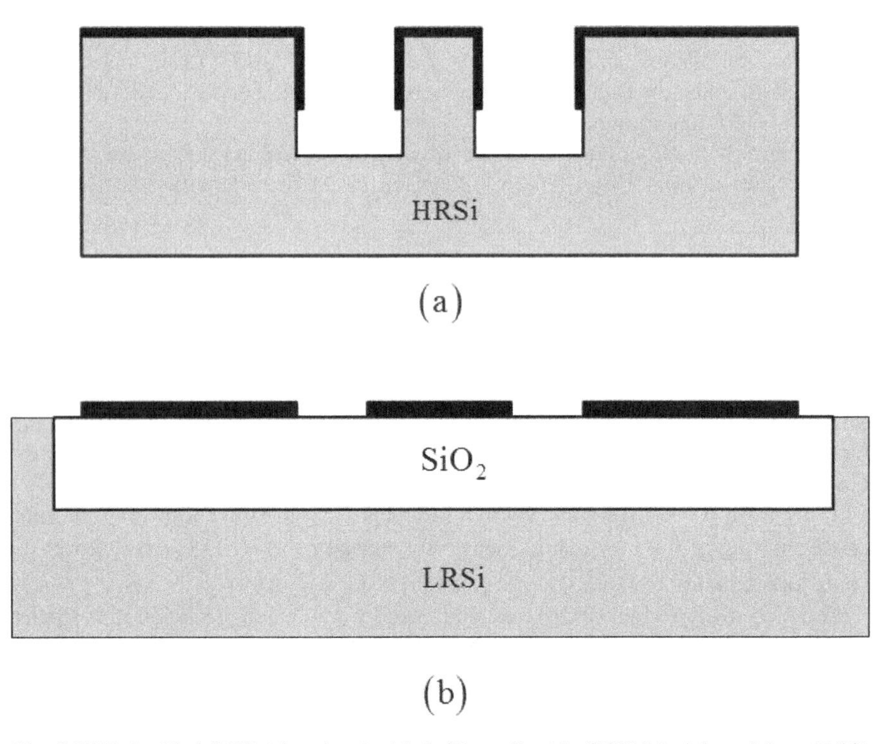

Fig. 6.33 Embedded CPW (a) and embedded silicon dioxide CPW (b). Adapted from [156], [158], correspondingly

To compare the proposed transmission line (Fig. 6.33a) and a conventional CPW, these waveguides have been manufactured using a high-resistivity silicon substrates ($\rho_s = 15$ KΩ-cm). It is found that the loss of the embedded CPW of the characteristic impedance 24.9-76.2 Ω is below 1.67 dB/cm at 50 GHz. In comparison, a CPW of conventional design of the characteristic impedance 30 Ω has loss about 7 dB/cm. This high-impedance CPW shows comparable loss regarding to the embedded waveguide but remains higher. The conclusion of the authors of this cited paper is that the proposed CPW is promising for many increased frequency and bandwidth applications which require an extended usable impedance range from 20 to 110 Ω.

Some results are with the etching of thick polyimide, BCB, or SU-8 organic layers. They, being of the low loss and permittivity, allow for increasing Q-factors of CPWs. Further improvements are with trenching of these thick layers which decreases the penetration of the electric field into semiconductor and polymer. For instance, in a polyimide-based CPW placed over a low-resistivity (LRSi) silicon, the etching of this dielectric layer allows for 28% loss reduction [121]. At the difference to the polyimide-based CPWs, the BCB etching is insufficient to reduce the loss, although it decreases the effective permittivity as shown in [157] in 5-65 GHz frequency band and which is interesting in high-speed applications.

For comparison, Fig 6.33b shows a CPW placed over embedded thick (h=50 μm) dioxide layer [158]. In this line, a low-resistivity silicon ($\rho < 0.01$ Ω-cm) is etched. In addition, a thick dioxide silicon island is formed inside the cavity. The CPW's 3-μm gold conductors are placed over this island, and this geometry provides 85-Ω characteristic impedance. A low attenuation close to 3.2 dB/cm is found at 40 GHz.

6.3 Micromachined Rectangular-coaxial and Rectangular Waveguides

The achievements of 3-D integrated technologies and a strong interest in the interconnects and passive components working at millimeter-wave and terahertz frequencies allow for implementing some known transmission lines manufactured using new techniques and integrated with semiconductor chips.

One of the lines of this kind is the micromachined rectangular-coaxial waveguide (Fig. 6.34) which bandwidth can be extended up to several hundreds of Gigahertz [159],[160].

Several technologies are available for manufacturing and integration of this line and components on its base. One of the first micron-size coaxial waveguides is proposed to be manufactured using the bulk etching of silicon (see Section 5.7) which provides the grooves of the triangle and trapezoidal shapes [46]. The walls of grooves are covered by a conducting layer using evaporating and electroplating techniques. The strip conductor is on the surface of a thin dielectric layer placed between two etched silicon substrates covered by conducting layers and bonded to each other. Taking into account that the supporting dielectric of a low permittivity

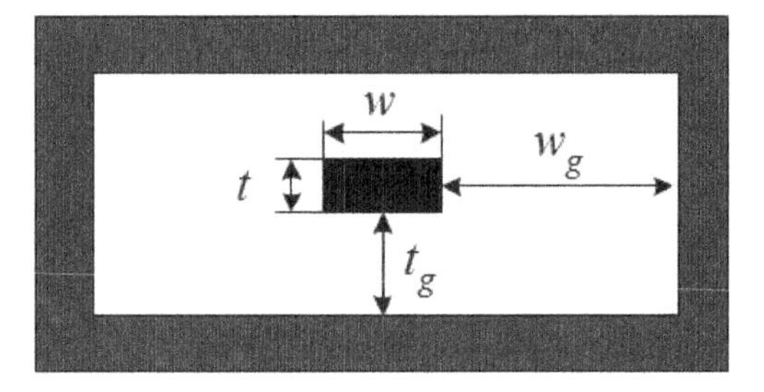

Fig. 6.34 Rectangular-coaxial waveguide cross-section

is thin, this coaxial waveguide has extremely low dielectric loss and increased cut-off frequencies of the higher-order modes, and, due to it, the line is workable at frequencies up to 300 GHz.

Another way is to use the EFAB technology [161],[162], which relates to the surface micromachining, and it is with the etching of copper (Cu) sacrificial layers through the windows made in nickel (Ni) ones (see Section 5.7). The line is filled by air, and the central conductor is supported by metallic (Ni) stubs or pedestals placed over the dielectric substrate. The typical size of components is several tens of microns. Although nickel is of increased loss in comparison to copper, the lines, studied in [162], show acceptable characteristics up to 60 GHz. For instance, the quality factor of a quarter-wavelength resonator is greater than 150. It is shown by modeling that similar resonator made of copper would have the quality factor about Q=300. A microstrip line made of gold has $Q \approx 250$. Although the Ni-made rectangular-coaxial waveguides show higher loss than the Au/Cu-microstrips, their advantage is in regulation of the bandwidth by varying the cross-section geometry and in possibility of stronger isolation of the central conductor from the aggressor lines. Additionally to regular waveguides, some components are studied at millimeter-wave frequencies. Among them are the bends, inductors, cross-overs, branch-line couplers, etc. It is mentioned that all such components can be placed over any dielectric or semiconductor substrate, and they are easily integrated with active devices.

Another technology known for manufacturing of coaxial waveguides is based on the intensive use of many-step lithography and electroplating [163]. Initially, on a surface of a substrate covered by a conducting layer, a photoresist mold is formed which is covered by copper or gold using the electroplating technique. Then the photoresist inside the mold is removed through the windows in metal layers using chemical etching. A coaxial transmission line of the rectangular cross-section was studied and measured at frequencies up 10 GHz, and the loss close to 2.5 dB/cm was found. The central Cu conductor of the 5-μm thickness embedded inside the 100-μm height channel is supported by contacting pads placed over silicon covered by its dioxide. It was found that the 10-μm Cu ground layer did not shield completely the substrate, and an over-glass-placed coaxial line showed reduced loss about 0.5 dB/cm at the same frequency.

To avoid the increased sensitivity of micromachined coaxial waveguides towards mechanical vibrations, the PolyStrata process was used in [164], for instance. It allows forming in the metallic (copper) cavities the supporting thin-dielectric straps while photoresist is etched through the windows in metallic shield.

It is found that the Q-factors of half-wave resonators are close to 400 in frequencies 37-39.5 GHz. The averaged measured loss is within 0.086-0.092 dB/cm. Several components have been designed and studied using these rectangular micro-coaxials which cross-section size is comparable with several hundreds of microns. Among them are the resonators, couplers, a socket assembled with active devices, and the impedance transformers [164]-[167]. The simulations performed using the HFSS (Ansoft Corp.) show good correspondence with measurements. This technology requires multiple technology windows in the metallic shield, and the influence of them is studied analytically, numerically, and by measurements in [168]. It is shown the coupling of two coaxials is close to 55 dB in frequencies 20-40 GHz for 1-cm long coupled coaxial waveguides having 27 technological holes of the 100-μm length made in their common wall.

An interesting technology that allows for avoiding the holes is with the micromachining of photoresist (SU-8) layers. Each SU-8 layer is covered by 2-μm gold and bonded to each other forming a 3-D hollow structure. Using this technology, the coaxial, rectangular, and ridge waveguides are manufactured, and several components are studied in [169],[170]. The central conductor of coaxial line is supported by contacting pads and/or conducting stubs. The calculated loss of rectangular coaxial lines of 2-μm width of the central conductor is around 0.07-0.125 dB/cm at 38 GHz depending on the geometry of formed shielded channel (full-wave simulation results).

Coaxial waveguides can be manufactured filled by a dielectric. In [171], a technology and the measurement results are described for a polyimide coaxial waveguide. A 0.75-μm thick gold conductor is inside a polyimide core covered by Ti/Al layer. A single and coupled coaxial waveguides are measured in frequencies up to 65 GHz, and the loss about 7.5 dB/cm is found. It is supposed that a better dielectric would provide reduced loss of these waveguides. They, being manufactured on the common ground layer, show parasitic coupling, and the isolation better than 30 dB is found for these transmission lines placed at the distance 175 μm from each other.

Rectungular-coaxial waveguide supports the TEM fundamental mode, and the line's bandwidth is restricted by excitation of the higher-order modes. To calculate the parameters of this waveguide working in the TEM regime, a set of formulas verified by full-wave simulations and measurements from [162] can be used. More formulas on this line are from [172],[173].

Taking into account the TEM nature of the fundamental mode, its phase velocity is equal to the speed of light in vacuum. The characteristic impedance is

$$Z_c = 1/cC. \tag{6.67}$$

The line capacitance C [162] is

$$C = 2\varepsilon_0\left(\frac{w}{t_g}+\frac{t}{w_g}\right)+\frac{4\varepsilon_0}{\pi}\left[\ln\left(\frac{w_g^2+t_g^2}{4t_g^2}\right)+2\frac{t_g}{w_g}\arctan\left(\frac{w_g}{t_g}\right)\right]+$$
$$+\frac{4\varepsilon_0}{\pi}\left[\ln\left(\frac{w_g^2+t_g^2}{4w_g^2}\right)+2\frac{w_g}{t_g}\arctan\left(\frac{t_g}{w_g}\right)\right],\ w>w_g;\ t>t_g. \tag{6.68}$$

For thin central conductor lines $(t<t_g)$, the capacitance C is

$$C = 2\frac{\varepsilon_0 w}{t_g}+\frac{4\varepsilon_0}{\pi\ln 2}\left[1+\coth\left(\frac{\pi w_g}{2t_g+t}\right)\right].$$
$$\cdot\left[\frac{t+2t_g}{2t_g}\ln\left(\frac{4t_g+t}{t}\right)+\ln\left(\frac{t\cdot(4t_g+t)}{4t_g^2}\right)\right],\ t<t_g,\ w<w_g. \tag{6.69}$$

These formulas for the characteristic impedance calculations are validated by comparison with full-wave HFSS simulations and measurements, and they are in good agreement with each other up to 65 GHz for the chosen geometries of the tested lines [162]. In overall, rectangular-coaxial waveguides are promising for the ultra-bandwidth communications, narrow-band filters, and oscillators.

The above-considered technologies can be used to manufacture the micron-sized hollow waveguides of the diamond, hexagonally-shaped, and rectangular cross-sections. These waveguides have the cut-off frequencies of their fundamental modes in millimeter-wave and Terahertz frequencies, and they are to be expected of a low loss. There are some papers on the use of bulk silicon etching to form the hollow waveguides of different shapes in the substrate. For instance, the wet etching is used in [174] to prepare the V-groves in two silicon substrates. After that, they are covered by a bilayer of titanium/gold. Then these two halves are bonded to each other forming a diamond-shape silicon waveguide. This manufactured waveguide is assigned for the frequencies over 87.1 GHz, and its cross-section has the diagonals $a\times b = 2418\times 1710\ \mu\text{m}^2$. The measurements are in good correspondence with the HFSS full-wave simulations, and the found loss is 0.08 dB/(waveguide wavelength) at 110 GHz if the conductivity of walls is $\sigma = 5\times 10^6\ (\Omega\text{-m})^{-1}$.

Similar technology is used in [175] for manufacturing of hexagonally-shaped waveguides, which inner walls are covered by gold/germanium layers of 800-nm thickness. The height of the waveguiding channel is 550 μm and the width is 2032 μm. The sidewalls tilt angles are 54.7°. The main mode of this waveguide is the TE_{10} one, and the frequency band is limited by the cut-off frequency of the TE_{20} mode. At the difference to the known rectangular waveguides, the parameters of the fundamental TE_{10} mode depend on the waveguide height, not only on its width. Some theoretical results on the eigenmodes of regular polygonal waveguides, which can be useful for the design of anisotropically etched waveguides, are in [176].

The measured attenuation for the considered waveguide together with the transitions is close to 0.04 dB/cm in frequencies 100-170 GHz. The loss can be reduced further using a waveguide with the ratio of its sides $b = a/2$ and a thicker gold metallization up to five skin-depths at the highest frequency [175].

The vertical-wall rectangular waveguides and their components are manufactured using the deep-reactive ion etching (DRIE) technology applied to the silicon material [176]-[180]. The results of measurements of straight and folded rectangular waveguides, resonators, filters, and directional couplers are known in frequencies up to 430 GHz. For instance, the loss in fabricated WR-10 waveguide, which walls are covered by a Ti /Cu/Au layer, is 0.41-0.69 dB/cm over 90-110 GHz [178]. Increased quality factors of micromachined resonators are reached using their non-propagating modes excited by capacitive probes. For instance, a DRIE manufactured resonator shows Q=400-1400 in 20-100 GHz [180].

The photolithography process is used by several research teams to prepare the rectangular waveguides on a chip and some components on their base. One of these waveguides is described in [181] where the measurements are considered in frequencies 75-110 GHz. It is noticed on the difficulties with proper calibration; nevertheless, the estimated loss is close to 0.2 dB per wavelength.

Silicon can be left at the waveguide channels with the price of the loss increase. Such integrated trapezoidal waveguides filled by a high-resistivity silicon $(\sigma = (0.7-1.3)\ \text{k}\Omega\text{-cm})$ are described in [182] where they have loss 2 dB/cm for waveguides working over 77 GHz cut-off frequency. It is less than of the traditional integrated lines using the same silicon as the substrate material. The next positive effect is the slow-wave factor 2.5 of these integrated waveguides, which increases the compactness of devices.

Additionally to the above-described technology, the rectangular waveguides can be manufactured by etching of the SU-8 photoresist layers and covering them by vacuum evaporated seed metal layers and electroplated gold. These layers are bonded to each other, and no any technology windows are needed [183]. A novel micromachined waveguide bend for frequencies 220-335 GHz is presented in [184]. In [185], a WR-3 waveguide filter with four resonators is designed and measured at the 293.2-GHz central frequency. The insertion loss is close to 3.3 dB in the frequency band 25.8 GHz.

It is found that while the frequency increases up to several THz, the accuracy of the technology and quality of processed surfaces start to be the major problem for the terahertz transmission lines and components. An example of this kind of research is published, for instance, in [186] where the regular and meandered waveguides and the horn antennas for 3-THz applications are studied. This frequency range is especially interesting in many civilian and military applications. The studied waveguides are of the size $a \times b = 75 \times 37.5\ \mu\text{m}^2$, and they work in 2-4-THz range. The waveguides are manufactured using the surface micromachining of photoresist and using a combination of vacuum evaporating and electroplating of conducting layers.

The photoresist is removed from the inner volume of waveguides by solvents, and the technological windows in the waveguide walls are needed. The final roughness of the walls is estimated below 20-nm RMS (root mean square), which is less than the 50-nm skin-depth at 3 THz. Theoretical modeling of this waveguide is performed using analytical formulas and the HFSS, and the necessity of the use of complex conductivity of gold at such increased frequencies is found. The results show the loss of studied waveguides is close to 13 dB/cm at 3.11 THz, which is acceptable for future integrated applications.

6.4 Substrate Integrated Waveguides (SIW)

6.4.1 Review on the SIWs, Simulation Methods and Main Results

An interesting solution for interconnects and passive components of microwave and millimeter wave frequencies is with the substrate integrated waveguides [187]-[189]. The predecessor of this interconnecting and component technology is the fence waveguide proposed in 1971 [190]. Fence waveguide consists of a dielectric slab and the inserted to it two rows of closely placed conducting pins. Thus the EM field is confined due to the dielectric surface effect and the reflections from the pin fences. This waveguide and multiple components were studied by modeling and measurements, and they demonstrated a reduced loss at millimeter-wave frequencies [191].

The substrate integrated waveguide confines the field by two conducting planes placed on two sides of a dielectric substrate and by two rows of conducting pins shorting these two plates. An example of this waveguide is shown in Fig. 6.35.

This waveguide unifies the advantages of planar and waveguide technologies, and it is distinguished by a relatively low loss and improved shielding properties. The cross-section size of this waveguide depends on the frequency range because the cut-off frequency of the first TE mode is defined by the width of the channel. Similarly to the waveguides, the modal cut-off frequencies are decreased with the permittivity of the substrate, and the monomodal bandwidth is similar to the one of rectangular waveguides. The loss of these waveguides is defined by the substrate, conductors, and radiation from the fence, and it is slightly higher of the loss of a rectangular waveguide filled by the same dielectric [192],[193], but it is essentially less than in microstrip lines at these frequencies. The SIW's loss is extremely high close to the cut-off frequency of the main mode. The predicted working frequencies of SIW today are up to several hundreds of GHz [194], and some estimations extend them up to the THz range [195],[196].

The design considerations of these waveguides can be found in [192], for instance. Today several types of SIWs have been known. Among them are the *Substrate Integrated Non-radiating Waveguide* with an increased permittivity dielectric insert at the center of the channel (Fig. 6.36a) [197], *Half Mode SIW* (Fig. 6.36b) [198], *Multichannel SIW*, (Fig. 6.36c) [188], and *Wideband Ridged SIW* (Fig. 6.36d) [195]. Further improvements are with the air-filled fenced waveguides. In [199], a technology and components are considered allowing for

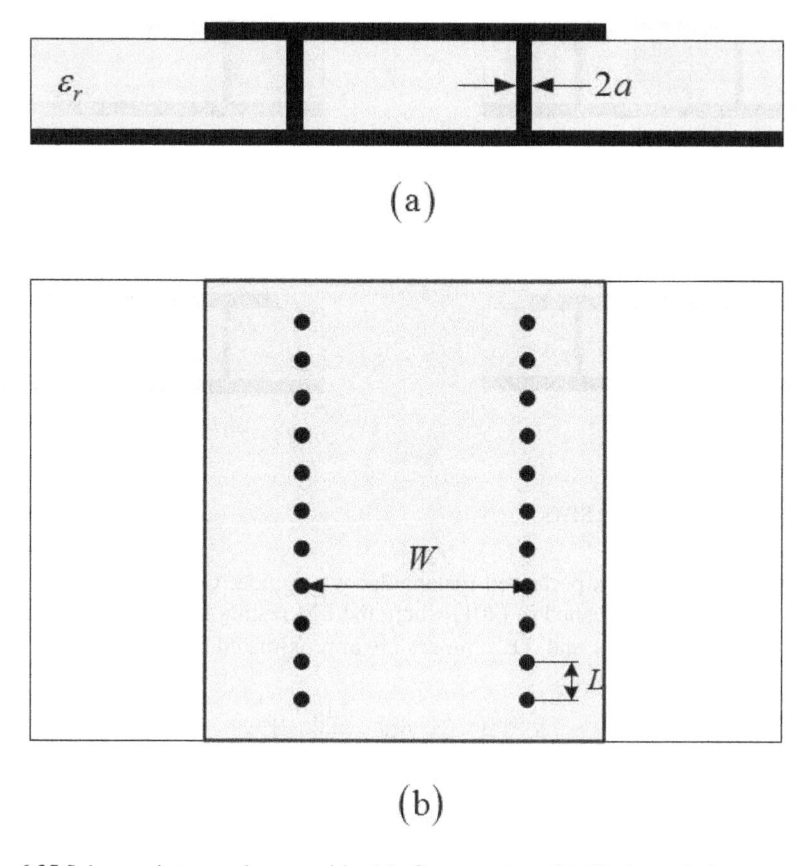

Fig. 6.35 Substrate integrated waveguide. (a)- Cross-section; (b)- Horizontal view

manufacturing the air-filled resonators and filters formed by the fences of metalized SU-8 pillars, which are covered by metalized silicon caps. The quality factor of these cavities can be of several hundreds of units at millimeter-wave frequencies. The SIW channel can be filled by a low-loss dielectric, and this type of waveguides is considered in [200] where a BCB SIW shows loss 0.8 dB/cm at 36 GHz. The SIW's passive components, including the transitions, resonators, filters, directional couplers, etc. are considered in many publications.

The most accurate modeling approach to the SIWs is with the full-wave simulations, and the EM solvers calculate these waveguides in the frequency and time domains. They allow for the modal phase and loss constants, characteristic impedances, and S-matrices of SIW's components.

For engineering practice, the analytical models and the design rules are in a great interest. There are two techniques allow for some approximate engineering formulas to calculate the SIW modal parameters. The first of them are with finding the equivalency between the rectangular waveguides and the SIWs. Both waveguides support the TE modes, and if we do not take into account the radiation, a SIW

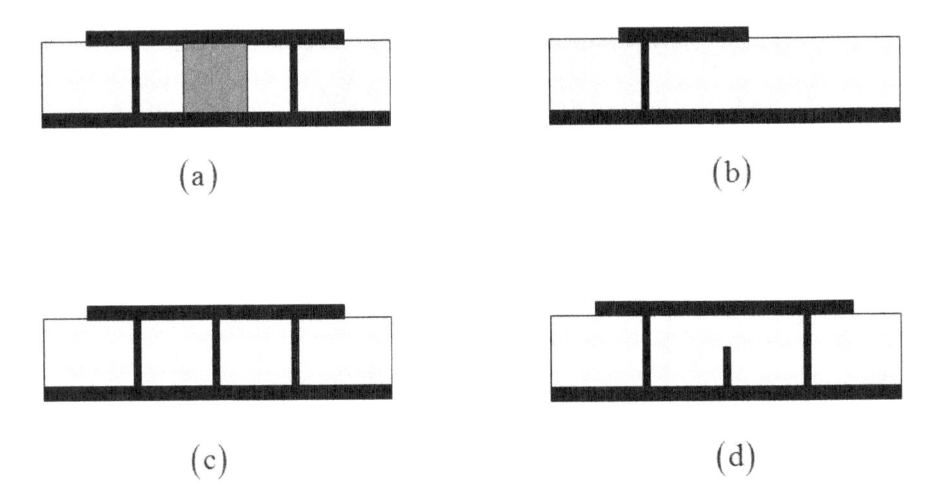

Fig. 6.36 Some modified SIWs

can be considered as a perturbed rectangular waveguide. One of the first equivalency formulas is published in [201] where the EM results for the cut-off frequencies of the SIW's TE_{10} and TE_{20} modes are approximated by

$$F_c^{(10)} = \frac{c}{2\sqrt{\varepsilon_r}}\left(W - \frac{4a^2}{0.95L}\right)^{-1}, \tag{6.70}$$

$$F_c^{(20)} = \frac{c}{\sqrt{\varepsilon_r}}\left(W - \frac{4a^2}{1.1L} - \frac{8a^3}{6.6L^2}\right)^{-1}. \tag{6.71}$$

Then for the fundamental mode TE_{10}, the equivalent rectangular waveguide width $W_{\text{eff}}^{(10)}$ is calculated as

$$W_{\text{eff}}^{(10)} = W - \frac{4a^2}{0.95L}. \tag{6.72}$$

It means, the rectangular waveguide of this width should have about the same modal parameters as the modeled SIW, and the known analytical formulas for propagation constants can be used.

Initially, this equivalency was validated comparing the full-wave simulations and the measurements of a SIW (up to 42 GHz) with the calculated phase coefficient of the equivalent rectangular waveguide. The error of calculation of the phase constant using (6.72) is within $\pm 5\%$. The second mode phase constant is calculated with the error $-9\% \ldots +4\%$ if the effective width is

$$W_{\text{eff}}^{(20)} = \left(W - \frac{4a^2}{1.1L} - \frac{8a^3}{6.6L^2}\right). \tag{6.73}$$

In [202], it is found that the dielectric and conductor losses in SIW and in the equivalent rectangular waveguide are close to each other. The known analytical formulas for the rectangular waveguides can be used to calculate the loss in SIW.

Another EM formula for the equivalent waveguides is [203]

$$W = \frac{2W_{\text{eff}}^{(10)}}{\pi} \operatorname{arc\,cot}\left(\frac{\pi L}{4W_{\text{eff}}^{(10)}} \ln \frac{L}{4a} \right) \tag{6.74}$$

which is valid for $L < 1/20\lambda_g$, with λ_g as the wavelength in the equivalent rectangular waveguide. This equivalency formula was tested by comparison with the measurements (up to 12 GHz) and the results derived with (6.72), and the found error in phase constant calculation was within 2% for studied geometries.

Further improvements of formulas for the effective width can be found in [204], for instance, where SIW is modeled by periodically cascaded joints of rectangular waveguides different from each other on the radius size of the shortening posts, and the proposed formula is

$$W_{\text{eff}}^{(10)} = \frac{W}{\sqrt{1 + 8\left(\frac{W-a}{L}\right)\left(\frac{a}{W-2a}\right)^2 - \frac{4W}{5L^4}\left(\frac{4a^2}{W-2a}\right)^3}}. \tag{6.75}$$

It demonstrates an increased accuracy of calculations of the propagation constants in frequencies up to 20 GHz in comparison to (6.72). One of the reason of this improving is that (6.75) takes into account the periodicity of the SIW's geometry in origin. This periodicity leads to the fence impedance effect and to the concentration of the field close to the fence, which cannot be occured in conventional rectangular waveguides. In this case, the surface impedance approach is prospective, and this idea is considered shortly below.

The walls of SIWs are composed of periodical grids of shorted cylindrical posts, and the techniques known from the grid theory can be applied to calculate the SIWs. The guiding and resonance properties of grids in open space started to be studied in the end of the 19th Century using different EM methods [205], which are very time-consuming even now. A convenient way is the use of the surface impedance approach which allows for substituting a grid by its equivalent impedance surface [206]-[208]. For many grid configurations, the surface impedance is expressed analytically if the grid geometrical parameters are essentially less than the wavelength in open space. For instance, M.I. Kontorovich introduced the averaged boundary conditions for periodical 1-D and 2-D grids in 1939. Simulations of many configurations using this technique were performed and validated by measurements and full-wave calculations in [206].

The idea of this and similar approaches is shown in Figs 6.37 and 6.38 where a fence of cylindrical posts is substituted by an impedance wall on which the following boundary condition is valid [206]

$$E_y^{(1)}(x=0,z) = Z_F' \cdot j_y(x=0,y,z) + Z_F'' \cdot \frac{\partial^2 j_y(x=0,y,z)}{\partial y^2} \tag{6.76}$$

where Z_F' and Z_F'' are the surface impedance components, $j_y(x=0,y,z)$ is the current density along the parallel conductors oriented in parallel to the y-axis. In a particular case, when $j_y = const$, only Z_F' is needed, and, for the electrically thin $(a << \lambda)$ cylindrical wires placed close to each other $(L < \lambda/2)$ it is calculated as [206]

$$Z_F' = j\frac{120\pi}{\sqrt{\varepsilon_r}}\frac{L}{\lambda}\ln\left(\frac{L}{2\pi a}\right). \tag{6.77}$$

To find the modal propagation constant, the general solutions of the wave equations are written with the following matching of the tangential field components on each boundary using (6.76) or conventional boundary conditions depending on the type of boundary. The obtained homogeneous system of linear algebraic equations regarding to the unknown field amplitudes gives an equation to derive the propagation constant k_z, as it was performed in [206] for a rectangular waveguide with the inserted grids. In this manner, in one of the first papers [190] on the fenced waveguides, the modal characteristics are computed.

Additionally to the direct matching approach that can be rather complicated for the multiple-fence waveguides, the transverse-resonance method (see Chpt. 2) is used taking into account that several formulas for surface impedance are known. In this case, the cross-section of a SIW is represented by an equivalent network, and, at an arbitrary plane of this cross-section, a transverse-resonance condition is written using known formulas for the input impedances of cross-sectional transmission lines and the surface impedances of the fences [209],[210]. As usually, the obtained equations for the modal propagation constants are the transcendental ones, and they need a root-searching algorithm to be applied. An example of the use of this technique is considered below for a shorted-by-a-fence waveguide intended for the differential signaling [96].

6.4.2 Modeling of Differential SIW

From the preceding review, it follows that the fenced waveguides can provide interesting solutions for the interconnecting of microwave signals and for the design of many components of the improved parameters.

Very often, to support the increased rate of transmitted signals, the differential interconnects are used. The simplest transmission line of this kind is the tightly coupled strips or microstrips lines. They allow for propagating two modes, and the information is carried by only the odd mode. The noise induced by the environment is of the common (even) mode nature, and it is cancelled by a differentially designed receiver. A variety of transmission lines can be used to support this type of signaling. One of the ideas has been published recently in [211] where the authors proposed to use the EH_1 mode of a microstrip for differential signaling. This mode has a variation of the electric field with its zero along the axis of the strip, and it is excited by two anti-phase signals (Fig. 6.37a).

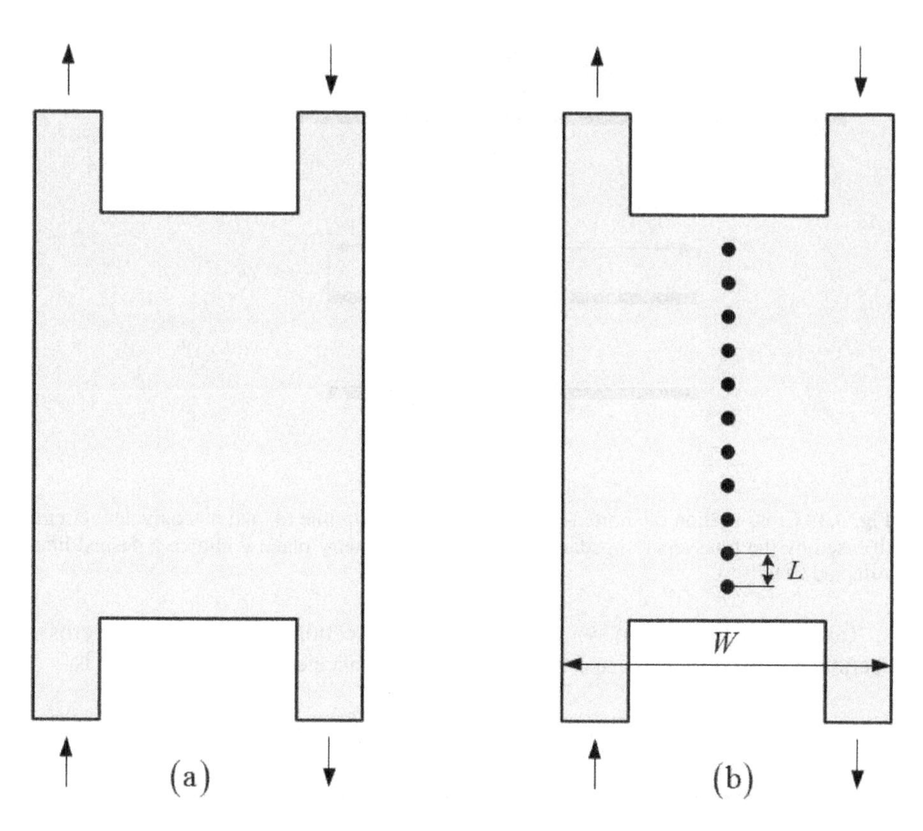

Fig. 6.37 Conventional higher-order mode (a) and shorted-by-the-fence (b) microstrips as the microwave differential transmission lines. Adapted from [96]

This interconnect can be improved further [96]. To avoid the transformation of the working EH_1 mode to the fundamental quasi-TEM one, a fence shorts the strip along its z-axis, and each branch should be excited differentially (Fig. 6.37b). Such a line, in fact, is used as an element of directional couplers considered in [212]. It is supposed that the lowest-critical-frequency mode of this line is of the quasi-TE type, and, in the first approximation, it can be modeled by the TE_x mode, which has no variation of its field along the y-axis. To calculate the modal longitudinal parameter k_z, the method of transverse resonance is chosen with the impedance model of the fence from [206], and the equivalent circuit for this model is shown in Fig. 6.38.

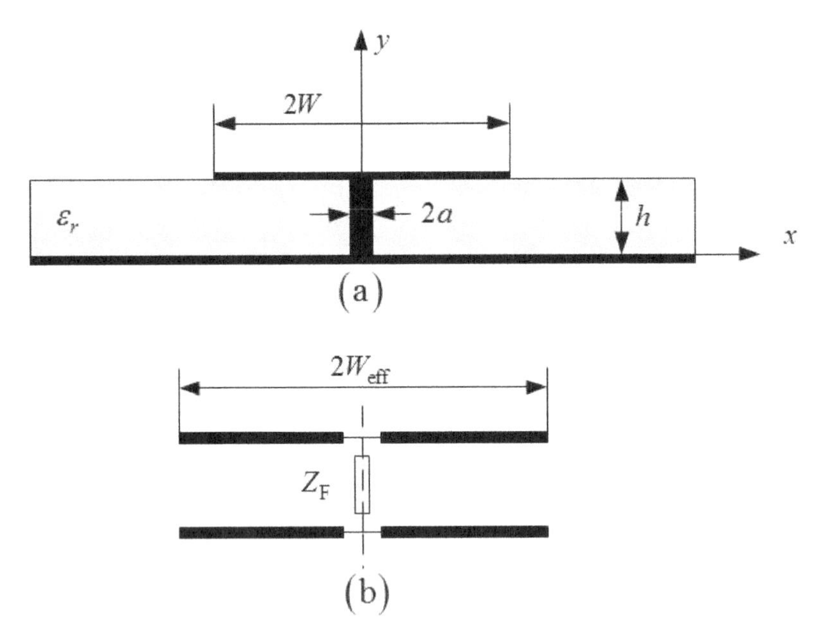

Fig. 6.38 Cross-section of shorted-by-the-fence microstrip line (a) and its equivalent circuit (b) used by the transverse impedance model. The symmetry plane is shown a dashed line. Adapted from [96]

Taking into account the symmetry of the cross-section, only a half of the cross-section can be considered, and the transverse resonance equation (see Part I) is

$$Z_{\text{F}}' + Z(W_{\text{eff}}) = Z_{\text{F}}' - j\frac{\omega\mu_0}{k_x}\cot(k_x W_{\text{eff}}) = 0 \tag{6.78}$$

where the surface impedance Z_{F} of the fence composed of conducting cylinders is calculated according to (6.77) and $k_x = \sqrt{k_0^2\varepsilon_r - k_z^2}$.

Supposing that $k_z = 0$, the equation for the modal cut-off frequency is obtained

$$Z_{\text{F}}' - j\frac{\omega\mu_0}{k_0\sqrt{\varepsilon_r}}\cot\left(k_0\sqrt{\varepsilon_r}W_{\text{eff}}\right) = 0. \tag{6.79}$$

In the above-considered formulas, only one geometrical parameter has been left undefined, and it is the effective width of the microstrip W_{eff}. Taking into account that the field distribution of this mode close to the edge of conductor is similar to the one of the first mode of a microstrip patch, a pertinent approximation is the empirical formula (6.80) carefully validated in [97]:

$$W_{\text{eff}} = W + \Delta W_e = W + 0.533h\,. \tag{6.80}$$

The results of simulations are shown in Fig. 6.39 where the normalized propagation constants k_z/k_0 of the first two modes are shown versus frequency F.

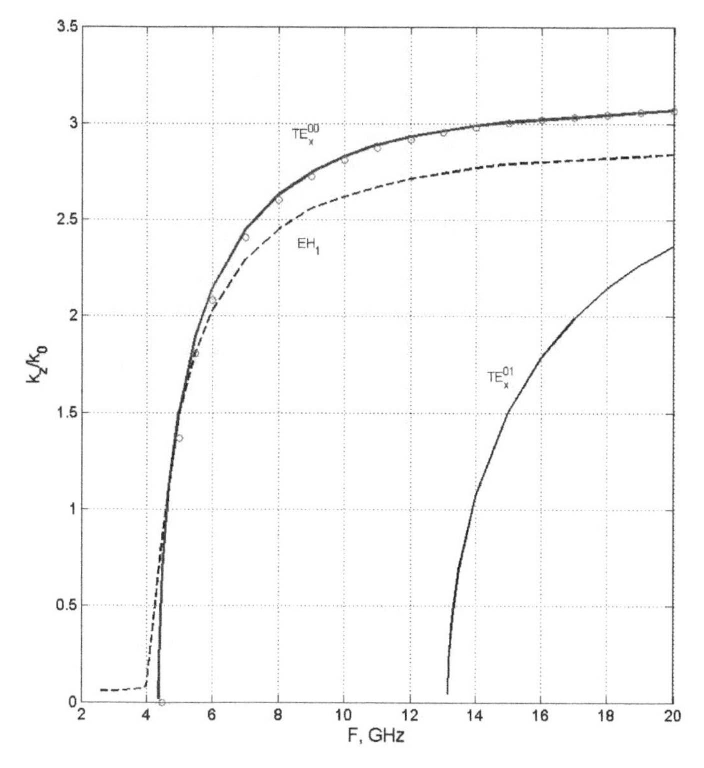

Fig. 6.39 Frequency dependence of the normalized modal propagation constants of two first TE_x^{00} and TE_x^{10} modes of the shorted-with-fence microstrip and of the first higher-order mode EH_1 of the microstrip line of the same size: $h = 1$ mm, $W = 5h$, $\varepsilon_r = 9.9$, $L = 1.5$ mm, and $a = 0.4$ mm. Adapted from [96]

It is supposed that the electric field $E_y(x)$ of the mode TE_x^{00} has no variation of its direction along the x-axis at low frequencies, but it depends on this spatial variable. The second index in the mode notation is with the field variations along the y-axis. For comparison, the frequency dependence of the first higher-order mode of a microstrip of the same width is shown calculated according to [213]. It is seen that the propagation constants of the modes of two waveguides are close to each other in spite of our calculations do not take into account the radiation effect started at frequencies of the effective coupling to the surface mode.

Although the eigenvalue transcendental equation (6.78) is simple, an analytical formula for propagation constant is attractive for this waveguide. To obtain it, let us follow the technique from [203]. It is shown that the use of the equivalent impedance condition allows for a simple formula for the extension of the equivalent rectangular waveguide, and it gives analytical expression for the propagation constant. The equivalent rectangular waveguide with one of the wall, which is the perfect magnetic one, is shown in Fig. 6.40. It corresponds to a half of the differential SIW shown in Fig. 6.37b.

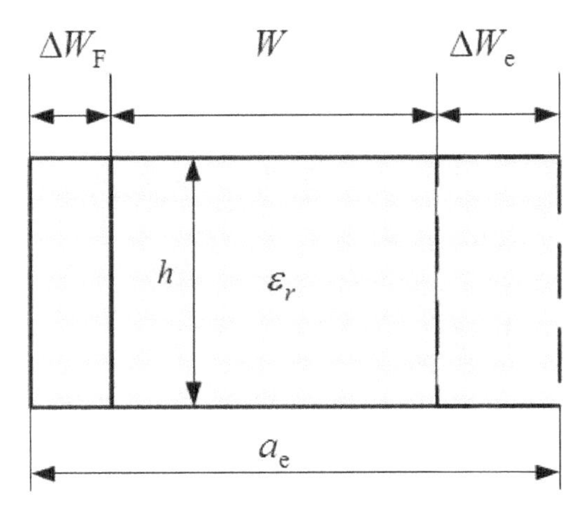

Fig. 6.40 Equivalent electric/magnetic wall waveguide for calculation of differential mode

In Fig. 6.40, the extention of the waveguide ΔW_F due to the fence inductive effect is found from the equation:

$$Z_\mathrm{F} = \frac{j\omega\mu_0}{\left(\frac{\pi}{W}\right)} \cot\left(\frac{\pi}{W}\Delta W_\mathrm{F}\right). \tag{6.81}$$

It allows an approximate analytical expression if the extention is small. Then the propagation constant k_z is

$$k_z = \sqrt{k_0^2 \varepsilon_r - \left(\frac{\pi}{2a_e}\right)^2}. \tag{6.82}$$

The results of calculation are shown in Fig. 6.39 by red circles, and they show good correspondence to the data derived by solution of the transcendental eigenvalue equation (6.78).

To calculate this differential interconnect for practical applications, the characteristic impedance is to be obtained

$$Z_\mathrm{c} = \int_0^h E_y\left(x = W_\mathrm{eff}\right) dy \bigg/ \int_0^{W_\mathrm{eff}} H_x(x)\, dx. \tag{6.83}$$

The modal fields are calculated as

$$\mathbf{E} = -j\omega\mu_0 \nabla \times \psi_h, \tag{6.84}$$

$$\mathbf{H} = \varepsilon_r k_0^2 \psi_h + \nabla \cdot \left(\nabla \cdot \psi_h\right) \tag{6.85}$$

where $\psi_h = \mathbf{x}_0 A \cos\left(k_x \left(W_{\text{eff}} - x\right)\right) e^{-jk_z z}$ with A as modal amplitude and the final formula for characteristic impedance is

$$Z_c = \frac{\omega\mu_0 k_z k_x h}{\varepsilon_r k_0^2 + k_x^2} \sin\left(k_x W_{\text{eff}}\right). \tag{6.86}$$

Some results of calculation of the characteristic impedance of the mode TE_x^{00} are shown in Fig. 6.41 for the same geometry as in Fig. 6.39.

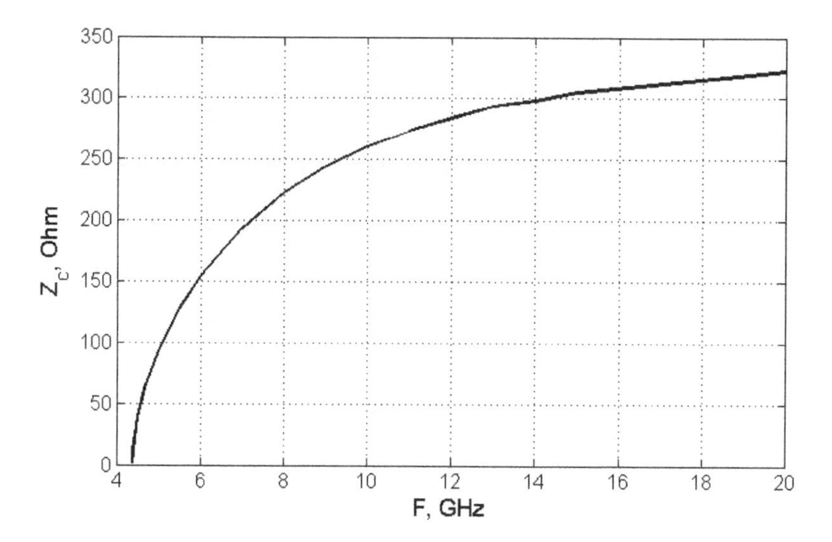

Fig. 6.41 Characteristic impedance Z_c of the TE_x^{00} mode of differential SIW versus frequency F. Adapted from [96]

6.4.3 EM Model of Fenced Strip Line

In many papers, it is shown that the fencing allows for effective isolation of the strip and microstrip lines from the electromagnetically induced noise [50]-[56]. Additionally, the strip lines embedded into SIW allows for multichannel communications, as it is shown in [194]. A part of the results on fenced strip and microstrip lines has been already mentioned in Section 6.1.5.4. Here an analytical model a strip line shielded by a fence is proposed and considered. Geometry of this line is shown in Fig. 6.42. A conductor strip is embedded into the SIW channel, and the via-hole walls can influence the parameters of the TEM mode if the distance between the conductor and the wall is less than three heights of the substrate.

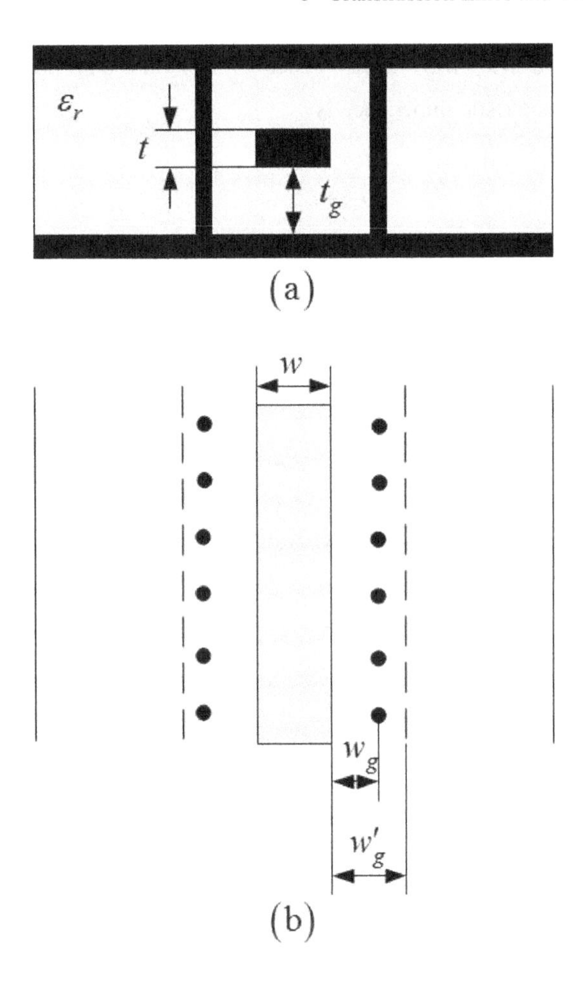

Fig. 6.42 Fenced strip line. (a)- Cross-section; (b)- Horizontal view

To calculate the parameters of this line, we assume that the fence, in the first approximation, does not change the TEM nature of the strip fundamental mode, and its propagation constant is still $k_z = k_0\sqrt{\varepsilon_r}$. The characteristic impedance Z_c is influenced by the fences, and it is calculated using the modified formulas for rectangular coaxial waveguide (6.67)-(6.69). We are supposing that, similarly to the conventional SIW [203], the fence inductive effect can be described by the effective channel widening, and it is

$$w'_g = w_g + \Delta W_F \tag{6.87}$$

where $\Delta W_F = 60\frac{\sqrt{\varepsilon_r}}{c\mu_0}\ln\left(\frac{L}{2\pi a}\right)$. Here L is the fence period, and a is the radius of the via-holes composing this fence. Then the characteristic impedance is calculated using (6.67)-(6.69) substituting there the distance between the central conductor w_g with w_g': $Z_c\left(w_g' \to w_g\right)$. It is interesting to validate this formula by full-wave computations and measurements. One of the challenges is with the minimal distance between the fence and the central conductor where the fence model is still providing correct results. Another question is with the frequencies where the frequency dependence of ΔW_F should be taken into account.

6.5 Comparative Analysis of Integrated Transmission Lines

Modern integrated technologies allow for manufacturing hundreds of transmission lines of different types. This "explosion" occured in the 80s of the last Century when the conception and the first results on microwave 3-D integrated circuits were proposed and published [91].

The used today transmission lines are different according to their design, functionality, and provided EM characteristics. Here the attention is paid to the comparison of the Si-based integrated transmission lines which are the most prospective for ultra-wide band or increased frequency commercial applications. Only one parameter is compared, and it is the loss per unit length of these transmission lines (Table 6.1). As well, short information on used technologies and materials is available from this Table with the references to the papers from which the data has been obtained.

Comparing this data, it has been found that not all needed information on the geometry of lines and other their parameters is available from the published papers due to the use of commercial technologies or the difficulties to measure the substrates and dielectric layers. Very often, the analytical models are not available due to the complexity of the cross-sections of the analyzed lines, and only full-wave simulation results are published. In some cases, the engineering models are developed which are valid only for the used technology and the employed IC stack. Another problem met in this comparative analysis, is that the loss is estimated, mainly, only for the 50-Ω lines. For some transmission lines, there are several geometries for this characteristic impedance, and not all of them have the same loss. In some cases, the thin-film microstrip lines are better than the CPWs especially if the CPW's dioxide isolation layer is thin and the substrate is very lossy. Besides, the CPW loss is the bias-voltage dependent, and, in many cases, this fact has not been messaged by the authors who studied, presumably, the zero-voltage-bias CPWs.

Anyway, the composed Table gives the figures which are typical for the different Si-technologies, and the data is useful for the initial estimation of loss in transmission lines and Si integrated circuits. Additionally to this Table and the comparative analysis of transmission lines for millimeter- and submillimeter-waves, some prospective interconnects for digital inter- and -intra-chip communication are considered in the Sections below.

Table 6.1 Loss Data for some Transmission Lines

	Transmission Line	Technology and Materials	Reference	Z_c,Ω	Loss, dB/cm	F, GHz	Measurements (M) or Simulations (S)
1	Thin-film MS	SiGe BiCMOS	[105]	50	11.5	110	M
2	Thin film MS	$\tan\delta = 0.015$	[9]	50	25	100	S
3	Thin-film MS	Ti/Au, Cyclic-olefin copolymer,	[29]	50	<17.5	220	M/S
4	Superstrated MS	GaAs/Au	[27], [28]	-	5-18	40	S/M
5	Stacked MS	32-nm STM CMOS, Cu/Al	[45]	70	16-17	110	S/M
6	Stacked MS	45-nm STM CMOS, Cu/Al	[45]	70	16-17	110	S/M
7	Stacked MS	65-nm STM CMOS, Cu/Al	[45]	70	<14	110	S/M
8	Stacked MS	45-nm CMOS, Al/Cu	[44]	-	14	110	M
9	Air-filled Strip	Multilayer Process	[50]	50	<3	40	M
10	Buried MS	Multilayer Process	[50]	50	<1.2	40	M
11	MS	Si/Au/ BCB/Au	[63], [64]	50	<6	220	M
12	CPW	SiGe HBT BiCMOS	[105]	50	80	110	M
13	CPW	Ion-implanted	[38]	-	6	120	M
14	CPW	30-μm BCB/LRSi	[86]	50	3	30	
15	Elevated Conductor-backed Air-filled CPW	Conductors are supported by pedestals	[50]	50	0.7	40	M
16	Conductor-backed COC filled CPW	Cyclic-olefin copolymer, Ti/Au	[29]	50	5-13.5	220	M
17	Elevated CPW	Si/Polyimide	[121]	65	3.5	40	M
18	CPS	SU-8/LRSi	[131]	100	8-9	40	M
19	Overlay CPW	Quartz 520-μm	[138]	36	2	110	M
20	3-D CPW	GaAs/Polyimide/Au	[142]	<50	0.5-8	15	M
21	Membrane Supported CPW	Si Micromachining	[145]-[149]	100	≈1	130-170	M
22	Trenched CPW	HRSi (10 k -cm)	[153]	50	1.6	30	M
23	Trenched CPW	HRSi (1k -cm)	[154]		35	300	M
24	Trenched Si, Bended CPW	HRSi (15 k -cm)	[156]	25-76	1.67	50	
25	CPW over Trenched LRSi 0.02 -cm	Filled with SiO_2	[158]	85	3.2	40	M
26	Integrated Rectangular Coax	Polystrata Techn., Cu	[164]	-	0.092	39.5	M

Table 6.1 (*continued*)

27	Integrated Rectangular Coax	SU-8/Au Techn.	[169]	50	0.07-0.125	38	S
28	Integrated Rectangular Waveguides	DRIE-Silicon, Ti/Cu/Au	[178]	-	0.41-0.69	90-110	M
29	Integrated Trapezoidal Waveguide	Micromachining, filled with HRSi	[182]	-	2	>77	M
30	Sub-mm Integrated Rectangular Waveguide	SU-8/Au/Air-filled	[184]	-	0.9-4.4	220-335	M
31	THz Integrated Rectangular Waveguide	Photoresist/Au/Air-filled	[186]	-	13	3311	M

6.6 Interconnects for Optoelectronics

6.6.1 Optical Interconnects

Taking into account that the conductor interconnects and passives are a bottleneck for many applications and they are the major energy consumers, the optical interconnects and optoelectronics are considered for some microwave and high-speed integrations [214],[215]. A prospective silicon/optoelectronic IC consists of traditional chip structure with placed over this chip several bonded optical layers where the optical waveguides and the necessary optoelectronics are embedded. It is supposed that these waveguides of decreased loss can provide the frequency bandwidth of several hundred GHz. A careful comparative analysis of optical and conductor interconnects including the energy consumption of laser sources [215],[216] shows that the optoelectronics has its advantage for signaling for a large distance of around several tens or hundreds of microns ([215], pp. 1166-1185). Then the best area of the optoelectronics is the global on-chip and off-chip applications.

Unfortunately, due to poor optical parameters of silicon, the waveguides are made of organic materials or SiON with n=1.46-2.0. The best transmission window of this material is close to λ=850 nm, and the outer laser sources are used to avoid the increased heating of chip ([215], pp. 1186-1198). According to this work, the inserted loss at this window of a 2-μm cross-section waveguide is 0.2-0.3 dB/cm. The increased optical density of this material allows for the 90°-bending without essential light radiation. To excite the conductor interconnects the plasmon nano-diodes are used. Although this one of the first optoelectronic silicon chip is working with the signal of the 5-GHz frequency, it is supposed that further progress might allow for the developments of new ultra-wide bandwidth chips. Some current problems on this way are analyzed in [215].

6.6.2 Plasmon-polariton Interconnects and Components

6.6.2.1 Plasmon-polariton Interconnects

As it was mentioned in the previous section, the co-integration of optical components and silicon electronics allows to overcome some limitations of electric interconnects regarding to the ultra-high-speed and multichannel signaling. Unfortunately, the large footprints of optical components, which are comparable with the several wavelengths, do not allow for reaching the ultimate advantage. In the ideal case, the optoelectronics parts should be comparable with the electronic ones, which size tends to several nanometers, and it is essentially smaller than the wavelength of the visible light.

The cross-section of optical waveguides can be decreased down to several tens of nanometers by the use of plasmons in conductors [214],[217]-[222]. The plasmons are the quantum pseudo particles of oscillating electronic plasma in metals. Close to a certain resonance frequency, the plasmons are coupled to the EM field, and these hybrids are called the plasmon-polaritons with their optical range propagating frequencies. Microwave plasmon-polaritons can be modeled by metamaterials [218]. In both cases, the conductors and metamaterials are described by the effective complex dielectric permittivity $\varepsilon_m = \varepsilon'_m + j\varepsilon''_m$:

$$\varepsilon_m = \left(\varepsilon_\infty - \frac{\omega_p}{\omega^2 + j\omega\omega_\tau} \right) \tag{6.88}$$

where the parameters $\varepsilon_\infty, \omega_p$, and ω_τ depend on the material. The EM waves in such media are described by Maxwell equations, and the well-known methods from optics and microwaves can be used to calculate the modal parameters.

The simplest guiding structure is just a boundary between a conductor and dielectric (Fig. 6.43). The wave propagates along the axis $0x$, which is normal to the picture plane. The field concentrates close to the surface $z = 0$, and it does not need additional reflection walls to guide the energy along this surface (Fig. 6.43).

The propagation constant k_{sp} of this wave is found analytically matching the fields of both domains at the boundary $z = 0$ [217]:

$$k_{\mathrm{sp}} = k_0 \left(\frac{\varepsilon_r \varepsilon_m}{\varepsilon_r + \varepsilon_m} \right)^{\frac{1}{2}}. \tag{6.89}$$

Unfortunately, there are many problems with the plasmonic waveguides. The increased loss of them is associated with the conductor, and it is caused by relaxation of plasmons in it. Due to the loss, the wave magnitude decreases exponentially while propagating along the x-axis. For instance, it decreases e-times on the distance δ_{sp}, which for the considered waveguide is [217]

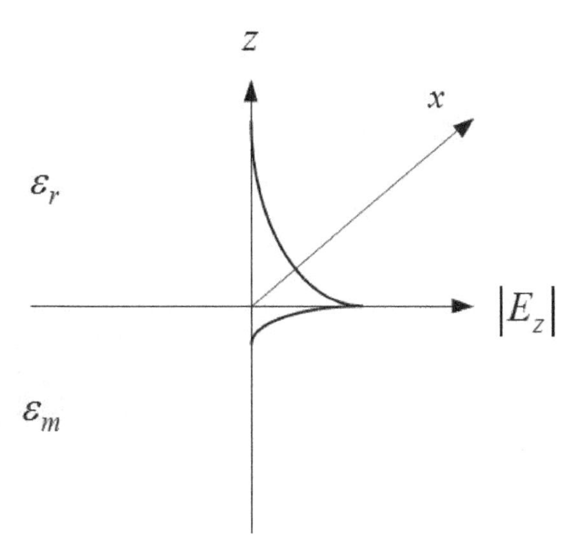

Fig. 6.43 Electric field distribution along the axis $0z$ which is normal to the boundary dielectric-conductor

$$\delta_{\mathrm{sp}} = \frac{1}{2k''_{\mathrm{sp}}} = \left[k_0 \frac{\varepsilon_r^{1.5} \varepsilon''_m}{\sqrt{\varepsilon'_m} \left(|\varepsilon'_m| - \varepsilon_r \right)^{1.5}} \right]^{-1} \tag{6.90}$$

where $\varepsilon'_m = \mathrm{Re}(\varepsilon_m)$, $\varepsilon''_m = \mathrm{Im}(\varepsilon_m)$, and $k''_{\mathrm{sp}} = \mathrm{Im}(k_{\mathrm{sp}})$. For different plasmonic waveguides, this parameter δ_{sp} is about of several tens or even hundreds of microns.

Another important parameter is the signals' cross-sections. Only ones, which field concentrates close to the nanoconductors, can be considered as the subwavelength signals. Unfortunately, the decreased cross-sections of plasmonic wires are with the essential loss.

Some studied guiding structures are shown in Fig. 6.44. The first three of them are the parallel-plate waveguides of different designs, and they are calculated analytically or semi-analytically in [214],[217]-[222], for instance. The modal field of plasmon-polariton waves concentrates close to the conductors and/or between them. The loss and phase constants are derived from the transcendental eigenvalue equations or using simplified analytical formulas. Additionally, some numerical computations are known, as well. The plasmon-polariton waveguides can support several propagating modes in a chosen frequency band. Besides, the modal spectrum has leaky modes, or the ones, which are leaky under certain conditions [220]. The considered lines are of the infinite cross-sections, and they are only theoretical abstractions.

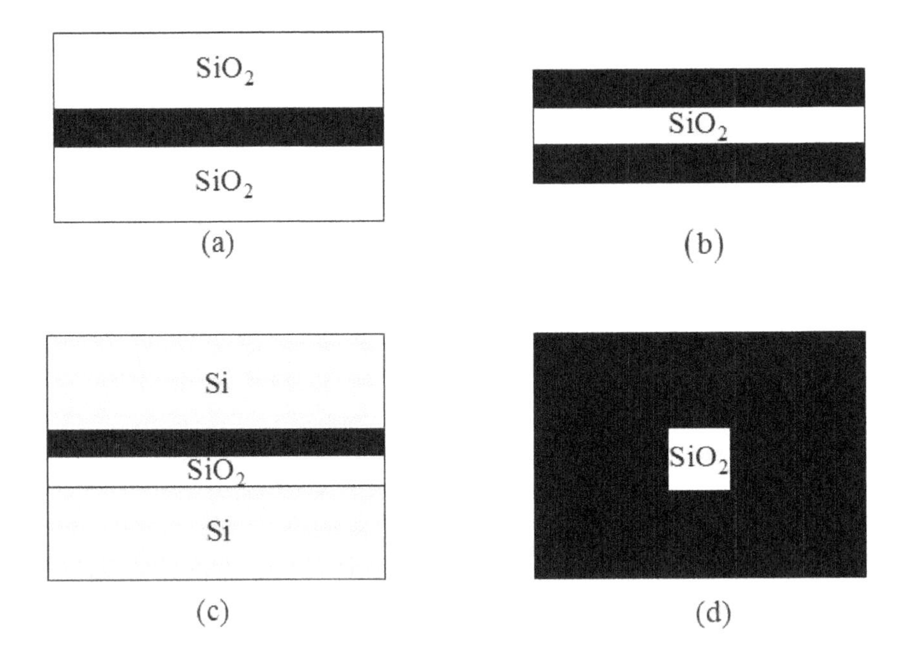

Fig. 6.44 Plasmon-polariton (plasmonic) waveguides. Conductors are shown in black

To confine the field, an analog of rectangular waveguide with lossy walls is proposed and studied in [223] (Fig. 6.44d). It allows for concentrating the field in its cross-section, which is only 10% of the wavelength. The surface mode field is "glued" to the inner walls, and such modes are interesting in subwavelength signaling. Numerical simulations allow for selecting the fundamental mode of the hybrid type and its unimodal frequency band. The used materials and geometry define the guiding parameters.

Some experimental results for coaxial nanowaveguides and resonators are given in [224]. The Ag-waveguides of 100-175-nm wide dielectric channel and length 265-485 nm were manufactured and measured for the light of the wavelength from the 450-nm range. A theory of plasmonic coaxial waveguides with complex shapes of cross-sections is given in [225],[226].

The integrated strip waveguides are shown in Fig. 6.45. They consist of a strip or coupled strips embedded into dielectric layers made of Si, SiO_2, or/and BCB. Both, numerical and experimental results are available given for micron-size conductors of the nanometric height (10-60 nm) [214],[227]-[231]. These studies show that the propagation loss decreases with the width (6-60 μm, 1-2 dB/cm), but these narrowed strips may have increased loss. Formally, they allow for the centimeter-long on-chip communication. Unfortunately, the typical cross-sectional field extension of the studied long-range waveguides is within 10-40 μm, then the increased cross-talks are expected in such integrations. Some experimentations with the silicon-gold-polymer strip waveguides (Fig. 6.45d) show the error-free transmission of 10- and 40-Gbit signals over 4-cm distance without essential distortions of their eye-patterns [229],[230].

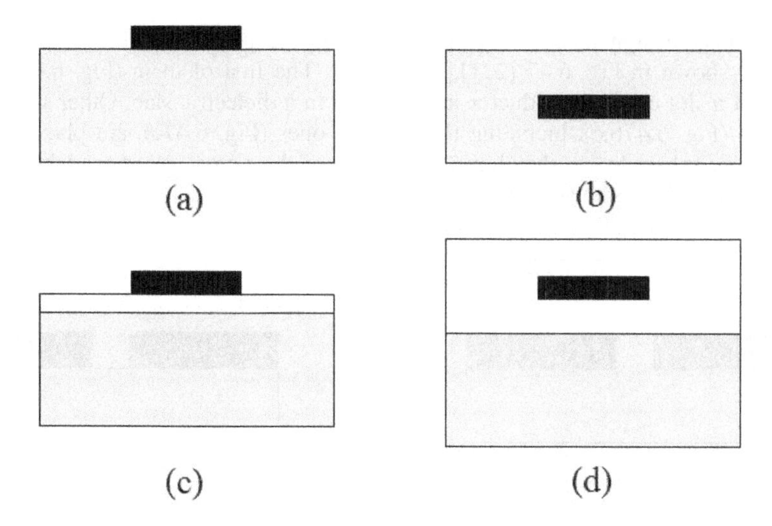

Fig. 6.45 Integrated strip plasmon-polariton waveguides

Another interesting type of waveguides is shown in Fig. 6.46a where a groove of a triangle shape guides the plasmon-polariton waves along the axis, which is normal to the figure plane [231]. The field of them is "glued" to the metallic surface, and these grooves provide increased concentration of the wave field in the channel.

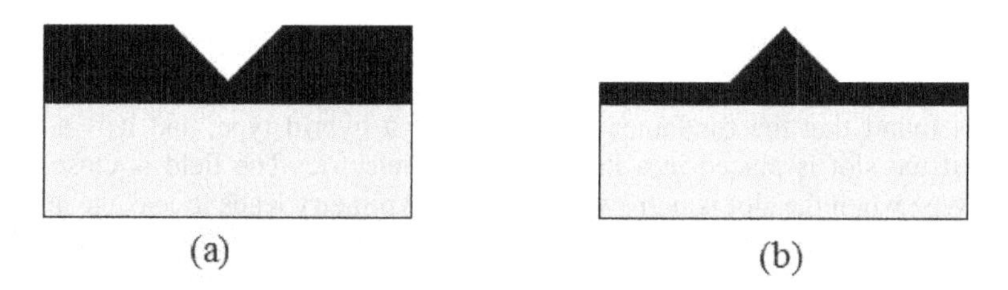

Fig. 6.46 Groove (a) and wedge (b) waveguides cut in a gold layer (in black) placed over a silica substrate

The measurements and approximate calculations, being in agreement, show the propagating length $\delta_{sp} \sim 100\ \mu m$ for micrometric grooves guiding light of the 1425-1620 nm range.

The wedge waveguide, shown in Fig 6.46b, provides more compactly "packaged" field around the wedge [232]. The performed numerical simulations show increased confinement of the field and a longer propagation length regarding to the surface mode of a groove waveguide (Fig. 6.46a). Typical data for $\lambda = 1.5\ \mu m$ is that $\delta_{sp} = 37\ \mu m$ if the gold wedge height is $0.2\ \mu m$ and the wedge angle 20º. Additionally, the circular bends of this waveguide show less radiation regarding to the groove ones.

Another design taken from microwaves is the plasmon-polariton slot waveguides shown in Fig. 6.47 [221],[233]-[235]. The first of them (Fig. 6.47a) consists of a slot cut in a conductor layer placed in a dielectric slab. Other slot waveguides (Fig. 6.47b,c), including the coupled ones (Fig. 6.47c), are placed over a dielectric substrate, i.e. they have asymmetry of their cross-sections. Additionally, the waveguides can be covered by a clad layer.

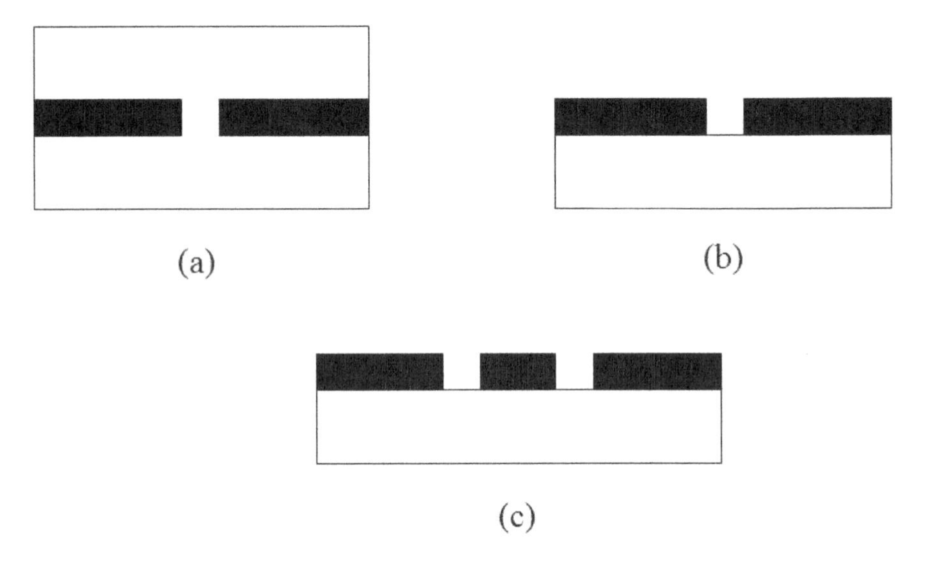

Fig. 6.47 Integrated plasmonic slot waveguides

It is found that the fundamental mode is of a hybrid type, and it is a bound mode if the slot is placed in a homogeneous dielectric. The field is close to the TEM type when the slot is narrow. Dielectric asymmetry leads to leakage at lower frequencies. The fundamental mode is highly concentrated around the slot, which can be of several tens of nanometers. The propagation length increases with the slot width, and it can be of several tens of micrometers. The modal area occupied by field decreases with the width. Additional loss increase is due to the field concentration near the conductor corners.

6.6.2.2 Components of Plasmon-polariton Integrations

Although the contemporary plasmonic guiding structures are rather far from to be perfect, some research concentrates on the discontinuities, components, and their theory. From general point of view, the modes of these waveguides diffracting at the discontinuities are transformed into an infinite set of modes of the connected waveguides, and they are radiated into the open space by the waves of the continuous frequency spectrum [236],[217]. This radiation, being parasitic, can be increased for antennas of visible light.

These discontinuities are studied by full-wave software tools and experimentally. For instance, a theoretical study of steps of metal-insulator-metal (MIM) waveguide (Fig. 6.44b) is given in [237]. The authors calculate the *S*-matrices of steps supposing the exponential decay of the high-order modes with the distance from a step. Taking into account the concentration of higher-order modal fields close to this junction, the equivalent circuits are obtained based on the known approach from the theory of perfect waveguides. Similarly to the microwave network theory, the *S*-matrices of several joints are calculated, and their accuracy is analyzed in [237]. This simplified model of MIM joints is in a good agreement with the full-wave simulations performed by the authors of that paper. A number of components on long-range strip waveguides have been proposed, simulated, and measured in other papers. They are the Y-joints, directional couplers, strip waveguides, and antennas [228],[238]-[240].

Some components on slot waveguides are described in [241]-[243]. Among them are the bends, 3-dB branchline coupler, Fabri-Perot resonator, 4-way joint, and a transition from an optical waveguide to a slot plasmonic one.

The components on V-shaped grooves and their measurements are published in [240]. The authors study the Y-joints, ring resonators, and interferometers. Measurements are in a good correspondence with the single mode theory. It is noticed that the designed optical circuitry has a decreased footprint, and the components are compatible with electronic integration.

A compact surface-plasmon-polariton microcavity with $Q \approx 1-2.4\times10^3$ is described in [244]. It consists of a silica disc covered by silver. It is shown that this increased quality factor is typical for whispering-gallery modes. It is stated on a possibility to create the surface-plasmon lasers based on this type of oscillations and cavities.

One of the smallest lasers has been developed recently, and it generates the plasmons in a gold spherical nano-core covered by silica and amplified by organic dye [245]. Although, it radiates coherent light isotropically, this development is a very important step towards practical nano-plasmonics. Another interesting work in this direction is a theoretical and experimental study of quantum dots coupled to nanowires [246]. It allows for exciting the quantized plasmon-polaritons for digital plasmonics.

6.6.3 On the EM Theory of Plasmon-polariton Interconnects and Components

The most theoretical research on plasmon-polariton integrations is based on the use of equivalent dielectric permittivity of real conductors or metamaterials at optical frequencies (6.88). The parameters in this formula depend on the conductor material or metamaterial, and they can be found in [247]. Considering these materials as the frequency-depending solids, the Maxwell or Helmholtz equations with the boundary conditions are solved to find the waveguiding parameters of plasmon-polariton transmission lines.

The simplest structures as the parallel-plate waveguides (Fig. 6.44b) are solved by matching the analytically found fields of partial domains at their boundaries.

Then the transcendental equations regarding to the complex longitudinal modal propagation constants are solved numerically. In some cases, the lossy conductor layers are substituted by equivalent impedance walls, and the Leontovitch boundary condition is used as the shortest way to obtain the eigenvalue equations [217].

Similarly to the parallel-plate waveguide analytical theory [22], the transcendental functions of small arguments in the eigenvalue equations can be substituted by the first-order approximations, and some analytical formulas for the modal propagation constants are derived [248]. In that work, the authors proofed the idea that the TEM-wave based plasmon-polariton waveguides could be calculated by perturbation method if the skin-depth is essentially smaller than the height of conductor in optical range. The criterion of TEM-like modes in multi-conductor plasmon-polariton waveguides is also given. Qualitatively, the TEM- and perturbation approaches are applicable if the conductors are close to each other, i.e. the field-decay distance in the substrate should be smaller than its height. The conductor slabs support practically independent surface modes outside of this region of frequencies and geometries,

For example, the complex propagation constant γ_z of a parallel-plate waveguide (Fig. 6.44b) calculated in the quasi-TEM approximation is [248]

$$\gamma_z = \frac{\sqrt{\varepsilon_r}\,|\varepsilon_m''|}{|\varepsilon_m'|^{3/2}\,2d} + j\left(k_0\sqrt{\varepsilon_r} + \frac{\sqrt{\varepsilon_r}}{d\sqrt{|\varepsilon_m'|}}\right) \tag{6.91}$$

where d is the dielectric height and the limited thickness of metallic slabs is not taken into account.

To understand the optic light phenomena in nanometric circuits the analogies with the equivalent electric network theory can be helpful [249]. Some practical applications of this and similar approaches have been already considered above for calculation of discontinuities and plasmonic components [237],[241].

The solutions, which are more accurate, are from the full-wave solvers based on different numerical methods like the FDTD or FEM. In these cases, the attention should be paid to the modeling of the field distribution inside the lossy thin conductors or tiny irregularities.

The high-memory cost is typical for solving the problems where the fields from small obstacles are calculated in a large surrounding space. In this case, the surface integral method is preferable [250]. The unknown currents are calculated only on the surface of lossy conductors; the fields inside and outside them are computed using analytical integral-based formulas. Some typical problems with this method as the singularities of the Green functions are solved analytically in [251].

6.7 Prospective Interconnects

As follows from the analysis of many authors on transmission lines and digital interconnects the main limitation of using the traditional transmission lines is their increasing loss and modal dispersion.

In digital electronics, a half of the loss is associated with the interconnects because the CMOS gates are consuming power only during their switching. Further improving of the transistors can lead to the extremely low power dissipation by these elements which is close to several $k_B T$ where k_B is the Boltzmann's constant and T is the absolute noise temperature. There are some ideas how to substitute the CMOS transistors by quantum-mechanical devices operating the signals of different types, and the loss and the time delay of traditional interconnects would seriously limit the performance of the future chips.

One of the alternatives has been already considered above in the Section 6.6. It is supposed to use the plasmon-polariton interconnects as the transmission lines for large bandwidth signals taking into account their optical frequencies. Unfortunately, these lines still have rather increased loss, and, additionally they need optical transistors or low-loss EM-optical transformers. Encouraging experimentations have been published recently on the developments of the plasmon-polariton lasers [245].

Besides the plasmon effects, several other ones can be used for transmission of signals in integrated chips of near of relative far future. Most of them are analyzed and compared in [252], for instance. They are:

- Diffusion effects
- Drift effects
- Ballistic effects
- Propagation of spin-waves
- Propagation of plasmonic waves
- Electric signal propagation along wire interconnects.

Diffusion is arisen in a system where a difference of concentration of carriers exists. Thus the motion of different carriers can be used for signaling, including the spins, phonons, and excitons. This signaling is described by corresponding diffusion equations written regarding to the concentration of carriers. *Drift* is with physical motion of a carrier from one point to another. For instance, spin can be transported together with electron. *Ballistic effects* are arisen in low-dimension conducturs, for instance, in graphene, and they are distinguished by the scattering-less motion. *Spin-wave interconnects* are based on the wave-like motion of atom spins from one point of space to another.

The above-considered effects were analyzed together in [252], and it is shown the fastest interconnects are of those which guiding plasmons, electric voltage waves (CMOS interconnects), and then spin waves. Other effects can be competitive with them regarding to the delay, only if the based on them interconnects are short. In [252], a conculusion is made that the future integration could be of the hybrid nature composed of elements and interconnects of different types. It will require the developments of new simulation and design tools which can seamlessly model these new integrations. One of the ideas of this sort is considered in the Chpt. 9.

References

[1] Barret, R.M.: Microwave printed circuits - the early years. IEEE Trans., Microw. Theory Tech. 32, 983–990 (1984)
[2] Marcuvitz, N.: Waveguide Handbook. Inst. of Eng. and Techn. Publ. (1986)
[3] Hoffman, R.K.: Handbook of Microwave Integrated Circuits. Artech House (1987)
[4] Kompa, G.: Practical Microstrip Design and Applications. Artech House (2005)
[5] Kovalev, I.S.: Theory and Calculation Strip Waveguides. Minsk (1967) (in Russian)
[6] Nefeodov, E.I., Fialkovskyi, A.T.: Strip Transmission Lines. Theory and Calculation of Discontinuities. Nauka, Moscow (1974) (in Russian)
[7] Nikolskyi, V.V.: Computer Aided Design of Microwave Devices. Nauka (1980) (in Russian)
[8] Kouzaev, G.A., Deen, M.J., Nikolova, N.K.: A parallel-plate waveguide model of lossy microstrip transmission line. IEEE Microw. Wireless Comp. Lett. 15, 27–29 (2005)
[9] Schnieder, F., Heinrich, W.: Model of thin-film microstrip line for circuit design. IEEE Trans., Microw. Theory Tech. 49, 104–110 (2001)
[10] Olyslager, F., De Zutter, D., Blomme, K.: Rigorous analysis of the propagation characteristics of general lossless and lossy multiconductor transmission lines in multilayered media. IEEE Trans., Microw. Theory Tech. 41, 79–88 (1993)
[11] Tzuang, C.-K.C., Tseng, J.-D.: A full-wave mixed potential mode-matching method for the analysis of planar or quasi-planar transmission lines. IEEE Trans., Microw. Theory Tech. 39, 1701–1711 (1991)
[12] Gong, X., Gao, B., Tian, Y., et al.: An analysis of characteristics of coplanar transmission lines with finite conductivity and finite strip thickness by the method of lines. J. Infrared Milli. Terahz Waves 30, 792–801 (2009)
[13] Pucel, R.A., Masse, D.J., Hartwig, C.P.: Losses in microstrip. IEEE Trans., Microw. Theory Tech. 16, 342–350 (1968)
[14] Lee, H.-Y., Itoh, T.: Phenomenological loss equivalence method for planar quasi-TEM transmission lines with a thin normal conductor or superconductor. IEEE Trans., Microw. Theory Tech. 37, 1904–1909 (1989)
[15] Verma, A.K., Bhupal, A.: Conductor loss of multilayer microstrip line using the single layer reduction formulation. Microw. Opt. Tech. Lett. 19, 20–24 (1998)
[16] Tuncer, E., Lee, B.-T., Islam, M.S., et al.: Quasi-static conductor loss calculations in transmission lines using a new conformal mapping technique. IEEE Trans., Microw. Theory Tech. 42, 1807–1815 (1994)
[17] Djordjevic, A.R., Sarkar, T.P.: Closed-form formulas for frequency-dependent resistance and inductance per unit length of microstrip and strip lines. IEEE Trans., Microw. Theory Tech. 42, 241–248 (1994)
[18] Goldfarb, M.E., Platzker, A.: Losses in GaAs microstrip. IEEE Trans., Microw. Theory Tech. 38, 1957–1963 (1990)
[19] Conn, D.R., Naguib, H.M., Anderson, C.M.: Mid-film for microwave integrated circuits. IEEE Trans., Comp., Hybrids, Manuf. Tech. 5, 185–191 (1982)
[20] Torres-Torres, R.: Extracting characteristic impedance in low-loss substrate. El. Lett. 47(3), 191–193 (2011)
[21] Crute, J.R., Davis, L.E.: Loss characteristics of high-ε_r microstrip lines fabricated by an etchable thick-film on ceramic MCM technology. IEEE Trans., Adv. Pack. 25, 393–396 (2002)

[22] Guckel, H., Brennan, P.A., Paloscz, I.: A parallel-plate waveguide approach to micro-miniaturized, planar transmission lines for integrated circuits. IEEE Trans., Microw. Theory Tech. 15, 468–476 (1967)

[23] Pond, J.M., Krowne, C.M., Carter, W.L.: On the application of complex resistive boundary conditions to model transmission line consisting of very thin superconductors. IEEE Trans., Microw. Theory Tech. 37, 181–190 (1967)

[24] Kouzaev, G.A.: Quasistatic model of ribbed nonsymmetrical slotted line. Radio Eng. Electron. Phys. 28, 137–138 (1983)

[25] Schneider, M.V.: Dielectric loss in integrated microwave circuits. Bell Syst. Tech. J. 48, 2325–2332 (1969)

[26] Glib, J.P.K., Balanis, C.A.: Transient analysis of distortion and coupling in lossy coupled microstrips. IEEE Trans., Microwave Theory Tech. 38, 1894–1899 (1990)

[27] Paleczny, E., Kinowski, D., Legier, J.F., et al.: Comparison of full wave approaches for determination of microstrip conductor losses for MMIC applications. El. Lett. 26(3), 2076–2077 (1990)

[28] Wang, E.-K., Tzuang, C.-K.C.: Full-wave analyses of composite-metal multidielectric lossy microstrips. IEEE Microw. Guided Wave Lett. 1, 97–99 (1991)

[29] Peytavit, E., Donche, C., Lepilliet, S., et al.: Thin-film transmission lines using cyclic olefin copolymer for millimetre-wave and terahertz integrated circuits. El. Lett. 47(7), 453–454 (2011)

[30] Mongia, R., Bahl, I., Bhartia, P.: RF and Microwave Coupled-Line Circuits. Artech House (1999)

[31] Svačina, W.J.: Analysis of multilayer microstrip lines by a conformal mapping method. IEEE Trans., Microw. Theory Tech. 40, 769–772 (1992)

[32] Lutz, R.D., Tripathi, V.K., Weisshaar, A.: Enhanced transmission characteristics of on-chip interconnects with orthogonal gridded shield. IEEE Trans., Adv. Pack. 24, 288–293 (2001)

[33] Chiang, M.-J., Wu, H.-S., Tzuang, C.-K.C.: Design of synthetic quasi-TEM transmission line for CMOS compact integrated circuit. IEEE Trans., Microw. Theory Tech. 55, 2512–2520 (2007)

[34] Chiang, M.-J.: Highly integrated three-dimensional synthetic transmission line design on silicon substrate. Microw. Opt. Tech. Lett. 53, 2604–2607 (2011)

[35] Williams, D.F.: Metal-insulator-semiconductor transmission lines. IEEE Trans., Microw. Theory Tech. 47, 176–181 (1999)

[36] Schellenberg, J.M.: CAD models for suspended and inverted microstrip. IEEE Trans., Microw. Theory Tech. 43, 1247–1252 (1995)

[37] Yamashita, E.: Variational method for the analysis of microstrip-like transmission line. IEEE Trans., Microw. Theory Tech. 16, 529–535 (1968)

[38] Chen, C.C., Hung, B.F., Chin, A., et al.: High-performance bulk and thin-film microstrip transmission lines on VLSI standard Si substrates. Microw. Optical Tech. Lett. 43, 148–151 (2004)

[39] Chin, A., Chan, K.T., Huang, C.H., et al.: RF passive devices on Si with excellent performance close to ideal devices designed by electro-magnetic simulations. In: Proc. IEDM 2003, pp. 375–378 (2003)

[40] Kouzaev, G.A., Deen, M.J., Nikolova, N.K., Rahal, A.: An approximate parallel-plate waveguide model of a lossy multilayered microstrip line. Microw. Opt. Tech. Lett. 45, 23–26 (2005)

[41] Hammerstad, E., Jensen, O.: Accurate models for microstrip computer-aided design. In: 1980 IEEE MTT-S Int. Microw. Symp. Dig., pp. 407–409 (1980)

[42] Lowther, R., Lee, S.-G.: On-chip interconnect lines with patterned ground shields. IEEE Microw. Guided Wave Lett. 10, 49–51 (2000)
[43] Dubuc, D., De Raedt, W., Carchon, G., et al.: MEMS-IC integration for RF and millimeterwave applications. In: Proc. 13th GAAS Symp., Paris, pp. 529–532 (2005)
[44] Shi, J., Kang, K., Xiong, Y.Z., et al.: Millimeter-wave passives in 45-nm digital CMOS. IEEE El. Device Lett. 31, 1080–1082 (2010)
[45] Quemerais, T., Moquillon, L., Fournier, J.-M., et al.: 65-, 45-, 32-nm aluminum and copper transmission line model at millimeter-wave frequencies. IEEE Trans., Microw. Theory Tech. 58, 2426–2433 (2010)
[46] Dib, N.I., Harokopus Jr., W.P., Katehi, L.P.B., et al.: Study of a novel planar transmission line. In: 1991 IEEE MTT-S Int. Microw. Symp. Dig., pp. 623–626 (1991)
[47] Cheng, K.-K.M., Robertson, I.D.: Quasi-TEM study of microshield lines with practical cavity sidewall profiles. IEEE Trans., Microw. Theory Tech. 43, 2689–2694 (1995)
[48] Yuan, N., Ruan, C., Lin, W.: Analytical analyses of V, elliptic and circular-shaped microshield transmission lines. IEEE Trans., Microw. Theory Tech. 42, 855–859 (1994)
[49] Du, Z., Ruan, C.: Analytical analysis of circular-shaped microshield and conductor-backed coplanar waveguide. Int. J. Infrared Milli. Waves 18, 165–171 (1997)
[50] Jeong, A., Shin, S.-H., Go, J.-H., et al.: High-performance air-gap transmission lines and inductors for millimeter-wave applications. IEEE Trans., Microw. Theory Tech. 50, 2850–2855 (2002)
[51] Ponchak, G.E., Chun, D., Yook, J.-G., et al.: The use of metal filled via holes for improving isolation in LTCC RF and wireless multichip packages. IEEE Trans., Advanced Pack. 23, 88–99 (2000)
[52] Zhang, R., Fang, D.G., Wu, K.L., et al.: Study of the elimination of surface wave by metal fences. In: Proc. Asia-Pacific Conf. Environmental Electromagnetics, CEEM 2000, Shanghai, China, May 3-7, pp. 174–178 (2000)
[53] Kim, J., Qian, Y., Feng, G., et al.: Millimeter–wave silicon MMIC interconnect and coupler using multilayer polyimide technology. IEEE Trans., Microw. Theory Tech. 48, 1482–1487 (2000)
[54] Ponchak, G.E., Tentzeris, E.M., Papapolymerou, J.: Coupling between microstrip line embedded in polyimide layers for 3D-MMICs on Si. In: IEE Proc., Microw. Antennas Propag., vol. 151, pp. 344–350 (October 2003)
[55] Gipprich, J., Stevens, D.: A new via fence structure for cross-talk reduction in high density strip line packages. In: 2001 IEEE MTT-S Microw. Symp. Dig., pp. 1719–1722 (2001)
[56] May, J.W., Rebeiz, G.M.: A 40-50-GHz SiGe 1:8 differential power divider using shielded broadside-coupled striplines. IEEE Trans., Microw. Theory Tech. 56, 1575–1581 (2008)
[57] Jin, J.-D., Hsu, S.S.H., Yang, M.-T., et al.: Low-loss differential semicoaxial interconnects in CMOS process. IEEE Trans., Microw. Theory Tech. 54, 4333–4340 (2006)
[58] Smith, C.E., Chang, R.-S.: Microstrip transmission line with finite-width dielectric. IEEE Trans., Microw. Theory Tech. 28, 90–94 (1980)
[59] Engel, A.G., Katehi, L.P.B.: Frequency and time domain characterization of microstrip-ridge structures. IEEE Trans., Microw. Theory Tech. 41, 1251–1262 (1993)
[60] Gvozdev, V.I., Kouzaev, G.A., Kulevatov, M.V.: Narrow band-pass microwave filter. Telecommun. Radio Eng. 49, 1–5 (1995)

[61] Chow, Y.L., Tang, W.C.: Formulas of microstrip with a truncated substrate by synthetic asymptotes-a novel analysis technique. IEEE Trans., Microw. Theory Tech. 49, 947–953 (2001)
[62] Smith, C.E., Chang, R.-S.: Microstrip transmission line with finite-width dielectric and ground plane. IEEE Trans., Microw. Theory Tech. 33, 835–839 (1985)
[63] Six, G., Prigent, G., Rius, E., et al.: Fabrication and characterization of low-loss TFMS on silicon up to 220 GHz. IEEE Trans., Microw. Theory Tech. 53, 301–305 (2005)
[64] Prigent, G., Rius, E., Happy, H., et al.: Design of branch-line coupler in the G- frequency band. In: Proc. 36th Eur. Microw. Conf., pp. 1296–1299 (2006)
[65] Yook, J.-M., Kim, K.-M., Kwon, Y.-S.: Air-cavity transmission lines on anodized aluminum for high-performance modules. IEEE Microw. Wireless Comp. Lett. 19, 623–625 (2009)
[66] Bang, Y.-S., Kim, N., Cheon, C., et al.: Fabrication of hybrid shielded-strip-line using half substrate integrated waveguide and half shielded-stripline structures. El. Lett. 47, 110–111 (2011)
[67] Mathaei, G., Young, L., Johnes, E.M.T.: Microwave Filters, Impedance-Matching Networks, and Coupling Structures. Artech House (1961)
[68] Wadell, B.C.: Transmission Line Design Handbook. Artech House (1991)
[69] Gvozdev, V.I., Kouzaev, G.A., Nefedov, E.I.: Filters on multilayered microwave integrated circuits for antennas applications. In: Proc. Conf. Design and Computation of Strip Transmission Line Antennas, Sverdlovsk, Russia, pp. 72–76 (1982) (in Russian)
[70] Stellary, F., Lacaita, A.L.: New formulas of interconnect capacitances based on results of conformal mapping method. IEEE Trans., El. Dev. 47, 222–231 (2000)
[71] Gvozdev, V.I., Kouzaev, G.A., Nefedov, E.I., et al.: Physical principles of the modeling of three-dimensional microwave and extremely high-frequency integrated circuits. Soviet Physics-Uspekhi 35, 212–230 (1992)
[72] Kim, J.P., Jeong, C.H., Kim, C.H.: Coupling characteristics of aperture-coupled vertically mounted strip transmission line. IEEE Trans., Microw. Theory Tech. 59, 561–567 (2011)
[73] Kim, C.S., Kim, Y.-T., Song, S.-H., et al.: A design of microstrip directional coupler for high directivity and tight coupling. In: Proc. 31st Eur. Microw. Conf., pp. 1–4 (2001)
[74] Malutin, N.D.: Multicoupled Strip Structures and Circuits on their Base. Tomsk State University (1990) (in Russian)
[75] Kuroki, F., Kimura, M., Yoneyma, T.: Analytical study on guided modes in vertical strip line embedded in NRD guide. El. Lett. 40(18), 1121–1122 (2004)
[76] Tokumitsu, T., Nishikawa, K., Kamogawa, K., et al.: Three-dimensional MMIC technology for multifunction integration and its possible application to masterslice MMIC. In: Proc. 1996 IEEE Microwave and Millimeter-Wave Monolithic Circ. Symp. Dig., pp. 85–88 (June 1996)
[77] Onodera, K., Hirano, M., Tokimutsu, M., et al.: Folded U-shaped microwave technology for ultra-compact three-dimensional MMIC's. IEEE Trans., Microw. Theory Tech. 44, 2347–2353 (1996)
[78] Toyoda, F., Nishikawa, K., Tokumitsu, T., et al.: Three-dimensional Masterslice MMIC on Si substrate. IEEE Trans., Microw. Theory Tech. 45, 2524–2530 (1997)

[79] Svačina, W.J.: A simple quasi-static determination of basic parameters of multilayer microstrip and coplanar waveguide. IEEE Microw. Guided Wave Lett. 2, 385–387 (1992)
[80] Ghione, G., Goano, M.: Revisiting the partial-capacitance approach to the analysis of coplanar transmission lines on multilayered substrates. IEEE Trans., Microw. Theory Tech. 51, 2007–2014 (2003)
[81] Iskander, M.F., Lind, T.S.: Electromagnetic coupling of coplanar waveguides and microstrip lines to highly lossy dielectric media. IEEE Trans., Microw. Theory Tech. 37, 1910–1917 (1989)
[82] Ponchak, G.E., Katehi, L.P.B.: Measured attenuation of coplanar waveguide on CMOS grade silicon substrates with a polyimide interface layer. El. Lett. 34(13), 1327–1329 (1998)
[83] Bouchriha, F., Grenier, K., Dubuc, D., et al.: Minimization of passive circuit losses realized on low resistivity silicon using micro-machining techniques and thick polymer layers. In: 2003 IEEE MTT-S Dig., pp. 959–962 (2003)
[84] Grenier, K., Lubecke, V., Bouchrihia, F., et al.: Polymers in RF and millimeter-wave applications. In: Proc. SPIE 1st Symp. Microtechnologies for New Millennium 2003, vol. 5116, pp. 502–513 (2003)
[85] Schoellhorn, C., Zhao, W., Morschbach, M., et al.: Attenuation mechanisms of aluminum millimeter-wave coplanar waveguides on silicon. IEEE Trans., El. Dev. 50, 740–746 (2003)
[86] Leung, L.L.W., Hon, W.-C., Chen, K.J.: Low-loss coplanar waveguides interconnects on low-resistivity silicon substrate. IEEE Trans., Comp. Pack., Techn. 27, 507–512 (2004)
[87] Ghione, G., Goano, M., Madonna, G., et al.: Microwave modeling and characterization of thick coplanar waveguides on oxide-coated lithium niobate substrates for electrooptical applications. IEEE Trans., Microw. Theory Tech. 47, 2287–2293 (1999)
[88] Gvozdev, V.I.: Use of the unbalanced slotted line in SHF microcircuits. Radioeng. Electron Physics (Radiotekhnika i Elektronika) 27(11), 42–47 (1981)
[89] Gvozdev, V.I., Nefedov, E.I.: Some possibilities of three-dimensional integrated UHF structures. Sov. Phys. Dokl. 27, 959–960 (1982)
[90] Gvozdev, V.I., Gulayev, Y.V., Nefedov, E.I.: Possible use of the principles of three-dimensional integrated microwave circuits in the design of ultra-fast digital computers. Sov. Phys. Dokl. 31, 760–761 (1986)
[91] Gvozdev, V.I., Nefedov, E.I.: Three-Dimensional Microwave Integrated Circuits. Nauka Publ., Moscow (1985) (in Russian)
[92] Gvozdev, V.I., Nefedov, E.I.: Volumetrical Microwave Integrated Circuits – Element Base of Analog and Digital Radioelectronics. Nauka Publ., Moscow (1987) (in Russian)
[93] Kouzaev, G.A.: Balanced slotted line. In: Gvozdev, V.I., Nefedov, E.I. (eds.) Microwave Three-Dimensional Integrated Circuits, pp. 45–50. Nauka Publ., Moscow (1985)
[94] Gvozdev, V.I., Kouzaev, G.A., et al.: Directional couplers based on transmission lines with corners. J. Commun. Techn. Electron (Radiotekhnika i Elektronika) 37, 37–40 (1992)
[95] Gvozdev, V.I., Kolosov, S.A., Kouzaev, G.A., et al.: Directional coupler. USSR Invention Certificate No 1786561 dated on, February 14 (1990)

[96] Kouzaev, G.A.: Electromagnetic model of differential substrate integrated waveguide. In: Proc. Eur. Computing Conf., Paris, France, April 27-29, pp. 282–284 (2011)
[97] Kouzaev, G.A., Deen, M.J., Nikolova, N.K., Rahal, A.: Cavity models of planar components grounded by via-holes and their experimental verification. IEEE Trans., Microwave Theory Tech. 54, 1033–1042 (2006)
[98] Williams, D.F., Janezic, M.D.: QuasiTEM model for coplanar waveguide on silicon. In: Proc. El. Performance of El. Packaging 1997, IEEE 6th Topical Meeting, October 27-29, pp. 225–228 (1997)
[99] Amaya, R.E., Li, M., Harrison, R.G., et al.: Coplanar waveguides in silicon with low attenuation and slow wave reduction. In: Proc. 37th Eur. Microw. Conf., pp. 508–511 (2007)
[100] Bedair, S.S., Wolff, I.: Fast, accurate and simple approximate analytic formulas for calculating the parameters of supported coplanar waveguides for (M)MIC's. IEEE Trans., Microw. Theory Tech. 40, 41–48 (1992)
[101] Seki, S., Hasegawa, H.: Cross-tie slow-wave coplanar waveguide on semi-insulating GaAs substrates. El. Lett. 17(25), 940–941 (1981)
[102] Wang, P., Kan, E.C.-C.: High-speed interconnects with underlayer orthogonal metal grids. IEEE Trans., Adv. Pack. 27, 497–507 (2004)
[103] Tiemeijer, L.F., Pijper, R.M.T., Havens, R.J., et al.: Low-loss patterned ground shield interconnect transmission lines in advanced IC processes. IEEE Trans., Microw. Theory Tech. 55, 561–570 (2007)
[104] Sayag, A., Ritter, D., Goren, D.: Compact modeling and comparative analysis of silicon-chip slow-wave transmission lines with slotted bottom metal ground planes. IEEE Trans., Microw. Theory Tech. 57, 840–847 (2009)
[105] Morton, M., Andrews, J., Lee, J., et al.: On the design and implementation of transmission lines in commercial SiGe HBT BiCMOS processes. In: Proc. 2004 Topical Meeting on Silicon Monolithic Integrated Circuits in RF Systems, pp. 53–56 (2004)
[106] Kaddour, D., Issa, H., Franc, A.-L., et al.: High-Q slow-wave coplanar transmission lines on 0.35 μm CMOS process. IEEE Microw. Wireless Comp. Lett. 19, 542–544 (2009)
[107] Levenets, V.V., Amaya, R.E., Tarr, N.G., et al.: Characterization of silver CPWs for applications in silicon MMICs. IEEE El. Device Lett. 26, 357–359 (2005)
[108] Tsuji, M., Shigesawa, H.: Packaging of printed-circuit lines: a dangerous cause for narrow pulse distortion. IEEE Trans., Microw. Theory Tech. 42, 1784–1790 (1994)
[109] Kim, S.-J., Yoon, H.-S., Lee, H.-Y.: Suppression of leakage resonance in coplanar MMIC packages using a Si sub-mount layer. IEEE Trans., Microw. Theory Tech. 48, 2664–2669 (2000)
[110] Horno, M., Mesa, F.L., Medina, F., et al.: Quasi-TEM analysis of multilayered, multiconductor coplanar structures with dielectric and magnetic anisotropy including substrate losses. IEEE Trans., Microw. Theory Tech. 38, 1059–1068 (1990)
[111] Mesa, F.L., Cano, G., Medina, F., et al.: On the quasi-TEM and full-wave approaches applied to coplanar microstrip on lossy dielectric layered media. IEEE Trans., Microw. Theory Tech. 40, 524–531 (1992)
[112] Collier, R.J.: Coupling between coplanar waveguides and substrate modes. In: Proc. 29th Eur. Microw. Conf., vol. 3, pp. 382–385 (1999)
[113] Yip, J.G.M., Collier, R.J., Jastrzebski, A.K., et al.: Substrate modes in double-layered coplanar waveguides. In: Proc. 31st Eur. Microw. Conf., pp. 1–4 (2001)

[114] Elgaid, K., Thayne, I.G., Whyte, G., et al.: Parasitic moding influences on coplanar waveguide passive components at G-band frequency. In: Proc. 36th Eur. Microw. Conf., pp. 486–488 (2006)
[115] Schnieder, F., Doerner, R., Heinrich, W.: High-impedance coplanar waveguides with low attenuation. IEEE Microw. Guided Wave Lett. 6, 117–119 (1996)
[116] Hofschen, S., Wolff, I.: Simulation of an elevated coplanar waveguide using 2-D FDTD. IEEE Microw. Guided Wave Lett. 6, 28–30 (1996)
[117] Hettak, K., Stubbs, M.G., Elgaid, K., et al.: Design and characterization of elevated CPW and thin film microstrip structures for millimeter-wave applications. In: Proc. 2005 EuMW Conf., vol. 2, p. 4 (2005)
[118] Agarwal, B., Schmitz, A.E., Brown, J.J., et al.: 112-GHz, 157-GHz, and 180-GHz InP HEMT traveling-wave amplifiers. IEEE Trans., Microw. Theory Tech. 46, 2553–2559 (1998)
[119] Yoon, S.-J., Jeong, S.-H., Yook, J.-G., et al.: A novel CPW structure for high-speed interconnects. In: 2001 IEEE MTT-S Microw. Symp. Dig., pp. 771–774 (2001)
[120] Arbabian, A., Niknejad, A.M.: A tapered cascaded multi-stage distributed amplifier with 370 GHz GBM in 90nm CMOS. In: 2008 RFIC Symp., pp. 57–60 (2008)
[121] Ponchak, G.E., Margomenos, A., Katehi, L.P.B.: Low-loss CPW on low-resistivity Si substrates with a micromachined interface layer for RFIC interconnects. IEEE Trans., Microw. Theory Tech. 49, 866–870 (2001)
[122] Sharma, R., Chakravarty, T., Bahattacharyya, A.B.: Analytical modeling of microstrip-like interconnections in presence of ground plane aperture. IET Microw. Antennas Propag. 3, 14–22 (2009)
[123] Ghione, G., Naldi, C.U.: Coplanar waveguides for MMIC applications: Effect of upper shielding, conductor backing, finite extent ground planes, and line-to-line coupling. IEEE Trans., Microw. Theory Tech. 35, 260–267 (1987)
[124] McGregor, F.A., Agharmoradi, F., Elgaid, K.: An approximate analytical model for the quasi-static parameters of elevated CPW lines. IEEE Trans., Microw. Theory Tech. 58, 3809–3814 (2010)
[125] Kneppo, A., Gotzman, J.: Basic parameters of nonsymmetrical coplanar line. IEEE Trans., Microw. Theory Tech. 25, 718 (1977)
[126] Ghione, G., Naldi, C.: Analytical formulas for coplanar lines in hybrid and monolithic MICs. El. Lett. 20(4), 179–181 (1984)
[127] Frankel, M.Y., Voelker, R.H., Hilfiker, J.N.: Coplanar transmission lines on thin substrates for high-speed low-loss propagation. IEEE Trans., Microw. Theory Tech. 42, 396–402 (1994)
[128] Frankel, M.Y., Gupta, S., Valdmanis, J., et al.: Terahertz attenuation and dispersion characteristics of coplanar lines. IEEE Trans., Microw. Theory Tech. 39, 910–916 (1991)
[129] Helliger, N.M., Pfeifer, T., Vosseburger, V., et al.: Influence of insulation layers on the high-frequency properties of coplanar waveguides on Si. In: Proc. CLEO 1996, p. 452 (1996)
[130] Arif, M.S., Peroulis, D.: Loss optimization of coplanar strips for CMOS RFICs. In: Proc. Microw. Conf. APMC 2009, pp. 2144–2147 (2009)
[131] Deng, T.Q., Leong, M.S., Kooi, P.S., et al.: Synthesis formulas for coplanar lines in hybrid and monolithic ICs. El. Lett. 32(13), 2253–2254 (1996)
[132] Kim, S., Leong, S., Lee, Y.T., et al.: Ultra-wideband (from DC 110 GHz) CPW to CPS transition. El. Lett. 38(13), 622–623 (2002)

[133] Anagnostou, D.E., Morton, M., Papapolymerou, J., et al.: A 0-55-GHz coplanar waveguide to coplanar strip transition. IEEE Trans., Microw. Theory Tech. 56, 1–6 (1999)

[134] Prieto, D., Cayrout, J.C., Cazaux, J.L., et al.: CPS/CPW structure potentialities for MMICs: a CPS/CPW transition and a bias network. In: 1998 IEEE MTT-S Symp. Dig., pp. 111–114 (1998)

[135] Kim, H.-T., Lee, S., Kim, S., et al.: Millimetre-wave CPS distributed analogue MMIC phase shifter. El. Lett. 39(23), 1660–1661 (2003)

[136] Nordquist, C.D., Muyhondt, A., Pack, M.V., et al.: An X-band to Ku-band RF MEMS switched coplanar strip filter. IEEE Microw. Wireless Comp. Lett. 14, 425–427 (2004)

[137] Fan, L., Chang, K.: Uniplanar power dividers using coupled CPW and asymmetrical CPS for MIC's and MMIC's. IEEE Trans., Microw. Theory Tech. 44, 2411–2420 (1996)

[138] Kwon, Y., Kim, H.T., Park, J.-H., et al.: Low-loss micromachined inverted overlay CPW lines with wide impedance ranges and inherent air bridge connection capability. IEEE Wireless Comp. Lett. 11, 59–61 (2001)

[139] Kolosov, S.A., Kouzaev, G.A., Skulakov, P.I., et al.: Slot transmission line. USSR Invention Certificate No 1683100 dated on, May 17 (1989)

[140] Gvozdev, V.I., Kouzaev, G.A., Tikhonov, A.N.: New transmission lines and electrodynamical models for three-dimensional microwave integrated circuits. Sov. Physics-Doklady 35, 675–677 (1990)

[141] Gillick, M., Robertson, I.D.: Ultra low impedance CPW transmission lines for multilayer NNIC's. In: 1993 IEEE MTT-S Microw. Symp. Dig., pp. 145–148 (1993)

[142] Vo, V.T., Krishnamurthy, L., Sun, Q., et al.: 3-D low-loss coplanar waveguide transmission lines in multilayer MMICs. IEEE Trans., Microw. Theory Tech. 54, 2864–2871 (2006)

[143] Dib, N.I., Katehi, L.P.B.: Impedance calculation for the microshield line. IEEE Microw. Guided Wave Lett. 2, 406–408 (1992)

[144] Kiang, J.-F.: Characteristic impedance of microshield lines with arbitrary shield cross section. IEEE Trans., Microw. Theory Tech. 46, 1328–1331 (1998)

[145] Weller, T.M., Katehi, L.P.B., Rebeiz, G.M.: High performance microshield line components. IEEE Trans., Microw. Theory Tech. 43, 534–543 (1995)

[146] Herrick, K.J., Yook, J.G., Katehi, L.P.B.: Microtechnology in the development of three-dimensional circuits. IEEE Trans., Microw. Theory Tech. 46, 1832–1844 (1998)

[147] Duwe, K., Hirch, S., Judaschke, R., et al.: Micromachined coplanar waveguides on thin HMDSN-membranes. In: Int. Conf. Infrared and Millimeterwaves 2000, Conf. Dig., pp. 299–300 (2000)

[148] Duwe, K., Hirch, S., Mueller, J.: Micromachined low pass filters and coplanar waveguides for D-band frequencies based on HMDSN-membranes. In: Proc. of MSMW 2001 Symp., Kharkov, Ukraine, June 4-9, pp. 675–677 (2001)

[149] Hirsh, S., Chen, Q., Duwe, K., et al.: Design and characterization of coplanar waveguides and filters on thin dielectric membranes at D-band frequencies. In: Proc. of MSMW 2001 Symp., Kharkov, Ukraine, June 4-9, pp. 678–680 (2001)

[150] Margomennos, A., Herrick, K.J., Herman, M.I., et al.: Isolation in three-dimensional integrated circuits. IEEE Trans., Microw. Theory Tech. 51, 25–32 (2003)

[151] Duwe, K., Mueller, J.: Realization of a 150 GHz to 450 GHz tripler circuit based on a thin dielectric HMDS-N-membrane. In: 2003 MTT-S Microw. Symp. Dig., pp. 755–757 (2003)
[152] Margomenos, A., Lee, Y., Katehi, L.P.B.: Wideband Si micromachined transitions for RF wafer-scale packages. In: Proc. Conf. Silicon Monolithic Integrated Circuits in RF Systems, pp. 183–186 (2007)
[153] Yang, S., Hu, Z., Buchanan, N.B., et al.: Characteristics of trenched coplanar waveguide for high-resistivity Si MMIC applications. IEEE Trans., Microw. Theory Tech. 46, 623–631 (1998)
[154] Sahri, N., Nagatsuma, T., Mashida, K., et al.: Characterization of micromachined coplanar waveguides on silicon up to 300 GHz. In: Proc. 29th Eur. Microw. Conf., pp. 254–257 (1999)
[155] Leung, L.L.W., Chen, K.J.: CAD equivalent-circuit modeling of attenuation and cross-coupling for edge-suspended coplanar waveguides on lossy silicon substrate. IEEE Trans., Microw. Theory Tech. 54, 2249–2255 (2006)
[156] Lin, C.-P., Jou, C.F.: New CMOS-compatible micromachined embedded coplanar waveguide. IEEE Trans., Microw. Theory Tech. 58, 2511–2516 (2010)
[157] Bouchriha, F., Grenier, K., Dubuc, D., et al.: Miniaturization of passive circuits losses realized on low resistivity silicon using micro-machining techniques and thick polymer layers. In: 2003 IEEE MTT-S Microw. Symp. Dig., pp. 959–962 (2003)
[158] Wang, G., Bacon, A., Abdolvand, R., et al.: Finite ground coplanar lines on CMOS grade silicon with a thick embedded silicon oxide layer using micromachining techniques. In: Proc. 33rd Eur. Microw. Conf., pp. 25–27 (2003)
[159] Reid, J.R., Marsh, E.D., Webster, R.T.: Micromachined rectangular-coaxial transmission lines. IEEE Trans., Microw. Theory Tech. 54, 3433–3442 (2006)
[160] Bishop, J.A., Hashemi, M.M., Kiziloglu, K., et al.: Monolithic coaxial transmission lines for mm-wave ICs. In: Proc. IEEE/Cornell Conference on Advanced Concepts in High Speed Semiconductor Devices and Circuits, pp. 252–260 (1991)
[161] Chen, R.T., Brown, E.R., Singh, R.S.: A compact 30 GHz low loss balanced hybrid coupler fabricated using micromachined integrated coax. In: Proc. RAWCON 2004, pp. 227–230 (2004)
[162] Reid, J.R., Marsh, E.D., Webster, R.T.: Micromachined rectangular-coaxial transmission lines. IEEE Trans., Microw. Theory Tech. 54, 3433–3442 (2006)
[163] Yoon, J.-B., Kim, B.-I., Choi, Y.-S., et al.: 3-D construction of monolithic passive components for RF and microwave ICs using thick-metal surface micromachining technology. IEEE Trans., Microw. Theory Tech. 51, 279–288 (2003)
[164] Filipovic, D.S., Lukic, M.V., Lee, Y., et al.: Monolithic rectangular coaxial lines and resonators with embedded dielectric support. IEEE Microw. Wireless Comp. Lett. 18, 740–742 (2008)
[165] Popovic, Z., Vanhille, K., Ehsan, N., et al.: Microfabricated micro-coaxial millimeter-wave components. In: Proc. 33rd Int. Conf. Infrared, Millimeter and Terahertz Waves, IRMMW-THz 2008, pp. 1–3 (2008)
[166] Ehsan, N., Cullens, E., Vanhille, K., et al.: Micro-coaxial lines for active hybrid-monolithic circuits. In: Proc. IMS 2009, pp. 465–468 (2009)
[167] Ehsan, N., Vanhille, K.J., Rondineau, S., et al.: Micro-coaxial impedance transformers. IEEE Trans., Microw. Theory Tech. 58, 2908–2914 (2010)
[168] Saito, Y., Filipovic, D.S.: Analysis and design of monolithic rectangular coaxial lines for minimum coupling. IEEE Trans., Microw. Theory Tech. 55, 2521–2530 (2007)

[169] Wang, Y., Ke, M., Lancaster, M.J., et al.: Micromachined millimeter-wave rectangular-coaxial branch-line coupler with enhanced bandwidth. IEEE Trans., Microw. Theory Tech. 57, 1655–1660 (2009)
[170] Murad, N.A., Lancaster, M.J., Wang, Y., et al.: Micromachined rectangular coaxial line to ridge waveguide transition. In: Proc. IEEE 10th Annual Wireless Microw. Techn. Conf., WAMICON 2009, pp. 1–5 (2009)
[171] Natarajan, S.P., Hoff, A.M., Weller, T.M.: Polyimide core 3D rectangular micro coaxial transmission lines. Microw. Opt. Techn. Lett., 1291–1253 (June 2010)
[172] Gunston, M.A.R.: Microwave Transmission Line Impedance Data. Van Nostrand Reinhold Comp. LTD. (1972)
[173] Lukic, M., Rondineau, S., Popovic, S., et al.: Modeling of realistic rectangular μ-coaxial lines. IEEE Trans., Microw. Theory Tech. 54, 2068–2076 (2006)
[174] Becker, J.P., East, J.R., Katehi, L.P.B.: Performance of silicon micromachined waveguide at W-band. El. Lett. 38(13), 638–639 (2002)
[175] Matvejev, V., De Tandt, C., Ranson, W., et al.: Wet silicon bulk micromachined THz waveguides for low-loss integrated sensor applications. In: Proc. 35th Int. Conf. Infrared Millimeter and Terahertz Waves, IRMMW-THz, pp. 1–2 (2010)
[176] Komarov, V.V.: Eigenmodes of regular polygonal waveguides. J. Infrared Milli. Terahz Waves 32, 40–46 (2011)
[177] Kirby, P.L., Pukala, D., Manohara, H., Mehdi, I., et al.: A micromachined 400 GHz rectangular waveguide and 3-pole bandpass filter on silicon substrate. In: 2004 MTT-S Microw. Symp. Dig., pp. 1185–1188 (2004)
[178] Li, Y., Kirby, P.L., Papapolymerou, J.: Silicon micromachined W-band folded and straight waveguides using DRIE technique. In: IEEE Int. Microw. Symp. Dig., pp. 1915–1918 (2006)
[179] Margomenos, A., Lee, Y., Kuo, A., et al.: K and Ka-band silicon micromachined evanescent mode resonators. In: Proc. 37th Eur. Microw. Conf., pp. 446–449 (2007)
[180] Li, Y., Kirby, P.L., Offranc, O., et al.: Silicon micromachined W-band hybrid coupler and power divider using DRIE technique. IEEE Microw. Wireless Comp. Lett. 18, 22–24 (2008)
[181] Digby, J.W., McIntosh, C.E., Parhurst, G.M., et al.: Fabrication and characterization of micromachined rectangular waveguide components for use at millimeter-wave and terahertz frequencies. IEEE Trans., Microw. Theory Tech. 48, 1293–1302 (2000)
[182] Gentile, G., Dekker, R., De Graaf, P., et al.: Silicon filled integrated waveguides. IEEE Microw. Wireless Comp. Lett. 20, 536–538 (2010)
[183] Smith, C.H., Skavonous, A., Barker, N.S.: SU-8 micromachining of millimeter and submillimeter waveguide circuits. In: IEEE MTT-S Int. Microw. Symp., vol. 3, pp. 961–964 (2009)
[184] Skaik, T., Wang, Y., Ke, M., et al.: A micromachined WR-3 waveguide with embedded bends for direct flange connections. In: Proc. 40th Eur. Microw. Conf., pp. 1225–1228 (2001)
[185] Shang, X., Ke, M.L., Wang, Y., et al.: Micromachined WR-3 waveguide filter with embedded bends. El. Lett. 47(9), 545–547 (2011)
[186] Nordquist, C.D., Wanke, M.C., Rowen, A.M., et al.: Properties of surface metal micromachined rectangular waveguide operating near 3 THz. IEEE J. Selected Topics in Quant. Electron. 17, 130–137 (2011)

[187] Hirokawa, J., Ando, M.: Single-layer feed waveguide consisting of posts for plane TEM wave excitation in parallel plates. IEEE Trans., Antennas Propag. 46, 625–630 (1998)
[188] Uchimura, H., Takenoshita, T., Fujii, M.: Development of a "laminated waveguide. IEEE Trans., Microw. Theory Tech. 46, 2438–2443 (1998)
[189] Wu, X.H., Kishk, A.A., Balanis, K. (ed.): Analysis and Design of Substrate Integrated Waveguide Using Efficient 2D Hybrid Method. Synthesis Lectures on Computational Electromagnetics, Series. Morgan and Claypool Publ. (2010)
[190] Tischer, F.J.: Fence guide for millimeter waves. Proc. IEEE 59, 1112–1113 (1971)
[191] Agarwal, K.K., Tischer, F.J.: Components using fence guide for millimeter-wave applications. El. Lett. 22(6), 330–331 (1986)
[192] Deslandes, D., Wu, K.: Accurate modeling, wave mechanisms and design considerations of a substrate integrated waveguide. IEEE Trans., Microw. Theory Tech. 54, 2516–2526 (2006)
[193] Bozzi, M., Perregrini, L., Wu, K.: Modeling of radiation, conductor, and dielectric losses in SIW components by the BI-RME method. In: Proc. 3rd Eur. Microw. Integrated Circ. Conf., pp. 230–233 (2008)
[194] Simpson, J.J., Taflove, A., Mix, J.A., et al.: Substrate integrated waveguides optimized for ultrahigh-speed digital interconnects. IEEE Trans., Microw. Theory Tech. 54, 1983–1989 (2006)
[195] Bozzy, M., Peregrini, L., Wu, K., et al.: Current and future research trends in substrate integrated circuit waveguide technology. Radioengineering 18, 201–209 (2009)
[196] Wu, K.: Towards the development of terahertz substrate integrated circuit technology. In: Proc. SiRF 2010, pp. 116–119 (2010)
[197] Cassivi, Y., Wu, K.: Substrate integrated nonradiative dielectric waveguide. IEEE Microw. Wireless Comp. Lett. 14, 89–91 (2004)
[198] Grioropoulos, N., Young, P.R.: Compact folded waveguides. In: Proc. 34th Eur. Microw. Conf., pp. 973–976 (2004)
[199] Pan, B., Li, Y., Tentzeris, M.M., et al.: Surface micromachining polymer-core-conductor approach for high-performance millimeter-wave air-cavity filters integration. IEEE Trans., Microw. Theory Tech. 56, 959–970 (2008)
[200] Hyeon, I.-J., Park, W.-Y., Lim, S., et al.: Fully micromachined, silicon-compatible substrate integrated waveguide for millimetre-wave applications. El. Lett. 47(3), 328–330 (2011)
[201] Cassivi, Y., Peregrini, L., Arcioni, P., et al.: Dispersion characteristics of substrate integrated rectangular waveguide. IEEE Microw. Wireless Comp. Lett. 12, 333–335 (2002)
[202] Bozzy, W., Pasian, M., Perregrini, L., et al.: On the losses in substrate integrated waveguides. In: Proc. 37th Eur. Microw., Conf., pp. 384–387 (2007)
[203] Che, W., Xu, L., Wang, D., et al.: Short-circuit equivalency between rectangular waveguides of regular sidewalls (rectangular waveguide) and sidewalls of cylinders (substrate-integrated rectangular waveguides), plus its extension to cavity. IET Microw. Antennas Propag. 1, 639–644 (2007)
[204] Salehi, M., Mehrshahi, E.: A closed-form formula for dispersion characteristics of fundamental SIW mode. IEEE Microw. Wireless Comp. Lett. 21, 4–6 (2011)
[205] Shestopalov, V.P., Kirilenko, A.A., Masalov, S.A., et al.: Resonant Scattering of Waves. In: Diffraction Grids, vol. 1. Naukova Dumka (1986) (in Russian)

[206] Kontorovitch, M.I., Astrakhan, M.I., Akimov, V.P., et al.: Electrodynamics of Grid Structures. Radio i Svayz, Moscow (1987) (in Russian)
[207] MacFarlane, G.G.: Surface impedance of an infinite wire grid, at oblique angles of incidence (parallel polarization). J. IEE 93(IIIA), 1523–1527 (1946)
[208] Wait, J.R.: The impedance of a wire grid parallel to a dielectric interface. IRE Trans., Microw. Theory Tech. 5, 99–102 (1957)
[209] Bray, J.R., Roy, L.: Resonant frequencies of post-wall waveguide cavities. IEE Proc. Microw. Antennas Propag. 150, 365–368 (2003)
[210] Che, W., Geng, L., Deng, K., et al.: Analysis and experiments of compact folded substrate-integrated waveguide. IEEE Trans., Microw. Theory Tech. 56, 88–93 (2008)
[211] Chiu, L.: Oversized microstrip line as differential guide-wave structure. El. Lett. 46(2), 144–145 (2010)
[212] Liu, B., Hong, W., Wang, Y.-Q., et al.: Half mode substrate integrated waveguide (HMSIW) 3-dB coupler. IEEE Microw. Wireless Comp. Lett. 17, 22–24 (2007)
[213] Liu, J., Long, Y.: Formulas for complex propagation constant of first higher mode of microstrip line. El. Lett. 44(4), 261–262 (2008)
[214] Deen, M.J., Basu, P.K.: Fundamentals and Devices. John Wiley and Sons (2012)
[215] Silicon Photonics. In: Tsubebeskov, L., Lockwood, D.J., Ichikawa, M. (eds.) Proc. IEEE, Special Issue, vol. 97, pp. 1159–1360 (2009)
[216] Noginov, M.A., Zhu, G., Belgrave, A.M., et al.: Demonstration of a spacer-based nanolaser. Nature, Online Publication (August 17, 2009), doi: 10.1038/nature08318
[217] Zayats, A.V., Smolayninov, I.I., Maradudin, A.A.: Nano-optics of surface plasmon polaritons. Physics Reports 408, 131–314 (2005)
[218] Barnes, W.L., Dereux, A., Ebbesen, T.W.: Surface plasmon subwavelength optics. Nature 424, 824–830 (2003)
[219] Furtak, T.E., Durfee, C.G., Sabbah, A.J., et al.: Toward silicon-compatible modulation of plasmonic waveguides. In: Proc. OSA/CLEO/QUELS 2008 Conf., Paper No. JThA131 (2008)
[220] Chen, J., Smolaykov, G.A., Bruek, S.R.J., et al.: Surface plasmon modes of finite, planar, metal-insulator-metal plasmonic waveguides. Optics Express 16(19), 14902–14909 (2008)
[221] Dionne, J.A., Lezec, H.J., Atwater, H.A.: Highly confined photon transport in subwavelength metallic slot waveguides. Nanoletters 6(9), 1928–1932 (2006)
[222] Shahvarpour, A., Gupta, S., Caloz, C.: Schroedinger solitons in left-handed SiO_2-Ag-SiO_2 and Ag-SiO_2-Ag plasmonic waveguides calculated with a nonlinear transmission line approach. J. Appl. Phys. 104(1-5), 124510
[223] Hosseini, A., Nieuwoudt, A., Massoud, Y.: Optimizing dielectric strip plasmonic waveguides for subwavelength on-chip optical communication. IEEE Trans., Nanotechn. 7, 189–196 (2008)
[224] de Waele, R., Burgos, S.P., Polman, A., et al.: Plasmon dispersion in coaxial waveguides from single-cavity optical transmission measurements. Nanoletters 209, 2832–2837 (2009)
[225] Kozina, O., Nefedov, I., Melnikov, L., et al.: Plasmonic coaxial waveguides with complex shapes of cross-sections. Materials 4, 104–116 (2011)
[226] Kozina, O.N., Mel'nikov, L.A., Nefedov, I.S.: Strong field localization in subwavelength metal-dielectric optical waveguides. Optics and Spectroscopy 111, 241–247 (2011)

[227] Park, S., Kim, M.-S., Kim, J.T., et al.: Long range surface plasmon polariton waveguides at 1.31 and 1.55 wavelength. Opt. Commun. 281, 2057–2061 (2008)
[228] Boltasseva, A., Nikolajesen, T., Leosson, K., et al.: Integrated optical components utilizing long-range surface plasmon polaritons. J. Ligthwave Techn. 23, 413–422 (2005)
[229] Ju, J.J., Kim, M.-S., Park, S., et al.: 10 Gbps optical signal transmission via long-range surface plasmon polariton waveguide. ETRI J. 29, 808–810 (2007)
[230] Ju, J.J., Park, S., Kim, M.-S., et al.: 40 Gbit/s light signal transmission in long-range surface plasmon waveguides. Appl. Phys. Lett. 91(1-3), 171117 (2007)
[231] Bozhevolnyi, S.I., Volkov, V.S., Devaux, E., et al.: Channel plasmon-polariton guiding by subwavelength metal grooves. Phys. Rev. Lett. 95(1-4), 046802 (2005)
[232] Moreno, E., Rodrigo, S.G., Bozhevolnyi, S.I., et al.: Guiding and focusing of electromagnetic fields with wedge plasmon polaritons. Phys. Rev. Lett. 100(1-4), 023901 (2008)
[233] Veronis, G., Fan, S.: Modes of subwavelength plasmonic slot waveguides. J. Lightwave Techn. 25, 25 (2007)
[234] Dionne, J.A., Sweatlock, L.A., Atwater, H.A.: Plasmon slot waveguides: Towards chip-scale propagation with subwavelength-scale localization. Phys. Rev. B 73(1-9), 035407 (2006)
[235] Wuenschell, J., Kim, H.K.: Excitation and propagation of surface plasmons in a metallic nanoslit structure. IEEE Nanotechn. 7, 229–236 (2008)
[236] Shestopalov, Y., Shestopalov, V.: Spectral Theory and Excitation of Open Structures. Peter Peregrinus Ltd. (1996)
[237] Kocabas, S.E., Veronis, G., Miller, D.A.B., et al.: Transmission line and equivalent circuit models for plasmonic waveguide components. IEEE J. Selected Topics Quant. El 14, 1462–1472 (2008)
[238] Boltasseva, A., Bozhevolni, S.I.: Directional couplers using long-range surface plasmon polariton waveguides. IEEE J. Selected Topics Quant. Electron 12(6), pt. 1, 1233–1241 (2005)
[239] Pakizeh, T., Käll, M.: Unidirectional ultracompact optical nanoantennas. Nanoletters 9, 2343–2349 (2009)
[240] Bozhevolnyi, S.I., Volkov, V.S., Devaux, E., et al.: Channel plasmon subwavelength waveguide components including interferometers and ring resonators. Nature 440, 508–511 (2006)
[241] Ly-Gagnon, D.-S., Kosabas, S.E., Miller, D.A.B.: Characteristic impedance model for plasmonic metal slot waveguides. IEEE J. Selected Topics Quant. Electron 14(6), 1473–1478 (2008)
[242] Feigenbaum, E., Orenstein, M.: Perfect 4-way plasmon splitting in cross gap waveguides intersection. IEEE Laser & Electro Optics Soc., 651–652 (October 2006)
[243] Ginzburg, P., Arbel, D., Orenstein, M.: Efficient coupling of nano-plasmonics to micro-photonic circuitry. In: Proc. 2005 Conf. Lasers & Electro-Optics, CLEO, pp. 1482–1484 (2005)
[244] Min, A., Ostby, E., Sorger, V., et al.: High-Q surface-plasmon-polariton whispering-gallery microcavity. Nature 457, 455–459 (2009)
[245] Noginov, M.A., Zhu, G., Belgrave, A.M., et al.: Demonstration of a spacer-based nanolaser. Nature Online (August 16, 2009)
[246] Akimov, A.V., Mukherjee, A., Yu, C.L., et al.: Generation of single optical plasmons in metallic nanowires coupled to quantum dots. Nature 450, 402–406 (2007)

[247] Johnson, P.B., Christy, R.W.: Optical constants for the noble metals. Phys. Rev. B 6, 4370–4379 (1972)

[248] Berglind, E., Thylen, L., Liu, L.: Microwave engineering approach to metallic based photonic waveguides and waveguide components. In: Proc. Int. Symp. Biophotonics, Nanophotonics and Metamaterials (2006)

[249] Nunes, F.D., Weiner, J.: Equivalent circuits and nanoplasmonics. IEEE Trans., Nanotechn. 8, 298–302 (2009)

[250] Kern, A.M., Martin, O.J.F.: Surface integral formulation for 3D simulations of plasmonic and high permittivity nanostructures. J. Opt. Soc. Am. A 26, 732–740 (2009)

[251] Bozhevolnyi, S.I. (ed.): Plasmonic Nanoguides and Circuits. Pan Stanford Publishing (2009)

[252] Rakheja, S., Naeemi, A.: Interconnects for novel state variables: Performance modeling and device and circuit implications. IEEE Trans., Electron. Dev. 57, 2711–2718 (2010)

7 Inter-component Transitions for Ultra-bandwidth Integrations

Abstract. In this Chapter, the planar and multi-level transitions between different transmission lines are considered. Among them are the ones realized using the via-holes, the EM coupling or combinations of both means. The most attention is paid to the millimeter-wave or ultra-wide bandwidth solutions for silicon technology. An EM theory of grounding via-holes for millimeter-wave packages is considered. Contemporary state of modeling of multilayer motherboards shorted by multiple through-via-holes is reviewed as well. References -104. Figures -16. Pages -34.

Contemporary technology, which allows for the planar and 3-D components embedded into a multilayered environment, is friendly to realize the principle of optimality of electronic system components [1]-[6]. It means that the transmission lines, used in these circuits, are adapted to each component to achieve its best characteristics, and the reflection-less transitions from one circuit to another are required. Additionally, these transitions can filter the signals, connect different signal layers in a 3-D module, cancel or provide the DC coupling, etc. The transitions are used for narrow- and ultra-wide bandwidth intra- and inter-chip communications, and they are the ones of the most important components of contemporary micro-, millimeter-wave, and high-speed electronics [7]-[9].

A large amount of work on the design of transitions was performed during the developments of the microwave hybrid 3-D integrated circuits, and these transitions can be re-scaled for millimeter-wave monolithic applications.

As an example, some of these transitions are shown in Fig. 7.1 [1],[4]-[6]. The elements of them are placed on different layers, and additionally to the matching the lines' characteristic impedances and fields, these transitions are for the EM connections of different layers and for the power distribution between the circuit's branches. To match the transition's lines, the open or short-end stubs are used. More information on the designs and their characteristics is from the above-cited publications.

The transitions are divided into two large classes. The first of them provide the DC and microwave connections, and they can be used in microwave and high-speed digital electronics. Others realize only the EM connections.

G.A. Kouzaev: Applications of Advanced Electromagnetics, LNEE 169, pp. 307–340.
springerlink.com

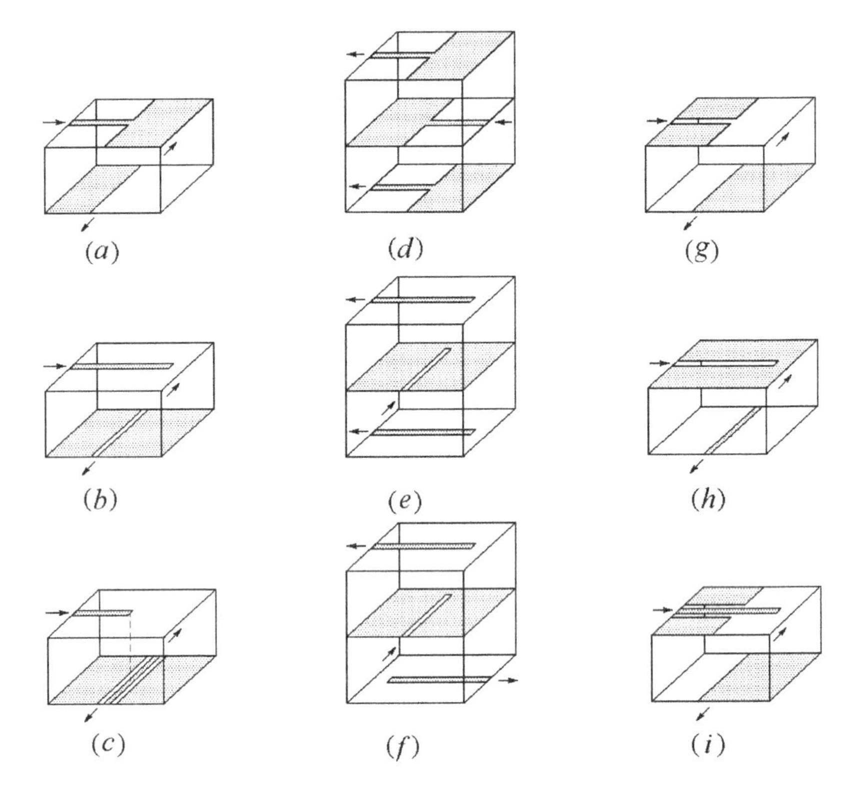

Fig. 7.1 Three-dimensional T-joints for inter-layer EM transitions. (a)- Microstrip-antipodal slot transmission line joint; (b)- Microstrip-slot line joint; (c)- Microstrip-CPW via-hole joint; (d)- Microstrip-microstrip power distribution tree; (e)- Slot-microstrip tree; (f)- Slot-microstrip tree; (g)- Slot-antipodal slot joint; (h)- Slot-microstrip joint; (i)- CPW-antipodal slot joint

The mentioned components can be of the planar, quasi-planar, or 3-D design depending on the needs. The most developed theory is known for the transitions of transmission lines of the same geometry but of different characteristic impedances. Their design solutions are close to the filters, and many matching circuits are known with the formulas for their synthesis. The most increased bandwidth solutions are provided by continuous transitions and circuits based on resistive networks. Some examples of these transitions are shown in Fig. 7.2.

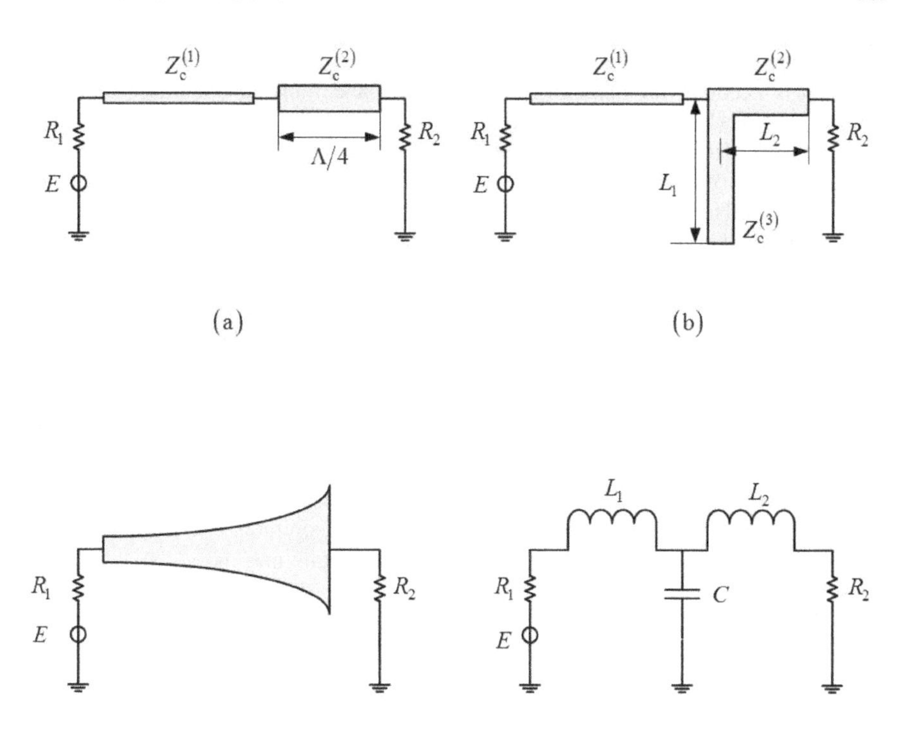

Fig. 7.2 Matching circuits. (a)- Quarter wavelength transformer; (b)- Open-stub transformer; (c)- Transformer on tapered microstrip line; (d)- LC transformer

7.1 Planar Line Transitions

Consider, initially, the planar lines as microstrips, coplanar waveguides, and coplanar strips placed on the same level of integration and connected to each other without using via-holes or multilayered components. These transitions are rather simple, and they can be attractive due to their low-cost. Connections of the same lines of different characteristic impedances have been already mentioned, and they are easy to be calculated. The most difficult problem is to design the reflectionless transitions for different types of lines, and no any well-developed theory is known, according to the best knowledge of the Author. In these transitions, the matching of fields and impedances should be provided, and it is still a problem, which is difficult to be formalized analytically.

Some transitions between the CPW and microstrip line having planar design are shown in (Fig. 7.3). The transitions between a conductor-backed CPW (GCPW) and a microstrip line are important for millimeter-wave monolithic integration because the probes used for the feed and measurements of monolithic circuits. The simplest design consists of CPW-like pads and a CPW-microstrip step-wise transition (Fig.7.3a). Taking into account the common ground for CPW and microstrip, the design can work starting at zero frequency, but the spatial orthogonality of the modal fields of the connected lines limits the bandwidth of this transition, and careful optimization is needed. For instance, an ultra-wide band transition of this type is described in [10] where it is realized on a grounded BCB substrate placed over silicon. By simulations, it was found possibility to extend the bandwidth of this GCPW-MS transition up to several tens of GHz, and the measurements verified a component for 200 MHz-40 GHz bandwidth were published. According to the authors of the cited paper, it is reached using a thin (10 μm) BCB layer, wide GCPW grounded strips, etc. Additionally, theory shows on possible increase of the upper frequency of these monolithically manufactured transitions up to 84 GHz.

The frequency band can be extended using the open-end stubs (Fig. 7.2b,c) [11],[12], and the bandwidth about 90% is found in [11] at the 20-GHz center frequency. A tapered GCPW-microstrip transition (Fig. 7.3d) realized over a high-resistivity $(\rho > 5000\ \Omega\text{-cm})$ silicon substrate is simulated and optimized by a full-wave software tool in [13] for 2-110 GHz. The transition loss, which is less than 0.4 dB, is registered for 40-100 GHz, and it was the widest via-less transition at the moment of the paper [13] publication according to authors' opinion. It was found a good correspondence between the Method-of-Moments' results and measurements in the considered frequency band. A theory of CPW-microstrip transitions based on the equivalent circuits and validated for frequencies up to 30 GHz is published in [14].

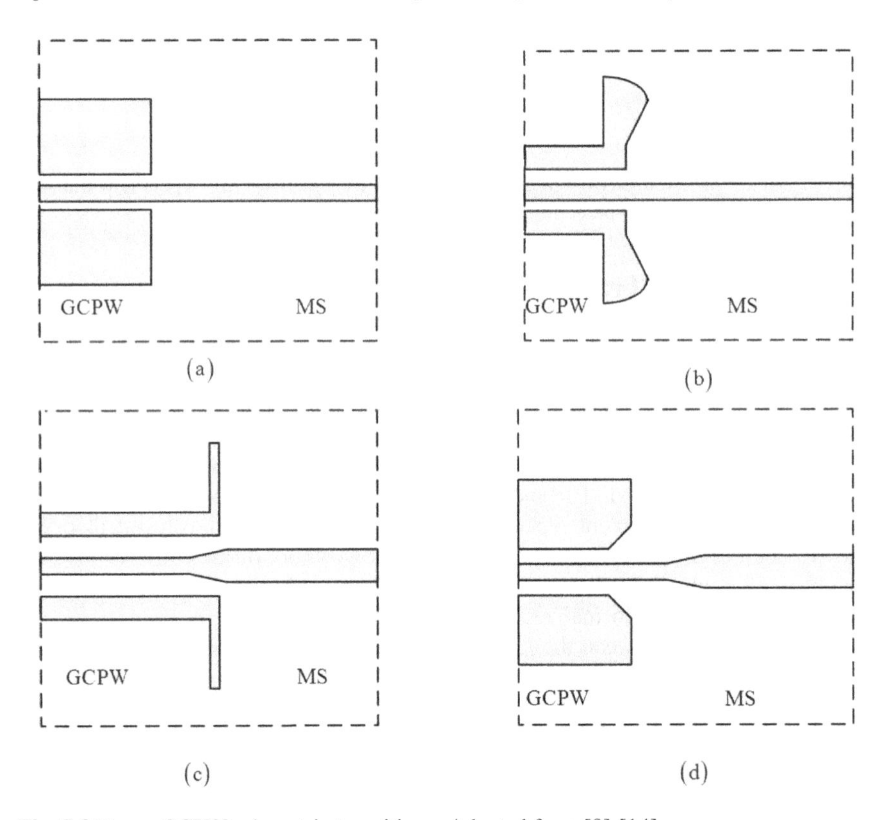

Fig. 7.3 Planar GCPW-microstrip transitions. Adapted from [9]-[14]

7.2 Quasi-planar Transitions

The quasi-planar transitions are the ones which connect the signal traces placed on one level of integration but employing the matching elements placed on both sides of a substrate [6],[4]. They use the via-holes, bridges, etched ground planes, and the EM couplings to provide the needed characteristics, and, as a rule, they have better parameters towards the completely planar analogs. Some examples of these transitions are studied in [15], for instance, where several CPW-slot line transitions are introduced in 1-40-GHz frequency band. Mostly, they use the air-bridges to equalize the potentials of conductors. Authors show that the quasi-planar transitions can provide more degree of freedom, less radiation loss, wide-band performance (up to 160%), compactness, etc. compared to the planar analogs.

7.2.1 CPW-microstrip Transitions

Several of these circuits are shown in Fig. 7.4 [16]-[23]. The first of them (Fig. 7.4a,b) are realized on a 100-μm thick high-resistive silicon substrate, and they connect a 50-Ω CPW and a microstrip line. In spite the equalizing of their characteristic impedances, the lines need additional circuit components, which are the quarter wavelength transformers, to eliminate the reflection due to difference of the modal fields of these two lines. The central frequency is chosen 94 GHz, and the bandwidth is close to 20% for 0.3-dB insertion loss.

Additionally to these CPW-MS transitions, several GCPW-MS ones are studied, and it is found that the etching of the silicon below GCPW conductors allows improving characteristics of these components. The measurements results are available in frequencies from 75 to 110 GHz. For instance, the CPW-microstrip transition from Fig. 7.4c uses a section of GCPW to transform the fundamental CPW mode into the microstrip one. In this length, in fact, two propagating GCPW modes are excited, and one of them has similar field structure with the quasi-TEM mode of the microstrip. Additionally, a step-wise tapering is used to provide the matching in wide frequency band. The transitions of this type are simulated and measured for Alumina and InP substrates in frequencies up to 25 GHz [17], and they show good performance with the bandwidth up to 9-15 GHz. An equivalent circuit model of this transition is published in [18].

Several designs are known which use the via-holes connecting ground layers of CPW and microstrips (Fig. 7.4d-f) [19]-[21]. The first figure of this set is for adaptation of coplanar probes, and two pads there are grounded to the common ground layer by via-holes. A semi-empirical equivalent circuit model of it is published in [19], and its accuracy is good enough at frequencies up to 30 GHz.

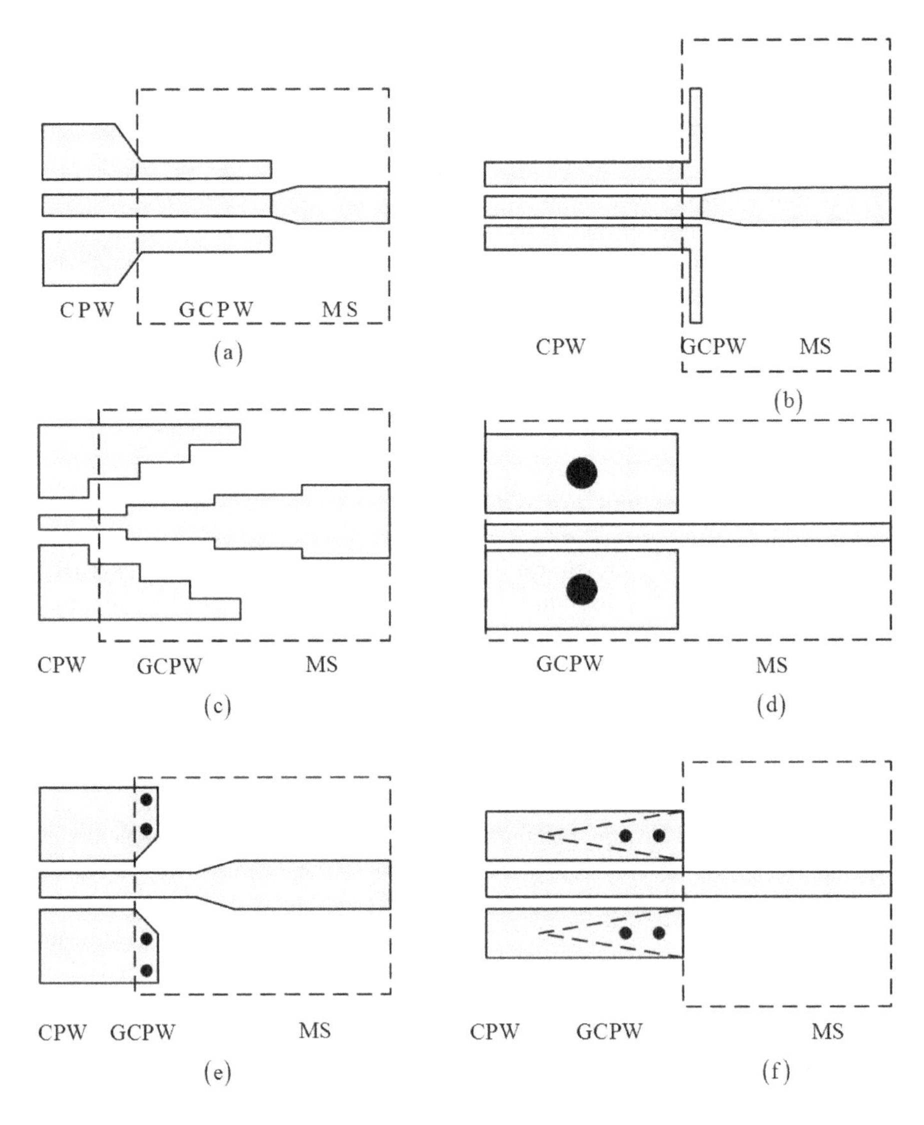

Fig. 7.4 Quasi-planar CPW-microstrip transitions. Back-side ground layer is shown by dashed lines. Adapted from [16]-[23]

Fig 7.4e shows another design where the via-holes connect the ground CPW strips to the microstrip ground layer [20]. The central conductor and the grounded CPW strips are slightly tapered for better matching of the transition. The transition is realized on several Duroid substrates and measured in frequencies up to 24 GHz. They show the insertion loss around 1.5 dB for a back-to-back transition.

More aggressive tapering and via-hole grounding are used in [21] (Fig. 7.4f). Here a microstrip is connected to GCPW, which ground layer is tapered, and the grounding strips of CPW are connected by multiple via-holes to the tapered ground layer. This circuit allows for the wide-bandwidth gradual transforming of the main modes of the connected lines. The transitions are manufactured using the low- and high-permittivity Duroid substrates, and they show only 0.7 dB loss in frequencies from 9 GHz to 40 GHz $(\varepsilon_r = 2.2)$ and from DC to 20 GHz $(\varepsilon_r = 10.2)$.

An interesting design is published in [22] where a CPW is connected to a suspended ground microstrip. The CPW's conductors are placed on a high-resistivity silicon covered by a layer of silicon dioxide. A gold tapered strip is connected to the central conductor of the CPW. Using photoresistor micromachining and electroplating, a suspended ground layer is placed over the strip with the spacing from 100 to 400 μm, depending on the sample. This suspended layer is connected to metallic strips, which are the continuations of grounded strips of the CPW. The spacing between them is large enough and does not influence the mode of the suspended microstrip. The transition shows a broad bandwidth up to 30 GHz and low insertion loss which is less than 0.36 dB in this frequency range. Additionally, different other Si-micromachined transitions studied in frequencies up to 40 GHz can be found in [23].

A novel wide-band transition between a finite-ground CPW and balanced stripline is proposed, simulated, and measured in [24]. The conductors of balanced strip line are placed on different sides of a dielectric substrate. The CPW signal conductor is tapered and prolonged behind the ends of the ground CPW strip conductors, and it is a strip of the balanced strip line. These ends are shorted by via holes to another tapered metallization of balanced strip line placed on other side of dielectric substrate. The overall length of this tapered transition is 3000 μm, and three layers of a Duroid dielectric are used in the transition design. Simulations and measurements show the insertion loss about 2.1 dB in frequencies up to 65 GHz. This transition can be used as a component of millimeter-wave packages.

7.2.2 Microstrip-CPS Transitions

As it was mentioned in the Section 6.2, the CPS transmission line is a popular solution for balanced low-loss applications at micro- and millimeter-waves. Taking into account the combined nature of contemporary integrations, the transitions between the microstrips and CPS need to be developed. Today many transitions are known for microwave frequencies including those assigned for the antenna applications [25]-[28]. They are designed using the tapering of the line geometry, the ground layer etching, and the via-hole connecting of conductors placed on different layers. Most results are obtained by full-wave simulations, optimization, and measurements. In Fig. 7.5, a transition is shown measured in frequencies from 4 GHz to 40 GHz [29]. It is seen that the field of microstrip mode is transformed gradually to the main CPS mode using smooth etching of the ground layer shown by dashed line. It is used the RT/Duroid 5880 material as the substrate. Transition exhibits the insertion loss about 1 dB from 6 GHz to 40 GHz. The EM simulations performed using the CST Microwave Studio and the Ansoft HFSS show the workability of this design in frequencies up to 100 GHz with the maximum insertion loss of 3 dB. Some circuits using this type of transition in millimeter-wave frequencies are published in [30]-[32].

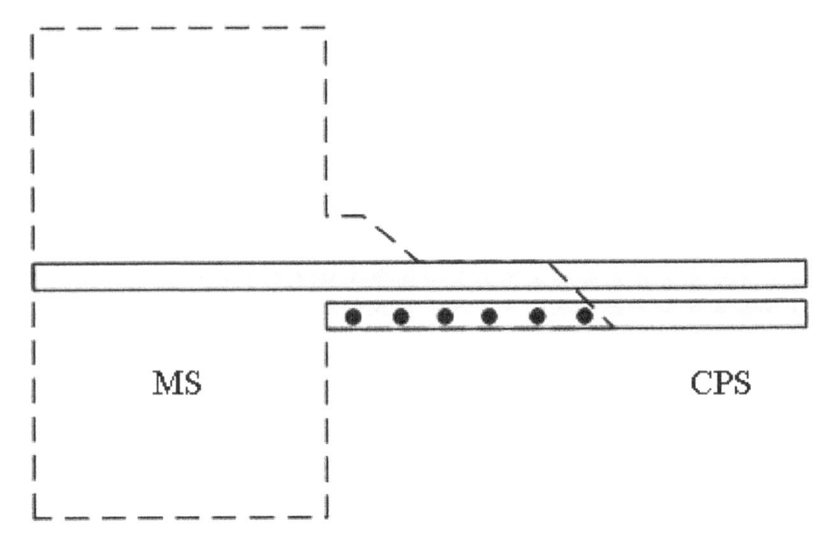

Fig. 7.5 Quasi-planar microstrip-coplanar strip line transition. Adapted from [29]

7.2.3 CPW-CPS Transitions

The CPW and CPS transmission lines are of the planar design, and they can have decreased dispersion and loss. At the difference to CPW, coplanar strips line can be used for balanced applications. Several designs have been known for frequencies up to 110 GHz, including the circuits on silicon substrates [33]-[36].

Fig. 7.6 shows a CPW-CPS transition designed for 0-55-GHz frequency band [37]. It is manufactured on a 400-μm silicon substrate covered by 1-μm silicon dioxide and 0.3-μm silicon nitride. Conductors are made of 2-μm gold. The total length of the transition is 8600 μm. Taking into account the wire-bonds used for equalization of the potentials and for preventing higher-order mode excitation, this transition relates to the quasi-planar ones.

The design uses the wire bonds to equal the potentials of the connected lines and tapering of conductor strips for smooth matching of the transition in a wide frequency band. The back-to-back configuration with two transitions and a small length of a CPS line gives the insertion loss less than 1.9 dB in 2-55 GHz, according to the measurements from the above-cited paper.

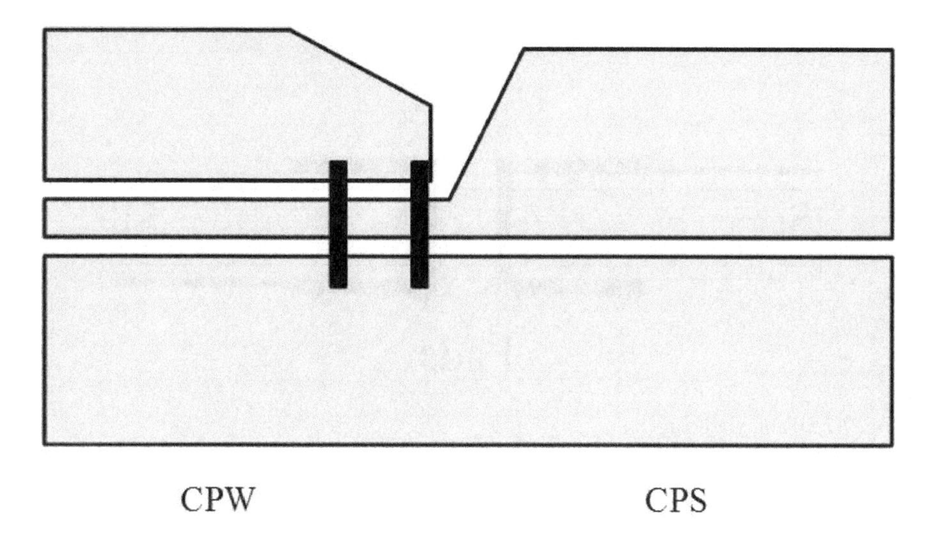

Fig. 7.6 Tapered CPW-CPS transition for 0-55-GHz applications. Wire-bonds are shown by black strips. Adapted from [37]

7.3 Interlayer Transitions for 3-D Integrations

Many contemporary monolithic and LTCC integrations are of the 3-D design, and the interlayer transitions between different transmission lines are needed. Some transitions were intensively studied during the research and developments of the first 3-D hybrid microwave integrations. The 3-D transitions are divided into three large classes: via-hole based, EM transitions, and of the hybrid design based on combinations of via-holes and EM circuit elements to match the characteristic impedances and fields to minimize the reflection coefficient of the transitions.

7.3.1 Silicon Via-holes and Transitions

Contemporary silicon integrations are of the three-dimensional design, and the circuit components, placed at different IC levels, need shorted vertical connections to each other and to the ground. The wide-frequency band connections are realized by via-holes, which are the vertical wires placed inside the holes etched in the silicon by dry or wet etching [38]. Some holes of the increased diameters are made mechanically or by laser light [39].

A via-hole consists of a vertical conductor made of a solid metal cylinder or a hollow microtube which inner wall is made of a well-conducting metal (Fig. 7.7). In the case of lossy silicon, the via-hole's wall is covered by silicon dioxide to isolate this conductor from the substrate. Then a seed-metal layer, distinguished by its increased adhesion towards the silicon or silicon dioxide, covers the inner surface of the hole. A metal to form the vertical well-conducting connection to the pads of a square or circular shape covers afterwards this surface.

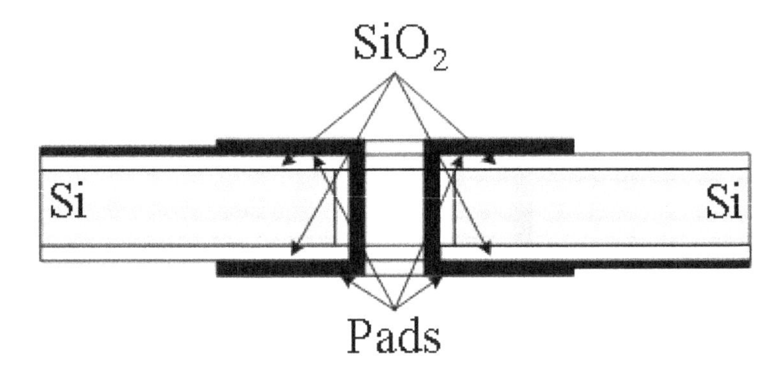

Fig. 7.7 Silicon via-hole

Due to the anisotropy of some etching processes, the via-holes are of the conical design that, practically, does not influence the via-hole performance even at millimeter-wave frequencies.

Via-hole has the parasitic dynamic capacitance of the pads and the inductance L_v of its rod. Additionally, the central conductor penetrating the ground layer has capacitance C_g to it. At low frequencies, the pads and rod's capacitances are negligible, and the inductance L_v of the central rod is dominant. It can be estimated as the inductance of a cylindrical or rectangular wire of the length *h*. The inductivity L_v of this inductor is calculated with a formula for cylindrical rod of the radius a [40]:

$$L_v = \frac{\mu_0}{2\pi}\left[h\ln\left(\frac{h+\sqrt{a^2+h^2}}{a}\right)+\frac{3}{2}\left(a-\sqrt{a^2+h^2}\right)\right]. \tag{7.1}$$

In [40], the calculated inductance L_v of a grounding via-hole was compared with the measurements and numerical simulations for a wide ratio a/h. Typically, the inductance of micron-sized via-holes is in the order of tens or hundreds of pico-Henry. At increased frequencies, the via-holes of large parasitic inductivity and capacitance are able to be resonating and radiating. Some of these frequency limitations can be overcome by the EM interlayer transitions.

In Fig. 7.8, the via-holes filled by copper and manufactured by the Allvia Inc. are shown. A typical size of these via-holes of the used technology is 30-500 μm. Due to the complete copper filling, these via-holes have reduced resistivity.

Fig. 7.8 Copper-filled via-hole. Reprinted with the permission of Allvia Inc. (www.allvia.com)

Some results on the study of via-holes for silicon integrations are published in [23],[41]. In the last paper, several through-silicon-substrate via-holes are studied. The thickness of the high-resistivity $(4\ \text{k}\Omega\text{-cm})$ silicon substrate is circa 400 μm, and the diameters of holes vary within 130-200 μm. The conical-shape holes are filled by gold in full, and the gold pads of circular, square, and octagonal shapes surround them. The measured via-hole DC resistance is less than 0.5 Ω for all studied items. Simulations and measurements show the via-hole inductance is about 54-384 pH, depending on the geometry. In addition, the dynamical inductance is influenced by the frequency due to the pads capacitances and the skin effect.

In the 0-40 GHz frequency range, the via-hole microstrip-to-microstrip transitions are studied in [23]. These transitions are designed to connect the 50-Ω microstrips with minimum reflection. They are fabricated on a high-resistivity substrate of the thickness 100 μm, and the conductors are made of gold. The design is optimized using an EM software tool, and the measured items show the de-embedded loss per via transition around 0.08-0.17 dB in frequencies 20-40 GHz. Some applications of via-holes for shielding of integrated inductors of increased quality-factor are described in [42].

Coupling of silicon via-holes is studied in [43] using the Foldy-Lax method, which allows for representation of the diffraction problem by a system of linear equations regarding to the amplitudes of cylindrical waves. The system can be reduced taking into account the strength of the via-hole interactions, and thousands of via-holes can be modeled by this approach. The published results are in a good accordance to the HFSS simulations up to 40 GHz.

Silicon multiwafer vertical interconnects are described in [44]. Two 100-μm-thick wavers of silicon $(\rho > 2\ \text{k}\Omega\text{-cm})$ are stacked to each other to avoid the increased loss of long microstrips in the case of planar design. The via-holes connect both signal microstrips and both ground layers of the two stacked wafers. The transition is simulated and measured in frequencies up to 20 GHz, and it shows the insertion loss better than 0.5 dB.

One limitation of via-hole transitions is that a via-hole is, in fact, a radiating wire, and intensive use of these transitions causes parasitic EM coupling of chip components. There are many works on EM simulations of via-holes and their radiation. One of them is on the via-hole microstrip-strip line transitions where, on the base of numerical simulations and measurements, the leakage phenomenon is studied [45]. It is shown that the support of equal characteristic impedance of all transition components is an effective way to decrease the reflection, but radiation dumping should be provided by additional design optimization. For instance, if a via-hole penetrating the ground plane, the smaller gaps between this plane and central conductor decrease the radiation. Asymmetry of strip line regarding its ground planes can decrease the leakage, too. Although these results are obtained for hybrid integrations, they can be useful in millimeter-wave monolithic circuits.

An interesting design is published in [46] where a CPW is connected to a vertically oriented parallel-plate transmission line which field is confined between the plates, and the reduced radiation and better matching of characteristic impedances can be achieved. Although this transition is studied in the 1-14-GHz frequency range, the similarly designed devices can be used in monolithic integrations of increased frequencies.

A vertical transition between two CPWs isolated by a ground is modeled in [47]. In fact, all conductors of these two CPWs are connected by vertical via-holes forming a CPW normally oriented to the connected lines. Due to that, reduced radiation is expected, and the transition can be of more improved characteristics. It is simulated by an FDTD method, and the parameters of simple equivalent circuit are extracted from full-wave simulation data.

A tapered wide-band transition between a microstrip line and CPW is considered in [48], for instance, where it allows for matching different characteristic impedances and for transferring the signals from one side of a dielectric substrate to another. Several designs are studied by simulations and measurements, including the 50-Ω-to-50-Ω and 50-Ω-to-16-Ω CPW- microstrip transitions, which show the 1.1-dB insertion loss in frequencies 40 MHz-60 GHz.

The use of via-holes is not limited by only transitions. As it has been mentioned above, the via-hole fences can shield the components and improve cross-talk issues [41],[49].

7.3.2 EM Interlayer Transitions

Typically, the EM transitions use the coupling elements and matching stubs of the size which is close to a quarter of the wavelength. They allow for avoiding some of the via-hole severe problems appearing in micro- and millimeter-wave hybrid integrations. As it was mentioned, a strong attention had been paid to these transitions during the developments of hybrid 3-D integrations [6],[50]-[53].

Being scaled, the size of EM transitions is close to several hundreds of microns even at 80 GHz, and it is comparable with the overall chip real-estate. Some of these transitions can be modified by tapering of their elements and the overall space can be decreased several times. Another area of applications of these large components is the communications between the chips and millimeter-wave motherboards. The EM components can avoid some typical problems of via-holes, wire-bonds, and flip-chip interconnects. Several microwave transitions, which are prospective for increased frequency integrations, are shown in Fig. 7.1 [6],[50] and Fig. 7.9 [54]-[56].

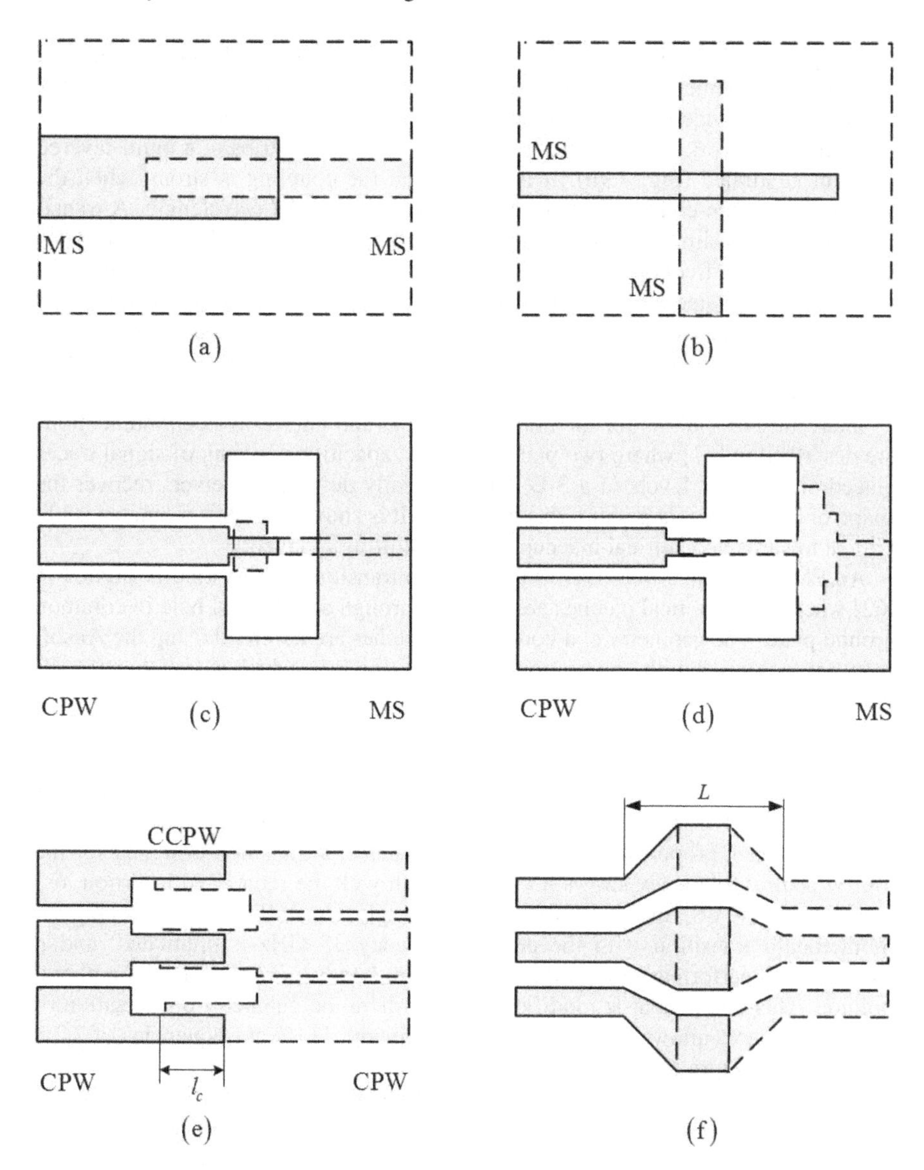

Fig. 7.9 EM vertical transitions. (a)- Microstrip-microstrip transition (ground plane shown by dashed line); (b)- Transversal crossing of two microstrips; (c)- CPW-microstrip transitions; (d)- CPW-tapered microstrip transition; (e)- CPW-CPW transition; (f)- Tapered CPW-CPW transition. Adapted from [54]-[56]

The transition shown in Fig. 7.9a is for two microstrip lines placed on different layers and having their common ground [56]. The strips and the ground are divided by the layers of dielectrics. The strips are overlapped, and they are electromagnetically coupled to each other. The coupling level depends on the overlapping area length. This transition is studied by the integral equation method. The *S*-matrix elements are calculated regarding to the geometry and frequency variation, and a wide-bandwidth geometry is found by simulations.

In the same work, a transversal transition of two microstrips in a multi-layered medium is studied (Fig. 7.9b). It is shown that the coupling is strong when the length of the open-end stubs is close to a half of the modal wavelength. A transition geometry is simulated and measured which provides the VSWR less than 1.8 in the 7-11-GHz frequency band. Some theory on skewed transmission lines for monolithic integrated circuits can be found in [57],[58]. Engineering formulas for broadside-coupled strip lines and directional couplers are published in [59]. Filters composed of stacked coupled microstrip and strip resonators are considered in [60], for instance.

Interesting transitions for the inter-chip digital and microwave communications are described in [61] where two plates provide capacitive coupling of signal traces placed on different levels of a 3-D chip. Specially designed receivers recover the shape of digital signals passing this transition. It is shown that this geometry is not critical towards geometrical inaccuracies and shifting of layers.

An EM microwave microstrip-to-microstrip transition of this kind is studied in [62] where two elliptical patches are coupled through an elliptical hole in common ground plate. The geometry and coupling of patches are modified using the Ansoft software tool to reach the minimum of the insertion loss, which is less than 0.6 dB in (3.1-10.6)-GHz frequency band.

Fig. 7.9c shows a microstrip-CPW transition [63]. The lines placed on different sides of a substrate, and they are coupled to each other in a small area which length and shape can be regulated to control the coupling and the transition bandwidth. A full-wave software tool simulates this component, and its geometry is optimized to reach proper characteristics. For instance, a transition designed for the 4-GHz central frequency shows 35% bandwidth with the return and insertion loss coefficients 20 dB and 0.7 dB, correspondingly. A millimeter-wave chip-to-motherboard transition with the center frequency 35 GHz is simulated, and it shows better performance regarding to the wire bonds, according to the authors' opinion. This component is modified in [64] where the geometry optimization of the coupling area allows for achieving the bandwidth 111% at frequencies of 7.2-8 GHz with the return loss better than 10 dB.

Other results on simulation of microstrip-CPW transitions are considered in [65]. It is shown that, in a high-frequency MMIC, the interlayer or inter-chip connecting by wire bonds can degrade the circuit performance. At the same time, the EM coupling can be realized because the length of a typical structure for EM coupling is about several hundreds of microns at high frequencies. Such a transition can be used for connection of a chip to motherboard. The authors analyzed several these transitions. It is found that they are not sensitive towards technologically caused displacements of the overlapped transmission lines. At high frequencies, these studied circuits have a wide bandwidth from 75 to 95 GHz realized by a coupling area about 160 μm.

A modified type of the microstrip-CPW transition is considered in [66] where the coupling area is tapered (Fig. 7.9d). The shape of the overlapped conductors is optimized numerically to reach the minimum of the reflection coefficient. The simulated transition has a bandwidth from 6.9 to 12.4 GHz at the level of 20-dB return loss.

An interesting transition is described in [67] where the CPW and microstrip are placed on different sides of a substrate to provide vertical connection. These lines are matched by circular shape stubs of the splitted input microstrip and of the slot lines formed in the ground sheets of CPW. It allows to match the transition in a multioctave frequency band. An 1.5-15-GHz transition was simulated and measured by the author of [67].

In some applications, it is preferable to use the transitions matched by transversal stubs. An example of these planar transitions is studied in [68] where its frequency band is about 8 GHz for the 20-GHz center frequency and the return loss is 10 dB.

Another design of this transition contains a non-uniform microstrip and the CPW stubs [54]. These lines are placed on different sides of a substrate, and they have no any galvanic contacts. In spite using the resonant stubs, the frequency band is rather wide, and the developed transition shows good performance in frequencies 2.8-7.5 GHz with the return loss better than 15 dB. These experimentally tested solutions, although being designed for low microwave frequencies, can be scaled up to millimeter-waves if a careful study would be performed.

A CPW-to-CPW vertical transition is shown in Fig. 7.9e. It is developed for those applications at millimeter-wave frequencies where the via-holes and the wire bonds distort severely the frequency performance of components. Especially, such effects are dangerous in phase array antennas containing multiple transitions of different types. In [55], a vertical EM transition between two CPWs placed on different levels of a module is studied. These open-end CPWs have an overlapping area of the length l_c where they are coupled to each other electromagnetically. The coupling length is described by the even $Z_c^{(e)}$ and the odd $Z_c^{(o)}$ characteristic impedances. The average phase constant of this length is

$$\beta^2 = \left(\frac{\omega}{c}\right)^2 \left(\frac{\varepsilon_{\text{eff}}^{(o)} + \varepsilon_{\text{eff}}^{(e)}}{2}\right) \tag{7.2}$$

where $\varepsilon_{\text{eff}}^{(e)}$ and $\varepsilon_{\text{eff}}^{(o)}$ are the effective dielectric constants of the even and odd modes of the coupled CPWs, correspondingly. The impedance parameters of the transition are

$$\begin{aligned} Z_{11} &= -j\frac{Z_c^{(e)} + Z_c^{(o)}}{2}\cot(\beta l_c), \\ Z_{12} &= -j\frac{Z_c^{(e)} - Z_c^{(o)}}{2}\csc(\beta l_c). \end{aligned} \tag{7.3}$$

To adjust this transition, the modal characteristic impedances should be equalized to each other at the central frequency, first. In [55], the derivation of these characteristic impedances is realized electromagnetically by solving an integral equation regarding to the currents on conductors. This transition is analyzed and measured at 5 GHz and 10 GHz. A 10-GHz single transition shows 15-dB return loss over 10% bandwidth. The maximum bandwidth of the simulated and measured transitions does not exceed 25%. Similar components are studied in 75-110-GHz frequency band, and the length of a transition is 210-350 μm. To match, each CPW is connected to the transition through a short length of a taper (Fig. 7.9f). The center of the bandwidth is placed in 96-100-GHz band. In frequencies from 85 to 110 GHz, the insertion loss is within 0.35-0.6 dB, and the return loss is about 20 dB.

7.4 Via-holes of Micro- and Millimeter-wave Motherboards

The millimeter-wave chips packaged to a motherboard need reliable grounding. Very frequently, it is achieved by the via-holes which connect the traces to the ground. If a motherboard consists of several layers of traces, the via-holes are to connect the chip or the traces in the vertical direction. The typical size of via-holes used in motherboards of the MCM type allows for a number of high-frequency effects. Among them are the resonances, radiation, and parasitic coupling of via-holes. To design the millimeter-wave motherboards, these effects should be calculated to avoid strong distortions of signals.

Below, the electromagnetics of grounding via-holes is considered in the details, and the developed theory allows for calculating the resonant frequencies and scattering parameters of grounding via-holes. This approach can be applied for the analysis of through via-holes connecting the traces placed on different conductor layers of the millimeter-wave motherboards.

7.4.1 Grounding Via-holes[1]

A grounding via-hole connects a signal trace to the ground. It consists of a pad and a shorting cylindrical rod. A circular grounding via-hole has a circular pad (Fig. 7.10).

Square-pad via-holes are of particular importance in the microwave hybrid planar and 3-D integrated circuits and packages, where lumped elements such as capacitors, inductors and monolithic microwave integrated circuits require square pads for attachment [6],[38],[40]. In addition, the rectangular and square shorted patches are used as printed antennas, resonators and filters, periodic high-impedance surfaces and fenced waveguides [69]-[81].

[1] Sections 7.4.1-7.4.4 are written together with M.J. Deen, N.K. Nikolova, and A. Rahal.

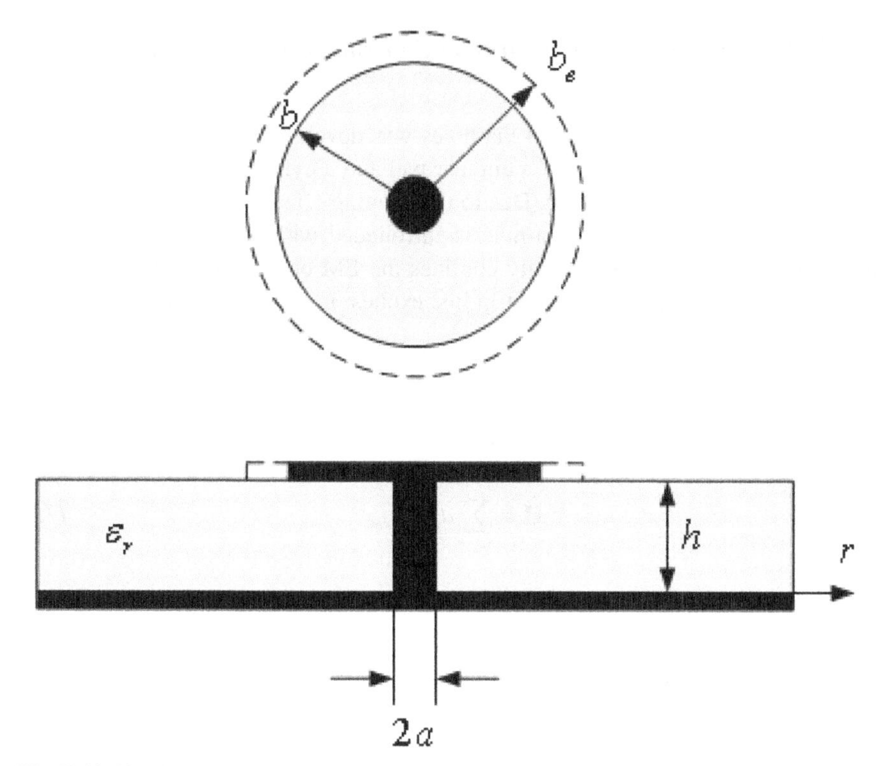

Fig. 7.10 Circular-pad via-hole. Adapted from [84]

The via-hole performance is excellent up to a certain frequency where parasitic resonances and radiation occur. For high-speed and microwave applications, this frequency must be calculated to define the maximum frequency band of operation [74]-[78]. The resonances and radiation are parasitic effects in via-holes; however, they are useful in antennas. In all these cases, rigorous EM analysis is required. Some of the general-purpose 3-D EM solvers can offer adequate modeling of the performance characteristics of the via-holes. However, they require long computation times, and the results are valid only for the specific substrate and dimensions.

In [82], a new and fast analytical model of a grounding circular-pad via-hole based on the cavity representation was developed. In [83], the eccentric distortions of circular-pad via-holes were considered semi-analytically. Square- and rectangular-pad via-holes are considered in [84]. All models are based on the cavity representation and on the assumption that the electric field is constant along the substrate height. Hence, the pad size b is greater than the substrate height h, and the substrate is electrically thin at the considered frequencies. The studied via-holes and antenna patches are verified by our own measurements as well as by other published experimental results. Additional confirmation is obtained from simulations with EM full-wave software tools. Overall, the errors of our models vary from sample to sample, and they are within 0.4-6.4% for the substrates whose relative permittivity fulfills $2.2 < \varepsilon_r < 10$, and whose relative pad radius b/h is within $1.3 < b/h < 32$. The normalized rod radius a/b must vary according to $0.018 < a/b < 0.46$.

7.4.2 Cavity Model of Circular-pad Grounding Via-hole

The cavity theory of circular-pad via-holes was developed in [82]. A grounding circular-pad via-hole consists of a circular pad and a cylindrical shorting pin connected to the ground (Fig. 7.10). Due to the fringing effect, the radius of the pad b is extended to b_e, where the via-hole is surrounded with an effective cylindrical magnetic wall. The via-hole cavity confines the EM energy and defines its resonant frequencies. The input microstrip line excites, in general, an infinite series of resonant modes [85],[86]:

$$\mathbf{E} = \sum_{n,m,k}^{\infty} c_{nmk} \mathbf{E}_{nmk}, \tag{7.4}$$

$$\mathbf{H} = \sum_{n,m,k}^{\infty} d_{nmk} \mathbf{H}_{nmk} \tag{7.5}$$

where c_{nmk} and d_{nmk} are the modal amplitudes, and the indices $n = 0,1,...,$ $m = 1,2,...,$ and $k = 0,1,...,$ correspond to the modal field variations along the angular, radial and normal-to-the-substrate axes, respectively. In addition to the resonant modes, the series (7.4) and (7.5) includes the magnetic zero-frequency mode. For thin substrates, the resonant modes are of the TM_z type: the normal-to-the-substrate magnetic field is zero, $H_{z_{nmk}} = 0$, and the $E_{z_{nmk}}$ field component does not vary with z ($k = 0$).

For practical applications, the series (7.4) and (7.5) is modeled with an equivalent circuit consisting of a finite number ($n = 1,\ldots,N$; $m = 1,\ldots,M$; $k = 0$) of resonant tanks serially connected to each other (Fig. 7.11). Each modal equivalent circuit or a tank consists of a capacitor $C_{\text{cir}}^{(nm0)}$, an inductor $L_{\text{cir}}^{(nm0)}$, and a number of equivalent resistors representing the losses in the via-hole.

In Fig. 7.11b, the resistance $R_r^{(nm0)}$ represents radiation loss [85], $R_d^{(nm0)}$ is the equivalent dielectric loss resistance [86], $R_c^{(nm0)}$ is the equivalent conduction resistance [87], and $R_{\text{sw}}^{(nm0)}$ is the equivalent surface-wave loss resistance [88]. The zero-frequency mode is described with the inductor $L^{(0)}$ [40],[82].

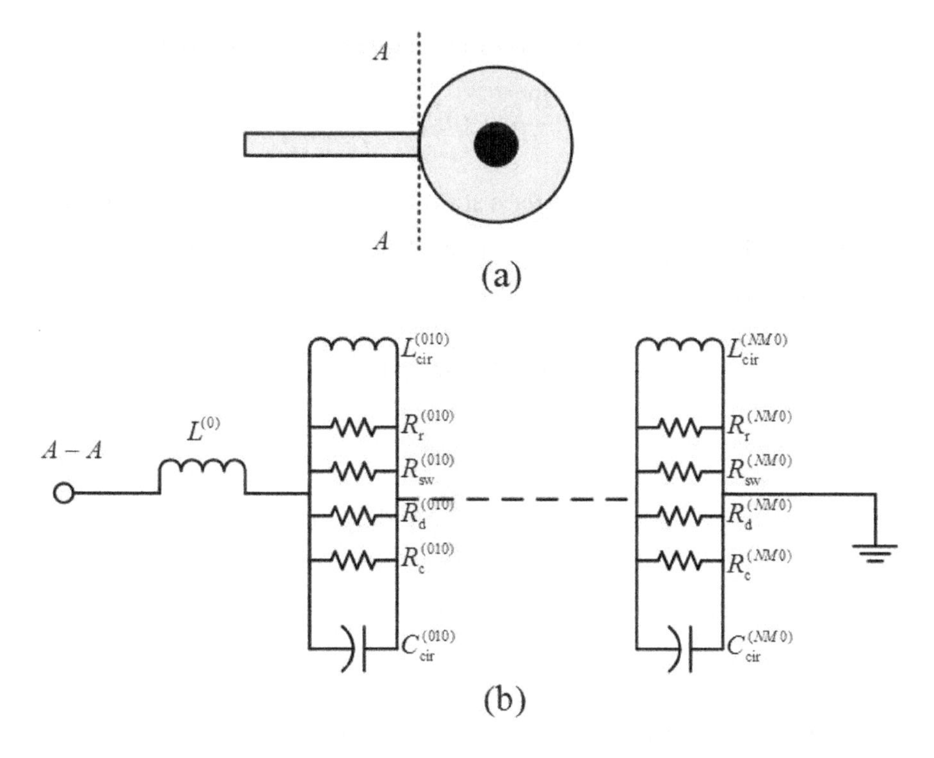

Fig. 7.11 Circular-pad via-hole directly excited by a microstrip line: (a)- Top view; (b)- Multimode equivalent circuit [84]. Reprinted with permission of the IEEE, license #2873190190705

To calculate the tank parameters, the modal resonant frequencies Ω_{nm0} are computed from an eigenvalue equation, which is derived by matching the modal fields on the rod's surface $(r = a)$ and on the outer magnetic wall $(r = b_e)$:

$$J_n(k_{nm0}a)Y_n'(k_{nm0}b_e) - J_n'(k_{nm0}b_e)Y_n(k_{nm0}a) = 0, \; n = 0,1,\ldots \qquad (7.6)$$

Here, J_n and Y_n are the Bessel and Neumann functions of order n ($n = 0,1,\ldots$), J_n' and Y_n' are their derivatives with respect to the argument $(k_{nm0}r)$, n is the number of field variations in the ϕ-direction, $k_{nm0} = (\Omega_{nm0}/c)\sqrt{\varepsilon_r}$, and ε_r is the substrate relative permittivity. The effective radius b_e is different for the dominant and for the higher-order modes and is given by [82]:

$$b_e^{(010)} = b + 0.553h, \; n = 0, \; m = 1, \qquad (7.7)$$

$$b_e^{(nm0)} = b + 0.450h, \; n \geq 1, \; m \geq 1. \qquad (7.8)$$

Then the modal equivalent capacitances and inductances are computed as

$$C_{\text{cir}}^{(nm0)} = \frac{2W_{e_{\text{cir}}}^{(nm0)}}{U_{\text{edge}}^2}; \; L_{\text{cir}}^{(nm0)} = \frac{1}{C_{\text{cir}}^{(nm0)} \Omega_{nm0}^2}. \tag{7.9}$$

Here, U_{edge} is the edge voltage defined at any point P_j where the mode magnitude is maximal. The energy of the modal electric field $W_{e_{\text{cir}}}^{(nm0)}$ stored in the equivalent volume $V_e = \pi h(b_e^2 - a^2)$ is computed as

$$W_{e_{\text{cir}}}^{(nm0)} = \frac{1}{4}\varepsilon_r \varepsilon_0 \int_{V_e} (E_{z_{nm0}}^{(I)})^2 dV_e \tag{7.10}$$

where

$$E_{z_{nm0}}^{(I)} = A_{nm0} \left[J_n(k_{nm0} r) - \frac{J_n(k_{nm0} a)}{Y_n(k_{nm0} a)} Y_n(k_{nm0} r) \right] \cos n\phi. \tag{7.11}$$

Here, A_{mn0} is the modal amplitude proportional to the edge voltage U_{edge} at $r = b_e$.

7.4.3 Eccentric Circular-pad Grounding Via-hole

The circular-pad via-holes can be distorted due to the technologically caused eccentricity (Figs 7.12 and 7.13). The eccentricity decreases the frequency band of the via-holes [83]. On the other hand, the eccentricity is exploited in large shorted patch resonators in order to shift the modal resonant frequency or to change the input impedance. This important geometry is analyzed with the cavity approach based on the theory of circular-pad via-holes and the method of transformation of coordinates [89],[90]. To compute the resonant frequencies of an eccentric via-hole, the distorted geometry is transformed into the concentric one, whose modes are calculated by the above theory. Then, the geometry is transformed back to the eccentric one, and the field of the modal series is matched at the outer effective magnetic wall. The derived infinite system of linear algebraic equations is reduced, and it is solved with a root-searching algorithm. The study shows that the small-pad via-holes are influenced significantly by the eccentricity: their frequency band decreases in the worst-case scenario by up to 20-30%. Such effects must be taken into account in manufacturing tolerances as well as when the via-hole models are verified by measurements.

The eccentric circular via-hole consists of a grounding rod and a circular pad (Fig. 7.12). The rod is misaligned with respect to the geometrical center of the circular-patch.

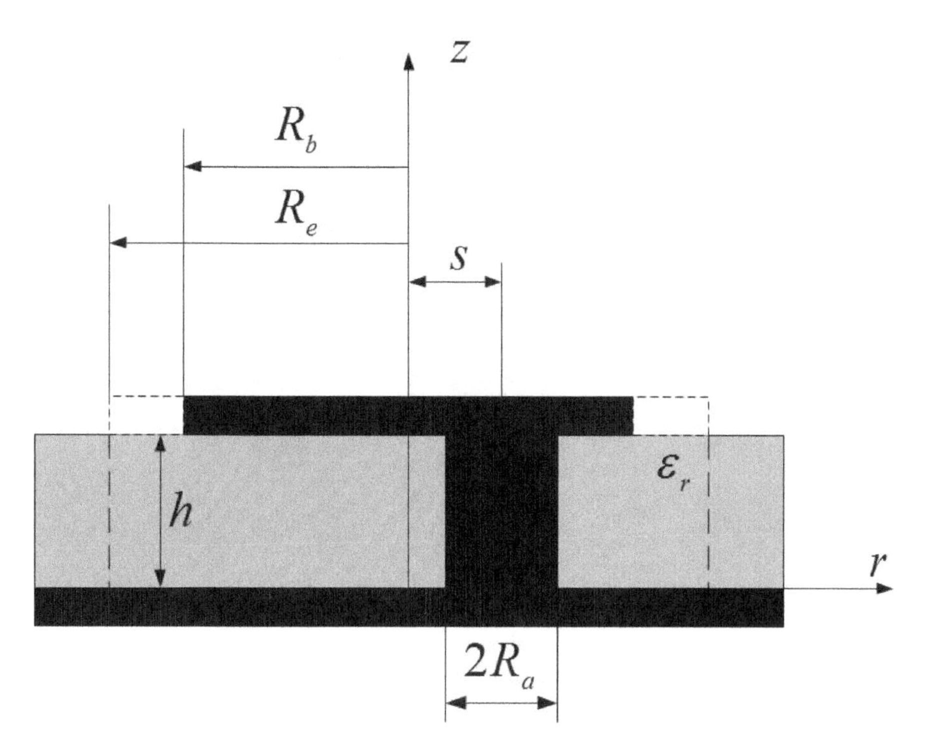

Fig. 7.12 Eccentric via-hole cavity [83]. Reprinted with permission of the IEEE, license # 2873210115411

In antennas, this misalignment is used for transforming the input characteristic impedance and shifting resonant frequencies. In grounding via-holes whose pad-radii are small and comparable with the substrate height, the technologically caused eccentricity leads to the shift of the fundamental modal frequency and, in the worst-case scenario, additionally reduces the via-hole frequency band.

The excited field in the eccentrically shorted cavity is expanded in series of modes:

$$\mathbf{E} = \sum_{\nu}^{\infty} a_{\nu} \mathbf{E}^{(\nu)}, \; \mathbf{H} = \sum_{\nu}^{\infty} b_{\nu} \mathbf{H}^{(\nu)} \tag{7.12}$$

where a_{ν} and b_{ν} are the modal amplitudes, and the index ν corresponds to the respective eigenvalue in the solution of the Helmholtz equation. Because of the eccentricity of the cavity, the analytical method of eigenfunction derivation applied to the concentrically grounded circular-patch resonator is not valid. Here, we use a polar coordinate transformation method. The polar coordinate system of the eccentric structure (Fig. 7.13a) is transformed linearly into the coordinate system, where the studied cavity is centered at the origin (Fig. 7.13b). Any point of the eccentric structure described by a vector $\boldsymbol{\rho}$ with the origin at O_{ρ} is shifted to a point with the new coordinate $\mathbf{r} = \boldsymbol{\rho} - \mathbf{s}$, where $\mathbf{s}$ is the shift-vector and $\mathbf{r}$ is the vector with the origin O_r (Fig. 7.13). It allows the use of separation of variables method for the general solution of the centrically shorted cavity.

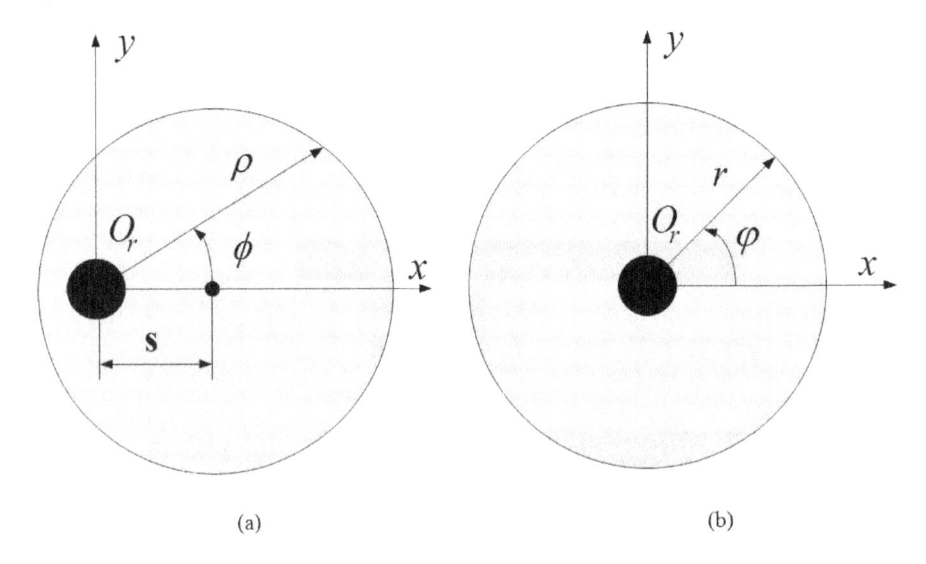

Fig. 7.13 Linear transformation of the polar coordinate system associated with the via-hole structure [83]. Reprinted with permission of the IEEE, license # 2873210115411

Consider only the TM_z-modes of the transformed concentric resonator (Fig. 7.13b). In this case, the normal-to-the-substrate magnetic field $H_z^{(v)}$ is zero, and $E_z^{(v)}$ does not vary with z. In general, the $E_z^{(v)}$ field component is expanded into a series of cylindrical functions in the concentric via cavity (Fig. 7.13b):

$$E_z^{(v)} = \sum_n E_{z_n}^{(v)} = \sum_n \left[A_n J_n(k_c r) + B_n Y_n(k_c r) \right] e^{jn\varphi} \tag{7.13}$$

where A_n and B_n are unknown coefficients.

Each term of the series (7.13) is set to zero at the rod surface, $E_{z_n}^{(v)}(|\mathbf{r}| = R_a) = 0$, which allows for the expression of the unknown coefficient B_n in terms of A_n. After which, we derive the angular component of the magnetic field $H_\varphi^{(v)}$ using Maxwell's equations and (7.13)

$$H_\varphi^{(v)} = -jc\varepsilon_0 \sum_n A_n \left[J_n'(k_c r) - \frac{J_n(k_c R_a)}{Y_n(k_c R_a)} Y_n'(k_c r) \right] e^{jn\varphi} \tag{7.14}$$

where J_n' and Y_n' are the derivatives of Bessel and Neumann cylindrical functions, respectively, with respect to the argument $(k_c r)$.

On the outer magnetic wall, the $H_\phi^{(\nu)}$ component is zero. This matching should be done in the original coordinates where the cavity is eccentric (Fig. 7.13a):

$$H_\phi^{(\nu)}\left(|\mathbf{s}+\boldsymbol{\rho}|=b_e\right)=0. \tag{7.15}$$

We apply the inverse transformation of the coordinates to (7.14) and (7.15), and change the arguments of the Bessel and Neumann cylindrical functions. Then, we rewrite them according to Graf's addition formulas (7.16) and (7.17):

$$Z_n\left(k_c r\right)e^{jn\varphi}=\sum_{p=-\infty}^{\infty}Z_p\left(k_c s\right)Z_{n+p}\left(k_c\rho\right)e^{j(n+p)\phi}, \tag{7.16}$$

$$Z'_n\left(k_c r\right)e^{jn\varphi}=\sum_{p=-\infty}^{\infty}Z_p\left(k_c s\right)Z'_{n+p}\left(k_c\rho\right)e^{j(n+p)\phi} \tag{7.17}$$

where Z and Z' are the cylindrical functions (J or Y) and its derivative, respectively. Then, the boundary condition (7.15) at the outer magnetic wall becomes

$$\sum_n\sum_{p=-\infty}^{p=\infty}A_nJ_p\left(k_c s\right)\left[J'_{n+p}\left(k_c b_e\right)-\frac{J_n\left(k_c R_a\right)}{Y_n\left(k_c R_a\right)}Y'_{n+p}\left(k_c b_e\right)\right]e^{j(n+p)\phi}=0\,\cdot \tag{7.18}$$

Since (7.15) is valid for all values of the angular variable ϕ, we rewrite this expression as a set of equations using $n+p=i$:

$$\sum_n A_nJ_{i-n}\left(k_c s\right)\left[J'_i\left(k_c b_e\right)-\frac{J_n\left(k_c R_a\right)}{Y_n\left(k_c R_a\right)}Y'_n\left(k_c b_e\right)\right]=0,\ i=0,\pm1,\pm2,\ldots,\pm\infty\ . \tag{7.19}$$

The infinite system of linear equations (7.19), where A_n are unknown coefficients, has a nontrivial solution if $D=0,$ where D is its determinant. It produces the modal resonant frequencies ω_ν. Similar to the concentric cavity, the number of the modes ν is infinite and each mode $\mathrm{TM}_{\varsigma lk}$ is classified according to the field variation described by the numbers: k – along the normal-to-the-substrate axis z, l – along the radial axis ρ, ς – along the azimuthal variable ϕ.

7.4.4 Modeling and Measurement Results for Grounding Via-holes

The via-holes' models were verified by the authors' own experiments, published measurements, and simulations by several full-wave software tools at frequencies up to 26 GHz [82]-[84]. The overall inaccuracy in the calculation of modal resonant frequencies of circular- and square-pad via-holes does not exceed 5.4%. Additionally, the reflection coefficient from the ground via-hole was calculated according to the equivalent network model (Fig. 7.11) and compared with the measurements. The overall error of our via-hole model is found to be no greater than 5.5 % when $2.2 < \varepsilon_r < 9.9$, $k_0 h \ll 1$, and $b/h \geq 1.3$.

The circular-pad via-hole computation results, which relate to the millimeter-wave range are shown in Fig. 7.14, where the modal resonant frequencies ω_{mn0} of the first three modes are plotted versus the pad radius *b* (solid lines). Additionally, the results are verified by Agilent Momentum simulations (circles). The first mode TM_{010} resonance can seriously distort the via-hole performance, and the via-hole geometry should be designed to avoid this resonance from the frequency band of millimeter-wave signals. It is reached by the use of low-permittivity materials and small-radius via-hole pads.

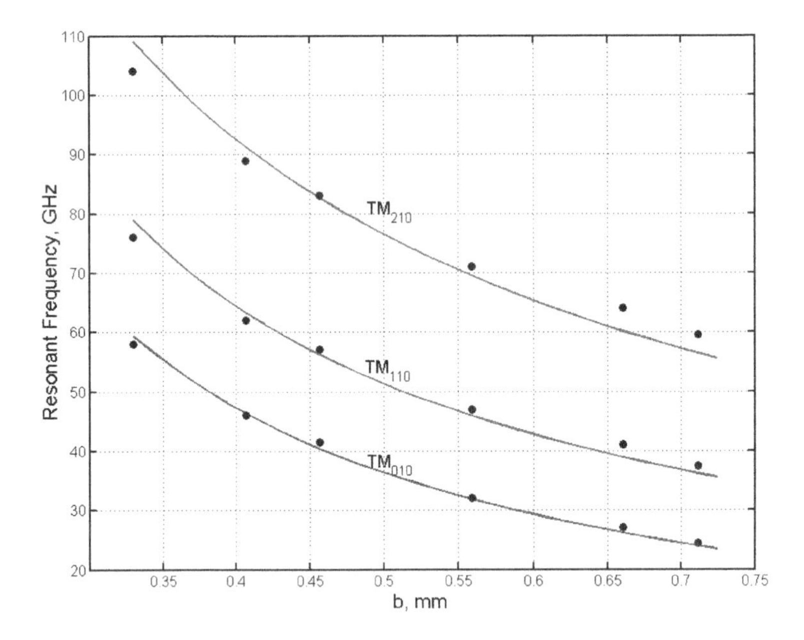

Fig. 7.14 Modal resonant frequencies of a circular-pad via-hole versus the pad radius: solid lines – our results, circles – Agilent Momentum simulations [84]. Via-hole parameters: $a = 0.152$ mm, $h = 0.254$ mm, $\varepsilon_r = 9.9$. Reprinted with permission of the IEEE, license # 2873191334118

A full-wave study of the eccentric via-hole is published in [83]. It was found that the eccentricity influence is especially strong for small via-hole pads, and where the technology inaccuracy is comparable with the pad diameters. In the worst case, their frequency band calculated from the zero frequency to the first resonant frequency can decrease by 20% if the eccentricity is comparable to the via-pad radius.

The validation of the eccentric grounding via-hole theory is performed by comparisons with the authors' own experiments, published measurements, and simulations by the Agilent Momentum and IE3D [83],[84]. Additionally, the convergence of the used numerical algorithm is tested. It is found that the error does not exceed several percent in the computations of the resonant frequencies.

Two modal resonant frequencies versus the eccentricity parameter s are shown for two via-holes (Fig. 7.15). The via-pad radius b is the parameter. It is seen that the resonances are in the millimeter frequency band for the studied via-hole with the typical size for this technology and application.

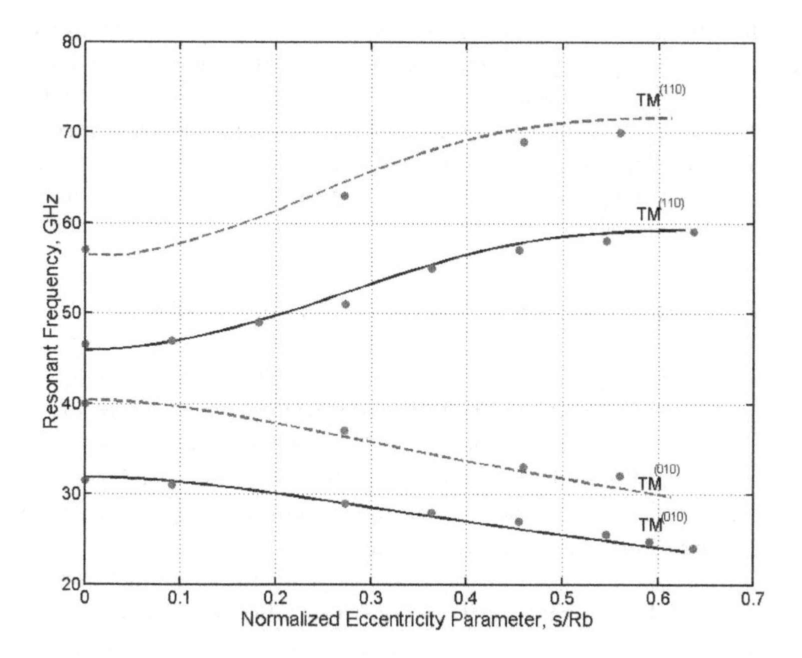

Fig. 7.15 Modal resonant frequencies of circular-pad via-holes versus the normalized eccentricity parameter s/b : solid and dashed lines – current approach, circles – Momentum [83]. Via-hole parameters: $a=0.152$ mm, $h=0.254$ mm, $\varepsilon_r=9.9$, b =0.46 mm (dashed lines), b =0.558 mm (solid lines). Reprinted with permission of the IEEE, license # 2873210115411

The $TM^{(010)}$ resonant frequencies of both via-holes decrease as the eccentricity parameter s increases, while the second modal frequencies increase (Fig. 7.15). The maximum shift of the modal resonant frequencies is within 22-30% with respect to the corresponding resonant frequency of the concentric grounding via. This occurs for large misalignments when s is close to (but smaller than) the via-pad radius.

The modes define the resonant region and the high-frequency limit of the via frequency band. In the worst-case scenario, the lowest modal $TM^{(010)}$ resonant frequency is the high frequency limit of the via-hole. Thus, the eccentricity may change the via frequency limit with about 22-30% in comparison with the frequency limit of the concentric via. The manufacturing misalignment of the via-hole pad – estimated at 50-100 μm – has strong influence on small pad via-holes used for ultra-wide-band applications. For the studied cases (see Fig. 7.15), the 100- μm misalignment decreases the first modal frequency with up to 8%. This influence increases to 20% for via-holes having a pad radius of about of 300 μm. The nature of this technologically caused eccentricity is random and thus the high-frequency limit may vary between the two resonant frequencies of the modes $TM^{(010)}$ and $TM^{(110)}$. Besides, the degree of the via-hole performance degradation depends on the position of the exciting port at the via pad edge with respect to the eccentricity. In summary, the worst-case frequency limit of a via can be estimated as the resonant frequency of the $TM^{(010)}$ mode corresponding to the largest manufacturing eccentricity.

7.5 Through-substrate Via-holes and Their Modeling

More complicated high-frequency effects are typical for via-holes connecting microstrip traces placed on different layers separated from each other by one or several ground and power planes. Although the numerical tools are rather fast today, the EM simulation of multiple via-holes is still very time-consuming, and the interest in analytical or semi-analytical modeling and enhancing the element library of circuit-based simulators is strong up to now.

Fig. 7.16 shows a couple of basic via-holes geometries for which approximate analytical models are available. They consist of a metallic cylindrical post and two pads to which microstrip lines are connected. The post intersects one (Fig. 7.16a) or several ground and power planes (Fig. 7.16b). In the first case, a via-hole, being resonating, excites the grounded-substrate modes and spatial waves from its pad.

In the second case, additionally to the mentioned waves, the parallel-plate modes are propagating away from a via-hole, and they are a source of severe cross-talks between the via-holes and/or other components of motherboard. Then the via-holes should be designed with the respect of minimization of the EM radiation.

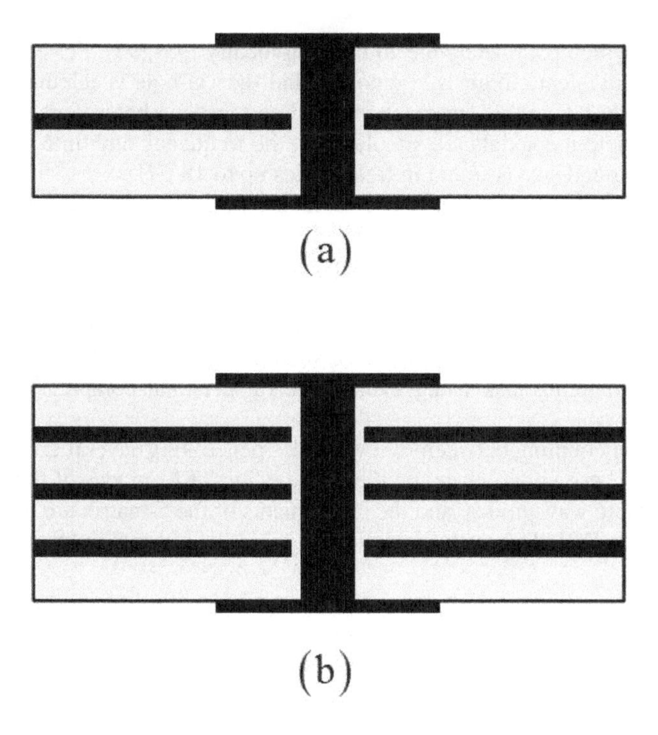

Fig. 7.16 Through-substrate via-holes

A full-wave study of the via-hole shown in Fig. 7.16a is published, for instance, in [91] where the resonances are modeled, and it is shown that the frequency response depends on the angular orientation of the microstrips joined to the via-hole pads. One of the reasons is that the cylindrical via-holes support the degenerated angular modes, and violation of the symmetry leads to formation of a two-mode cavity with different resonance frequencies, and this difference is defined by coupling of modes and the asymmetry of the geometry. The full-wave results are confirmed by measurements, and a good correspondence is found in the mentioned paper in frequencies up 24 GHz.

An interesting solution to improve the characteristics of a microstrip-microstrip transition is proposed in [92] where a multiple of via-holes, instead of a single one, connect two microstrips through a thick ground layer. It improves the transition performance and shields partly the radiation of via-holes. The measurements and full-wave simulations are compared in 1-10 GHz band.

As it has been already mentioned above, simulation of via-holes penetrating several ground and power planes (Fig. 7.16b) is a complicated task due to the needs to take into account the parallel-plate modes spreading from via-holes. One the first analytical models of these via-holes is available from [93],[94] where a decomposition approach is used.

A via-hole of this type shown in Fig. 7.16b is separated into several elements. For instance, it is modeled by a coaxial line if the central rod passing a metallic plate. Out of these plates, a via-hole is represented as a radial wave antenna or waveguide, which parameters are found analytically [93],[95]. For each such of a length, an equivalent circuit is composed, and the via-hole is calculated as a cascaded network. Using this approach, several pad-less via-holes connected directly to the input/output coaxials are simulated in the frequency and time domains, and a good correspondence is found in frequencies up to 18 GHz.

The via-holes can be placed close to each other, and they are coupled having mutual inductance and capacitance. Via-holes excite the traveling parallel-plate modes and waves of microstrip substrates, and this is another reason for cross-talk. The effect must be simulated, taken into account, and a microwave board should be optimized to decrease this EM cross-talk. Besides, this radiation can interact with the interconnects, and simulation of such boards with the EM full-wave tools is a very consuming task using even powerful personal computers or designer workstations.

In [96], the coupling between two via-holes penetrating several grounded plates is modeled. These via-holes excite the propagating TEM modes of the radial type in parallel-plate waveguides, and the components of the S-matrix are derived using the ideas from [93]. The model is validated by measurements in frequencies up to 18 GHz.

Analytical models of via-holes can be combined with the numerical ones, and this approach is effective in many cases. For example, via-holes can be modeled analytically using different methods, including the quasistatic ones [97], but interactions of via-holes and excitation of them by microstrips can be described electromagnetically using, for instance, the method of moments. This combination of methods and models allows for increasing the effectiveness of algorithms and decreasing the required computer resources and simulation time [94]-[100].

Interesting results are derived in [101]-[103] where an effective algorithm is described. Being applied to the motherboards with thousands of via-holes, it gives accurate results in short time. The idea of this method consists of the known fact that the problem of diffraction of a plane TEM wave on a perfectly conducting cylinder is solved analytically, and the field is expressed by an infinite series. The convergence of the series depends on the cylinder wave radius, and it is fast for electrically thin cylinders. In many cases, only a few series members can represent the main part of the field. In the above-cited papers, the field, excited and scattered by multiple via-holes, is represented by summation of such a reduced series, and a system of linear algebraic equations regarding to the scattering coefficients is derived. This matrix, named after Foldy and Lax, can be sparse due to decreasing of interactions of via-holes with the distance between them. The coupling of via-holes to outer layers is modeled by a moment method. The method is validated by comparison with the results obtained using known commercial full-wave software tools and with the measurements.

Besides the computational aspects, the via-hole implementation issues are very important. For instance, the ground via-holes or even via-hole fences are widely used to prevent the cross-talks between the integrated components of microwave and high-speed boards [104]. Meanwhile, these effects highly depend on the location of ground via-holes assigned for suppressing of propagating parallel-plate modes. Close distance of the ground via-holes to the signal ones leads to the strong magnetic coupling and distortions. Large distance between the ground suppressing via-holes is the cause of low-frequency resonances of parallel-plate modes excited by signals of these via-holes. The EM consideration of these problems and proper design of circuit boards can improve their parameters.

Although the performed here review touches only several typical ideas on the via-hole transition simulations, it shows that further attention should be paid on this 3-D problem of practical electromagnetics for effective modeling of micro- and millimeter-wave boards.

References

[1] Gvozdev, V.I.: Use of the unbalanced slotted line in SHF microcircuits. Radioeng. Electron Physics (Radiotekhnika i Elektronika) 27(11), 42–47 (1981)

[2] Gvozdev, V.I., Nefedov, E.I.: Some possibilities of three-dimensional integrated UHF structures. Sov. Phys. Dokl. 27, 959–960 (1982)

[3] Gvozdev, V.I., Gulayev, Y.V., Nefedov, E.I.: Possible use of the principles of three-dimensional integrated microwave circuits in the design of ultra-fast digital computers. Sov. Phys. Dokl. 31, 760–761 (1986)

[4] Gvozdev, V.I., Nefedov, E.I.: Three-Dimensional Microwave Integrated Circuits. Nauka Publ., Moscow (1985) (in Russian)

[5] Gvozdev, V.I., Nefedov, E.I.: Volumetrical Microwave Integrated Circuits. Element Base of Analog and Digital Radioelectronics. Nauka Publ., Moscow (1987) (in Russian)

[6] Gvozdev, V.I., Kouzaev, G.A., Nefedov, E.I., et al.: Physical principles of the modeling of three-dimensional microwave and extremely high-frequency integrated circuits. Soviet Physics-Uspekhi 35, 212–230 (1992)

[7] Izadian, J.S., Izadian, S.M.: Microwave Transition Design. Artech House, Norwood (1988)

[8] Gvozdev, V.I., Kouzaev, G.A., Kulevatov, M.V., et al.: Transitions and Matching Circuits for Microwave Three-dimensional Integrated Circuits. MSIEM Publ., Moscow (1999) (in Russian)

[9] Gvozdev, V.I., Kouzaev, G.A., Podkovyrin, S.I.: Element base and functional components of three-dimensional integrated circuits of the microwave range. In: Modeling and Design of Devices and Systems of Micro- and Nanoelectronics, pp. 5–10. MIET Publ., Moscow (1994) (in Russian)

[10] El-Gibari, M., Averty, D., Lupi, C., et al.: Ultra-wideband GCPW-MS transitions for characterizing microwave and photonic components based on thin polymer. El. Lett. 47(9), 553–555 (2011)

[11] Strauss, G., Ehret, P., Menzel, W.: On-wafer measurement of microstrip-based MIMICs without via holes. In: 1996 MTT-S Microw. Conf. Dig., pp. 1399–1402 (1996)

[12] Liu, D., Floyd, B.: Microstrip to CPW transitions for package applications. In: Proc. APSURSI Conf., pp. 1–4 (2010)

[13] Zheng, G., Kirby, P., Rodriguez, A., et al.: Design and on-wafer measurement of a W-band via-less CPW RF probe pad to microstrip transition. In: Proc. 33rd Eur. Microw. Conf., pp. 443–446 (2003)

[14] Torres-Torres, R., Hernandes-Sosa, G., Romo, G., et al.: Characterization of electrical transitions using transmission line measurement. IEEE Trans., Adv. Pack. 32, 45–52 (2009)
[15] Hettak, K., Dib, N., Sheta, A., et al.: New miniature broad-band CPW-to-slotline transitions. IEEE Trans., Microw. Theory Tech. 48, 138–146 (2000)
[16] Raskin, J.P., Gauthier, G.P., Katehi, L.P.B., et al.: Mode conservation at GCPW-to-microstrip line transitions. IEEE Trans., Microw. Theory Tech. 48, 158–161 (2000)
[17] Safwat, A.M.E., Zaki, K.A., Johnson, W., et al.: Novel transition between different configurations of planar transmission lines. IEEE Microw. Wireless Comp. Lett. 12, 128–130 (2002)
[18] Hong, J., Wang, B., Zhang, Y.: Novel distributed circuit parameters model for coplanar waveguide to microstrip transition. In: Proc. APSI 2004 Symp., vol. 1, pp. 994–997 (2004)
[19] Wiatr, W., Walker, D.K., Williams, D.F.: Coplanar-waveguide-to-microstrip transition model. In: 2000 IEEE MTT-S Dig., pp. 1797–1800 (2000)
[20] Han, L., Wu, K., Hong, W., et al.: Compact and broadband transition of microstrip line to finite-ground coplanar waveguide. In: Proc. 38th Eur. Microw. Conf., pp. 480–483 (2008)
[21] Kim, Y.-G., Kim, K.W., Cho, Y.-K.: An ultra-wideband microstrip-to-CPW transition. In: IEEE MTT-S Int. Symp. Dig., pp. 1079–1082 (2008)
[22] Hsu, H.-H., Peroulis, D.: A micromachined high-Q microstrip line with a broadband microstrip-to-CPW transition. In: Proc. IEEE Topical Meeting Silicon Monolithic Integr. Circuits in RF Systems, SiRF 2009, pp. 1–4 (2009)
[23] Margomenos, A., Lee, Y., Katehi, L.P.B.: Wideband Si micromachined transitions for RF wafer-scale packages. In: Proc. 2007 Topical Meeting Silicon Monolithic Integ. Circ. in RF Systems, pp. 183–186 (2007)
[24] Bulja, S., Mirshekar-Syahkal, D., Yazdanpanahi, M.: Novel wide-band transition between finite ground coplanar waveguide (FGCPW) and balanced stripline. In: Proc. 4th Eur. Microw. Integrated Circ. Conf., pp. 301–303 (2009)
[25] Simons, R.N., Dib, N.I., Katehi, L.P.B.: Coplanar stripline transition. El. Lett. 31(20), 1725–1726 (1995)
[26] Suh, Y.-H., Chang, K.: A wideband coplanar stripline to microstrip transition. IEEE Microw. Wireless Comp. Lett. 11, 28–29 (2001)
[27] Tu, W.-H., Chang, K.: Wide-band microstrip-to-coplanar stripline/slotline transitions. IEEE Trans., Microw. Theory Tech. 54, 1084–1089 (2006)
[28] Lim, T.B., Zhu, L.: Compact microstrip-to-CPS Transition for UWB application. In: Proc. 2008 IEEE MTT-S Int. Microw. Workshop Series on Art of Miniaturizing RF and Microw. Passive Comp., pp. 153–156 (2008)
[29] Kim, Y.-G., Woo, D.-S., Kim, K.W., et al.: A new ultra-wideband microstrip-to-CPS transition. In: Proc. IEEE/MTT-S Int. Symp., pp. 1563–1566 (2007)
[30] Kim, Y.-G., Kim, K.W., Cho, Y.-K.: A planar ultra-wideband balanced doubler. In: 2008 MTT-S Int. Microw. Symp. Dig., pp. 1243–1246 (2008)
[31] Kim, Y.-G., Kim, I.-B., Song, S.-Y., et al.: Ultra-wideband surface-mountable double-balanced multiplier. In: Proc. 40th Eur. Microw. Conf., pp. 308–311 (2010)
[32] Wu, P., Wang, Z., Zhang, Y.: Wideband planar balun using microstrip to CPW and microstrip to CPS transitions. El. Lett. 46(24), 1611–1613 (2010)
[33] Prieto, D., Cayrou, J.C., Cazaux, J.L., et al.: CPS structure potentialities for MMICS: a CPS/CPW transition and a bias network. In: 1998 IEEE MTT-S Symp. Dig., pp. 111–114 (1998)

[34] Mao, S.-G., Hwang, C.-T., Wu, R.-B., et al.: Analysis of coplanar waveguide-to coplanar stripline transitions. IEEE Trans., Microw. Theory Tech. 48, 23–29 (2000)

[35] Lin, Y.-S., Chen, C.H.: Novel lumped-element uniplanar transitions. IEEE Trans., Microw. Theory Tech. 49, 2322–2330 (2001)

[36] Kim, S., Jeong, S., Lee, Y.-T., et al.: Ultra-wideband (from DC to 110 GHz) CPW to CPS transition. El. Lett. 38(13), 622–623 (2002)

[37] Anagnostou, D.E., Morton, M., Papapolymerou, J., et al.: A 0-55-GHz coplanar waveguide to coplanar strip transition. IEEE Trans., Microw. Theory Tech. 56, 1–6 (2008)

[38] Ryzhii, M., Ryzhii, V. (eds.): Physics and Modeling of Tera- and Nanodevices. Imperial College Press (2008)

[39] Lau, J.H., Lee, S.W.R.: Microvias for Low Cost, High-density Interconnects. McGraw-Hill (2001)

[40] Goldfarb, M.E., Pucel, R.A.: Modeling via hole grounds in microstrip. IEEE Microw. Guided Wave Lett. 1, 135–137 (1991)

[41] Zhu, J., Yu, Y., Yang, N., et al.: Micromachined silicon via-holes and interdigital band-pass filters. Microsyst. Technol. 12, 913–917 (2006)

[42] Kretly, L.C., Barbin, S.E.: Planar inductor for RFICs surrounded by metallic vias forming a cavity-backed structure improving isolation from circuitry. PIERS Abstracts, Moscow, Russia, August 18-21 (2009)

[43] Wu, B., Gu, X., Tsang, L., et al.: Electromagnetic modeling of massively coupled through silicon vias for 3D interconnects. Microw. Opt. Techn. Lett. 53, 1204–1206 (2011)

[44] Lahiji, R.R., Herrick, K.H., Lee, Y., et al.: Multiwafer vertical interconnects for three-dimensional integrated circuits. IEEE Trans., Microw. Theory Tech. 54, 2699–2706 (2006)

[45] Kim, J., Lee, H.-Y., Itoh, T.: Novel microstrip-to-stripline transitions for leakage suppression in multilayer microwave circuits. In: Proc. IEEE 7th Topical Meeting El. Performance of El. Packaging, pp. 252–255 (1998)

[46] Kwon, D.-H., Kim, Y.: A wideband vertical transition between co-planar waveguide and parallel-strip transmission line. IEEE Microw. Wireless Comp. Lett. 15, 591–593 (2005)

[47] Zhong, X., Wang, B.-Z., Wang, H.: Full-wave analyses for vertical interconnections of shielded coplanar waveguide. Int. J. Infrared Milli. Waves 22, 1683–1694 (2001)

[48] Bulja, S., Mirshekar-Syahkal, D.: Novel wideband transition between coplanar waveguide and microstrip line. IEEE Trans., Microw. Theory Tech. 58, 1851–1857 (2010)

[49] Dubuc, D., De Raedt, W., Carchon, G., et al.: MEMS-IC integration for RF and millimeterwave applications. In: Proc. 13th GAAS Symp., Paris, pp. 529–532 (2005)

[50] Kouzaev, G.A., Deen, M.J.: 3D-integrated circuits. Part. 1. High-frequency 3D-hybrid ICs. An analytical review. Microelectronics Res. Lab. Report, McMaster University, Canada (2002)

[51] Kouzaev, G.A.: Balanced slotted line. In: Gvozdev, V.I., Nefedov, E.I. (eds.) Microwave Three-Dimensional Integrated Circuits, pp. 45–50. Nauka Publ., Moscow (1985) (Invited Chapter, in Russian)

[52] Gvozdev, V.I., Kouzaev, G.A., Nefedov, E.I.: Balanced slotted line. Theory and experiment. Radio Eng. Electron. Physics (Radiotekhnika i Elektronika) 30, 1050–1057 (1985)

[53] Gvozdev, V.I., Kouzaev, G.A., Nefedov, E.I., Utkin, M.I.: Electrodynamical calculation of microwave volume integrated circuit components based on a balanced slotted line. J. Commun. Techn. Electronics (Radiotekhnika i Electronika) 33, 39–43 (1989)
[54] Lin, T.-H.: Via-free broadband microstrip to CPW transition. El. Lett. 37, 960–961 (2001)
[55] Jackson, R.W., Matolak, D.W.: Surface-to-surface transitions via electromagnetic coupling of coplanar waveguides. IEEE Trans., Microw. Theory Tech. 35, 1027–1032 (1987)
[56] Yang, H.-Y., Alexopoulos, N.G.: Basic blocks for high-frequency interconnects: theory and experiment. IEEE Trans., Microw. Theory Tech. 36, 1258–1264 (1988)
[57] Carpentier, J.F., Pribetich, P., Kennis, P.: Quasi-static analysis of skewed transmission lines for MMIC applications. IEE Proc., Microw. Antennas Propag. 141, 246–252 (1994)
[58] Mehdipour, A., Kamarei, M.: Optimization of the crosstalk of crossing microstrips in multilayered media using lumped circuit model. In: Proc. 11th Int. Conf. Math. Methods in Electromagnetic Theory, Karkiv Ukraine, June 26-29, pp. 279–281 (2006)
[59] Shelton, P.J.: Impedances of offset parallel coupled strip transmission line. IEEE Trans., Microw. Theory Tech. 14, 7–15 (1966)
[60] Gvozdev, V.I., Kouzaev, G.A., Nefedov, E.I.: Filters on multilayered microwave integrated circuits for antennas applications. In: Proc. Conf. Design and Computation of Strip Transmission Line Antennas, Sverdlovsk, Russia, pp. 72–76 (1982) (in Russian)
[61] Kuhn, S.A., Kleiner, M.B., Thewes, R., et al.: Vertical signal transmission in three-dimensional integrated circuits by capacitive coupling. In: Proc. ISCAS 1995, 1995 IEEE Int. Symp. Circuits and Systems, April 28 - May 3, vol. 1, pp. 37–40 (1995)
[62] Abbosh, A.M.: Ultra wideband vertical microstrip-microstrip transition. IET Microw. Antennas Propag. 1, 968–972 (2007)
[63] Burke, J.J., Jackson, R.W.: Surface-to-surface transition via electromagnetic coupling of microstrip and coplanar waveguide. IEEE Trans., Microw. Theory Tech. 37, 519–525 (1989)
[64] Zhu, L., Menzel, W.: Broad-band microstrip-to-CPW transition via frequency-dependent electromagnetic coupling. IEEE Trans., Microw. Theory Tech. 52, 1517–1522 (2004)
[65] Strauss, G., Menzel, W.S.: Millimeter-wave monolithic integrated circuit interconnects using electromagnetic field coupling. IEEE Trans., Comp., Pack., Manuf. Techn., Part B 19, 278–282 (1996)
[66] Chiu, T., Shen, Y.-S.: A broadband transition between microstrip and coplanar stripline. IEEE Microw. Wireless Comp. Lett. 13, 66–68 (2003)
[67] Abbosh, A.M.: Multioctave microstrip-to-coplanar waveguide vertical transition. Microw. Opt. Techn. Lett. 53, 187–188 (2011)
[68] Menzel, W., Kassner, J.: Novel techniques for packaging and interconnects in mm-wave communication front-ends. In: Proc. Packaging and Interconnects at Microwave and mm-Wave Frequencies (Ref. No. 2000/083) IEE Seminar, June 26, pp. 1/1–1/18 (2000)
[69] Vaughan, R.G.: Two-port higher mode circular microstrip antennas. IEEE Trans. Antennas Propag. 36, 309–321 (1988)

[70] Waterhouse, R.B., Targonski, S.D.: Performance of microstrip patches incorporating a single shorting post. In: Proc. Antennas Propag. Soc. Int. Symp., 1996, AP-S Dig., July 21-261, vol. 1, pp. 29–32 (1996)

[71] Clavijo, S., Dias, R.E., McKinzie, W.E.: Design methodology for Sievenpiper high-impedance surfaces: an artificial magnetic conductor for positive gain electrically small antennas. IEEE Trans. Antennas Propag. 51, 2678–2690 (2003)

[72] Lin, S.-C., Wang, C.-H., Chen, C.H.: Novel patch-via-spiral resonators for the development of miniaturized bandpass filters with transmission zeros. IEEE Trans., Microw. Theory Tech. 55, 137–146 (2007)

[73] Razalli, M.S., Ismail, A., Mahdi, M.A., et al.: Compact configuration ultra-wideband microwave filter using quarter-wave lengthy short-circuited stub. In: Proc. APMC 2007, pp. 1–4 (2007)

[74] Razalli, M.S., Ismail, A., Mahdi, M.A., et al.: Compact ultra-wide band microwave filter utilizing quarter-wave length short-circuited stubs with reduced number of vias. Microw. Opt. Techn. Lett. 51, 2116–2119 (2009)

[75] Razalli, M.S., Ismail, A., Mahdi, M.A.: Novel compact "via-less" ultra-wide band filter utilizing capacitive microstrip patch. PIER 91, 213–227 (2009)

[76] Nath, J., Soora, S.: Design and optimization of coax-to-microstrip transition and through-hole signal via on multilayer printed circuit boards. In: Proc. 37th Eur. Microw. Conf., pp. 134–137 (2007)

[77] Xu, J., Sun, C., Xiong, B., et al.: Resonance suppression of grounded coplanar waveguide in submount for 40Gb/s optoelectronic modules. J. Infrared Milli Terahz Waves 30, 103–108 (2009)

[78] Luna-Rodríguez, J.-J., Martín-Díaz, R., Hernández-Igueño, M., Varo-Martínez, et al.: Simul-EMI II: An Application to Simulate Electric and Magnetic Phenomena in PCB Designs. In: García-Pedrajas, N., Herrera, F., Fyfe, C., Benítez, J.M., Ali, M. (eds.) IEA/AIE 2010. LNCS, vol. 6096, pp. 489–498. Springer, Heidelberg (2010)

[79] Mahajan, M., Khah, S.K., Chakarvarty, T., et al.: Computation of resonant frequency of annular microstrip antenna loaded with multiple shorting posts. IET Microw. Antennas Propag. 2, 1–5 (2008)

[80] Kamgaing, T., Ramahi, O.M.: Multiband electromagnetic-bandgap structures for applications in small form-factor multichip module packages. IEEE Trans., Microw. Theory Tech. 56, 2293–2300 (2008)

[81] Hettak, K., Dib, N., Sheta, A., et al.: New miniature broad-band CPW-to-slotline transitions. IEEE Trans., Microw. Theory Tech. 48, 131–146 (2000)

[82] Kouzaev, G.A., Nikolova, N.K., Deen, M.J.: Circular-pad via model based on cavity field analysis. IEEE Microw. Wireless Lett. 13, 481–483 (2003)

[83] Kouzaev, G.A., Deen, M.J., Nikolova, N.K., Rahal, A.: Influence of eccentricity on the frequency limitations of circular-pad via-holes. IEEE Microw. Wireless Lett. 14, 265–267 (2004)

[84] Kouzaev, G.A., Deen, M.J., Nikolova, N.K., Rahal, A.: Cavity models of planar components grounded by via-holes and their experimental verification. IEEE Trans., Microwave Theory Tech. 54, 1033–1042 (2006)

[85] Derneryd, A.G.: Analysis of the microstrip disk antenna element. IEEE Trans., Antennas Propag. 27, 660–664 (1979)

[86] Okoshi, T.: Planar Circuits for Microwaves and Light Waves. Springer (1985)

[87] Lin, Y., Shafai, L.: Characteristics of concentrically shorted circular patches. IEE Proc., Part H 137, 18–24 (1990)

[88] Bhattacharyya, A.K.: Characteristics of space and surface-waves in a multilayered structure. IEEE Trans., Antennas Propag. 38, 1231–1238 (1990)
[89] Zhang, S.-C., Thumm, M.: Eigenvalue equations and numerical analysis of a coaxial cavity with misaligned inner rod. IEEE Trans., Microw. Theory Tech. 48, 8–14 (2000)
[90] Zhang, H.-B., Lai, Y.-X., Zhang, S.-C.: Eigenfrequency analysis of a coaxial cavity with large eccentricity. In: Proc of APMC 2005, vol. 5 (2005)
[91] Maeda, S., Kashiwa, T., Fukai, I.: Full wave analysis of propagation characteristics of a through hole using the finite-difference time-domain method. IEEE Trans., Microw. Theory Tech. 39, 2154–2159 (1991)
[92] Casares-Miranda, F.P., Viereck, C., Camacho-Penalosa, C., et al.: Vertical microstrip transition for multilayer microwave circuits with decoupled passive and active layers. IEEE Microw. Wireless Comp. Lett. 16, 401–403 (2006)
[93] Gu, Q., Yang, Y.E., Tassoudji, M.A.: Modeling and analysis of vias in multilayered integrated circuits. IEEE Trans., Microw. Theory Tech. 41, 206–214 (1993)
[94] Abhari, R., Eleftheriades, G.V., van Deventer-Perkins, E.: Physics-based CAD models for the analysis of vias in parallel-plate environments. IEEE Trans., Microw. Theory Tech. 49, 1697–1707 (2001)
[95] Otto, D.V.: The admittance of cylindrical antennas driven from a coaxial line. Radio Sci. 2, 1031–1042 (1967)
[96] Gu, Q., Tassoudji, A., Poh, S.Y., et al.: Coupled noise analysis for adjacent vias in multilayered digital circuits. IEEE Trans., Circ. Systems-I: Fundamental Theory Appl. 41, 796–804 (1994)
[97] Wang, T., Harrington, R.F., Mautz, J.R.: Quasi-static analysis of a microstrip via through a hole in a ground plane. IEEE Trans., Microw. Theory Tech. 36, 1008–1013 (1988)
[98] Oo, Z.Z., Liu, E.-X., Li, E.-P., et al.: A semianalytical approach for system-level electrical modeling of electronic packages with large number of vias. IEEE Trans., Adv. Pack. 31, 267–274 (2008)
[99] Rimolo-Donaldo, R., Gu, X., Kwark, Y.H., et al.: Physics-based via and trace models for efficient link simulation on multilayer structures up to 40 GHz. IEEE Trans., Microw. Theory Tech. 57, 2072–2083 (2009)
[100] Zhang, Y.-J., Fan, J.: An intrinsic circuit model for multiple vias in an irregular plate pair through rigorous electromagnetic analysis. IEEE Trans., Microw. Theory Tech. 58, 2251–2265 (2010)
[101] Tsang, L., Chen, H., Huang, C.C., et al.: Modeling of multiple scattering among vias in planar waveguides using Foldy-Lax equations. Microw. Opt. Techn. Lett. 31, 201–208 (2001)
[102] Chen, H., Li, Q., Tsang, L.T., et al.: Analysis of a large number of vias and differential signaling in multilayered structures. IEEE Trans., Microw. Theory Tech. 51, 818–829 (2003)
[103] Huang, C.-C., Tsang, L., Chan, C.H., et al.: Multiple scattering among vias in planar waveguides using preconditioned SMGG method. IEEE Trans., Microw. Theory Tech. 52, 20–28 (2004)
[104] Seijas-Garcia, S.C., Romo, G., Torres-Torres, R.: Impact of the configuration of ground vias on the performance of vertical transitions used in electronic packages. In: Proc. 2009 IEEE MTT-S Int. Microw. Workshop Series on Signal Integrity and High-Speed Interconnects (IMWS2009-R9), Guadalajara, Mexico, February 19-20, pp. 17–20 (2009)

8 Integrated Filters and Power Distribution Circuits

Abstract. In this Chapter, the monolithic integrated filters, power dividers, directional couplers and baluns for millimeter-wave frequencies are reviewed. It is shown that the increased real-estate of them can be partly resolved by the 3-D integration, meandering of conductors, and using periodically patterned ground planes as the photonic band gap structures for reducing the effective wavelength. The increased loss and poor selectivity of the monolithic distributed filters can be improved by using better dielectrics, copper conductors of increased thickness, multilayered ground planes, etc. References -82. Figures -3. Pages -17.

8.1 Monolithic Integrated Filters

One of the important problems in millimeter-wave silicon integration is the development of miniature frequency selective circuits [1]-[5]. Filters reject or absorb signals of certain frequencies; while others are passing them with minimal loss. Most of them consist of resonant sections, and the low-frequency analogs of them are the resonant loops. In microwave electronics, filters are for the frequency band selection, channel separation, and selective matching of devices or sub-circuits. There are several types of filters, and they are the band-pass, stop-band, low- and high-frequency ones. Many of them are the two-port devices, but circuits arranging signals into several outputs according their frequencies are in use, and they are called the multiplexors. For characterization of the filters, the S-matrix formalism is used and the additional engineering parameters:

- Specific frequencies for each type of filters (center frequency for pass-band and stop-band filters, relative frequency band, and cut-off frequencies)
- VSWR estimated at the each port
- Insertion loss $A = 20\lg\left(\frac{1}{|S_{21}|}\right)$, dB
- Maximal loss in the pass-band
- Minimal loss in the stop band
- Loss pulsation in the pass-band.

Design of microwave filters is calculation of geometry of the filter elements and their connections and couplings providing the characteristics which are close to

G.A. Kouzaev: Applications of Advanced Electromagnetics, LNEE 169, pp. 341–357.
springerlink.com

those given by a customer. Unfortunately, any well-developed EM theory of the filter design is not known, and the similarity with the low-frequency resonant loop filters is used. The idea is to find a prototype filter composed of capacitors and inductors which provide the frequency response close to the wanted. Then the geometries of distributed components are found which provide the same response at a narrow frequency band near the filter's center frequency. As a rule, such technique needs following EM optimization or tuning of the designed filters. More detailed information on the design of distributed microwave filters is from [1]-[5], for instance.

There are many obstacles to reach the needed frequency characteristics in monolithically integrated filters. Among them are the increased loss of silicon substrates and thin-film conductors, insufficient accuracy of technology, and parasitic coupling of high-dense integration elements. Today the activity of researchers and engineers is concentrated on several ways of solving the above-mentioned problems. One study is on the design of resonators and filters of increased frequencies using integrated capacitors and inductors, and the main problem here is the low parasitic resonant frequencies of inductors and their low Q-factors [6]. Many techniques have been developed, including the 3-D shaping of inductors, the isolating components from the lossy substrate by the solid and patterned grounding planes, applications of ferrite layers to increase the inductivity, etc. The design of these filters can be performed by techniques known for decades [1]-[3], and many commercial circuit simulators have the packages or detailed instructions on the calculation of filters of this type. Unfortunately, the parasitics of capacitors, inductors, and interconnects can seriously distort the filter characteristics, and optimization of their layout is required using realistic models of these integrated components. In some cases, the EM optimization can be performed which allows to calculate the filters with improved accuracy.

At increased frequencies, the length of distributed resonators starts to be comparable with the one of the millimeter-wave chips and the developments of resonators and distributed filters of increased selectivity are the main direction of the research in this field. Here the known topologies of the distributed printed filters can be used [1],[4]-[5]. Their design consists in the calculation of idealized lumped component prototype filters providing the needed frequency response, and finding the filter distributed topology which have similar characteristics with the earlier calculated prototype.

Many papers are for the design of distributed filters using standard CMOS technologies dealing with the low-resistivity silicon substrates $(\rho = 5-15\ \Omega\text{-cm})$. In this case, the lossy substrate is shielded by a solid or patterned conducting layer serving as the ground layer for thin-film microstrips. Very often, the thin layer M1 is used which height is 0.5 μm only, and it causes increased resistivity and some effects associated with this non-ideal grounding.

For instance, a meander line pass-band filter is realized in [7] using standard 180-nm CMOS technology. A meandered strip (M6 level, thickness 2 μm) is coupled to another meandered resonator (M5 level) through a silicon dioxide layer of the thickness 6.7 μm, and the structure is shielded from the lossy silicon by a ground layer (M1, thickness 0.5 μm). The designed circuit shows filtering properties at 41.5 GHz with the insertion loss 2.5 dB. Suppression loss reaches 36 dB at 100 GHz. The estimated quality factor of this resonator is about Q=19.5, and it depends highly which metallization layer is used for the resonator conductors.

In [8], the same technology is used to manufacture the 60- and 65-GHz 2-resonator filters. Each resonator has a shape of a rectangular sine function. They are coupled to each other being placed close in the area of their ¼ length arms. The input/output arms are on the M5 level, and they are tightly coupled to the resonators which are exactly below their arms. Additionally to meandering of resonators, the stacking allows for decreasing the overall length of these two-resonator filters, which is only 790 μm. The designed filters show 51.8% and 51.3% bandwidth, correspondingly, with the minimum insertion loss 3.9 and 2.7 dB at the center frequencies.

Further improvements of selectivity of filters manufactured by standard CMOS technology are available, using, for instance, the dual-mode resonators known from the waveguide and printed strip technologies. Some circuits are designed and studied in [9],[10] where a dual-mode square-shape microstrip resonator is folded further into a Π -like shape to decrease its space. These resonators are analyzed using the equivalent circuit theory and a full-wave software tool, and the results of simulations are compared to the measurement data. The 3-dB bandwidth of one of them is of 18 GHz at 70 GHz [10]. The insertion loss is 3.6 dB. The rejection level is greater than 35 dB within 2-53 GHz and 95-110 GHz. The proposed filter can be used in the 70-GHz RF transceivers.

Several other dual-mode bandpass filters are considered in [11]. They are realized using a low-loss $\left(\tan\delta = 2\cdot 10^{-3}\right)$ 25-μm BCB substrate placed on a thick gold layer for shielding the filters from the lossy silicon. Depending on the geometry of microstrip line, the filters show the bandwidth of 2.4-11.27% at the center frequency 60 GHz. The maximum size of the filters is circa 1494×1280 μm^2, and it is reduced due to the dual mode filtering. In [11] a CPW filter for the same frequency is designed, studied, and compared as well.

The selectivity of bandpass filters can be improved further staying with the standard 180-nm CMOS technology. For instance, some filters on coupled parallel microstrip resonators are considered in [12] where again the signal conductors are placed on the M6 level. The lossy silicon is shielded by a grounding layer placed on the M1 level. The designed filters are modeled using the Agilent Momentum and Sonnet (Sonnet Software Inc.), which show similar results close to the measurements. The designed 60-GHz filter has insertion loss of 9.3 dB with a 3-dB bandwidth of 6.2 GHz. The designed 77-GHz filter has a 3-dB bandwidth of 7.7 GHz with the insertion loss of 9.3 dB. The thin conductors and thin ground layer according to the opinion of the authors of the cited paper cause this increased loss.

The loss can be reduced using copper as a conductor material. For instance, in [13], an interdigital 5-resonator filter is described where the ground layer is made of patterned copper. The strip conductors made of aluminum are placed on the highest metallization level and their length at 55 GHz is less than 600 μm. Unfortunately, the technology-caused patterned ground allows for penetration of the field into the lossy silicon substrate, and the resonator projection areas in silicon are made of increased impedance, which prevents loss and parasitic coupling of resonators due to the eddy currents. A design methodology of these filters is considered, and the results of the full-wave simulations and measurements are given. A 5-order symmetric interdigital bandpass filter manufactured using the IBM 130-nm CMOS technology shows the mid-band frequency of 55.3 GHz with a fractional bandwidth of 3.25%. The insertion loss is close to 4.5 dB, while the return loss is 13 dB.

Several filters for 60-GHz applications are designed in [14] by the same authors where the 130-nm CMOS technology is used. Similarly to the above-cited paper, the ground layer is of copper, and the used standard technology does not allow for solid copper layer. Instead, these authors use a patterned one. The filters are designed as the coupled open-loop resonators, and these two-loop filters show an insertion loss 1 dB with the bandwidth 9 GHz. Good correspondence of simulation results is found with the measurements in 1-110-GHz band.

A 60-GHz filter manufactured using a 90-nm Al/Cu CMOS technology is considered in [15] where miniaturization of the filter is performed by meandering of a dual-mode rectangular ring microstrip resonator and using additional capacitors connected to the corners of this resonator. The loss decreasing is achieved by using the thickest M9 aluminum layer and patterned copper ground. The filter bandwidth on the level of 3 dB is close to 15 GHz at 60-GHz center frequency.

Additionally to the thin-film microstrips, the CPWs are used in millimeter-wave filters, and they require high-resistivity $(\rho > 1\ \text{k}\Omega\text{-cm})$ or selectively ion-implanted substrates with $\rho \sim 10^6\ \Omega\text{-cm}$. In this case, the results on dielectric-substrate or GaAs CPWs filters can be used taking into account the specifics of the CMOS technology [16]. For instance, the size of resonators can be decreased using the high-quality-factor inductors on elevated conductors showing the increased resonance frequencies and reduced loss [17].

For instance, some results on the CPW filter designs are given in [18] where a 130-nm SOI CMOS technology is used to manufacture the CPW filters. One of the filters consists of a CPW with the connected symmetrically shorted CPW stubs. It shows a band-pass response in 60-80 GHz with the transmission coefficient about -2.01 dB, which is close to the one (-1.8 dB) provided by the same-geometry GaAs filter. In this same work, a technique on the design of a narrow-band filter is considered, too. Two factors are important in this case, namely, the conductor loss and technology accuracy. It is proposed to stack all six layers of integration to form a multilayer CPW. One layer is assigned for the realization of bridges, which prevents the excitation of the unwanted higher-order quasi-TEM mode in CPW. The studied filter consists of three coupled CPW resonators. Two of them are connected at their centers to the input/output CPWs. The designed structure provides a 5% bandwidth at the frequency which is close to 60 GHz. Due to technology inaccuracy, the shift of the frequency band is close to 5% which leads to some return loss degradation. In spite of it, the insertion loss is about 6.6 dB, which is the encouraging result for further research and design of silicon-based narrow-bandwidth filters.

In 2001, a new technology was developed allowing selective increase of the silicon resistivity up to $\rho \sim 10^6$ Ω-cm by proton bombarding of a substrate through windows in a mask [19],[20]. It allows for the essential decrease of the loss of transmission lines and resonators. For instance, the 50-Ω-CPW's loss at 100 GHz is about 6 dB/cm, according to the performed measurements. The transmission line is realized using the 4-μm thick Al conductors placed over the 1.5-μm silicon dioxide layer covering the silicon substrate.

Several CPW bandpass and stop-band filters are manufactured, modeled, and measured using the mentioned technology. For example, a 40-GHz filter shows -3.4 dB loss with the bandwidth of 9 GHz, and it is an essential improvement regarding to the filter on conventional Si/SiO_2 substrate [19].

A bandpass filter for 91 GHz is designed and studied in [20] using the same technology and substrate as in [19]. The overall length of the CPW based filter is close to 420 μm. It shows only 1.6 dB transmission loss at 91 GHz. In the same paper, a bandstop filter is studied, too. Additionally to the measurements, the IE3D (Mentor Graphics Corp.) simulations are performed, and the equivalent circuits for these filters are considered. The overall results show that the ion bombarding allows for improving of the filter parameters, as it was validated by measurements and simulations in frequencies of 22-91 GHz.

Many results are published on micromachined millimeter-wave filters. As was mentioned, the technology, born in 1991 [21], allows for the cavities in silicon covered by a conducting layer to shield the signal conductors from the lossy substrate. These cavities can be the air- or a low-loss dielectric filled, and the CPWs and microstrip lines are realized, including the dielectric thin-membrane supported ones.

Ones of the first filters on a membrane supported CPW are described in [22] where the CPW conductors are supported by a 1.5-μm thick dielectric layer. The used waveguide shows only 0.03-dB/cm loss at 35 GHz, and a designed bandpass filter consisting of three open-end series stubs demonstrates the insertion loss around 1 dB/cm in 22-32 GHz, which is comparable with the best waveguide bandpass filters on suspended striplines. Practically at the same time, this technology is used for the design of a 250-GHz bandpass filter consisting of four sections of the open-end series stubs [23]. These filters are simulated by a full-wave software tool, and they are measured in 130-360 GHz. The filter bandwidth is 58% with the insertion loss 1.5 dB. Up to now, many designs of filters based on the mentioned technology are known, and one of them, for instance, is a low-pass filter with the corner frequency 120 GHz published in [24]. A narrow-bandwidth 5.3% filter on low-loss membrane CPW (0.05-1 dB/cm) is described in [25], where several shunt-stubs filters are modeled and measured. The registered loss of a 94-GHz filter is about 6.46 dB with the return loss better than 20 dB. Further reading on these components based on different membranes and at frequencies 53, 95, and 150 GHz are from [26]-[27].

The ideas of bulk and surface micromachining are very fruitful in millimeter-wave technology, and many variants of micromachining technology have been proposed which provide low cost of components and their increased resistance towards mechanical vibrations and shocks. One of them is the surface micromachining of BCB layers placed over a thick gold layer covering a lossy silicon substrate, [30]. The cross-sections of these lines are trapezoidal due to the used anisotropic etching. As was mentioned, the regular microstrip and CPW lines of this type demonstrate a reduced loss which is close to 3-6 dB/cm in 94-220 GHz. Some millimeter-wave filters on these lines are described in [28],[31] where it is demonstrated that the 5% 3-dB bandwidth filters show the loss 4.64 dB and 6.968 dB at 50 and 94 GHz, correspondingly. These filters consist of four symmetrical microstrip open stubs connected to each other by microstrip lines. The line geometry is found using the equivalent network theory and simulations performed using the Agilent ADS and its Momentum simulation package. For instance, the thickness of BCB layers 20 μm is defined to be optimal for the microstrips to decrease the insertion loss.

8.2 Power Dividers, Directional Couplers, and Baluns

As known, many components of waveguide devices and hybrid integrated circuits are based on the interference and diffraction effects, and they employ the elements which size is comparable with the wavelength [1],[4],[32]. At low microwave frequencies, the wavelength is in the centimeter or millimeter range, and the size of such microwave components is large. Practically, semiconductor active circuits can perform all functions of these distributed components, but they increase the noise and require additional power. The lumped equivalent circuits composed of integrated elements can substitute the distributed components. We review only some works on this design, touching only the low frequency components, and we suppose that this idea is still applicable at increased frequencies.

8.2.1 Power Dividers and Directional Couplers on Discrete Integrated Components

At low frequencies of the considered range 30-100 GHz, the distributed power dividers (Fig 8.1a), directional couplers (Fig. 8.2a), and hybrids can be realized using discrete integrated circuits [33],[34]. Each $\lambda/4$ -length of a transmission line is substituted by its equivalent π-circuit. Then a device equivalent circuit is modified to decrease the number of reactive components and its connections to the ground.

Fig. 8.1a shows an equivalent circuit of Wilkinson power divider. Here $C = 1/\sqrt{2}\,\omega_0 Z_c$, $L = \sqrt{2}Z_c/\omega_0$, and ω_0 is the center cyclic frequency [33]. Its idealized S-matrix (8.1) defined at this center frequency shows that this 3-port circuit is matched, and the incoming power from the input 1 is divided between the ports 2 and 3 at an equal portion. The equal phase signals coming from these ports are summed at the output 1. The signals, which are out of phase, loose a part of their power in the balancing load R:

$$S = \begin{pmatrix} 0 & -j/\sqrt{2} & -j/\sqrt{2} \\ -j/\sqrt{2} & 0 & 0 \\ -j/\sqrt{2} & 0 & 0 \end{pmatrix}. \quad (8.1)$$

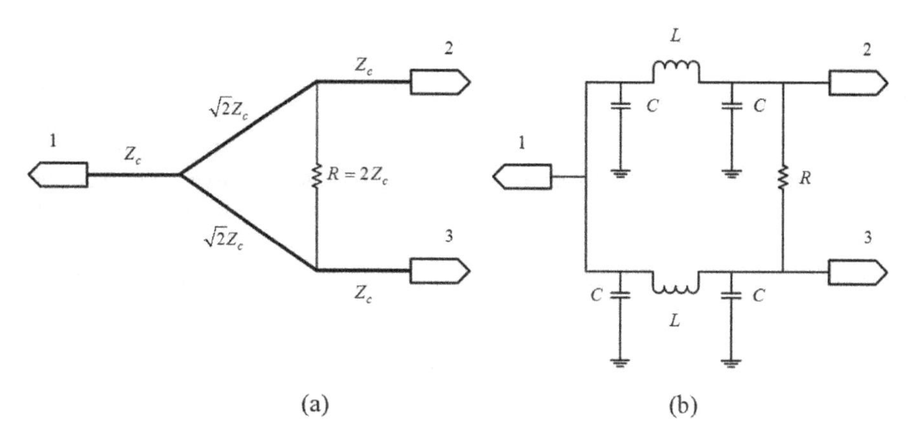

Fig. 8.1 Wilkinson power divider (a) and its narrow-band equivalent circuit (b)

Directional couplers are the devices to route a microwave signal to the prescribed ports while others stay isolated. One of them, called the branch-line directional coupler, is composed of $\lambda/4$-length lines, and it is shown in Fig. 8.2a with its equivalent circuit (Fig. 8.2b).

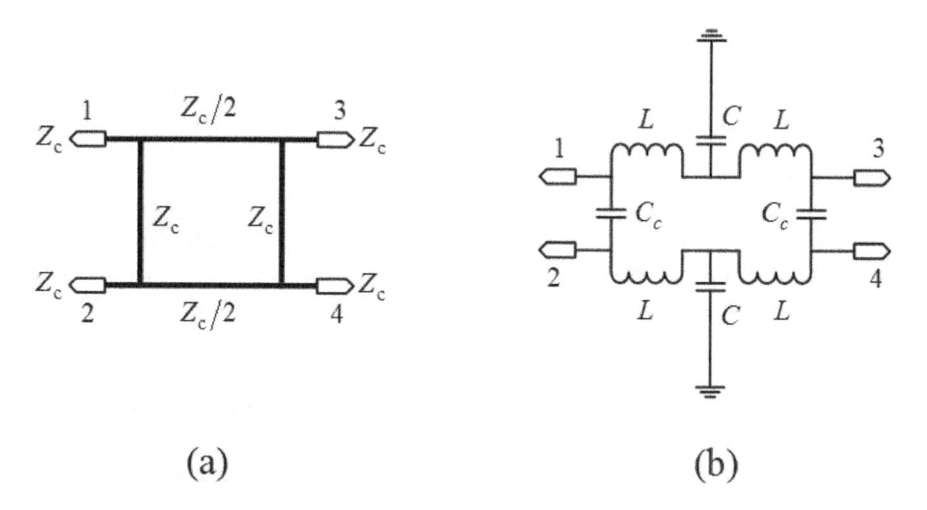

Fig. 8.2 Branch-line directional coupler (a) and its narrow-band equivalent circuit (b)

In this figure, $C = 1/\omega_0 Z_c$, $L = Z_c/\omega_0$ and $C_c \approx (1/\omega Z_c)\left[10^{-C_F/20}\right]$. Here C_F is the coupling factor of the coupler, expressed in dB. The equivalent circuit can be modified to decrease the number of space-consuming inductors, and a network consisting of two inductors and four capacitors is available. Its idealized S-matrix written at the center frequency is

$$S = -\frac{1}{\sqrt{2}}\begin{pmatrix} 0 & j & 1 & 0 \\ j & 0 & 0 & 1 \\ 0 & 1 & j & 0 \end{pmatrix} \tag{8.2}$$

It shows that the directional coupler is matched from all their four ports. The signals, coming from an input port to other two outputs, are divided in an equal portion, and they have a 90° – phase difference.

Additionally to the S-matrix, the following engineering parameters are used for characterization of the directional couplers:

- Coupling loss $A = 10\lg(P_1/P_4)$, dB
- Direct loss $L = 10\lg(P_1/P_2)$, dB
- Directivity $C = 10\lg(P_4/P_3)$, dB
- Isolation $B = 10\lg(P_3/P_1)$, dB
- Each port's VSWR
- Center frequency
- Relative frequency band.

The directional couplers are the analogue devices, and they can provide arbitrary parameters. For convenience, they are classed according to the coupling coefficient A, and the following couplers can be designed with $A = 3,\ 6,\ 10,\ 20,\ 30\ \text{dB}$. For instance, in the case of 3 dB, the input power is divided between the outputs 2 and 4 in the equal ratio. More knowledge on the directional coupler's design can be found from many books on microwave techniques [1],[4].

The considered-above equivalent circuits of the power divider and the branch-line directional coupler (Figs 8.1b and 8.2b) have decreased bandwidth regarding to known distributed prototypes due to rough approximation of transmission lines by their equivalent circuits. Besides, the integrated discrete components have their own parasitics limiting the bandwidth of the designed devices.

8.2.2 *Distributed Power Dividers and Directional Couplers*

The increased millimeter-wave frequencies allow for realizing the distributed power dividers and directional couplers, although their overall size might be comparable with one of the chip. The size reduction is available using the aggressive planar and 3-D meandering of their branches.

In [35],[36], a couple monolithic designs of Wilkinson power dividers are considered. They are using the coplanar waveguides placed on a surface of a polyimide layer covering a low-resistivity $(20\ \Omega\text{-cm})$ silicon. The first of them [35] shows 6.8 dB insertion loss and 14 dB return loss at 20 GHz. The output isolation is about 16 dB at the same frequency. The measurement results are available up to 30 GHz. This design is modified in [36] using a finite ground CPW showing better isolation of the signal conductor from the lossy silicon. Due to that and optimization of the overall geometry, the insertion loss of the designed Wilkinson power divider is only 0.57 dB at the center frequency 12.5 GHz. The isolation of the outputs is better than 15 dB. This relatively low center frequencies cause the length of both dividers greater than 3 mm.

One of the examples of millimeter-wave distributed power dividers manufactured using a commercial SiGe BiCMOS process is considered in [37]. A 1:8 differential power divider is designed for the 45-GHz center frequency using the shielded broadside-coupled striplines.

A single-ended input signal is transformed into a differential one using a differential amplifier. Then it is divided using a passive 1:2 Wilkinson power divider. Each signal from its two outputs is divided again into four streams by a couple of 1:8 dividers equipped by differential amplifiers. It is shown by measurements that the circuits have a 3-dB bandwidth from 37 to 45 GHz. Due to the differential design, it allows for canceling the common mode noise, which is strong at increased frequencies. The considered power divider occupies the space circa $1.12\times1.5\ \text{mm}^2$, and it can be used in couplers, baluns, and matching networks of the millimeter-wave silicon integration.

Another monolithic realization of the integrated distributed Wilkinson power divider is published in [38] where the standard 180-nm SiGe technology is used to manufacture these circuits as parts of a four-element phased-array front-end receiver for the 30-50 GHz frequency band. The center frequency of the designed and tested divider for this *Q*-band phased array front-end is close to 45 GHz. The direct loss is 0.5-0.6 dB, and the isolation of the ports is about 15-20 dB in the frequency band 30-50 GHz. This power divider is manufactured using microstrips in their thin-film realizations, i.e. the signal conductor is placed on the highest and thickest (2.81 μm) metallization level, and it is isolated from the lossy silicon $(\rho = 8\text{-}10\ \Omega\text{-cm})$ by a ground metal layer of the thickness 0.62 μm. The height of the silicon dioxide substrate for this microstrip is 5.59 μm.

An integrated power combiner is described in [39] where four CMOS amplifiers feed one load. This circuit consists of four quarter-wavelength coupled transmission lines, and they form the voltage on a common load. It is noticed that this power combiner is less lossy than the Wilkinson circuits which have matching resistors inside. The circuit is designed for 60 GHz frequency, and it has only 1.09-dB insertion loss in frequencies 55-65 dB with variation of this loss within 0.16 dB, only. These combining circuits allow for employing the low-power CMOS amplifiers for forming the signals with the needed increased energy parameters.

The increased size of integrated power dividers stimulates researchers to find some ways to reduce the real-estate of them. For instance, in [40], a power divider is described where the meandering, patterning of ground layer, and periodical underneath conductor bridges reduce the overall size of a 20-GHz Wilkinson power divider. The signal conductor of this circuit manufactured by a standard 180-nm CMOS technology is placed on the thickest M6 layer. The ground M1 layer is patterned. The bridges are made using the M2-M5 layers providing periodical loads of strip line by capacitors. It improves the slow wave factor of 8.2% compared with the same power divider with the meandered line.

A microrectangular waveguide realization of a matched T-junction power divider is published in [41] for 75-110-GHz frequency band. To manufacture the microwaveguides, the cavities are dry-etched in two blocks of silicon. The surfaces of these two etched blocks are covered by a Ti/Cu/Au layer, and they are fixed to each other forming the waveguides and power dividers. Measurements and simulations show the reflection coefficient from the input -18... - 20 dB in 95-100-GHz frequency band.

Many non-silicon realizations of integrated power dividers are known. One of them is realized on a low-loss laminated liquid crystal polymer substrate of the thickness 50 μm [42]. The geometry of Wilkinson power divider is modified to avoid the sharp bends, and additional short lines are added to the thin-film resistors to compensate the parasitic reactivities of the components. Good correspondence was found between the HFSS simulations and measurements. The power divider occupies the space circa $556\times253\ \mu m^2$. It shows the insertion loss between 3.5-4 dB in 85-95 GHz, and its isolation is about 20 dB in 85-95 GHz.

The micro-coaxial Gysel power dividers manufactured by the Polystrata technology are described in [43] for 30 GHz. In these micro-coaxials, the central conductors are supported by thin dielectric membranes, and the measured loss of this divider is close to 0.22 dB with the reflection coefficient and isolation better than 20 dB in the Ka-band. Thermal analysis performed by the authors of the cited paper shows that the Gysel dividers allow for handling more power than the Wilkinson one.

The directional couplers are very important components of contemporary millimeter-wave circuits. Unfortunately, being distributed, the elements occupy a large space, and the miniaturization of them is a very important problem. An example of a modified branch-line directional coupler shown in Fig. 8.2 is considered in [44] where the quarter-wavelength microstrip lines composing this directional coupler are meandered to decrease the overall length. The conductors of this circuit are placed on the top of a silicon dioxide layer, and ground conductor of the thickness 1 μm isolates the design from a low-resistivity silicon. Unfortunately, the non-zero surface resistance of the ground layer causes additional distortion, and the geometry of grounding layers was optimized to provide the potential which is close to the homogeneous one along the meandered lines. Several designs are simulated using the Ansoft HFSS and the Cadence, and these EM simulations give the best topology showing $|S_{21}|_{dB}=-4$ and -15-dB reflection coefficient at 30 GHz.

A broadband surface-micromachined 15-45-GHz microstrip coupler is designed and studied in [45]. The gold 1.5-μm conductors of coupled microstrips are supported in the air by the metal covered the SU-8 posts of the height 200 μm.

It allows for avoiding the substrate-caused loss and minimization of radiation. The coupler is simulated and optimized using the Ansoft HFSS. The circuit shows acceptable characteristics in the frequency range 15-45 GHz, and it demonstrates 10-dB return loss and 12.5-dB coupling. The transmission loss is within 0.015-1.85 dB in these frequencies. Although the design is built over a soda-lime glass, the used technology is completely compatible with the CMOS one, and the air-lifted directional couplers are promising for the silicon integrations.

Another broadside coupled directional coupler is described in [46]. The design is based on the vertically placed CPWs embedded into a silicon dioxide layer. The ground strips are shorted to each other; meanwhile the signal conductors are placed on different adjacent metal layers to realize the strong coupling in a wide frequency band. All design is over a 10-Ω-cm silicon covered by a layer of dioxide. The directional coupler was manufactured using the National Semiconductor's 180-nm standard digital CMOS technology, measured, and simulated up to 40 GHz. Measurements show 7-10-dB coupling and greater than 10 dB directivity within 10-40 GHz.

A 3-D integration approach is used in [47] where the aggressive planar and 3-D meandering reduces the real-estate of the Lange couplers and rat-race hybrid rings down to acceptable space for the frequencies 20-40 GHz. The first of these studied components is a Lange coupler realized using the broadside coupled strips placed on two layers of metallization and meandered to reduce the required real-estate of the silicon substrate down to 217x185 μm^2. The strip conductors are shielded from the lossy silicon by a ground layer. The design is simulated and optimized using the IE3D EM software tool (Mentor Graphics Corp.). The circuit demonstrates coupling of 3.3-3.7 dB and 12-dB isolation of the ports in frequencies 25-35 GHz.

Another component, described in the same paper, is a multilayered ring hybrid rat-race coupler. Its conventional design is distinguished by its increased real-estate due to the overall length 3/2λ of its ring-like geometry. To decrease the size, its meandered conductors are placed on different layers. This hybrid shows the center frequency about 30 GHz and the isolation of the ports around 17 dB in frequencies 25-35 GHz. It has been shown that further space reducing is possible implementing the multilayered slow-wave substrates.

At the difference to the above-mentioned papers, the authors of [48] use a proposed by them Si-BCB microstrip line [29]. This line is realized on a thick 20-μm BCB $\left(\tan\delta \approx 1.2 \cdot 10^{-2}\right)$ plate placed over a thick (10 μm) gold ground layer isolating the line from the lossy silicon. The designed branch-line couplers have the center frequency close to 180 GHz, and they show loss which is less than 4 dB, and isolation of the corresponding ports is greater than 45 dB at the center frequency.

Another design uses the broad-side coupled strips placed inside a thick BCB layer covering the etched silicon [49],[50]. The first of the coupled lines is a CPW, and the second one is a broadside coupled strip. This design allows for adjusting the modal phase velocities of the odd and even modes with the aim at increasing the isolation of the ports. A 20-GHz design is simulated and measured in frequencies 1-40 GHz, and it demonstrates the 28-dB isolation. The length of 470 μm is reduced twice by the capacitive coupling at the ends of the directional coupler. Unfortunately, it leads to reducing of the isolation better than 14 dB at the center frequency.

Interesting results on miniaturization of hybrid couplers are published in [51] where the use of different transmission lines placed on different layers of integration reduces the size of 81%, as compared to a conventional design. The coupler is built using the meandered CPW and microstrips. The shortening of the branches is realized using capacitive stubs, which are the open-end rectangular coaxial lines. They are buried inside the silicon dioxide layers. The vertical walls of these waveguides are made of the via-hole fences. In the paper, the properties of all lines are analyzed, and their advantages are used properly in the circuit design for 60-GHz applications. This coupler is simulated using the Agilent Momentum, and the apparatus of the equivalent circuits is used for preliminary analysis. At 60 GHz, the return loss and isolation are better than 18 dB. The insertion loss is about of 5.1 dB. It is stated that combinations of different lines and 3-D integration allow for overcoming the most serious problems associated with the silicon technology.

8.3 Integrated Baluns

Baluns are the circuits for conversation of a single-end signal to a pair one of the same magnitude but opposite in sign. Baluns are used for bipolar antennas, differential amplifiers, mixers, filters, etc. There are several circuits performing this function, and some of them have been already considered in Sections 8.2.2 and 8.2.3, namely, the microstrip-CPS and CPW-CPS transitions [52]-[66]. They are realized using the tapered transformations of modes, or they cannot contain any resonant elements being the electrically small [67].

Many other designs are based on the quarter-wavelength $(\lambda/4)$ Marchand balun [68] which is shown in Fig. 8.3.

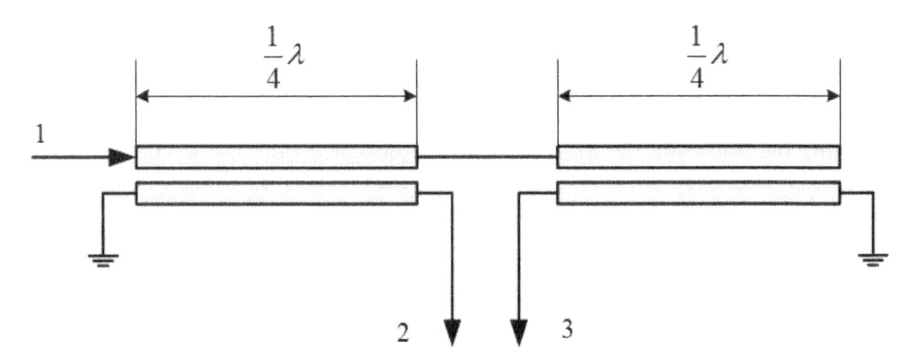

Fig. 8.3 Marchand balun

It consists of two quarter-wavelength sections of coupled lines, and its work is based on the interference of modes excited in coupled transmission lines. In general, this balloon is matched only at the port 1, and, if the coupling coefficient of quarter-wavelength lines is -4.8 dB, its S-matrix [69] is

$$S = \begin{pmatrix} 0 & \frac{j}{\sqrt{2}} & -\frac{j}{\sqrt{2}} \\ \frac{j}{\sqrt{2}} & \frac{1}{2} & \frac{1}{2} \\ -\frac{j}{\sqrt{2}} & \frac{1}{2} & \frac{1}{2} \end{pmatrix}. \tag{8.3}$$

It is seen that $S_{22} = S_{33} \neq 0$, and this Marchand balun should be additionally matched to the outputs 2 and 3 [69],[70]. The main research directions are with the size reduction of baluns and their bandwidth increase. Essential concerns are with the accuracy of balancing in a wide frequency band.

Consider only the main techniques found to be used in monolithic millimeter-wave integrations. Taking into account the size of the Marchand balun, its distributed variant can be used only at increased millimeter-wave frequencies, and its size is comparable with the one of the integrated chip. Very often, the meandering and 3-D topology reduces the overall real-estate of baluns [69]-[74]. The bandwidth increase is achieved by broadside coupling of strips placed on different layers of 3-D integrations [69]-[75], or by using the planar multi-conductor coupled lines [76].

The baluns can be realized using lumped analogs of coupled lines [77], and the integrated transformers can substitute them. They can have their parasitic resonant frequencies shifted towards 100 GHz [78]-[81], and these transformers can be used in millimeter-wave circuits now. Some papers are on the combinations of distributed elements and lumped capacitors for miniaturization of the impedance transforming baluns [82], and this idea can be implemented in monolithic integrations of increased frequencies.

References

[1] Mathaei, G., Young, L., Johnes, E.M.T.: Microwave Filters, Impedance-Matching Networks, and Coupling Structures. Artech House (1961)

[2] Zverev, A.I.: Handbook of Filter Synthesis. John Wiley & Sons (1969)

[3] Williams, B., Taylor, F.J.: Electronic Filter Design Handbook. McGraw-Hill (1995)

[4] Matsumoto, A. (ed.): Microwave Filters and Circuits. Academic, New York (1970)

[5] Jarry, P., Beneat, J.: Advanced Design Techniques and Realizations of Microwave and RF Filters. J. Wiley & Sons, IEEE Press (2008)

[6] Aguilera, J., Berenguer, R.: Design and Test of Integrated Inductors for RF Applications. Kluwer Academic Press (2003)

[7] Sun, S., Shi, J., Zhu, L., et al.: 40 GHz compact TFMS meander-line bandpass filter on silicon substrate. El. Lett. 43(25), 1433–1434 (2007)

[8] Sun, S., Sni, J., Zhu, L., et al.: Millimeter-wave bandpass filters by standard 0.18-μm CMOS technology. IEEE El. Device Lett. 28, 220–222 (2007)
[9] Hsu, C.-Y., Chen, C.Y., Chuang, H.-R.: A 60-GHz millimeter-wave bandpass filter using 0.18 μm CMOS technology. IEEE El. Dev. Lett. 29, 246–248 (2008)
[10] Hsu, C.Y., Chen, C.-Y., Chuang, H.-R.: 70 GHz folded loop dual-mode bandpass filter fabricated using 0.18 μm standard CMOS technology. IEEE Microw. Wireless Comp. Lett. 18, 587–589 (2008)
[11] Salleh, M.K.M., Prigent, G., Crampagne, R., et al.: Quarter wavelength side-coupled ring filters for V-band applications. In: Proc. of 2008 IEEE Int. RF Microw. Conf., Kuala Lumpur, Malaysia, December 2-4, pp. 487–490 (2008)
[12] Nan, L., Mouthaan, K., Xiong, Y.Z., et al.: Design of 60- and 77-GHz narrow-bandpass filters in CMOS technology. IEEE Trans., Circ. Systems - II: Express Briefs 5, 738–742 (2008)
[13] Yang, B., Skafidas, E., Evans, R.: Design of integrated millimeter wave microstrip interdigital bandpass filters on CMOS technology. In: Proc. 37th Eur. Microw. Conf., pp. 680–683 (2007)
[14] Yang, B., Skafidas, E., Evans, R.: Design of 60 GHz millimeter-wave bandpass filter on bulk CMOS. IET Microw. Antennas Propag. 3, 943–949 (2009)
[15] Lu, H.-C., Ye, C.-S., Wei, S.-A.: Miniaturized 60 GHz rectangular ring bandpass filter in 90 nm CMOS technology. El. Lett. 47(7), 448–450 (2011)
[16] Rius, E., Prigent, G., Happy, H., et al.: Wide- and narrow-band bandpass coplanar filters in the W-frequency band. IEEE Trans., Microw. Theory Tech. 51, 784–791 (2003)
[17] Hettak, K., Elgaid, K., Thayne, I.G., et al.: 3D MMIC compact semi-lumped loaded CPW stubs for spurious suppression fabricated with a standard air bridge process. In: IEEE MTT-S Microw. Symp. Dig., pp. 1033–1036 (2009)
[18] Prigent, G., Gianesello, F., Gloria, D., et al.: Bandpass filter for millimeter-wave applications up to 220 GHz integrated in advanced thin SOI CMOS technology on high resistivity substrate. In: Proc. 37th Eur. Microw. Conf., pp. 676–679 (2007)
[19] Chan, K.T., Chen, C.Y., Chin, A., et al.: 40-GHz coplanar waveguide bandpass filters on silicon substrate. IEEE Microw. Wireless Comp. Lett. 12, 429–431 (2002)
[20] Chan, K.T., Chin, A., Li, M.-F., et al.: High-performance microwave coplanar bandpass and bandstop filters on Si substrates. IEEE Trans., Microw. Theory Tech. 51, 2036–2040 (2003)
[21] Dib, N.I., Harokopus Jr., W.P., Katehi, L.P.B., et al.: Study of a novel planar transmission line. In: 1991 IEEE MTT-S Int. Microw. Symp. Dig., pp. 623–626 (1991)
[22] Weller, T.M., Katehi, Rebeiz, G.M.: High performance microshield line components. IEEE Trans., Microw. Theory Tech. 43, 534–543 (1995)
[23] Weller, T.M., Katehi, L.P., Rebeiz, G.M.: A 250-GHz microshield bandpass filter. IEEE Microw. Guided Wave Lett. 5, 153–155 (1995)
[24] Hirsch, S., Chen, Q., Duwe, K., et al.: Design and characterization of coplanar waveguides and filters on thin dielectric membranes at D-band frequencies. In: Proc. of MSMW 2001 Symp., Kharkov, Ukraine, June 4-9, pp. 678–680 (2001)
[25] Vu, T.M., Prigent, G., Plana, R.: Membrane technology for band-pass filter in W-band. Microw. Opt. Techn. Lett. 52, 1393–1397 (2010)
[26] Bouchirha, F., Dubuc, D., Grenier, K., et al.: IC-compatible low loss passive circuits for millimeter-wave applications. In: Proc. EuMC, pp. 380–383 (2006)
[27] Vu, T.-M., Prigent, G., Mazenq, L., et al.: Design of bandpass filter in W-band on a silicon membrane. In: Proc. APMC 2008, pp. 1–4 (2008)

[28] Prigent, G., Rius, E., LePennec, F.L., et al.: Design of narrow-band DBR planar filters in Si-BCB technology for millimeter-wave applications. IEEE Trans., Microw. Theory Tech. 52, 1045–1051 (2004)

[29] Six, G., Prigent, G., Rius, E., et al.: Fabrication and characterization of low-loss TFMS on silicon up to 220 GHz. IEEE Trans., Microw. Theory Tech. 53, 301–305 (2005)

[30] Prigent, G., Rius, E., Happy, H., et al.: Design of branch-line coupler in the G- frequency band. In: Proc. 36th Eur. Microw. Conf., pp. 1296–1299 (2006)

[31] Prigent, G., Rius, E., Blary, K., et al.: Design of narrow band-pass planar filters for millimeter-wave applications up to 220 GHz. In: IEEE MTT-S Symp. Dig., pp. 1491–1494 (2005)

[32] Altman, J.L.: Microwave Circuits. D. Van Nostrand. Comp. (1964)

[33] Lee, T.H.: Planar Microwave Engineering. Cambridge University Press (2004)

[34] Sakagami, S., Fujii, M., Wuren, T.: Impedance-transforming lumped element two-branch 90° couplers in case of type C. In: Proc. 10th WSEAS Int. Conf. Circuits, Vouliagmeni, Athens, Greece, July 10-12, pp. 271–274 (2006)

[35] Papapolymerou, J., Ponchak, G.E., Tentzeris, E.M.: A Wilkinson power divider on a low resistivity Si substrate with a polyimide layer for wireless circuits. In: 2002 IEEE MTT-S Dig., pp. 593–596 (2002)

[36] Ponchak, G.E., Bacon, A., Papapolymerou, J.: Monolithic Wilkinson power divider on CMOS grade silicon with a polyimide interface layer for antenna distribution networks. IEEE Antennas Wireless Prop. Lett. 2, 167–169 (2003)

[37] May, J.W., Rebeiz, G.M.: A 40-50-GHz SiGe 1:8 differential power divider using shielded broadside-coupled striplines. IEEE Trans., Microw. Theory Tech. 56, 1575–1581 (2008)

[38] Koh, K.-J., Rebeiz, G.M.: A Q-band four element phased-array front-end receiver with integrated Wilkinson power combiners in 0.18-μm SiGe BiCMOS technology. IEEE Trans., Microw. Theory Tech. 56, 2046–2053 (2008)

[39] Niknejad, A.M., Bohsali, M., Adabi, E., et al.: Integrated circuit transmission-line transformer power combiner for millimeter-wave applications. El. Lett. 43(5), 47–48 (2007)

[40] Yu, H.Y., Choi, S.-S., Kim, Y.-H.: K-band power divider with metal bridge structures for size reduction using CMOS technology. Microw. Opt., Techn. Lett. 53, 379–381 (2011)

[41] Li, Y., Kirby, P.L., Offranc, O., et al.: Silicon micromachined W-band hybrid coupler and power divider using DRIE technique. IEEE Microw. Wireless Comp. Lett. 18, 22–24 (2008)

[42] Horst, S., Bairavasubramanian, R., Tentzeris, M.N., et al.: Modified Wilkinson power dividers for millimeter-wave integrated circuits. IEEE Trans., Microw. Theory Tech. 55, 2439–2446 (2007)

[43] Saito, Y., Fontaine, D., Rollin, J.-M., et al.: Micro-coaxial Ka-band Gysel power divider. Microw. Opt. Techn. Lett. 52, 474–478 (2011)

[44] Lee, J., Tretiakov, Y.V., Cressler, J.D., et al.: Design of a monolithic 30 GHz Branch line coupler in SiGe HBT technology using 3-D EM simulation. In: Proc. 2004 Topical Meeting on Silicon Monolithic Integrated Circuits in RF Systems, pp. 274–277 (2004)

[45] Pan, B., Yoon, Y., Zhao, Y.Z., et al.: A broadband surface-micromachined 15-45 GHz microstrip coupler. In: 2005 IEEE MTT-S Symp. Dig., pp. 989–992 (2005)

[46] Zhu, Y., Wu, H.: A 10-40 GHz 7 dB directional coupler in digital CMOS technology. In: Proc. 2006 Int. Microw. Symp., pp. 1551–1554 (2006)
[47] Chirala, M.K., Nguen, C.: Multilayer design techniques for extremely miniaturized CMOS and millimeter-wave distributed passive circuits. IEEE Trans., Microw. Theory Tech. 54, 4218–4224 (2006)
[48] Prigent, G., Rius, E., Happy, H., et al.: Design of branch-line coupler in the G-frequency band. In: Proc. 36th Eur. Microw. Conf., pp. 1296–1299 (2006)
[49] Do, M.-N., Dubuc, D., Grenier, K., et al.: High compactness/high isolation 3D-broadside coupler design methodology. In: Proc. APMC 2006, pp. 799–802 (2006)
[50] Do, M.-N., Dubuc, D., Grenier, K., et al.: 3D-BCB based branchline coupler for K-band integrated microsystem. In: Proc. 36th Eur. Microw. Conf., pp. 498–501 (2006)
[51] Hettak, K., Amaya, R.E., Morin, G.A.: A novel compact three-dimensional CMOS branch-line coupler using the meandering ECPW, TFMS, and buried micro coaxial technologies at 60 GHz. In: Proc. IMS 2010, pp. 1576–1579 (2010)
[52] Bulja, S., Mirshekar-Syahkal, D., Yazdanpanahi, M.: Novel wide-band transition between finite ground coplanar waveguide (FGCPW) and balanced stripline. In: Proc. 4th Eur. Microw. Integrated Circ. Conf., pp. 301–303 (2009)
[53] Simons, R.N., Dib, N.I., Katehi, L.P.B.: Coplanar stripline transition. El. Lett. 31(20), 1725–1726 (1995)
[54] Suh, Y.-H., Chang, K.: A wideband coplanar stripline to microstrip transition. IEEE Microw. Wireless Comp. Lett. 11, 28–29 (2001)
[55] Tu, W.-H., Chang, K.: Wide-band microstrip-to-coplanar stripline/slotline transitions. IEEE Trans., Microw. Theory Tech. 54, 1084–1089 (2006)
[56] Lim, T.B., Zhu, L.: Compact microstrip-to-CPS transition for UWB application. In: Proc. 2008 IEEE MTT-S Int. Microw. Workshop Series on Art of Miniaturizing RF and Microw. Passive Comp., pp. 153–156 (2008)
[57] Kim, Y.-G., Woo, D.-S., Kim, K.W., et al.: A new ultra-wideband microstrip-to-CPS transition. In: Proc. IEEE/MTT-S Int. Symp., pp. 1563–1566 (2007)
[58] Kim, Y.-G., Kim, K.W., Cho, Y.-K.: A planar ultra-wideband balanced doubler. In: 2008 MTT-S Int. Microw. Symp. Dig., pp. 1243–1246 (2008)
[59] Kim, Y.-G., Kim, I.-B., Song, S.-Y., et al.: Ultra-wideband surface-mountable double-balanced multiplier. In: Proc. 40th Eur. Microw. Conf., pp. 308–311 (2010)
[60] Wu, P., Wang, Z., Zhang, Y.: Wideband planar balun using microstrip to CPW and microstrip to CPS transitions. El. Lett. 46(24), 1611–1613 (2010)
[61] Prieto, D., Cayrou, J.C., Cazaux, J.L., et al.: CPS structure potentialities for MMICS: a CPS/CPW transition and a bias network. In: 1998 IEEE MTT-S Symp. Dig., pp. 111–114 (1998)
[62] Mao, S.-G., Hwang, C.-T., Wu, R.-B., et al.: Analysis of coplanar waveguide-to coplanar stripline transitions. IEEE Trans., Microw. Theory Tech. 48, 23–29 (2000)
[63] Lin, Y.-S., Chen, C.H.: Novel lumped-element uniplanar transitions. IEEE Trans., Microw. Theory Tech. 49, 2322–2330 (2001)
[64] Kim, S., Jeong, S., Lee, Y.-T., et al.: Ultra-wideband (from DC to 110 GHz) CPW to CPS transition. El. Lett. 38(13), 622–623 (2002)
[65] Anagnostou, D.E., Morton, M., Papapolymerou, J., et al.: A 0-55-GHz coplanar waveguide to coplanar strip transition. IEEE Trans., Microw. Theory Tech. 56, 1–6 (2008)
[66] Herrick, K.J., Yook, J.-G., Katehi, L.P.B.: Microtechnology in the development of three-dimensional circuits. IEEE Trans., Microw. Theory Tech. 46, 1832–1844 (1998)

[67] Kim, H.-T., Lee, S., Park, J.-H., et al.: Ultra-wideband uniplanar MMIC balun using field transformations. El. Lett. 42(6), 359–361 (2006)

[68] Marchand, N.: Transmission-line conversation transformers. Electronics 17, 142–145 (1944)

[69] Liu, J.-X., Hsu, C.-Y., Chuang, H.-R., et al.: A 60 GHz millimeter-wave CMOS Marchand balun. In: Proc. 2007 IEEE RFIC Symp., pp. 445–448 (2007)

[70] Chongcheawchamnan, M., Ng, C.Y., Robertson, I.D.: Miniaturized and multilayer Wilkinson divider and balun for microwave and millimeter-wave applications. In: Proc. 6th IEEE HF Postgraduate Student Colloquium, pp. 174–179 (2001)

[71] Chiou, H.-K., Yang, T.-Y., Hsu, Y.-C., et al.: 15-60 GHz asymmetric broadside coupled balun in 0.18 μm CMOS technology. El. Lett. 43(19), 1028–1030 (2007)

[72] Chiou, H.-K., Yang, T.-Y.: Low-loss and broadband asymmetric broadside-coupled balun for mixer design in 0.18-μm CMOS technology. IEEE Trans., Microw. Theory Tech. 56, 835–848 (2008)

[73] Ding, H., Lam, K., Wang, G., et al.: On-chip millimeter wave rat-race hybrid and Marchand balun in IBM 0.13μm BiCMOS technology. In: Proc. APMC 2008, pp. 1–4 (2008)

[74] Wang, G., Bavisi, A., Woods, W., et al.: A 77-GHz Marchand balun for antenna applications in BiCMOS technology. Microw. Opt. Techn. Lett. 53, 664–666 (2011)

[75] Dawn, D., Sen, P., Sarkar, S., et al.: 60-GHz integrated transmitter development in 90-nm CMOS. IEEE Trans., Microw. Theory Tech. 57, 2354–2366 (2009)

[76] Lin, C.-S., Wu, P.-S., Yeh, M.C., et al.: Analysis of multiconductor coupled-line Marchand baluns for miniature MMIC design. IEEE Trans., Microw. Theory Tech. 55, 1190–1199 (2007)

[77] Johansen, T., Krozer, V.: Analysis and design of lumped element Marchand baluns. In: Proc. 17th Int. Conf. MIKON 2008, pp. 1–4 (2008)

[78] Huang, D., Wong, R., Chien, C., et al.: 1.2 V and 8.6 mW CMOS differential receiver front-end with 24 dB gain and -11 dBm IRCP. El. Lett. 42(19), 1449–1450 (2006)

[79] Ragonese, E., Sapone, G., Palmisano, G.: High-performance interstacked transformers for mm-wave ICs. Microw. Opt. Techn. Lett. 52, 2160–2163 (2010)

[80] Felic, G., Skafidas, E.: An integrated transformer balun for 60 GHz silicon RF IC design. In: Proc. Int. Symp. Signals, Systems and Electronics, ISSSE 2007, pp. 541–542 (2007)

[81] Wei, H.-J., Meng, C., Wu, P.-Y., et al.: K-band CMOS sub-harmonic resistive mixer with a miniature Marchand balun on lossy silicon substrate. IEEE Microw. Wireless Comp. Lett. 18, 40–42 (2008)

[82] Ang, K.S., Leong, Y.C., Lee, C.H.: Analysis and design of miniaturized lumped-distributed impedance-transforming baluns. IEEE Trans., Microw. Theory Tech. 51, 1009–1017 (2003)

9 Circuit Approach for Simulation of EM-quantum Components

Abstract. This Chapter is on the circuit approach to describe the quantum-mechanical phenomena. Being proposed by G. Kron many years ago, this technique is now a very powerful tool for modeling and design of hybrid electronics integrating the classical and quantum-mechanical components. The linear and non-linear Schrödinger equations are transformed into the first-order partial differential equations with respect to currents and voltages, and the obtained equivalent circuits are modeled using a commercially available simulator. The approach is pertinent for seamless simulation of the future-generation integration, although the main attention in this Chapter is paid to the modeling of trapped Bose-Einstein condensates. References -108. Figures -34. Pages -54.

Quantum-mechanical effects have been used for creation of many electronic components. The focus of today's physics is the control of single atoms, molecules, and charge carriers or their clusters by the EM field [1],[2]. This individual control supposes the use of EM signals which level is comparable with the one induced or radiated by a single particle. The problem is described by the Maxwell and Schrödinger equations, which are solved jointly, and it is a complicated problem due to the nonlinearity of large systems of these partial differential equations [3]-[8].

The quantum effects relate to the wave phenomena, and the equivalent electric circuits [9]-[20] can model the governing EM/Schrödinger equations. Then these circuits can be the subject of modeling by the Electronic Design Automation (EDA) tools. Unfortunately, for a long period, the equivalent quantum circuits could not be simulated directly by known EDA systems operating the real time-dependent currents and voltages. The probability density wave function is a complex one, and only recently, some software packages have been modified to calculate the complex signals [21],[22]. Splitting the Schrödinger equation into the real and imaginary parts [23] leads to doubling the unknown variables, which causes an essential increase of simulation time and required computational resources, especially, in the case of multi-dimensional and nonlinear equations.

G.A. Kouzaev: Applications of Advanced Electromagnetics, LNEE 169, pp. 359–412.
springerlink.com

The goal of this Chapter is to develop such a technique for computations of EM/quantum equations by EDA tools operating complex signals and imaginary circuit components, which are typical for modeling of quantum-mechanical phenomena by the circuit approach. An additional goal of this contribution is to estimate creation of an approach for joint co-design of electronic and quantum devices and circuits by commercially available circuit simulators.

9.1 Circuit Models of Linear Schrödinger Equation

Starting with the works of G. Kron, there are many papers on quantum equations in their circuit form. Consider how to obtain them in the case of commonly known linear Schrödinger equation describing the motion of a particle of the mass m in the external time-independent potential $V(\mathbf{r})$:

$$-\left(\frac{\hbar^2}{2m}\right)\nabla^2\cdot\Psi(\mathbf{r},t)+V(\mathbf{r})\Psi(\mathbf{r},t)=j\hbar\frac{\partial\Psi}{\partial t}(\mathbf{r},t) \tag{9.1}$$

where $\Psi(\mathbf{r},t)$ is the probability density wave function. Supposing that the quantum voltage $u(\mathbf{r},t)$ and current $\mathbf{i}(\mathbf{r},t)$ are connected with the wave function $\Psi(\mathbf{r},t)$ and with each other as

$$\begin{aligned} u(\mathbf{r},t)&=\Psi(\mathbf{r},t),\\ \mathbf{i}(\mathbf{r},t)&=j\frac{\hbar}{m}\nabla\cdot u(\mathbf{r},t). \end{aligned} \tag{9.2}$$

The circuit equations written regarding to the introduced voltage and current (9.2) are obtained substituting the above expressions into (9.1):

$$\begin{aligned} \nabla\cdot u(\mathbf{r},t)&=-j\frac{m}{\hbar}\mathbf{i}(\mathbf{r},t),\\ \nabla\cdot\mathbf{i}(\mathbf{r},t)&=2\frac{\partial u(\mathbf{r},t)}{\partial t}+\frac{2j}{\hbar}V(\mathbf{r})u(\mathbf{r},t). \end{aligned} \tag{9.3}$$

The attractiveness of this form of Schrödinger equation is that an equivalent circuit can be introduced describing an infinitesimal volume of space, and the whole spatial domain is described by connections of these infinitesimal circuits. Calculations of this network can be performed using some circuit simulators, and these simulations are distinguished by their increased stability in contrast to the conventional FDTD methods [10]. Besides, the known techniques of circuit analysis can be used which is more convenient for electronic engineers, and it is important for creation of future CAD tools for simulation and design of hybrid quantum/electronic integrations.

9.2 Circuit Models of Nonlinear Schrödinger Equations

Nonlinear Schrödinger equations are very important to describe many micro- and macro-phenomena, and they are used as a theoretical tool for physics of condensed matter. An example is the Gross-Pitaevskii equation obtained using an averaging method applied to many-body interactions of ultra-cold bosonic atoms:

$$j\hbar\frac{\partial\Phi(\mathbf{r},t)}{\partial t}=\left\{-\frac{\hbar^2}{2m}\nabla^2+V(\mathbf{r})+g\left|\Phi(\mathbf{r},t)\right|^2\right\}\Phi(\mathbf{r},t) \tag{9.4}$$

where $\Phi(\mathbf{r},t)$ is the unknown function describing the condensate density, m is the effective mass of boson, $V(\mathbf{r})$ is the trapping potential, and g is a constant.

To obtain the circuit equation for Gross-Pitaevskii one, the equivalent voltage $u(\mathbf{r},t)$ and current $\mathbf{i}(\mathbf{r},t)$ are introduced similarly to (9.2):

$$\begin{aligned} u(\mathbf{r},t)&=\Phi(\mathbf{r},t),\\ \mathbf{i}(\mathbf{r},t)&=j\frac{\hbar}{m}\nabla\cdot u(\mathbf{r},t). \end{aligned} \tag{9.5}$$

Then the circuit equations regarding to the equivalent voltage and current are

$$\begin{aligned} \nabla\cdot u(\mathbf{r},t)&=-j\frac{m}{\hbar}\mathbf{i}(\mathbf{r},t),\\ \nabla\cdot\mathbf{i}(\mathbf{r},t)&=2\frac{\partial u(\mathbf{r},t)}{\partial t}+\frac{2j}{\hbar}\left(V(\mathbf{r})+g\left|u(\mathbf{r},t)\right|^2\right)u(\mathbf{r},t). \end{aligned} \tag{9.6}$$

Similarly to the linear cases, these equations can be solved using equivalent circuits and some circuit simulators.

Another interesting nonlinear equation is employed in statistical mechanics. Some effects are described by generalized Fokker-Plank equation [24],[25]:

$$j\hbar\frac{\partial\Psi}{\partial t}=-\frac{1}{2-q}\frac{\hbar^2}{2m}\nabla^2\cdot\left[(\Psi)^{2-q}\right] \tag{9.7}$$

where Ψ is the normalized unknown function, and q is the nonlinearity order. It can be transformed into the circuit equations although they are supposed to be more complicated.

Thus the circuit approach proposed by G. Kron can describe many physical effects from the classical EM ones to the complicated nonlinear phenomena arisen in many branches of physics.

9.3 Circuit Models of EM-quantum Equations

Interaction of particles with the EM field is described by the Maxwell and corresponding quantum equations. In Section 1.6 the compact Hertz-quantum equations are obtained, and they can be represented by equivalent circuit equations and models. A couple of them are considered below.

9.3.1 Circuit Model for Schrödinger-Hertz Equation

Following to known procedure, let us to introduce the quantum voltage $u_\psi = \Psi$ and current $\mathbf{j}_\psi = \frac{j\hbar}{m}\nabla \cdot u_\psi$, Then the Schrödinger-Hertz equation (1.65) is transformed into a system of partial differential equations of the first order:

$$\nabla \cdot \mathbf{j}_\psi + \frac{2jq}{\hbar c^2}\frac{\partial \mathbf{Z}}{\partial t}\mathbf{j}_\psi = 2\left\{\frac{\partial}{\partial t} + \frac{j}{\hbar}\left[\frac{q}{2mc^2}\left(q\left|\frac{\partial \mathbf{Z}}{\partial t}\right|^2 - j\hbar\left(\nabla \cdot \frac{\partial \mathbf{Z}}{\partial t}\right)\right) - q(\nabla \cdot \mathbf{Z})\right]\right\}u_\psi,$$
$$\nabla \cdot u_\psi = -j\frac{m}{\hbar}\mathbf{j}_\psi. \tag{9.8}$$

These equations allow to put into correspondence an equivalent circuit for an infinitesimally small domain $\Delta x \Delta y \Delta z$ at $\mathbf{r} = \mathbf{r}_k$ (Fig. 9.1) where $\mathbf{n} = \frac{2jq}{\hbar c^2}\frac{\partial \mathbf{Z}(\mathbf{r}_k)}{\partial t}$,

$$X_x^{(k)} = \frac{m}{2\hbar}\Delta x_k, X_y^{(k)} = \frac{m}{2\hbar}\Delta y_k, X_z^{(k)} = \frac{m}{2\hbar}\Delta z_k, C^{(k)} = 2\left(\Delta x_k + \Delta y_k + \Delta z_k\right),$$

$$Y^{(k)} = \frac{2}{\hbar}\left[\frac{q}{2mc^2}\left(q\left|\frac{\partial \mathbf{Z}}{\partial t}\right|^2 - j\hbar\left(\nabla \cdot \frac{\partial \mathbf{Z}}{\partial t}\right)\right) - q(\nabla \cdot \mathbf{Z})\right](\Delta x + \Delta y + \Delta z).$$

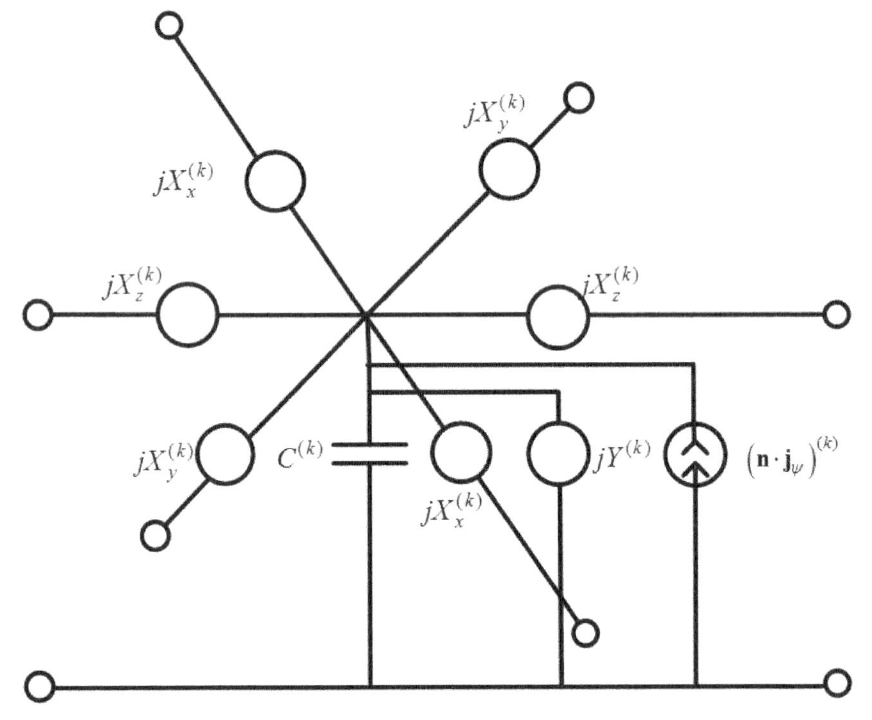

Fig. 9.1 Infinitesimal equivalent circuit for the Schrödinger-Hertz equation. Reprinted from [14] with permission of the World Scientific Publ. Co.

A particular case of this equation is the Schrödinger one, and its circuit equation system is

$$\begin{aligned} \nabla \cdot \mathbf{j}_\psi &= 2\left\{\frac{\partial}{\partial t} + \frac{jq}{\hbar}\Phi_0\right\}u_\psi, \\ \nabla \cdot u_\psi &= -j\frac{m}{\hbar}\mathbf{j}_\psi. \end{aligned} \tag{9.9}$$

Taking into account that $\mathbf{n} = 0$, $X_x^{(k)} = \frac{m}{2\hbar}\Delta x_k, X_y^{(k)} = \frac{m}{2\hbar}\Delta y_k, X_z^{(k)} = \frac{m}{2\hbar}\Delta z_k$, $C^{(k)} = 2(\Delta x_k + \Delta y_k + \Delta z_k)$, and $\Upsilon^{(k)} = \frac{2q\Phi_0}{\hbar}(\Delta x + \Delta y + \Delta z)$ with Φ_0 as the external potential. The equivalent electric circuit of a larger domain consists of a number of the infinitesimal ones, which is enough to provide the necessary convergence of results. Numerical simulations of some of these circuits and quantum phenomena, including the nonlinear ones, are considered and verified in [14],[15], for instance. Similarly, the equations and infinitesimal equivalent circuits can be derived for the Pauli-Hertz (1.72), Dirac-Hertz (1.75) and Klein-Gordon-Hertz (1.77) equations.

The above-considered approach dealing with the equivalent circuits defined for the EM field and the wave functions opens a way for co-joint simulation or even co-design of the EM-quantum components.

9.4 Verification of Circuit Model of Tunneling Problem

As was stated above, some contemporary simulators allow calculating the complex currents and voltages in the linear and nonlinear circuits composed of conventional and imaginary components.

Our simulations of the linear and nonlinear quantum-mechanical equations are performed using the Agilent Advanced Design System (ADS) [21],[22], and the results are verified by comparison with the analytical data derived for some geometries and potentials [14],[15]. For the first case, the quantum tunneling of an electron through a rectangular barrier of the limited height $U_0 = \text{const}$ and the width d_z (Fig. 9.2) is modeled by the above-mentioned simulator and analytically.

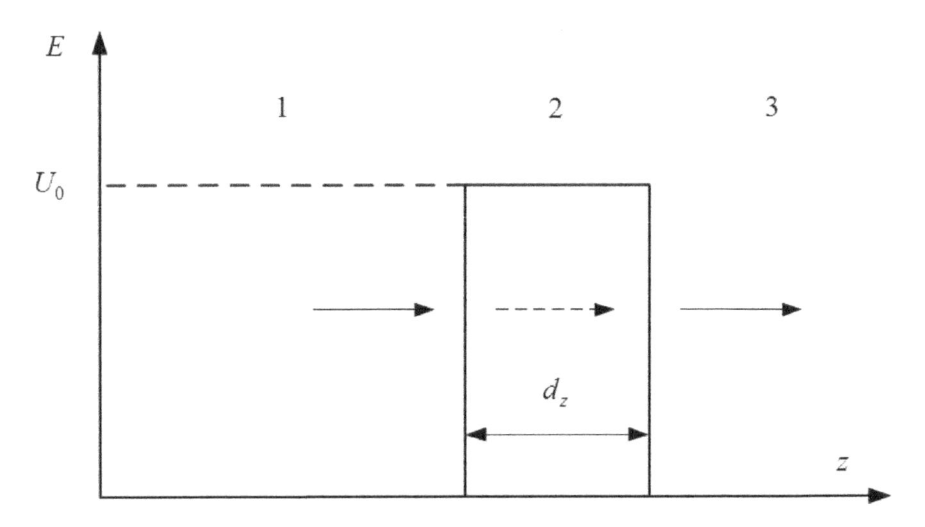

Fig. 9.2 Tunneling of an electron through the barrier of the height U_0 and the width d_z. Reprinted from [15] with permission of the World Scientific Publ. Co.

The governing Schrödinger equation for electron is

$$-j\hbar\frac{\partial}{\partial t}\Psi(z,t)=-\frac{\hbar^2}{2}\frac{\partial}{\partial z}\left\{\frac{1}{m_{\text{eff}}(z)}\frac{\partial}{\partial z}\right\}\Psi(z,t)+U_0(z)\Psi(z,t) \tag{9.10}$$

where m_{eff} is the effective mass of electron and U_0 is the potential, as usually.

The derived transmission line equation is

$$\begin{aligned}-\frac{\partial u_\psi}{\partial z}&=j\frac{m_{\text{eff}}(z)}{\hbar}j_\psi.\\ -\frac{\partial j_\psi}{\partial z}&=-\frac{2j}{\hbar}U_0(z)u_\psi+2\frac{\partial u_\psi}{\partial t}\end{aligned} \tag{9.11}$$

where $u_\psi=\Psi$ is the equivalent voltage, and $j_\psi=j\dfrac{\hbar}{m_{\text{eff}}(z)}\dfrac{\partial u_\psi}{\partial z}$ is the equivalent current.

The problem shown in Fig. 9.2 and described by the equations (9.11) can be modeled by two different equivalent networks. Supposing that the effective mass of electron $m_{\text{eff}}(z)=\text{const}$ and the potential $U_0(z)=\text{const}$, then the first of the circuits is just a length of a transmission line (Fig. 9.3) with the propagation constant $\gamma^{(2)}=\sqrt{-2m_{\text{eff}}(\hbar\omega-U_0)}/\hbar$ and loaded by the characteristic impedances $Z_c^{(1,3)}=jm_{\text{eff}}/\hbar\gamma^{(1,3)}$ where $\omega=E_0/\hbar$, E_0 is the initial kinetic energy of electron, and $\gamma^{(1,3)} = \sqrt{-2m_{\text{eff}}\hbar\omega}$. All parameters of this circuit are the frequency-dependent ones, and special means should be applied in the ADS environment to calculate its *S*-matrix or transients.

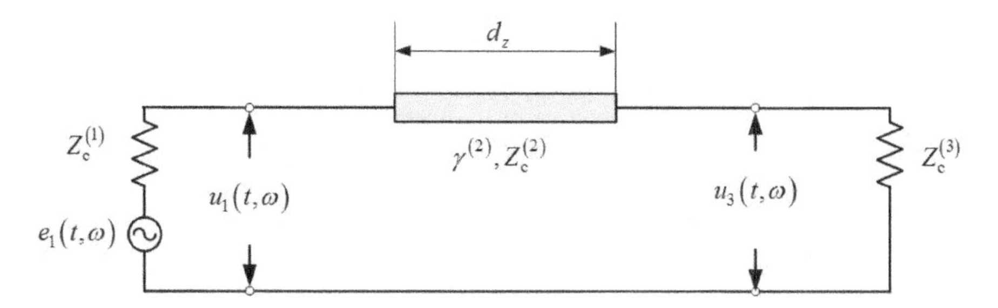

Fig. 9.3 Equivalent transmission line model for a tunnel problem. Reprinted from [15] with permission of the World Scientific Publ. Co.

In general, the effective electron mass m_{eff} can depend on the coordinate *z*, and the potential $U_0(z,t)$ is a function of time and space. Usually, the circuit simulators have no any embedded models for such equivalent lines, and the equivalent infinitesimal T-networks should be used to simulate the quantum phenomena. From the equations (9.11) it follows an equivalent infinitesimal T-network given for the *k-th* length Δz_k (Fig. 9.4).

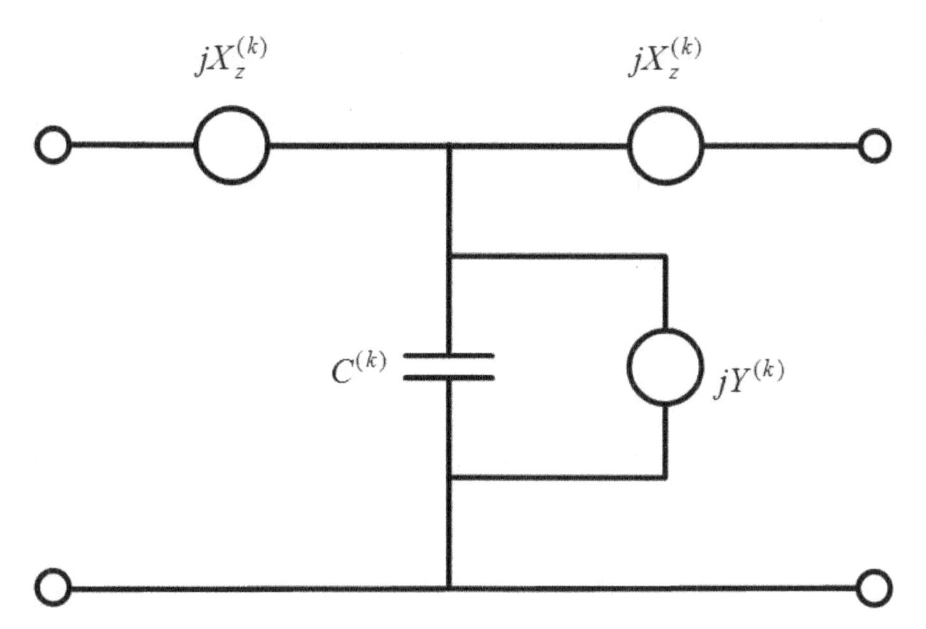

Fig. 9.4 Infinitesimal equivalent T-network. Reprinted from [15] with permission of the World Scientific Publ. Co.

In this figure, $C_k = 2\Delta z_k$, $X_z^{(k)} = \frac{m_{\text{eff}}(z_k)}{2\hbar}\Delta z_k$, and $Y_k = -\frac{2U_0(z_k)}{\hbar}\Delta z_k$. Now, the transmission line from Fig. 9.3 is modeled by a ladder circuit (Fig. 9.5) composed of *N* circuits from Fig. 9.4.

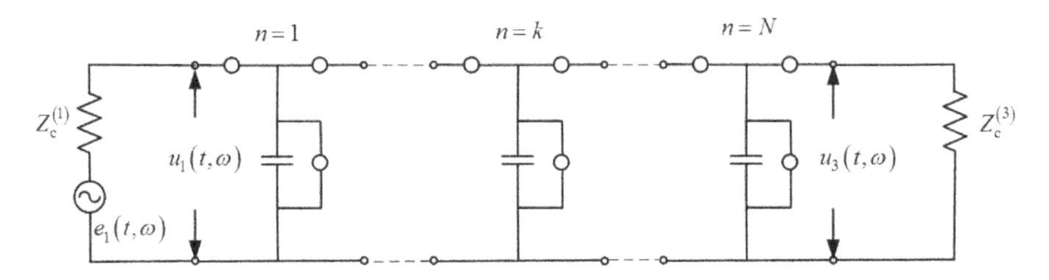

Fig. 9.5 Ladder network model of a generalized tunnel problem. Reprinted from [15] with permission of the World Scientific Publ. Co.

Here, only the impedances $Z_c^{(1,3)}$ are frequency dependent. The circuit is composed of imaginary resistors shown by circles in Fig. 9.5. Additionally, the source or launching signal $e_1(t,\omega)$ can be a complex frequency-time dependent function. Such circuits and problems are common in the signal processing area, and they are simulated by circuit envelope tools [21],[22]. They allow modeling not only conventional circuits, but also those composed of the imaginary, nonlinear, frequency- and time-dependent components. The results are on the signal envelops and their spectral content. Anyway, taking into account the numerical nature of the calculations, they should be carefully tested, especially, in such unusual for the circuit simulators area as the calculations of quantum phenomena. For this purpose, the above-considered circuits (Figs 9.3 and 9.5) are assembled in the ADS Schematic Window, and the envelope simulations are performed to calculate the output voltage $u_3(t,\omega)$ at the frequency corresponding to the energy E_0. The *S*-matrices of these linear circuits are calculated according to a formula from [26]:

$$S_{kl} = \frac{2u_k\sqrt{Z_c^{(k)}}}{e_l\sqrt{Z_c^{(l)}}}, k,l = 1,3. \tag{9.12}$$

For the problem shown in Fig. 9.2, the analytical formulas of the transmission *T* and reflection *R* coefficients are given in [27], and they are used for verification of our numerical simulations:

$$T = \frac{2jk\kappa}{(k^2+\kappa^2)\sin(\kappa d_z) + 2j\kappa k\cos(\kappa d_z)}, \tag{9.13}$$

$$R = \frac{\left(k^2 - \kappa^2\right)\sin\left(\kappa d_z\right)}{\left(\kappa^2 - k^2\right)\sin\left(\kappa d_z\right) + 2jk\kappa\cos\left(\kappa d_z\right)} \tag{9.14}$$

where $k = \sqrt{2m_{\text{eff}}E_0}/\hbar$, $\kappa = \sqrt{2m_{\text{eff}}\left(E-U\right)}/\hbar$, and $m_{\text{eff}} = 0.067m_e$.

In the case of $E_0 > U_0$, the comparison of analytical results and numerical ones obtained according to the network models (Figs 9.3 and 9.5) are shown in Table 9.1.

Table 9.1 Comparison of analytical and numerical calculations of the transmission probability for the tunneling problem shown in Fig. 9.2 ($E_0 = 1.1$ eV and $U_0 = 0.956$ eV)

d_z,Å	10	20	30	40	50
$\|T\|^2$, (9.13)	0.7490	0.4928	0.4104	0.4588	0.6683
$\|S_{31}\|^2$, Fig. 9.3	0.7482	0.4928	0.4109	0.4583	0.6675
$\|S_{31}\|^2$, Fig. 9.5	0.7586 (N=5)	0.4970 (N=10)	0.4104 (N=15)	0.4597 (N=20)	0.6691 (N=25)

Here, the number of the elementary T-networks N is chosen according to the criterion of the results stabilization. It shows the satisfactory accuracy of the envelope calculations of tunneling problem in this case.

The real tunneling takes its place when $E_0 < U_0$, i.e. the electron energy is not high enough to overcome the barrier. In this case, the equivalent transmission line corresponding to the barrier's region has the imaginary frequency-dependent propagation constant $\gamma^{(2)}$ and characteristic impedance $Z_c^{(2)}$. Unfortunately, the standard ADS library has no such a component, and only the ladder circuit from Fig. 9.5 is able to provide the results shown in Table 9.2 together with the ones calculated analytically. Again, the numerical simulations are of good accuracy.

Table 9.2 Comparison of analytical and numerical calculations of the transmission probability for the tunneling problem shown in Fig. 9.2 ($E_0 = 0.8$ eV and $U_0 = 0.956$ eV).

d_z,Å	10	20	30	40	50
$\|T\|^2$, (9.13)	0.6455	0.2593	0.0936	0.0331	0.0116
$\|S_{31}\|^2$, Fig. 9.5	0.6448 (N=5)	0.2591 (N=10)	0.0936 (N=15)	0.0331 (N=20)	0.0117 (N=25)

Additionally to the linear Schrödinger equation, a nonlinear one was simulated by this approach in [15], and good agreement of analytical and numerical data was found again.

The performed here initial research on the applicability of an available EDA tool shows good accuracy. The derived new EM-quantum equations and their circuit models are convenient for joint EM and quantum-mechanical simulations by the same already available EDA tool. It opens a way towards direct simulation and design of components and circuits integrating the classical electronics parts and nano-based quantum mechanical components.

9.5 Solution of 2-D Nonlinear Schrödinger Equations by Circuit Simulators[1]

The Bose-Einstein condensate is one of the interesting states of matter arising at temperature of several tens of nano-Kelvin. The thermal motion effects are negligible in this case, and the quantum interactions define the properties of ultra-cold gaseous matter. The condensate is confined in a trap, and it is described by the many-body quantum theory.

An averaged equation for this trapped matter is obtained by Gross [28] and Pitaevskii [29]. Today the condensates are interesting in the study of quantum effects [30],[31], for the developments of ultra-high-sensitivity interferometers and sensors [32],[33], cold-atom analogs of transistors [34],[35], superconductive atomic circuit [36],[37], prospective registers for quantum computers [38], etc.

The condensate can be described by the space- and time-dependent global wave function $\Psi(\mathbf{r},t)$. The time evolution of the condensate wave function in the presence of an external trapping potential V_{ext} is described by the mentioned Gross-Pitaevskii nonlinear equation:

$$j\hbar\frac{\partial\Psi}{\partial t}=\left[-\frac{\hbar^2}{2m}\nabla^2+V_{\text{ext}}+NU_0\,|\Psi|^2\right]\Psi \tag{9.15}$$

where j is the imaginary unit, $\hbar$ is the normalized Planck constant, m is the mass of atom, ∇^2 is the Laplace operator, N is the number of condensate atoms, $U_0=4\pi\hbar^2 a/m$ is the self-interacting coupling, and a is the scattering length of bosons. The last term in the right side of (9.15) describes the averaged interactions of trapped atoms.

The analytical and numerical simulations of the time-dependent Gross-Pitaevskii equation are great challenges due to its nonlinearity, multi-dimension, and time-dependence. Various kinds of techniques have been developed, and most of them are the numerical methods [39]. Among them is the Visscher's scheme [23], which is an explicit method for solving the time-dependent Schrödinger equations. Another explicit finite-difference scheme, which is an extension of the Visscher's, is proposed in [40]. Furthermore, a numerical method, which directly minimizes the energy functional via the finite element approximation, is used in [41].

---

[1] Written by G. Ying and G.A. Kouzaev.

In addition, the method of equivalent circuits [42] can be applied to solve the Gross-Pitaevskii equation. The initial idea of this method belongs to G. Kron, who built the circuit networks for EM problems [43] and linear Schrödinger equation [9]. Today, many software tools for the EM initial boundary value problems are available in the market.

The applications of this circuit technique to the quantum-mechanical problems are less known, and only the literature studies are available. For instance, the electron wave propagation in quantum-well devices is analyzed in [12], and the same method is applied to the resonant tunneling effect in multilayered structure [10].

A new interest in the equivalent circuit representation is within the hybrid integration [16],[44] of classical electronics and quantum-mechanical circuits, which is preferable to simulate and design using the same software tool and numerical methods.

This idea and technique are proofed in [14],[15],[45] for the solution of the linear and non-linear one-dimensional Schrödinger equations by means of the circuit simulator ADS [21]. It is shown that the adaptation of circuit simulators to the quantum-mechanical calculations is not a simple task, and much effort should be made to the techniques of such simulations.

The purpose of this Section is the development of the verified circuit-based simulation techniques for the 2-D trapped Bose-Einstein condensates, which can be used by engineers for simulation and design of traps and circuits of ultra-cold matter.

9.5.1 Equivalent Circuit Model for Gross-Pitaevskii Equation

9.5.1.1 Telegraph Equation for Two-dimensional Condensates Trapped in a Cylindrical Domain

For verification of the circuit-based technique, the normalized Gross-Pitaevskii equation from [40] is used. It is supposed that cold atoms are trapped in a cylindrical angularly symmetrical trap, and the external potential is

$$V_{\text{ext}} = \frac{1}{2} m\Omega^2 \left(r^2 + \varepsilon z^2 \right) \tag{9.16}$$

where the angular dependence is ommited, r and z are the radial and axial axes, correspondingly, Ω is the spatial frequency of the harmonic potential V_{ext}, and ε is the aspect ratio of this cylindrical trap. In this case, the Gross-Pitaevskii equation is solved in the cylindrical system of coordinates.

Initially, the authors of [40] applied a mathematical scaling technique to (9.15) by introducing the dimensionless variables ρ, ζ, α, and τ:

$$\rho \equiv \frac{r}{S_l},\ \zeta \equiv \frac{z}{S_l},\ \alpha \equiv \frac{a}{S_l},\ \tau \equiv \frac{t}{S_t} \tag{9.17}$$

where S_l is the unit length (μm) and S_t is the unit time (ms), respectively.

Once the wave function Ψ is scaled, a dimensionless form of the Gross-Pitaevskii equation is obtained

$$j\frac{\partial}{\partial\tau}\Phi = [T+V]\Phi \tag{9.18}$$

where the terms T, V, and the normalized wave function Φ are

$$T = -\frac{\partial^2}{\partial\zeta^2} - \frac{\partial^2}{\partial\rho^2} + \frac{1}{\rho}\frac{\partial}{\partial\rho} - \frac{1}{\rho^2},\ V = \frac{\rho^2+\varepsilon\zeta^2}{4} + \frac{8\pi N\alpha|\Phi|^2}{\rho^2},\ \Phi = \rho S_l^{3/2}\Psi.$$

The Eq. (9.18) is solved in [40] for an unbounded space domain. The FDTD simulations of (9.18) are performed in [40] for a domain of a limited size. The theoretical results are compared with the measurements in [40] and [46].

In our work, the nonlinear 2-D time-dependent Eq. (9.18) is analyzed by the equivalent circuit approach and solved by a circuit simulator, which have a verified integrated tool for calculations of transients in nonlinear circuit networks. It avoids spending time on writing complicated codes for nonlinear simulations.

By analogy with the modeling of wave propagation in a transmission-line, the complex voltage $u=\Phi$ and current $\mathbf{i}=j\nabla'\cdot u$ are introduced. The operator ∇' is defined as $\nabla'\cdot f = \partial f/\partial\rho\hat{\mathbf{e}}_\rho + \partial f/\partial\zeta\hat{\mathbf{e}}_\zeta$ when acting on the scalars with $\hat{\mathbf{e}}_\rho$, $\hat{\mathbf{e}}_\zeta$ as the unit vectors along the radial and axial axes. This operator acting on a vector is $\nabla'\cdot\mathbf{A} = \rho\left(\partial\left(\rho^{-1}A_\rho\right)/\partial\rho\right) + \partial A_\zeta/\partial\zeta$, with A_ρ, A_ζ are the vector's radial and axial components, respectively.

The telegraph equation set is written in analogy with [14], and it is a system of partial differential equations of the first order:

$$\nabla'\cdot u = -j\,\mathbf{i}, \tag{9.19}$$

$$\nabla'\cdot\mathbf{i} = \frac{\partial u}{\partial\tau} - \frac{ju}{\rho^2} + \frac{j}{4}\left(\rho^2+\varepsilon\zeta^2\right)u + \frac{8j\pi N\alpha}{\rho^2}|u|^2 u. \tag{9.20}$$

In the coordinate form these equations are

$$\frac{\partial u}{\partial\rho}\hat{\mathbf{e}}_\rho + \frac{\partial u}{\partial\zeta}\hat{\mathbf{e}}_\zeta = -ji_\rho\hat{\mathbf{e}}_\rho - ji_\zeta\hat{\mathbf{e}}_\zeta \tag{9.21}$$

$$\frac{\partial i_\rho}{\partial\rho} + \frac{\partial i_\zeta}{\partial\zeta} - \frac{i_\rho}{\rho} = \frac{\partial u}{\partial\tau} + j\left[\frac{1}{4}\left(\rho^2+\varepsilon\zeta^2\right) - \frac{1}{\rho^2} + \frac{8\pi N\alpha}{\rho^2}|u|^2\right]u. \tag{9.22}$$

9.5.1.2 Domain Discretization and Boundary Conditions

To solve these equations, a 2-D mesh is introduced in the considered domain $[\rho_1\le\rho\le\rho_2]\times[\zeta_1\le\zeta\le\zeta_2]$ (Fig. 9.6, left part) where (9.19)-(9.22) are valid. A uniform isotropic grid is used particularly, and the space steps are equal in the radial and axial normalized coordinates, i.e., $\Delta\rho=\Delta\zeta=\Delta l$.

The corresponding radial and axial coordinates are easy to be calculated. For example, one can derive that $\rho=\rho_1+p\Delta l$ and $\zeta=\zeta_1+q\Delta l$ for a node indexed (p,q).

The grid consists of the sequentially connected equivalent circuits corresponding to a portion Δl^2. The right part of Fig. 9.6 illustrates the corresponding schematic view of nodes (p,q), $(p+1,q)$, $(p,q-1)$, and $(p+1,q-1)$ in the considered 2-D domain.

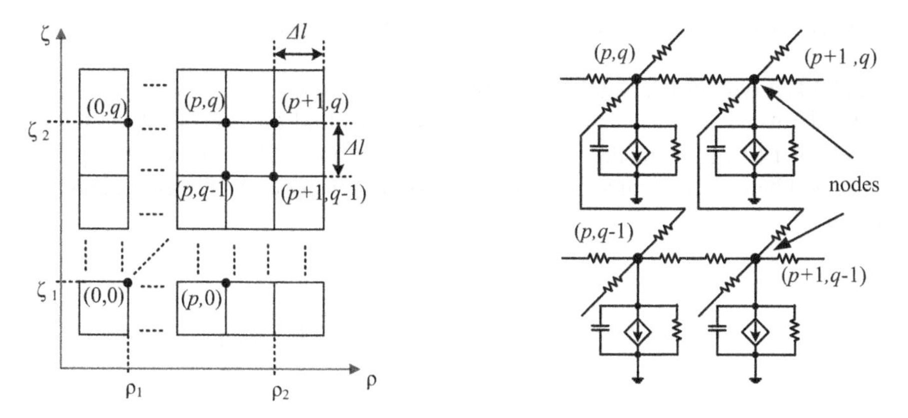

Fig. 9.6 Two-dimensional grid of considered domain. Left: computation grid for considered domain $[\rho_1 \le \rho \le \rho_2] \times [\zeta_1 \le \zeta \le \zeta_2]$; Right: corresponding schematic view of nodes (p,q), $(p+1,q)$, $(p,q-1)$, and $(p+1,q-1)$ in the considered domain

An attention should be paid to the nodes near the boundaries of the considered domain. Two nodes are shown in the left part of Fig. 9.7 near the boundary, and the schematic views of the two different boundary conditions are illustrated.

The middle part of Fig. 9.7 shows the view of the Type-1 boundary condition, which is similar to the Dirichlet one: $u = \Phi = 0|_{\text{on the boundary}}$. In this case, all connections to the boundary are grounded.

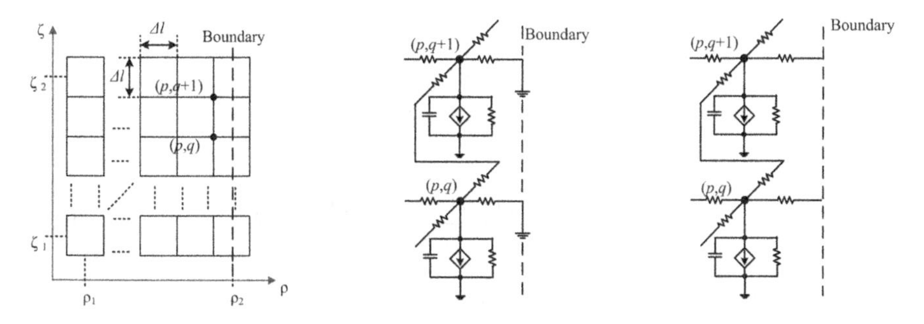

Fig. 9.7 Illustration of boundary conditions. Left: illustration of two nodes located near the boundary of the considered domain; Middle: schematic view of the Type-1 boundary condition for corresponding nodes; Right: schematic view of the Type-2 boundary condition for corresponding nodes

The right part of Fig. 9.7 shows the schematic view of the Type-2 boundary condition, which is similar to the Neumann one:

$$i_\rho = \frac{\partial u}{\partial \rho} = \frac{\partial \Phi}{\partial \rho} = 0 \,\Big|_{\text{on the boundary}} . \tag{9.23}$$

It is seen that all boundary connections of circuits are floating in this case.

9.5.1.3 Equivalent Circuit for an Infinitesimal Domain

Considering an elemental portion, one intends to find an equivalent circuit for (9.21) and (9.22). A model of an infinitesimal portion Δl^2 is shown in Fig. 9.8. Here and below, all parameters in the circuits are given per the dimensionless unit length.

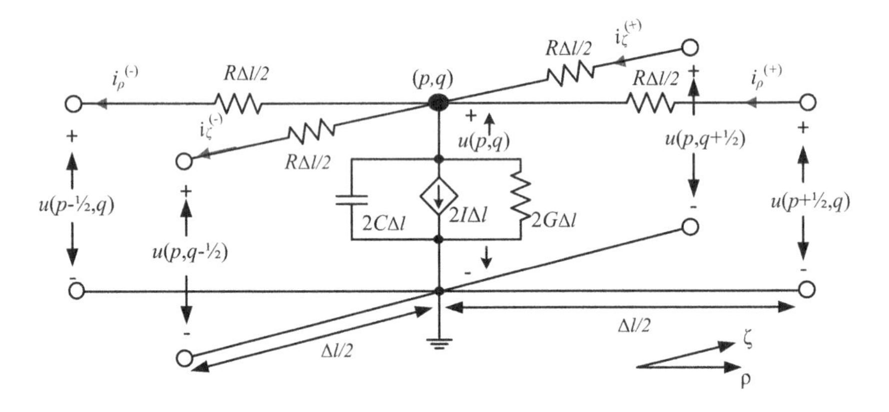

Fig. 9.8 Complex equivalent circuit for an infinitesimal portion Δl^2

The Kirchhoff's voltage law applied to the left loop in the radial branch of the circuit from Fig. 9.8 gives:

$$u(p,q) = i_\rho^{(-)} R \frac{\Delta l}{2} + u(p-1/2,q), \tag{9.24}$$

or

$$\frac{u(p,q) - u(p-1/2,q)}{\Delta l/2} = i_\rho^{(-)} R. \tag{9.25}$$

Taking the limit $\Delta l \to 0$ of (9.25) leads to

$$\frac{\partial u}{\partial \rho} = i_\rho R. \tag{9.26}$$

Similarly, applying the Kirchhoff's voltage law to the left loop in the axial branch of this circuit, one obtains

$$\frac{\partial u}{\partial \zeta} = i_\zeta R. \tag{9.27}$$

Comparing the corresponding terms in (9.26) and (9.27) with (9.21), one easily finds that $R = -j$ for the resistance.

Applying the Kirchhoff's current law to the node (p,q) gives

$$i_\rho^{(+)} - i_\rho^{(-)} + i_\zeta^{(+)} - i_\zeta^{(-)} = 2C\Delta l \frac{\partial u(p,q)}{\partial \tau} + 2I\Delta l + 2G\Delta l u(p,q), \tag{9.28}$$

or

$$\frac{i_\rho^{(+)} - i_\rho^{(-)}}{\Delta l} + \frac{i_\zeta^{(+)} - i_\zeta^{(-)}}{\Delta l} = 2C \frac{\partial u(p,q)}{\partial \tau} + 2I + 2Gu(p,q). \tag{9.29}$$

When $\Delta l \to 0$, one obtains from (9.29)

$$\frac{\partial i_\rho}{\partial \rho} + \frac{\partial i_\zeta}{\partial \zeta} = 2C \frac{\partial u}{\partial \tau} + 2I + 2Gu. \tag{9.30}$$

Comparing the corresponding terms in (9.30) with (9.22), it is easy to find that $C = 1/2$ for capacitance, $I = i_\rho / 2\rho = \left(i_\rho^{(+)} + i_\rho^{(-)}\right)/4\rho$ for the ideal current source, and $G = j\left[\left(\rho^2 + \varepsilon\zeta^2\right)/4 - \rho^{-2} + 8\pi N\alpha|u|^2 \rho^{-2}\right]/2$ for the conductance. The corresponding radial and axial coordinates are obtained according to the domain described in Section 9.5.1.2, namely, $\rho = \rho_1 + p\Delta l$ and $\zeta = \zeta_1 + q\Delta l$.

The ideal current source I is controlled by the average currents in the radial branch. The reason for the approximation $i_\rho \approx \left(i_\rho^{(+)} + i_\rho^{(-)}\right)/2$ is as follows. Imagine that three infinitesimal equivalent circuits are radially connected to each other. Here, $u(p-1,q)$, $u(p,q)$, and $u(p+1,q)$ denote the voltages for corresponding nodes as shown in Fig. 9.9.

Applying the Kirchhoff's voltage law to the loop, indicated by the dashed arrow in Fig. 9.9, we obtain

$$u(p-1,q) + i_\rho^{(-)} R\Delta l + i_\rho^{(+)} R\Delta l = u(p+1,q), \tag{9.31}$$

or

$$\frac{u(p+1,q) - u(p-1,q)}{2\Delta l} = R\frac{i_\rho^{(+)} + i_\rho^{(-)}}{2}. \tag{9.32}$$

Taking the limit $\Delta l \to 0$ and using $R = -j$, (9.32) is then written as

$$\frac{\partial u}{\partial \rho} = -j\frac{i_\rho^{(+)} + i_\rho^{(-)}}{2}. \tag{9.33}$$

Comparing (9.33) with the definition $\partial u / \partial \rho = -ji_\rho$, one obtains that $i_\rho = \left(i_\rho^{(+)} + i_\rho^{(-)}\right)/2$.

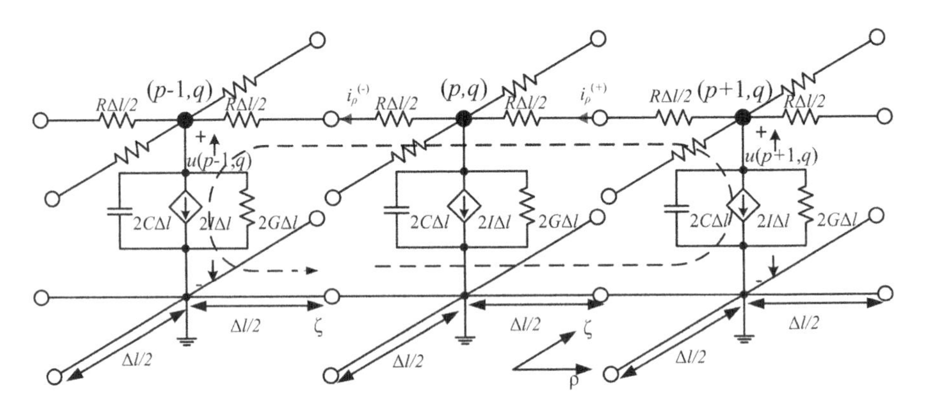

Fig. 9.9 Explanation of the approximation of the radial component of complex current

Due to the use of the imaginary components such a model is not compatible directly with many circuit tools, and our goal is to obtain the circuits and techniques, which are allowed for simulations of nonlinear Schrödinger equations.

In the following Sections, three different types of models based on the equivalent circuit approach are introduced. The mesh grid and the boundary conditions for each model are not repeated again since they are similar to those described in the previous Section.

9.5.1.4 Cross-coupled Equivalent Circuit

The first idea is to avoid the imaginary resistors, and one option is to separate the normalized wave function Φ into the real and imaginary parts: $\Phi = \mathrm{Re}(\Phi) + j\,\mathrm{Im}(\Phi)$.

By defining the real $\mathrm{Re}(u) = \mathrm{Re}(\Phi)$ and imaginary $\mathrm{Im}(u) = \mathrm{Im}(\Phi)$ voltages, (9.18) is written as

$$\frac{\partial}{\partial \tau}\mathrm{Im}(u) = \left[-\frac{\partial^2}{\partial \rho^2} + \frac{1}{\rho}\frac{\partial}{\partial \rho} - \frac{\partial^2}{\partial \zeta^2} - \frac{1}{\rho^2} + \frac{1}{4}\left(\rho^2 + \varepsilon\zeta^2\right) + \frac{8\pi N\alpha}{\rho^2}\left(\mathrm{Re}^2(u) + \mathrm{Im}^2(u)\right)\right]\mathrm{Re}(u), \tag{9.34}$$

$$\frac{\partial}{\partial \tau}\mathrm{Re}(u)=\left[-\frac{\partial^2}{\partial \rho^2}+\frac{1}{\rho}\frac{\partial}{\partial \rho}-\frac{\partial^2}{\partial \zeta^2}-\frac{1}{\rho^2}+\frac{1}{4}\left(\rho^2+\varepsilon\zeta^2\right)+\frac{8\pi N\alpha}{\rho^2}\left(\mathrm{Re}^2(u)+\mathrm{Im}^2(u)\right)\right]\mathrm{Im}(u). \quad (9.35)$$

The real and imaginary currents are defined as $\mathrm{Re}(\mathbf{i})=\nabla'\cdot\mathrm{Re}(u)$ and $\mathrm{Im}(\mathbf{i})=\nabla'\cdot\mathrm{Im}(u)$, respectively, and the resulting telegraph equations for (9.34) are

$$\nabla'\cdot\mathrm{Re}(u)=\mathrm{Re}(\mathbf{i}), \quad (9.36)$$

$$\nabla'\cdot\mathrm{Re}(\mathbf{i})=\frac{\partial\,\mathrm{Im}(u)}{\partial \tau}-\frac{\mathrm{Re}(u)}{\rho^2}+\frac{1}{4}\left(\rho^2+\varepsilon\zeta^2\right)\mathrm{Re}(u)+\frac{8\pi N\alpha}{\rho^2}\left(\mathrm{Re}^2(u)+\mathrm{Im}^2(u)\right)\mathrm{Re}(u). \quad (9.37)$$

Similarly, the telegraph equations for (9.35) are written as

$$\nabla'\cdot\mathrm{Im}(u)=\mathrm{Im}(\mathbf{i}), \quad (9.38)$$

$$\nabla'\cdot\mathrm{Im}(\mathbf{i})=-\frac{\partial\,\mathrm{Re}(u)}{\partial \tau}-\frac{\mathrm{Im}(u)}{\rho^2}+\frac{1}{4}\left(\rho^2+\varepsilon\zeta^2\right)\mathrm{Im}(u)+\frac{8\pi N\alpha}{\rho^2}\left(\mathrm{Re}^2(u)+\mathrm{Im}^2(u)\right)\mathrm{Im}(u). \quad (9.39)$$

It is seen that these telegraph equations are coupled with each other after this separation of the real and imaginary parts of the normalized wave function Φ.

Some simple mathematical derivations allow to rewrite (9.36)-(9.39) as

$$\frac{\partial\,\mathrm{Re}(u)}{\partial \rho}\hat{\mathbf{e}}_\rho+\frac{\partial\,\mathrm{Re}(u)}{\partial \zeta}\hat{\mathbf{e}}_\zeta=\mathrm{Re}(i_\rho)\hat{\mathbf{e}}_\rho+\mathrm{Re}(i_\zeta)\hat{\mathbf{e}}_\zeta, \quad (9.40)$$

$$\begin{aligned}\frac{\partial\,\mathrm{Re}(i_\rho)}{\partial \rho}+\frac{\partial\,\mathrm{Re}(i_\zeta)}{\partial \zeta}&=\\ =\frac{\partial\,\mathrm{Re}(u)}{\partial \tau}&+\left[\frac{1}{4}\left(\rho^2+\varepsilon\zeta^2\right)-\frac{1}{\rho^2}+\frac{8\pi N\alpha}{\rho^2}\left(\mathrm{Re}^2(u)+\mathrm{Im}^2(u)\right)\right]\mathrm{Re}(u)+ \quad (9.41)\\ +\frac{\mathrm{Re}(i_\rho)}{\rho}&+\frac{\partial\,\mathrm{Im}(u)}{\partial \tau}-\frac{\partial\,\mathrm{Re}(u)}{\partial \tau},\end{aligned}$$

$$\frac{\partial\,\mathrm{Im}(u)}{\partial \rho}\hat{\mathbf{e}}_\rho+\frac{\partial\,\mathrm{Im}(u)}{\partial \zeta}\hat{\mathbf{e}}_\zeta=\mathrm{Im}(i_\rho)\hat{\mathbf{e}}_\rho+\mathrm{Im}(i_\zeta)\hat{\mathbf{e}}_\zeta, \quad (9.42)$$

and

$$\frac{\partial \operatorname{Im}(i_\rho)}{\partial \rho}+\frac{\partial \operatorname{Im}(i_\zeta)}{\partial \zeta}=$$
$$=\frac{\partial \operatorname{Im}(u)}{\partial \tau}+\left[\frac{1}{4}\left(\rho^2+\varepsilon\zeta^2\right)-\frac{1}{\rho^2}+\frac{8\pi N\alpha}{\rho^2}\left(\operatorname{Re}^2(u)+\operatorname{Im}^2(u)\right)\right]\operatorname{Im}(u)+ \tag{9.43}$$
$$+\frac{\operatorname{Im}(i_\rho)}{\rho}-\frac{\partial \operatorname{Im}(u)}{\partial \tau}-\frac{\partial \operatorname{Re}(u)}{\partial \tau}.$$

Considering an infinitesimal portion Δl^2, we intend to find two equivalent circuits for the real and imaginary parts. Thanks to these systems of the telegraph equations, the two resulting circuits are coupled with each other.

The equivalent circuit model for the real part is shown in the left part of Fig. 9.10.

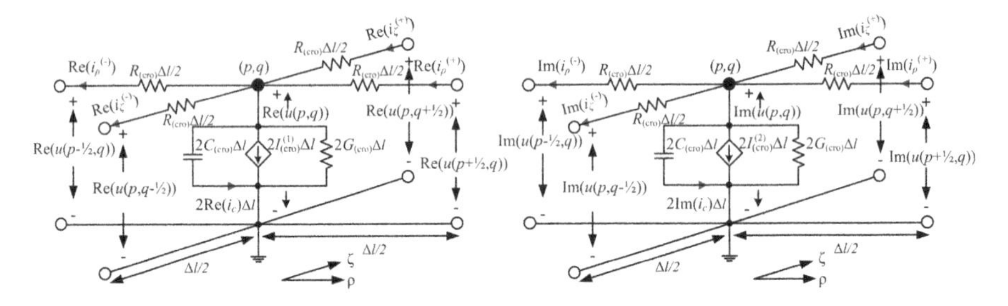

Fig. 9.10 Real (left) and imaginary (right) parts of a cross-coupled equivalent circuit for an infinitesimal portion Δl^2

Applying the Kirchhoff's voltage law to the left loop in the radial branch of the real part (Fig. 9.10, left), we obtain

$$\operatorname{Re}(u(p,q))=\operatorname{Re}\left(i_\rho^{(-)}\right)R_{(\text{cro})}\Delta l/2+\operatorname{Re}(u(p-1/2,q)), \tag{9.44}$$

or

$$\frac{\operatorname{Re}(u(p,q))-\operatorname{Re}(u(p-1/2,q))}{\Delta l/2}=\operatorname{Re}\left(i_\rho^{(-)}\right)R_{(\text{cro})}. \tag{9.45}$$

Taking the limit $\Delta l \to 0$ of (9.45) leads to

$$\frac{\partial \operatorname{Re}(u)}{\partial \rho}=\operatorname{Re}(i_\rho)R_{(\text{cro})}. \tag{9.46}$$

Similarly, applying the Kirchhoff's law to the left loop of the axial branch of this circuit yields

$$\frac{\partial \operatorname{Re}(u)}{\partial \zeta}=\operatorname{Re}(i_\zeta)R_{(\text{cro})}. \tag{9.47}$$

Comparing the corresponding terms in (9.46) and (9.47) with (9.40), one finds that $R_{(\text{cro})} = 1$ for the resistance parameter. Applying the Kirchhoff's current law to the node (p,q) in the left part of Fig. 9.10 yields

$$\begin{aligned}&\operatorname{Re}\left(i_\rho^{(+)}\right)-\operatorname{Re}\left(i_\rho^{(-)}\right)+\operatorname{Re}\left(i_\zeta^{(+)}\right)-\operatorname{Re}\left(i_\zeta^{(-)}\right)=\\&=2C_{(\text{cro})}\Delta l\frac{\partial\operatorname{Re}\left(u(p,q)\right)}{\partial\tau}+2I_{(\text{cro})}^{(1)}\Delta l+2G_{(\text{cro})}\Delta l\operatorname{Re}\left(u(p,q)\right),\end{aligned}\tag{9.48}$$

or

$$\begin{aligned}&\frac{\operatorname{Re}\left(i_\rho^{(+)}\right)-\operatorname{Re}\left(i_\rho^{(-)}\right)}{\Delta l}+\frac{\operatorname{Re}\left(i_\zeta^{(+)}\right)-\operatorname{Re}\left(i_\zeta^{(-)}\right)}{\Delta l}=\\&=2C_{(\text{cro})}\frac{\partial\operatorname{Re}\left(u(p,q)\right)}{\partial\tau}+2I_{(\text{cro})}^{(1)}+2G_{(\text{cro})}\operatorname{Re}\left(u(p,q)\right).\end{aligned}\tag{9.49}$$

Taking the limit $\Delta l \to 0$ of (9.49) leads to

$$\frac{\partial\operatorname{Re}\left(i_\rho\right)}{\partial\rho}+\frac{\partial\operatorname{Re}\left(i_\zeta\right)}{\partial\zeta}=2C_{(\text{cro})}\frac{\partial\operatorname{Re}(u)}{\partial\tau}+2I_{(\text{cro})}^{(1)}+2G_{(\text{cro})}\operatorname{Re}(u).\tag{9.50}$$

Comparing the corresponding terms in (9.50) with (9.41), one easily finds the capacitance $C_{(\text{cro})} = 1/2$, the ideal current source $I_{(\text{cro})}^{(1)} = \left[\operatorname{Re}\left(i_\rho\right)/\rho + \partial\operatorname{Im}(u)/\partial\tau - \partial\operatorname{Re}(u)/\partial\tau\right]/2$, and the conductance $G_{(\text{cro})} = \left[\left(\rho^2+\varepsilon\zeta^2\right)/4 - \rho^{-2} + 8\pi N\alpha\rho^{-2}\left(\operatorname{Re}^2(u)+\operatorname{Im}^2(u)\right)\right]/2$, respectively.

The current owing to the capacitance is denoted by $\operatorname{Re}(i_c)$ per dimensionless unit length (Fig. 9.10, left), therefore it is easy to find that $2\operatorname{Re}(i_c)\Delta l = 2C_{(\text{cro})}\Delta l\partial\operatorname{Re}(u)/\partial\tau$. Using $C_{(\text{cro})} = 1/2$, one obtains $\operatorname{Re}(i_c) = \partial\operatorname{Re}(u)/2\partial\tau$.

Similarly, the current due to the capacitance for the imaginary part of the equivalent circuit is denoted by $\operatorname{Im}(i_c)$, and it satisfies $\operatorname{Im}(i_c) = \partial\operatorname{Im}(u)/2\partial\tau$. Thus, the ideal current source can be simplified as $I_{(\text{cro})}^{(1)} = \operatorname{Re}\left(i_\rho\right)/2\rho + \operatorname{Im}(i_c) - \operatorname{Re}(i_c)$.

This procedure can be applied to the right part of Fig. 9.10 to derive the circuit parameters. The resistance $R_{(\text{cro})} = 1$, capacitance $C_{(\text{cro})} = 1/2$, and conductance $G_{(\text{cro})} = \left[\left(\rho^2+\varepsilon\zeta^2\right)/4 - \rho^{-2} + 8\pi N\alpha\rho^{-2}\left(\operatorname{Re}^2(u)+\operatorname{Im}^2(u)\right)\right]/2$ are of the same value as those in the real part (Fig. 9.10, left). The ideal current source for the imaginary part is given by $I_{(\text{cro})}^{(2)} = \operatorname{Im}\left(i_\rho\right)/2\rho - \operatorname{Im}(i_c) - \operatorname{Re}(i_c)$. The corresponding radial and axial coordinates are obtained according to the domain described in Section 9.5.1.2.

In summary, the proposed cross-coupled equivalent circuit is a general model, and it can be used to simulate the arbitrary time dependences. Furthermore, it is compatible for most circuit simulators, since it does not consist of imaginary resistors. Unfortunately, because the circuit topology is doubled, the computational complexity also increases, and some techniques on faster simulations are considered below.

9.5.1.5 Envelope Technique

It is possible to use the envelope technique [15] provided by the Agilent ADS when the solution of Gross-Pitaevskii equation is written in the form of an envelope function:

$$\Phi = \sum_k f_k(\mathbf{r},\tau)e^{-j\omega_k\tau}. \tag{9.51}$$

Similarly to the theory of the modulated signals, this time-domain waveform is considered as a sum of different k-th carriers. Then $f_k(\mathbf{r},\tau)$ is the modulation function in (9.51), and ω_k is the angular frequency of the k-th carriers.

In our particular case, only one carrier frequency is considered, and (9.51) is simplified as

$$\Phi = f(\mathbf{r},\tau)e^{-j\omega\tau} \tag{9.52}$$

The idea of the envelope simulation is to transform the original solution with high-frequency modulated carrier into an equivalent one of the low-frequency modulation. This brings benefits when performing simulation.

Substituting (9.52) into (9.18) the normalized Gross-Pitaevskii equation can be written as

$$j\frac{\partial}{\partial\tau}f(\mathbf{r},\tau)+\omega f(\mathbf{r},\tau)=[T+V]f(\mathbf{r},\tau), \tag{9.53}$$

or

$$j\frac{\partial}{\partial\tau}f(\mathbf{r},\tau)=[T+V_{(\mathrm{env})}]f(\mathbf{r},\tau), \tag{9.54}$$

where the coefficient $V_{(\mathrm{env})}$ is defined as

$$V_{(\mathrm{env})}=\frac{1}{4}\left(\rho^2+\varepsilon\zeta^2\right)+\frac{8\pi N\alpha}{\rho^2}\left|f(\mathbf{r},\tau)\right|^2-\omega. \tag{9.55}$$

The problem is now solving the modulation function $f(\mathbf{r},\tau)$ which usually varies more slowly with time than Φ does. The telegraph equations for (9.54) can be easily obtained by defining the complex voltage $u_{(\mathrm{env})} = f(\mathbf{r},\tau)$ and the complex current $\mathbf{i}_{(\mathrm{env})} = j\nabla' \cdot u_{(\mathrm{env})}$

$$\nabla' \cdot u_{(\mathrm{env})} = -j\,\mathbf{i}_{(\mathrm{env})}, \tag{9.56}$$

$$\nabla' \cdot \mathbf{i}_{(\mathrm{env})} = \frac{\partial u_{(\mathrm{env})}}{\partial \tau} + j\left[\frac{1}{4}\left(\rho^2 + \varepsilon\zeta^2\right) - \frac{1}{\rho^2} + \frac{8\pi N\alpha}{\rho^2}\left|u_{(\mathrm{env})}\right|^2 - \omega\right]u_{(\mathrm{env})}. \tag{9.57}$$

The equivalent circuit model for an infinitesimal portion Δl^2 is shown in Fig. 9.11.

Using the same method as in Section 9.5.1.3, the values of the circuit parameters are obtained, i.e., $R_{(\mathrm{env})} = -j$, $C_{(\mathrm{env})} = 1/2$, $I_{(\mathrm{env})} = \left(i^{(+)}_{\rho\,(\mathrm{env})} + i^{(-)}_{\rho\,(\mathrm{env})}\right)/4\rho$, and $G_{(\mathrm{env})} = j\left[\left(\rho^2 + \varepsilon\zeta^2\right)/4 - \rho^{-2} + 8\pi N\alpha\left|u_{(\mathrm{env})}\right|^2\rho^{-2} - \omega\right]/2$, respectively. The corresponding discrete radial and axial coordinates are calculated according to the domain described in Section 9.5.1.2.

The proposed model contains complex resistors, and the Agilent ADS circuit envelope simulator, which supports the imaginary ones, can emulate them. For other software tools that do not support these components, one can also separate the real and imaginary parts of (9.54) to obtain almost the same cross-coupled topology shown in Fig. 9.10. The derivation process is similar to that described in Section 9.5.1.4.

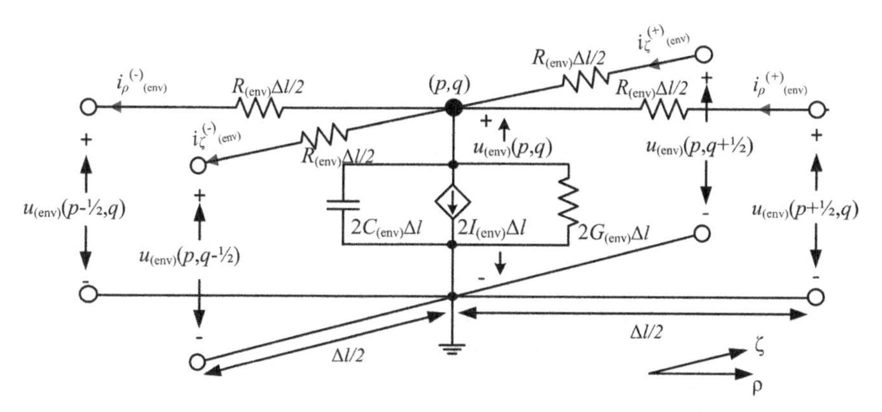

Fig. 9.11 Equivalent circuit of an infinitesimal portion Δl^2 for envelope simulations

9.5.1.6 Time-dependent Component Approach

Another technique to speed-up the calculations is with the time-depending resistors. By applying the Euler's formula, the normalized wave function (9.52) can be written as

$$\begin{aligned}\mathrm{Re}(\Phi) &= f(\mathbf{r},\tau)\cos\omega\tau,\\ \mathrm{Im}(\Phi) &= -f(\mathbf{r},\tau)\sin\omega\tau.\end{aligned} \tag{9.58}$$

Accordingly, the imaginary part can be written as a function of the real part:

$$\mathrm{Im}(\Phi) = -\mathrm{Re}(\Phi)\tan\omega\tau. \tag{9.59}$$

Using the relationship between the real and imaginary parts in (9.59), eq. (9.35) is transformed to:

$$\frac{\partial}{\partial\tau}\mathrm{Re}(\Phi) = -\tan\omega\tau\left[T + V_{(\mathrm{td})}\right]\mathrm{Re}(\Phi), \tag{9.60}$$

where the coefficient $V_{(\mathrm{td})}$ is

$$V_{(\mathrm{td})} = \frac{1}{4}\left(\rho^2 + \varepsilon\zeta^2\right) + \frac{8\pi N\alpha}{\rho^2}\left(1 + \tan^2(\omega\tau)\right)\mathrm{Re}(\Phi)^2. \tag{9.61}$$

By defining the real voltage $u_{(\mathrm{td})} = \mathrm{Re}(\Phi)$ and the real current $\mathbf{i}_{(\mathrm{td})} = \tan(\omega\tau)\nabla'\cdot u_{(\mathrm{td})}$, the telegraph equations for (9.60) are written as

$$\nabla'\cdot u_{(\mathrm{td})} = \frac{1}{\tan(\omega\tau)}\mathbf{i}_{(\mathrm{td})}, \tag{9.62}$$

$$\nabla'\cdot\mathbf{i}_{(\mathrm{td})} = \frac{\partial u_{(\mathrm{td})}}{\partial\tau} + \tan(\omega\tau)\left[\frac{1}{4}\left(\rho^2 + \epsilon\zeta^2\right) - \frac{1}{\rho^2} + \frac{8\pi N\alpha}{\rho^2}\left(1 + \tan^2\omega\tau\right)u_{(\mathrm{td})}{}^2\right]u_{(\mathrm{td})} \tag{9.63}$$

The resulting equivalent circuit for an infinitesimal portion Δl^2 is shown in Fig. 9.12.

Using the same method in Section 9.5.1.3, the values of circuit parameters are obtained. They are given by $R_{(\mathrm{td})} = 1/\tan(\omega\tau)$, $C_{(\mathrm{td})} = 1/2$, $I_{(\mathrm{td})} = \left(i^{(+)}_{\rho\,(\mathrm{td})} + i^{(-)}_{\rho\,(\mathrm{td})}\right)/4\rho$, and $G_{(\mathrm{td})} = \tan(\omega\tau)\left[\left(\rho^2 + \varepsilon\zeta^2\right)/4 - \rho^{-2} + 8\pi N\alpha u^2_{(\mathrm{td})}\rho^{-2}\left(1 + \tan^2(\omega\tau)\right)\right]/2$, respectively. The corresponding discrete radial and axial coordinates are calculated according to the domain described in the Section 9.5.1.2.

Note that this topology consists of no imaginary resistors and it is as simple as the circuit presented in Fig. 9.8. The time-dependent components are easy to be implemented in most circuit simulators. The performance is expected to be twice faster than the cross-coupled equivalent circuit approach since the topology is simplified by half.

The resistors and conductance parameters in this model are the functions of $\tan(\omega\tau)$, therefore $R_{(\mathrm{td})}$ and $G_{(\mathrm{td})}$ grow to infinity when $\omega\tau = k\pi/2$, where k is a positive integer. Thus the dimensionless time step needs to be carefully chosen to avoid these time points.

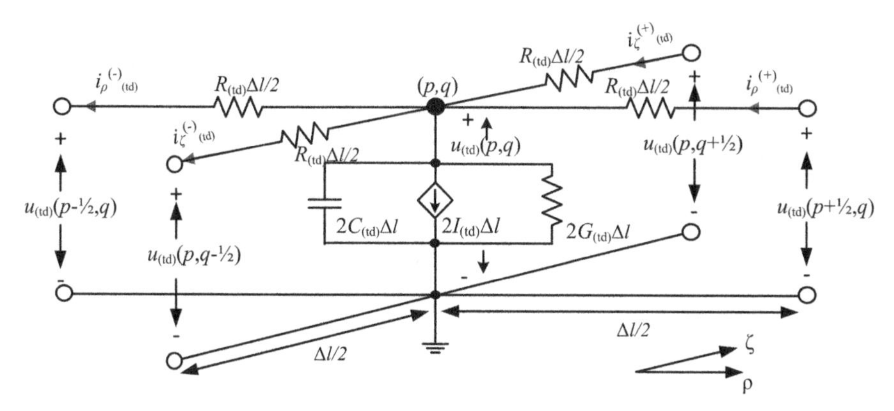

Fig. 9.12 Equivalent circuit with the time-dependent components for an infinitesimal portion Δl^2

9.5.1.7 Stability and Convergence of the Solution

The equivalent circuit approach is a discretization process similar to other numerical methods. The difference is that the equivalent circuit approach is only a spatial discretization process, and the time discretization is left for a circuit simulator. This results in the question how one specifies the time step in the circuit solver to keep the stability.

It is very difficult to derive a general stability condition in this work because it is largely dependent on a used software tool. Our simulations are performed by the Agilent ADS. According to our observation, the stability condition for the complex equivalent circuits (Fig. 9.8) and the cross-coupled equivalent circuits (Fig. 9.10) generally comply with the same stability condition in [40], namely, $\Delta\tau\left|4/\Delta^2+V_{\mathrm{M}}\right|<2$ where $\Delta\tau$ is the dimensionless time step, Δ is the dimensionless space step, and V_{M} is the maximum potential of V in Eq. (9.18).

The envelope equivalent circuit method is a larger-time-marching model compared with the cross-coupled one and with the numerical algorithm from [40]. If the critical time steps (largest acceptable time step to keep stability) for the cross-coupled and envelope equivalent circuits are $\Delta\tau_{\mathrm{c}}$ and $\Delta\tau_{\mathrm{c(env)}}$, respectively, it is proved that the envelope approach has larger critical time steps (i.e., $\Delta\tau_{\mathrm{c(env)}}>\Delta\tau_{\mathrm{c}}$). A proof of it is given in [24].

The convergence of the solution can be a problem if the mesh is too coarse. An easy way is to use a finer mesh ($\Delta l\to 0$), but this increases the memory usage. The installed memory on a computer also puts a limit on the mesh size. In our simulations presented below, a mesh of minimum 25 by 25 cells is recommended for all cases as shown by the study of calculations of the convergence.

9.5.2 Numerical Validation

The models proposed in Section 9.5.1 are validated by comparisons with the published theoretical and experimental data from [40],[46], where the analytical solution of ground-state condensate (no self-interaction), the analytical solution of free self-interacting condensates (no external potential), and the experimental data are given. Our simulations are carried out on a typical desktop computer, and its operative memory restricts the grid resolution to only a few thousands of elementary equivalent circuits.

9.5.2.1 Ground-state Condensate

In this case, the self-interacting term is neglected, and the harmonic potential is described by

$$V_{\text{ext}}(\rho,\zeta)=\frac{\rho^2+\varepsilon\zeta^2}{4}. \tag{9.64}$$

The analytical solution of the nonlinear Schrödinger equation (9.18) for ground-state condensate is given in [40]:

$$\Phi(\rho,\zeta,\tau)=\rho\frac{\varepsilon^{1/8}}{(2\pi)^{3/4}}e^{-(\rho^2+\varepsilon\zeta^2)/4}e^{-j[(2+\varepsilon)/2]\tau}. \tag{9.65}$$

Since (9.65) is written in the form of $\Phi=f(\mathbf{r},\tau)\exp(-j\omega\tau)$, this case can be tested by all models proposed in Section 9.5.1.

For simplicity, the aspect ratio of the trapping potential is set to $\varepsilon=0.5$. The analytical solution shows that the initial voltage of all nodes within the considered domain follows a space distribution of $\rho\varepsilon^{1/8}(2\pi)^{-3/4}\exp\left(-(\rho^2+\varepsilon\zeta^2)/4\right)$.

The simulation is performed on a 33×33 grid with the integration domain $[-10\le\rho\le 10]\times[-10\le\zeta\le 10]$ and the time step is set to $\Delta\tau=10^{-2}$. The simulation lasts until $T=10$ time units. The absolute value of the analytical solution on the boundaries $\rho=\pm 10$ and $\zeta=\pm 10$ are approximately zero, and the small reflections from the boundaries can be neglected. Therefore, the Type-1 boundary condition (Fig. 9.7, middle) is used to match the analytical solution.

The time correspondence is checked for the node located at $(\rho,\zeta)=(1.25,0)$, where the absolute value of the analytical solution almost reaches the maximum. The space distribution of the numerical solution at $(\zeta,\tau)=(0,10)$ along the radial axis and at $(\rho,\tau)=(1.25,10)$ along the axial axis is then compared with the analytical one.

The cross-coupled equivalent circuit method is used first to analyze this case. The results are shown in Fig. 9.13.

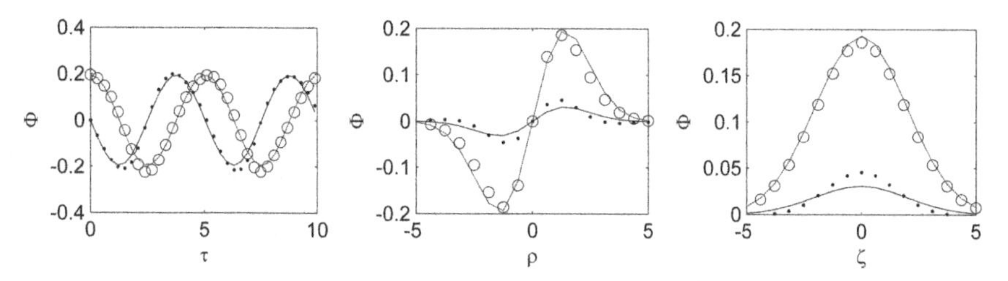

Fig. 9.13 Ground-state condensates analyzed with the cross-coupled equivalent circuit. Dashed line: real part of analytical solution; Solid line: imaginary part of analytical solution; Circles: real part of numerical solution; Points: imaginary part of numerical solution. Left: time evolution of wave function at $\rho = 1.25$, $\zeta = 0$; Middle: space distribution at $\zeta = 0$, $\tau = 10$ as a function of ρ; Right: space distribution at $\rho = 1.25$, $\tau = 10$ as a function of ζ

Both the time and space correspondences are investigated. Visual inspection of the time distribution leads to the conclusion that the numerical solutions show good agreement with the analytical ones (Fig. 9.13, left). The space distribution along the radial and axial axis also shows a good accordance with the analytical solutions (middle and right parts of Fig. 9.13, correspondingly).

The envelope technique is applied to this case. It should be mentioned that the numerical results are compared with the analytical ones of the modulation function $f(\mathbf{r},\tau) = \rho\varepsilon^{1/8}(2\pi)^{-3/4}\exp\left(-\left(\rho^2+\varepsilon\zeta^2\right)/4\right)$ with a single carrier frequency $\omega = (2+\varepsilon)/2$. Note that $f(\mathbf{r},\tau)$ is a complex function when solving Eq. (9.54), but, in this special case, the modulation function has only real part, and it is not time-dependent.

Starting from $\tau = 0$, the real part of the modulation function remains unchanged, and the imaginary part is always zero. The numerical solutions and the analytical ones are plotted and compared in Fig. 9.14. The numerical curves have some slight fluctuations in the time domain compared to the analytical ones (Fig. 9.14, left). The space distribution of numerical solutions has tiny fluctuation in both radial and axial coordinates regarding to the imaginary part (middle and right parts of Fig. 9.14). These tiny fluctuations are due to the use of coarse grid with a larger space step. In summary, the envelope approach provides good accuracy for the ground-state harmonic potential calculations.

Finally, the equivalent circuits with time-dependent resistors are tested. It is necessary to set the angular frequency $\omega = (2+\varepsilon)/2$ so that the time-dependent resistors and conductors can be evaluated. The results are shown in Fig. 9.15. The time step is carefully chosen for $\Delta\tau = 0.0087$ so that the time-dependent resistors and conductors do not reach the infinity.

A conclusion is made analyzing the results in Fig. 9.15 that the numerical solutions in this case show good correspondence to the analytical ones in both time and space domains.

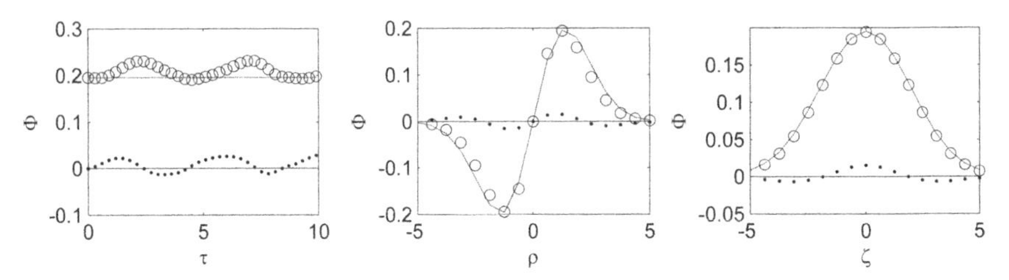

Fig. 9.14 Ground-state condensates analyzed with the envelope approach. Dashed line: real part of analytical solution; Solid line: imaginary part of analytical solution; Circles: real part of numerical solution; Points: imaginary part of numerical solution. Left: time evolution of modulation function at $\rho = 1.25$, $\zeta = 0$; Middle: space distribution at $\zeta = 0$, $\tau = 0$ as a function of ρ ; Right: space distribution at $\rho = 1.25$, $\tau = 10$ as a function of ζ

It should be pointed out that $\tan(\omega\tau)$ approaches zero or infinity when $\omega\tau = k\pi/2$, where k is an integer number. The time-dependent components attain large values periodically with the time, and, consequently, additional error is introduced due to the numerical approximation of derivatives. Therefore, our data periodically suffer from large errors close to these points (Fig. 9.15, left).

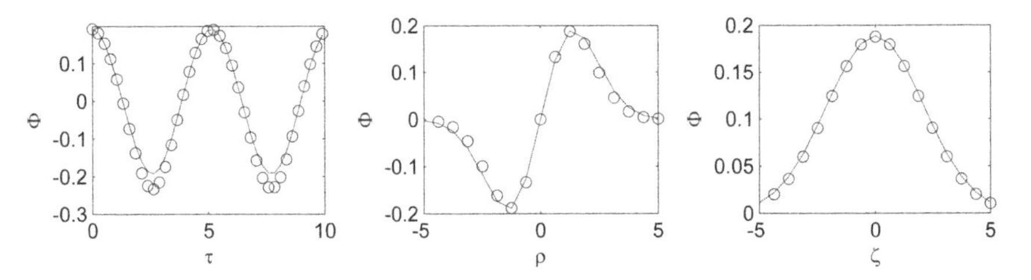

Fig. 9.15 Ground-state condensates analyzed with the equivalent circuits with time-dependent resistors. Dashed line: analytical solution; Circles: numerical solution. Left: Time evolution of modulation function at $\rho = 1.25$, $\zeta = 0$; Middle: Space distribution at $\zeta = 0$, $\tau = 10$ as a function of ρ ; Right: Space distribution at $\rho = 1.25$, $\tau = 10$ as a function of ζ

9.5.2.2 Free Self-interacting Condensate

It is time to turn to the free self-interacting condensate without external potential. The simulation is performed by a 33x33 grid with the integration domain $[1 \le \rho \le 2] \times [1 \le \zeta \le 2]$. The simulation goes up to $T = 0.1$ time unit with the time step $\Delta\tau = 10^{-4}$.

Since the analytical solution is a continuous function, the boundary conditions are set to the Type-2 (Fig. 9.7, right) on all boundaries so that the continuity is preserved. Although setting the boundary conditions to the Type-2 leads to some reflections of condensate, they are so small that can be neglected according to our study.

The analytical solution of this case is given in [40]:

$$\Phi(\rho,\zeta,\tau)=\rho\frac{e^{\zeta\sqrt{16\pi\alpha N}}+1}{e^{\zeta\sqrt{16\pi\alpha N}}-1}e^{-8j\pi\alpha N\tau}. \tag{9.66}$$

The constant parameters of the self-interacting term in (9.66) are chosen for the same value as in [40], which corresponds to $8\pi N\alpha=210$.

The time evolution of numerical solutions for the elemental equivalent circuit at $(\rho,\zeta)=(1.6,1.5)$ is compared with the analytical one. The space distribution of the numerical solutions at $(\zeta,\tau)=(1.5,0.1)$ along the radial axis and the space distribution at $(\rho,\tau)=(1.6,0.1)$ along the axial axis are compared with the analytical ones. The cross-coupled equivalent circuit is used in this case, and the time and space distribution of the numerical solutions are plotted and compared with the analytical ones (Fig 9.16, left). The space distribution also shows good correspondence (middle and right parts of Fig. 9.16).

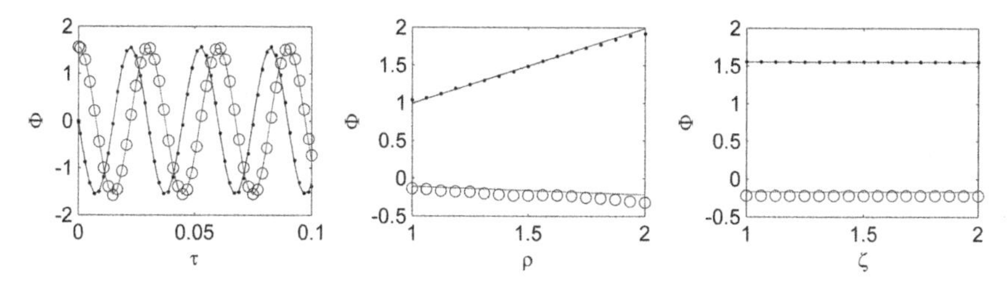

Fig. 9.16 Self-interacting condensate analyzed with the cross-coupled equivalent circuit approach. Dashed line: real part of analytical solution; Solid line: imaginary part of analytical solution; Circles: real part of numerical solution; Points: imaginary part of numerical solution. Left: time evolution of wave function at $\rho=1.6$, $\zeta=1.5$; Middle: space distribution at $\zeta=1.5$, $\tau=0.1$ as a function of ρ; Right: space distribution at $\rho=1.6$, $\tau=0.1$ as a function of ζ

The envelope approach analyzes this case. The numerical solutions are compared with the analytical modulation function $f(\mathbf{r},\tau)=\rho\cdot(\exp(\zeta\sqrt{16\pi\alpha N})+1)(\exp(\zeta\sqrt{16\pi\alpha N})-1)^{-1}$ at the angular frequency $\omega=8\pi N\alpha=210$. The numerical results in Fig. 9.17 show good correspondence with the analytical solutions in both time and space domains. It has to be mentioned that Fig.9.17 shows much less fluctuation in both time and space domain compared with Fig.9.14, and this is due to the use of smaller space step.

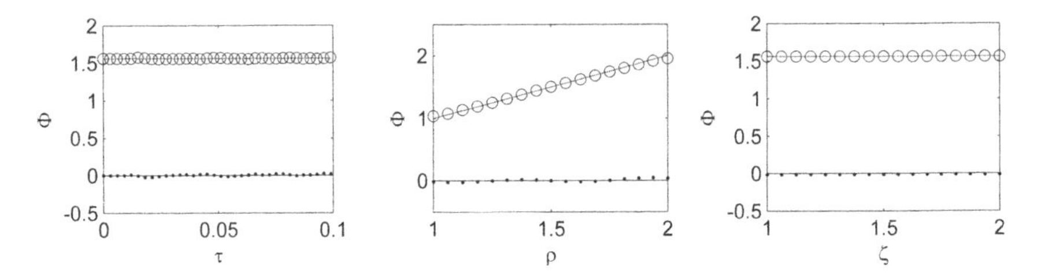

Fig. 9.17 Self-interacting condensate analyzed with the envelope approach. Dashed line: real part of analytical solution; Solid line: imaginary part of analytical solution; Circles: real part of numerical solution; Points: imaginary part of numerical solution; Left: time evolution of modulation function at $\rho = 1.6$, $\zeta = 1.5$; Middle: space distribution at $\zeta = 1.5$, $\tau = 0.1$ as a function of ρ; Right: space distribution at $\rho = 1.6$, $\tau = 0.1$ as a function of ζ

Finally, the equivalent circuit model with the time-dependent resistors is applied to this case. The time step is chosen for $\Delta\tau = 9.8\times10^{-4}$, so that the time-dependent resistors and conductors do not grow to infinity. The time and space distributions of the numerical solutions still show good correspondence with the analytical ones (Fig. 9.18). The time distribution of the numerical solutions does not suffer from obvious periodic error. This is mainly because the space step used here is smaller than that in the ground-state harmonic potential case (Section 9.5.2.1).

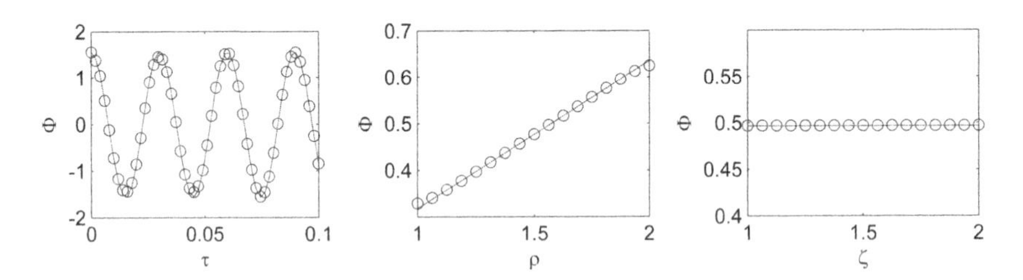

Fig. 9.18 Self-interacting condensate analyzed with equivalent circuits with time-dependent resistors. Dashed line: analytical solution; Circles: numerical solution. Left: time evolution of modulation function at $\rho = 1.6$, $\zeta = 1.5$; Middle: space distribution at $\zeta = 1.5$, $\tau = 0.1$ as a function of ρ; Right: space distribution at $\rho = 1.6$, $\tau = 0.1$ as a function of ζ

9.5.2.3 Comparison with the Experimental Data

Additionally to comparisons with the analytical and numerical data, our approach is verified by published experimental results from [40],[46]. Unfortunately, only the cross-coupled equivalent circuit can be used to simulate this experimental case, since there is no any analytical solution available and no information of oscillation frequency is known.

The experiment refers to 4000 Rb^{87} atoms, which are initially confined in a harmonic trap of frequency 56.25 Hz. This leads to the unit length $S_l = 1\,\mu m$ and the unit time $S_t = 2.8$ ms. The condensate is allowed to expand freely by suddenly removing trap at $t = 0$.

The scattering length of Rb^{87} atoms is $a = 110a_0$ with a_0 as the Bohr radius. In order to reduce the computation time, only a quarter of the condensate is computed since it is symmetric about both radial and axial axes. The integration domain is $[0 \le \rho \le 20] \times [0 \le \zeta \le 20]$ with 31x31 grid resolution. The time step is chosen as $\Delta\tau = 10^{-3}$. Since the condensate is symmetric, the boundaries along $\rho = 0$ and $\zeta = 0$ are set to the Type-2 condition (Fig. 9.7, right) in order to preserve continuity.

Since the condensate is expanding in a very slow rate, and the wave function is almost zero during entire simulation along $\rho = 20$ and $\zeta = 20$, the boundary condition of the Type-1 (Fig. 9.7, middle part) is therefore introduced to these boundaries. The initial distribution of the condensate density is chosen in the form of a Gaussian function $\Phi(\rho, \zeta, 0) = A\exp\left(-\beta(\rho^2 + \varepsilon\zeta^2)\right)$ according to [40]. Other parameters are $\varepsilon = 8$ and $\beta = 0.08$, which are specially chosen so that the initial distribution is close to the one of the measurements [46]. The scaling factor A should be small enough in order to maintain the stability for nonlinearity, and its typical value is $A = 10^{-3}$.

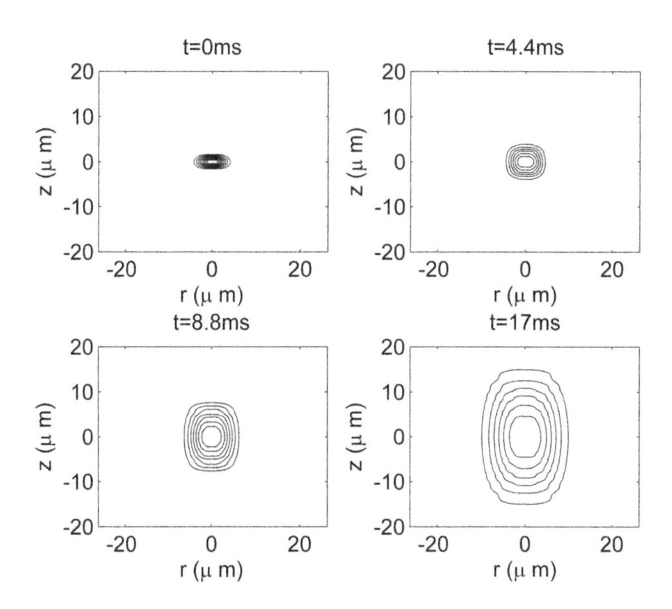

Fig. 9.19 Numerical simulations of experiment described in [40],[46] for free expanded condensate shown by contour plots of condensate density at t = 0, 4.4, 8.8, and 17 ms

The previous studies in [40] and [46] show that the condensate turns from an initial radial-elongated shape to the axial-elongated one. Our simulations, as shown in Fig. 9.19, also confirm this behavior. The contour lines of condensate density is plotted at t=0, 4.4, 8.8, and 17 ms with respect to the upper left, upper right, lower left, and lower right parts. Note the negative radial part in Fig. 9.19 has an angular shift of π regarding to the positive radial one.

To verify the results in a systematical way, one needs to compare the condensate width. Detailed description of how to calculate the condensate widths σ_r and σ_z along the radial and axial axes can be found in [40] and [46].

Fig. 9.20 shows the condensate width as a function of time. The results are consistent with the measurements (crosses and circles in Fig. 9.20), although a slightly underestimation of the expansion rate on σ_z is found.

Our calculations (solid and dashed lines in Fig. 9.20) are also compared with the numerical scheme (star marks and points in Fig. 9.20) proposed in [40], and good agreement of the two proposed methods is found.

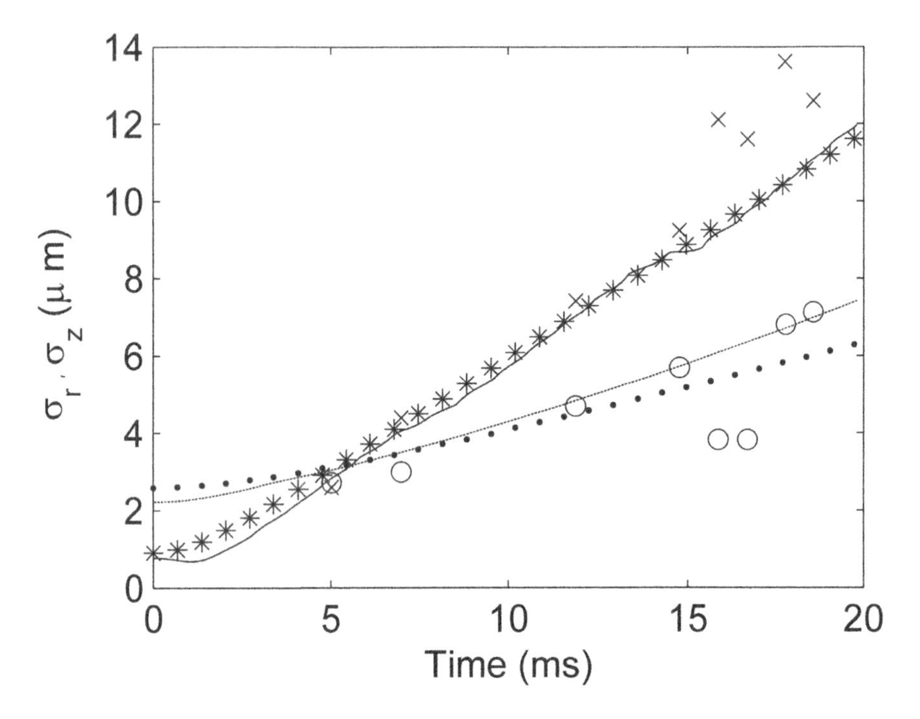

Fig. 9.20 Simulation results and comparisons with the experimental and numerical data from [40] given for the radial σ_r and the axial σ_z widths plotted as functions of time. Solid line: axial width calculated with the cross-coupled equivalent network approach; Dashed line: radial width calculated with cross-coupled equivalent network; Star marks: axial width calculated by numerical scheme proposed by [40] (redrawn from Fig. 5 in [40]); Points: radial width calculated by numerical scheme proposed by [40] (redrawn from Fig. 5 in [40]); Crosses: axial width (measurements redrawn from Fig. 5 in [40]); Circles: radial width (measurements redrawn from Fig. 11.10 in [40])

9.5.3 Comparison of the Proposed Techniques

In this paper, three different techniques are proposed in Section 9.5.1, and they are used for simulations of several cases. The numerical solutions are compared with the analytical ones in Section 9.5.2.

The cross-coupled equivalent circuit is a general model whereas the rest of other ones can be considered as the more specialized techniques. They bring their own advantages and they speed up the simulations when some a priory data on enveloped wave function is known.

It was mentioned in Section 9.5.1.7, the envelope approach allows for larger critical time steps in comparison to the cross-coupled one, which can be unstable for these cases. For comparisons, the error estimate is calculated

$$\text{error}=\frac{\sum_{\text{all time steps}}\sum_{\text{all nodes}}\left|\Phi_{A}-\Phi_{N}\right|^{2}\Delta l^{2}}{\text{number of nodes}\times\text{number of time steps}} \tag{9.67}$$

where the subscripts A and N denote the analytical and numerical solutions. It is a simple way to show the stability by checking Eq. (9.67), although it does not give physical meaning. The envelope approach is still accurate even if the time step increases to $\Delta\tau = 10^{-2}$. However, the cross-coupled equivalent circuit suffers from the instability when the time step is above $\Delta\tau = 2\times10^{-4}$.

The conclusion is made that one can use the envelope approach with the time step up to 10^{-2}, whereas one can use the time steps no larger than 2×10^{-4} only, for the cross-coupled model in order to keep stability.

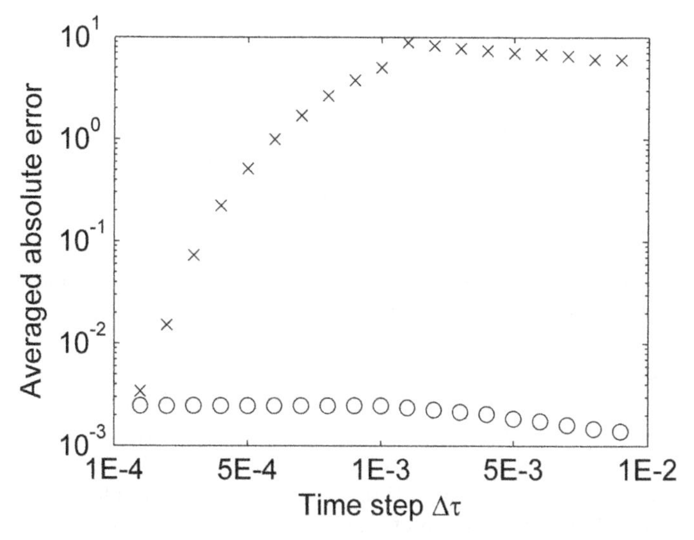

Fig. 9.21 Comparison of the averaged absolute error regarding different time steps. Cross marks: cross-coupled equivalent circuit approach; Circles: envelope approach

The equivalent circuit with the time-dependent resistors has simpler circuit topology than the cross-coupled one. It is expected that the former model speeds up the simulations. To verify this, both models are applied to the free self-interacting condensate (Section 9.5.2.2) using the Agilent ADS, and the simulations are carried out on an HP workstation with an Intel Quard Core Prossesor (2.8 GHz). The tests are performed on $N \times N$ grids, whereas N is chosen to be 8, 16, 24, and 32, respecively. The time step is specially chosen as $\Delta\tau = 9.8\times10^{-4}$. The boundary conditions and simulation domain are the same as in Section 9.5.2.2. The simulation time as a function of the grid size is plotted in Fig. 9.22.

It follows that these two models have approximately the same performance when the grid size is small. This is because the time-dependent resistors need extra computation, and they cancel out the benefit brought by the simpler circuitry. When the grid size grows large, the simplier circuit topology saves more computation time, and the performance is therefore improved by more than twice.

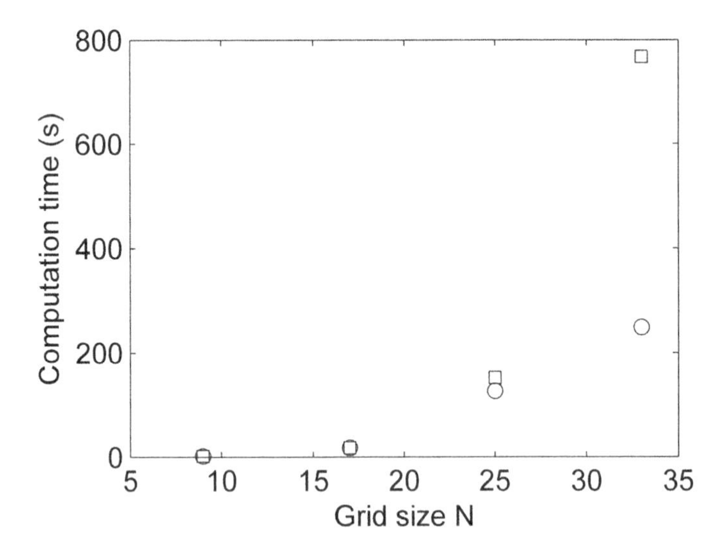

Fig. 9.22 Comparison of simulation time between cross-coupled equivalent circuit and equivalent circuit with time-dependent resistors. Square marks: cross-coupled equivalent circuit. Circles: equivalent circuit with time-dependent resistors

In this Section, the enhancements of the equivalent circuit approach for simulation of Bose-Einstein condensate described by the Gross-Pitaevskii equation have been proposed, considered, and verified. The mentioned equation has been transformed into a system of partial differential equations of the first order regarding to the equivalent voltage and current. It has been solved by three techniques realizable by the available circuit simulators. These approaches have been carefully verified by comparisons with the available analytical, numerical, and experimental results, and good correspondence has been found. The comparative analysis of the proposed three simulation techniques has also been performed.

It has been found that the equivalent circuit approach is a powerful tool for simulations of quantum-mechanical equations. The main advantage of the equivalent circuit approach is its great flexibility and versatility. With the help of modern circuit software tools, it is possible to perform the frequency-domain analysis, S-parameter analysis or even harmonic balance simulation [21]. Although they are not considered in this work, these calculations may provide a new insight on the condensate. The use of circuit simulators also allows reducing the time needed for coding and calculations. Finally, it opens a way for seamless simulation and design not only the EM traps and Bose-Einstein condensates, but also the hybrid electronic/EM/quantum integrations.

9.6 Developments and Design of Hardware for Cold Matter

9.6.1 Current State of Circuits for Trapping of Cold Matter

Interesting effects can be studied at extremely low temperatures below 1 K. Here, the thermal motion of atoms is slow and more quantum effects arisen obvious. Especially interesting effects occur below certain temperatures where the distance between the slowly moving atoms is comparable with the de Broglie wavelength and a special quantum state of matter or a quantum liquid, called the Bose-Einstein condensate, arises which was predicted and described by these scientists in 1924-1925. The initial theoretical findings in this field are with the theory of Bogolubov [47], who proposed to separate the discrete and continuous parts of the bosonic field operator and derived a nonlinear integro-differential equation for condensates. One of the means to describe this state of matter is a theory named after E.P. Gross [28] and L.V. Pitaevskii [29],[30] with their non-linear Schrödinger equation (9.4) on the condensate with locally interacting bosons.

The temperature T, where the condensate emerges, depends on the density of matter ρ and the mass m of a particle, and it can be found from [49], for instance:

$$\rho\Lambda_{dB} \simeq 2.6 \tag{9.68}$$

where $\Lambda_{dB} = \hbar\sqrt{2\pi}/\sqrt{mk_BT}$ is the thermal wavelength, and k_B is the Boltzmann constant. Depending on the mentioned parameters, this critical temperature of the matter transition to a new quantum state is around several tens or hundreds of nanoKelvin. Today multiple modifications of this equation are known, including those which take into account the nonlocal nature of condensates [50]. For special cases of trapping potential and spatial domains, the analytical solutions of this equation are known [30],[48]. Simulation of realistic traps requires numerical modeling [39]-[41]. Some of these simulations have been already considered above.

The gaseous Bose-Einstein condensate was discovered experimentally in 1995 by two groups of scientists; at Boulder - E. Cornel and C. Wieman [51] and at MIT - W. Ketterle [52], which were awarded later by the 2001 Nobel Prize.

9.6.1.1 Laser Cooling of Neutral Atoms

To reach the Bose-Einstein condensate state, the gaseous matter should be cooled enough. It is performed by using the laser cooling and evaporating the fast moving atoms in traps. Theory and practice of these processes are described in many papers, reviews and books.

An atom, placed in laser light, is moved under two forces - the radiation pressure $\mathbf{F}_{\mathrm{rp}}$ and the gradient $\mathbf{F}_{gr}$ one. The first of them is the result of scattering of photons when they transfer their momentums to atoms. Under some conditions, the averaged momentums received by atoms are aligned along the photon flow, and this effect can be used for cooling or accelerating atoms. The gradient force appears in spatially non-homogeneous fields acting on atoms having the induced or fixed electric dipolar moments, and the direction of this force can be away from or towards the field maximum location depending on the frequency of wave.

These forces can be calculated according to the semi-classical formulas [53]:

$$\mathbf{F}_{\mathrm{rp}}(\mathbf{r}) = \hbar\mathbf{k}\gamma \frac{G(\mathbf{r})}{1+G(\mathbf{r})+(\delta-\mathbf{kv})^2/\gamma^2}, \tag{9.69}$$

$$\mathbf{F}_{\mathrm{gr}}(\mathbf{r}) = -\frac{1}{2}\hbar(\delta-\mathbf{kv})\frac{\nabla\cdot G(\mathbf{r})}{1+G(\mathbf{r})+(\delta-\mathbf{kv})^2/\gamma^2} \tag{9.70}$$

where $\mathbf{k}$ is the vector light wave constant, $\mathbf{v}$ is the atom velocity, $\gamma = 0.5A$ with A as the spontaneous decay probability (Einstein coefficient), $G(\mathbf{r}) = I(\mathbf{r})/I_S$ is the dimensionless saturation parameter with $I(\mathbf{r})$ as the intensity of the laser beam at a point $\mathbf{r}$, and I_S as the saturation intensity. The formulas (9.69) and (9.70) are valid for two-state atoms, and $\delta = \omega - \omega_0$ is the difference between the wave frequency ω and the quantum inter-state frequency ω_0 defined by the energy difference between these two quantum states of this atom.

It is seen that the radiation pressure force $\mathbf{F}_{\mathrm{rp}}$ is oriented along the wave direction of propagation, and it can act against the vector of motion of atoms. The gradient force orientation depends on the sign of detuning $(\delta-\mathbf{kv})$ and $\nabla\cdot G(\mathbf{r})$. It is seen that variation of the laser light parameters can de-accelerate atoms, and this effect is used to cool and trap them.

Using this idea, several techniques have been developed on the laser cooling. One of them applies the red-detuned laser light to atoms. It means that the frequency ω of this laser is lower than the transition frequency ω_0 of atoms. The most effectively cooling atoms have velocity which is close to $|\delta|/|k|$. Applying three pairs of counter-propagating laser beams, a standing wave field is formed, and atoms are cooled in this area. A semi-classical theory of this cooling can be found in [53], for instance, and it says that these atoms can be cooled up to only certain temperature $T < \alpha E_0^2/k_B$, where α is the atom polarizability, E_0 is the light field magnitude, k_B is the Boltzmann constant. There are some more effective cooling processes, and the reviews of them can be found in [53]-[56].

In 1997, several scientists (S. Chu [54], C. Cohen-Tannoudji [55], and W.D. Fillips [56]) were awarded by the 1997 Nobel Prize for the developments of the methods and techniques of laser cooling and trapping of atoms.

9.6.1.2 Optical Trapping of Neutral Atoms

As it follows from the previous paragraph, atoms can be cooled and placed in areas where the potential is minimal. Depending of the frequency de-tuning, these minima can be associated with the light increased (red-detuned) or decreased intensities (blue-detuned). Close to a potential minimum, the gradient force is prevailing in comparison to the radiation pressure force if the far-detuned light is used. An expression of this force in the case of a 3-D red-detuned trap is shown below [53]:

$$\mathbf{F}_{\text{gr}}(\mathbf{r}) = \sum_{i=1}^{3} \hbar \mathbf{k}_i \gamma G(\mathbf{r}) \left\{ \frac{1}{1+(\delta - kv_i)^2/\gamma^2} - \frac{1}{1+(\delta + kv_i)^2/\gamma^2} \right\}. \tag{9.71}$$

Here, the trap area of increased field intensity is surrounded by the space of low intensity, and it prevents escaping these atoms from the trap.

For instance, the 3-D traps are known to be realized in the focus of a Gaussian laser beam. Using different approaches to trapping, a cold matter with a long lifetime can be obtained, and this parameter is around several hundreds of seconds limited, for instance, by the quantum-mechanical effects as the fluctuation of laser light composed of photons. This effect is reduced in the blue-detuned or dark optical traps [57]. The trapping potential $U(\mathbf{r})$ is proportional to the light intensity, and the two-level atoms are placed at minima of this potential:

$$U(\mathbf{r}) \approx \frac{3\pi c^2}{2\omega_0^2} \frac{\gamma}{\delta} I(\mathbf{r}). \tag{9.72}$$

The acting force $\mathbf{F}$ can be found as the gradient of this potential. The walls of these traps are the areas of increased field intensity, and different designs of dark optical traps are considered in the mentioned review [57]. Additional information on trapping atoms by hollow optical systems can be found in [58].

An especial interest is in the optical lattices produced by standing waves. In this case, cold atoms are in a periodic potential, and they can be trapped at maxima or minima of the light intensity depending on the frequency detuning δ. A potential $U(z)$ formula for 1-D lattice case is [53]

$$U(z) = \hbar\delta \frac{G(z)}{1+2G(z)+\delta^2/\gamma^2} \cos 2kz. \tag{9.73}$$

The first experimentations with the lattice trapping relate to the end of the 80s and to the beginning of the 90s (see [53],[57]). It was found possibility to trap individual atoms in each potential minimum of standing light field. Strong field allows isolating these atoms from each other and realizing the Mott insulator state that is used for initialization of quantum registers. Lower magnitude fields lead to the correlated quantum state of trapped atoms or to the Bose-Einstein condensates. Joint applications of DC magnetic fields and optical trapping potentials allows to split the quantum states due to the Zeeman effect, and atoms can be controlled by microwave field which transfers them from one state to another. The stored atoms can be moved in a deterministic way to realize an optical belt [59]. To upload the atoms into a qubit state, they can be collided in a coherent way [60], and massive quantum computation can be performed using the entanglement of atoms in these periodical traps. In [61], the first experimentations with a 5-atom quantum linear register are published. It is shown a possibility on the individual control of optically trapped cesium atoms placed in nonhomogeneous DC magnetic field by impulses of a microwave field. More information of this technology and some theoretical and experimental results is in [62],[63], for instance.

9.6.1.3 Magnetic and EM Traps[2]

Magnetic field can be used to trap neutral atoms having magnetic moments $\mathbf{p}_m$. These atoms interact with the external magnetic field $\mathbf{B}$, and the potential U and the force $\mathbf{F}$ acting on a dipolar magnetic atom are:

$$U = -\mathbf{p}_m \mathbf{B}, \tag{9.74}$$

$$\mathbf{F} = \mathbf{p}_m \nabla \cdot \mathbf{B}. \tag{9.75}$$

The atoms, which magnetic moments are aligned against the magnetic field, are attracted to the minima of magnetic field. In these sites, their potential energy is minimal. It allows isolating them from the conductors carrying currents on which surface the magnetic field is strong. Similarly to the force acting on the electric dipoles, the magnetic one is very weak, hence, the atoms should be cooled preliminary to be trapped, and the relatively strong magnetic fields are used. The trapping force (9.75) depends on the magnetic field gradient, and a given configuration can trap the atoms having temperature T less than a certain value [53]:

$$T < p_m |\Delta B| / k_B \tag{9.76}$$

where ΔB is the potential depth.

Magnetic trapping can be combined with the optical one, and even gravitation is used to realize some trapping configurations. Today many magnetic traps are known, and only several of them are considered here to highlight the principles and the main results in this field. The history of magnetic trapping of particles can be found in [53].

A quadrupole trap proposed by W. Paul consists of two wire rings placed at a certain distance from each other and carrying the static opposite currents. It produces zero minimum of magnetic field at the center of the design where the atoms

[2] Written by G.A. Kouzaev and K.J. Sand.

can lose their orientation due to the Majorana effect. For practically used configurations, this trap keeps atoms of the temperature around 10-20 μK during no longer than one minute.

Another trap consists of four bars with opposite DC currents and two rings in which the currents have the same direction. These rings are to create confining field along the axis, and the bars are to confine the atoms in the radial direction. Analytical formulas for calculating this Ioffe-Pritchard trap are from [64]. The traps can be simulated numerically with increased accuracy using commercially available tools, and one of them, Amperes™ [65], was used in our calculations [66],[67].

A modified quadrupole trap is shown in Fig. 9.23. It consists of two rings with the currents opposite to each other. To increase the steepness of the minimum walls, these four Ioffe-Pritchard bars are installed inside the rings. They carry the opposite DC currents to trap atoms in the radial direction.

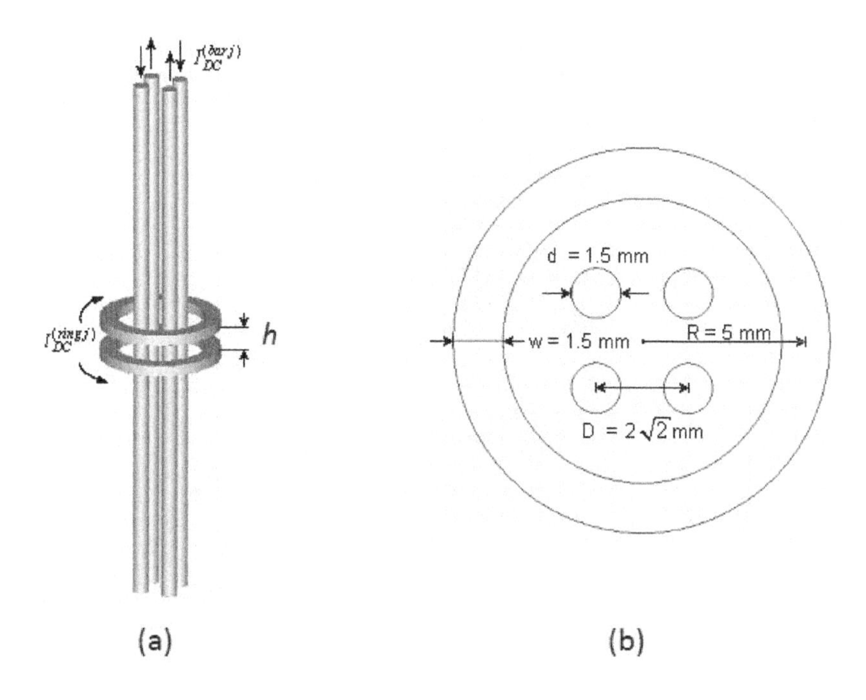

Fig. 9.23 Modified quadrupole trap. (a)- General view; (b)- Cross-section

In Fig. 9.24, the contour plots of magnetic field $|\mathbf{B}|$ are shown in this quadrupole trap. Here the field minimum is shown in dark-blue color at the center of the trapping structure. To prevent the zero potential at the center, an additional biasing magnetic field along the *z*-axis should be applied, caused, for instance, by asymmetry of rings or by additional rings with parallelly flowing currents [66]. More modifications of the Ioffe-Pritchard trap are known from [68],[69], for instance.

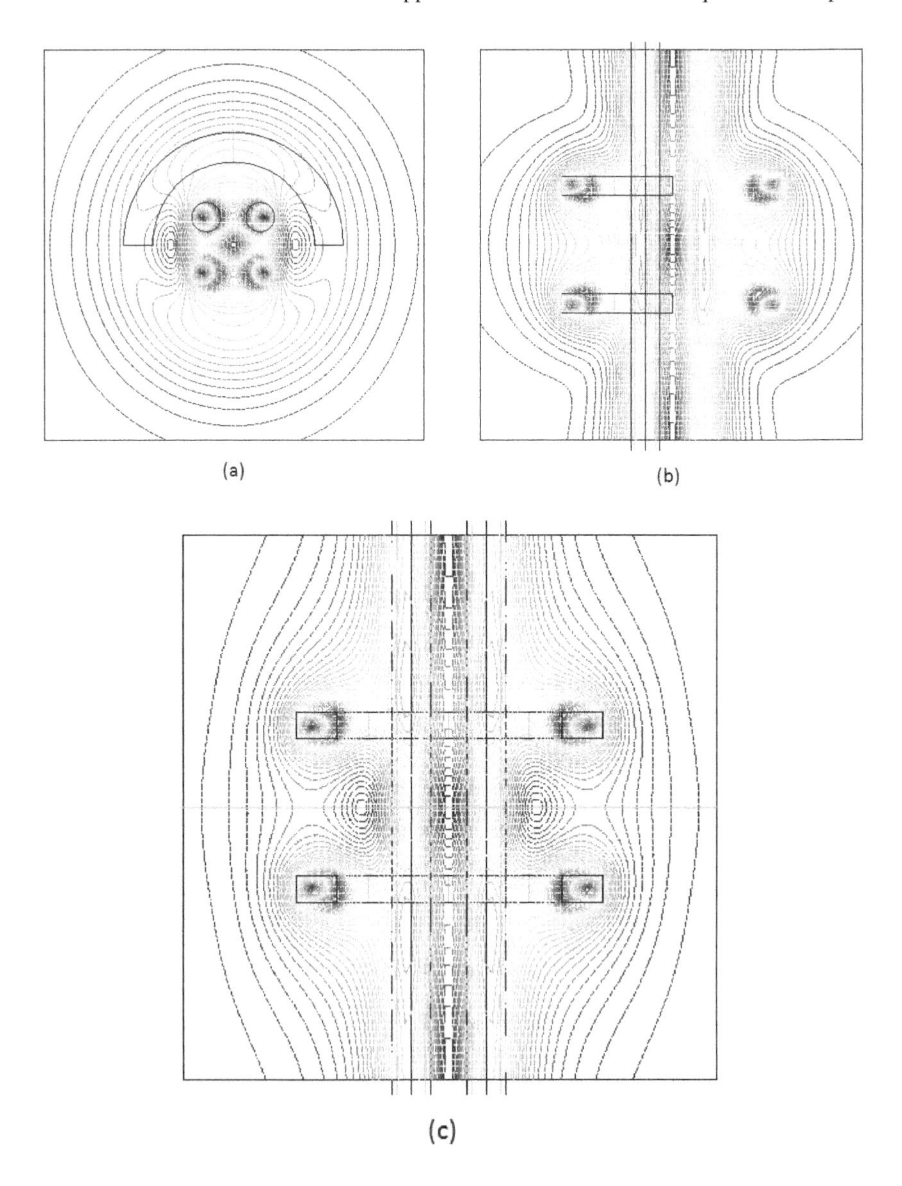

Fig. 9.24 Contour plots of the magnetic field $|\mathbf{B}|$ for a gap size $h = 5\,\text{mm}$ and $I_{\text{DC}}^{(\text{bar})} = I_{\text{DC}}^{(\text{ring})} = 1$ A. (a)- xy -plane for $z = 0$; (b)- yz -plane for $y = 0$; (c)- xz -plane for $x = 0$

Although these considered traps can be scaled down to several millimeters with the decreasing of currents [66], they are still bulky, and they are not always convenient for practical experimentations. In 2001, the integrated magnetic micro-traps, called now the atomic microchips, were proposed, and the first measurements were performed with these traps [70].

A trap of this sort consists of a dielectric substrate on the surface of which a topology of thick conductors is placed. Currents moving along the conductors generate the magnetic field of certain geometry confining atoms at their minima. The conductors can be placed on several layers of integration using a multilayered technology [71],[72]. Taking into account the increased currents used, the thickness of conductors can be of several tens of microns and their conductivity should be high to avoid increased heating and damaging of these conductors by strong currents. The used dielectrics should be resistant to the electric discharges and be suitable to high vacuum. An additional advantage of these traps is the consumption of decreased currents needed to trap atoms in a smaller area in comparison to the bulky Paul and Joffe-Pritchard traps. Due to decreased currents, the integrated traps do not need cooling by water, and it diminishes mechanical vibration, which is important for study of coherent Bose-Einstein condensates. At the same time, further trap miniaturization is limited due to several effects. For instance, sputtering technology provides conductors of increased granularity (800 nm) and surface roughness (16 nm). It leads to the inhomogeneous magnetic field, which is unwanted for trapping and handling of condensates [73]. In that paper, some other technological questions for micron-sized traps are considered, and the used electroplating technique allows improving the quality of conductors about 10 times in comparison to the mentioned sputtering technology.

Usually, the traps are of the room temperature. The warmed electrons moving along the grained conductors are scattered at the grain boundaries, and it increases the magnetic field fluctuations. It is shown that atoms having magnetic moments can lose their spin orientation due to it, and this effect influences the magnetic field close to a few microns from the conductor surface [74]. It can be decreased, if carbon nanotubes, which are highly homogeneous, are used as the trap conductors. As it is shown in that cited paper, the life-time of trapped bosons is around of a few seconds if they are placed at 20-nm distance from the nanotube surface. Trapping distances over 120 nm give the life-time over one minute.

Unfortunately, another effect harms the trapping by nanometric circuits [75],[76]. Close to the body surface, the EM field of vacuum oscillations is changed, and the attracting Casimir-Polder force acts. The atoms can tunnel the barrier wall, and they tend to the surface of carbon nanotube. It reduces the life-time of trapped bosons. To reach several-seconds life-time of trapped particles, they should be placed at the distance over 100-160 nm [75].

Today many miniaturized integrated traps are known, and they are the most convenient hardware for experimentations with cold matter and condensates confined by magnetic field.

An example of an integrated trap proposed by us is shown in Fig. 9.25 [77].

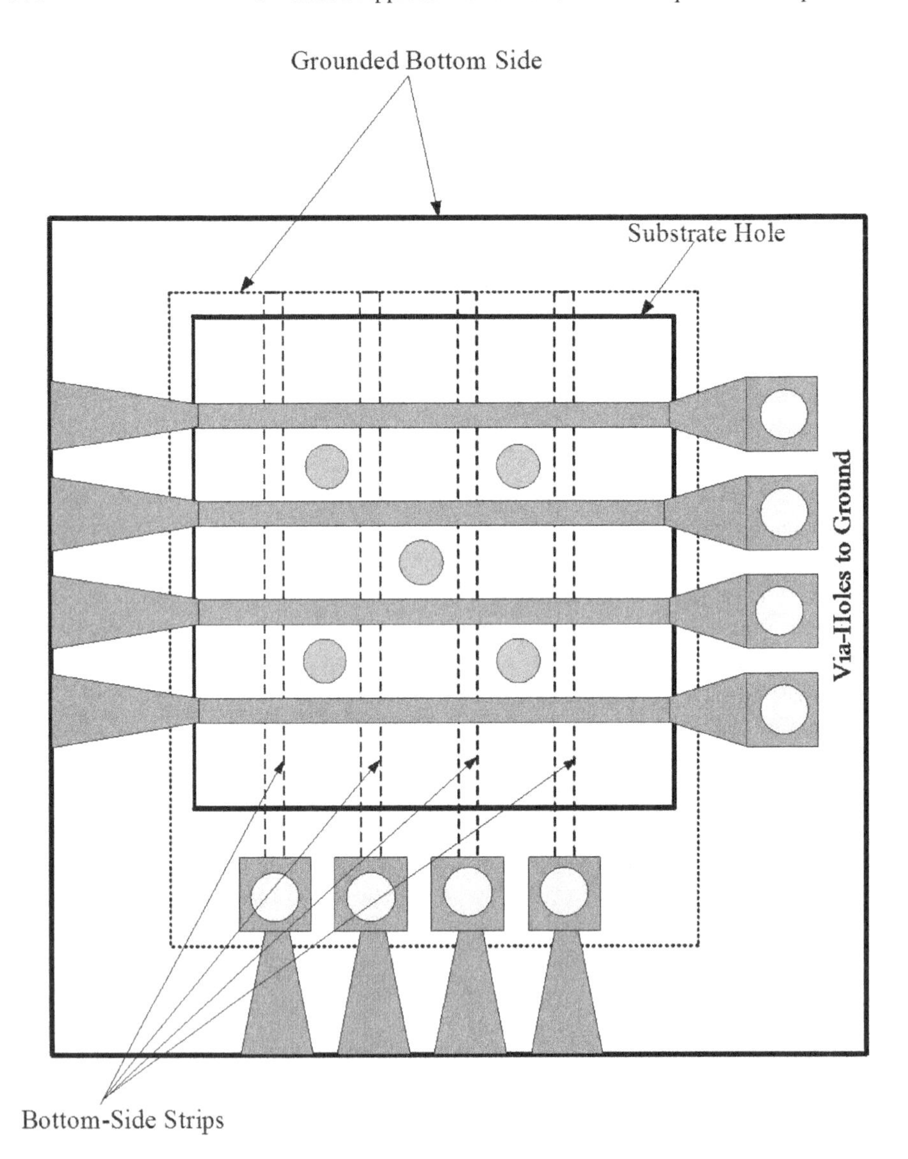

Fig. 9.25 Five-cell integrated trap of could atom matter. The trapped cold matter is shown by circles in cyan color

It consists of two crossed grids of strip-like conductors placed on different levels. The 3-D trapping domains are in the cells formed by crossed strip layers (circles shown in cyan color). To manufacture this integration, an available 3-D micromachining technology should be used. The strip conductors are placed photolithographically on the both surfaces of a substrate. To support short distance between the trapping cells, the strips can be made very narrow, while their height is increased to support strong currents. The hole in the substrate can be realized using an etching technology. It is possible to consider stacking of these layers, and a 3-D lattice can be manufactured. This integrated chip can be fed by DC or DC/RF currents to obtain a variety of trapping potential shapes [66].

Our interest in the design of these multi-cell structures was inspired by [38], [78],[79] where the Josephson Effect between two Bose-Einstein condensate drops was studied theoretically. It was predicted that coupling of these pairs could lead to the entanglement of these microscopic drops, and it could be promising for ultra-cold quantum computing. In [80],[81], this effect was found experimentally between the light-trapped Bose-Einstein condensate drops.

To this moment, only the potential shape formation has been studied in our one- and -multi-layer trapping structures [66],[77]. By numerical and analytical modeling and for a variety of geometries, it is shown a possibility to generate the traps with the depth around of 30-80 μK, which is enough to trap cold matter. The shape of trapping potential can be controlled by the DC and RF currents, and even merging of wells can be performed.

In Fig. 9.26, a single cell constructed of two pairs of crossed cylindrical wires is given, and the DC field minimum is at the center of this cell shown by a dark-blue colored circle.

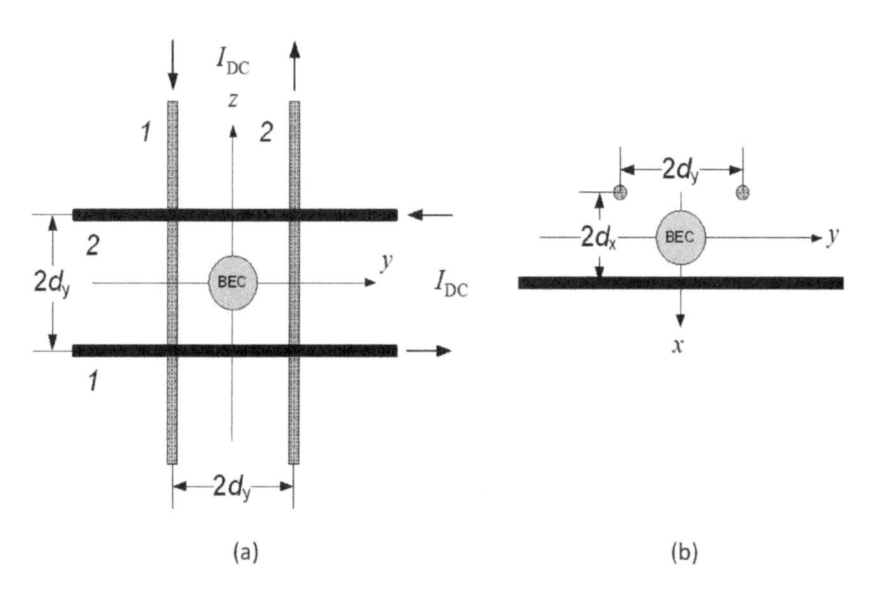

Fig. 9.26 A trapping cell made of four wires fed by DC currents I_{DC}

The potential U distribution (Fig. 9.27) is calculated according to a known formula:

$$U = m_F \mu_B g_F |\mathbf{B}_{DC}| \tag{9.77}$$

where $m_F = 2$ is the quantum number of the state of atom, μ_B is the Bohr magneton, and g_F is the Lande factor.

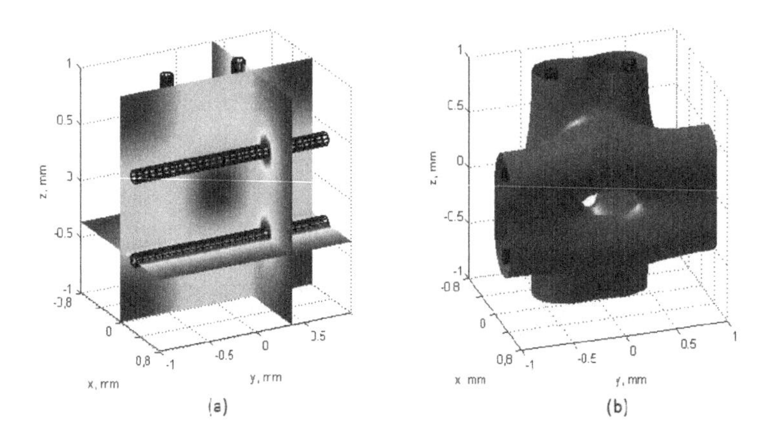

Fig. 9.27 (a)- Potential distribution in a four-wire trapping cell fed by DC currents $I_{DC} = 92.5$ mA . (b)- Potential surface for this four-wire cell calculated for the potential level $U = 8 \cdot 10^{-29}$ J . Cell geometry: $d_y = 0.37$ mm and $d_x = 0.267$ mm

To estimate the trap depth, the potential was recalculated in Kelvin, and its profile along the z -axis is shown in Fig. 9.28.

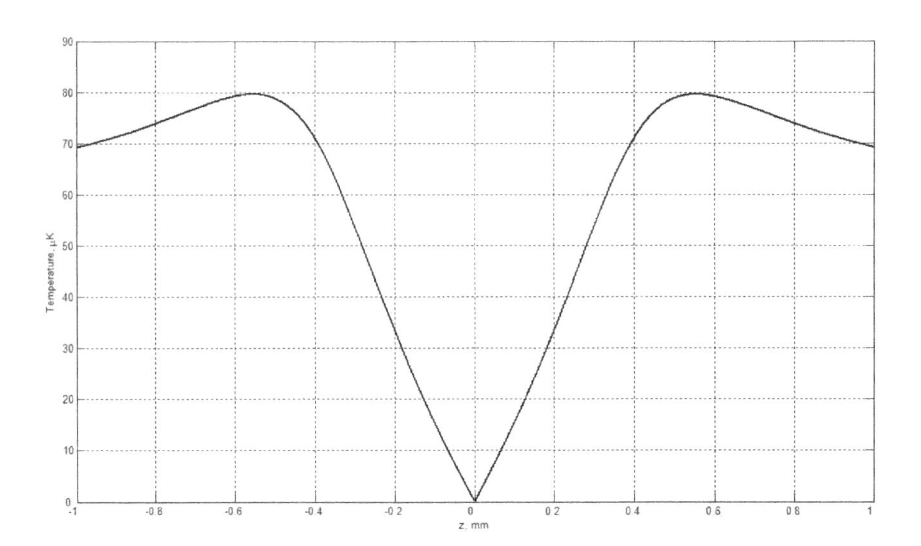

Fig. 9.28 Effective potential distribution $T = U/k_B$ along the z -axis $(x = 0,\ y = 0)$

It is seen that the potential reaches zero at the cell center. To avoid an increased rate of the atom spin-off at this center, this cell should be biased by additional DC or RF field. In a multi-cell structure, this biasing is realized by the field from neighboring cells. An example of the potential distribution in a multi-cell trapping structure interesting in study of coupling of Bose-Einstein condensate drops is shown in Fig. 9.29.

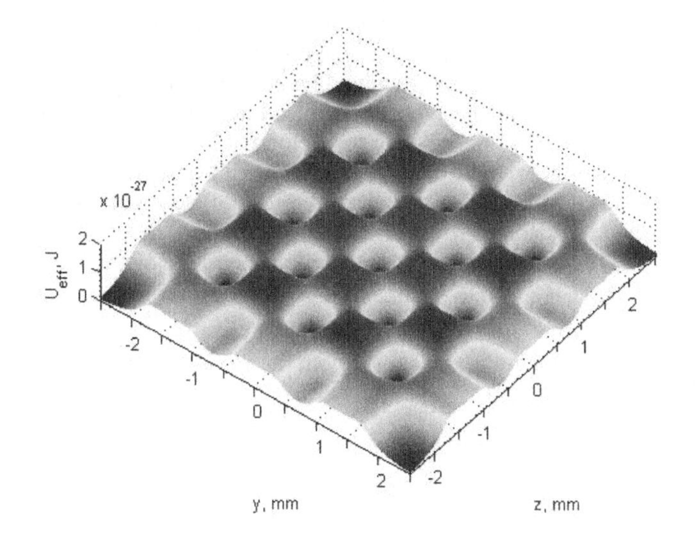

Fig. 9.29 Potential distribution (in Joule) in a multi-cell structure composed of 6×6 wires. Reproduced from [77] with permision of the World Scientific Publ. Co.

Additionally, the stacked or multilayered trapping structures can be designed. The potential shapes can be modified using the combined DC and RF excitations [66]. We evaluated the nanometric trapping multi-cell structures. In this case, to create a trapping well of a needed depth, only the currents of around several tens of μA are needed. The most atoms should be placed far enough from the carbon nanotubes to avoid the areas with strong attracting Casimir-Polder force, and the traps of 400 nm size are modeled, which may provide several seconds life-time of trapped bosons according to the estimations from [75],[76].

Our results presented here are of the preliminary character, and they can be modified further taking into account our near future simulation of condensates using the equivalent circuit approach.

Interesting results are obtained for the traps fed by combination of strong DC and RF fields. The effective potential formulas are derived using the dressed formalism obtained initially for the slow atoms trapped by laser light [55]. In contrast to the semi-classical theory, the outer EM field is quantized in this case. The atomic Hamiltonian is represented by a sum of the EM, atom, and dipole-interaction ones. In many cases, it allows to derive approximate analytical formulas for the spatially dependent effective potentials [82]-[87]. Additionally, combinations of DC and RF fields allow realizing a variety of potential shapes unavailable for the

DC magnetic traps, including those having the minima and maxima in space, which can trap atoms of different quantum states [66],[77],[82]-[87]. In these circuits, the quantum states of atoms can be controlled by a microwave field, which is interesting in quantum computing [88].

9.6.1.4 Transportation of Cold Matter[3]

Transportation of cold matter is interesting in study of dynamics of cooled atoms and condensates. For instance, the atoms can be trapped and cooled in a special magneto-optical trap; meanwhile, a study of cold effects can be performed using integrated traps and devices. In this case, the cooled matter should be transported to another place or even into a new vacuum chamber. Besides, many effects as the quantum interferometry and the prescribed interactions of atoms or condensates take place in motion of these particles or their clouds.

The history of the theory and practice of cold atom transportation from the 90s to 2000 is considered in [53],[58], for instance. Several atom guides are shown in Fig. 9.30. It seems that the first experimental works were described in [88],[89] where a hollow optical waveguide was used to transport cold atoms (Fig. 9.30a). In this waveguide, the EH_{11} mode has maximum at the center of the hole. The red-detuning allows isolating atoms from the waveguide wall, and the radiation pressure transports atoms along this waveguide with the velocity around 40 cm/s [89].

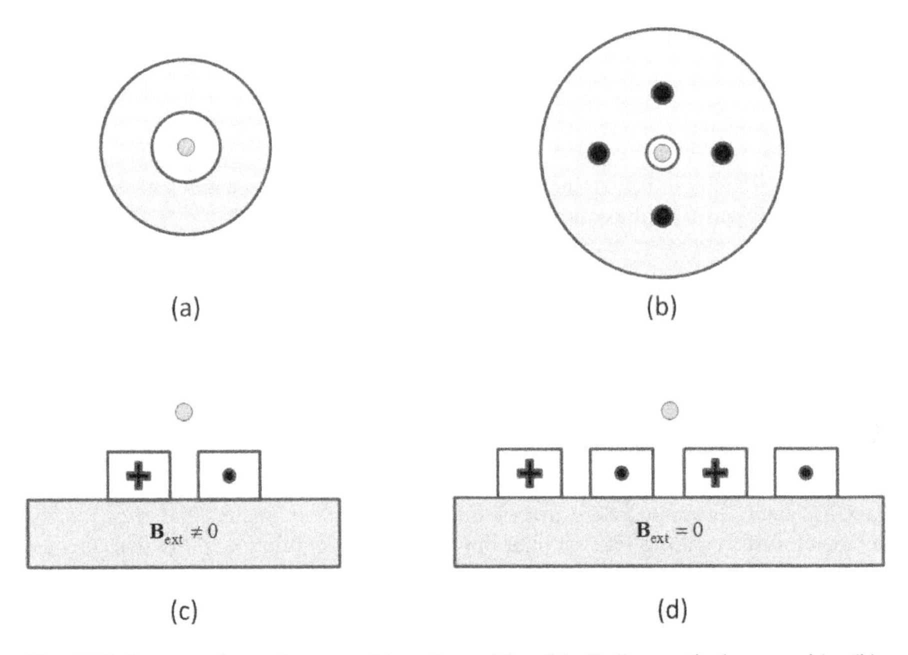

Fig. 9.30 Cross-sections of some cold matter guides. (a)- Hollow optical waveguide; (b)- Hollow optical waveguide equipped by four wires; (c)- Integrated two-strip guide with external biasing magnetic field $\mathbf{B}_{ext}$; (d)- Integrated four-strip guide. Cold matter areas are shown in cyan color

[3] Written by G.A. Kouzaev and K.J. Sand.

The increased heating of atoms in the light can be avoided using the blue-detuned laser light and the modes having minimum of the field along the axis of a hollow optical waveguide [53],[58]. In experiments of M.J. Renn and co-authors [90], atoms were trapped close to the axis of a hollow optical waveguide by blue-detuned laser light, and they were escorted along a 6-cm waveguide by the red-detuned light from another laser which radiation pressure moves the atoms.

Cold atoms can be moved through open space using hollow beams of blue-detuned laser light being confined inside the light tunnel [91]. The first experimentations allowed transportation of around of 10^8 Cs atoms through an 18-cm-long light tunnel of 1-mm diameter.

Magnetic field confining atoms in a 2-D channel can be used for guiding cold matter. For instance, in [92], a hexagonal guide is proposed. It is built of six wires each carrying 300-A current. They confine ^{87}Rb atoms in the cross-sectional directions and guide these atoms to another trap placed below of the atom vapor source. In [93], a curved beam guide of the radius 300 mm is described which is made of permanent magnets creating a quadrupole field. An additional weak longitudinal field is applied to prevent the spin-offs of atoms propagating along the curved axis of this guide. It demonstrates effective directing properties for atoms moving with the velocity from 30 $\mathrm{m/s}$ to 60 $\mathrm{m/s}$.

In [94],[95], four wires are placed inside a silica tube for confining atoms moving inside a hole made in the center of this cylinder (Fig. 9.30b). The wires carry static electric currents which magnetic field confines the atoms close to the axis of this sealed hole. Additional pinch coils are used to avoid the spin-offs of atoms and to pump them along guide [95]. In [94], cold atoms move under the gravitational force.

A sealed guide for cold atoms proposed by us is shown in Fig. 9.31. It consists of a non-magnetic conducting cylinder with a hole inside. Close to the surface of the cylinder, a wire is placed. Both conductors carry currents I_{cyl} and I_{wire}. The magnetic field inside the hole is composed of two fields from the cylinder and the wire, and zero magnetic field can be realized inside the hole. Being biased by a longitudinal magnetic field varying in the space and time, this waveguide can be used for transportation of cold matter along its $z-$axis.

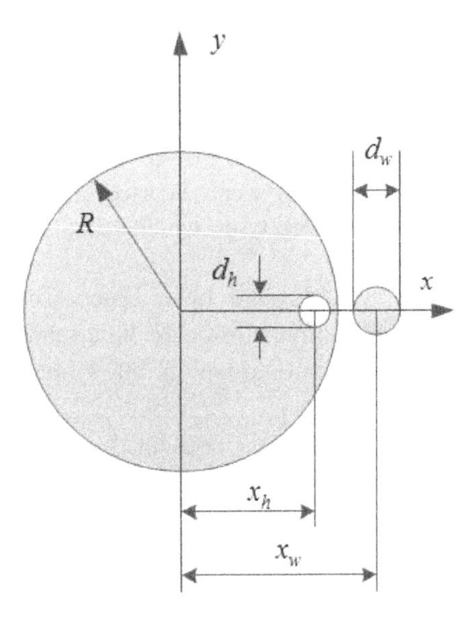

Fig. 9.31 Cross-section of a cylinder guide for cold atoms placed inside the hole

Practical calculations of this guide and optimization of its geometry and currents were performed in [66] both analytically - using some formulas from [96], and numerically by the software tool AMPERES™ [65].

Fig. 9.32 shows distributions of the magnetic field $|\mathbf{B}(x, y=0)|$ for Fig. 9.31 along the axis x and given for different currents I_{cyl} and I_{wire}. It is seen that the magnetic field has a zero at the center of the cylinder $(x=1.1 \text{ mm})$. Adjusting the currents on conductors, it is possible to place the second magnetic field zero close to the center of the hole $(x=1.6 \text{ mm})$. It is found that the depth of the trapping channel is varied from 14 μK to 93 μK, and the guide should be biased by a longitudinal magnetic field to avoid the Majorana effect destroying the orientations of atom spins along the hole axis. It is reached by using the biasing rings around the guide (Fig. 9.33) on which the same-direction currents are flowing. These rings are centered around the hole to avoid the influence of radial components of the ring magnetic field on the geometry of trapping channel.

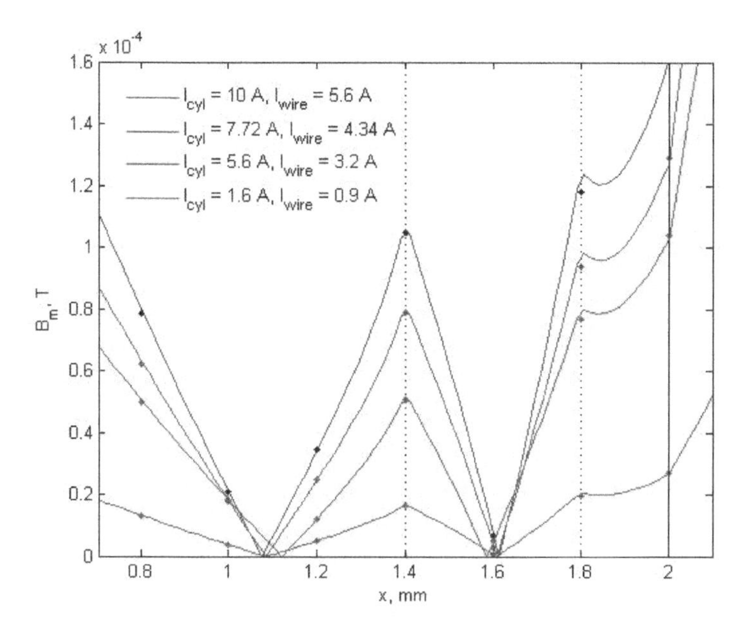

Fig. 9.32 Magnetic field distribution $B_m = |\mathbf{B}|$ along the x -axis $(y=0)$ in cylindrical cold-atom guide given for different cylinder's and wire's currents. Geometrical parameters are: $R = 2$ mm , $d_w = 1$ mm , $x_h = 1.6$ mm , and $x_w = 3$ mm . Hole is shown by dashed lines. AMPERES simulations

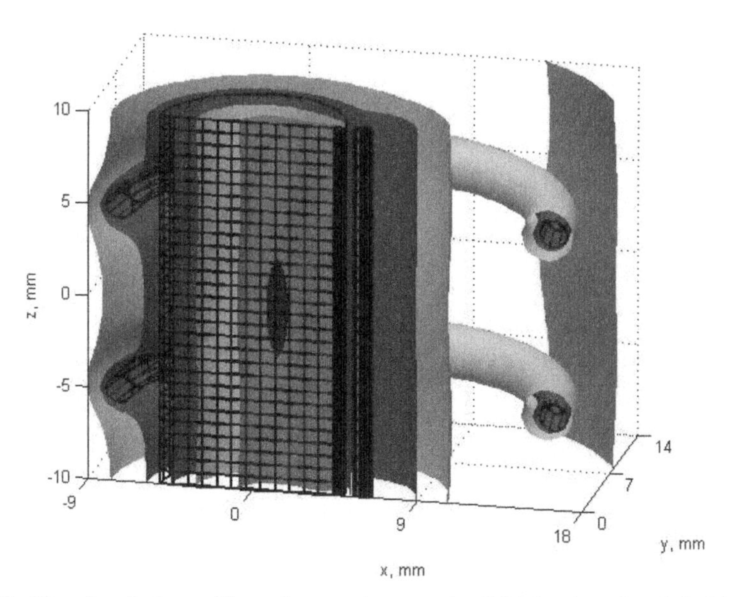

Fig. 9.33. Biased cylinder guide and several magnetic field level surfaced inside and outside the guide. AMPERES simulation

Fig. 9.34 illustrates the biasing effect showing the trapping potential in μK along the z-axis of the hole. The low-field seeking atoms can be trapped at the potential minimum $(z=0)$. The geometry and currents are optimized to reach confining of atoms inside the hole (see Table 9.3).

Table 9.3 Geometry and guide currents

Cylinder radius R, mm	Cylinder current, I_{cyl}, A	Wire current, I_{wire}, A	Hole coordinate x_h, mm	Ring radius, mm
2	5.6	3.17	1.6	5.85
3	3.6	1.46	2.6	7.85
5	5	1.29	4.6	11.85
7.5	5.4	0.955	7.1	16.85
10	6	0.81	9.6	21.85

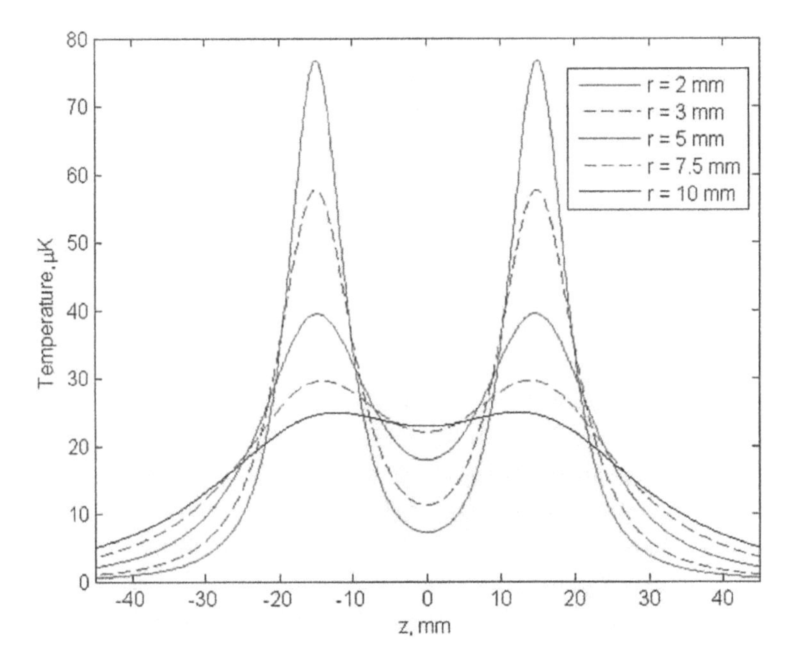

Fig. 9.34. Longitudinal dependence of the trapping potential $T = U/k_B$ along the guide channel biased by two currents 0.8 A flowing along two rings of the radii 1.75 mm. The rings are placed at the distance 30 mm from each other. The curves are given for 5 geometries and cylinder and wire currents (see Table 9.3). AMPERES simulations

It is supposed that the slowly changing ring fields applied, it is possible to move the cold matter along the guide axis similarly to known solutions [95].

Cold matter can be moved not only along the straight channels, but also along the ones, which change the geometry and direction. For instance, in [97], it was proposed a convergent hollow optical fiber for increasing the atomic-phase density. The surface-wave mode field isolates the cold atoms from the wall and transports them from 500-μm side to 10-μm output, and the increase of 5 orders of phase space density occurs after 1-cm distance.

The cold atom flow can be turned, and one of these transportation elements is considered in [98] where atoms are moved at a distance around 30 cm through a 90° corner of a guide composed of the transfer coils with slowly varying currents. It allows moving the trapping potential minimum along the transportation trace.

Atoms can be transported in the vertical direction between the crossed channels. Some simulations of possible potential transformation are considered in [99] where two crossed wires carrying DC and RF currents are biased by DC rings. Slowly-varying ring fields allow touching the potential minima of these two crossed wires, and atoms can wander from one wire to another.

A number of integrated guides and components for handling cold matter and condensates have been proposed and studied in [70],[100]-[104]. For instance, atoms are confined in the minimum of the magnetic field created by DC currents and a permanent magnet. This minimum is placed at a distance from the substrate surface (Fig. 9.30c). In multi-strip guides, the non-zero trap minimum is created by four strips (Fig. 9.30d). These mentioned printed guides are manufactured using combination of sputtering and electroplating techniques. The confined cold matter moves along the waveguides under gravitational force in [103].

A slowly varied biasing magnetic field can be used to transport the cold atom clouds along an integration (conveyor belt) as it was proposed and shown experimentally in [70]. It can be used for prescribed interaction of cold atom clouds and condensates, which is interesting in quantum interferometry and quantum computing.

The trapping of cold atoms close to the domains of ferromagnetic nanowires was proposed and simulated in [102]. The nano-domain walls are moved together with atoms, and this effect can be used for transportation of atoms and for quantum processing.

The above-considered results are mostly on the transportation of cold matter as clouds of classical and semi-classical particles. Cooling the gaseous matter down to several tens or hundreds of nanoKelvins allows the Bose-Einstein condensates which particle wave functions are coherent. For many applications as the atomic clocks, interferometers and quantum gates and their integrations, the coherent handling of condensates is very important, and some theoretical and experimental results for integrated components can be found, for instance, in [80],[81],[86]-[88],[104]-[107].

In this Section only a couple areas have been touched upon in brief where the circuit models and classical circuit simulators can be used for simulation of the quantum-mechanical phenomena. Besides this review, many other problems are worthy of attention. Among them are the prospective elements of nano-integrations where classical electromagnetics meets its limitations and new methods of advanced semi-classical electromagnetics should be created and applied. An interesting area is the imitation of quantum effects by classical circuits and EM structures which can lead to new components and systems [18],[44],[109],[110]. We suppose that the equivalent circuit approach which is ready to be used in these fields can deliver many interesting results in the future.

References

[1] Tour, J.M.: Molecular Electronics. World Sci. (2003)
[2] Deleonibus, S. (ed.): Electronic Device Architectures for the Nano-CMOS Era. World Sci. (2008)
[3] Benci, V., Fortunato, D.: An eigenvalue problem for the Schrödinger-Maxwell equations. Topological Methods in Nonlinear Analysis 11, 283–293 (1998)
[4] Ginbre, B., Velo, G.: Long range scattering for the Maxwell-Schrödinger system with large magnetic field data and small Schrödinger data. Publ. RIMS, Kyoto Univ. 42, 421–459 (2006)
[5] Yang, J., Sui, W.: Solving Maxwell-Schrödinger equations for analyses of nanoscale devices. In: Proc. 37th Europ. Microw. Conf., pp. 154–157 (2007)
[6] Pieratoni, B., Mencarelli, D., Rozzi, T.: A new 3-D transmission line matrix scheme for the combined Schrödinger-Maxwell problem in the electronic/electromagnetic characterization of nanodevices. IEEE Trans., Microwave Theory Tech. 56, 654–662 (2008)
[7] Pieratoni, B., Mencarelli, D., Rozzi, T.: Boundary immitance operators for the Schrödinger-Maxwell problem of carrier dynamics in nanodevices. IEEE Trans., Microwave Theory Tech. 57, 1147–1155 (2009)
[8] Mastorakis, N.E.: Solution of the Schrödinger-Maxwell equations via finite elements and genetic algorithms with Nelder-Mead. WSEAS Trans. Math. 8, 169–176 (2009)
[9] Kron, G.: Electric circuit model of the Schrödinger equation. Phys. Rev. (1&2) (1945)
[10] Sanada, H., Suzuki, M., Nagai, N.: Analysis of resonant tunneling using the equivalent transmission-line model. IEEE J. Q. Electron. 33, 731–741 (1977)
[11] Anwar, A.F.M., Khondker, A.N., et al.: Calculation of the transversal time in resonant tunneling devices. J. Appl. Phys. 65, 2761–2765 (1989)
[12] Kaji, R., Koshiba, M.: Equivalent network approach for guided electron waves in quantum-well structures and its application to electron-wave directional couplers. IEEE J. Quant. Electron 31, 1036–1043 (1994)
[13] Civalleri, P.P., Gilli, M., Bonnin, M.: Equivalent circuits for two-state quantum systems. Int. J. Circ. Theory Appl. 35, 265–280 (2007)
[14] Kouzaev, G.A.: Hertz vectors and the electromagnetic-quantum equations. Mod. Phys. Lett. B 24(24), 2117–2212 (2010)
[15] Kouzaev, G.A.: Calculation of linear and non-linear Schrödinger equations by the equivalent network approach and envelope technique. Modern Phys. Lett. B 24, 29–38 (2010)
[16] Matyas, A., Jirauschek, C., Perretti, P., et al.: Linear circuit models for on-chip quantum electrodynamics. IEEE Trans., Microw. Theory Tech. 59, 65–71 (2011)
[17] Kouzaev, G.A., Nazarov, I.V., Kalita, A.V.: Unconventional logic elements on the base of topologically modulated signals. El. Archive, `http://xxx.arXiv.org/abs/physics/9911065`

[18] Kouzaev, G.A., Lebedeva, T.A.: Multivalued and quantum logic modeling by mode physics and topologically modulated signals. In: Proc. Int. Conf. Modelling and Simulation, Las Palmas de Grand Canaria, Spain, September 25-27 (2000), http://www.dma.ulpgc.es/ms2000

[19] Kouzaev, G.A.: Predicate and pseudoquantum gates for amplitude-spatially modulated electromagnetic signals. In: Proc. 2001 IEEE Int. Symp. Intelligent Signal Processing and Commun. Systems, Nashville, Tennessee, USA, November 20-23 (2001)

[20] Kouzaev, G.A.: Qubit logic modeling by electronic gates and electromagnetic signals. El. Archive (2001), http://xxx.arXiv.org/abs/quant-ph/0108012

[21] Advanced Design System 2008. Agilent Corp. (2008)

[22] A User Guide to Envelope Following Analysis Using Spectre RF. Cadence Corp. (2007)

[23] Visscher, P.B.: A fast explicit algorithm for the time-dependent Schrödinger equation. Comp. Phys. 5/6, 596–598 (1991)

[24] Frank, T.D.: Nonlinear Fokker-Planck Equations: Fundamentals and Applications. Springer, Berlin (2005)

[25] Norbe, F.D., Rego-Monteiro, M.A., Tsallis, C.: A generalized nonlinear Schroedinger equation: Classical field-theoretic approach. Eur. Phys. Lett. 97(1-5), 41001 (2012)

[26] Belevitch, V.: Classical Network Theory. Holden-Day (1968)

[27] Galizkyi, V.M., Kornakov, B.M., Kogan, V.I.: Tasks to Solve in Quantum Mechanics (Zadachi po Kvantovoy Mekhanike), Nauka (1981) (in Russian)

[28] Gross, E.P.: Structure of a quantized vortex in boson systems II. Nouvo Cimento 20, 454–457 (1961)

[29] Pitaevskii, L.V.: Vortex lines in an imperfect Bose gas. Soviet Phys. JETP 13, 451–454 (1961)

[30] Pitaevskii, L.P., Stringari, S.: Bose-Einstein Condensation. Clareton Press (2003)

[31] Ueda, M.: Fundamentals and New Frontiers of Bose-Einstein Condensation. World Scientific (2010)

[32] Vengalattore, M., Higbie, J.M., Leslie, S.R., et al.: High-Resolution Magnetometry with a spinor Bose-Einstein Condensate. Phys. Rev. Lett. 98, 200801 (2007)

[33] Simmonds, R.W., Marchenkov, A., Hoskinson, E., et al.: Quantum interference of super fluid ^{3}He. Nature 412, 55–58 (2001)

[34] Seaman, T., Krämer, M., Anderson, D.Z., et al.: Atomtronics: ultracold-atom analogs of electronic devices. Phys. Rev. A 75, 023615 (2007)

[35] Stickney, J.A., Anderson, D.Z., Zozulya, A.A.: Transistorlike behavior of a Bose-Einstein condensate in a triple-well potential. Phys. Rev. A 75, 013608 (2007)

[36] Ramanathan, A., Wright, K.C., Muniz, S.R., et al.: Superflow in a toroidal Bose-Einstein condensate: an atom circuit with a tunable weak link. Phys. Rev. Lett. 106, 13041 (2001)

[37] Farkas, M., Hudek, K.M., Salim, E.A., et al.: A compact, transportable, microchip-based system for high repetition rate production of Bose-Einstein condensates. App. Phys. Lett. 96, 093102 (2001)

[38] Cataliotti, F., Burger, S., Fort, C., et al.: Josephson junction arrays with Bose-Einstein condensates. Science 293, 843–846 (2001)

[39] Succi, S., Toschi, F., Tosi, M.P., et al.: Bose-Einstein condensates and the numerical solution of Gross-Pitaevskii equation. IEEE Comput. Sci. Eng. 7, 48–57 (2005)

[40] Cerimele, M.M., Chiofalo, M.L., Pistella, F., et al.: Numerical solution of the Gross-Pitaevskii equation using an explicit finite-difference scheme: an application to trapped Bose-Einstein condensates. Phys. Rev. E 62, 1382–1389 (2000)
[41] Bao, W., Tang, W.: Ground-state solution of Bose-Einstein condensate by directly minimizing the energy functional. J. Comput. Phys. 187, 230–254 (2003)
[42] Kron, G.: Numerical solution of ordinary and partial differential equations by means of equivalent circuits. J. Appl. Phys. 16, 172–186 (1945)
[43] Kron, G.: Equivalent circuit of the field equations of Maxwell. In: Proc. I. R. E., pp. 289–299 (1944)
[44] Dragoman, D., Dragoman, M.: Quantum-classical Analogies. Springer (2004)
[45] Kouzaev, G.A.: Co-design of quantum and electronic integrations by available circuit simulators. In: Proc. 13th Int. Conf. Circuits, Rodos, Greece, pp. 152–156 (2009)
[46] Holland, M.J., Jin, D.S., Chiofalo, M.L., et al.: Emergence of interaction effects in Bose-Einstein condensation. Phys. Rev. Lett. 78, 3801–3805 (1997)
[47] Bogolubov, N.: J. Phys 11, 23 (1947) (in Russian)
[48] Pethick, C.J., Smith, H.: Bose-Einstein Condensation in Dilute Gases. Cambridge Press (2003)
[49] Chevy, F., Dalibard, J.: Rotating Bose-Einstein condensates. Europhysicsnews 37, 12–16 (2006)
[50] Rozanov, N.N., Rozhdestvenkyi, Y.V., Smirnov, V.A., et al.: Atomic "Needles" and "Bullets" of the Bose-Einstein condensate and forming of nano-size structures. Pisma v ZHETF- Lett. J. Exper. Theor. Phys. 77, 89–92 (2003) (in Russian)
[51] Anderson, M.H., Ensher, J.R., Matthews, M.R., Wieman, C.E., Cornell, E.A.: Observation of Bose–Einstein condensation in a dilute atomic vapor. Science 269(5221), 198–201 (1995)
[52] Davis, K.B., Mewes, M.-O., Andrews, M.R., van Druten, N.J., Durfee, D.S., Kurn, D.M., Ketterle, W.: Bose–Einstein condensation in a gas of sodium atoms. Phys. Rev. Lett. 75, 3969–3973 (1995)
[53] Balykin, V.I., Minogin, V.G., Letokhov, V.S.: Electromagnetic trapping of cold atoms. Rep. Prog. Phys. 61, 1429–1510 (2000)
[54] Chu, S.: Laser manipulations of atoms and particles. Science 253, 861–866 (1991)
[55] Cohen-Tannoudji, C., Guerry-Odelin, D.: Advances in Atomic Physics: an Overview. World Scientific (2011)
[56] Phillips, W.D.: Nobel lecture: Laser cooling and trapping of neutral atoms. Rev. Mod. Phys. 70, 721–741 (1998)
[57] Friedman, N., Kaplan, A., Davidson, N.: Dark optical traps for cold atoms. Adv. Atomic, Molec., Opt. Phys. 48, 99–151 (2002)
[58] Noh, H.-R., Jhe, W.: Atom optics with hollow optical systems. Phys. Reports 372, 269–317 (2002)
[59] Kuhr, S., Alt, W., Schrader, D., et al.: Deterministic delivery of a single atom. Science 293, 278–280
[60] Mandel, A., Greiner, M., Widera, A., et al.: Controlled collisions for multi-particle entanglement of optically trapped atoms. Nature 425, 937–940 (2003)
[61] Schrader, D., Dotsenko, I., Khudaverdyan, M., et al.: Neutral atom quantum register. Phys. Rev. Lett. 93(1-4), 150501
[62] Bloch, I.: Exploring quantum matter with ultracold atoms in optical lattices. J. Phys. B 38, S629–S643 (2005)

[63] Bloch, I., Dalibard, J., Zwerger, W.: Many-body physics with ultra-cold gases. Rev. Mod. Phys., 885–964 (2008)
[64] Bergman, T., Erez, G., Metcalf, H.J.: Magnetostatic trapping fields for neutral atoms. Phys. Rev. A 35, 1535–1546 (1987)
[65] AMPERES Program Guide. Integrated Software Eng. Inc. (2006)
[66] Sand, K.J.: On the design and simulation of electromagnetic traps and guides for ultra-cold matter. PhD Thesis, NTNU, Trondheim, Norway, 252 p (2010)
[67] Kouzaev, G.A., Sand, K.J.: RF controllable Ioffe-Pritchard trap for cold dressed atoms. Modern Phys. Lett. B 21, 59–68 (2007)
[68] Thomas, N.R., Foot, C.J., Wilson, A.C.: Double-well magnetic trap for Bose-Einstein condensates. ArXiv: cond-mat/01108169 (2001)
[69] Tiecke, T.G., Kemmann, M., Buggle, C., et al.: Bose-Einstein condensation in a magnetic double-well potential. ArXiv: cond-mat/0211604 (2002)
[70] Rechel, J., Hansel, W., Hommelholf, P., et al.: Applications of integrated magnetic microtraps. Appl. Phys. B 72, 81–89 (2001)
[71] Jones, M.P.A., Vale, C.J., Sahagun, D., et al.: Cold atoms probe the magnetic field near a wire. J. Phys. B 37, L15–L20 (2004)
[72] Crookston, M.B., Baker, P.M., Robinson, M.P.: A microstrip ring trap for cold atoms. J. Phys. B 38, 3227–3289 (2005)
[73] Koukharenko, E., Mktadir, Z., Kraft, M., et al.: Microfabrication of gold wires for atom guides. Sensors and Actuators A 115, 600–607
[74] Henkel, Wilkens, M.: Heating of trapped atoms near thermal surfaces. Europhys. Lett. 47, 414–420 (1999)
[75] Fermani, R., Scheel, S., Knight, P.L.: Trapping cold atoms near carbon nanotubes: Thermal spin flips and Casimir-Polder potential. Phys. Rev. A 75(1-7), 062905 (2007)
[76] Bostroem, M., Sernelius, B.E., Brevik, I., et al.: Retardation turns the van der Waals attraction into a Casimir repulsion as close as 3 nm. Phys. Rev. A 85(1-4), 010701
[77] Kouzaev, G.A., Sand, K.J.: 3D multicell designs for registering of Bose-Einstein condensate clouds. Modern Phys. Lett. 22(25), 2469–2479 (2008)
[78] Shi, Y.: Entanglement between Bose-Einstein condensates. Int. J. Modern. Phys. B 15, 3007–3030 (2001)
[79] Yalabik, M.C.: Nonlinear Schrödinger equation for quantum computation. Modern Physics Lett. B 20, 1099–1106 (2006)
[80] Albiez, M., Gati, R., Foeling, J., et al.: Direct observation of tunneling and nonlinear self-trapping in a single bosonic Josephson junction. Phys. Rev. Lett. 95(1-4), 010402 (2005)
[81] Levy, S., Lahoud, E., Shomroni, I., et al.: The a.c. and d.c. Josephson effects in a Bose-Einstein condensate. Nature 449, 579–583 (2007)
[82] Muskat, E.: Dressed neutrons. Phys. Rev. Lett. 58, 2047–2050 (1987)
[83] Zobay, O., Garraway, B.M.: Two-dimensional atom trapping in field-induced adiabatic potentials. Phys. Rev. Lett. 86, 1195–1198 (2001)
[84] Colombe, Y., Knyazchyan, E., Morizot, O., et al.: Ultracold atoms confined in rf-induced two-dimensional trapping potentials. arXiv:quant-ph/0403006
[85] Courteille, P.W., Deh, B., Fortag, J., et al.: Highly versatile atomic micro traps generated by multifrequency magnetic field modulation. arXiv:quant-ph/0512061
[86] Schumm, T., Hofferberth, S., Andersson, L.M., et al.: Matter-wave interferometry in a double well on an atom chip. Nature Physics 1, 57–62 (2005)

[87] Lesanovsky, I., Schumm, T., Hofferberth, S., et al.: Adiabatic radio frequency potentials for coherent manipulation of matter waves. ArXiv:physics/0510076
[88] Ol'shanii, M.A., Ovchinnikov, Y.V., Letokhov, V.S.: Laser guiding of atoms in a hollow optical fiber. Opt. Commun. 98, 77–79 (1993)
[89] Renn, M.J., Mongomery, D., Vdovin, O., et al.: Laser-Guided atoms in hollow-core optical fibers. Phys. Rev. Lett. 75, 3253–3256 (1995)
[90] Renn, M.J., Donley, E.A., Cornell, E.A., et al.: Evanescent-wave guiding of atoms in hollow optical fibers. Phys. Rev. A 53, R648–R651 (1996)
[91] Song, Y., Milam, D., Hill III, W.T.: Long, narrow all-light atom guide. Opt. Lett. 24, 1805–1807 (1999)
[92] Myatt, C.J., Newbury, N.R., Ghrist, R.W., et al.: Multiply loaded magneto-optical trap. Opt. Lett. 21, 290–292
[93] Goepfert, A., Lison, F., Schutze, R., et al.: Efficient magnetic guiding and deflection of atomic beams with moderate velocities. Appl. Phys. B 69, 217–222
[94] Key, M., Hughes, I.G., Rooijakkers, W., et al.: Propagation of cold atoms along a miniature magnetic guide. Phys. Rev. Lett. 84, 1371–1373 (2000)
[95] Teo, B.K., Raithel, G.: Loading mechanism for atomic guides. Phys. Rev. A 63(1-4), 031402 (2001)
[96] Yung-Kuo, L.: Problems and Solutions on Electromagnetism. World Scientific (1993)
[97] Subbotin, M.V., Balykin, V.I., Laryushin, D.L., et al.: Laser controlled atom waveguide as a source of ultracold atoms. Opt. Commun. 139, 107 (1997)
[98] Greiner, M., Bloch, I., Haensh, T.W., et al.: Magnetic transport of trapped cold atoms over a large distance. Phys. Rev. A 63(1-4), 0131401 (2001)
[99] Kouzaev, G.A., Sand, K.J.: Inter-wire transfer of cold dressed atoms. Modern Phys. Lett. B 21, 1653–1665 (2007)
[100] Weinstein, J.D., Librecht, K.G.: Microscopic magnetic traps for neutral particles. Phys. Rev. A 52, 4004–4009 (1995)
[101] Thywissen, J.J., Olshanii, M., Zabow, G., et al.: Microfabricated magnetic waveguides for neutral atoms. Eur. Phys. J. D 7, 361–367 (1999)
[102] Allwood, D.A., Schrefl, T., Hrkac, G., et al.: Mobile atom traps using nanowires. Appl. Phys. Lett. 89(1-3), 014102 (2006)
[103] Dekker, N.H., Lee, C.S., Lorent, V., et al.: Guiding neutral atoms on a chip, vol. 84, pp. 1124–1127 (2000)
[104] Tonyshkin, A., Prentiss, M.: Straight macroscopic magnetic guide for cold atom interferometer. J. Appl. Phys. 108(1-5), 094904 (2010)
[105] Bongs, K., Burger, S., Dettmer, S., et al.: Waveguide for Bose-Einstein condensates. Phys. Rev. A 63(1-4), 031602 (2001)
[106] Treutlein, P., Hommelhoff, P., Steinmetz, T., et al.: Coherence in microstrip traps. Phys. Rev. Lett. 92(1-4), 203005 (2004)
[107] Treutlein, P., Steinmetz, T., Colombe, Y., et al.: Quantum information processing in optical lattices and magnetic microtraps. Fortschr. Phys. 54, 702–718 (2006)
[108] Boehi, P., Riedel, M.F., Hoffrogge, J., et al.: Coherent manipulation of Bose-Einstein condensates with state-dependent microwave potentials on an atom chip. Nature Physics 5, 592–597 (2009)
[109] Sun, Y., Tan, W., Jiang, H.-T. et al.: Metamaterial analog of quantum interference: From electromagnetically induced transparency to absorbtion. EPLA 98, 6407 (1-6) (2012)
[110] Rangelow, A.A., Suchowski, H., Silberberg, Y., et al.: Wireless adiabatic power transfer. Annals of Phys. 326, 626–633 (2011)

10 EM Topological Signaling and Computing

Abstract. In the Chapter 4, a topological approach to the theory of EM boundary value problems has been already considered. Additionally to the computational aspects, the ideas of topology are applicable to the noise-tolerant signaling, computing, microwave imaging, and this application area and some related results are studied below. References -191. Figures -49. Pages -82.

10.1 Introduction to Topological Signaling and Computing

Our world is interacting by shape-modulated signals, and an origin of this communications might be from the atomic level where the geometry of orbits of electrons around nuclei is connected to their quantum numbers. The molecular shape is an essential factor in chemical and bio-chemical reactions and in the cell membrane activity [1],[2].

The information in brain is represented by spatio-temporal patterns kept in cortical columns and processed by sophisticated spatially distributed neural networks [3]-[7]. Image recognition and processing are the most important application areas of the computer science and today's computers, and the powerful video co-processors are operating giga- and teraflops of information represented by pixels. The nature operates with whole images rather than with the detailed and digitized pictures, and a research on how it happens in vivo is crucial in the study of molecular, cellular, and brain activity. This study requires joint efforts of the scientists from mathematics, physics, computer engineering, neural physiology, and biology. Many scientists suppose that the most important information in the brain is with the topology of objects, and the detailed geometrical characteristics play only the secondary role. Additionally to the spatial characteristics, the patterned signals have their intensity or amplitude, and the interplay of these parameters opens a wide area of new logic circuit developments.

McKinsey and Tarski [8] proposed to assign a logical unit to the figure topology and to create the "shape logic". The most important problem in the implementation of this idea is to find a carrier of "topological" information compatible with contemporary electronic technology and design methodology.

Below some results on the use of these ideas are considered starting with the theory and simplified logic gates up to the recently developed predicate logic processor. Related data of other authors are shown as well.

G.A. Kouzaev: Applications of Advanced Electromagnetics, LNEE 169, pp. 413–494.
springerlink.com

10.2 Topology and EM Signaling

According to mathematics, topology studies the spatial properties of figures preserved by bi-continous deformations. Two figures, which cannot be transformed to each other by these transformations, are topologically different. Since the work of H. J. Poincaré on topology, it has been developed several mathematical techniques on the proof of topological identity of spatial forms in the 2-D, 3-D and multi-dimensional spaces, excepting the 4-dimensional one. It means that the spatial forms can be selected in such spaces and numbered by symbols.

Our world has four dimensions; the three of them are assigned to space and the fourth one is for time. Unfortunately, to find a way to distinguish the topological shapes in the 4-dimensional world was the most difficult task during many decades. Only recently, G. Perelman has solved this "Problem of the Millennium" and derived a proof for the Poincaré conjecture on the characterization of 3-D sphere amongst all 3-D manifolds [9],[10]. Another work related to this problem is a paper of A. Thomson who found a proof that an algorithm exists, and it works for detecting the 3-D sphere in the 4-D space [11]. Roughly speaking, the topologists are on the way to find the methods of digitalizing the shapes in the 4-D world.

In spite of these solved problems are from a highly abstract mathematics branch, such methods can find applications in digital signal processing for optical communications, radar engineering, and microwave communications where the signals have complicated space-time distributions of their EM fields. The field-force lines and their differential equations describe the spatial forms of these signals. The phase space of these equations has four dimensions, and a topological theory of these spatio-temporal forms is in a high demand.

Our research on topological analysis and solutions of boundary problems of electromagnetism started in 1988 [12],[13]. Traditionally, since the Faraday's time, the electric and magnetic fields have been described by their force lines. Each line is a trajectory of a hypothetical electric or "magnetic" charge moving in the electric or magnetic field, correspondingly. These trajectories are the solutions of ordinary differential equations - dynamical systems. A set of trajectories composes a field-force line map qualitatively described by its skeleton - topological scheme T that is a set of the field separatrices and the equilibrium field positions. The first ones are the vector manifolds in the phase space of dynamical systems. The topological solution of an EM problem consists of analytical or semi-analytical composing of topological schemes according to the given boundary. Later, this theory allowed proposing the topological modulation of the field, a theory of these signals, the first digital components and a couple of variants of a predicate logic processor [14]-[72].

Logical variables are assigned to the topological field schemes, and the mentioned signal is a series of field impulses with their discretely modulated spatio-temporal forms. These impulses propagate along the multimodal transmission lines. Each mode has a unique shape of its field, i.e. the modes are distinguished from each other by their topological schemes of their field-force line maps. A binary system requires two impulses with the topologically different spatio-temporal contents assigned for the logical units "0" and "1", and some transmission lines supporting the bi-modal regime were studied [18],[19],[34],[36]. For instance, a microstrip-slot transmission line (Fig. 10.1, a) guides the quasi-TEM microstrip mode and quasi-TE slot mode, and it is for microwave signaling. The coupled strips (Fig. 10.1b) and microstrips (Fig. 10.1c) lines are for digital and microwave signals. The ridged waveguide (Fig. 10.1d) supports two modes (TE_{10} and TE_{20}) with the equalized modal velocities and topologically different microwave fields.

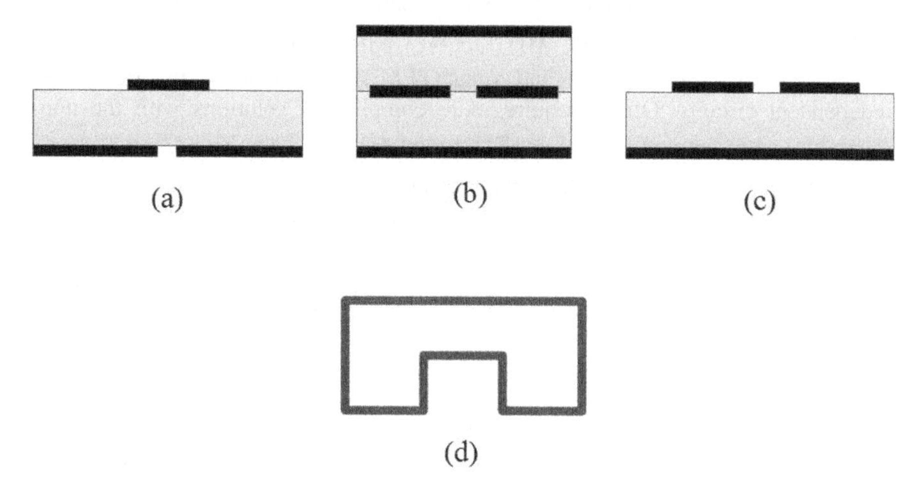

Fig. 10.1 Cross-sections of bi-modal transmission lines. (a)- Microstrip-slot transmission line; (b)- Coupled strips line; (c)- Coupled microstrips line; (d)- Ridged rectangular waveguide for TE_{10} and TE_{20} modes.

The main requirement is that the modal impulses having certain spatial topology should not be transformed to each other while propagating along the interconnects from one gate to another. Strong restrictions are for preventing the inter-symbol interference. The switching and energy recovering of each impulse should not be followed by parasitic excitation of an unwanted mode. From the EM point of view, these restrictions are severe for high-speed signaling and switching, and these cases should be studied carefully. Below some ideas of this kind are considered.

10.3 Topological Description of the Non-stationary EM Field

Geometrical description of the non-harmonic signals by field-force lines supposes solving the non-stationary wave equations with the boundary and initial conditions for the electric **E** and magnetic **H** fields [73]:

$$\Delta \cdot \mathbf{E}(\mathbf{r},t) - \frac{\varepsilon_r \mu_r}{c^2} \frac{\partial^2 \mathbf{E}(\mathbf{r},t)}{\partial t^2} = \frac{1}{\varepsilon_0 \varepsilon_r} \nabla \cdot \rho(\mathbf{r},t) + \mu_0 \mu_r \frac{\partial \mathbf{J}(\mathbf{r},t)}{\partial t}, \tag{10.1}$$

$$\Delta \cdot \mathbf{H}(\mathbf{r},t) - \frac{\varepsilon_r \mu_r}{c^2} \frac{\partial^2 \mathbf{H}(\mathbf{r},t)}{\partial t^2} = -\nabla \times \mathbf{J}(\mathbf{r},t) \tag{10.2}$$

where $\rho(\mathbf{r},t)$ is the time-dependent electric charge, and $\mathbf{J}(\mathbf{r},t)$ is the time-dependent electric current.

There are two types of the above-mentioned non-stationary phenomena. The first of them is described by the separable spatio-temporal functions $\mathbf{E}(\mathbf{r},t) = \mathbf{E}(x,y)e(z)e(t)$ and $\mathbf{H}(\mathbf{r},t) = \mathbf{H}(x,y)h(z)h(t)$. Typically, these are the effects in regular TEM and quasi-TEM lines modeled by frequency-independent circuits. Others require more complicated solutions with the non-separable spatio-temporal functions $\mathbf{E}(\mathbf{r},t)$ and $\mathbf{H}(\mathbf{r},t)$.

In general, the field-force line maps of non-stationary fields can be obtained solving the non-autonomous dynamical system considered in Chpt. 4:

$$\frac{d\mathbf{r}_{e,h}}{dt_{e,h}} = \left[\frac{\mathbf{E}(\mathbf{r}_e,t_e)}{|\mathbf{E}(\mathbf{r}_e,t_e)|} \left(\frac{\partial s_e}{\partial t_e} \right), \frac{\mathbf{H}(\mathbf{r}_h,t_h)}{|\mathbf{H}(\mathbf{r}_h,t_h)|} \left(\frac{\partial s_h}{\partial t_h} \right) \right]. \tag{10.3}$$

There are several ways how to study the topology of non-stationary fields. One way is to consider the time as a bifurcation parameter and to solve the field force-line equations (10.3). Another way is introducing a quasi-spatial variable τ, and the transformation of the non-autonomous system (10.3) into the autonomous one:

$$\begin{aligned} \frac{d\mathbf{r}_{e,h}}{d\tau_{e,h}} &= \left[\frac{\mathbf{E}(\mathbf{r}_e,\tau_e)}{|\mathbf{E}(\mathbf{r}_e,\tau_e)|} \left(\frac{\partial s_e}{\partial \tau_e} \right), \frac{\mathbf{H}(\mathbf{r}_h,\tau_h)}{|\mathbf{H}(\mathbf{r}_h,\tau_h)|} \left(\frac{\partial s_h}{\partial \tau_h} \right) \right], \\ \frac{d\tau_{e,h}}{dt_{e,h}} &= 1. \end{aligned} \tag{10.4}$$

This system requires a special description of its topology of the phase space, because it has no zeroes of the right part of the extended system (10.4). Anyway, the topological schemes in both cases (10.3) and (10.4) are composed of hyper planes of their 4-D spaces, and signaling using these fields can be considered as the communications by 3-D manifolds [71].

10.4 Topological Modulation of the Field

The carrier of topological information is a certain spatio-time structure of the EM field. To define the field characteristics, which are able carrying discrete information, consider the spatial field maps of the time-dependent 3-D electric and magnetic fields.

A field-force line map is described with its topological scheme. According to the Andronov's definition [74], it is an arranged set of equilibrium points and separatrices of the field force-line maps. For the electric or magnetic field, it is written each own topological scheme $T_{e,h}$, correspondingly. Taking into account the time dependence of fields, the considered schemes are the objects that have certain shapes in 4-D phase space.

The schemes of the *i-th* and *j-th* fields can be non-homeomorphic to each other, i.e. $T_{e,h}^{(i)} \not\sim T_{e,h}^{(j)}$, and a natural number i or j can be assigned to each of these schemes. For example, the modes of a transmission line can have non-homeomorphic topological skeletons of their fields, and they are identified by modal numbers. Additionally, the non-homeomorphic schemes can be composed of combinations of modes, as well. In general the number of modes and their combinations is infinite for a given transmission line, but for a certain frequency band only a few of them are propagating. A set of two non-homeomorphic topological schemes of the propagating modes can correspond to the binary system $\{i = 0, j = 1\}$. Manipulation of topological schemes according to a digital signal is the topological field modulation [18]. The digital operations with these signals are the topological computing [14].

10.5 Studied Topologically Modulated Field Shapes

Several types of signals have been studied. Among them are the impulses of the TE (transverse electrical) –modes in rectangular waveguides, hybrid modes of strip-transmission line, TEM, and quasi-TEM (transverse EM) modes in coupled triplate and microstrip lines (Fig.10.1).

10.5.1 Quasi-TEM Signaling

For the TEM or quasi-TEM signaling, a chosen transmission line should consist of several signaling strips and a ground. The coupled strips line or triplate (Fig.10.2b) is one of the best candidates for TEM communications. The even and odd modes of this line are of the TEM type and the hybridization of their fields is low which is caused only by imperfect conductors. These modes have equal modal velocities, and they have very low dispersion, which is very important for high-speed signaling.

The odd mode's electric field is marked by "0" in Fig. 10.2a. The even mode's electric field is marked by "1" in the same figure. Magnetic fields of both modes are shown in Fig. 10.2b.

The modal field topological schemes $T_{e,h}^{(\mathrm{odd})}$ and $T_{e,h}^{(\mathrm{even})}$ of these modes are not homeomorphic to each other, and the logical levels "1" and "0" are assigned to them. Then a topologically modulated signal is a series of modal impulses as shown in Fig. 10.2a,b, and the signal "topology" is changed from impulse to impulse.

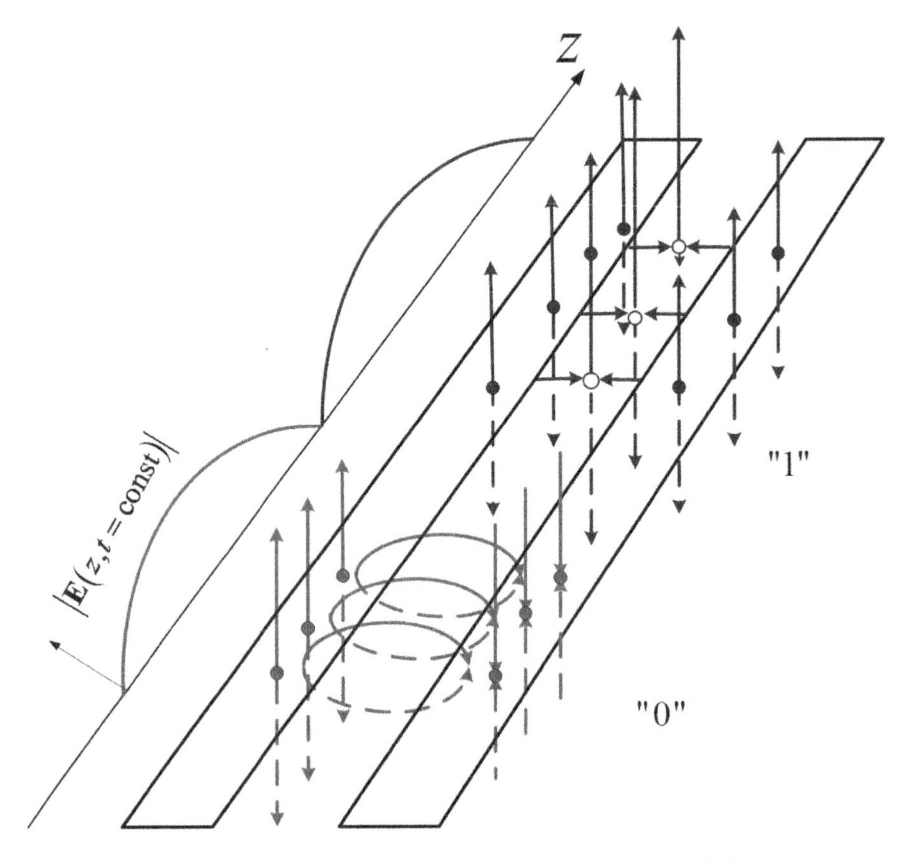

Fig. 10.2a Electric field of topologically modulated impulses in a coupled strips transmission line at a fixed moment of time. The ground plates are not shown

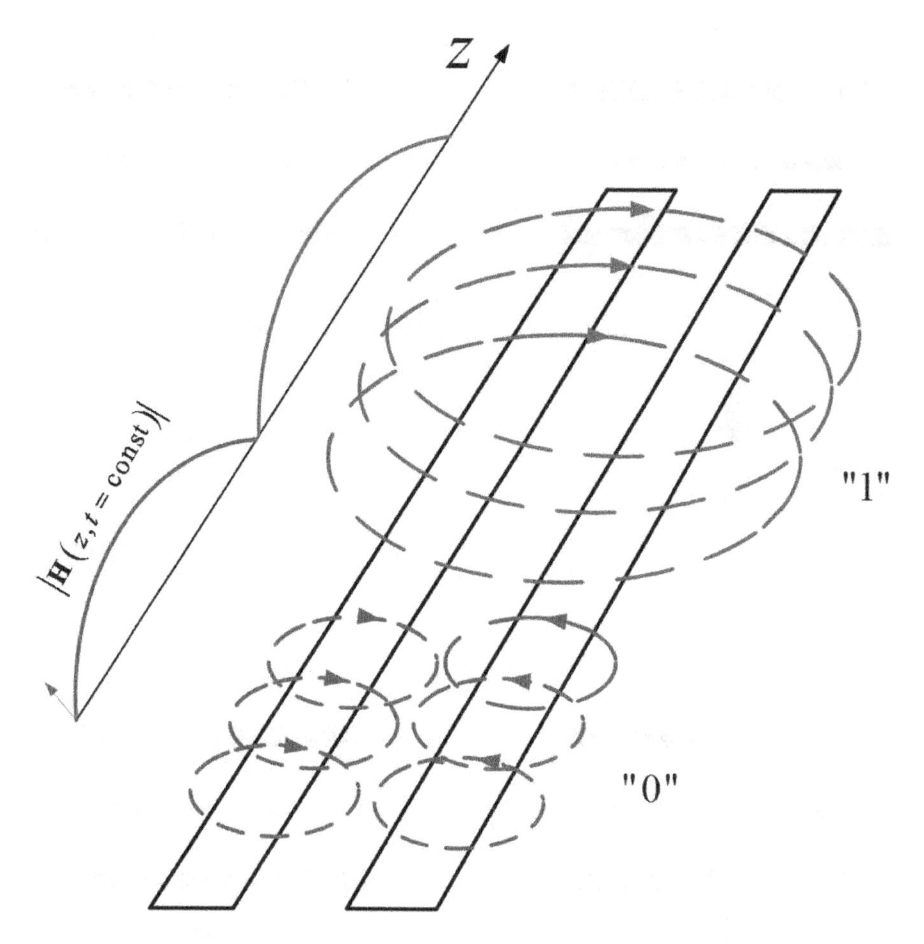

Fig. 10.2b Magnetic field of topologically modulated impulses in a coupled strips transmission line at a fixed moment of time. The ground plates are not shown

These TEM modes have negligible longitudinal field components, and their topological schemes are described by a projection of the field on the plane which is normal to the propagation direction. The modal field-force line pictures of the electric field at a fixed moment of time at an arbitrary point of the z-axis are shown in Fig.10.3 where the topological schemes of the odd $T_e^{(\text{odd})}$ and even $T_e^{(\text{even})}$ correspond to the logical "0" and "1", respectively. The modal magnetic field-force line pictures are shown in Fig. 10.4.

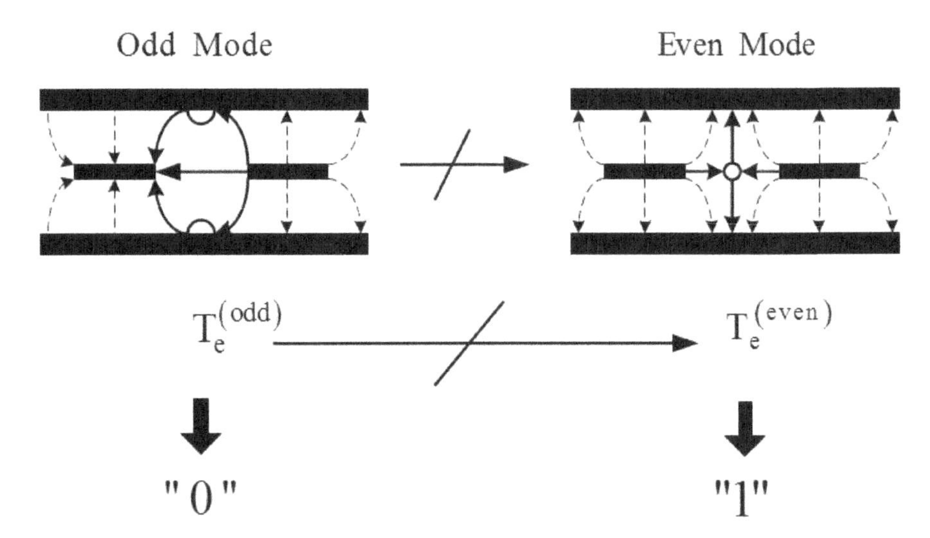

Fig. 10.3 Spatial structures the electric fields of the odd and even mode impulses at a fixed moment of time at an arbitrary coordinate z. The separatrices and equilibrium points are shown by solid lines

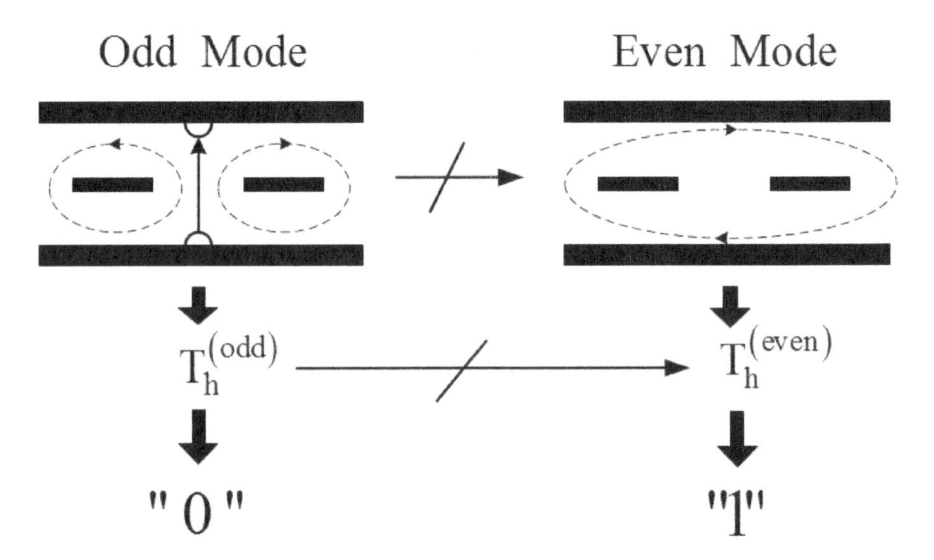

Fig. 10.4 Spatial structures of the magnetic fields of the odd and even mode impulses at a fixed moment of time at an arbitrary coordinate z. The separatrices and equilibrium points are shown by solid lines

During propagation along this transmission line, these modal schemes are unchanged, i.e. at any moment of time and at any point of the longitudinal coordinate z, the field-force line maps of impulses are identical to each other due to the negligible dispersion and loss. In this case, a signal impulse is described by the following formula:

$$\mathbf{S}(\mathbf{r},t) \sim \begin{Bmatrix} E(t)\left\{\dfrac{\mathbf{E}(\mathbf{r})}{|\mathbf{E}(\mathbf{r})|}\right\}|\mathbf{E}(\mathbf{r})| \\ H(t)\left\{\dfrac{\mathbf{H}(\mathbf{r})}{|\mathbf{H}(\mathbf{r})|}\right\}|\mathbf{H}(\mathbf{r})| \end{Bmatrix} \tag{10.5}$$

where $E(t)$ and $H(t)$ are the envelope functions, $\left\{\dfrac{\mathbf{E}(\mathbf{r})}{|\mathbf{E}(\mathbf{r})|}\right\}$ and $\left\{\dfrac{\mathbf{H}(\mathbf{r})}{|\mathbf{H}(\mathbf{r})|}\right\}$ are the space-dependent coefficients responsible for the spatial content of the signal.

It follows, that for such signals the envelope functions $E(t)$ and $H(t)$ and the vector-spatial coefficients in (10.5) can be manipulated independently. Then this space-time signal is a series of impulses of the even and odd modes propagating along the transmission line as shown in Fig. 10.2. It is supposed that the independent manipulation of the envelope and spatial contents is allowed in the case of non-TEM modes if only some restrictions are fulfilled.

Taking into account this possibility of independent manipulations of the impulse magnitude and its topological scheme, the introduced impulses are the two-place signals S:

$$S = (T, a) \tag{10.6}$$

where T corresponds to discrete spatial content or topological scheme, and a is the impulse magnitude. This expression relates to a predicate formula where T is the predicate variable, and a is the quantifier, for instance. Then the most pertinent logic for topologically modulated signals is the predicate one [75], and some circuitry can be designed using the rules of this logic.

10.5.2 Signaling by Non-separable Field Impulses

In general, the EM signals are described by the systems (10.3) and (10.4) for non-separable fields and with the topological schemes $T_{e,h}(\mathbf{r},\tau)$ defined at the (3+1)-D phase space $(\mathbf{r},\tau)$. Topological signaling means launching the EM impulses of topologically different contents defined in the (3+1)-D phase space of this system, and to which the Boolean variables are corresponding. This case is less studied, and more efforts should be applied to research the advantages of this signaling and the areas of its applications.

10.5.3 Noise Immunity

An attractive feature of this communication is that the increased noise immunity can be reached using these signals. Intuitively, it follows from the coarseness of topological schemes of the stable dynamical systems [74]. An accurate estimation of the noise immunity is derived from the general theory of space-time modulated signals [18].

Consider $u_{1,0}(\mathrm{r},t)$ are the binary space-time modulated impulses defined for the spatial volume *V* and the time interval *T*. The lower limit of the conditional probability error P_{err} is estimated according to a formula from [76]:

$$P_{\mathrm{err}} \geq \frac{1}{2}\left(1-\Phi\sqrt{\frac{1}{N_0}\int_0^T\int_0^V\left[u_0(\mathbf{r},t)-u_1(\mathbf{r},t)\right]^2 dvdt}\right) \tag{10.7}$$

where N_0 is the signal norm, and Φ is the *Krampf* function. It follows that P_{err} is the minimal one if the impulses corresponding to the binary "0" and "1" are orthogonal to each other:

$$\int_0^T\int_0^V u_0(\mathbf{r},t)u_1(\mathbf{r},t)dvdt = 0. \tag{10.8}$$

Particularly, such signals can be the impulses of the waveguide eigenmodes having topologically different maps of field-force lines. Thus, the topologically modulation can be based on the modal manipulation, and it can provide increased noise immunity. To establish a more accurate theory of these signals, the general theory space-time signaling should be applied [76],[77].

10.6 Passive Gating of Microwave Topologically Modulated Signals

Digital hardware for topologically modulated signals should detect the impulses with different time-varying 3-D digital shapes and compare them according to a certain logic. By the detection we mean the development of a signal which confirms the arriving of an impulse of a certain topological scheme. Additionally, the topologically modulated signals can be compared with each other or be interpreted regarding to a spatial hardware structure similarly to optical holography, for instance. Formally, this hardware needs an established theory of the non-autonomous 3-D dynamical systems or (3+1)-D dynamical systems and a topological theory of 3-D manifolds and oriented graphs in the (3+1)-D phase space. Unfortunately, these theories are only in the beginning of their developments. For instance, only recently the famous Poincaré conjecture regarding to 3-D manifolds from the 4-D space has been proofed by G. Perelman.

Our research started in 1988, and, intially, the 2-D and 3-D EM fields were studied from the topological point of view [12]. Then the main attention was paid to a qualitative theory of the boundary problems of electromagnetism (See Chpt. 4). These techniques allow for the analytical or semi-analytical composition of topological schemes of EM fields according to the given boundary conditions. The bifurcation parameters for these systems are the frequency or the given boundary conditions and the domain geometry. It is found that the field-force line maps nonlinearly depend on the boundary conditions, which smooth variation can change the topological schemes of the excited fields.

Later, in 1993, the modal diffraction was considered in a wave transformer from the viewpoint of the geometry of diffracted field [20],[21],[36]. It is found that the topology of the output signals discretely depends on the input signal magnitudes and their fields. Then this effect can be used to develop the circuits for processing of topologically modulated signals.

The following example illustrates the effect of passive switching of topologically modulated microwave signals. For this purpose, a circuit arranging the impulses of different topological schemes into different outputs is represented as a generalized wave transformer.

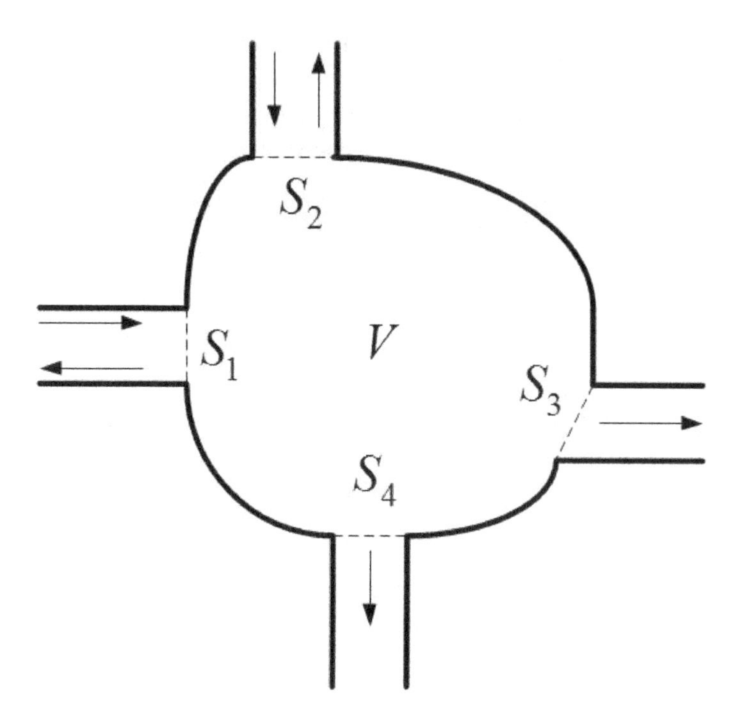

Fig. 10.5 Wave transformer

It consists of the central cavity V and the N arms. The input fields excite K arms. An example of a four-arm (N=4, K=2) wave transformer is shown in Fig. 10. 5.

The EM power conservation law equation in this cavity is written as

$$\frac{dW}{dt} = \sum_{k=1}^{K} p_{in}^{(k)} - \sum_{n=1}^{N} p_{sc}^{(n)} \tag{10.9}$$

where W is the EM field energy in the transformer volume V, $p_{in}^{(k)}$ is the vector Poynting flow of the incident fields from the k-*th* arm, and $p_{sc}^{(n)}$ is the vector Poynting flow of the scattered field to the n-*th* arm.

This expression is written using the field vectors $\mathbf{E}^{(V)}$ and $\mathbf{H}^{(V)}$ in the central cavity and in the *k-th* $(\mathbf{E}_{in}^{(k)},\mathbf{H}_{in}^{(k)})$ and *n-th* arms $(\mathbf{E}^{(n)},\mathbf{H}^{(n)})$:

$$\begin{aligned}
&\frac{1}{2}\frac{d}{dt}\int_V \begin{pmatrix} \varepsilon_a\left(\mathbf{E}^{(V)}(\mathbf{r},t)\cdot\mathbf{E}^{(V)}(\mathbf{r},t)\right)+ \\ +\mu_a\left(\mathbf{H}^{(V)}(\mathbf{r},t)\cdot\mathbf{H}^{(V)}(\mathbf{r},t)\right)\end{pmatrix} dV = \\
&=\sum_{k=1}^{K}\int_{s_k}\left[\mathbf{E}_{\text{in}}^{(k)}(\mathbf{r},t)\times\mathbf{H}_{\text{in}}^{(k)}(\mathbf{r},t)\right]\mathbf{v}_k ds- \\
&-\sum_{n=1}^{N}\int_{s_n}\left[\mathbf{E}^{(n)}(\mathbf{r},t)\times\mathbf{H}^{(n)}(\mathbf{r},t)\right]\mathbf{v}_n ds.
\end{aligned} \tag{10.10}$$

Substituting the electric and magnetic field vectors from (10.3), the power conservation law (10.10) is written using the geometrical field characteristics:

$$\begin{aligned}
&\frac{1}{2}\frac{d}{dt}\int_V \left(\varepsilon_a\left(\frac{d\mathbf{r}_e^{(V)}(t)}{ds_e}\right)^2+\mu_a\left(\frac{d\mathbf{r}_h^{(V)}(t)}{ds_h}\right)^2\right) dV = \\
&=\sum_{k=1}^{K}\int_{s_k}\left[\mathbf{E}_{\text{in}}^{(k)}(\mathbf{r},t)\times\mathbf{H}_{\text{in}}^{(k)}(\mathbf{r},t)\right]\mathbf{v}_k ds- \\
&-\sum_{n=1}^{N}\int_{s_n}\left[\frac{d\mathbf{r}_e^{(n)}(t)}{ds_e}\times\frac{d\mathbf{r}_h^{(n)}(t)}{ds_h}\right]\mathbf{v}_n ds
\end{aligned} \tag{10.11}$$

where ε_a and μ_a are the absolute permittivity and permeability, respectively, of the medium inside the transformer volume, and $\mathbf{v}_k$, $\mathbf{v}_n$ are the normal unit vectors to the opening surfaces of the *k-th* and *n-th* arms, respectively (Fig. 10.5). This equation shows the evolution of the geometry of the excited fields to a steady state if a transient happens with the incident signal.

Taking into account that a field-force line picture is a particular case of images, let us compare the proposed method with the techniques used in signal processing and topology. For example, the images can be enhanced using the partial differential diffusion equation [78]. For a 2-D image, an equivalent stationary energy functional is written, and the image enhancing is associated with the minimization of the "image energy." Unfortunately, this method increases the geometrical entropy of the processed image, and it has limitation of its use.

The energy approach was used by G. Perelman for his proof of the Poincare conjecture [9],[10]. The used Ricci flow is similar to the nonlinear partial differential diffusion equation. He shows that the time evolution of a closed 3-D manifold of an arbitrary geometry and associated with the Ricci-flow leads to the 3-D sphere, and this process is equivalent with reaching the maximum entropy because of the fundamental simplicity of the sphere.

Signal processing is with increase of information, and the above-mentioned equations have limited applications in logical processing of image-like objects. Here, the time evolution of a geometry should lead to a certain figure, which shape is not always coupled with the maximum of geometrical entropy.

The equations (10.11), derived in 1993, are the energy flow equations, but geometrical solutions of them are not with the maximum of geometrical entropy. The transients in the above-considered wave transformer lead to excitation of the cavity and the portal modes. Their fields have a certain space-time geometry which depends on the boundary, initial conditions, and transformer geometry. The resonances and the non-zero excitations do not allow degrading the field to the noise-like distribution of maximal geometrical entropy. Then the above-considered effects and the equations (10.10) and (10.11) can be used to describe the information dynamics of such a sort of devices instead of the methods associated with the entropy maximum.

The equ. (10.11) shows that the geometry of the field-force line maps inside the transformer and in the output arms depends nonlinearly on the input fields and the relationships of their amplitudes. The topological maps can be changed discretely by a smooth variation of the incident field parameters under a certain condition. Particularly, it allows controlling the field distribution of the output fields. If a field is close to a propagating mode of the *n-th* output, then the matching conditions allow transmitting an increased portion of power of the incident signal to this port. Other terminals can be isolated from the input if this signal excites the evanescent modes in them. Then switching of the input signal from one output to another can be realized by choosing the input signal parameters.

Considering the (3+1)-D dynamical systems (10.3) and (10.4), it follows that it is possible to design the devices handling the signals composed of 3-D manifolds and having their different space-time topology. Another conclusion is that the discrete dynamics of the vector maps is caused by wave diffraction and interference instead of switching mechanisms of semiconductors. Although the transistor is the best switching device ever made by engineers, the Nature can realize the digital-like mechanisms using other less-energy-consuming means.

10.7 Theoretical and Experimental Validations of the Proposed Signals and Their Processing

10.7.1 Passive Switches and Their Experimental and Theoretical Studies

Our initial research was more aimed to the validation of the proposed logical handling of microwave and digital topologically modulated signals instead of the direct developments for commercial applications. These designs and simulations proved the idea of logical handling of such signals. Later, these results inspired the developments which were more relevant for market needs, and a couple of variants of a predicate logic processor were proposed for artificial intelligence applications [63]-[72].

A couple of the developed and tested passive gates of topologically modulated signals are shown in Figs. 10.6 and 10.7, and they are to switch the even and odd modes of coupled microstrip lines (input I) to the separated outputs II (slotline) and III (two-conductor line). Taking into account the reversibility of the circuits, the modes of coupled microstrips can be excited by signals from corresponding ports II and III.

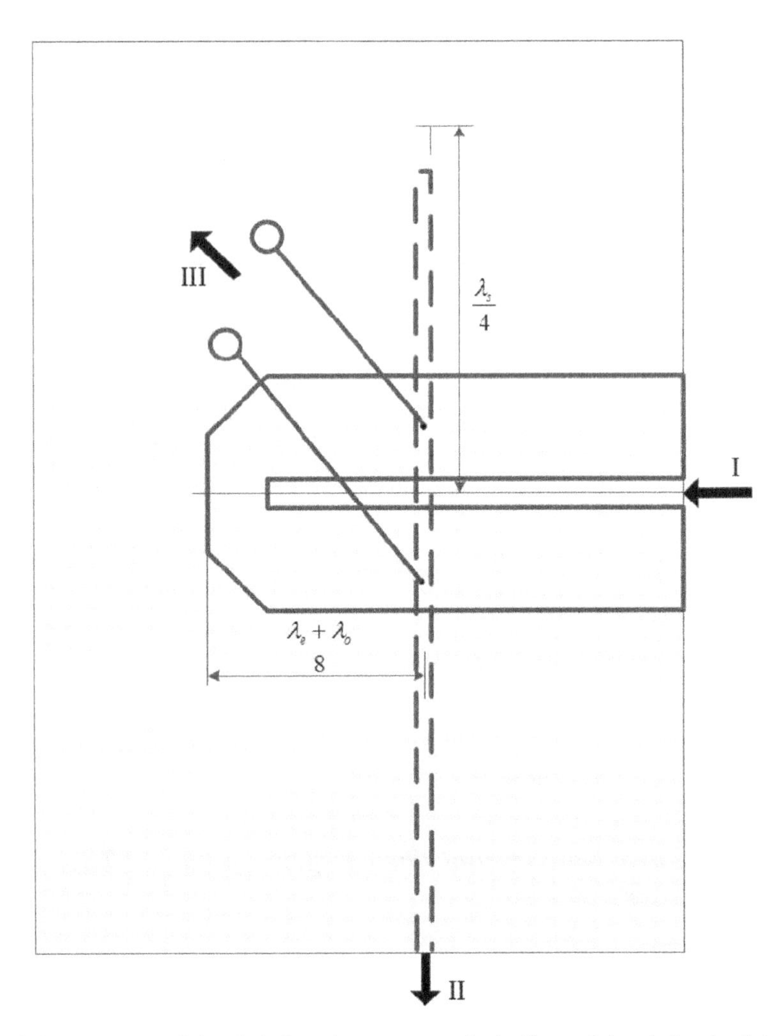

Fig. 10.6 Resonant spatial switch for microwave topologically modulated signals. Adapted from [16]

This originally proposed gate is based on the resonant coupling of the slot mode and the even mode of coupled microstrips. The differential signal of the odd mode can be taken by a two-conductor line or a coaxial waveguide. The signal from the slot line is taken by a coaxial waveguide in this case. Measurements show the isolation of the outputs over 20-30 dB at the resonant frequency 3 GHz, depending on design. The loss is about several dB measured for whole integration together with the transitions to the coaxial lines [16],[18],[28].

Another circuit, containing matching resistors, is assigned for digital applications where the signals can be restored by transistor gates (Figs 10.7 and 10.8). It demonstrates large bandwidth and, unfortunately, increased loss [21],[25],[26],[28],[36]. It consists of the input coupled microstrip transmission lines I shorted through the resistors *R* to the microstrip line II, the two-strip or two-wire line III orthogonally connected to this joint of the lines I-II, and the resistors *R*.

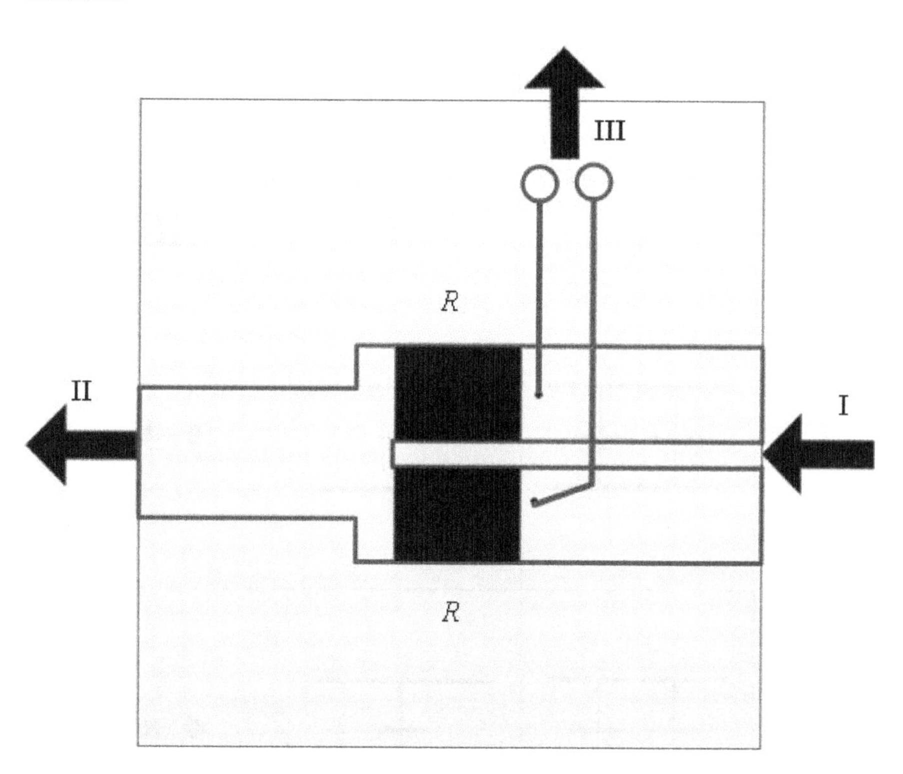

Fig. 10.7 Passive spatial switch for topologically modulated signals. Adapted from [36]

Its equivalent circuit and truth-table are shown in Figs. 10.8 and 10.9, correspondingly.

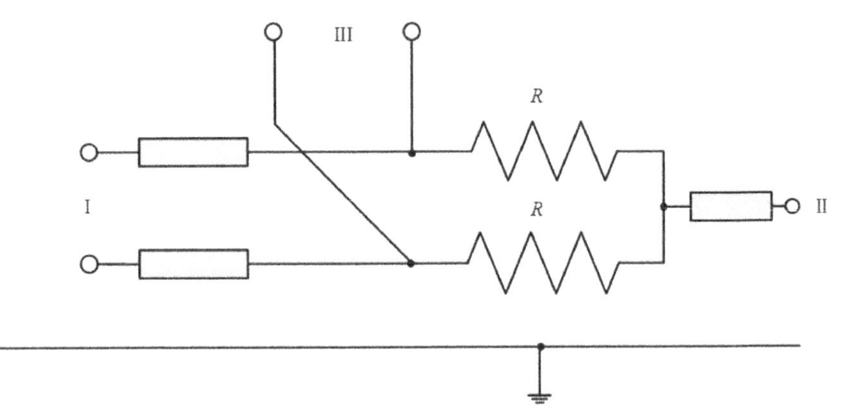

Fig. 10.8 Equivalent circuit of passive spatial switch. I- input on the coupled strip line, II- output of the topological "1", and III- output of the topological "0". Adapted from [36]

The input signal (Port I) is a series of video or sinusoidal impulses of the even and the odd modes of coupled strips line. Their field topological schemes are not homeomorphic to each other, and the even and the odd ones correspond to the logical levels "1" and "0", respectively.

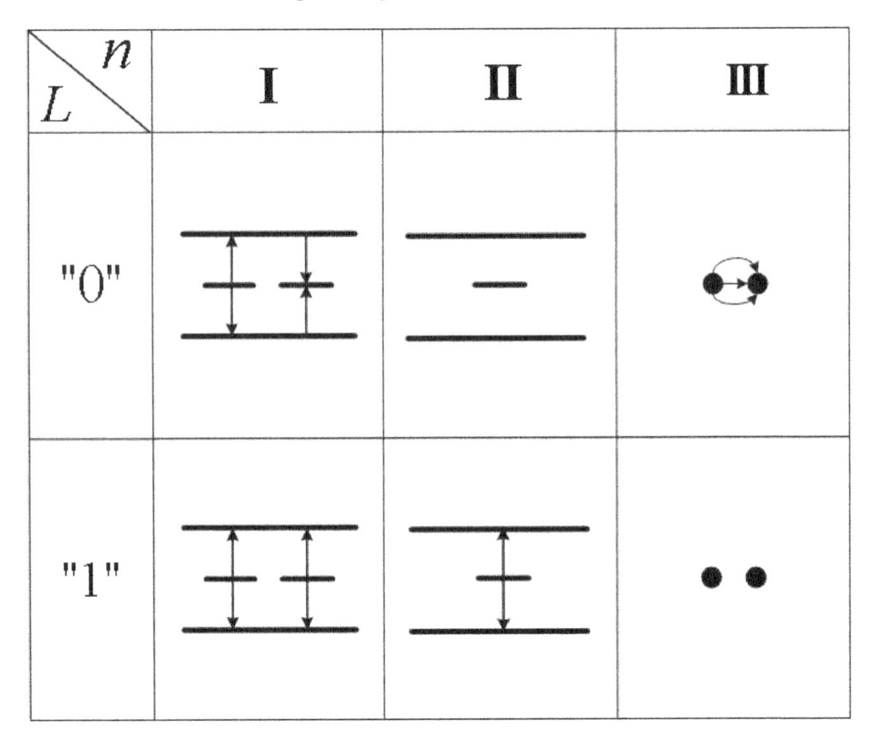

Fig. 10.9 Truth-table for the passive switch of topologically modulated signals realized on strip transmission lines. Adapted from [36]

These input signals are switched to the outputs II and III according to the truth-table shown in Fig. 10.9 where L is assigned for the logical levels, and n is for the input/output numbers. This circuit is logically reversible, and the signals from the outputs can excite corresponding modes separately in time or at the same moment.

This passive gate was studied experimentally for low-rate digital signals only [40]. A measurement setup is shown in Fig. 10.10.

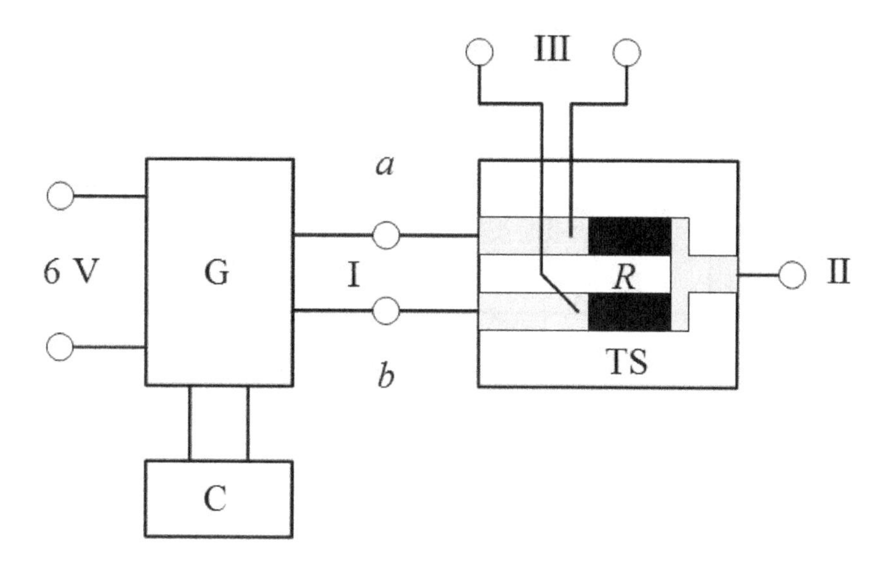

Fig. 10.10 Measurement setup for a passive switch of topologically modulated signals. G- Generator of topologically modulated signals, C- Controlling circuit, $R = 50\ \Omega$ –Thin-film resistors, and TS- Plate with the switch. Adapted from [40]

In this design, a generator of topologically modulated signals G and the plate with the switch TS are placed separately from each other but connected to the common ground. Some results are shown in Figs. 10.11 and 10.12. The input voltage is measured at the 2 points marked by "a" and "b" in Fig. 10.10. The output voltage of the logical "1"is at the output II loaded by a high-ohmic resistor to the ground. The output voltage of the logical level "0" is measured as the differential signal at the port III. The time diagrams show the switching of impulses to different outputs according to the truth-table of Fig. 10.9. The isolation of the outputs is measured at frequencies below 1 MHz, and it is around 30 dB.

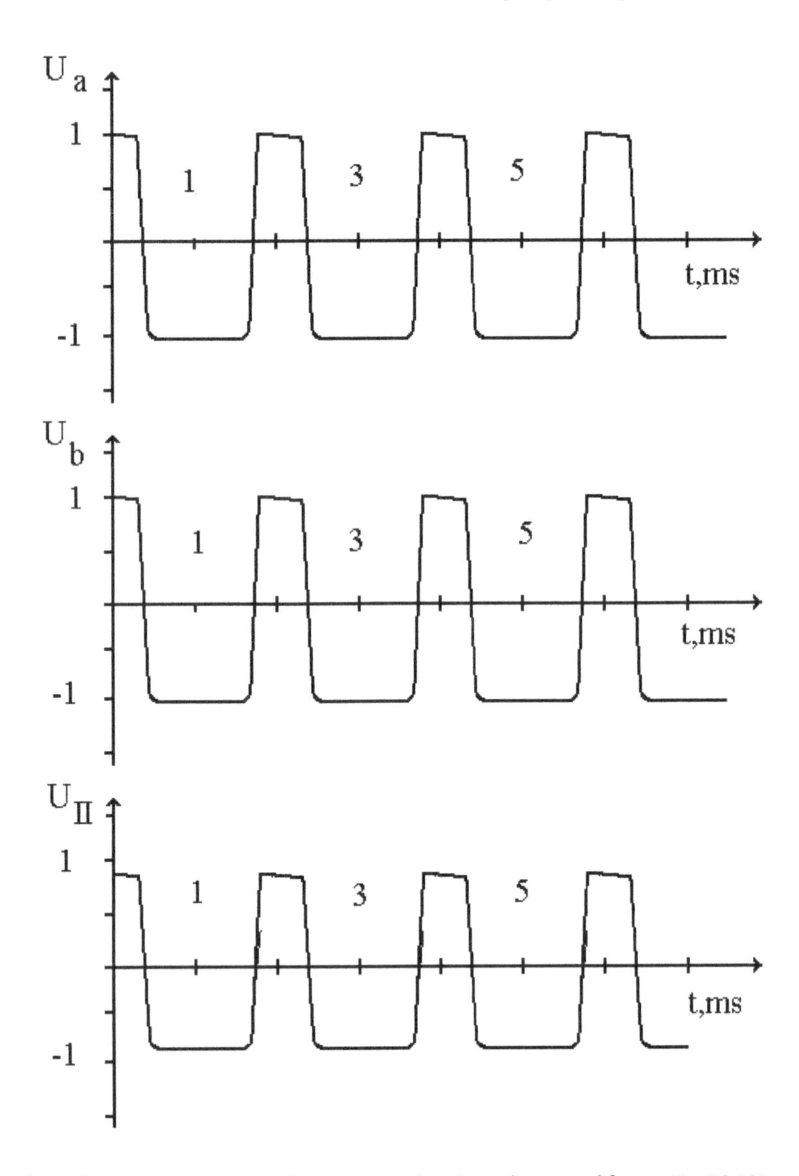

Fig. 10.11 Input even mode impulses measured at the points *a* and *b* (see Fig. 10.10) and the switched output signal U_{II}

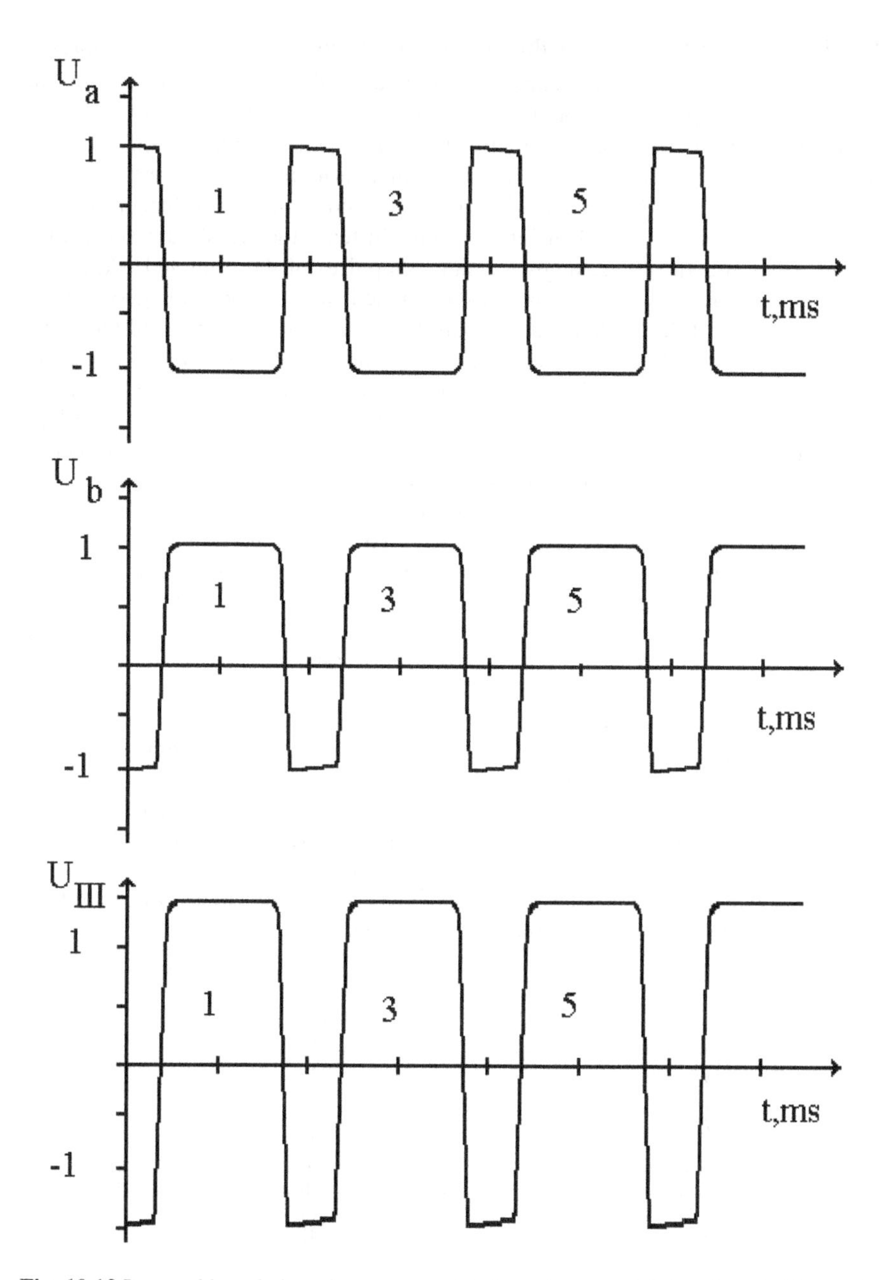

Fig. 10.12 Input odd mode impulses measured at the points *a* and *b* (see Fig. 10.10) and the switched output signal U_{III}

The transients of this passive switch of a monolithic design are estimated roughly in [36],[41] by theoretical modeling. It was supposed that the switch was manufactured on the substrates with $\varepsilon_r = 3.5$ and 9.6 and height $h = 3$ μm.

The thin-film resistors of the width w are considered as the transmission lines of the length Δl placed at the distance s from each other. The resistor length, the end and shunt capacitances of thin-film resistors are taken into account by the used equivalent circuit. The coupling of resistors is not considered, and the parasitic reactivities of junctions to a single microstrip and to a "vertical" two-conductor line are neglected. The dispersion of modes and the influence of skin-effect in resistors are not taken into account. Instead, the worst-case scenario is studied when practically rectangular voltage impulses of the even and odd modes are applied to the input *I* consisting of the coupled strips of the same geometry as the resistor part of the switch. The outputs *II* and *III* are loaded by 50-Ohms instead of using separate loads adapted to the common and differential excitations. Variation of parameters of the resistors allows to find those of them which minimally distort the signals of both common and differential types. The results of estimations of transients are given in Fig. 10.13a-c and Fig. 10.14a-c for the rectangular even and odd input signals, respectively.

These modeling results show that the time delay of the switched impulse is roughly determined by the length of resistors. To estimate its influence, the phase constant is approximately derived from the model of an ideal microstrip transmission line of the same geometry. More serious signal distortion is with mismatching and ringing. In the case of even mode, the transients of this kind can be decreased by proper matching and decreasing of parasitics of joints.

In the case of the odd-mode excitation, the switch resistors are shorted at their ends, and the signal is reflected multiple times from this joint. Proper tuning of the resistor value adapted to the input and output loads decreases this ringing. The performed simulations in both cases demonstrate the duration of transients of the order of several tenths of a picosecond in spite of the worst-case scenario excitation.

The transients can be decreased further using more realistic signals which time shape is close to the Gaussian function. The modeling results of this kind are in Fig. 10.15 where the Gaussian input signals are shown together with the output signals, and the ringing is not strong enough, as it is seen.

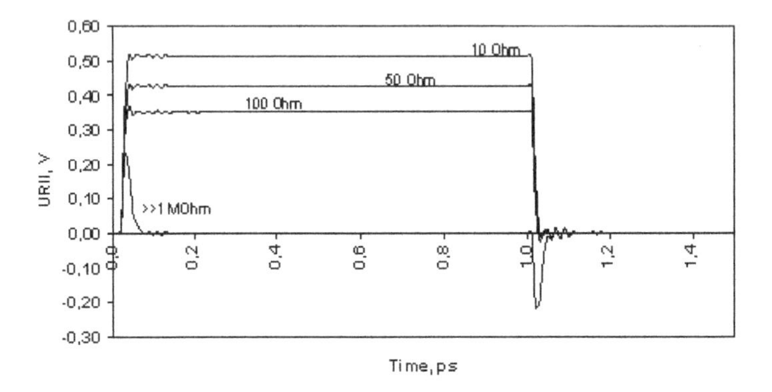

Fig. 10.13a Rectangular even mode impulses of the 1 V magnitude at the output *II* calculated for *R*=10, 50, 100 Ohm, and 1 MOhm. The output port *II* load is $Z^{(II)} = 50$ Ohm, $\varepsilon_r = 3.5$, *h*=3 µm, *Δl*=1 µm, *w*=1 µm, *s*=1 µm

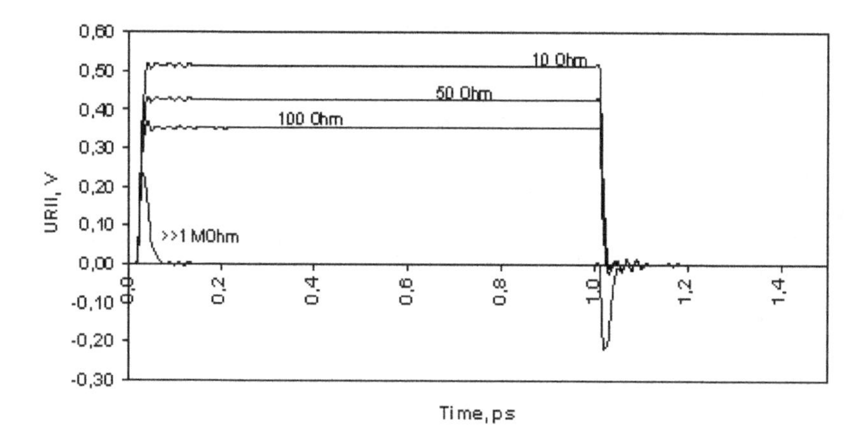

Fig. 10.13b Rectangular even mode impulses of the 1 V magnitude at the output *II* calculated for *R*=10 Ohm, 50 Ohm, 100 Ohm, and 1 MOhm. The output port *II* load is $Z^{(II)} = 50$ Ohm . $\varepsilon_r = 3.5$, *h*=3 μm, Δl =1 μm, *w*=4 μm, *s*=1 μm

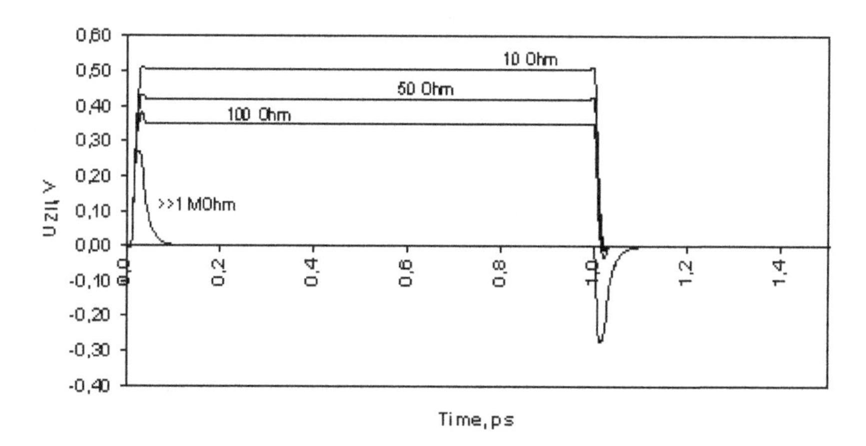

Fig. 10.13c. Rectangular even mode impulses of the 1 V magnitude at the output *II* calculated for *R*=10 Ohm, 50 Ohm, 100 Ohm, and 1 MOhm. The output port *II* load is $Z^{(II)} = 50$ Ohm , $\varepsilon_r = 9.6$, *h*=3 μm, *Δl*=1.5 μm, *w*=1.5 μm, *s*=1 μm

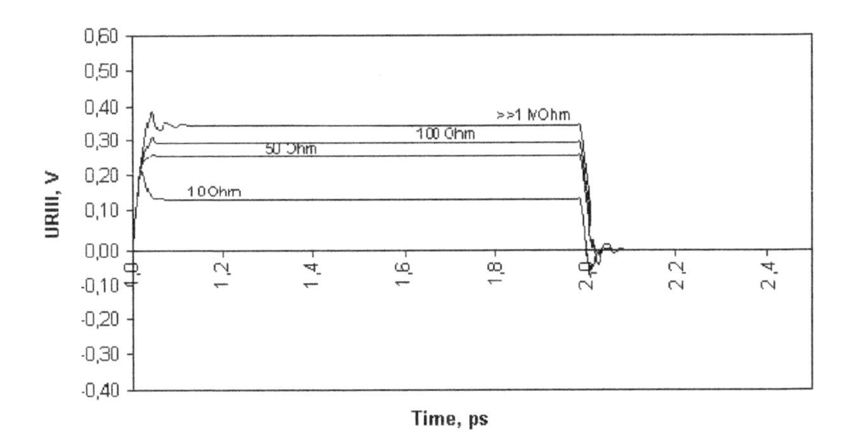

Fig. 10.14a Rectangular odd mode impulses of the 1 V magnitudes at the output calculated for *R*=10 Ohm, 50 Ohm, 100 Ohm, and 1 MOhm. The output port *III* load is $Z^{(III)} = 50$ Ohm, $\varepsilon_r = 3.5$, *h*=3 µm, *Δl*=1 µm, *w*=1 µm, *s*=1 µm

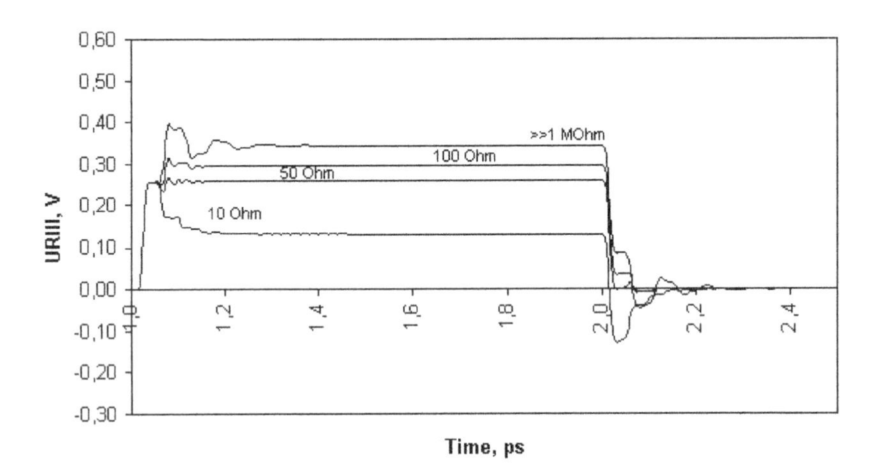

Fig. 10.14b Rectangular odd mode impulses of the 1 V magnitudes at the output calculated for *R*=10 Ohm, 50 Ohm, 100 Ohm, and 1 MOhm. The output port *III* load is $Z^{(III)} = 50$ Ohm, $\varepsilon_r = 3.5$, *h*=3 µm, Δ*l* =1 µm, *w*=4 µm, *s*=1 µm

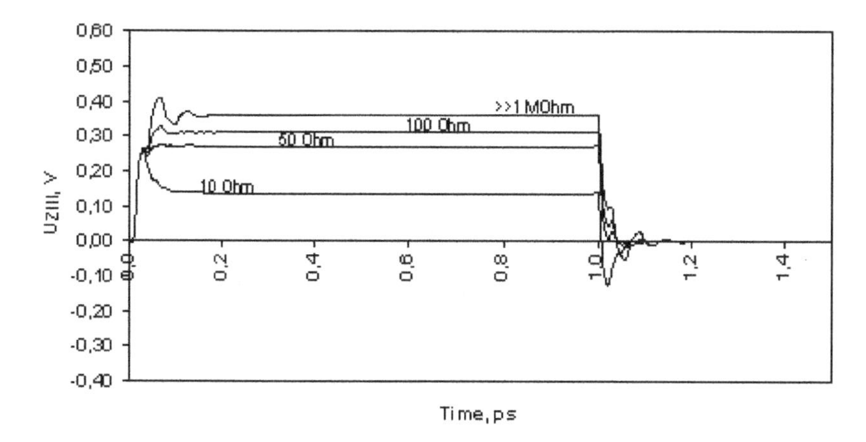

Fig. 10.14c Rectangular odd mode impulses of the 1 V magnitudes at the output calculated for R=10 Ohm, 50 Ohm, 100 Ohm, and 1 MOhm. The output port *III* load is $Z^{(III)} = 50$ Ohm, $\varepsilon_r = 9.6$, h=3 μm, Δl=1.5 μm, w=1.5 μm, s=1 μm

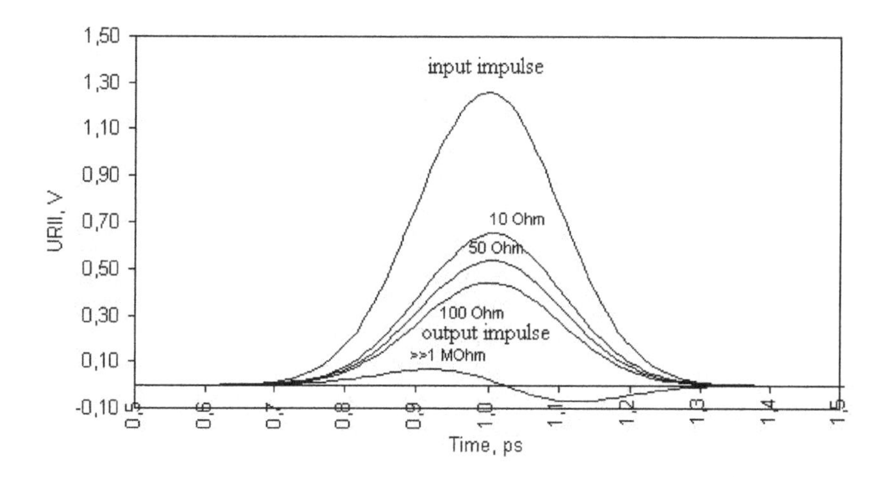

Fig. 10.15a Gaussian even mode impulses transmitted through the switch. R=10 Ohm, 50 Ohm, 100 Ohm, and 1 MOhm, $Z^{(II)} = Z^{(III)} = 50$ Ohm, $\varepsilon_r = 3.5$, h=3 μm, Δl=1 μm, w=1 μm, s=1 μm

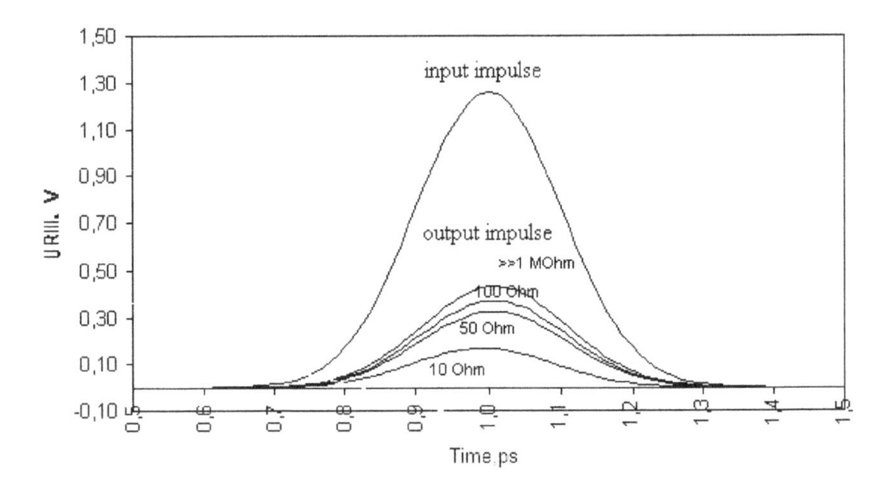

Fig. 10.15b Gaussian odd mode impulses transmitted through the switch. *R*=10 Ohm, 50 Ohm, 100 Ohm, and 1 MOhm, $Z^{(II)} = Z^{(III)} = 50\,\text{Ohm}$: $\varepsilon_r = 3.5$, *h*=3 µm, *Δl*=1 µm, *w*=1 µm, *s*=1 µm

A comparative energy analysis of this switch is performed in [28], where it is shown that even the picosecond delay-time passive switches of a micrometer size had the energy factor around 0.02 fJ, which is calculated as the thermal loss power multiplied by the time delay. This evaluation was performed in 1994, and according to the available data at that moment on the semiconductor gates consisting of a couple of transistors, it was a couple of orders better than those for other analogs. Contemporary technology allows manufacturing more enhanced integrated elements of the size comparable with a couple of tens of nanometers, and, according to the elementary scaling, the delay time of passive switches can be improved, again. Of course, the designed switch destorts the signals due to the loss and transients, and the switched impulses need periodical restoration.

Concluding the transient modeling, it is necessary to underline that the final estimation of the time delays and transients durations can be obtained from experimentations, only. Indirect confirmation of increased speed of the passive switching is from a later work [79] where an AND gate on coupled microstrips line is studied experimentally. It is shown the workability of these circuits for 80-Gps signals. It is predicted by simulations that the 200-Gps signals can be switched by passive gates with the length of coupled lines *l*=800 µm and width *w*=15 µm. Nowadays technologies allow manufacturing on the several-tens-of-nanometers scale, and the possibilities of passives and ultra-high speed transistors are estimated more optimistically for realization of space-time logic and computing.

10.7.2 Hardware for Measurements of Picosecond Circuits

For measurements of the picosecond even/common mode signals and the digital circuits, a custom-made generator was designed [57].

It consists of several building blocks (Fig. 10.16). The *Main Generator* 1 is the source of impulses with the front width around 1 ns, and they are used as clocks as well. The *Switched Delay Line Section* 2 allows regulation of the delay for 25, 50, and 100 ns. Smooth regulation in the limits of 0-25 ns is controlled by a separate handle. The sections of *Shaping of Impulses* 3 and 4 allow forming a series of impulses with the fronts around 100 ± 5 ps and duration of 1 μs. This long duration allows studying the transients of measured circuits for each front of an impulse. The common and differential output signals with the controlled amplitudes within 3-5 V are taken from the 50-Ω coaxial connectors.

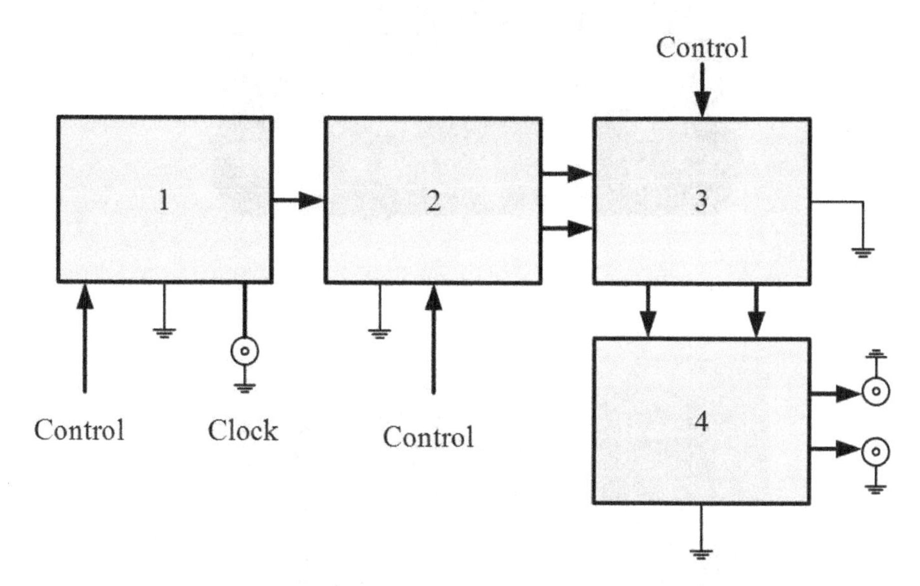

Fig. 10.16 Picosecond generator. 1- Generator; 2- Switched delay line section; 3- Main impulse shaper; 4- Impulse shaper

Fig. 10.17 shows the inside of the device. In Fig. 10.18, the front panel of the picosecond generator is given.

The generator characteristics are measured using some simple passive circuits and a high-speed oscilloscope, and the measurements show that the positive and negative impulses have the shape error no more than $\pm 3\%$, and they can be used for generation of differential/common modes in coupled microstrips transmission lines. Some results of the measurements of 3-port circuits (terminated and non-terminated Y-joints of microstrip lines) are published in [45],[57]. The obtained results and developed hardware were a part of an initial experimental R&D on the study of picosecond switches of topologically modulated signals.

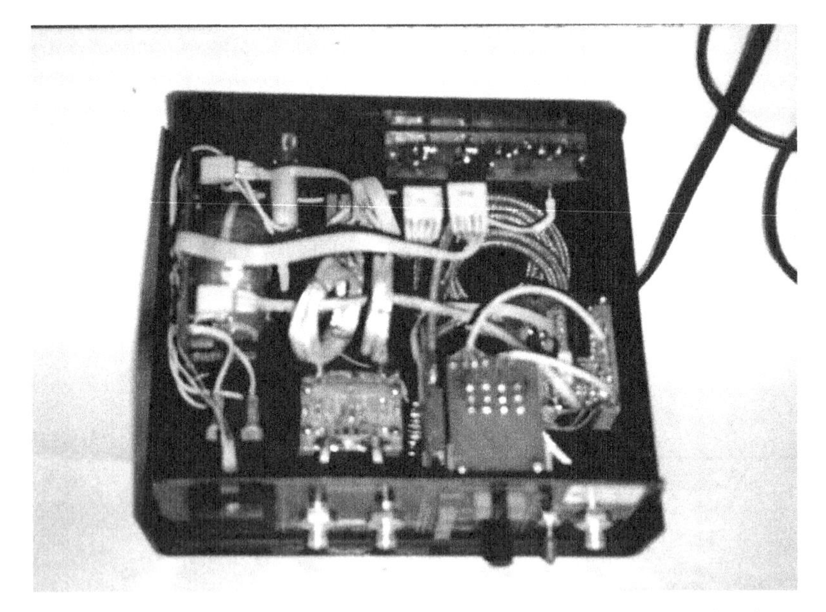

Fig. 10.17 The inside view of the picosecond generator of common/differential picosecond signals (Author's archive)

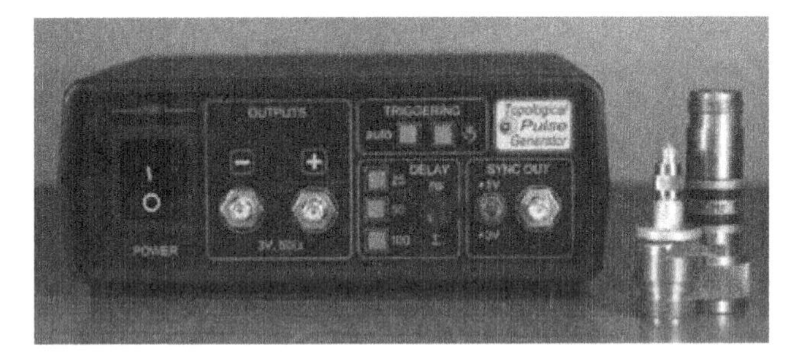

Fig. 10.18 Picosecond generator of common/differential signals (front view, Author's archive)

10.7.3 Diode Equipped Switch

Another part of our attention was paid to the digital circuits based on a mixed technology and which designs consisted of passive and semiconductor components to switch the signals according to their magnitudes and topological field schemes. One of the simplest gates of this kind and its truth-table are shown in Figs 10.19 and 10.20, respectively, and it is the above-considered resistive switch equipped additionally by diodes D_2 and D_3.

The input signal is the series of the even and odd modes of a coupled strips line I, and it is switched according to their topologically modulated spatial contents and the magnitudes of these impulses (Fig. 10.20). The even mode impulses are transmitted to the output II connected by the diode D_2 to the ground. The signal is at this output if the impulse has a large negative magnitude to keep the diode closed.

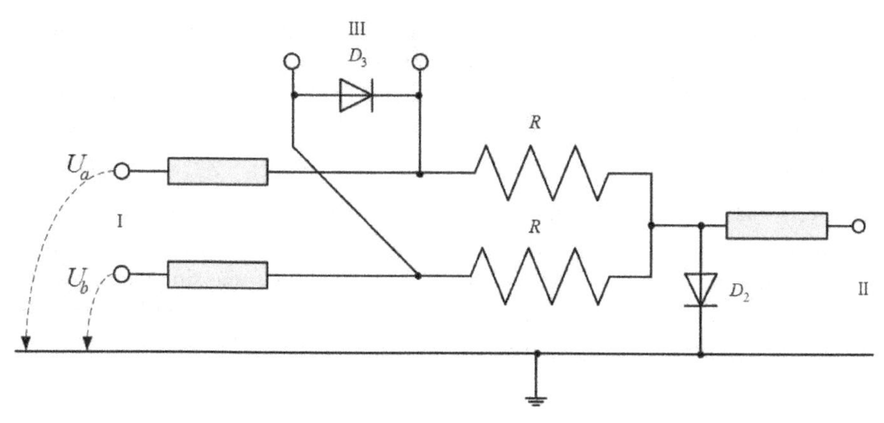

Fig. 10.19 Gate for space-magnitude switching of topologically modulated signals. I- input on the coupled strip lines, II- output of the topological "1", and III- output of the topological "0". Adapted from [39]

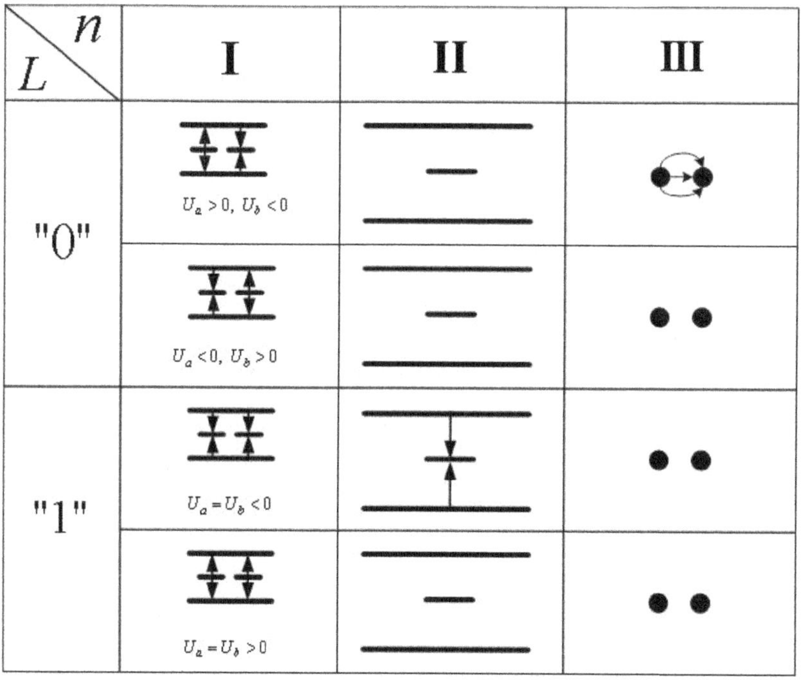

L \ n	I	II	III
"0"	$U_a > 0,\ U_b < 0$		
	$U_a < 0,\ U_b > 0$		
"1"	$U_a = U_b < 0$		
	$U_a = U_b > 0$		

Fig. 10.20 Truth-table for a diode switch of topologically modulated signals. Adapted from [39]

The odd signal should have a certain polarity and strength to close the diode D_3 to support a large differential signal corresponding to the logical "1". Thus, one information parameter can control the signal flow, and this gate is an example of the reconfigurable logic circuitry.

Different components of this type were simulated and measured in [39]. One of the studied switches was designed according to the microwave hybrid IC technology with the following parameters of the circuit [23]. The substrate of the height *h*=0.5 mm is of the permittivity $\varepsilon_r = 3.5$. The coupled microstrips line has the characteristic impedances of the even and odd modes $Z_c^{(e)} = 111.9\ \Omega$ and $Z_c^{(o)} = 37.1\ \Omega$, respectively. The switch's resistors are of $R = 24.5\ \Omega$.

The transients were simulated during the signal switching by diodes according to the truth-table from Fig. 10.20. Figs 10.21-10.24 show some simulation results. Each of these plots shows the input (dashed lines) and output (solid line) signals. They are distorted due to the parasitics of microstrip layout and lumped diodes.

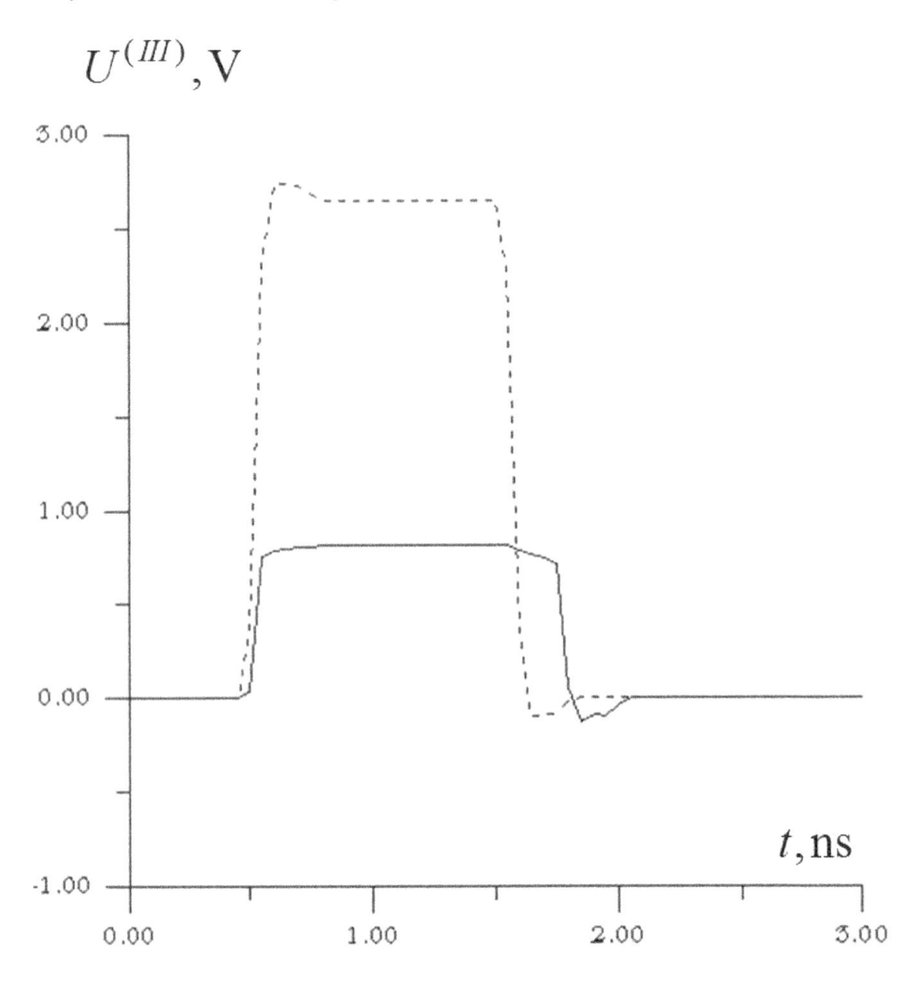

Fig. 10.21 Transients for the odd mode excitation. Diode D_3 is open and $U_a < 0,\ U_b > 0,\ |U_a| = |U_b|$. See the 2nd row in Fig. 10.20. Adapted from [23],[39]

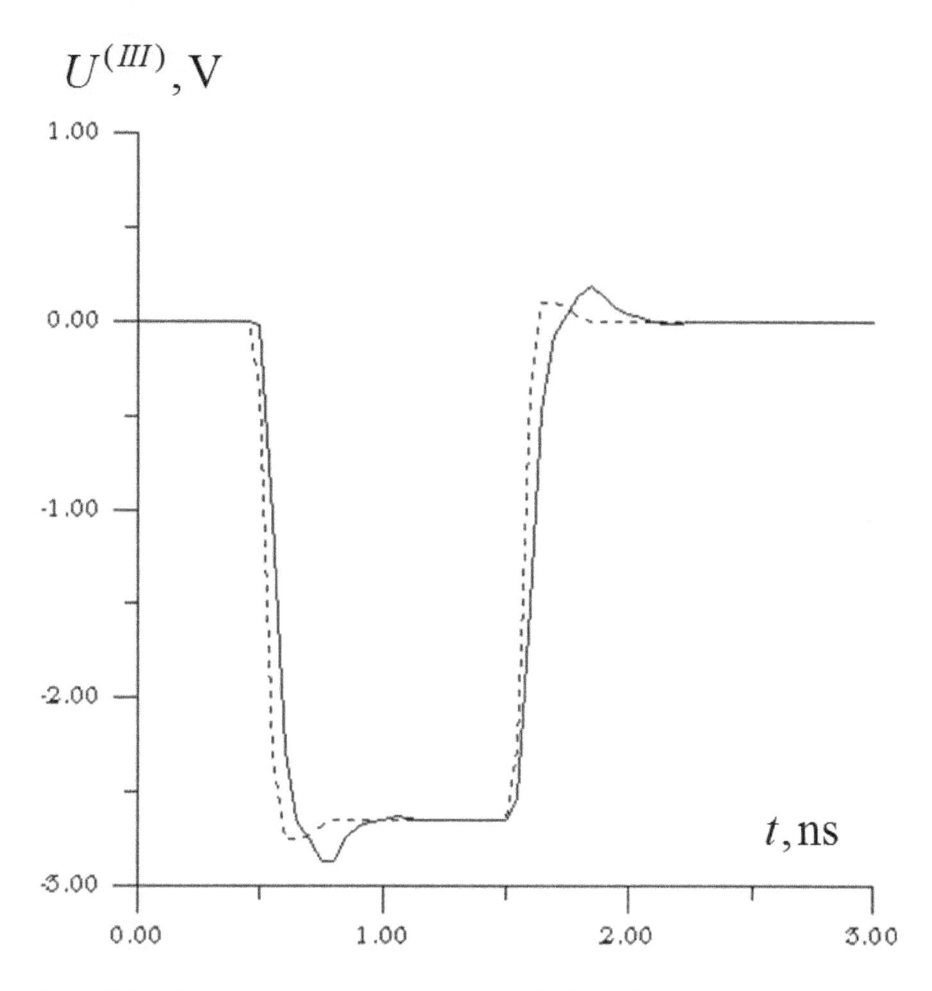

Fig. 10.22 Transients for the odd mode excitation. Diode D_3 is closed and $U_a > 0$, $U_b < 0$, $|U_a| = |U_b|$. See the 1st row in Fig. 10.20. Adapted from [23],[39]

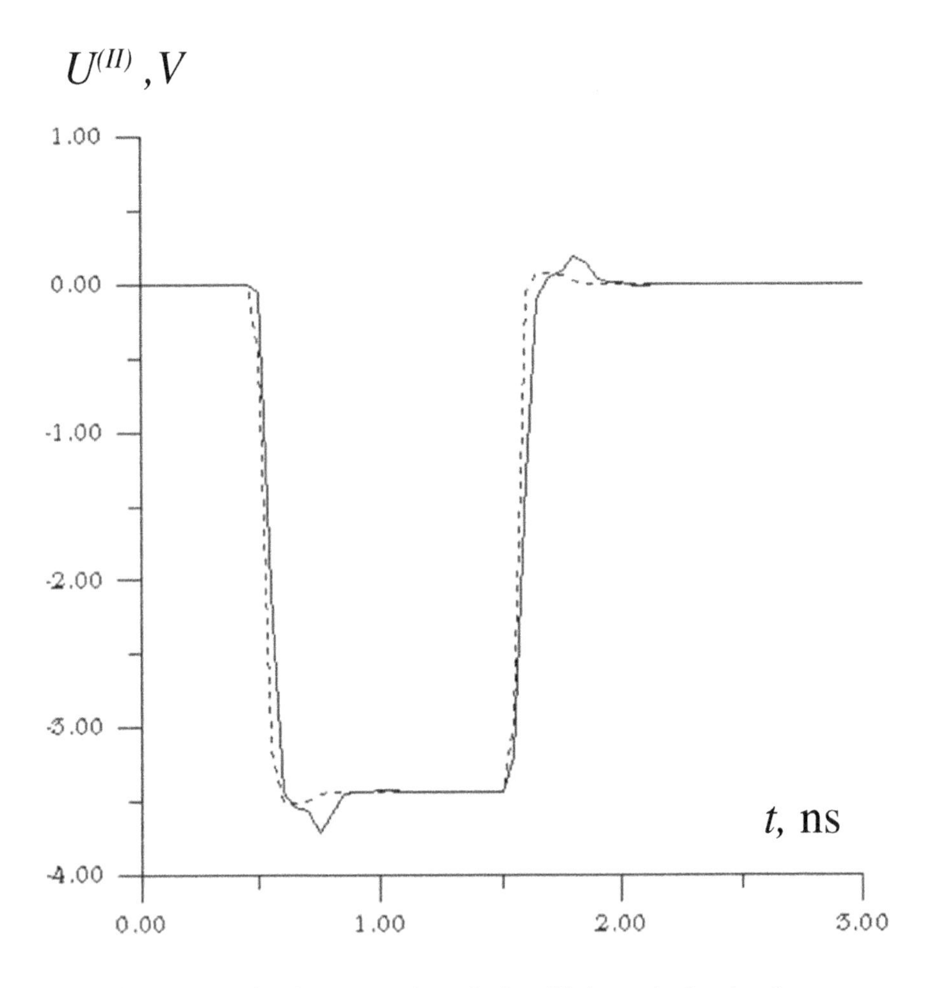

Fig. 10.23 Transients for the even mode excitation. Diode D_2 is closed and $U_a = U_b < 0$. See the 3rd row in Fig. 10.20. Adapted from [23],[39]

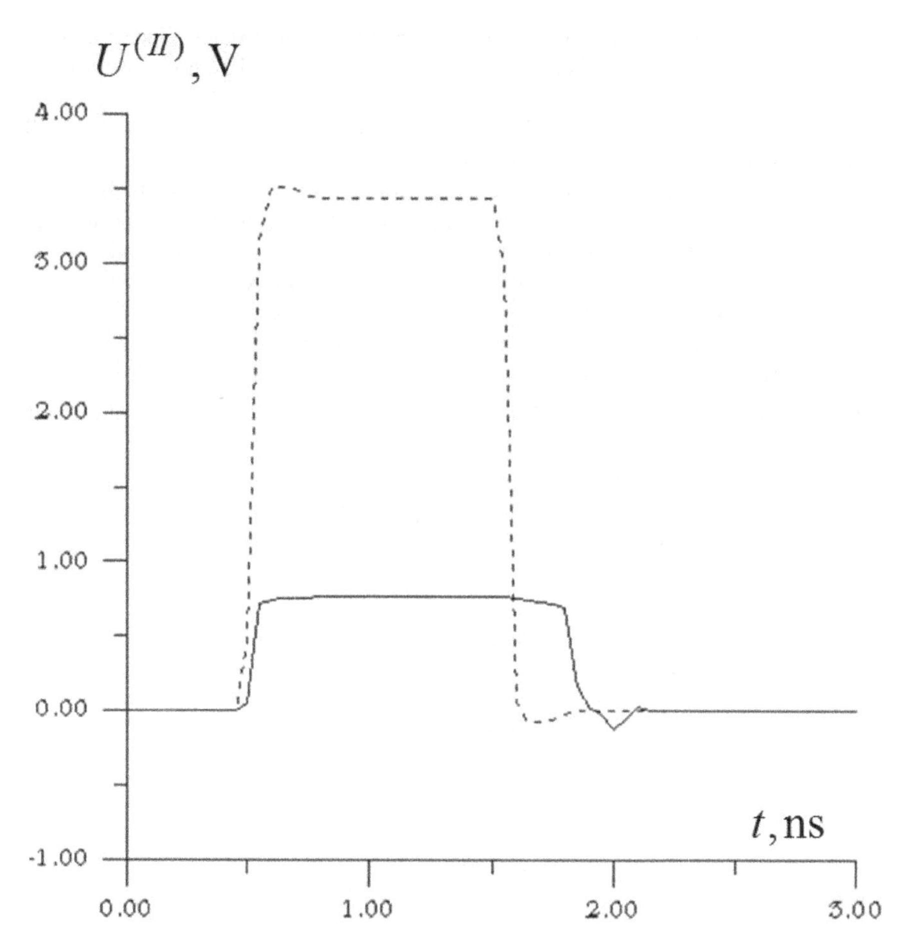

Fig. 10.24 Transients for the even mode excitation. Diode D_2 is open and $U_a = U_b > 0$. See the 4th row in Fig. 10.20. Adapted from [23],[39]

It is seen that the switching of topologically modulated signals occurs according to their field maps and magnitudes. This developed circuit is one of the first gates handling the two-place signals of this kind, and it can be related to the predicate logic gates [23],[24],[58],[60]. Additionally, it transforms the topologically modulated impulse series into the conventional, magnitude-modulated signals for further processing by Boolean logic circuitry.

10.7.4 Pseudo-quantum Gates for Topologically Modulated Signals

The pseudo-quantum logic [56],[58],[59] is based on the similarity of quantum and classical EM wave physics. A number of quantum gates can represent a quantum computation process based on the entanglement of the particles. Each gate is based on the handling of a qubit that has three logical states. One of them is for the logical "0", and another one is for the logical "1". The third one is specified for the intermediate state "*q*", which is the superposition of the two quantum states. The overall quantum parallelism is realized due to the entanglement of all particles taking part in the computing process.

Unfortunately, classical macro-effects are not able to provide this entanglement in spite of possibility of the imitation of the qubit-effect by electronic circuits, and the gates and hardware based on this imitation realize only classical parallelism of space-time computations. Some quantum algorithms can be realized using even classical circuits, and the results presented below are of a special interest.

In our case, the qubit effect can be associated with a superposition of the even and odd modes of a coupled strip transmission line, and three pseudo-quantum gates – NOT, CNOT, and "$\sqrt{\text{CNOT}}$ -gate" composing a full set of logical circuits were designed [58],[59].

Here the experimental results for a "$\sqrt{\text{CNOT}}$ -gate" (Fig. 10.25) are considered. The gate's truth-table is shown in Fig. 10.26. The two first logical states ("1") and ("0") are represented by the odd mode impulses of the high and low magnitudes, and the gate performs the NOT operation similarly to the classical NOT (the first two lines of Fig. 10.26).

The second set of impulses is represented by an algebraic sum of the odd and even modes, and this gate, instead of transforming it to the odd mode impulses, only exchanging the position of the excited wire, i.e. it performs the semi-negation of the input impulse (the last two lines of Fig. 10.26).

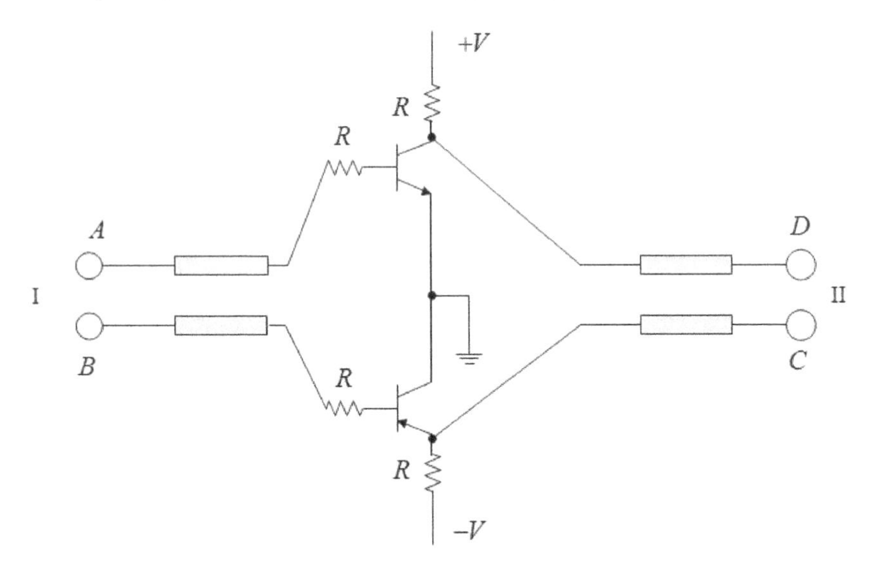

Fig. 10.25 $\sqrt{\text{NOT}}$ -gate" for pseudoquantum logic. Port I is the input of signals. Port II is the output of the gate. Adapted from [56]

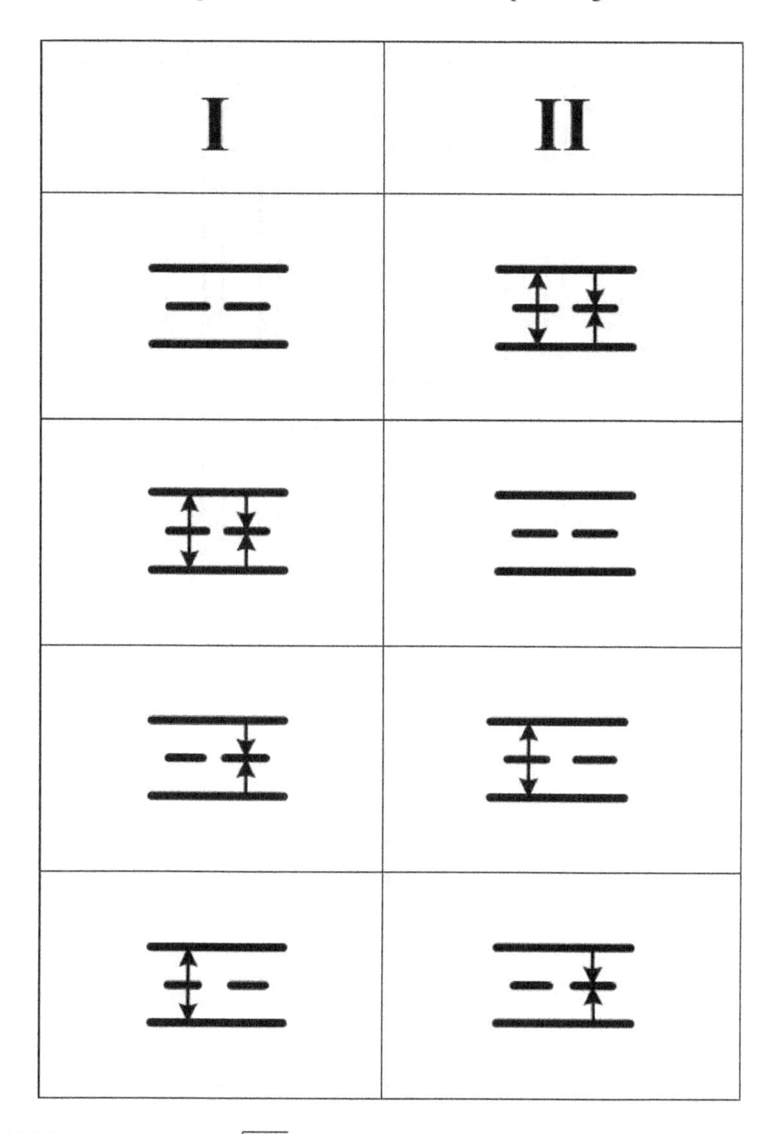

Fig. 10.26 Truth –table for "$\sqrt{\text{NOT}}$ -gate". Adapted from [56].

Some measurements for a low-rate signal are shown in Figs. 10.27-10.30. The input impulses are measured at the points A and B regarding to the ground (see Fig. 10.25). The output signals are from the points D and C, and they are measured similarly. Fig. 10.27 shows a set of the input odd mode impulses of high and low magnitudes. The signal is switched according to the Fig. 10.26, and the output impulses are demonstrated in Fig. 10.28.

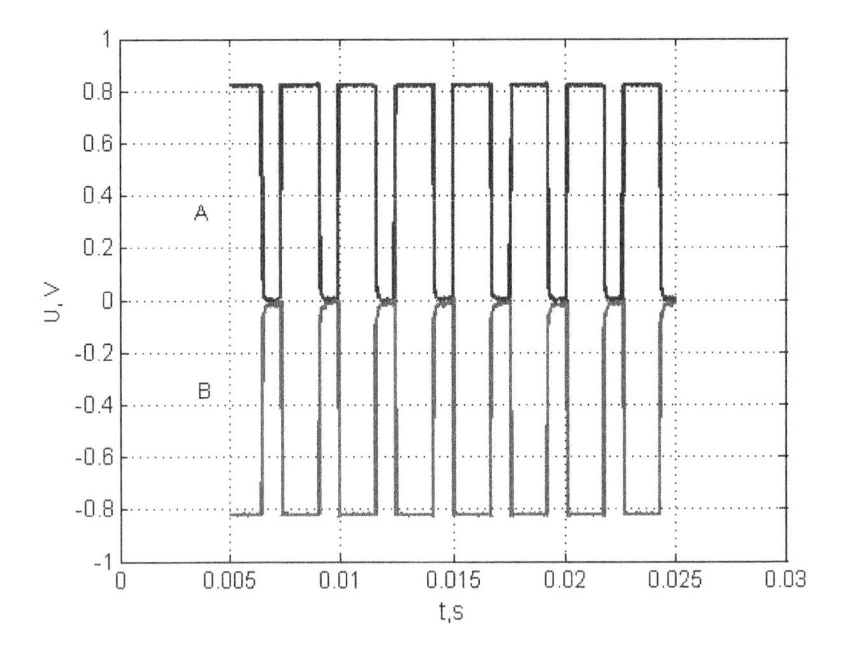

Fig. 10.27 Input odd mode impulses (2nd row of Fig. 10.26)

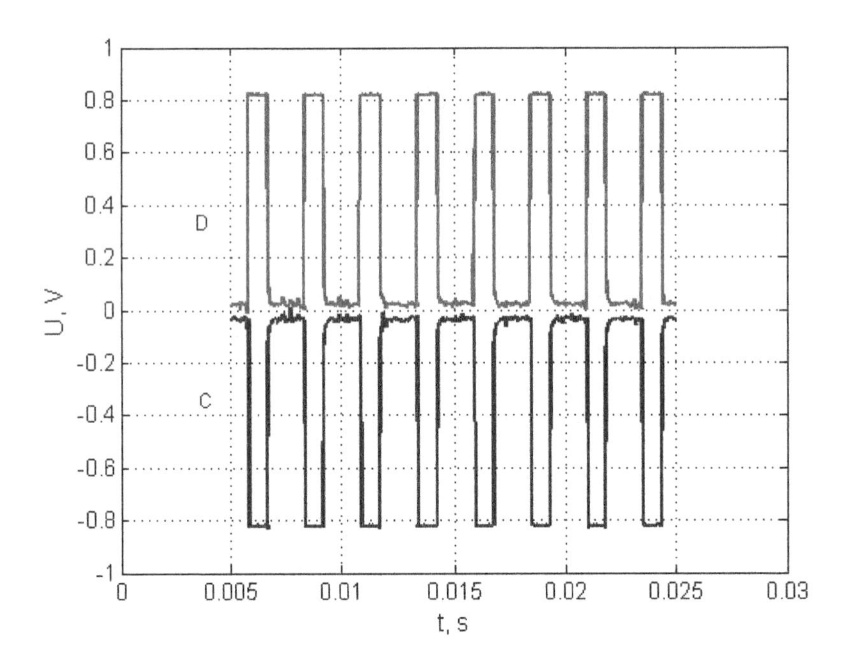

Fig. 10.28 Switched odd mode impulses (2nd row of Fig. 10.26)

The input signals, which model the superposition state, are shown in Fig. 10.29. The output switched signals are shown in Fig. 10.30.

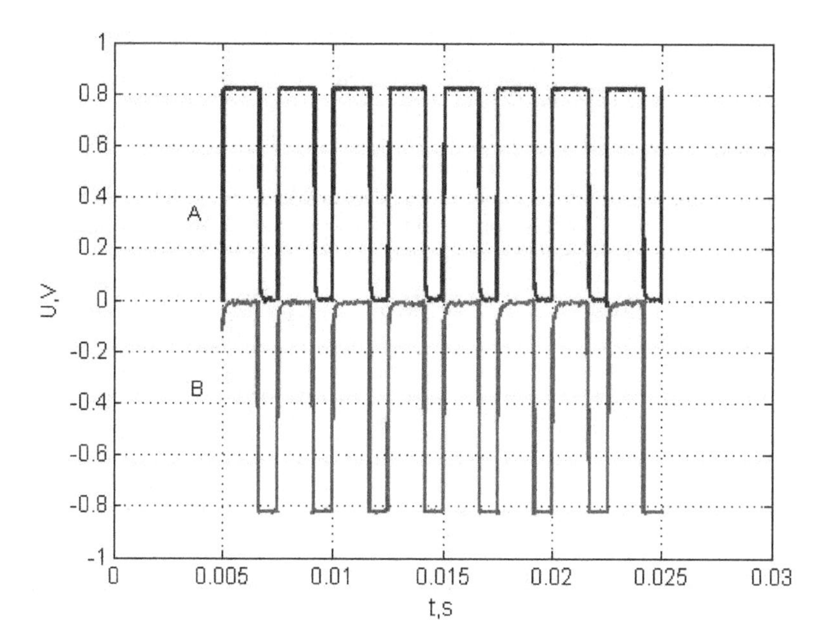

Fig. 10.29 Input "superposed" impulses (4th row of Fig. 10.6)

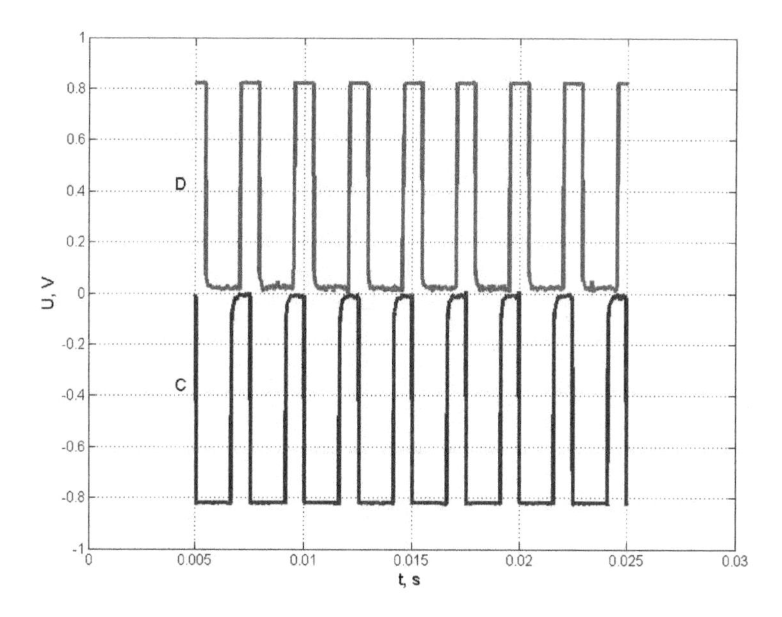

Fig. 10.30 Switched "superposed" impulses (4th row of Fig. 10.6)

10.7.5 Passive OR/AND Gate for Microwave Topologically Modulated Signals

An interesting question is on the possibility to perform the logical operations NOT, OR, AND by passive circuits designed for topologically modulated signals. It was the subject of serious debates in the beginning of the 90s when this idea was born. Several circuits were developed and described in [14],[20],[26],[39],[66], and one of them (Fig. 10.31), which had the minimum number of components, was simulated at microwaves using a PSPICE software tool and equivalent circuit models. The logic operations are performed by this component with the spatial contents of microwave signals. The main mechanism is the wave interference effect in a directional coupler.

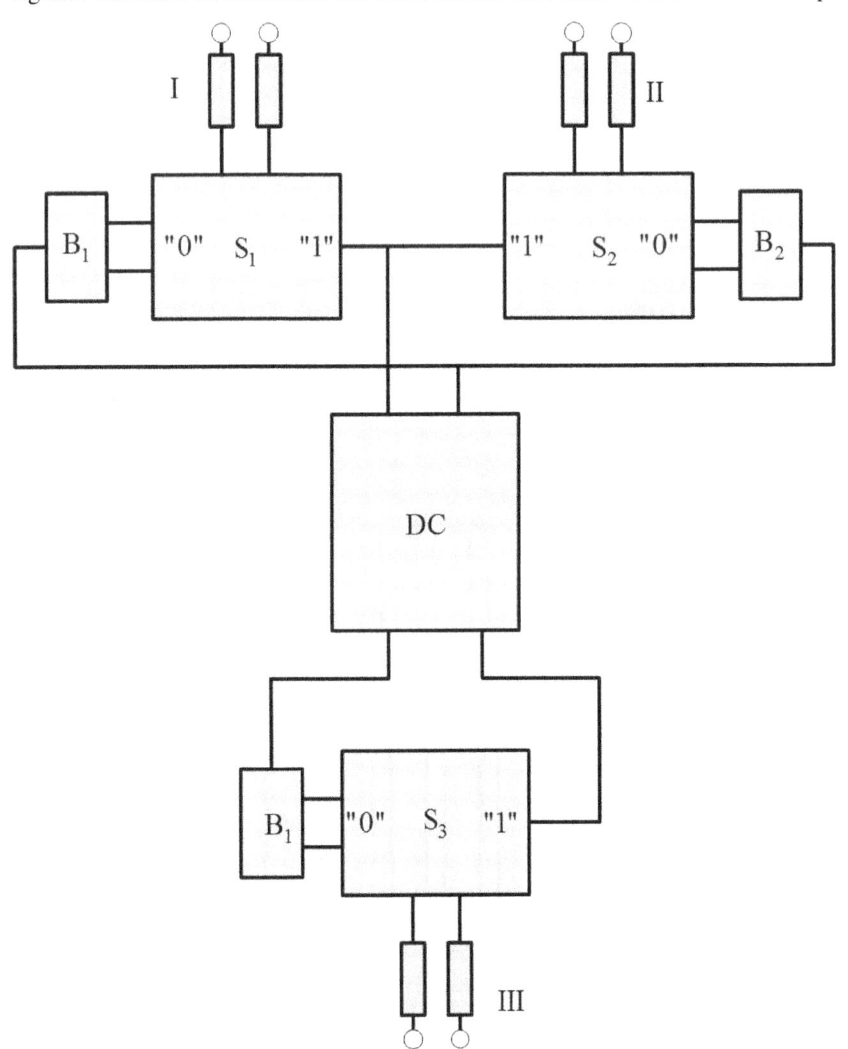

Fig. 10.31. Passive reconfigurable OR/AND gate for microwave topologically modulated signals. Adapted from [66]

This OR/AND gate consists of three switches $S_{1\text{-}3}$ of a microstrip design, three baluns $B_{1\text{-}3}$ and a branchline directional coupler (DC). The input signals I and II are the microwave impulses of the even (logical "1") or odd (logical "0") modes of coupled microstrips which have different topological charts of their fields.

The switches analyze the spatial spectrum of incoming impulses. The directional coupler mixes the spatial spectral contents. The switch S_3 forms a processed image, which depends on the spatio-temporal contents of incoming impulses and their magnitudes as shown below.

The express-simulations of this gate were performed by I.V. Nazarov (MSIEM, Moscow) using a SPICE-program, and the results are shown in Figs. 12.32 and 12.33 for signals of 10-GHz frequency. The gate was loaded by coupled microstrips lines I-III of the characteristic impedances Z_e=112 Ω and Z_0=37.1 Ω. The thin-film resistors of a microstrip switch (Fig. 10.7) are of the values R=24.5 Ω, and its outputs II and III are calculated for 50 Ω and 100 Ω loads, respectively. The models of resistors have been already considered in Section 10.7. The idealized baluns $B_{1\text{-}3}$ are to transform the differential and common voltages to each other. In the case of the identical incoming modes, the gate repeats them at the output III. The logical results of different incoming modes are described below.

Fig. 10.32 is for the incoming even mode (Port I) with the impulse magnitude 1 V and the incoming odd mode (Port II) of the magnitude 0.18 V measured regarding to ground. In this case, the curve of the increased magnitude is the voltage of the even mode on the output III, and the gate is working in the regime of logical OR. The logical signal "1" appears at the output if any or both inputs are excited by the even mode signal (s).

Fig. 10.33 is in the case of the excitation of the input I by a low-level even mode (logical "1") with the magnitude 0.18 V. The odd mode (logical "0") voltage is now 1 V measured regarding to ground. In this case, this mode suppresses the even one due to the increased magnitude and interference, and the magnitude of the output odd mode is higher than the magnitude of the even mode. Now this circuit is working in the regime of the AND gate. The logical "1" or the even mode appears at the output of this gate if both inputs *I* and *II* are excited by the even modes only.

The edges of the output impulses are distorted due to transient effects caused by mismatching and parasitics of resistors R in the switches $S_{1\text{-}3}$. These effects can be reduced by proper circuit optimization and using monolithic design of decreased size.

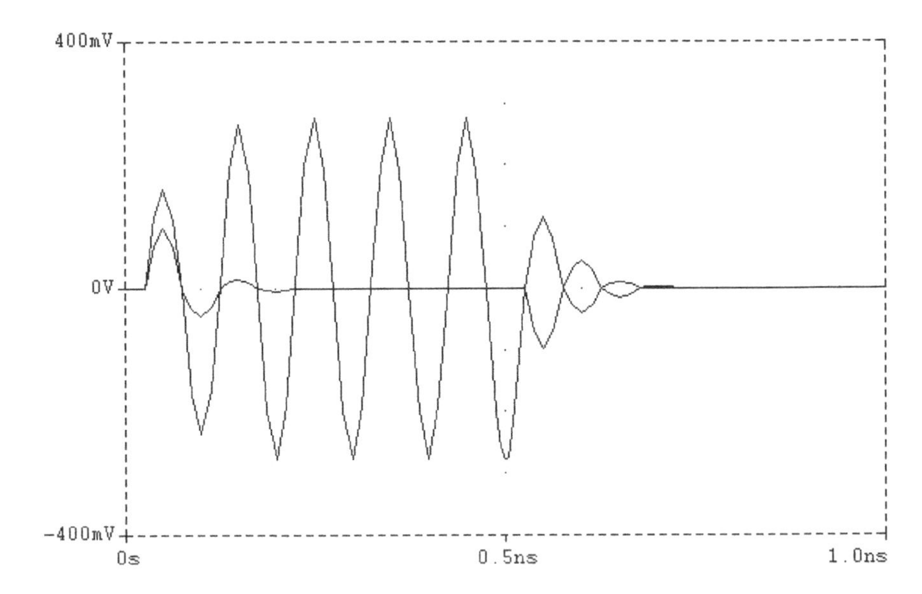

Fig. 10.32 OR-gate effect. The even mode signal is of the increased magnitude. Adapted from [66]

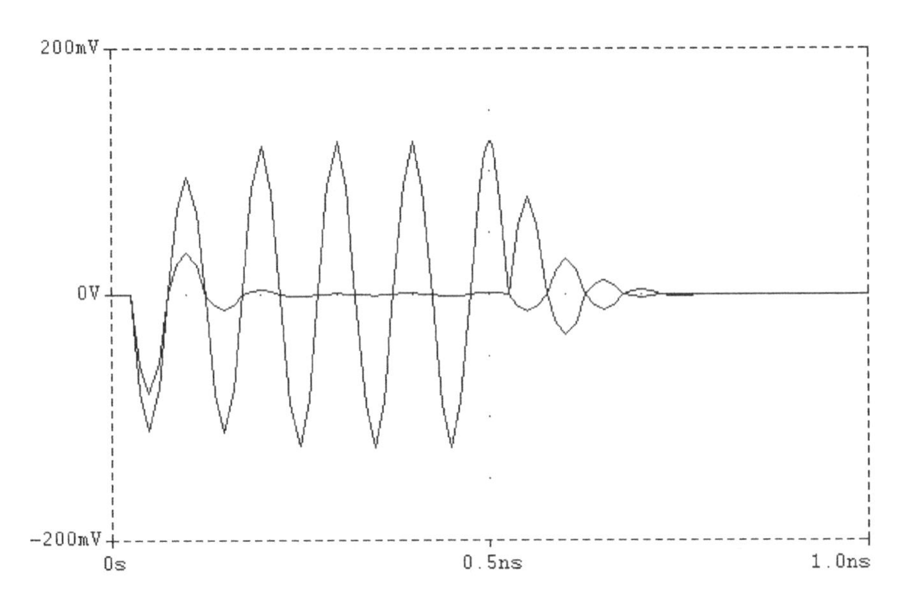

Fig. 10.33 AND-gate effect. Odd mode signal is of the increased magnitude. Adapted from [66]

Taking into account the dependence of this logical operation on the magnitude relationship, the above-considered circuit OR/AND gate belongs to the class of reconfigurable logic. Additionally, this gate can work as a controlled-NOT gate or a Follower depending on the control signal. Composition of larger microwave logic with the above-mentioned gates requires a more detailed study of degradation of signals due to the reflections, transients, loss, and parasitic modal transformations.

It is interesting to notice that the resistive switches used in the modeled design can be substituted by the resonant gates (Fig. 10.6). In this case, the microwave OR/AND gate is described by a symmetrical and reciprocal scattering matrix. It means, this gate is reversible, and it can be related to the class of the "zero-energy" logical circuitry. The proposed gates can be used for a passive flip-flop [14].

Of course, due to the use of directional couplers and baluns, the microwave gates are large even in monolithic implementation, and they were designed only to demonstrate the theoretical possibility of logical operations using the wave interference effects. Besides, the problem of degradation of signals cannot be solved without restoration of them by active circuitry. Later, some other passive logical circuitry for microwave and digital circuitry are studied by simulation and measurements in [79],[80].

At this moment, the interference-based gates are considered to be promising for all-optical computers [81] which are supposed to be superior of the electronic ones due to reduced loss of passive optical components.

10.8 Predicate Logic, Gates and Processor for Topologically Modulated Signals (TMS)[1]

10.8.1 Predicate Logic Theory

Although the predicate logic is well-known [75], the text below is for better explaining the proposed signals and hardware [63]-[70],[71]. Some initial results on the use of predicate logic (first-order logic, first-order predicate calculus, the lower predicate calculus, etc.) formalism and topologically modulated signals are from [23],[26],[58],[60].

Compared to the propositional logic, the predicate one additionally uses quantifiers, i.e. it deals with the logical expressions instead of declarative propositions:

$$L=(p,q). \tag{10.12}$$

Here, p is for predicate variable and q is for quantifier, i.e. q is a measure of p. The predicate L expresses the truth of this statement, i.e. it can be of a binary value. Such a structure is common for natural languages and mathematics ("*John* (p) is *tall* (q)"), and it allows comparing properties shared by many objects (*John*, *Peter*,...). It is known that this logic is complete and able to derive logically valid implications.

---

[1] Written together with A.N. Kostadinov

Generally, predicate logic uses an extended set of logical and non-logical symbols. Among them are the quantifier ones, conjunction (AND), disjunction (OR), negation (NOT), and implications (*if-then*). The reduced predicate logic uses only the AND, OR, and NOT logical units applied to a predicate expression L:

$$L = \bar{L} \text{ (NOT)}, \; L = L_1 \wedge L_2 \text{ (AND)}, \; L = L_1 \vee L_2 \text{ (OR)}. \tag{10.13}$$

The initial hardware realization can be introduced for these operators only.

As was mentioned, the topologically modulated signals are available for this reduced predicate logic, and the following correspondence of formal predicate expression (10.12) to topologically modulated signals can be stated as

$$L = (p, q) \rightarrow \left(T^{(1,0)}, a^{(1,0)}\right). \tag{10.14}$$

The system of predicate operations consists of a set of three operators, which truth-tables (10.1-10.3) are shown here.

Table 10.1 Truth-table for the predicate logical operation NOT (TMS)

#	Input (p_1, q_1)	Output (p_2, q_2)
1	$(T^{(0)}, a^{(0)})$	$(T^{(1)}, a^{(1)})$
2	$(T^{(0)}, a^{(0)})$	$(T^{(0)}, a^{(1)})$
3	$(T^{(0)}, a^{(0)})$	$(T^{(1)}, a^{(0)})$
4	$(T^{(1)}, a^{(1)})$	$(T^{(0)}, a^{(0)})$
5	$(T^{(1)}, a^{(1)})$	$(T^{(0)}, a^{(1)})$
6	$(T^{(1)}, a^{(1)})$	$(T^{(1)}, a^{(0)})$
7	$(T^{(0)}, a^{(1)})$	$(T^{(1)}, a^{(1)})$
8	$(T^{(0)}, a^{(1)})$	$(T^{(0)}, a^{(0)})$
9	$(T^{(0)}, a^{(1)})$	$(T^{(1)}, a^{(0)})$
10	$(T^{(1)}, a^{(0)})$	$(T^{(1)}, a^{(1)})$
11	$(T^{(1)}, a^{(0)})$	$(T^{(0)}, a^{(0)})$
12	$(T^{(1)}, a^{(0)})$	$(T^{(0)}, a^{(1)})$

Table 10.2 Truth-table for the predicate logical operation OR (TMS)

#	Input 1 (p_1,q_1)	Input 2 (p_2,q_2)	Output (p_3,q_3)
1	$(T^{(0)},a^{(0)})$	$(T^{(0)},a^{(0)})$	$(T^{(0)},a^{(0)})$
2	$(T^{(0)},a^{(1)})$	$(T^{(0)},a^{(0)})$	$(T^{(0)},a^{(1)})$
3	$(T^{(1)},a^{(0)})$	$(T^{(0)},a^{(0)})$	$(T^{(1)},a^{(0)})$
4	$(T^{(1)},a^{(1)})$	$(T^{(0)},a^{(0)})$	$(T^{(1)},a^{(1)})$
5	$(T^{(0)},a^{(0)})$	$(T^{(1)},a^{(0)})$	$(T^{(1)},a^{(0)})$
6	$(T^{(0)},a^{(1)})$	$(T^{(1)},a^{(0)})$	$(T^{(1)},a^{(1)})$
7	$(T^{(1)},a^{(0)})$	$(T^{(1)},a^{(0)})$	$(T^{(1)},a^{(0)})$
8	$(T^{(1)},a^{(1)})$	$(T^{(1)},a^{(0)})$	$(T^{(1)},a^{(1)})$
9	$(T^{(0)},a^{(0)})$	$(T^{(0)},a^{(1)})$	$(T^{(0)},a^{(1)})$
10	$(T^{(0)},a^{(1)})$	$(T^{(0)},a^{(1)})$	$(T^{(0)},a^{(1)})$
11	$(T^{(1)},a^{(0)})$	$(T^{(0)},a^{(1)})$	$(T^{(1)},a^{(1)})$
12	$(T^{(1)},a^{(1)})$	$(T^{(0)},a^{(1)})$	$(T^{(1)},a^{(1)})$
13	$(T^{(0)},a^{(0)})$	$(T^{(1)},a^{(1)})$	$(T^{(1)},a^{(1)})$
14	$(T^{(0)},a^{(1)})$	$(T^{(1)},a^{(1)})$	$(T^{(1)},a^{(1)})$
15	$(T^{(1)},a^{(0)})$	$(T^{(1)},a^{(1)})$	$(T^{(1)},a^{(1)})$
16	$(T^{(1)},a^{(1)})$	$(T^{(1)},a^{(1)})$	$(T^{(1)},a^{(1)})$

Table 10.3 Truth-table for the predicate logical operation AND (TMS)

#	Input 1 (p_1,q_1)	Input 2 (p_2,q_2)	Output (p_3,q_3)
1	$(T^{(0)},a^{(0)})$	$(T^{(0)},a^{(0)})$	$(T^{(0)},a^{(0)})$
2	$(T^{(0)},a^{(1)})$	$(T^{(0)},a^{(0)})$	$(T^{(0)},a^{(0)})$
3	$(T^{(1)},a^{(0)})$	$(T^{(0)},a^{(0)})$	$(T^{(0)},a^{(0)})$
4	$(T^{(1)},a^{(1)})$	$(T^{(0)},a^{(0)})$	$(T^{(0)},a^{(0)})$
5	$(T^{(0)},a^{(0)})$	$(T^{(1)},a^{(0)})$	$(T^{(0)},a^{(0)})$
6	$(T^{(0)},a^{(1)})$	$(T^{(1)},a^{(0)})$	$(T^{(0)},a^{(0)})$
7	$(T^{(1)},a^{(0)})$	$(T^{(1)},a^{(0)})$	$(T^{(1)},a^{(0)})$
8	$(T^{(1)},a^{(1)})$	$(T^{(1)},a^{(0)})$	$(T^{(1)},a^{(0)})$
9	$(T^{(0)},a^{(0)})$	$(T^{(0)},a^{(1)})$	$(T^{(0)},a^{(0)})$
10	$(T^{(0)},a^{(1)})$	$(T^{(0)},a^{(1)})$	$(T^{(0)},a^{(1)})$
11	$(T^{(1)},a^{(0)})$	$(T^{(0)},a^{(1)})$	$(T^{(0)},a^{(0)})$

Table 10.3 (*continued*)

12	$(T^{(1)},a^{(1)})$	$(T^{(0)},a^{(1)})$	$(T^{(0)},a^{(1)})$
13	$(T^{(0)},a^{(0)})$	$(T^{(1)},a^{(1)})$	$(T^{(0)},a^{(0)})$
14	$(T^{(0)},a^{(1)})$	$(T^{(1)},a^{(1)})$	$(T^{(0)},a^{(1)})$
15	$(T^{(1)},a^{(0)})$	$(T^{(1)},a^{(1)})$	$(T^{(1)},a^{(0)})$
16	$(T^{(1)},a^{(1)})$	$(T^{(1)},a^{(1)})$	$(T^{(1)},a^{(1)})$

10.8.2 Modeling of Predicate Logic by Unipolar Spatial Signals and Single-ended Gates

The considered 2-place signals provide many interesting possibilities. They are of the 2-polar type, and they require the circuits handling large positive and negative voltages in the same design. Although such circuitry is known in high-speed electronics, the use of gates working with one sign of voltage (unipolar logic) simplifies the design and manufacturing. For this purpose, our common/differential mode signals or bipolar ones are mapped into a set of word-like single-ended signals in which logical level "1" very often is associated with relatively large range of positive voltage levels in digital electronics. Fortunately, this mapping does not require any new additional wires (conducting strips).

Table 10.4 shows the correspondence of the bipolar signals to a set of unipolar ones. Again, the signals propagating along two wires are considered as spatial patterns for which logical circuits can be designed.

Table 10.4 Mapping of 2-polar TMS signals to a unipolar spatial set

#	2-polar signal (T,a)	1-polar signal (p,q)
1	$(T^{(0)},a^{(0)})$	(0,0)
2	$(T^{(1)},a^{(0)})$	(1,0)
3	$(T^{(0)},a^{(1)})$	(0,1)
4	$(T^{(1)},a^{(1)})$	(1,1)

This mapping allows introducing the truth-tables for logical operations with the mapped single-ended signals (Tables 10.5-10.7).

Table 10.5 Truth-table for the predicate logical operation NOT (mapped spatial signals)

#	p_1	q_1	p_2	q_2
	0	0	1	1
1	0	0	0	1
2	0	0	1	0
3	1	1	0	0
4	1	1	0	1
5	1	1	1	0
6	0	1	1	1
7	0	1	0	0
8	0	1	1	0
9	1	0	1	1
10	1	0	0	0
11	1	0	0	1
12	1	0	0	1

Table 10.6 Truth-table for the mapped predicate logical operation OR (mapped spatial signals)

#	p_1	q_1	p_2	q_2	p_3	q_3
1	0	0	0	0	0	0
2	0	1	0	0	0	1
3	1	0	0	0	1	0
4	1	1	0	0	1	1
5	0	0	1	0	1	0
6	0	1	1	0	1	1
7	1	0	1	0	1	0
8	1	1	1	0	1	1
9	0	0	0	1	0	1
10	0	1	0	1	0	1
11	1	0	0	1	1	1
12	1	1	0	1	1	1
13	0	0	1	1	1	1
14	0	1	1	1	1	1
15	1	0	1	1	1	1
16	1	1	1	1	1	1

Table 10.7 Truth-table for the mapped predicate logical operation AND (mapped spatial signals)

#	p_1	q_1	p_2	q_2	p_3	q_3
1	0	0	0	0	0	0
2	0	1	0	0	0	0
3	1	0	0	0	0	0
4	1	1	0	0	0	0
5	0	0	1	0	0	0
6	0	1	1	0	0	0
7	1	0	1	0	1	0
8	1	1	1	0	1	0
9	0	0	0	1	0	0
10	0	1	0	1	0	1
11	1	0	0	1	0	0
12	1	1	0	1	0	1
13	0	0	1	1	0	0
14	0	1	1	1	0	1
15	1	0	1	1	1	0
16	1	1	1	1	1	1

A signal on one wire corresponds to the logical value of the topological variable T (predicate variable p). The second conductor carries the logical level corresponding to the magnitude of topologically modulated signal (quantifier q).

It is well-known that logical functions can be modeled in different ways. The main requirements are the minimum of complexity, decreased propagation delay, and low energy consumption. In our case, the predicate gates proposed according to Tables 10.5-10.7 are the simplest ones, and they are completely realizable using the one type logical gates only.

For example, to satisfy the truth-table of NOT operation (Table 10.5), four types of NOT gates were designed. The first of these gates inverts only quantifier q, and it is called the Q-NOT gate (Table 10.8).

Table 10.8 Truth-table for the predicate Q-NOT gate with mapped spatial signals

#	p_1	q_1	p_2	q_2
1	0	0	0	1
2	0	1	0	0
3	1	0	1	1
4	1	1	1	0

The second designed logic gate (P-NOT) is negated p variable only (Table 10.9).

Table 10.9 Truth-table for the predicate P-NOT gate (mapped spatial signals

#	p	q_1	p	q_2
1	0	0	1	0
2	0	1	1	1
3	1	0	0	0
4	1	1	0	1

The predicate logic gate L-NOT inverts both variables p and q (Table 10.10).

Table 10.10 Truth-table for the predicate L-NOT gate (mapped spatial signals

#	p	q_1	p	q_2
1	0	0	1	1
2	0	1	1	0
3	1	0	0	1
4	1	1	0	0

The last of the proposed NOT gates is a universal one (U-NOT), and it performs all of the above mentioned NOT operations (Table 10.5). It consists of P-NOT, Q-NOT, L-NOT gates together with a multiplexor MUX controlled by a 2-bit signal S_c (Fig. 10.34). The logic gates OR (Table 10.6) and AND (Table 10.7) are very simple, and they consist of conventional AND and OR gates (Fig. 10.35 and 10.36).

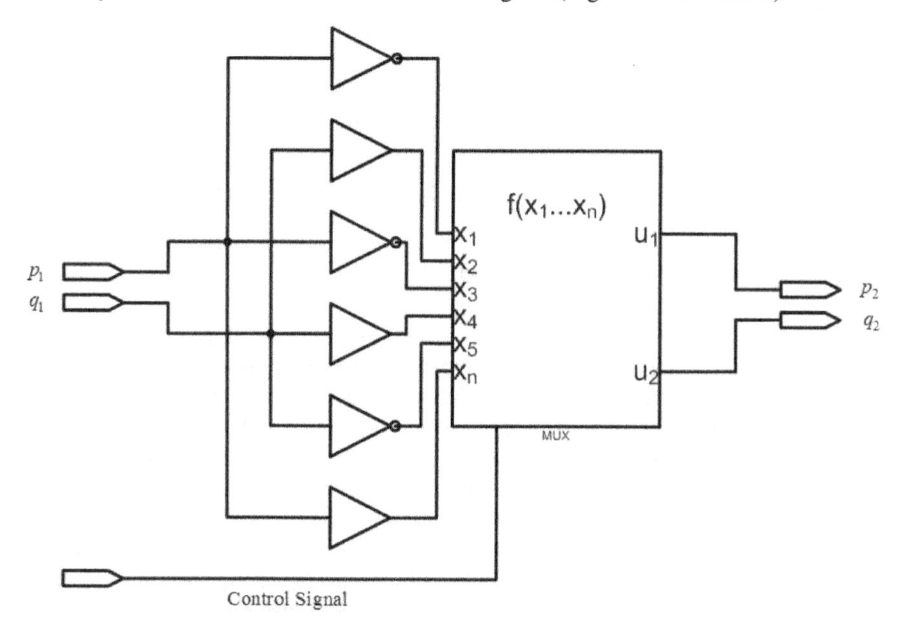

Fig. 10.34 Designed universal predicate (U-NOT) gate. Adapted from [72]

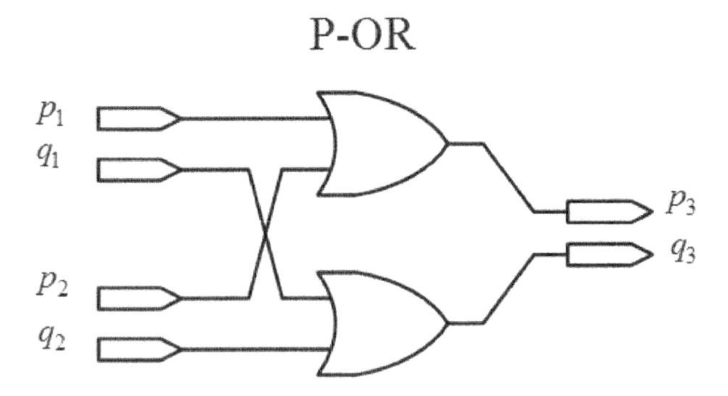

Fig. 10.35 Predicate OR-gate

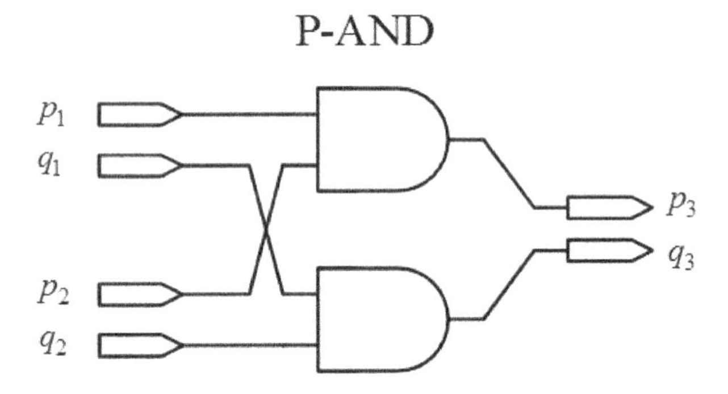

Fig. 10.36 Predicate gate AND

Additionally to the one-bit predicate gates, the multi-bit ones are designed in [67],[70] by unifying the above-mentioned logic elementary blocks. Then an arithmetic-logic unit called the predicate-logic unit (PLU) can be composed for operations with the predicate multi-bit signals. The design methodology, the gates, and the measurement results are presented in more detail in [71].

10.8.3 Predicate Logic Processor

The existing processors can be divided into three different groups – general-purpose (GPPs), single-purpose (SPPs) and application-specific instruction set processors (ASIPs). A GPP is suitable for a variety of applications to maximize the number of devices sold. This processor consists of a datapath and program memory. The first one is general enough to handle a variety of computations, and it has a large register file and one or more general-purpose ALUs.

A SPP is a digital circuit designed to execute exactly one program. The datapath contains only the essential components required for this program. Since the processor is only for one program, it is possible to hardwire the program's instructions directly into the control logic and to use a state register to step through those instructions, so no program memory is necessary.

ASIP is a compromise between the above-mentioned processor types. An ASIP is a programmable processor optimized for a particular class of applications having common characteristics. The designer of such a processor can optimize the datapath for an application class. The proposed PLP belongs to the ASIP class of processors, and it consists of a datapath and control unit (Fig. 10.37).

The datapath is responsible for the manipulation of data. It has a register named accumulator and denoted as ACCA in Fig. 10.37. Another part is a predicate logic unit (PLU) which performs the previously mentioned predicate logic operations AND, OR, and NOT. Additionally, the RAM module (Program Memory and Data Memory in Fig. 10.37) is used to keep the program code (instructions) and corresponding data.

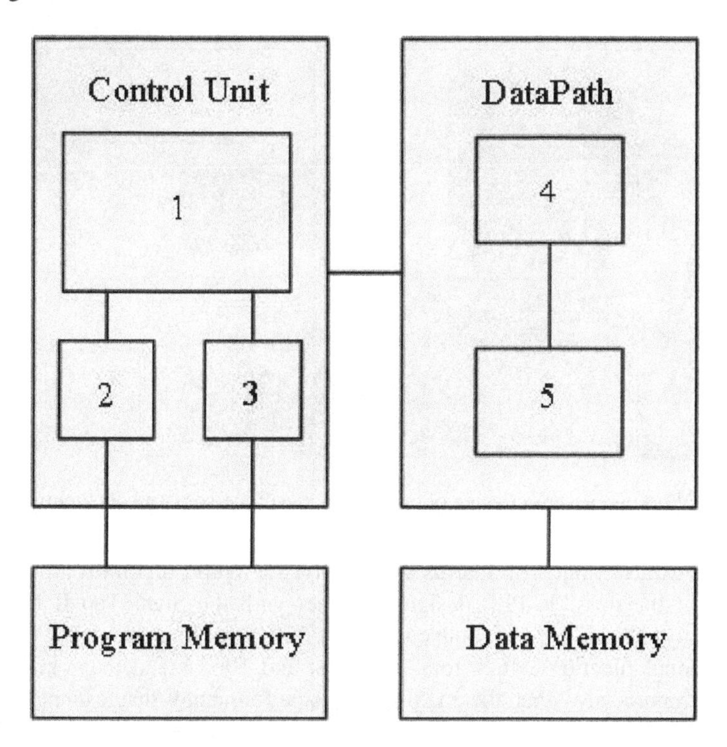

Fig. 10.37 Simplified PLP block diagram. 1- Control logic and state register; 2 - Instruction register (IR); 3- Program Counter (PC); 4- Data accumulator (ACCA); 5- Program logic unit (PLU). Adapted from [72]

The control unit of the microprocessor is a finite state machine (FSM). By stepping through a sequence of states, the control unit controls the operations of the datapath. For each state that the control unit is in, the control unit output logic will generate the entire appropriate control signal for datapath to perform a data operation. It consists of control logic and state register together with the instruction register (IR) and program counter (PC). The IR contains the code of the currently executed operation, and the PC is used to access to the memory cells in RAM.

An 8-bit prototype processor was designed. The synthesis of the whole circuit is performed with a DE2 (Development and Education 2) FPGA (Field Programmable Gate Array) board manufactured by Altera Corporation (Fig. 10.38).

Fig. 12.38 Working place for design of predicate logic processor (Authors archive)

The hardware compiler (Quartus® II version 7.1 Web Edition Program) reports that one of the possible PLP designs together with the SignalTap II Embedded Logic Analyzer and RAM module consists of 5868 total logic elements, 3482 total combinational functions, 4628 total registers, and 10624 total memory bits. The compiler reports also that the maximum clock frequency that can be achieved without violating the internal setup and hold-time requirements is 130.28 MHz. To verify the processor, sets of all implemented instructions were written into the memory. The results of execution of these operations (the data obtained from the FPGA board with help of the embedded logic analyzer) were compared with the preliminary calculated ones.

The developed hardware and used logic signals are completely compatible with contemporary technology. The design is prospective for artificial intelligence applications, especially, for linguistic calculations, database machines, and SQL (Structural Query Language) servers, and it needs further developments for realistic multi-bit calculations.

10.9 Related Results on the Space-time Signaling and Computing

In this Chapter, some results have been described on the spatial and topological signaling and computing. The further review performed here shows that the use of the signal's spatial contents is an effective way to design many components and systems of improved parameters, and these ideas are popular in analog and digital electronics and computer engineering.

10.9.1 Open-space EM Spatial Signaling

One of the first books written for space-time signaling in open space belongs to D.D. Klovskyi and V.A. Soeifer who published it in 1976 [76]. The theoretical questions on the multichannel propagation of space-time modulated signals $s(t,\mathbf{r})$ through stochastic medium were considered there. A theory of optimal and quasi-optimal signal processing by space-time matched filters was developed, and some hardware realizations were studied.

Later, it was found that an additional means to improve this type of communication was the space-time coded signaling and receiving by multiple sources and receivers [82], and the method has been known today as the multi-input/multi-output (MIMO) communication which has shown its effectiveness in practice.

The microwave waveguides and transmission lines allow the multimodal communications, and a two-eigenmode signaling was mentioned in [14]. In fact, our proposed spatial communication can be considered as a particular case of MIMO systems based on the signaling along the multi-wire environment.

Recently, a paper of F. Tamburiny and his colleagues has been published in which the two-channel communications by the field impulses of different topologies was realized through open space and a conclusion on multi-channel communications has been made [83].

10.9.2 Differential and Combined Differential/Common Mode Signaling

Differential signaling is communicating by two signals, and the information is carried by the difference between their individual voltages applied to wires placed close to each other, i.e. it relates to a particular case of spatial signaling. The origin of this spatial signaling is difficult to track, and it has been used for decades in analog telephony, see, for instance [84]. It is found that the noise, which is induced in a couple of wires placed in proximity to each other, is of the common nature, and it is canceled at the end of transmission line by the receiver's differential amplifier (Fig. 10.39). The value of balancing resistor R is equal to the characteristic impedance of differential line.

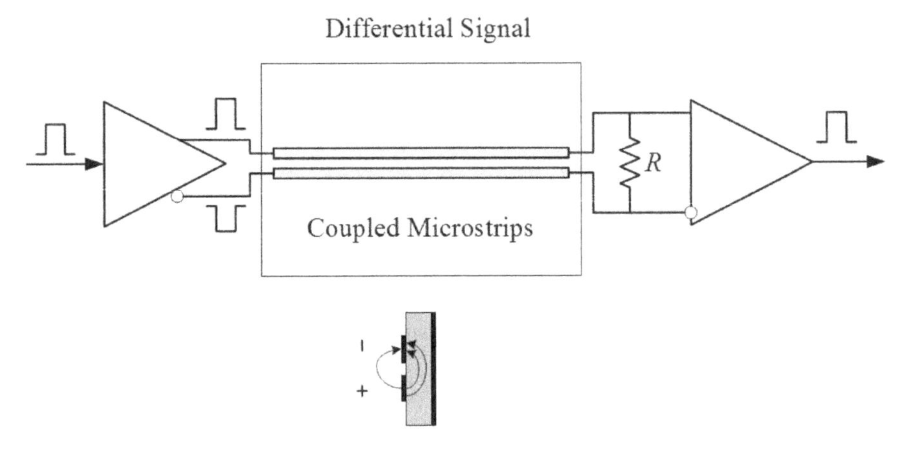

Fig. 10.39 Differential signaling with the zero offset of signals

Additionally to the differential amplifiers, the common and differential signals can be separated by passive [84] and enhanced active [85] baluns. For multi-wire systems, the differential signaling is introduced in [86], and it is for improving the bandwidth of signaling systems composed of more than two wires. The signal levels on each wire are adjusted to provide a voltage difference between all pairs at the same moment of time.

In digital electronics, the differential signaling is to provide increased bandwidth of communications through cables, PCB traces, chip and inter-chip connections, transmission lines, etc. Several standards are known which are according to the used hardware and the signals of communication interfaces. For instance, the differential signals can be shifted by a constant voltage, and this offset relates to a used standard. Usually, this voltage shift is between -2 V and +3.5 V. The voltage span of standards can be different, too.

One of the first types of differential digital signaling is the *Emitter Coupled Logic* (ECL). Several types of the ECL interfaces have been developed and used since that time. One of them was standardized by the ANSI, and it was the *High-Performance Parallel Interface* (HPPI) standard, for instance. More information on the ECL differential signaling is from [87].

Another standard of this type is the *Low-voltage Differential Signaling* (LVDS) introduced by the National Semiconductor (USA). Drivers of this standard provide the 350-mV differential voltage centered at +2.4 V for a multigigabit rate. Although the differential signaling provides increased bandwidth due to its improved noise immunity, the signaling rate limitations are with the dispersion and frequency-dependent loss of cables and PCB traces [88]. The differentially designed equalizers compensating the frequency-dependent loss should be used to increase the transmission rate further.

Additionally to differential signaling, the common mode can be simultaneously transmitted. Different modal switches, including the designs similar to Fig. 12.38, are considered in [89]. Another found paper [90] is on the combined common and differential mode signaling along the FR4 PCB's traces.

This type of signaling was found applications in a high-rate bus line for Si integrations, which was composed of multiple stacked wires [91]. The signals are sent through six stacked conductors of the length 1 mm, and this 12-Gbps link shows acceptable measured performance. Additionally to the measurements, the EM simulations were carried out using the Ansoft's 2-D Extractor Software Tool. Due to the 3-D stacking, this multi-wire interconnects shows a 30% smaller wiring area regarding to the coplanar design.

Some new findings are announced by Alcatel-Lucent for combined differential and common mode signaling along the DSL cables with the signal rate up to 600 Mbps [92]. The information on the enhancements of differential signaling can be found in [93], for instance.

In general, the use of differential, and, over the last years, combined differential/common mode signaling allows increasing the bandwidth of communications, but it needs increased number of components and it leads to a larger real-estate of integration and interconnection areas. Additionally, differential signaling needs special routing of traces and increased requirements for symmetry of all components of high-speed differential channels. Many of today's software tools allow such a routing.

Although the differential/common signaling is considered using the currents and voltages, in fact, it is realized by eigenmodes of transmission lines, i.e. it is a case of topological signaling.

10.9.3 Differential or Dual-rail Logic

During a long period, the analog and digital circuits designed for differential signals have been used only in the interfaces for the inter-system or inter-chip communications. The increased signal rate and high density of contemporary and future integrations dictate the use of differential signaling even for intra-chip communications.

For instance, the analog parts of mixed-signal ICs are suffering from increased noise induced by switching of digital components and by clock signals. Many years ago, the differential style of analog and digital circuits was proposed as the means to improve the noisy environment, and the most important gates were designed and tested [94],[95]. Recently, a differential logic-compatible multiple-time programmable memory cell has been created distinguished by its reliability towards multiple switching [96].

The differential logic is common in multiple-valued circuits where this style allows reducing the noise, which is very dangerous for small-span multi-level signals. A number of works are known in this area, and only a few of them are referenced here [97],[98]. Another field of applications of dual-rail traces and logic is the asynchronous computing where the components are communicating with each other by a pair of handshake wires [99]. Further improvements of differential logic is with its speed increase, diminishing of parasitics, differential routing and shortening traces, supporting the symmetry of traces and components, etc. [100],[101].

New applications of the differential or dual-rail logic style are with the secure integrated circuits. They need decreased EM radiation from traces, and the differential/common mode signaling and a special coding allow enhancing the security parameters of the embedded integrated circuits. Then even logical circuitry should be designed using the idea of differential/common mode signaling. Several approaches have been developed, and only some of them are considered here. An additional application area is the asynchronous processors where the dual-rail signaling is now common. Only the EM aspects of these problems are touched an increased attention here. For more information on the logical matter, the readers should follow the references to this section.

To solve the problems with the leakage of information from smart cards and embedded security ICs, the signaling on a pair of traces is used, as a rule. Additionally to this signaling, the most circuitry is re-designed in the differential style. Very often, the dual-rail name of this logic is used to avoid the misunderstanding with the commonly used "differential logic" as the "logic of change" which is, generally speaking, is applicable even for single-ended circuits.

One of the design styles is published in [102] where the wave dynamic differential logic (WDDL) is proposed together with a layout technique allowing differential routing. The problems encountered by the authors of [102] were that the commercially available design tools not allowed the differential-style routing of silicon integration, and this problem was resolved by creation of a special design flow for the secure chips with minimal changes of the standard CMOS cell design. It was shown that the proposed WDDL allowed masking the data-dependent energy consumption, and the differential routing decreased the EM radiation. The increase of the data rate using this type of signaling and gating was possible due to its improved noise immunity.

More enhanced dual-rail coding is proposed in [103] where to mask the data-driven power consumption, the dual-rail data (0,1) or (1,0) is followed by the spacers (0,0) or (1,1). It was shown that this coding and dual-rail design style allowed balancing the power consumption. Analyzing these signals and protocol, it is clear that they relate to a particular case of topological or common/differential mode signaling. Additionally to their circuits, the authors of [103] developed a software tool "Verimap Design Kit" compatible with the Cadence and allowing transforming the single-ended logic to the dual-rail one.

The dual-rail logic was found applications in the programmable logic arrays (PLA) [104]. The cells of the proposed PLA realize a 2-input logic. A new PLA is composed of these cells, and more complicated Boolean operations can be realized using it. Taking into account that these two-input gates are at the knots of the proposed PLA, the slow input decoders, usually connected to PLAs, are excluded, and the design is several times faster than the conventional single-rail PLA. It is noticed that the needed area is only 50% larger with respect to a similar single-rail PLA, and the time-delay and power consumption are reduced essentially, too.

More applications of dual-rail logic or even multiple-wire logic can be found in high-speed cross-bar architects [105]. The nano-conductors in these prospective circuits are close to each other, and an excited electric voltage impulse can induce a strong response in neighboring wires. In the beginning of the 90s, when the nano-arrays were not known, it was proposed an idea of collectively excited wire arrays, when a single impulse excited a number of modes in an array, and the logical outcome was a response of the whole circuit [106],[107]. A particular case was a two-wire circuit where the common and differential modes were excited and separated by spatial switches.

Unfortunately, not so many contributions are known on the study of high-speed effects in nano-arrays. For instance, the EM coupling is considered in [108] as one of the types of noise source in nano-integrations. Taking into account the faulty manufacturing and the noise of the thermal and quantum origins, a conclusion of the cited book is that the design principles of nanocircuits should be radically changed. One of the pertinent ideas to detect the noise-induced faults is on the use of dual-rail logic for the fault-tolerant computing by nano-arrays [109]. The approach is that two neighboring wires are allowed to support only the signals and their compliments. The electric or technology faults can lead to violation of this rule, and the defect would be detected by checking the output 2-bit spatial words. The authors of the mentioned paper simulated the typical elementary array operations, and it was found that the most faults, including the multiple ones, could be registered with that proposed dual-rail protocol.

10.9.4 Ultrafast Gating Circuits Using Coupled Waveguides

An interesting work was published in Japan on the time-division demultiplexors based on the management of propagation modes in traveling-wave transistors and coupled lines in the end of the 90s of the last Century. The developed circuits of hybrid (transistor/passive) logic were aimed at substitution of the slow flip-flops at that time [79],[110],[111].

A key component of the proposed demultiplexor is a section of coupled coplanar strips loaded by the resistors which values are equal to the characteristic impedance of the even mode (Fig. 10.40).

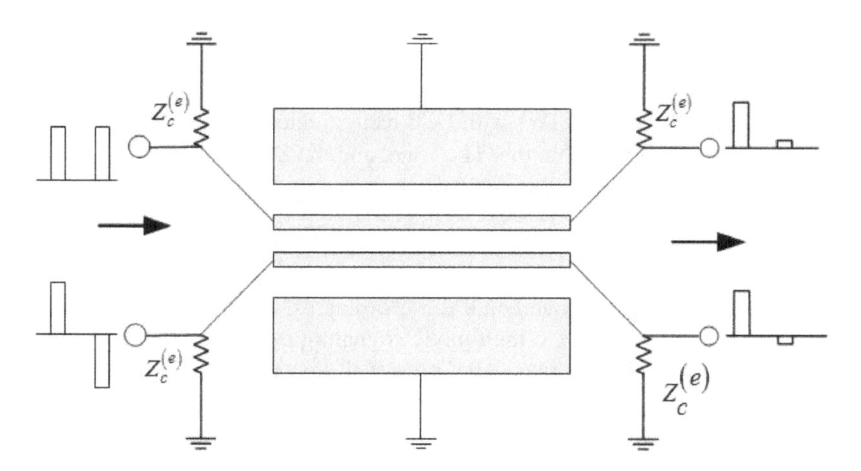

Fig. 10.40 Gating circuit on coupled CPW waveguides. Adapted from [79]

The incoming signals of the same sign excite the even modes, and they are positively interfering at the outputs, and the large voltages are on them. The odd modes excited by the impulses of different polarity are reflected from the outputs tuned to the even mode, and the level of these signals is lower than in the previous (even mode) case. The distinction rate of voltages K is calculated as

$$K = \frac{4Z_c^{(e)}/Z_c^{(o)}}{\left(1+Z_c^{(e)}/Z_c^{(o)}\right)^2}. \tag{10.15}$$

To increase this rate, the coupling of strips should be chosen stronger. Then this passive circuitry, together with a reshaping block, can act as an AND gate. The OR function can be realized, too, according to the opinion of the authors of the cited papers. An experimental study of this AND circuit is performed for a gate length of 800 μm. Each strip of the width 15 μm is placed at the distance 3 μm from each other on a polyimide substrate of the height 50 μm. The maximum frequency of the sample is estimated around 200 GHz.

The gating effect is studied for 20-, 40-, and 80-Gbps signals, and the predicted effect is completely confirmed. Additionally, it is noticed that the gate is able to work with the signals of the rate beyond 100 Gbps. The main factor, which limits the bandwidth, is the ringing of the odd-mode signals, which is decreased with the loss of strips. Additional distortion effects are the frequency-dependent loss due to the skin-effect in conductors, inequality of the modal velocities, and the modal dispersion. In these cited papers more logic circuits of improved characteristics are considered which are composed of passive spatial gates and digital high-speed semiconductor components.

10.9.5 Multimodal Data Transmission Using Hybrid Substrate Integrated Waveguides

The substrate integrated waveguides or SIWs have been already described in the Chpt. 6 of this book. The most attractive feature of them is a rather large bandwidth, decreased loss comparable with the one provided by rectangular waveguide, and compatibility of SIWs with PCB technologies.

The main mode of a SIW is the TE_{10} one, and the SIW's bandwidth is defined by the cut-off frequency of the second TE_{20} mode. It is estimated roughly as $\Delta F = F_c^{(20)} - F_c^{(10)}$. Another limitation is with rather large footprints of these lines, especially, at frequencies below 20-30 GHz in comparison with the printed strip or microstrip lines. To partly overcome these problems, an interesting idea is proposed in [112],[113] where the multimodal signaling is used to increase the overall aggregate transmission rate. The proposed interconnect is shown in Fig. 10.41a,b.

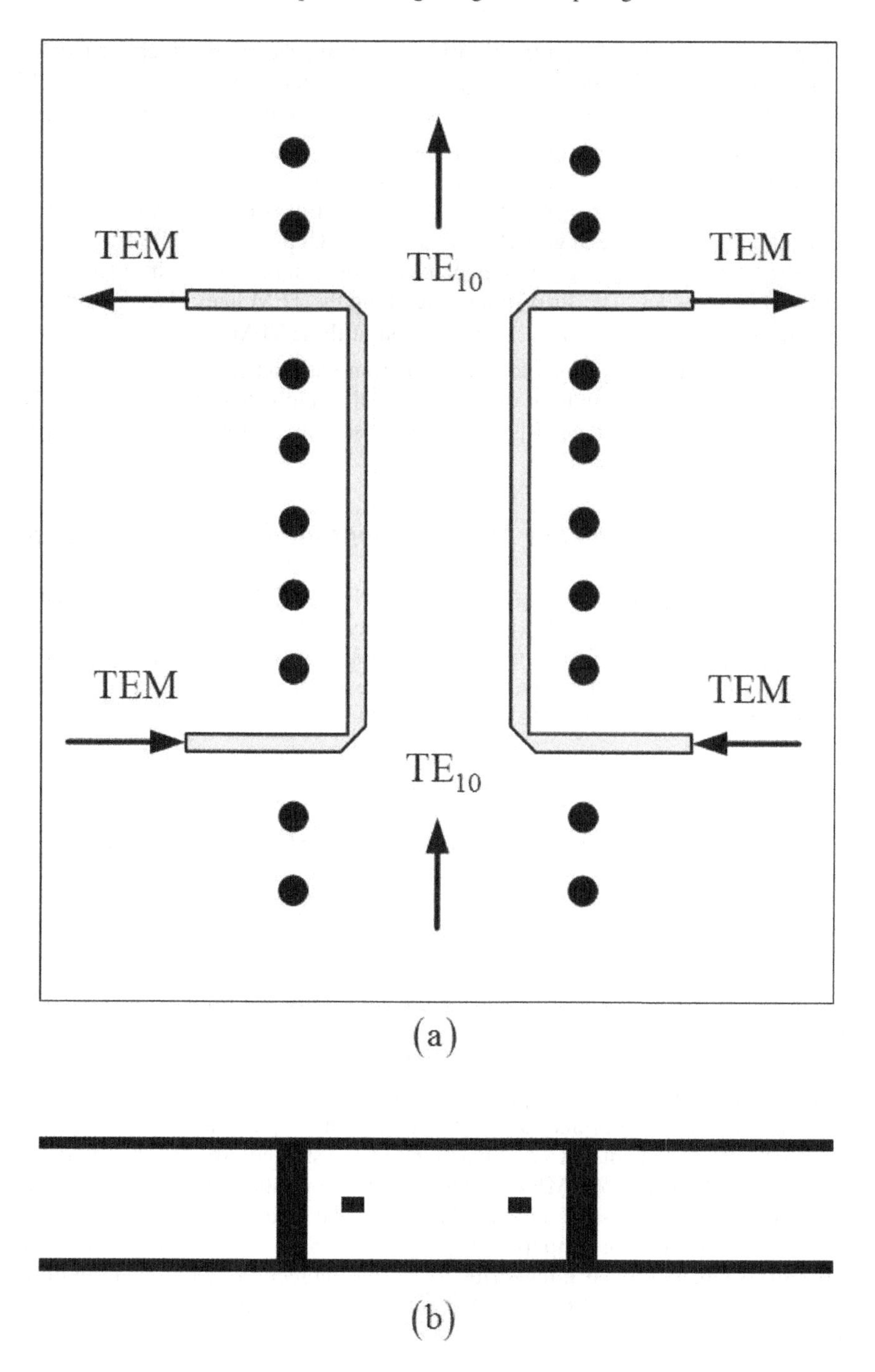

Fig. 10.41 Hybrid SIW/strip interconnect for multi-gigabit modal signaling. (a)- Frontal view; (b)- Cross-section. Adapted from [113]

It consists of a SIW with a placed inside two strip conductors which are excited independently on each other providing the two-channel signaling using the TEM modes. The SIW waveguide supports additional communications by its TE_{10} mode. All channels, being triggered, aggregated at the receiver side, and the measurements show an aggregate rate around 15 Gbps for a trace of the length 48 mm realized on the substrate Rogers 4003C. Theoretical full-wave simulations show at the potentially achievable aggregate rate around 47 Gbps. One of the factors providing such characteristics is the independency of channels on each other due to using different modes. The strip signals are along the TEM lines placed inside the SIW channel with its TE_{10} mode, and isolation of these TEM and TE_{10} signals is around 70 dB. The strip lines are isolated from each other with the cross-talks better than 20 dB. The authors of [113] note that more than two strips can be placed inside this SIW channel [114] without essential increase of the inter-wire cross-talks.

10.9.6 Passive Frequency Multiplier on Coupled Microstrips Line

One of the first papers on this circuitry was published in 1981 where a frequency multiplier was introduced and a technique on its calculation was given [115]. A particular circuit is shown in Fig. 10.42. It consists of three sections of coupled microstrips loaded by a resistive network. The input periodical impulse signal comes to port 1, and the output is the second port.

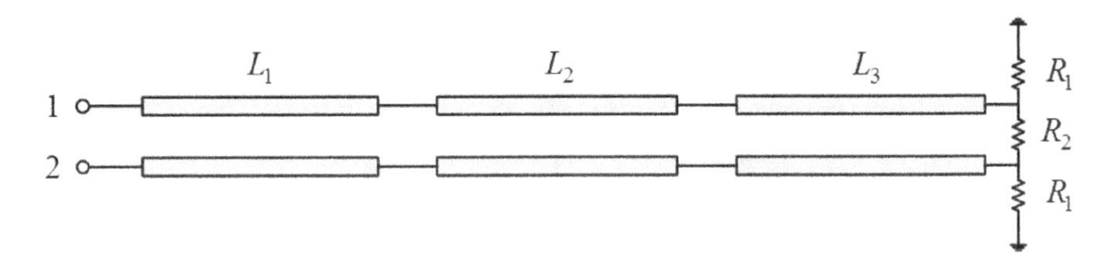

Fig. 10.42 Three-section frequency multiplier using two-strip coupled lines. Adapted from [115]

It is shown that due to multiple reflections and interference, the input clocked signal frequency is multiplied, and the doublers and triplers are realized without using any active elements. A good matching of the studied prototypes at frequencies from 100 MHz to 1600 MHz is found. More information on the use of passive and passive/active ring-like circuits based on the coupled uniform and non-uniform microstrips for impulse shaping is from [116], for instance. Non-uniformly shaped coupled microstrips lines and interaction of the even and odd modes in them can be used for the design of components with producing the linear group-delay variation along the operation band [117].

10.9.7 Microwave Passive Logic for Phase-modulated Signals

Some interesting results are presented in [80] where its authors designed the microwave gates using the interference of modes in passive circuitry – directional couplers or matched power dividers. The carrier of digital information is the phase of microwave signals. It is introduced a set of signals consisting of the zero-shift ('A'), 180°-shift ('B') and the zero-magnitude ('C') signals. i.e., they can represent a 3-level logical system. Additionally to the compared signals, the designed gates OR /AND use the phased reference signals of the logical levels 'A' or 'B', and the logical operations are realized by constructive/destructive interference of waves.

Five logic circuits are introduced, and among them there are the NOT, OR, and AND gates operating the 'A' and 'B' signals (Fig. 10.43). An additional couple of gates, which are not shown here, perform the logical operations with the zero, 'A' and 'B' signals.

The NOT gate is a phase shifter realized by transmission line of the 180° electrical length, and its truth-table is shown in Fig 10.43 (upper part). The OR gate works according to its truth-table shown in the inset of the center of Fig. 10.43. The input signal from the port 1 is compared with the "B" signal from the reference generator. If the phase difference between the reference signal 'B' and signal ('A') of the port 1 is 180°, then all power is absorbed at the balancing load of the first power coupler. The output signal 3 is defined by logical level of the signal at the input port 2 – see the two first lines of the OR truth-table. The input signal of type 'B' at the port 1 is summed with the reference signal of the 'B' type, and it is larger than the signal 'A' appearing at the port 2. Then the result of comparing of two signals in this case always is 'B'. Two signals of the type 'B' excite the output 'B' signal.

The gate AND works similarly, and, there, instead of the reference signal 'B', the one of the type 'A' is used. All 'A' -type signals coming at the inputs 1 and 2 are positively interfering with the reference generator signal 'A', and the 'A' one appears in this case at the output 3 of the gate AND. Similarly to the OR gate, the 'B' signal and the reference signal 'A' are absorbed in the balancing resistor of the power coupler 1, and the logical state of the output 3 is determined by the signal 2, which is 'A' or 'B', -see the truth-table shown at the bottom of Fig. 10.39 (right part).

These ideas and designed circuits are tested by simulations up to 300 GHz. Measurements are performed for the circuits of 300 MHz. It is shown that these circuits can provide extremely high speed of logical signal processing, but several disadvantages exist. Among them are the large power dissipation at balancing resistors and the signal distortions due to parasitics of circuit elements. The performed entropy analysis shows that only NOT gate is reversible, but others require essential power dissipation.

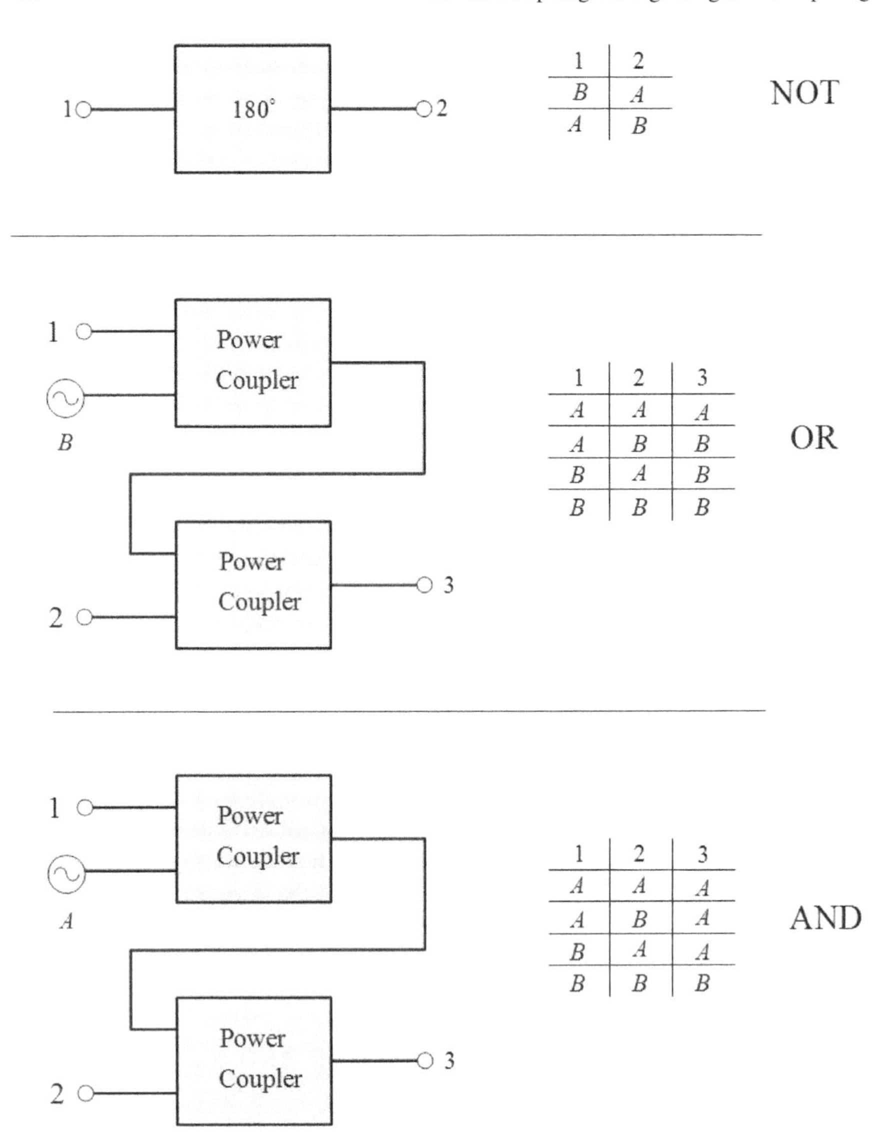

Fig. 10.43 Passive microwave logical gates for phase-modulated signals. Adapted from [80]

Although the research was conducted for microwave frequencies, the ideas of [80] can be implemented at optical range for the telecommunication subsystems or all-optical computers, which are prospective to overcome the principal limitations of electronic technologies and architectures. Additionally, the passive components performing sets of logical operations can be used in radiation-resistant processors to improve their realibility towards particle radiation.

10.9.8 Microwave Mode Selective Devices, Converters, and Multimodal Frequency Filters

Several advanced circuits have been considered above for the spatial computing and signaling. It is interesting to review and study the roots and contemporary state of these circuits assigned for modal processing in the known classical areas of microwave techniques and high-speed electronics. The main attention is paid to the modal selective devices or modal filters, transformers of modes, and frequency filters based on multimodal physics.

Since the end of the 19th Century, it has been known that the EM field is excited in waveguides as the eigenmodes which have certain spatial structures of their fields. The propagation properties of these modes can be regulated by the geometry and dielectric filling of waveguides, and the simplest modal filter is just a waveguide which allows propagating only one mode in a certain frequency band. Some devices are based on combination of multimodal and monomodal effects and on the components allowing selecting the excited modes.

The technique of selective excitation can be explained using Fig. 10.44 and some formulas from [118].

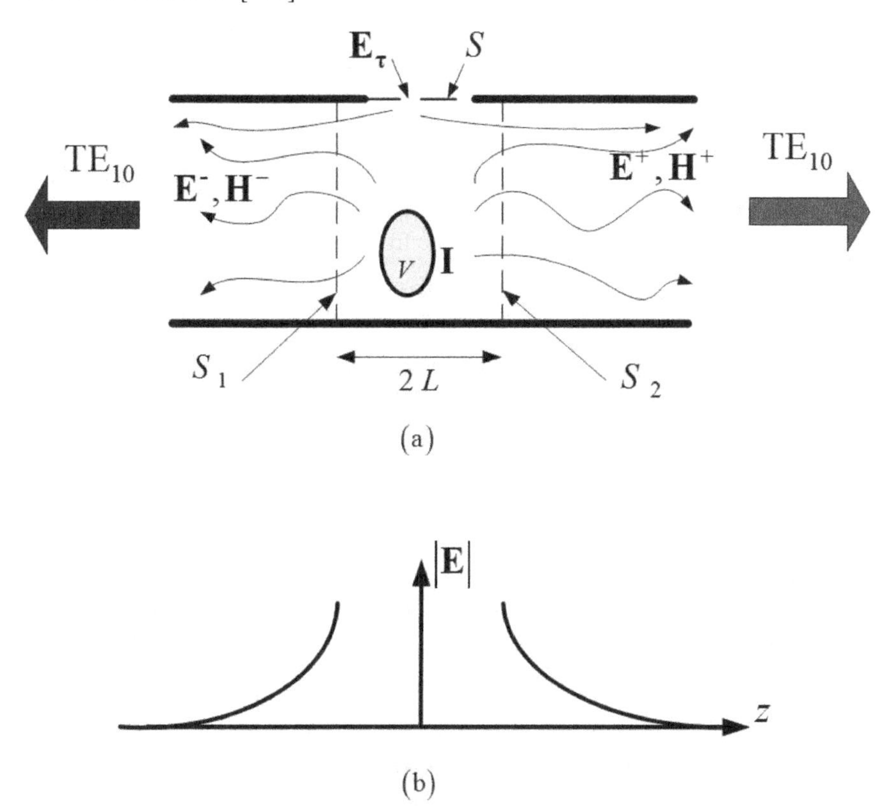

Fig. 10.44 Excitation of the TE_{10} mode of a rectangular waveguide. (a)- General view; (b)- Higher-order mode electric field distribution along the z-axis

Consider a rectangular waveguide excited by an electric current of the volume density $\mathbf{I}(r' \in V)$ and/or electric field $\mathbf{E}_\tau(r' \in S)$ defined on a slot S in the waveguide shield. According to the EM theory [118]-[120], it is excited infinite number of *n,m-th* modes propagating away from the source region of the length $2L$ limited by the surfaces S_1 and S_2:

$$\mathbf{E}^- = \sum_{m,n}^{\infty} c_{mn}^- \mathbf{E}_{mn}^-(x,y,z); \quad \mathbf{H}^- = \sum_{m,n}^{\infty} c_{mn}^- \mathbf{H}_{mn}^-(x,y,z) \; ; z<-L;$$
$$\mathbf{E}^+ = \sum_{m,n}^{\infty} c_{mn}^+ \mathbf{E}_{mn}^+(x,y,z); \quad \mathbf{H}^+ = \sum_{m,n}^{\infty} c_{mn}^+ \mathbf{H}_{mn}^+(x,y,z) \; ; z>L. \tag{10.16}$$

In (10.16) the excitation coefficients $c_{mn}^{\pm}$ are written separately for the propagating (10.17) and non-propagating (10.18) modes:

$$c_{mn}^{\pm} = -\frac{1}{2}\frac{W_{mn}}{|W_{mn}|}\left[\int_V \mathbf{I}(r')\cdot\mathbf{E}^{*\pm}_{mn}(r')dv' + \int_{S_\tau}\left[\mathbf{E}_\tau(r')\times\mathbf{H}^{*\pm}_{mn}(r')\right]d\mathbf{s}'\right], f > f_c^{(mn)}, \tag{10.17}$$

$$c_{mn}^{\mp} = -\frac{1}{2}\left[\int_V \mathbf{I}(r')\cdot\mathbf{E}^{*\mp}_{mn}(r')dv' + \int_{S_\tau}\left[\mathbf{E}_\tau(r')\times\mathbf{H}^{*\mp}_{mn}(r')\right]d\mathbf{s}'\right], f < f_c^{(mn)}. \tag{10.18}$$

They depend on the spatial shapes of the excitation regions V and S and the current and field distributions in them. It allows regulating the amplitudes of the excited modes in a certain range. Additionally, the fields of the non-propagating modes exponentially decay with the distance from the source region:

$$\mathbf{E}_{mn}^{\pm} = \mathbf{E}_{mn}^{\pm}(x,y)\exp\left(\mp\left|k_z^{(mn)}\right|\cdot(z\mp L)\right),$$
$$\mathbf{H}_{mn}^{\pm} = \mathbf{H}_{mn}^{\pm}(x,y)\exp\left(\mp\left|k_z^{(mn)}\right|\cdot(z\mp L)\right). \tag{10.19}$$

This theory allows to understand the excitation of waveguides and to design the transitions between different waveguides [120].

In some cases, the microwave and high-speed circuits work in the multi-modal regime, or propagation of several modes is allowed and the controlled excitation and propagation are needed. Among them are the high-power or high-Q oversized waveguides and resonators, microwave ovens and applicators, microwave and millimeter-wave tubes, multimodal filters and antennas, microwave imaging devices, the above-considered multi-wire TEM or quasi-TEM interconnects, etc.

Some of the components of these systems are the *modal filters* preventing propagation of unwanted modes along the potentially multimodal waveguides. Ones of the first components of this type are the *polarization filters* which reject a mode of a certain polarization while being transparent for a mode of other orientation of the field.

The simplest polarization filter is a parallel-rod grid placed in a square cross-section waveguide (Fig. 10.45a,b) which is transparent only for one-polarization field.

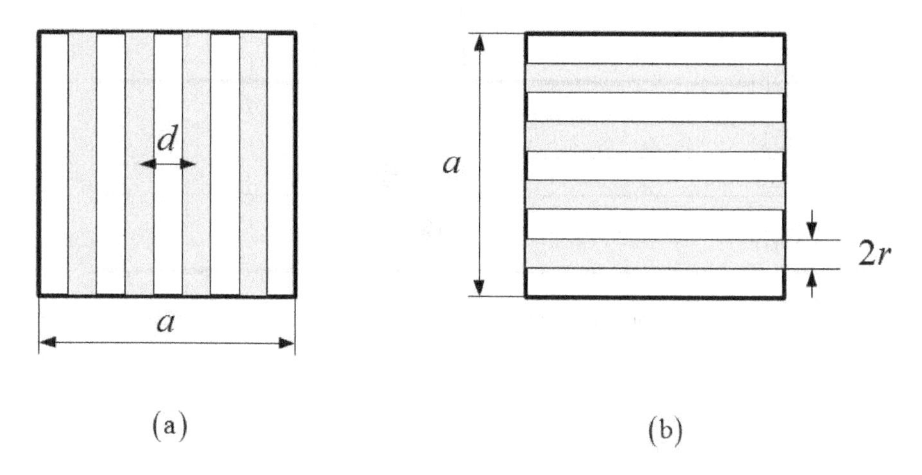

Fig. 10.45 Rejection filters for the TE_{10} (a) and TE_{01} (b) modes of a squared cross-section waveguide

For instance, the reflection and transmission coefficients for TE_{10} mode diffracted at the grid (Fig. 10.45a) composed of n conducting cylinders of the radius r and of the period d are estimated as [121]

$$\begin{aligned} |S_{11}|^2 &\approx 900\left(\frac{r^2}{d\lambda_0}\right)^2, \\ |S_{21}|^2 &\approx \left(\frac{a}{\Lambda_{10}}\right)^2 \frac{4}{(n+1)^2}\left[\ln\frac{2\pi(n+1)r}{a} - \frac{2(n+1)r}{a}\right]^2. \end{aligned} \tag{10.20}$$

In the first approximation, the mode of the orthogonal polarization TE_{01} is reflected completely from this grid. Combinations of these grids and T-junctions allow creating the *duplexers of polarized modes* for terrestrial and satellite telecommunication systems which use the polarization separation of up- and downstreams. One of them is designed in [121] for 3.4-3.9 GHz frequency band and is shown in Fig. 10.46 where the TE_{10} and TE_{01} modes are separated into different branches of a T-junction.

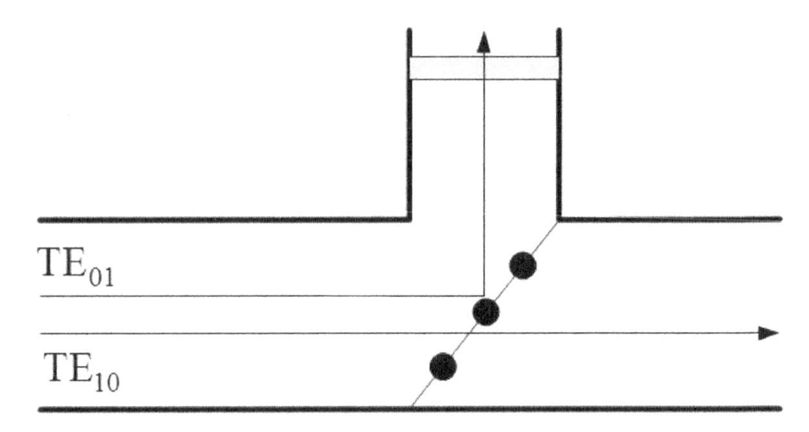

Fig. 10.46. Duplexer for the orthogonally polarized TE_{10} and TE_{01} modes. Adapted from [121]

The isolation of the output rectangular waveguides is better than 40 dB due to the reflection of unwanted mode by corresponding grids and waveguides. The messaged insertion loss is better than 0.98 dB in this frequency band. More advanced waveguide components for polarized signals, the design techniques and measurement results are considered in [121]-[124].

The modal filters and duplexers, built on the wire grids, have relatively narrow frequency bands. Better performance is shown by the polarization selective devices designed using distributed coupling between a two-mode waveguide and a one-mode sub-waveguide [121].

An interesting device, which allows the mode selective taking of modal power from an oversized cylindrical waveguide, is published in [125]. This waveguide is connected in parallel to the rectangular sub-waveguides through multiple circular holes in their common wall. The modal wavelengths of analyzed modes of cylindrical waveguides are adjusted to be close to the ones of the TE_{10} modes of the sub-waveguides. In each coupled arm, the selective constructive interference is adjusted additionally by proper choosing of the distance between the coupling holes. The designs allowing analyzing several modes of oversized cylindrical waveguides are considered for the 35- and 70-GHz gyrotrons.

The modal filters can be used to suppress the propagating higher-order modes excited at waveguide discontinuities and giving spurious responses at their cut-off frequencies. In [126], a low-loss filter is described in which a metallic septum is used to suppress the parasitic responses caused by the unwanted higher-order modes down to the level -50 dB.

Integrated modal filters are used in circuits designed on the transmission lines which allow multimodal propagation. For instance, the differential traces based on coupled strips or microstrips transmission lines suffer from parasitic excitation of the even (common) mode at the discontinuities [127]-[135]. Although the receivers are designed only for the differential signals, this parasitic excitation leads to the loss of power, distortions, and increased noise. To prevent this propagation of the even mode signals, the traces can be equipped by these common mode filters. For instance, one can consider splitting the ground plane and reflecting this parasitic mode by a slot in this ground plane [131]. Some measures should be applied to prevent parasitic slot radiation in this case. The slots can be transformed into advanced patterning of the ground plane. These defects form the resonant loops coupled only to common mode, and this mode is rejected in a certain frequency band [132]. Multiple defects of the ground plane form a photonic bandgap structure, and the isolation of the common and differential channels can be reached around 15- 20 dB in a several Gigahertz range [133],[134].

Some components of differential circuits can be designed with the reduced mode conversation. For instance, the authors of [135] optimized a microstrip 90°-bend to reach the noise suppression up to 14 dB in the DC to 6 GHz frequency band. The coupled microstrips are narrowed in a continuous manner, and they are tightly coupled at the bend area decreasing the length difference between the bent strips. Additionally, a large difference of the characteristic impedances of the differential and common modes is reached mostly due to increasing of the common mode impedance that causes reflection of this mode from this enhanced bend.

Many other passive components for microwave and high-speed electronics can be designed in a differential manner without using baluns but with the suppression of unwanted common modes. For instance, in [136]-[139], the designed devices, additionally to the frequency filtering, can suppress the common modes of circuits, and they can be directly connected to amplifiers or to antennas.

The best-known troublemakers in high-speed differential signaling are the via-holes, which should be designed in a differential manner with low common-noise radiation and reduced modal conversation. Some geometries of differential via-holes are shown in Fig. 10.47.

Taking into account the pads, the coupled microstrips are bent (Fig. 10.47a), or via-holes are shifted from each other (Fig. 10.47b). Both geometries may distort the differential mode, and this distortion should be compensated or minimized by optimization of the geometry of via-holes. In the case of close proximity to the via-hole resonances, the technologically caused eccentricity can influence the symmetry of the via-hole design, and it can lead to excitation of common mode. In general, the conversationless via-hole transitions should be symmetrical to provide matched connections to the output lines with the smooth field transformation along these transitions.

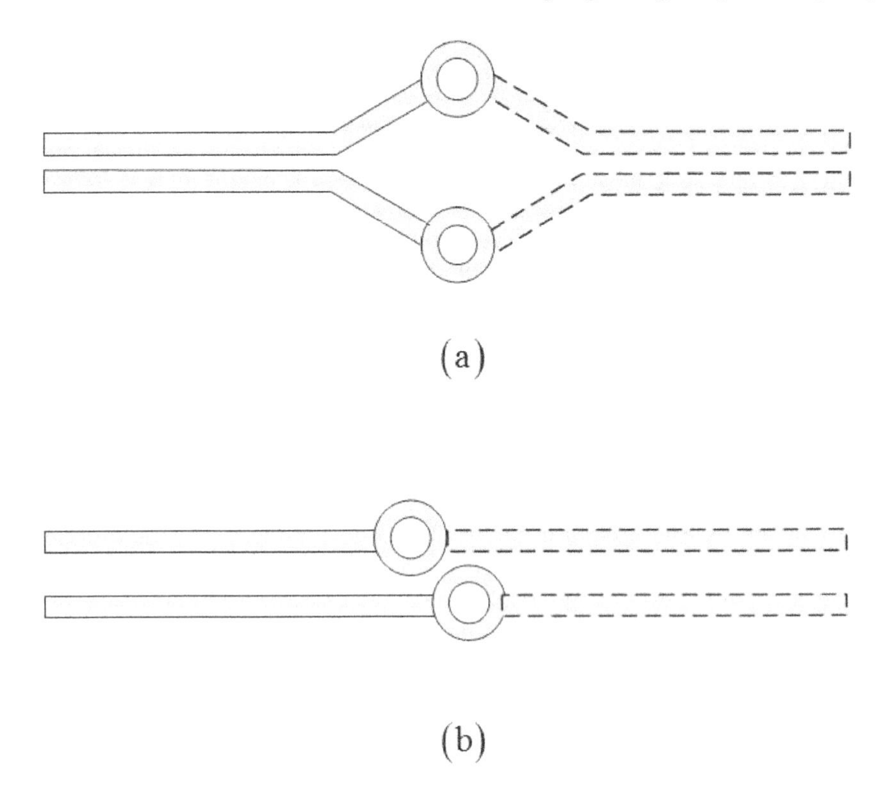

Fig. 10.47 Differential interlayer transition

Unfortunately, the differential via-holes are difficult to be simulated analytically, and only some rough estimates can be obtained using simple formulas. For instance, the characteristic impedance Z_c of a vertical two-cylinder via-hole in differential regime can be calculated using a formula given for a bifilar line [140]:

$$Z_c = \frac{119.904}{\sqrt{\varepsilon_r}} \ln\left(\frac{2H}{d} + \sqrt{\left(\frac{2H}{d}\right)^2 - 1}\right) \tag{10.21}$$

where d is the via-hole diameter, and H is the distance between the via-hole centers. Inductivity L per unit length of this line is [141]

$$L = \frac{\mu_0 \mu_r}{\pi} \operatorname{arccosh}\left(\frac{H}{d}\right). \tag{10.22}$$

Additionally, the capacitance between the differentially driven circular pads should be included into the equivalent circuits of via-holes.

Many papers are on the full-wave simulations and the extracted from the measurements and simulations equivalent circuit models. More information on these results is from [141]-[146], for instance, where the computational and signal integrity issues are considered.

Close to the modal filters there are the *suppressors of unwanted modes* which do not allow propagating them along the waveguides. For instance, in [147], a suppressor of the LSE_{01} mode of a non-radiating dielectric waveguide is described. It is reached by installing into its dielectric rod a metallic strip. It is normally oriented towards the conducting plates of this waveguide and along the electric field-force lines of this mode. The only working mode of the type LSM_{01} is propagating in this case if additional suppressing of the excited quasi-TEM mode is realized. It is reached by longitudinal patterning of the installed strip which allows filtering of the quasi-TEM mode in a wide frequency band.

In [148], a circular waveguide is considered where the radially oriented wires or metallic strips are connected to the waveguide shield. It allows suppressing all lossy modes which are different from $H_{0n}\left(TE_{0n}\right)$ ones.

Suppressing of unwanted modes is important in the frequency filtering where they are a source of parasitic outband transmission. In [149], the modes of dielectric cylindrical resonators are analyzed, and the higher-order modes are suppressed by holes in the places where the modal electric fields are strong. The results are supported by computations of fields and characteristics of a two-resonator filter. Additionally, a review and original results on suppressing spurious responses of dielectric filters are published in [150].

There are several techniques, applications of which allow less radical consequences for the modal spectrum of resonators. For instance, in the considered in Chpt. 7 shorted patch resonators, the eccentricity allows to increase the distance between the main and the higher-order modes resonances, and it can reduce the level of spurious response of filters based on these resonators.

In [151], a triangular microstrip resonator is studied, which patch is defected in a fractal manner. It is found that the defected resonators realized on a low-dielectric substrate have increased frequency bandwidth (up to 20.7%) due to this defection. It is explained by the distortion of the spectrum of resonances of a triangular patch resonator and the decreased due to that parasitics in the bandpass. Hollow waveguides and resonators allow to deform their shapes to increase the distance between the cut-off or resonance frequencies, and the Π- and H-like waveguides having increased mono-modal bandwidth have been known for decades.

Other components dealing with the discrete spatial properties of fields are the *mode converters* (Fig. 10.48). They are to transform a set of propagating modes to another one of the same or another waveguide.

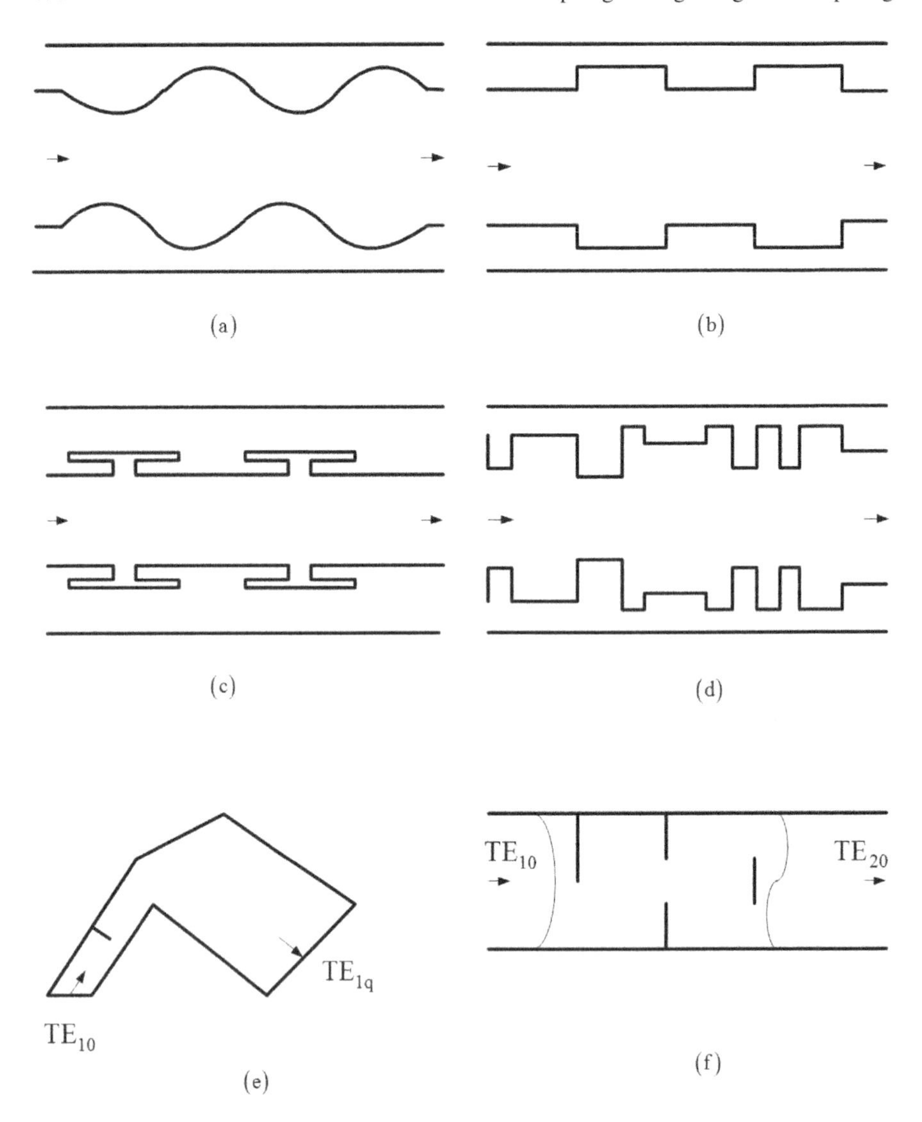

Fig. 12.48 Waveguide mode converter

The effect of transformation of modes is based on their coupling at a waveguide discontinuity (see Chpt. 3). Modifying its geometry, it is possible to reach the maximum of coupling of the input mode to the output one of a certain type.

One of the popular solutions is with the continuously varying cross-section waveguides allowing multimodal propagation (Fig. 10.48a). An incident wave of a regular waveguide is transformed into the coupled propagating and non-propagating modes in the converter. The geometry variation allows regulating the coupling coefficients of these modes, and the maximum of it can be reached for a selected output mode. Some of the periodical converters are considered, for instance, in [152]. Additionally, the waveguide geometry can be tapered along the longitudinal coordinate [153] or even be curved to provide the modal filtration or coupling of modes [154]-[156].

A theory of this converters is based on the method of cross-sections known from the 60s [157], and it gives a system of ordinary differential equations for the modal amplitudes in continuously varying waveguides. It allows finding in a fast way a geometry providing the needed conversation. Then these semi-analytical calculations are followed by full-wave EM simulations of designed converters [80]-[82]. Another simulation approach is with approximation of the boundary by steps along the longitudinal axis, and the use of the mode matching method at each step. The design goal is reached using an optimization procedure providing a geometry which gives the necessary conversation.

For instance, a converter of modes $H_{01} \rightarrow H_{02}$ is considered in [81], which is a length of a circular waveguide. Its diameter is varied periodically along the longitudinal coordinate as a cosine function. Eight periods allow reaching nearly full transformation of the input mode into the output one, according to calculations from [81].

The continuously varying waveguides allow rather wide bandwidth (10-30%) and high efficiency of modal transformations, although the converters of this type have increased length $l \sim (8-16)\lambda$.

The shorter length can be achieved using combinations of tapering and aperiodic radius variation, and several 1- and 1½-period designs are considered in [153] for different frequencies. One of them, which is of the length $l = 19$ cm, is designed and tested for 60 GHz frequency, and it shows the measured conversation efficiency around 96.6% for the modes TE_{02}-TE_{01} of a circular waveguide. Unfortunately, it demonstrates a narrower bandwidth regarding to longer converters. Additional information on the short continuous aperiodic converters and used design technique can be found, for instance, in [158].

The mode conversation can be reached using the corrugated waveguides (Fig. 10.48b,c). They are convenient to excite the corrugated wide-band horn antennas, and several geometries of corrugation of millimeter-wave circular waveguides are considered, for instance, in [159],[160].

Further improvement of characteristics of modal converters is obtained using the irregular discrete scatterers in waveguides (Fig.10.48d, [161]). It allows reducing the length of converters and improving their modal conversation efficiency regarding to their periodic counterparts up to 3-5 times [161]. For instance, a 60-GHz TE_{02}-to-TE_{01} circular waveguide converter has 4.8-cm length, in comparison with the periodical one of the length 18 cm [153]. These TE_{10} converters are designed using the mode matching method and optimization techniques to find a step-wise geometry of a scatterer providing the maximum conversation efficiency for a given couple of modes. Additionally to these converters, the authors of [161] consider some applications of irregular structures for power combiners/splitters, waveguide transitions, optically controlled phase shifters, and microwave switches allowing spatial separation of signals (modes) of different frequencies.

Further decreasing of the size of modal converters is with the *non-symmetrical corners* [162],[163] and the resonant diaphragm converters [164],[165]. The first of them is just an H-plane corner composed of two rectangular waveguides of different size (Fig. 10.48e). This corner is truncated, and the truncation angles can be regulated to reach the maximum efficiency of the modal conversation. The incident TE_{10} mode is diffracted at the corner, and it excites the propagating modes in the second waveguide. To provide better characteristics, an inductive diaphragm is used in the input waveguide. The maximum efficiency of transformation of the TE_{01} mode to the TE_{0q} one is found using an optimization procedure and varying the geometry of the corner. To design the converters, the EM method of semi-inversion [162] is used, and the results are compared with the Ansoft HFSS simulations. It is found that the mode conversation is realized, practically, at the corner area, which is confirmed by the field simulations. The conversation of the main mode to different 27 higher-order modes is simulated, and stability of this process with certain frequency variation is confirmed.

The use of the waveguide resonant volumes limited by two impedance walls was proposed in [164]. Later, the capacitive membranes substituted these walls and several mode converters were calculated using an EM full-wave method [165]. These converters consist of two 2-mode waveguide lengths separated from each other and from the input/output waveguides by capacitive diaphragms of a certain geometry. The length of each waveguide's volume is comparable with the height of the used rectangular waveguides, and it can be reduced further by filling the resonators with dielectrics of increased permittivity. In a designed TE_{10} - TE_{20} converter (Fig. 10.48f), these diaphragms are to transform the incident TE_{10} mode into the TE_{20} one. As the result of diffraction and superposition of these modes inside these two waveguide volumes, the input mode is transformed with the calculated efficiency 90-98% if the geometry of the structure is properly found by means of optimization and the EM methods of the diaphragm scattering calculations. The frequency band of the considered converters depends on the type of the converted modes, and it is varied within 0.5-27.5% regarding to the center frequency and for the 90% efficiency.

Longitudinal diaphragms can be used for mode converters, and one of these devices and its modeling results are studied in [166] where it is shown that the low-cost and high-precision diaphragm technology allows obtaining the parameters comparable with those of the tapered waveguide converters.

General principles of the waveguide converters are considered in [167], and several components are described there. Among them are a TM_{01}/TE_{31} converter for 9.75 GHz, 95-GHz taper transforming the TE_{11} of a circular waveguide to a Gaussian beam, and other components used for plasma heating and diagnostics, spectroscopy, and material technology applications and designed for the power of 10 W-1 GW, depending on the needs.

It is interesting to review *the methods of measurement of parameters of modes* in oversized waveguides. One of the first devices and a technique used until now are published in [168], where a test unit consists of a length of a waveguide connected to a load matched to several modes. The signals are taken from the launchers installed in this probe section, and the modal parameters are re-calculated using a system of linear algebraic equations regarding to the measured complex amplitudes of the launchers' signals. Another variant is the *k*-spectrometer [153],[169], which is a length of a regular multi-modal waveguide with a number of apertures drilled uniformly along the waveguide [153]. Modes show radiation aimed at different directions and the re-calculation of their parameters are available from this data. For measurements of parameters of modes, the above-mentioned mode-selective coupler [125] can be used, as well.

Other characterization method used in multimodal technique is with the measurements of radiation pattern of an open-end waveguide or from a horn antenna excited by this multimode waveguide. Then these measurements are compared with the calculations, and the modal parameters are obtained for each mode.

The above-considered mode converters and the measurement hardware are designed for microwave and millimeter-wave frequencies, and they are manufactured using conventional waveguide technology. A new wave of interests in mono- and overmoded waveguide components is with the sub-millimeter and terahertz frequencies where the integrated circuits are based on micromachined waveguides (Chpt. 5).

One of the components handling the spatial characteristics of modes in these micromachined waveguides is described in [170]. The device rotates the polarization plane of the main mode of a rectangular waveguide. It has ultra-bandwidth characteristics due to the 3-D tapering, and its parameters are measured in 200-340 GHz. A scaled variant is studied in 500-700 GHz frequency band. In these frequencies, the return loss of the measured samples is registered better than 20 dB. Taking into account the main interesting applications of terahertz frequencies, there are many needs in the developments of integrated components handling spatial properties of modes or their modal superpositions.

The knowledge of the spatial field characteristics is important for the developments of *compact multimodal frequency filters* used in satellite and wireless telecommunications. One of the first attempts to employ several modes excited in a waveguide volume relates, according to the best knowledge of the Author, to the end of the 40s when the single cylindrical cavity filters employing 2, 3, and 5 modes were designed and measured [171]. In these filters, the screws enable coupling of modes, and its level is defined by the positions of these screws and their geometry. Some formulas for the modal coupling coefficients are derived using a perturbation technique. The multimodal cavity equivalent circuits are found as the serially coupled resonant loops, and the correspondence between them and the prototype lumped element circuits allows designing these filters. The importance of understanding the modal spatial properties of cylindrical cavity and input/output rectangular waveguides to provide the prescribed coupling of modes is emphasized.

A much more intensive research started in the beginning of the 70s with the needs of the developments of compact narrow bandwidth filters for satellite telecommunications [172],[173]. The circular and squared cross-section waveguides are considered appropriate for filters, and some research was conducted regarding to the filter architectures, coupling elements, EM design models to improve the waveguide filter characteristics. The best-known designs use the dual-mode cavities serially connected to each other and employing corresponding degenerate orthogonal modes. Coupling elements allow splitting the resonance curves of cavities and providing a band-pass response. The intermodal couplings realize the Tchebyshev, quasi-, and elliptical filter responses.

Additionally to the screws allowing the tuned modal coupling, the cut corners which mix the modes in square-shaped waveguides are considered in [174]-[176]. An interesting idea is proposed in [177] where the coupling of orthogonal modes is realized just by perturbation of the shape of a squared waveguide. For cylindrical cavities, the sections of ridged or grooved waveguide sections are proposed in [178],[179] to avoid the screws worsening the electrical properties of filters. The design methodology of these filters can be found in many papers [171]-[174],[180].

The idea of multimodal filtering proposed more than 60 years ago is realizable using any type of waveguides if they provide the electric and technological advantages. For instance, the substrate integrated waveguide technology distinguished by its simplicity is used in [181] for X-band frequencies, and the designed multimodal filters demonstrate the insertion loss 1.4-1.8 dB and stopband rejection around 30-35 dB.

Many papers are on the design of *dual-mode dielectric resonator filters* having essentially decreased size regarding to the hollow cavity counterparts. The first design is considered in [182] where the dual-mode resonators are placed inside the cylindrical cavities which resonant mode frequency is essentially higher than the filter frequencies. The resonators are coupled to each other using the evanescent modes of cavities. The couplings are regulated by the geometry of cross-like slots in conducting membranes dividing each resonator section and by the length of screws mixing the modes of a single shielded dielectric resonator. The filter's input/output ports are realized using coaxial probes. Simple formulas for coupling coefficients are given in [182], and they can be used for calculations. The filters implementing this idea show the essential mass reduction in comparison with the hollow dual-mode counterparts.

In the dielectric dual-mode filters, the higher-order modes, which can distort the device characteristics, are a matter of serious concern. The modal suppressors are in use, including the metallization of resonators [183], drilling cylindrical holes in them [149],[150], shape grooving, and splitting of resonators [184]. All these methods require intensive study of the modal field topologies.

Today the development aims, mostly, at the *integrated dual-mode filters* manufactured using the hybrid integrated, LTCC, and monolithic technologies. Some of them are shown in Fig. 10.49. One of the first papers is on a dual-mode microstrip ring resonator (Fig. 10.49a [185]. Similarly to the circular waveguide, the modes of this microstrip resonator are degenerated, and there are two eigenmodes of the same resonant frequency, which are different from each other by the 90°-spatial shift of their modal fields. For instance, the normal-to-the-substrate electric field of these modes is [185]

$$E_z = \left[AJ_n(kr) + BN_n(kr)\right]\begin{Bmatrix}\sin n\varphi \\ \cos n\varphi\end{Bmatrix} \tag{10.23}$$

where A, B are unknown amplitude coefficients, $J_n(kr)$ is the Bessel function, $N_n(kr)$ is the Neumann function, n is the angular number, k is the wave number in substrate, and r is the radial variable.

Perturbation of symmetry leads to the coupling of these modes, and new ones of different resonant frequencies are formed. Variation of the perturbation controls the distance between the resonant frequencies and the shape of the frequency response in the bandpath area. Any microstrip resonator of a certain symmetry allows designing such dual-mode resonators. A dual-mode circular patch (Fig. 10.49b) resonators and filters are studied, for instance, in [186]. The triangle resonators (Fig. 10.49c) are considered in [187],[188]. A square microstrip ring (Fig. 10.49d) [186] and its modifications, including the meandered one [189], are studied in many papers, and the shape distortion in them regulates the modal coupling and the shift of resonant frequencies of initially degenerated modes. The resonators are placed on one substrate surface, or they are stacked and coupled through an aperture in a grounded shield [186]. Due to the additional vertical couplings, a quasi- or elliptical response is formed. Typical problems with the increased radiation from the resonator edges can be partly resolved by slotting the patches [190] or making them in a fractal manner [191]. Dual-modality can be realized by coupling of two identical resonators through a dielectric layer, and a couple of such resonators, which allow a wide bandwidth of their frequency responses (Fig. 10.49f), is described in Chpt. 2 of this book.

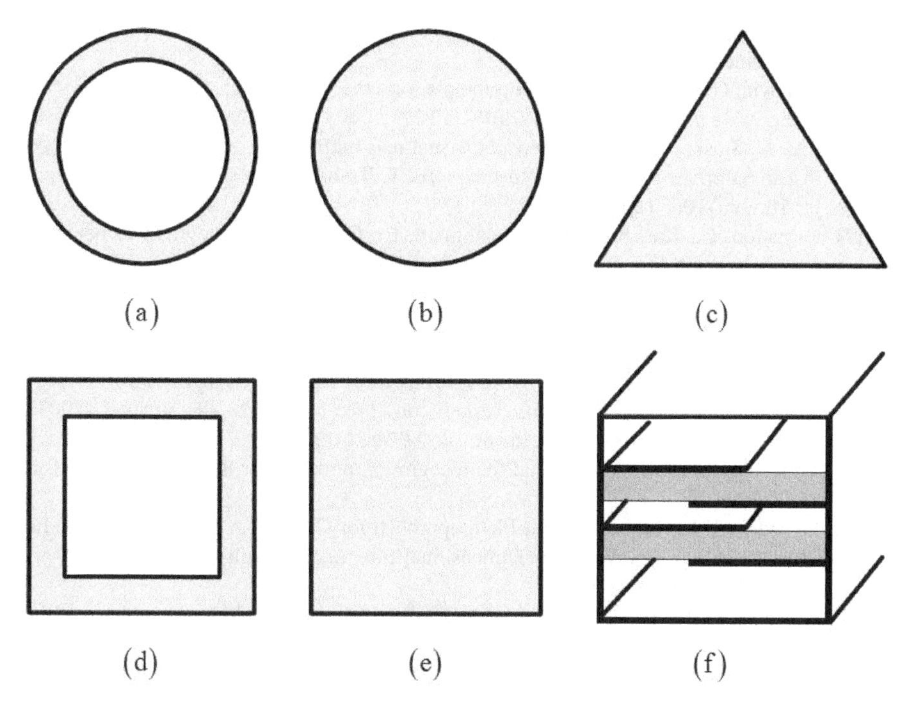

Fig. 10.49 Dual-mode microstrip patch resonators (a-e) and a shorted-end two-mode resonator on coupled antipodal slot lines (f)

Analysis of the results in the design of dual-mode filters and their design techniques indicates the importance of the EM models of resonators and topological representations of modes, which allow building new microwave filters distinguished by their increased compactness and essentially improved electric characteristics.

Concluding this Chapter, it is worthwhile to mention that the use of spatial properties of waves opens many interesting and fruitful possibilities in signaling, computing and design of novel electronic components and subsystems.

Additionally to this traditional electronics, the ideas of topology are used in fault-tolerant topological quantum computing [192], and they can find applications in magnetic ice solids for imitation of neural networks by excitation of topologically modulated impulses of magnetization of honey-comb structures [193].

References

[1] Conrad, M., Zauner, K.-P.: Molecular Computing: From Conformational Pattern Recognition to Complex Processing Networks. In: Hofestädt, R., Löffler, M., Schomburg, D., Lengauer, T. (eds.) GCB 1996. LNCS, vol. 1278, pp. 1–10. Springer, Heidelberg (1997)

[2] Păun, G., Rozenberg, G., Salomaa, A., Zandron, C. (eds.): WMC 2002. LNCS, vol. 2597. Springer, Heidelberg (2003)

[3] Freeman, W.J.: The neurobiological infrastructure of natural computing: intentionality. New Mathematics and Natural Computing (NMNC) 1(1), 19–29 (2009)

[4] Osterberg, B.: The Information Processing Mechanism of the Brain, E-Book, bertil@icon.co.za.

[5] de Pavia, G.: Pattern recognition principle for a theory of mind, http://arxiv.org/abs/0907.4509

[6] Liu, Z., Stasko, J.T.: Mental models, visual reasoning and interaction in information visualization: a top-down perspective. IEEE Trans. Visualization Comp. Graphics 16, 999–1008 (2010)

[7] Fernando, C., Karishma, K.K., Szathmary, E.: Copying and evolution of neuronal topology. PLoS One 3(11), e3775, 1–17 (2008)

[8] McKinsey, J.C.C., Tarski, A.: The algebra of topology. Annals of Mathematics 45(1), 141–191 (1944)

[9] Perelman, G.: The entropy formula for the Ricci flow and its geometric applications. El. Archive (2002), http://arxiv.org/abs/math.DG/0211159

[10] Perelman, G.: Ricci flow with surgery on three-manifolds. El. Archive (2003), http://arxiv.org/abs/math.DG/0303109

[11] Thomson, A.: Thin position and the recognition problem for S^3. Math. Res. Lett. 1, 613–630 (1994)

[12] Gvozdev, V.I., Kouzaev, G.A.: Field approach for CAD of microwave 3-D ICs. In: Proc. Conf. Microwave Three-Dimensional Integrated Circuits, Tbilisy, USSR, pp. 67–73 (1988) (in Russian)

[13] Kouzaev, G.A.: Mathematical fundamentals of topological electrodynamics and the three-dimensional microwave integrated circuits' simulation. In: Electrodynamics and Techniques of Microwaves and EHF, MIEM, pp. 37–44 (1991) (in Russian)

[14] Gvozdev, V.I., Kouzaev, G.A.: Microwave flip-flop. Russian Federation Patent, No 2054794 (February 26, 1992)

[15] Gvozdev, V.I., Kouzaev, G.A.: Topological computer. Computers and People (1), 2–5 (1992) (in Russian)
[16] Gvozdev, V.I., Kouzaev, G.A., Chernaykov, G.M., et al.: Topological demodulator. Telecommun. and Radio-Engineering 48, 26–28 (1993)
[17] Kouzaev, G.A., Nazarov, I.V.: Topological impulse modulation of the fields and the hybrid logic devices. In: Proc. Conf. and Exhibition on Microw. Techn. Satellite Commun., Sevastopol, Ukraine, vol. 4, pp. 443–446 (1993) (in Russian)
[18] Bykov, D.V., Gvozdev, V.I., Kouzaev, G.A.: Contribution to the theory of topological modulation of electromagnetic field. Russian Physics Doklady 38, 512–514 (1993)
[19] Gvozdev, V.I., Kouzaev, G.A.: A new technology of signal processing for super high-speed microwave circuits. Russian Microelectronics 22, 37–50 (1993)
[20] Kouzaev, G.A., Nazarov, I.V.: Quasineural effects for topologically modulated microwave field signals. Electrodynamics and Technique of Microwave and EHF 3, 17–18 (1993) (in Russian)
[21] Kouzaev, G.A., Nazarov, I.V.: On the theory of hybrid-logic devices. J. Commun. Technology and Electronics (Radiotekhnika i Elektronika) 39, 130–136 (1994)
[22] Kouzaev, G.A.: On the optimal design of super high-speed ICs for topologically modulated signals. Electrodynamics and Technique of Microwave and EHF 1, 70–73 (1994) (in Russian)
[23] Kouzaev, G.A., Kalita, A.V.: 4-valued gate for topologically modulated signals. Electrodynamics and Techniques of Microwave and EHF 3, 5 (1995) (in Russian)
[24] Kouzaev, G.A.: Information processing of field signals. Electrodynamics and Techniques of Microwave and EHF 4, 46–49 (1995) (in Russian)
[25] Kouzaev, G.A.: Topological pulse modulation of the electromagnetic field and super high-speed logical circuits of microwave range. In: Proc. Int. URSI Symp. Electromagnetic Theory, St.-Petersburg, Russia, May 23-26, pp. 584–586 (1995)
[26] Kouzaev, G.A.: Information properties of electromagnetic field superposition. J. Commun. Technology and Electronics (Radiotekhnika i Elektronika) 40, 39–47 (1995)
[27] Kouzaev, G.A.: Super high-speed switching of signals with discrete modulation of electromagnetic field structures. J. Techn. Physics 65, 205–207 (1995)
[28] Gvozdev, V.I., Kouzaev, G.A., Nazarov, I.V.: Topological switches for picosecond digital signal processing. Russian Microelectronics 24, 16–24 (1995) (in Russian)
[29] Kouzaev, G.A., Gvozdev, V.I.: Topological pulse modulation of the field and new microwave circuits designs for super-speed operating computers. In: Proc. Symp. Signals, Systems and Electronics, San Francisco, USA, October 25-27, pp. 383–384 (1995)
[30] Gvozdev, V.I., Kouzaev, G.A., Nazarov, I.V.: Topological pulse modulation of fields and new microwave circuits design for super-speed operating devices. In: Proc. Trans. Black Sea Region Symposium Applied Electromagnetism, Metsovo, Epirus-Hellas. Athens, Greece, April 17-19, pp. 174–175 (1996)
[31] Kouzaev, G.A., Nazarov, I.V.: Logical circuits for super high-speed processing of field impulses with topologically modulated structures. In: Proc. Int. Conf. Intelligent Technologies in Human-Related Sciences, incl. The 1996 System and Signals Symp., Leon, Spain, July 5-7 (1996)
[32] Kouzaev, G.A., Nazarov, I.V.: Theoretical and experimental estimations of the time delay of switches for topologically modulated electromagnetic field signals. In: Proc. AMSE Sci. Int. Conf. Commun., Signals and Systems, Brno, Czech Republic, September 10-12, pp. 181–183 (1996)

[33] Kuzaev, G.A.: Theoretical aspects of measurements of the topology of the electromagnetic field. Measurement Techniques 39, 186–191 (1996)
[34] Gvozdev, V.I., Kouzaev, G.A., Nazarov, I.V.: Problems of speed-increasing of the digital information processing. Zarubezhnaya Radioelektronika (Foreign Radio Electronics) (2), 19–30 (1996) (in Russian)
[35] Gvozdev, V.I., Kouzaev, G.A., Linev, A.A., et al.: Sensor for measurements of the permittivity of a medium in closed systems. Measurement Techniques 39, 81–83 (1996)
[36] Kouzaev, G.A., Nazarov, I.V., Tchernyi, V.V.: Circuits for ultra high-speed processing spatially modulated electromagnetic field signals. Int. J. Microcircuits and Electron. Packaging 20, 501–515 (1997)
[37] Kouzaev, G.A., Nazarov, I.V., Tcherkasov, A.S.: A physical view on broadband passive components for signal processing. In: Proc. 2nd Int. Sci. Conf. ELEKTRO 1997, Zilina, Slovak Republic, June 23-24, pp. 208–213 (1997)
[38] Kouzaev, G.A.: An active VLSI hologram for super high-speed processing of electromagnetic field signals. In: Proc. 3rd Int. Conf. Theory and Technique for Transmission, Reception, and Processing Digital Information, Kharkov, Ukraine, September 16-18, pp. 135–136 (1997) (in Russian)
[39] Kouzaev, G.A.: High-speed Signal Processing Circuits on the Principles of the Topological Modulation of the Electromagnetic Field. Doctoral Thesis, Moscow, MSIEM (1997) (in Russian)
[40] Kuzaev, G.A.: Experimental study of the transient characteristics of a switch for topologically modulated signals. J. Techn. Physics 40, 573–575 (1998)
[41] Kouzaev, G.A., Nazarov, I.V., Cherny, V.V.: Super broadband passive components for integrated circuits signal processing. In: Proc. SPIE, vol. 3465, pp. 483–490 (1998)
[42] Kouzaev, G.A., Nazarov, I.V., Tcherkasov, A.S.: Physical fundamentals for super-high speed processing spatially-modulated field signals. In: Proc. 28th Eur. Microw. Conf., Amsterdam, October 5-8, vol. 2, pp. 152–156 (1998)
[43] Kouzaev, G.A., Nazarov, I.V., Tchernyi, V.V.: The super broadband passive components for integrated circuits signal processing. In: Proc. 4th Conf. Millimeter and Submillimeter Waves and Applications, Digest, San Diego, July 20-24, pp. 161–163 (1998)
[44] Kouzaev, G.A., Nazarov, I.V., Tcherkasov, A.S.: Principles of processing of spatially modulated field signals. In: Proc. Int. AMSE Conf. Contribution of Cognition to Modelling, Lyon, France, July 6-8, Paper No 10.1 (1998)
[45] Kouzaev, G.A., Romanenkov, A.V., Smirnov, P.S.: Study of picosecond transients of microstrip components. Physics of Wave Process and Radiotechnical Systems (1) (1998) (in Russian)
[46] Kouzaev, G.A., Tcherkasov, A.S.: Circuit modeling for super high-speed processing spatially modulated field signals. In: Proc. 1998 Int. Conf. Math. Methods in Electromag. Theory, Kharkov, Ukraine, June 2-5, pp. 421–423 (1998)
[47] Kouzaev, G.A., Al-Shedifat, F., Smirnov, P.S.: Physical limitations of passive component speed-action. In: Proc. Int. Conf. Problems of Electronic Instrument Making, Saratov, Russia, September 7-9, vol. 2, pp. 117–121 (1998) (in Russian)
[48] Kouzaev, G.A.: Theoretical and experimental estimations of switching delay for topologically modulated signals. J. Commun. Technology and Electronics (Radiotekhnika i Elektronika) 43(1), 76–82 (1999)

[49] Kouzaev, G.A., Tchernyi, V.V., Al-Shedifat, F.: Subpicosecond components for quasioptical spatial electromagnetic signal processing. In: Proc. SPIE, vol. 3795, pp. 40–49 (1999)
[50] Kouzaev, G.A., Al-Shedifat, F., Kalita, A.V.: Currents and the frequency performance of modal filters on coupled microstrip transmission lines for microwave signals. Physics of Wave Process and Radiotechn. Systems 2, 42–43 (1999) (in Russian)
[51] Kouzaev, G.A., Nazarov, I.V., Kalita, A.V.: Unconventional logic elements on the base of topologically modulated signals. El. Archive, `http://xxx.arXiv.org/abs/physics/9911065`
[52] Kouzaev, G.A., Cherny, V.V., Lebedeva, T.A.: Multivalued processing spatially modulated discrete electromagnetic signals. In: Proc. 30th Eur. Microw. Conf., Paris, pp. 209–213 (October 2000)
[53] Kouzaev, G.A., Lebedeva, T.A.: New logic components for processing complex measurement data. Measurement Techniques 43, 1070–1073 (2000)
[54] Kouzaev, G.A., Lebedeva, T.A.: Multivalued and quantum logic modeling by mode physics and topologically modulated signals. In: Proc. Int. Conf. Modelling and Simulation, Las Palmas de Grand Canaria, Spain, September 25-27 (2000), `http://www.dma.ulpgc.es/ms2000`
[55] Kouzaev, G.A., Cherny, V.V., Lebedeva, T.A.: Multi-valued processing spatially modulated discrete electromagnetic signals (Invited paper). In: Proc. Int. Conf. Systems, Cybernetics, Informatics, Orlando, USA, vol. VI (July 2000)
[56] Kouzaev, G.A., Ermakov, A.: Multivalued electronic components for digital processing of discrete spatially-modulated field signals. In: Proc. Int. Conf. Systems, Analysis and Synthesis SCI200/ISAS 2000, vol. XI (2000)
[57] Kouzaev, G.A., Domashenko, G.D., Al-Shedifat, F., Potapova, T.A.: Picosecond generator for experimental studies of circuits for spatial processing of electromagnetic signals. Wave Processes and Radiotechn. Systems 3(1), 49–53 (2000) (in Russian)
[58] Kouzaev, G.A.: Predicate and pseudoquantum gates for amplitude-spatially modulated electromagnetic signals. In: Proc. 2001 IEEE Int. Symp. Intelligent Signal Processing and Commun. Systems, Nashville, Tennessee, USA, November 20-23 (2001)
[59] Kouzaev, G.A.: Qubit logic modeling by electronic gates and electromagnetic signals. El. Archive (2001), `http://xxx.arXiv.org/abs/quant-ph/0108012`
[60] Kouzaev, G.A.: Topologically modulated signals and predicate logic gates for their processing. El. Archive (2001), `http://xxx.arXiv.org/abs/physics/0107002`
[61] Kouzaev, G.A., Nazarov, I.V.: Discrete space-time modulated electromagnetic signals. In: Proc. 4th Int. Conf. Physics Techn. Appl. Wave Processes, Nizhny Novgorod, Russia, pp. 74–75 (October 2005)
[62] Kouzaev, G.A.: Space-time modulated signals. Noosphere (2005), `http://atss.brinkster.net/Noosphere/En/Default.asp`
[63] Kouzaev, G.A.: Topological computing (Invited paper). WSEAS Trans. Comp. 5, 1247–1250 (2006)
[64] Kouzaev, G.A.: Spatio-temporal electromagnetic field shapes and their logical processing. El. Archive (2007), `http://arXiv.org/physics/0701081`

[65] Kouzaev, G.A., Kostadinov, A.N.: Predicate gates for spatial logic. In: Proc. 11th Int. Multiconference Computer Science and Techn., CSCC, Agious Nikolaos, Crete Island, Greece, July 23-28, vol. 4, pp. 151–156 (2007)
[66] Kouzaev, G.A.: Spatial quasineural circuits for electromagnetic signals (Invited paper). In: Proc. 12th Int. Conf. Circuits, Heraklion, Greece, July 2-24, pp. 218–223 (2008), http://arxiv.org/abs/0805.4600
[67] Kouzaev, G.A., Kostadinov, A.N.: Predicate logic processor for space-time signals (invited paper). In: Proc. 7th Int. Conf. Physics and Techn. Wave Processes, Samara, Russia, September 15-21, Paper # 24 (2008)
[68] Kouzaev, G.A., Kostadinov, A.N.: Predicate logic processor. In: Innovation Forum 2008, 2009, Toronto, Canada (2008, 2009); (Booklet; Electronic version: Internet J. Noosphere, http://atss.brinkster.net/Noosphere/En/Magazine/Default.asp?File=20081207_Kouzaev_Kostadinov.htm)
[69] Kostadinov, A.N., Kouzaev, G.A.: Predicate logic processor of spatially patterned signals. In: Proc. Recent Advances in Systems Engineering and Applied Mathematics, pp. 94–96 (2008)
[70] Kouzaev, G.A., Kostadinov, A.N.: Predicate and Boolean operations processor. In: Proc. 8th Int. Conf. Applications of Electrical Eng., Houston, USA, April 5-May 2, pp. 199–203 (2009)
[71] Kouzaev, G.A.: Communications by vector manifolds (Invited paper). In: Mastorakis, M., Mladenov, V., Kontargry, V.T. (eds.) Proc. European Computing Conf. LNEE, vol. 1, 27, ch. 6, pp. 617–624. Springer (2009)
[72] Kouzaev, G.A., Kostadinov, A.N.: Predicate gates, components and a processor for spatial logic. J. Circuits, System, Computers 19(7), 1517–1547 (2010)
[73] Stewart, J.V.: Intermediate Electromagnetic Theory. World Scientific (2001)
[74] Andronov, A.A., Leontovich, E.A., Gordon, I.I., et al.: Qualitative Theory of Second Order Dynamical Systems. Halsted Press (1973)
[75] Stolyar, A.A.: Introduction to Elementary Mathematical Logic. Dover Publishing (1983)
[76] Klovskyi, D.D., Soifer, V.A.: Space-Time Signal Processing. Svyaz Publ., Moscow (1976) (in Russian)
[77] Migliore, M.D.: On electromagnetics and information theory. IEEE Trans., Microw. Theory Tech. 56, 3188–3200 (2008)
[78] Sapiro, G.: Geometric Partial Differential Equations and Image Processing. Cambridge University Press (2001)
[79] Narahara, K., Otsuji, T.: Ultrafast gating circuit using coupled waveguides. IEICE Trans. Electron E83-C, 98–108 (2000)
[80] Krishnamachari, B., Lok, S., Gracia, C., Abraham, S.: Ultra high speed digital processing for wireless systems using passive microwave logic. In: Proc. 1998 IEEE Int. Radio and Wireless Conf., RAWCON 1998, Colorado Springs, Colorado, pp. 43–46 (August 1998)
[81] Caufield, H.J., Dolev, S.: Why future supercomputing requires optics. Nature Photonics 4, 261–263 (2010)
[82] Raleigh, G.G., Cioffi, J.M.: Spatio-temporal coding for wireless communication. IEEE Trans. Commun. 46, 357–366 (1998)
[83] Tamburini, E., Mari, E., Sponselli, A., et al.: Encoding many channels on the same frequency through radio vorcity, First experimental test. New Phys. J. 14, 1–17 (2012)

[84] Eisenstad, W.R., Stengel, B., Thompson, B.M.: Microwave Differential Circuit Design Using Mixed-mode S-parameters. Artech House, Inc. (2006)

[85] Centurelli, F., Luzi, R., Marietti, P., et al.: An active balun for high-CMRR IC design. In: Proc. 13th GAAS Symp., Paris, pp. 621–624 (2005)

[86] Poulton, J.W., Tell, S., Palmer, R.: Multiwire differential signaling. White Paper on the US Pat. #6556628, April 29 (2003)

[87] Goldie, J.: LVDS, CML, ECL-differential interfaces with odd voltages. Planet Analog (January 21, 2003)

[88] Vega-Gonzales, V.H., Torres-Torres, R., Sanchez, A.S.: Analysis of the electrical performance of multi-coupled high-speed interconnects for SoP. In: Proc. 52nd MWSCAS 2009, pp. 1030–1033 (2009)

[89] Gabara, T.: Phantom mode signaling in VLSI systems. In: Proc. 2001 Conf. Advanced Research in VLSI, pp. 88–100 (2001)

[90] Ho, A., Stojanovic, V., Chen, F., et al.: Common-mode back-channel signaling system for differential high-speed links. VLSI Circuits. Dig. Tech. Papers, pp. 352–355 (2004)

[91] Kimura, M., Ito, H., Sugita, H., et al.: Zero-crosstalk bus line structure for global interconnects in Si ultra large scale integration. Jpn. J. Appl. Phys. 45, 4977–4981 (2006)

[92] Alcatel-Lucent Bell Labs achieves industry first: 300 Megabits per second over just two traditional DSL lines, `http://www.alcatel-lucent.com`

[93] Poulton, J.W., Palmer, R.: Multiwire differential signaling. White Paper on US Patent #6556628 (April 29, 2003)

[94] Allstot, D.J., Chee, S.-H., Kiaei, S., et al.: Folded source-coupled logic vs. CMOS static logic for low-noise mixed-signal ICs. IEEE Trans., Circuits and Systems-1 40, 553–563 (1993)

[95] Ng, P., Balsara, P.T., Steisiss, D.: Performance of CMOS differential circuits. IEEE J. Solid-State Circ. 31, 841–846 (1996)

[96] Tsai, Y.-H., Yang, H.-L., Lin, W.-J., et al.: A new differential logic-compatible multiple-time programmable memory cell. Jpn. J. Appl. Phys. 49, 04DD13-1-4 (2010)

[97] Hanyu, T., Mochizuki, A., Kameyama, M.: Design and evaluation of a multiple-valued arithmetic integrated circuit based on differential logic. In: IEE Proc. Circuits Devices Syst., vol. 143, pp. 331–336 (1996)

[98] Mochizuki, Hanui, T.: Highly reliable multiple-valued circuit based on dual-rail differential logic. In: Proc. 36th Int. Symp. Multiple-valued Logic, ISMVL 2006 (2006)

[99] Martin, A.J., Nystroem, M.: Asynchronous techniques for system-on-chip design. Proc. IEEE 94, 1089–1120 (2006)

[100] Azaga, M., Othman, M.: Source couple logic (SCL): Theory and physical design. Am. J. Eng. Appl. Sci. 1, 24–32 (2008)

[101] Zhang, L., Liu, J., Zhu, H., et al.: High performance current-mode differential logic. In: Proc. ASPDAC 2008, pp. 720–725 (2008)

[102] Tiri, K., Verbauwhede, I.: A digital design flow for secure integrated circuits. IEEE Trans., Computer-aided Design of Integrated Circuits and Systems 25, 1197–1208 (2006)

[103] Sokolov, D., Murphy, J., Bystrov, A., et al.: Design and analysis of dual-rail circuits for security applications. IEEE Trans., Computers 54, 449–460 (2005)

[104] Yamaoka, H., Yoshida, H., Ikeda, M., et al.: A dual-rail PLA with 2-input logic cells. In: Proc. ESSCIRC 2002, pp. 203–206 (2002)

[105] Stan, M.R., Franzon, P.D., Goldstein, S.C., et al.: Molecular electronics: from devices and interconnect to circuits and architecture. Proc. IEEE 91, 1940–1957 (2003)
[106] Kouzaev, G.A.: Research and development of ultra high-speed quasineural logical circuits for the field signals in VLSI. Techn. Report to the Russian Foundation on Basic Research, Grant No 94-02-04979a. Information Bulletin of RFBR 4(2), 488 (1996) (in Russian)
[107] Kouzaev, G.A.: Development of physical fundamentals of new high-dense integrated circuits on collective effects for topologically modulated signals. Techn. Report to the Russian Foundation on Basic Research, Grant No 96-02-1744a. Information Bulletin of RFBR 6(2), 355 (1998) (in Russian)
[108] S.N. Yanushkevich, V.P. Shmerko, and S. E. Lyshevski, Logic Design of NanoICs. CRC Press (2005)
[109] Farazmand, N., Tahoori, M.B.: Online detection of multiple faults in crossbar nanoarchitectures using dual rail implementations. In: Proc. 2009 IEEE/ACM Int. Symp. Nanoscale Archtectures, pp. 79–82 (2009)
[110] Narahara, K., Otsuji, T.: A traveling-wave time-division demultiplexer. Jpn. J. Appl. Phys. 38, 4021–4026 (1999)
[111] Narahara, K., Otsuji, T.: Characterization of wave propagation on traveling-wave field effect transistors. Jpn. J. Appl. Phys. 37, 6328–6339 (1998)
[112] Suntives, A., Abhari, R.: Dual-mode high-speed data transmission using substrate integrated waveguide interconnects. In: Proc. 16th IEEE Elect. Performance Electron. Packag., Atlanta, GA, October 29-31, pp. 215–218 (2007)
[113] Suntives, A., Abhari, R.: Ultra-high speed multichannel data transmission using hybrid substrate integrated waveguides. IEEE Trans., Microw. Theory Tech. 56, 1973–1984 (2008)
[114] Guckenberger, D., Schuster, C., Kwark, Y., et al.: On-chip crosstalk mitigation for densely packed differential striplines using via fence enclosures. El. Lett. 41(7), 412–414 (2005)
[115] Sakagami, I., Miki, N., Nagai, N., et al.: Digital frequency multipliers using multisection two-strip coupled line. IEEE Trans., Microw. Theory Tech. 29, 118–122 (1981)
[116] Iluishenko, V.N., Avdochenko, B.I., Baranov, V.Y., et al.: Picosecond Impulse Techniques. Energoatomizdat Publ., Moscow (1993) (in Russian)
[117] Lujambio, A., Arnedo, I., Chudzik, M., et al.: Dispersive delay line with effective transmission-type operation in coupled-line technology. IEEE Trans., Microw. Wireless Comp. Lett. 21, 459–461 (2011)
[118] Nikolskyi, V.V., Nikolskaya, T.I.: Electrodynamics and Wave Propagation. Nauka, Moscow (1987) (in Russian)
[119] Mashkovzev, B.M., Zibisov, K.N., Emelin, B.F.: Theory of Waveguides. Nauka, Moscow (1966) (in Russian)
[120] Collin, R.E.: Foundation of Microwave Engineering. John Wiley & Sons (2001)
[121] Model, A.M.: Microwave Filters in Radio Relay Systems. Svyaz Publ., Moscow (1967) (in Russian)
[122] Behe, R., Brachat, P.: Compact duplexer-polarizer with semicircular waveguide. IEEE Trans., Antennas Prop. 39, 1222–1224 (1991)
[123] Tuzbekov, A.R., Goldberg, B.K.: Wide bandwidth waveguide duplexer of a small cross-section for the G-frequencies. In: Proc. 4th All-Russia Conf. Radiolocation and Radio Telecommunication, Moscow, IRE RAS, November 29- December 3, pp. 887–895 (2010) (in Russian)

[124] Pisano, G., Melhuish, S., Savini, G., et al.: A broadband W-band polarization rotator with very low cross polarization. IEEE Microw. Wireless Comp. Lett. 21, 127–129 (2011)

[125] Wang, W., Gong, Y., Yu, G., et al.: Mode discriminator based on mode-selective coupling. IEEE Trans., Microw. Theory Tech. 51, 55–63 (2003)

[126] Alessandri, F., Comparini, M., Vitulli, F.: Low-loss filters in rectangular waveguide with rigorous control of spurious responses through a smart modal filter. In: 2001 IEEE MTT-S Microw. Symp. Dig., pp. 1615–1617 (2001)

[127] Harms, P.H., Mittra, R.: Equivalent circuits for multiconductor microstrip bend discontinuities. IEEE Trans., Microw. Theory Tech. 41, 62–69 (1993)

[128] Bockelman, D.E., Eisenstadt, W.R.: Combined differential and common-mode scattering parameters: theory and simulation. IEEE Trans., Microw. Theory Tech. 43, 1530–1539 (1995)

[129] Shiue, G.-H., Guo, W.-D., Liu, L.-S., et al.: Circuit modeling and noise reduction for bent differential transmission lines. In: Proc. IEEE 13th Topical Meeting El. Performance of Electron. Pack., pp. 143–145 (2004)

[130] Hagmann, J.H., Dickmann, S.: Determination of mode conversation on differential lines. In: Proc. Int. Symp. EMC Europe 2008, pp. 1–5 (2008)

[131] Chuang, H.-H., Wu, T.-L.: A novel ground resonator technique to reduce common-mode radiation on slot-crossing differential signals. IEEE Microw. Wireless Comp. Lett. 20, 660–662 (2010)

[132] Wu, S.-J., Tsai, C.-H., Wu, T.-L., et al.: A novel wideband common-mode suppression filter for gigahertz differential signals using coupled patterned ground structure. IEEE Trans., Microw. Theory Tech. 57, 848–855 (2009)

[133] De Paulis, F., Orlandi, A., Raimondo, L., et al.: Common mode filtering performances of planar EBG structures. In: Proc. Int. Symp. EMC Europe 2009, pp. 86–90 (2009)

[134] Tsai, C.-H., Wu, T.-L.: A broadband and miniaturized common-mode filter for gigahertz differential signals based on negative-permittivity metamaterials. IEEE Trans., Microw. Theory Tech. 58, 195–202 (2010)

[135] Gazda, C., Ginste, D.V., Rodiger, H., et al.: A wideband common-mode suppression filter for bend discontinuities in differential signaling using tightly coupled microstrips. IEEE Trans., Microw. Theory Tech. 43, 969–978 (2010)

[136] Wu, C.-H., Wang, C.-H., Chen, C.H.: Balanced coupled-resonator bandpass filters using multisection resonators for common mode suppression and stopband extension. IEEE Trans., Microw. Theory Tech. 55, 1756–1763 (2007)

[137] Saitou, A., Ahn, K.P., Aoki, H., et al.: Differential mode bandpass filters with four coupled lines embedded in self-complementary antennas. IEICE Trans., Electron. E90-C(7), 1524–1532 (2007)

[138] Lim, T.B., Zhu, L.: A differential-mode wideband bandpass filter on microstrip line for UWB application. IEEE Microw. Wireless Comp. Lett. 19, 632–634 (2009)

[139] Lim, T.B., Zhu, L.: Highly selective differential-mode wideband bandpass filter for UWB application. IEEE Microw. Wireless Comp. Lett. 21, 133–135 (2011)

[140] Gunston, M.A.R.: Microwave Transmission Line Impedance Data. Van Nostrand Reinhold Company Ltd (1972)

[141] Laermans, E., De Geest, J., De Zutter, D., et al.: Modeling differential via holes. IEEE Trans., Adv. Pack. 24, 357–363 (2001)

[142] Wang, C., Drewniak, J.L., Fan, J., et al.: Transmission lines modeling of vias in differential signals. In: Int. Symp. Electromag. Compatibility, EMC 2002, pp. 249–252 (2002)
[143] Antonini, G., Scogna, A.C., Orlandi, A.: *S*-parameters characterization of through, blind, and buried via holes. IEEE Trans., Mob. Comp. 2, 174–184 (2003)
[144] Wang, C.-C., Kuo, C.-W., Kuo, C.-C.: A time-domain approach for extracting broadband macro-π models of differential via-holes. IEEE Trans., Adv. Pack., 789–797 (2006)
[145] Cao, Y., Simonovich, L., Zhang, Q.-J.: A broadband and parametric model of differential via holes using space-mapping neural network. IEEE Microw. Wireless Comp. Lett. 19, 533–535 (2009)
[146] Rimolo-Donadio, R., Duan, X., Bruns, H.-D., et al.: Differential to common mode conversation due to asymmetric ground via configuration. In: Proc. SPI 2009, pp. 1–4 (2009)
[147] Huang, J., Wu, K., Kuroki, F., et al.: Computer-aided design and optimization of NRF-guide mode suppressors. IEEE Trans., Microw. Theory Tech. 44, 905–910 (1996)
[148] Baum, C.E.: Use of the $H_{0,1}$ mode in circular waveguide for microwave pulse compression. Circuit and Electromagnetic System Design Notes, Note 64 (October 30, 2009)
[149] Moraes, M.O., Borges, F.R., Hernandez-Figueroa, H.E.: Efficient technique for suppression of undesirable modes in dielectric resonator filters. In: Proc. Microw. Optoelectronics Conf., pp. 775–777 (2009)
[150] Weily, A.R., Mohan, A.S.: Microwave filters with improved spurious performance based on sandwiched conductor dielectric resonators. IEEE Trans., Microw. Theory Tech. 49, 1501–1507 (2001)
[151] Xiao, J.K., Chu, Q.-X., Huang, H.-F.: New wideband microwave bandpass filter using single triangular patch resonator with low permittivity substrate. In: Proc. ICCS 2008, pp. 608–612 (2008)
[152] Thumm, M.: High-power millimeter-wave mode converters in overmoded circular waveguides using periodic wall perturbation. Int. J. Electron. 57, 1225–1246 (1984)
[153] Stein, D.A., Vernon, R.J.: A single period $TE_{02} - TE_{01}$ mode converters in a highly overmoded circular waveguide. IEEE Trans., Microw. Theory Tech. 39, 1301–1306 (1991)
[154] Li, H., Thumm, M.: Mode conversation due to curvature in corrugated waveguides. Int. J. Electronics 71(2), 333–347 (1991)
[155] Yang, S., Li, H.: Optimization of novel high-power millimeter-wave $TM_{01} - TE_{01}$ mode converters. IEEE Trans., Microw. Theory Tech. 45, 552–554 (1997)
[156] Zemlaykov, V.V., Zargano, G.F., Sinaykovskyi, G.P.: Mode transformation due to curvature and diameter variations in smooth-wall circular waveguides. In: Proc. MSMW 2004 Symp., pp. 647–649 (2004)
[157] Katsenelenbaum, B.Z., Del Rio, L.M., Pereyaslavetz, M., Thumm, M.: Theory of Nonuniform Waveguides: the Cross-section Method, Inst. of Engineering and Technology (1999)
[158] Luneville, E., Krieg, J.-M., Giguet, E.: An original approach to mode converter optimum design. IEEE Trans., Microw. Theory Tech. 46, 1–9 (1998)
[159] James, G.L.: Analysis and design of $TE_{11} - HE_{11}$ corrugated cylindrical waveguide mode converters. IEEE Trans., Microw. Theory Tech. 29, 1059–1066 (1981)

[160] James, G.L., Thomas, B.M.: TE_{11} to HE_{11} cylindrical waveguide mode converters using ring-loaded slots. IEEE Trans., Microw. Theory Tech. 30, 278–285 (1982)

[161] Haq, T.U., Webb, K.J., Gallagher, N.C.: Optimized irregular structures for spatial- and temporal-field transformation. IEEE Trans., Microw. Theory Tech. 46, 1856–1867 (1998)

[162] Shestopalov, V.P., Kirilenko, A.A., Rud, L.A.: Resonant Scattering of Waves. Waveguide Discontinuities, vol. 2. Naukova Dumka (1986)

[163] Kirilenko, A.A., Rud, L.A., Tkachenko, V.I.: Nonsymmetrical H-plane corners for TE_{01} – TE_{q0}-mode conversion in rectangular waveguides. IEEE Trans., Microw. Theory Tech. 54, 2471–2477 (2006)

[164] Katsenelenbaum, B.Z., Korshunova, E.N., Pangonis, L.I., et al.: Synthesis of a converter for guided wave fields. Radiotekhnika i Elektronika 27, 2373–2380 (1982)

[165] Shcherbak, V.V.: Broadband regime of a conversation for $TE_{n,0}$ -modes on the cascade of three strip diaphragms. In: Proc. 2010 Int. Kharkov Symp. Physics and Engineering of Microwaves, Millimeter and Submillimeter-waves (MSMW), pp. 1–3 (2010)

[166] Zemlaykov, V.V., Zargano, G.F.: Mode transformers of longitudinal diaphragms in waveguides of complex cross-section. In: Proc. MSMW 2007 Symp., Kharkov, Ukraine, June 25-30, pp. 657–659 (2007)

[167] Denisov, G.G., Chirkov, A.V., Belousov, V.I., et al.: Millimeter wave multi-mode transmission line components. J. Infrared Milli. Terahz Waves 32, 343–357 (2011)

[168] Levinson, D.S., Rubinstein, I.: A technique for measuring individual modes propagating in overmoded waveguide. IEEE Trans., Microw. Theory Tech. 14, 310–322 (1966)

[169] Kasparek, W., Muller, G.A.: The wavenumber spectrometer - An alternative to the directional coupler for multimode analysis in oversized waveguide. Int. J. Electron. 65, 5–20 (1988)

[170] Chattopadhyay, C., Ward, J.S., Llombert, N., et al.: Submillimeter-wave 90° polarization twists for integrated waveguide circuits. IEEE Microw. Wireless Comp. Lett. 20, 592–594 (2010)

[171] Lin, W.: Microwave filters employing a single cavity excited in more than one mode. J. Appl. Phys. 22, 989–1011 (1951)

[172] Atia, A.E., Williams, A.E.: New types of bandpass filters for satellite transponders. COMSAT Tech. Rev. 1, 21–43 (1971)

[173] Williams, A.E., Atia, A.E.: Dual-mode canonical waveguide filters. IEEE Trans., Microw. Theory Tech. 25, 1021–1026 (1977)

[174] Liang, X.-P., Zaki, K.A., Atia, A.E.: Dual mode coupling by square corner cut in resonators and filters. IEEE Trans., Microw. Theory Tech. 40, 2294–2302 (1992)

[175] Levy, R.: The relationship between dual mode cavity cross-coupling and waveguide polarizers. IEEE Trans., Microw. Theory Tech. 43, 2614–2620 (1995)

[176] Ruiz-Cruz, J.A., Zhang, Y., Monteo-Garai, J.R., et al.: Longitudinal dual-mode filters in rectangular waveguide. In: 2008 IEEE Int. Microw. Symp. Dig., pp. 631–634 (2008)

[177] Orta, R., Savi, P., Tascone, R., et al.: Rectangular waveguide dual-mode filters without discontinuities inside the resonators. IEEE Microw. Guided Lett. 5, 302–304 (1995)

[178] Guglielmi, M., Molina, R.C., Melcon, A.A.: Dual-mode circular waveguide filters without tuning screws. IEEE Microw. Guided Lett. 2, 457–458 (1992)

[179] Yoneda, N., Miyazaki, M.: Analysis and design of grooved waveguide dual-mode filters. In: 2001 IEEE MTT-S Dig., pp. 1791–1794 (2001)
[180] Amari, S., Rosenberg, U.: Characteristics of cross (bypass) coupling through higher/lower order modes and their applications in elliptic filter design. IEEE Trans., Microw. Theory Tech. 53, 3135–3141 (2005)
[181] Chen, X., Hao, Z., Hong, W., et al.: Planar asymmetric dual-mode filters based on substrate integrated waveguide (SIW). In: 2005 IEEE MTT-S Int. Microw. Symp. Dig., pp. 949–952 (2005)
[182] Fedziusko, S.J.: Dual-mode dielectric resonator loaded cavity filters. IEEE Trans., Microw. Theory Tech. 30, 1311–1316 (1982)
[183] Fumagalli, M., Macchiarella, G., Resnati, G.: Dual-mode filters for cellular base station using metallized dielectric resonators. In: IEEE MTT-S Int. Microw. Dig., vol. 3, pp. 1799–1802 (2001)
[184] Accationo, L., Bertin, G., Mongiardo, M., et al.: Dual-mode filters with grooved/splitted dielectric resonators for cellular-radio base stations. IEEE Trans., Microw. Theory Tech. 50, 2882–2889 (2002)
[185] Wolff, I.: Microstrip passband filter using degenerate modes of a microstrip ring resonator. El. Lett. 8(12), 302–303 (1972)
[186] Curtis, J.A., Fiedziusko, S.J.: Multi-layered planar filters based on aperture coupled, dual mode microstrip or stripline resonators. In: 1992 IEEE MTT-S Dig., pp. 1203–1206 (1992)
[187] Hong, J.-S., Li, S.: Theory and experiment of dual-mode microstrip triangular patch resonators and filters. IEEE Trans., Microw. Theory Tech. 50, 1237–1243 (2004)
[188] Lugo, A., Papapolymerou, J.: Bandpass filter design using a microstrip triangular loop resonator with dual-mode operation. IEEE Microw. Wireless Comp. Lett. 15, 475–477 (2005)
[189] Gorur, A., Karpuz, C.: Miniature dual-mode microstrip filters. IEEE Microw. Wireless Comp. Lett. 17, 37–39 (2007)
[190] Zhu, L., Wecowski, P.M., Wu, K.: New planar dual-mode filter using cross-slotted patch resonator for simultaneous size and loss reduction. IEEE Trans., Microw. Theory Tech. 47, 650–654 (1999)
[191] Chen, W.-Y., Chang, S.-J., Weng, M.-H., et al.: A novel miniature dual-mode filter based on modified Sierpinski fractal resonator. In: Proc. APMC 2008, pp. 1–4 (2008)
[192] Freedman, M.H., Kitaev, A., Larsen, M.J., et al.: Topological quantum computation. Bull. Amer. Math. Soc. 40, 31–38 (2003)
[193] Branford, W.R., Ladak, S., Read, D.E., et al.: Emerging chirality in artificial spin ice. Science 335, 1597–1600 (2012)

11 EM Radiometry and Imaging

Abstract. This Chapter is on the basics of microwave radiometry used for registration of weak thermal signals. Additionally to the principles of radiometry, some original results are considered. Among them is the application of the methods of stochastic dynamics to the analysis of radiation and a technique developed to separate the parasitic deterministic and human-body thermal signals. A millimeter wave imager of a novel design is described allowing working in the radiometric, in scattering, and in holographic regimes. References -74. Figures -18. Pages -39.

11.1 Principles of the EM Radiometry

Microwave radiometry is the registration of the EM thermally caused radiation of objects in natural condition. Usually, the power of thermal signals is below the noise level of conventional receivers, and a special kind of them, called the radiometers, is designed for reliable signal registration. The material and its spatial structure influence this radiation, allowing the remote sensing of natural or artificial objects to define their spatial shapes and inner material structure [1]-[4]. Many applications of radiometry are within medicine where it allows to obtain the images from the inside of body or profile of the body temperature at the difference to the infrared imaging [4],[5]-[7]. Now some developments are with the security applications, and several radiometric vision systems have been developed for detecting the weaponry hidden beneath the clothes without additional illumination of the investigated individuals [8]-[15].

A pre-World-War-II history of the attempts of creation of high-sensitive hardware can be found in [16]. Today the circuits for registration of even single microwave photons are known [17].

The most used radiometers are based on the first design invented by R. H. Dicke in 1946 [18], and one of them, designed by Yu. Turygin from the Special Design Bureau at the Institute of Radioengineering and Electronics, Russian Academy of Sciences, is shown below (Fig. 11.1).

This radiometer eliminates the resulting hardware noise by periodical deduction of a power portion, equivalent to the noise caused by electronics, from the input signal. A square-law detector is used to produce an outcome signal which is proportional to the input power. Then it is integrated over a rather long period to smooth the fluctuations, digitized by an A/D (analog-to-digital) converter, and registered as a time series in the memory of a computer. A radiometer can be cooled to reduce the noise or placed into a thermostat to support the stable characteristics.

G.A. Kouzaev: Applications of Advanced Electromagnetics, LNEE 169, pp. 495–531.
springerlink.com

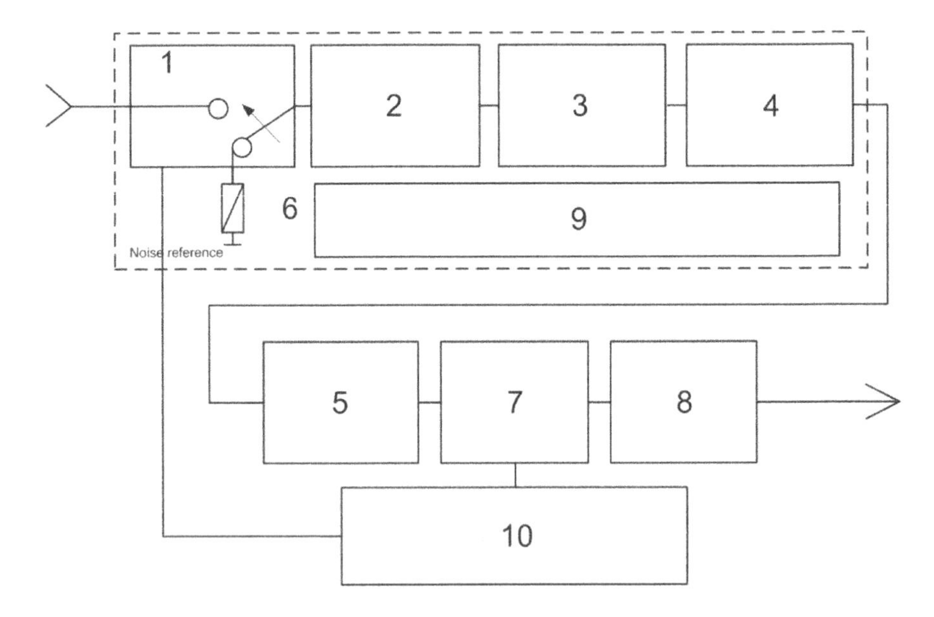

Fig. 11.1 A Dicke-based radiometer R-166 used for our experimentations. 1- Switch; 2- Ferrite gate; 3- Amplifier; 4- Square-law detector; 5- Low frequency amplifier; 6 – Embedded matched load as a noise source; 7- Synchronous detector; 8- Integrator; 9- Thermostat; 10- Clock oscillator. All blocks in the dashed box are stabilized thermostatically

Unfortunately, any of these radiometers is not free of the residual time-dependent noise if the practically interesting integration times are used. As a rule, the radiometer output signals show rather complicated time dynamics. The thermal and digital noise and the incoming thermal signals are mixed with each other in the hardware, and the algorithmic means to increase the sensitivity are preferable instead of expensive cooling of radiometers [19].

11.2 Natural Microwave Radiation and Its Characterization

The origin of natural microwave radiation is warmth. Any heated object radiates EM waves in a wide frequency region, and its qualitative parameters can be calculated using the Planck radiation law. The theory of this radiation is based on the phenomenon of EM power absorption. A real body is substituted by black body which absorbs all radiation. The radiation properties are calculated according to the mentioned absorption properties. At microwaves, Planck's formula originated from quantum mechanics is substituted by its low-frequency approximation or the Rayleigh-Jeans formula. It connects the spectral radiance L [20] measured in $\mathrm{W{\cdot}m^{-2}{\cdot}sr^{-1}{\cdot}Hz^{-1}}$ with the frequency f and absolute temperature T measured in Kelvin:

$$L \approx \frac{2f^2}{c^2} kT \tag{11.1}$$

where $k = 1.38 \cdot 10^{-23}$ J/K is the Boltzmann constant and c is the light velocity. From (11.1), the brightness temperature is defined:

$$T_L = \frac{\lambda^2}{2k} L \tag{11.2}$$

which is used for comparisons of any transmitters.

The power P_B radiated by black body in a certain frequency band is

$$P_B = kT\Delta f \; . \tag{11.3}$$

Thus, the proportionality between the radiance and temperature allows measuring the emitted signal in temperature units or measuring temperature of a distant object in terms of received radiation power.

Taking into account that the radiated power can be not only of the thermal nature, the radiation temperature is introduced from (11.3) as

$$T = \frac{P_B}{k\Delta f} \; ; \tag{11.4}$$

moreover, it can be used for characterization of any EM sources and antennas.

A real-body radiation differs from the above-mentioned P_B, being less than the power of ideal black body. The emissivity coefficient depends on the material, surface quality, and inner structure of the physical object. Furthermore, in real-life conditions, the antennas register the sum of the brightness temperature and the temperature of background EM noise in the frequency band concerned. This means that the radiance of an object is not proportional to the measured power but rather concealed in it.

The EM radiation is registered by non-ideal receivers which components are also inherently noisy. This noise can again be re-calculated in the temperature units. In microwaves, the temperature of natural radiation is essentially less than the temperature of uncooled receiver's noise, and special hardware or radiometers are used to register such very weak signals as it was mentioned above [20]. The statistical characteristics of natural radiation are close to those of Gaussian noise. Due to that, it has not so much to do mathematically with improving the parameters of radiometers, in contrast to the telecommunication hardware dealing with certain type of signals the statistics of which is known.

Averaging of a number N of independent noise samples results in the fractional uncertainty $(N/2)^{-1/2}$ of ideal radiometer's average noise power. This enables the low-level detection of a signal, which raises the antenna temperature by a small fraction of the total noise power.

One of the first types of radiometers developed by R. Dicke in 1946 [9] exploits an idea of periodical switching of the receiver from an antenna to the equivalent noise source which has temperature equal to that of the receiver (Fig. 11.1).

11.3 Calibration of Radiometers and Calculation of Physical Temperature

Very often, radiometers are used to measure the object's radio-brightness expressed in the temperature units. In the case of an ideally linear radiometer, the brightness temperature T_A is proportional to the output voltage V_{out}, and the coefficient of proportionality can be established during a calibration procedure measuring the objects with a priori known temperatures. Among them are the noise sources, liquid nitrogen, sky, the Moon, etc. Different techniques of calibration of linear and nonlinear radiometers are considered in many publications, for instance, in [4].

Besides the radio-brightness measurements, the physical temperature is of great interest in medicine. For instance, many health problems are with local deviation of the tissue temperature. Among them are tumors, blood flow disturbance, etc. Very often, even normal function of the human brain is accompanied by the redistribution of the blood flow in the brain, which can be detected using a system of antennas placed on the surface of the human head [5]. Some health problems can be treated by local heating or cooling of human tissues, and it is very important to monitor the absolute value of the temperature inside the human body. Unfortunately, the radio-brightness temperature depends on the inner structure of the source, and, to calculate the physical temperature 3-D distribution, the permittivity and conductivity of the human tissue layers need to be known.

The measured brightness temperature $T_{B_{measured}}$ corresponds to the physical one $T(\mathbf{r})$ at the $i-th$ frequency as [21]:

$$T_{B_{measured,i}} = \frac{1}{1-R_i} \iiint_V W_i(\mathbf{r}) T(\mathbf{r}) dV \tag{11.5}$$

where R_i is the reflection coefficient of the antenna placed at the body surface and operating at the particular frequency f_i, V is the tissue volume from which this applicator collects the radiation, and $W_i(\mathbf{r})$ is the tissue-dependent radiometric weighting function. If this function is known, the temperature $T(\mathbf{r})$ is calculated from a series of measurements at different frequencies. To calculate $W_i(\mathbf{r})$, a direct EM problem should be solved given the conductivity of the tissue. This weighting function is calculated as the normalized power absorption rate when the antenna-applicator is radiating [21]:

$$\frac{W_i(\mathbf{r})}{1-R_i} = \frac{0.5\sigma_i |\mathbf{E}_i(\mathbf{r})|^2}{0.5 \iiint_V \sigma_i |\mathbf{E}_i(\mathbf{r})|^2 dV} \tag{11.6}$$

Taking into account the complexity of this EM problem, the numerical calculations are performed to obtain $\mathbf{E}_i(\mathbf{r})$. Additionally, this formula requires heavy numerical calculations of the EM field radiated by an antenna. Today different algorithms and measurements schemes allow determining the temperature distribution inside the body within the error 0.01-0.05°C.

11.4 Sensitivity of Radiometers

The sensitivity of an ideal Dicke radiometer expressed in the equivalent temperature units is [4]:

$$\Delta T = \frac{T_A + T_N}{2\sqrt{\Delta f \, \Delta\tau}} \tag{11.7}$$

where T_A is the effective temperature at the output of the radiometer antenna, T_N is the radiometer noise, Δf is the radiometer bandwidth, and $\Delta\tau$ is the time constant of the receiver's integrator averaging the random fluctuations. This parameter is the standard deviation of the output signal, and it is improved with the increase of the bandwidth Δf , integration time τ , and decrease of the noise of the radiometer T_N .

It is seen that the Dicke's radiometer allows for registration of these weak signals, but it does not eliminate completely the noise influence. Another interesting fact on the Dicke's radiometers is that the output voltage V_{out} is proportional to the difference of the temperatures of the incoming signals T_A and the hardware T_R :

$$V_{out} = C\left(T_A - T_R\right)G \tag{11.8}$$

where C is a constant and G is the radiometer gain.

One of the effective means to increase the sensitivity is the cooling of radiometers up to the temperatures of liquid nitrogen or even liquid helium, which decreases T_N . In practice, more factors influence the sensitivity. Among them are the slow thermal drift of parameters of radiometers, the residue of thermal fluctuating noise, the digital noise caused by the high-frequency switches used in radiometers, the nonlinearity of receivers, the noise of the A/D convertors, the mismatch of microwave components and antennas, the induced EM noise due to non-ideal grounding of all equipment, etc. It leads to the worsening of the radiometer parameters and fluctuation of signals.

A part of this noise can be reduced by modification of hardware and application of special numerical procedures. To develop them, the characterization of radiometers is needed, and one of them is considered in this book with the aim to study the hidden digital noise in radiometric signals using the state-space reconstruction algorithms [22]-[24].

11.5 Radiometric Studies of Thermal Human-body EM Radiation[1]

11.5.1 Measurements and Statistical Analysis

Human body is a source of thermal radiation, and, depending on the frequency, it comes from the inside of tissue (microwaves) or from the body skin (short millimeter waves). In general, the radiation has rather complicated space-time distribution, and it can be modulated additionally by the heart beating, body tremor, etc. More information on the use of microwaves in medicine is from [5],[17], for instance.

Here the objects under investigation are considered not just as individuals, but generally as the sources radiating the thermal EM waves. These signals are compared with each other and with those originating from a matched $50-\Omega$ load, connected to the radiometer instead of antenna. Other comparisons are made with the measured signals in the laboratory room, which is normally polluted by parasitic radiation of used computer and thermal radiation of walls and human bodies. The goal of this research is to study the statistical characteristics of these signals, to find their differences, and to develop some recommendations for further improvements of radiometers. Our initial results on experimentations of this sort were already considered in [23],[24].

For our measurements, a variant of the Dicke radiometer is used (Fig. 11.1). It is designed for 1.88 GHz with the input frequency bandwidth around 100 MHz. The sensitivity announced by the manufacturer is about of 0.05 K for $\Delta\tau = 0.1$ s .

The device is placed in a thermostat (9) with a fixed temperature to avoid slow thermal drift of its parameters. The signal is collected by an 8-element microstrip antenna with VSWR $\approx$ 1.05-1.1 in the mentioned frequency band. The radiometer consists of many components in block, which are shortly described in the legend to Fig. 11.1.

The radiometer temperature 40°C is maintained by the thermostat 9 during all the measurements. According to the Dicke's radiometer principle [4], the objects, which have effective radio brightness temperatures close to this reference thermostat temperature, produce the voltages near the zero level.

The following measurements were performed. Eleven individuals were chosen for the measurements of their EM radiation. The signals are taken from the chest and head areas of the individuals standing at the 3-m distance from the antenna during one minute. Additionally, a part of the received power was taken by the antenna from the surrounding space due to its wide main lobe and its multiple side lobes.

An analog signal from the radiometer is digitalized by an embedded 12-bit A/D converter with the time step $\Delta t = 0.5$ s , and for each measurement, $N = 121$ samples form a time series $s(t)$. An example of the received signals as a function of time is shown in Fig. 11.2, and it looks as a noise-like curve.

---

[1] Written by S. V. Kapranov and G. A. Kouzaev.

Additionally to these individuals, the noise from a 50-Ohm load and the signal from the environment were registered for the comparison purpose. All these signals have about the same level, and they are in the linear zone of the radiometer, which was confirmed by the manufacturer.

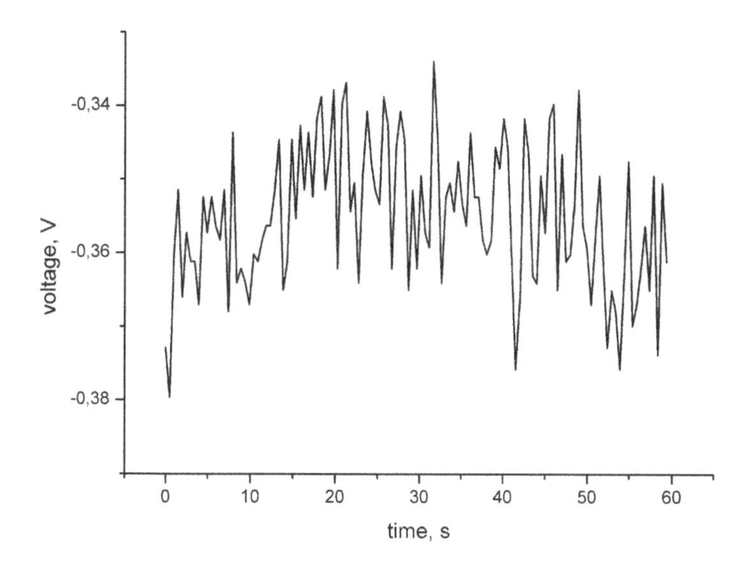

Fig. 11.2 EM signal of the individual marked as U

Spectra of the signals obtained by means of the fast Fourier transformation show that the most part of energy is concentrated close to zero frequency, and it corresponds to the sought signals. An example of such a spectrum is presented in Fig. 11.3 for a signal from the person denoted as U . The higher frequency lobes can correlate to the residue of the thermal noise of the receiver, transients, the EM induced noise due to the non-ideal grounding of the radiometer and computer, and the digital noise of the A/D converter the board of which is embedded inside the computer. However, the fine structure of the spectra is particular for each signal.

Fig. 11.4, in which the signal mean values $\overline{s}$ are plotted versus the standard deviations σ, illustrates these signals further, and it shows that no any artifacts are produced by the radiometer. It is evident that the outer standard 50-Ohm load heated by the warmth drift from the thermostat, generates the lowest voltage, marked "Load", and its temperature is closer to the reference (40°C) according to (11.8). Its standard deviation is smallest, and it corresponds to the thermal nature of this signal.

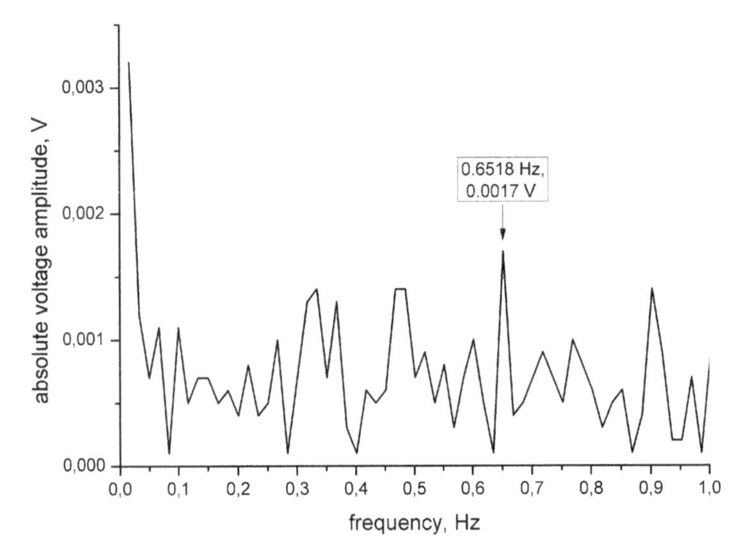

Fig. 11.3 Spectral representation of the signal from the U person obtained by the fast Fourier transformation. The coordinate of the maximum used for calculation of the time delay in the phase space reconstruction is given over the peak point

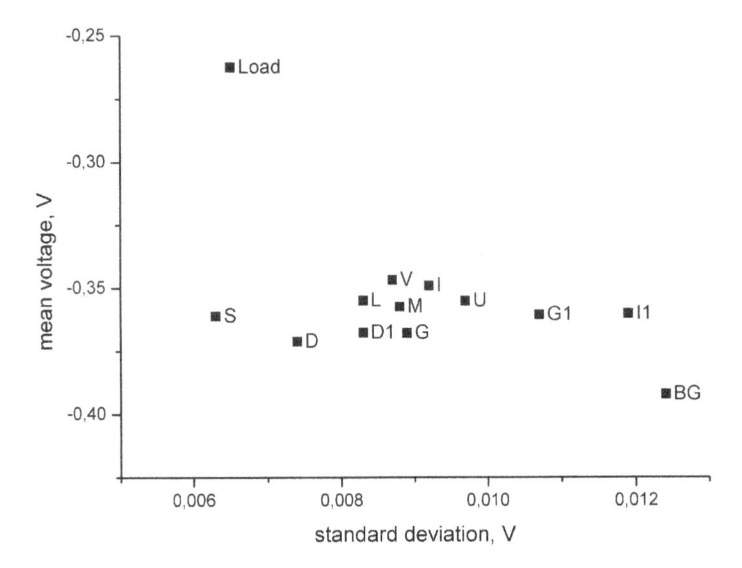

Fig. 11.4 Signal mean voltage $\bar{s}$ versus standard deviation σ for the individuals, matched load, (Load) and background EM noise (BG)

The background noise ("BG") has the largest standard deviation, which is caused by the radiation of computing hardware, according to our opinion. Anyway, the radiation temperature in the laboratory, taking into account the background signal level and (11.8), is rather low at the time of measurements.

It is apparent that the mean voltage $\overline{s}$ and the standard deviation σ of the matched load signal (Load) and background EM noise (BG) are beyond the main group of the individuals' signals marked as D, G, D1, G1, I1, I, L, M, S, U, and V. The difference between the background noise and human-generated signals is observed in the time trends of the autocorrelation functions. The autocorrelation function of a time lag k in a discrete time series $s(t) = s_1(t_1),\ s_2(t_2), \ldots, s_N(t_N)$ with N samples is defined [25] as

$$R(k) = \frac{1}{(N-k)\sigma^2} \sum_{i=1}^{N-k} (s_i - \overline{s})(s_{i+k} - \overline{s}). \tag{11.9}$$

The time coordinate of the first zero in the autocorrelation functions $R(n)$, termed the linear decorrelation time, is shown in Fig. 11.5. It is seen that the decorrelation times of all data sets excepting one are shorter than that of the background noise. This suggests that the human-body radiation has more in common with white noise than the background radiation does. This difference implies dissimilar nature of the human-body radiation and the EM noise patterns.

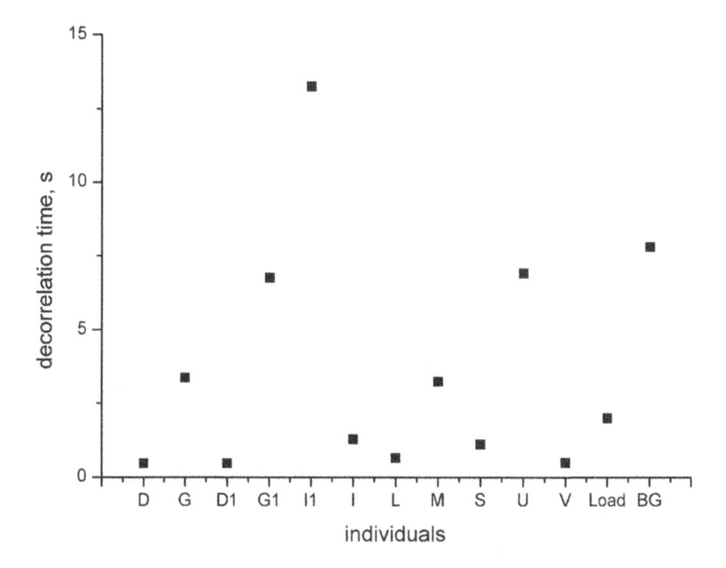

Fig. 11.5 Time coordinates of the first zeroes of the autocorrelation functions obtained for signals of individuals, of the matched load (Load), and of the background noise (BG)

The human-body radiation in the given frequency range is of the Gaussian noise shape, and the corresponding signals should be weakly correlated to each other in the ideal measurement environment. Fig. 11.6 shows the dependence of the statistical linear correlation factors [25] of all individuals, and one can see that they are different from zero. This is explained by the correlated digital noise inside the radiometer and the EM-polluted environment in the lab, although some measures were taken to reduce this EM noise. The full shielding of the used computer was not realized to bring the method to the conditions, which are closer to the reality.

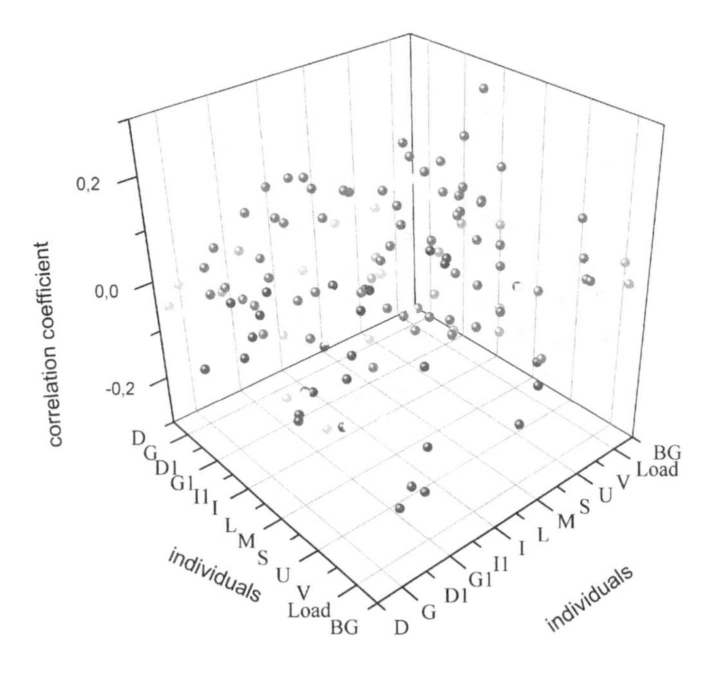

Fig. 11.6 Correlation coefficients of the registered time series of signals of the measured individuals, matched load and background noise

Taking into account the influence of the correlated digital noise, these signals are studied by an algorithm allowing detection of the deterministic noise in the output signals. The noise nature can be discovered by comparing the calculated parameters of the signals from the individuals, the load, and the environment. It will allow improving the radiometer characteristics in the future. This research and techniques are described below.

11.5.2 Technique of Calculation of the Attractor Dimension

A formal look at the registered signals shows their chaotic time-dependence, whereas the elementary statistical analysis has detected some correlated contents with its, presumably, digital nature due to the clocked radiometer, A/D converter, and computer radiation and its induced noise [24]. This periodicity is nonlinearly mixed with the Gaussian thermal noise of hardware and the Gaussian-like human-body signals. It is interesting to study this non-thermal hidden deterministic noise with its subsequent separation from the radiometric signals for improving the overall sensitivity of radiometers. There are several algorithms to find the quasi-periodicity or oscillatory contents in noisy signals, and the Grassberger-Procaccia and sphere-counting methods are chosen for calculation of the fractal dimension in this contribution.

Many natural signals are very complicated in their origin, and they may contain hidden oscillations. For instance, even deterministic processes can be chaotic due to their instability, but traces of this determinism can be detected using the Grassberger-Procaccia algorithm [22]. It allows to consider a one-dimensional time series in an extended phase space, and to find an attracting subset of this space, or an attractor, around which the trajectories are concentrated and where they display an oscillatory character. The low-dimensional attractors were found in a variety of processes, e.g. the weather and climate change [28]-[31], turbulence in flows [32], chaos in unipolar ion injection [33], economic time series [34], atmospheric turbulence [35], and predicting epileptic seizures from the electroencephalography (EEG) data [36],[37].

If the deterministic chaos (stochasticity) exists in a process in question, this procedure also enables calculating the fractal dimension of a stochastic system's attractor. This dimension is an indication of how completely a fractal appears to fill the phase space, as one zooms down to finer and finer scales. The nearest integer above the attractor dimension indicates the minimum number of degrees of freedom which determine the system dynamics. The attractor dimension is a valuable measure since it allows finding the minimal number of variables which fully describe the stochastic dynamics of complex random processes.

There are several types of attractor dimension used to describe the deterministic chaos [26],[38],[39]. One of them is the correlation dimension d_c introduced by Grassberger and Procaccia.

Consider the time series of N samples $s(t)$ registered by the radiometer. It is formed by many spacio-temporal processes $\mathbf{X}(t)$ which are unknown due to the complexity of the source dynamics. The idea of the algorithm is to obtain a minimal set of independent or orthogonal components of this vector $\mathbf{X}(t)$.

The normalized orthogonal components of this vector are the orts of a phase space where an attractor can be found. Then the state-space reconstruction procedure consists in constructing an m-dimensional vector time series $\mathbf{X}(t)$ corresponding to the initial series $s(t)$. In some cases, the vectors, being still chaotic, concentrate with time close to a compact space domain, and this domain of the phase space is called an *attractor*. Finding an attractor will give information on the influence of the deterministic component on the time behavior of the initial signal $s(t)$.

To find these vectors, a technique based on a rigorous proof [41] is used. It replaces the original time series $s(t)$ by its $(m-1)$ time lags τ, which compose the vector time series in an m-dimensional state space [22]-[24],[26]-[31],[33]-[40],[42]-[45]:

$$\begin{aligned} &X_0(t):\ s(t_1),\dots,s(t_n) \\ &X_1(t):\ s(t_1+\tau),\dots,s(t_n+\tau) \\ &\vdots \\ &X_{m-1}(t):\ s\left[t_1+(m-1)\tau\right],\dots,s\left[t_n+(m-1)\tau\right] \end{aligned} \tag{11.10}$$

where $n = N - [(m-1)\tau/\Delta t]$.

This vector series is further processed to yield the attractor dimension estimates. The reconstructed-space dimension m is referred to as the embedding dimension.

The choice of the time delay τ is an important issue since the correct state-space reconstruction requires that the vector components be decorrelated from each other. The simplest solution of this problem is construction of the autocorrelation function $R(t)$ from (11.9) and finding the coordinate of its first zero [26]-[31],[36],[38]. However, the autocorrelation measures the linear dependence among the points, and it might be inappropriate for the nonlinear analysis, as pointed out by Fraser and Swinney [42]. They propose that τ should be found as the local minimum of mutual information among the successive points. Another suggestion was made to take $\tau = T/m$ where T is the period of the dominant peak in the Fourier transformation of the signal [31] (see Fig. 11.3).

Next step in this study is to find an attractor of the time evolution of the vector $\mathbf{X}(t)$ and the attractor dimension. Taking into account the discrete nature of the object in question, the dimension can be non-integer and different from the commonly understood bulk dimension.

Several types of calculation of the attractor dimension by the means of the time-delay state space reconstruction are proposed [26],[38]. Among them one can distinguish two measures, the correlation dimension d_c and the Hausdorff dimension d_H. In general, the latter, also termed as the limit capacity, is the upper limit of the fractal dimension:

$$d_c \le d_H. \tag{11.11}$$

For calculating d_c, the correlation integral $C(N,r)$ introduced by Grassberger and Procaccia is required:

$$C(n,r) = \frac{1}{n(n-1)} \sum_{\substack{k,j=1 \\ k\neq j}}^{n} \theta\left(r - \left\|\mathbf{X}_k - \mathbf{X}_j\right\|\right) \tag{11.12}$$

where $\theta(x)$ is the Heaviside step function: $\theta(x) = 0$ for $x < 0$ and $\theta(x)=1$ for $x \ge 0$. The absolute difference between the vectors is determined in the Euclidean norm as

$$\left\|\mathbf{X}_k - \mathbf{X}_j\right\| = \sqrt{\sum_{i=1}^{m}\left(X_{k+(i-1)\tau/\Delta t} - X_{j+(i-1)\tau/\Delta t}\right)^2}. \tag{11.13}$$

This correlation integral is shown to be a power-law function of the correlation dimension [22],[26]:

$$C(r) \sim r^{d_c} \ (r \to 0). \tag{11.14}$$

Consequently, it can be determined as the slope of the logarithms ratio at very small r, once the fractal is completely embedded in the state space:

$$d_c = \lim_{r\to 0}\frac{\ln C(r)}{\ln r}. \tag{11.15}$$

The Hausdorff dimension dealing with the infinite number of points is usually expressed as the box-counting dimension for a finite set of vectors. In the box-counting procedure, one assesses how many m-dimensional hypercubes with the edge ε are sufficient to cover all points specified by vectors $\mathbf{X}(t)$ in the state space. Sometimes, it is more convenient to count hyperspheres of radius ε instead of the hyperboxes [26].

Let $n(\varepsilon)$ be the minimum number of the hyperspheres of radius ε, which is necessary to cover all mentioned points of the m-dimensional phase space:

$$N(\varepsilon) \sim \varepsilon^{-d_H} \ (\varepsilon \to 0). \tag{11.16}$$

This number of these spheres is then determined as the number of unique integer part combinations of $\mathbf{X}/\varepsilon$ [43]. Then under the stipulation that the fractal is completely embedded in the phase space, the Hausdorff dimension in its box-counting variant is [41],[43]

$$d_H = \lim_{\varepsilon\to 0}\frac{\ln n(\varepsilon)}{\ln(1/\varepsilon)}. \tag{11.17}$$

According to (11.15) or (11.16) the fractal dimension is found as the slopes of straight lines plotted by the least-squares method in the coordinates $\log C(r) - \log(r)$ and $\log n(\varepsilon) - \log(1/\varepsilon)$ in the region of small r and ε, respectively.

An idealized example of finding a single fractal dimension d_c is shown in Fig. 11.7. The slopes of lowermost linear sections of the logarithmic graphs (Fig. 11.7a) are plotted against the embedding dimension m (Fig. 11.7b). The convergence value $d_c = 1.87$ is the correlation dimension of the fractal.

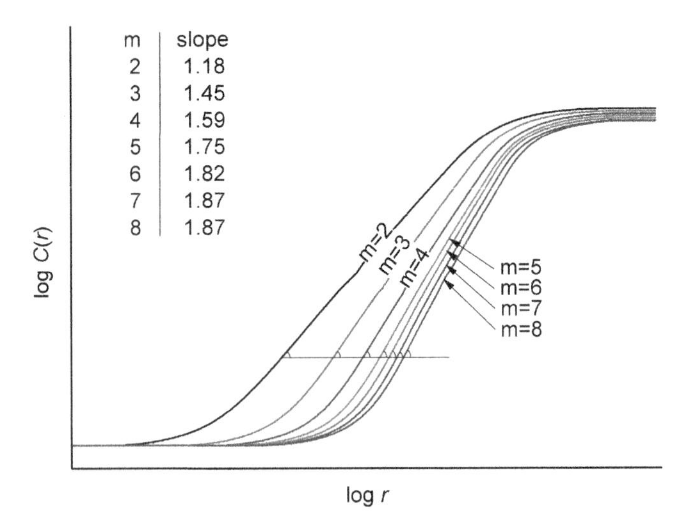

Fig. 11.7a Correlation integrals of a single-fractal system against the size parameter r given in the logarithmic coordinates for the embedding dimension m ranging from 2 to 8. The onset of linearity is marked with a horizontal line relative to which the plot slopes are determined

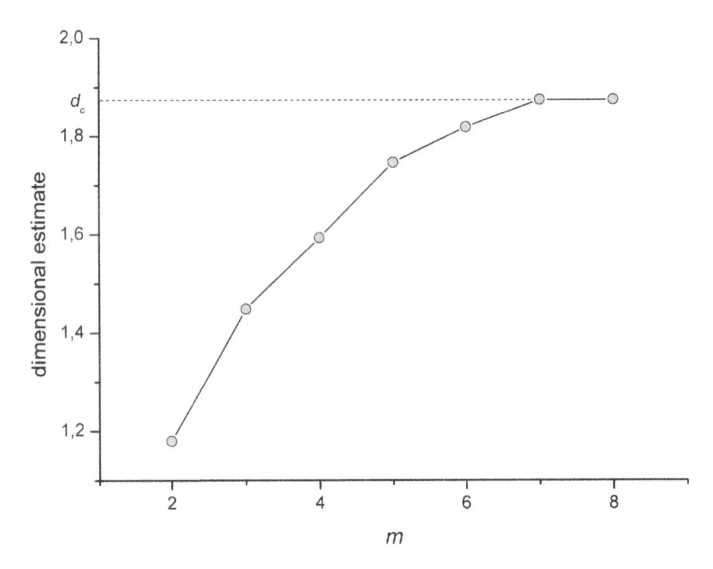

Fig. 11.7b Slopes (dimensional estimates) plotted against the embedding dimension. The convergence, indicated by a dashed line, is attained at d_c=1.87

The procedure of finding the single-fractal Hausdorff dimension is very similar with the exception that the slopes are found for the uppermost linear part of the plots because ε, by definition of the Hausdorff dimension (11.17), must approach zero.

The above-mentioned dimensions can be calculated using a formula given below [22],[26],[39]. For its derivation, an interval of the phase space is divided into n subintervals. Then the probability of the trajectory visiting a particular i-th subinterval is defined as p_i. Taking this into account, the definitions of dimension (11.15) and (11.17) can be combined into a generalized dimension of the order q:

$$d^{(q)} = \frac{1}{q-1} \lim_{\varepsilon \to 0} \frac{\ln \sum_{i=1}^{n} p_i^q}{\ln \varepsilon}. \tag{11.18}$$

It was shown in [26],[39] that the generalized dimension acquires the form of the box-counting dimension (11.17) when $q=0$, and it turns into the correlation dimension (11.15) when $q=2$. The equality $d_c = d_H$ in (11.11) is fulfilled only when trajectory points have uniform distribution over the phase space interval considered $(p_i = 1/n)$, in accordance with (11.18). Noteworthy, the box-counting dimension is a purely geometrical measure, which estimates the total covering of the phase space, and it has no bearing on the probability of finding a trajectory point in a certain range of the phase space. This probability, however, is allowed for in the correlation dimension.

It is possible to find the Hausdorff dimension by means of another procedure, which is somewhat different from counting the minimum number of hyperspheres covering $\mathbf{X}$. Consider a set of vector differences $\mathbf{X}_i - \mathbf{X}_j$. Similar to finding whether a phase space vector belongs to a certain hypersphere [43], the number G_i of unique combinations of integer parts of $(\mathbf{X}_i - \mathbf{X}_j)/\varepsilon$ determines the minimum number of hyperspheres sufficient to cover these differences. If $\mathbf{X}_i$ is a reference vector and $\mathbf{X}_j$ vectors are conditionally clustered in groups smaller than ε, then G_i determines how many ε-sized groups exist with respect to $\mathbf{X}_i$. Hence, the average number of vectors in a hypersphere associated with the $i-th$ vector is n/G_i.

According to the interpretation set forth in [39], the correlation integral ensues from (11.18) with $q=2$:

$$\begin{gathered} C(\varepsilon) = \sum_{i=1}^{n} p_i^2 = \langle p_i \rangle = \frac{1}{n} \sum_{i=1}^{n} p_i \approx \\ \approx \frac{1}{n} \sum_{i=1}^{n} \frac{(\text{average number of vectors in } i\text{th hypersphere})}{n}. \end{gathered} \tag{11.19}$$

Then

$$C(\varepsilon)=\langle p_i\rangle \approx \frac{1}{n}\sum_{i=1}^{n}\frac{n}{G_i}=\left\langle \frac{n}{G_i}\right\rangle \tag{11.20}$$

and

$$1/p_i \approx G_i/n . \tag{11.21}$$

Hence, the box-counting dimension derived from (11.18) with $q=0$ and finite n can be rewritten as

$$\begin{aligned} d_{\mathrm{H}} &= -\lim_{\varepsilon\to 0}\frac{\ln\sum_{i=1}^{n} p_i\,(1/p_i)}{\ln\varepsilon} = -\lim_{\varepsilon\to 0}\frac{\ln\langle G_i/n\rangle}{\ln\varepsilon} = \\ &= -\lim_{\varepsilon\to 0}\frac{\ln\langle G_i\rangle-\ln n}{\ln\varepsilon} \approx -\lim_{\varepsilon\to 0}\frac{\ln\langle G_i\rangle}{\ln\varepsilon}. \end{aligned} \tag{11.22}$$

The box-counting dimension defined according to (11.17) will be referred in this work to as the Method-1 Hausdorff dimension ($d_{\mathrm{H},1}$) and that found from (11.22) will be termed the Method-2 Hausdorff dimension ($d_{\mathrm{H},2}$).

Consider practical applications of the above formulas with the goal of verification of the developed codes by the example of the Lorenz attractor, for which the numerical results are well-known [46]. The Lorenz equations represent the derivatives of dimensionless convective motion functions with respect to dimensionless time t with σ=10, g=28, b=8/3 as in [30]:

$$\begin{aligned} \frac{dx}{dt} &= \sigma(y-x), \\ \frac{dy}{dt} &= x(g-z)-y, \\ \frac{dz}{dt} &= xy-bz. \end{aligned} \tag{11.23}$$

The Lorenz equations are integrated numerically in Matlab using the fourth-order Runge-Kutta algorithm [47] with Δt =0.005 and the number of samples $N=2048$. Using this data, the first zero $\tau=0.79$ of the time-dependent autocorrelation function $R(t)$ is found. It allows calculating the vectors $\mathbf{X}(t)$ according to (11.10), the correlation integral $C(r)$, and the numbers of hyperspheres $n(\varepsilon)$ and $\langle G\rangle(\varepsilon)$.

The range of r, in which the correlation dimension can be determined, is limited by the values of the correlation integral varying theoretically from $2/n^2$ to 1 [26]. However, the correlation integral (11.12) and the number of hypersphere groups G_i related to the $i-th$ hypersphere in (11.22) are meaningful when r or ε lie in the range between the smallest and the largest absolute vector differences. In practice, the ranges of these arguments used for finding the dimensions depend on n and m and the range of data set variation.

The Hausdorff dimension $d_{\mathrm{H},1}$ calculated according to (11.17) has an upper limit for the argument ε, too. This limit is also related to the largest absolute vector difference in the series. However, there is no definable lower limit of ε because the dimensional estimate logarithmically tends to zero, as it follows from (11.17). The lower boundary of the slope linearity shifts towards smaller values of ε as the number of points in the series increases and embedding dimension decreases. An empirical method of finding the lower boundary of ε and r used in determining the dimensions of an attractor is described below in the section on data processing method.

All the three dimensions are calculated for $m=2,3,4,...,10$ and d_{c}, $d_{\mathrm{H},1}$, and $d_{\mathrm{H},2}$ are found as the saturation levels of the dimensional estimate curves (Fig. 11.8). Takens suggested [41] that a functional series dimension converges when

$$2m+1 \geq d. \tag{11.24}$$

As it is seen in Fig. 11.8, the convergence of the correlation dimension plot (squares) and Method-2 box-counting dimension plot (circles) is attained when the condition (11.24) is satisfied. The Method-1 box-counting dimension plot (triangles) comes most sluggishly to saturation, converging only at $m=8$. The slower convergence of the Method-1 and Method-2 box-counting dimension plots is due to the relatively small number of data samples, which is most crucial for finding the Hausdorff dimension [26].

In theory, the slopes of the logarithmic plots will not change after the convergence has been reached. In practice, due to numerical errors and certain arbitrariness of choosing the linearity interval in the log-log plots (Fig. 11.7), the dimensional estimates fluctuate around mean dimension values. Thus, the dimension is assumed to be the average of the dimensional estimates, which are obtained for the embedding dimensions above the convergence. The uncertainty of calculating the dimensions is represented by the standard deviation from the mean value.

It is seen in Fig. 11.8 that the Lorenz attractor in our case converges to the correlation dimensions d_{c}=2.09 with the standard deviations 0.06 and to the Hausdorff dimensions $d_{\mathrm{H},1}=d_{\mathrm{H},2}$=2.05 with the standard deviations 0.02. Thus, in the case of the Lorenz attractor the correlation dimension estimation yields lower precision than that of the Hausdorff dimension. Nearly equal precisions are found in both hypersphere-counting procedures. The found dimensions of the Lorenz attractor are in agreement with the results obtained by other authors [22],[30],[43].

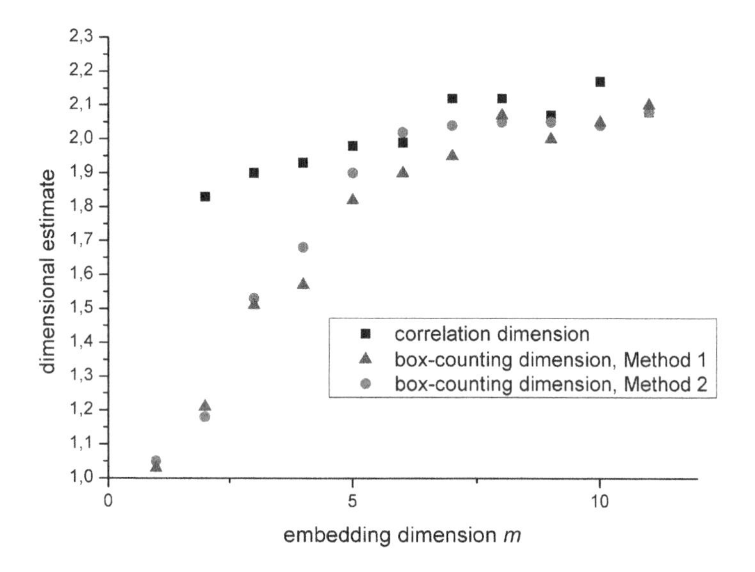

Fig. 11.8 Convergence of the Lorenz attractor as found from both the box-counting and correlation dimension estimates

Although both the Grassberger-Procaccia and the box-counting methods have demonstrated their effectiveness in many applications, there are some problems in calculation of the attractor dimensions of signals polluted by the Gaussian noise which alone produces the slopes equal to the embedding dimension in the region of small r [26],[27],[30], and thus it can affect the correlation dimension [26]. A simplified picture of the Gaussian noise influence is given in Fig. 11.9. It is seen that adding moderate levels of noise to the Lorentz attractor results in appearance of spurious higher correlation dimensions not observed in the absence of the noise. The box-counting dimension appears to be more noise-immune, according to [30], and it can be utilized in some cases for verifying correctness of correlation dimensions found for noisy systems.

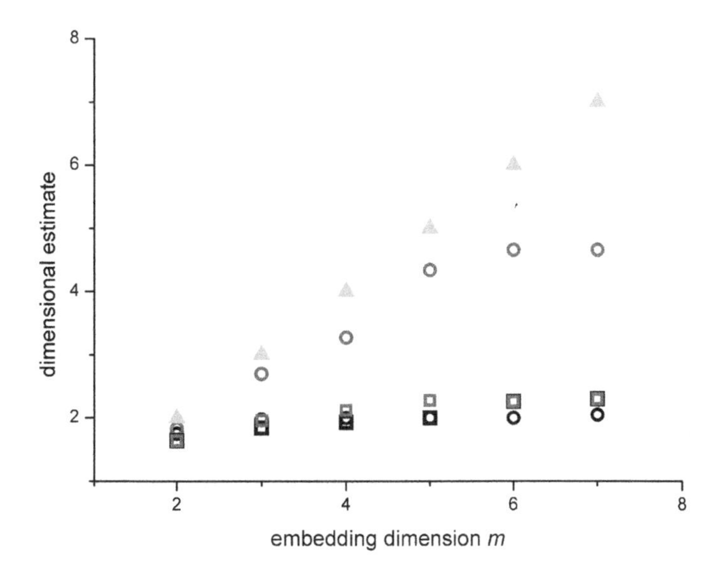

Fig. 11.9 Estimates of the correlation dimension (red symbols) and box-counting dimension (black symbols) of the Lorenz attractor with the level of Gaussian noise 1% (squares) and 10% (circles) of the maximum value of the Lorentz attractor coordinate. Cyan triangles represent the dimensional estimates of pure Gaussian noise. Adapted from [30]

As it has been already mentioned, the number of data points is critical for measuring the fractal dimension. According to different approaches, a 5% error in the dimension estimation probability is attained for a data set with a minimum number of points 5^d [26]. In [31], this error is obtained for $10^{2+0.4d}$ points, and even the number of points 42^d is suggested in [44]. The methods which can essentially reduce the minimum number of data points were reported, e.g. in [45]. Takens' theory of best estimator results in the 5% error level for as few as 1.25^d data points [26]. Theiler's estimate indicates that an acceptable accuracy would be attained for the attractor dimensions not exceeding 3 in our data sets [26].

Although the ampler data sets are highly desirable in the measurements, we show below that even 121 points are sufficient to perform the task of calculation of dimension estimates in some cases. To verify these results, both the correlation dimension and the Method-2 Hausdorff dimension have been calculated. The Method-1 of the hypersphere-counting appears to be unworkable in this case because of slow convergence and insufficient number of samples.

Another problem arises in calculation of dimensions in multi-attractor processes. When a multiple-attractor pattern exists in the time series, it is helpful to reveal the full spectrum of local slopes over the whole range of r or ε, although a failure to detect some dimensions is not excluded in this case, either. Below, some algorithmic ideas are considered to calculate the parameters of multi-attractor time series, which is typical for the signals of complicated nature as the radiometric ones polluted by external and internal noise of digital and continuous nature.

11.5.3 Calculation of Parameters of Multi-attractor Signals

The above procedures describe finding of a single attractor dimension from the time-delay reconstruction of data series' phase space. However, the time-dependent signals radiated by human body can contain more than one fractal set. The resulting signal will include traces of these fractals on different size scales. It will be referred to as the *multi-attractor* signal. Properly speaking, this signal contains only one attractor with the corresponding dimensions defined according to (11.15), (11.17), (11.18), but it is considered as if it were composed of several attractors isolated in delimited size ranges of the phase space. To put it another way, multi-attractors can have more than one dimension of a fixed order in different areas of the size parameter variation. The introduced concept of a multi-attractor should not be confused with the generally-accepted term *multiple fractal*, or *multifractal*. Multiple fractals emerge when probability of a trajectory visiting different areas of the phase space is not uniform. The existence of multifractals is well documented in dynamics of a number of mathematical and physical systems (see, for example, [26],[48]-[50] and references therein, [51]-[54])

The singularity scaling formalism introduced in [49] yields analytical spectra of generalized dimensions (11.18) versus q varying from $-\infty$ to $+\infty$ for several fractal sets. This method, however, is inapplicable in gaining dimension spectra of signals with originally unknown probability distribution in reconstructed phase space, as in our case. The procedure of calculating these probabilities utilized in [48],[51] solves this problem but complicates the dimension analysis to a considerable extent.

We propose a facile multi-attractor approach to the estimation of the fixed-order dimension patterns of the radiometric signals. As it is shown below, these spectra manifest themselves as imprecise, but they can be easily obtained and used for distinguishing the signals of individuals from those of the matched load and background noise.

To understand the origination of several local dimension measures in the correlation integral of a radiated signal, consider two cases of overlapping and non-overlapping fractal sets with different dimensions. If one fractal of the correlation dimension d_1 overlaps with another one of the correlation dimension d_2 starting from r_0, then their contribution C_o to the correlation integral at distance $r > r_0$ is

$$C_o = c_1 r^{d_1} + c_2\left(r^{d_2} - r_0^{d_2}\right) = c_2 r^{d_2}\left[1 + (c_1/c_2)\,r^{d_1 - d_2} - (r_0/r)^{d_2}\right] \tag{11.25}$$

and

$$\ln C_o = \ln c_2 + d_2 \ln r + \ln\left[1 + (c_1/c_2)\,r^{d_1 - d_2} - (r_0/r)^{d_2}\right] \tag{11.26}$$

where c_1 and c_2 are the proportionality factors.

The last term in the right-hand part of (11.26) is negligible if $1 < d_1 < d_2$ and $c_1 \sim c_2$. Then $\ln C_o$ has the slope d_2 for $r \gg r_0$ and the slope d_1 at distances below r_0. This case is represented by the correlation integral curves of Fig. 11.10a with the sharp bends between their linear sections corresponding to r_0-values. If, on the other hand, $1 < d_2 < d_1$ (11.26) can be represented as

$$\ln C_o = \ln c_1 + d_1 \ln r + \ln\left[1 + (c_2/c_1)\left(r^{d_2} - r_0^{d_2}\right)/r^{d_1}\right] \qquad (11.27)$$

in which the last term can again be neglected. Here, the correlation integral has slope d_1 over the whole range of distances except a narrow area near r_0.

In the case of two non-overlapping fractals with the same parameters as above and the boundary lying at r_0, the correlation integral is

$$C_{no} = c_1 r_0^{d_1} + c_2\left(r^{d_2} - r_0^{d_2}\right) = c_2 r^{d_2}\left[1 + (c_1/c_2)\left(r_0^{d_1}/r^{d_2}\right) - (r_0/r)^{d_2}\right] \qquad (11.28)$$

and

$$\ln C_{no} = \ln c_2 + d_2 \ln r + \ln\left[1 + (c_1/c_2)\left(r_0^{d_1}/r^{d_2}\right) - (r_0/r)^{d_2}\right] \qquad (11.29)$$

in which the last term is of the order of zero provided $c_1 \sim c_2$. Then $\ln C_{no}$ has slope d_2 at distances above r_0 and slope d_1 at distances below r_0.

Thus, finding the local slopes at the correlation integral curve against r in the log-log coordinates allows identifying the individual dimensions of a multi-attractor data set when the condition of the size parameter tending to zero is relaxed.

An ideal picture of the correlation integral plots describing two overlapping fractals are shown in Fig. 11.10a. In contrast to Fig. 11.7a, two distinct local slopes in these plots are discernable. As explicitly shown in Fig. 11.10b, the lower fractal dimension converges at $d_{1,c} = 0.97$ and the higher dimension converges at $d_{2,c} = 1.88$.

The similar reasoning of finding spectrum of fractal dimensions from the local slopes also applies to the Hausdorff dimension curves calculated using (11.22). Idealized bi-attractor box-counting plots resulting from the same hypothetic data set are given in Fig. 11.11a, and the convergence of the slopes is shown in Fig. 11.11b.

To obtain the spectrum of attractor dimensions of the human-radiated signals, analysis resting upon extracting local slopes of $C(r)$ and $\langle G\rangle(\varepsilon)$ in logarithmic coordinates in the whole region of $r-$ and $\varepsilon-$variation has been performed. It is discussed in the next Section.

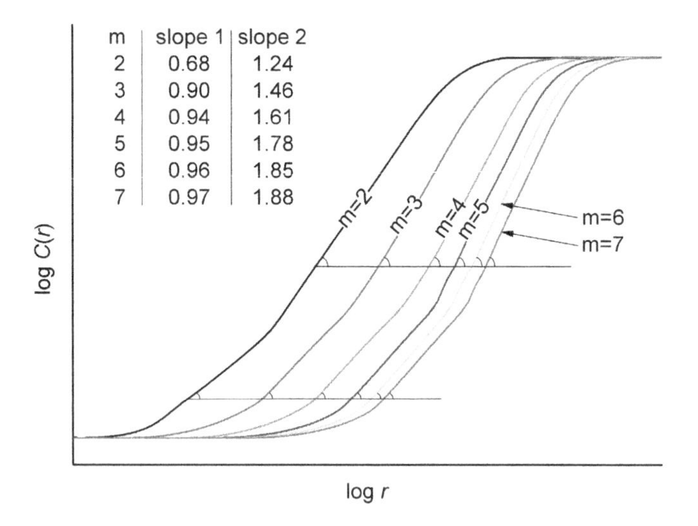

Fig. 11.10a Correlation integrals of an idealized bifractal system against the size parameter r in logarithmic coordinates for the embedding dimension m ranging from 2 to 7. The slope estimation sites are marked with the horizontal lines

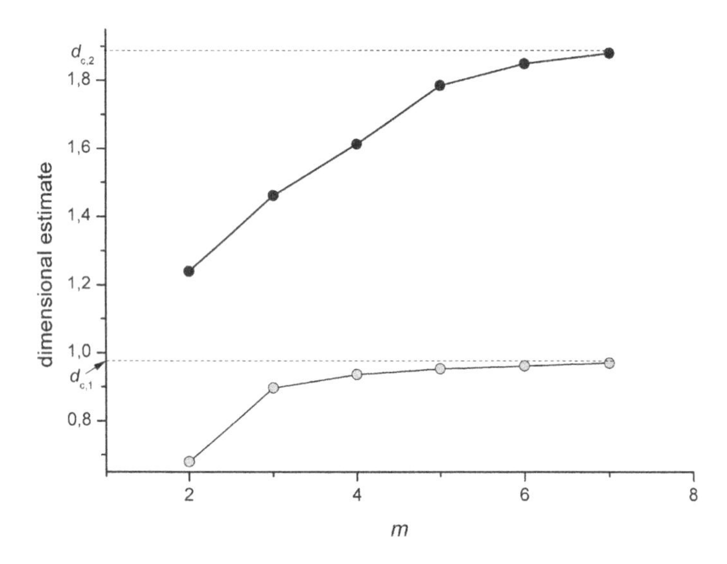

Fig. 11.10b Slopes (dimensional estimates) plotted against the embedding dimension. The convergence, indicated by the dashed lines, is attained at $d_{1,c}$=0.97 and $d_{2,c}$=1.88

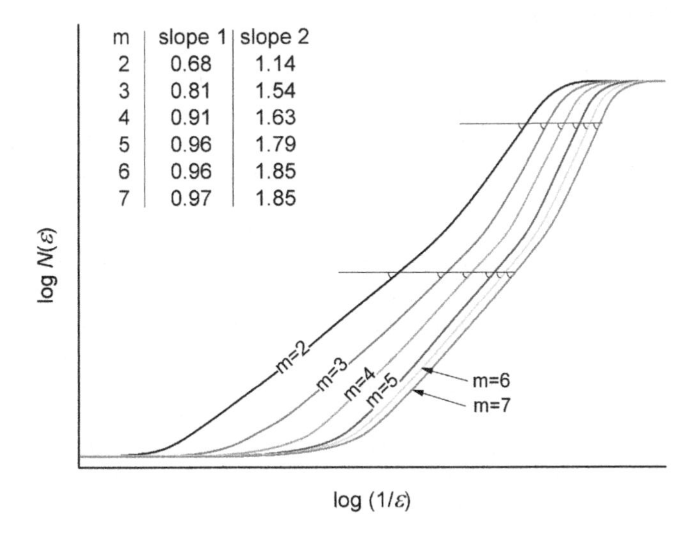

Fig. 11.11a Box-counting plots of the same idealized bifractal system as in Fig. 11.10 against $1/\varepsilon$ in logarithmic coordinates for the embedding dimension m ranging from 2 to 7. The slope estimation sites are marked with horizontal lines

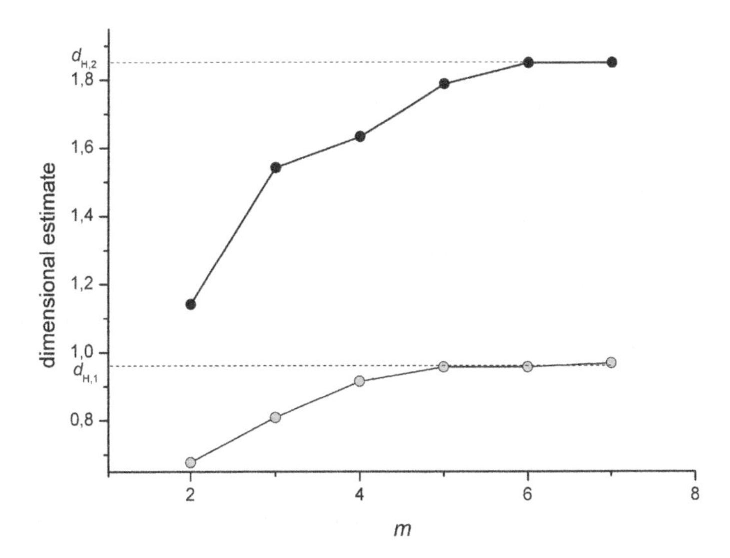

Fig. 11.11b Slopes (dimensional estimates) plotted against the embedding dimension. The convergence, indicated by the dashed lines, is attained at $d_{1,c}$=0.97 and $d_{2,c}$=1.89

11.5.4 Data Processing

The Grassberger-Procaccia algorithm was first applied to microwave human-body radiometric signals in [23],[24]. It was found that the shape of the curves of fractal dimension was different from linear, and this shape depends on the measured object. A preliminary conclusion on some deterministic contents in the registered radiometric signals was made there. In this Chapter, the results of application of more advanced signal processing algorithms are considered, and more knowledge on the nature of the deterministic contents is obtained. The integrations of the emitted microwave signals were recorded and further processed using the time-delay state space reconstruction algorithms. The integration data were registered every 0.495055 s. The data accumulation was restricted to 121 readings for each individual, so the total measurement time was 60.39671 s for each person. The data fewness and diversity of fractal structures concealed in the body-emitted signal complicate the analysis and interpretation of the reconstructed time series largely. Therefore, one cannot distinguish between the human and device/background attractors if their dimensions coincide within the measurement error.

The time of the first zero of the autocorrelation function was adopted as the time delay in processing all data sets, excepting four cases (persons G1, I1, U, and background noise). The decorrelation time in these cases was too great for the state-space reconstruction at high embedding dimensions so that the higher-order dimensions could be overlooked. The time delay for these data sets was calculated from the period of the first peak in the Fourier spectrum of the signals (an example of spectrum is shown in Fig. 11.3).

It is seen in Figs. 11.10, and 11.11 that the linear sections corresponding to certain dimensional estimates are separated from each other by bends, i.e. the sections with the considerable curvature. This fact is used in this work in the slope-finding procedure.

The following algorithm for finding local slopes of the correlation integral and number of spheres in the logarithmic coordinates for a constant embedding dimension has been used. The lower limit of both of the correlation integral and hypersphere-counting slopes is determined as the boundary where the curvature of the line connecting the adjacent points falls to the near-zero values. The curvature of a function $y(x)$ is determined as

$$\kappa = \frac{d^2y/dx^2}{\left[1+(dy/dx)^2\right]^{3/2}}. \tag{11.30}$$

The assumption that $y_1(x_1) \equiv \log C(r)$ and $y_2(x_2) \equiv \log \langle G \rangle (\varepsilon)$ are continuous functions of $x_1 \equiv \log r$ and $x_2 \equiv \log(1/\varepsilon)$, respectively, allows applying this formula to the dimension plots. By analogy drawn to the discrete data series, one can replace the first and second derivatives in (11.30) by the finite increment ratios, i.e. dy_i/dx_i by $\Delta y_i/\Delta x_i$ and d^2y_i/dx_i^2 by $\Delta(\Delta y_i/\Delta x_i)/\Delta x_i$ where i=1, 2.

The area of the linear sections is bounded on the side of the small values of $\log r$ and the great values of $\log(1/\varepsilon)$ by the finite resolution of the phase space portraits. At the opposite sides of the arguments, it is restricted by the finite size of the fractals in the phase space. For demarcating the linearity region, the minimum value of slopes equal to unity is fixed. The first points from both sides fitted by a linear equation satisfying this criterion will designate the limits of the working area. Other points are discarded.

In fact, the points in the working areas never lie on smooth curves like those in Figs. 11.10, and 11.11. The twisting pattern of lines connecting the points in real profiles can be accounted for not only by traces of multiple dimensions, but also by the presence of errors, which are discussed in [26]. To detect the individual linearity areas, one should find the intervals of the arguments, in which the near-zero curvatures are continuous. Their deviations from zero should be described by a certain distribution function. Since these deviations are assumed to be random, the linear area curvatures, obtained using κ; are expected to obey the normal distribution with a maximum close to zero.

Once the parameters of this Gaussian distribution function are found, one can determine the full range of curvatures satisfying the linearity conditions by means of the three-sigma rule. However, the curvature range, in which fitting by the Gaussian function is valid, is unknown. Furthermore, the normal distribution approximation is recognized to be not always justified in the statistical investigations. Hence, the statistical samplings require special tests for the normality. The Gaussian fittings are performed over various curvature extension ranges (from ±0.1 to ±15 with step 0.1). In each step, the 95%-probable validity of the normal distribution utilization is tested using the Shapiro-Wilk [55]-[57] and the D'Agostino K-squared tests [55],[58] realized in Matlab codes [59],[60]. Only those distributions are used for finding the zero-curvature scattering, which satisfy both tests for the normality.

After each step of the validated zero-curvature range determination, the slopes of linear sections in the working areas of the log-log plots are found and accumulated. A section is supposed to be linear if the curvatures in this section are within the found range of the zero-curvature scattering. The points, corresponding to the curvatures outside this range, separate the linear sections from each other.

The data sets are thus processed with the embedding dimensions varying from 2 to 21 for all the individual signals except the two ones (G and M), which display moderately long decorrelation times (over 3 s). Consequently, the phase space of the signals of these two persons could be reconstructed by means of the time delays only up to the embedding dimension 17. As already mentioned, the decorrelation times for three individuals (G1, I1, U) and background noise prove to be so long that the fractions of the main period of oscillation are employed instead in the state space reconstruction.

Due to the restrictions imposed by the finite time delays, the phase spaces can be embedded in no more than 21 dimensions. Furthermore, the saturation of the attractor dimensions sets in approximately in the last third of this embedding dimension range. It is the interval in which the observed dimensional estimates will be statistically processed. The slope values have been accumulated in the embedding dimension range from 14 to 17 for G's and M's data sets and from 15 to 21 for the other signal series. Some examples of the dimension convergence plots are available in [61].

These slopes in arrays are not equally distributed over the dimension scale, but rather they are cumulated in groups (clusters). The clusters are separated from each other by large gaps. Because there is an uncertainty in finding the fractal dimensions, each of the dimensions is assumed to lie within a cluster. For the sake of simplicity, the uncertainty of locating the dimension is assumed to be the standard deviation from the mean in the group of slopes. Thus, the task of finding the dimension spectrum is reduced to the determination of the optimum number of the slope clusters, the mean values, and the standard deviations in each of them. For this purpose, the use of a special branch of mathematical statistics, namely, the cluster analysis and its partitioning method is used here.

The standard algorithm of this method, called the k-means algorithm, attempts to minimize the within-cluster sum of squares:

$$\sum_{i=1}^{n_c} \sum_{x_j \in S_i} \left\| x_j - \mu_i \right\|^2 \tag{11.31}$$

where **S** is the set of the mutually exclusive spherical clusters $\left\{S_1, S_2, \ldots, S_{n_c}\right\}$, μ_i is the mean, or the centre of the points x_j in the S_i cluster.

The algorithm tries the random starting points and eventually converges at local minima of the sum of the squared Euclidean distances. There is no guarantee, however, that the found minimum is global. Another drawback of the algorithm is that one has to assign the number of clusters in advance.

The global minimum of (11.31) can be secured by repeatedly applying the k-means algorithm to the arrays of attractor dimensions when a number of slope clusters has been predefined. A minimum of the residual sum of squares, found among the all trials, is taken as the global minimum.

In this work, the k-means algorithm, as a built-in function of Matlab 7, is applied 300 times to each of the correlation dimension and the Hausdorff dimension arrays. After the minimum of the squared distances has been found, the coordinates of the centers in this trial represent the mean attractor dimension.

Belonging of each slope to a certain cluster is stored and used for finding the standard deviations for each dimension in the spectrum after the optimum number of clusters has been found.

Finding the optimum number of slope clusters is less straightforward. It is shown in [61] for the groups of slopes of the same hierarchical level that the global minimum of the sum of distances varies exponentially with the assumed number of clusters. Then the optimum number of clusters is found as the first breakpoint, in which the hierarchical level of the clustering dramatically changes, i.e. this is the intersection point of two best linear fits of logarithms of the minimal sums plotted against the assumed number of clusters.

Summing up, the proposed method provides a reliable estimate of the number of attractors and their fractal dimensions. Like any time-delay phase space reconstruction procedure, it has certain restrictions [26]. In particular, the method can yield the distorted or spurious dimensions if there is

- Insufficient amount of initial data
- Lacunarity in the phase space filling close to the fractal
- High level of the Gaussian noise in measured signals
- Correlation of the reconstructed vectors.

However, the influence of these errors is smoothed away in this method owing to

- Accumulation of slopes
- Detection of curvature scattering limits
- Statistical averaging of the slopes within a cluster.

In general, the above-mentioned errors can be further reduced by using the minimum mutual information method in the determination of the time delays [42] and utilizing more extensive data time series.

11.5.5 Results and Discussion

The data processing method described in Section 11.5.4 allows calculating the multi-attractor dimension spectra from the slopes clustering. Each dimension in the spectrum lies within a certain interval, specified by the standard deviation found for an individual group of slopes.

The mean values and spans of all the revealed clusters are shown in Fig. 11.12. The number of clusters for each kind of the dimensions varies from 3 to 10. Because of the method's restrictions shown above, the slope spans like those in Fig. 11.12 may not quite accurately reproduce the location of specific attractors in the total pattern. However, the method's advantages minimize the influence of these errors.

The analysis of the results plotted in Fig. 11.12 shows that 34 of 81 intervals in the correlation dimensions do not intersect with the corresponding Hausdorff dimension intervals. On the other hand, 37 of 77 intervals in the Hausdorff dimension chart have no correspondence in the correlation dimension results. This means that the data set processing produces, respectively, 42% and 48% results not confirmed by the alternative estimation. In the case of the Hausdorff dimensions, most of these intervals fall on the upper part of the chart. Nonetheless, these unconfirmed dimensions cannot be rejected either because their area of existence can be very narrow ($r \sim r_0$) and they can be masked with overlapping dimensions (see (11.25)), or because of the very prolate generalized dimension spectra [26],[48],[49]. The joint pattern from the superposition of the correlation dimensions and the Hausdorff dimensions is demonstrated in Fig. 11.13.

Poorly resolved low-dimensional attractors in the signals of individuals may originate from the receiver's own attractors and the induced EM noise from other equipment (Load and BG in Figs. 11.12 and 11.13). Thus, the human attractors with dimensions up to about 5.5, if they exist, cannot be discerned by this method. Some of the high-dimensional attractors (5.86–8.36, 9.64–11.56, and 14.80–15.71) can be ascribed to the background noise and equipment, too.

There are some empty spaces (8.36–9.64 and 11.56–14.80) in the combined dimension spectra of the background noise and matched load. One can notice that some of the dimension spans of certain individuals fall into these "windows".

In particular, the lower space contains the dimension intervals of the persons S and V, and, in the higher "window", one can find the dimension intervals of the persons S, D1, G1, and U. Although all these spans, except for the person U, mutually intersect, and, therefore, cannot be used as the personal "fingerprints", one can implement this peculiarity for separation of human-generated signals from those of the hardware and background EM noise.

Because the range of embedding dimensions covered in this work does not exceed 21, it is worthwhile to mention that the fractal dimensions above 10 may not attain the saturation, according to the Takens suggestion (11.24). Thus, the higher-order attractor dimensions in the signals are still disputable, and further research is required to determine them more accurately and to find the origins of these attractors. One can speculate that these attractors correlate to human-body tremor, heart beating, or even correspond to similar attractors registered in the normal state of human-brain activity.

More general point of view at the arising of fractals and attractors in signals or time-depending processes involves the projective approach. In fact, antenna collects the EM power from a larger area than the human-body torso, and these spatio-temporal signals are converted into the time-dependent ones processed by noisy hardware. This projection may lead to the fractal-like time dependence [27],[62].

Unfortunately, the available data does not allow any final conclusion, and more experimentation is needed. In particular, it should include the study of human-body radiation in more ideal conditions to avoid the intensive background EM noise, and longer expositions of individuals to obtain larger amount of time-dependent data are required.

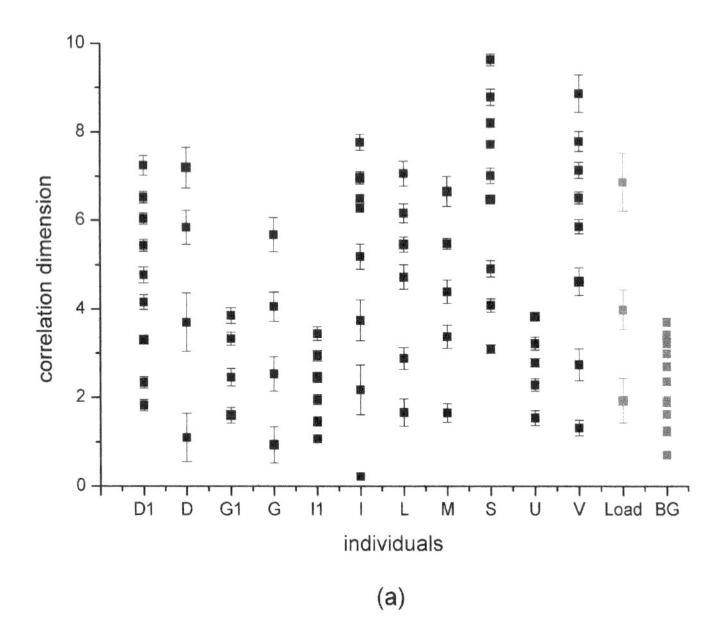

Fig. 11.12a Correlation dimensions of found attractors in microwave signals of several individuals, matched load, and background noise

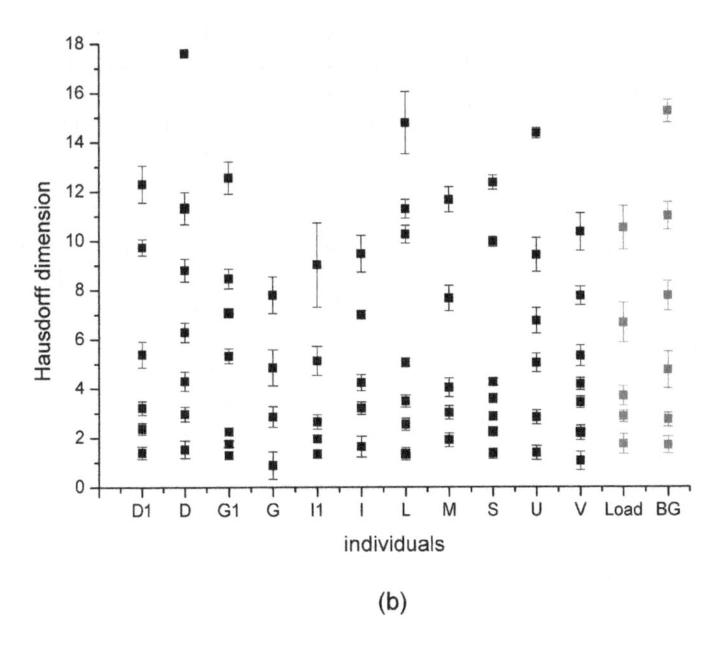

Fig. 11.12b Hausdorff dimensions of found attractors in microwave signals of several individuals, matched load, and background noise

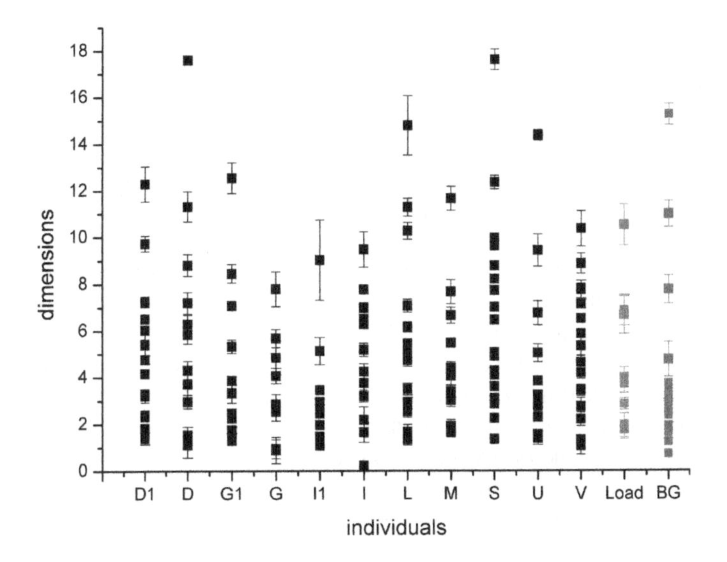

Fig. 11.13 Combination of spectra of both dimensions from Figs 11.12a and 11.12b

Here the time- and frequency-domain characterization of a 1.88 GHz microwave radiometer has been performed. To obtain the reliable data, different sources of thermal microwave radiation have been used. Among them are the matched load, background noise in the laboratory, and thermal radiation from eleven individuals. Comparing the time- and frequency-dependent data, some traces of deterministic noise caused by non-ideal design have been found in the time series. To provide more details, the obtained data have been studied by several state-space reconstruction algorithms. They allow detection of the deterministic component in the noise-like signals. An original procedure of finding the Hausdorff dimension has been utilized in this work. In contrast to the well-known box-counting method, it is based on averaging the number of the unique integer parts of ratios between vector differences and the size parameter.

Multi-attractors have been found in the signals, and their dimensions have been calculated and compared with each other. It has been allowed to assign the majority of attractors to the hardware and EM noise in the laboratory or to associate them with the calculation errors. Some higher-order attractors detected only in the signals of the individuals have been unidentified and further experimentation has been suggested in the future.

The obtained results and developed techniques on time- and frequency-domain characterization of radiometers are of special interest for increasing the sensitivity and speed of the radiometric hardware used for observing large scenes where the exposition time must be as small as possible.

11.6 Radiometric Imagers

The studied radiometers collect the signals from an area of an observed object and they provide a pixel signal corresponding to this square [3],[4],[63]. Mechanically moved antennas can scan the surrounding space, and a radio image is created from the registered multiple pixels. The antenna or radiometer moving velocity must be adapted to the time constant of the used hardware. To improve this parameter, cooling of radiometers is recommended which increases their sensitivity.

The radiometers with their individual antennas can be placed in a line with the parallelization of the image registration, and moving of this linear array allows synthesizing a radio image along the direction of the movement. The obtained array of pixels is processed, stored in the computer memory, and it can be transformed into an optical image for visual processing [64]. These systems are known in the air-born and satellite applications where the scanning hardware is moved on the board of an airplane or a satellite, and they allow observing large areas of the Earth surface to define the effective temperature of the ground, soil moisture, salinity of water, ice, snow, cloudiness, atmosphere vapor, etc. [1]-[4],[63],[65].

An example [12],[13] of such systems working at 8-mm wavelength is shown in Fig. 11.14.

Fig. 11.14 Linear array radiometric system consisting of 10 millimeter-wave radiometers. Photo was provided by V.I. Krivoruchko [13]

It consists of 10 radiometers equipped with horn antennas separated by 14.3-mm distance from each other with the isolation of them around 50 dB. The rays are focused at the antenna line by a PTFE dielectric lens of the 200-mm size placed at the 3-m distance from this line. The sensitivity of this 8-mm radiometric system is around 0.07 K , and the integration time constant is close to 1 s. Each channel is adjustable, and this system is calibrated by a computer and a calibration setup or gas-discharge noise generators. To avoid the thermal drift, this hardware is placed on a massive metallic plate heated up to 48°C and all sensitive hardware is thermally isolated from the environment. The received signals are digitized by 11-bit AC/DC converters, memorized by a computer, and this data can be processed further.

Additionally to linear array scanning vision systems, many results are known for the squared arrays composed of radiometers. They have miniaturized individual antennas receiving signals directly or being placed in a focus plane of a parabolic or an ellipsoidal antenna. The quality of these EM images can be improved using different means, including the neural networks for signal processing [64]. Today these radiometric systems can be manufactured using the discrete and monolithic technologies. In the first case, each individual radiometer consists of several integrations, and it allows its tuning by proper choice of components and integrated circuits. The main disadvantage is the increased cost and technologically caused variation of electric parameters. In fact, each channel needs individual calibration and tuning before the measurements.

Monolithic radio-imagers can integrate thousands of radiometers if an antenna system is not large [66]. Additionally, each radiometer has similar electric characteristics and it makes easier to tune such systems as a whole.

Tuning is realized using the surfaces with known radio-temperature, for instance. The controlling computer should regulate the amplification coefficients of channels to obtain a smooth distribution of received pixels. Further, depending on the amplification characteristics, the calibration of radiometers can be performed to define the radio temperature in Kelvin, if necessary.

To prevent thermal drift of amplifiers, the system should be thermostated, and cooling down to the nitrogen temperatures is recommended to increase the sensitivity of used radiometers, although it makes the exploitation cost higher.

Additional problem arising in exploitation of large number radiometric systems is the parasitic coupling of individual antennas. The each-antenna's main and parasitic lobes can be overlapped with those of a neighboring sensing element, and the pixels are correlated with each other. To prevent it, the antennas should be placed at an increased distance from each other, or special algorithms of improved directivity should be used for processing these signals [67].

The square-array radio imagers can be used for real-time observation of large areas [1]-[3], human presence and hidden weaponry detection [10],[11],[68], medicine [5],[6], landmine detection [11], etc. Depending on the needs, the frequency of the radio imagers can be of several hundreds Gigahertz. They can perform spectral analysis [69], produce the stereoscopic or near-field 3-D images [70]-[72], combine passive and scattered radiation pictures [73], etc.

As an example of these combined imagers, consider one of the radiometric systems developed by us in the end of the 90s of the last Century [13]-[15]. The imager consists of 16 radiometers arranged on a square plane (Fig. 11.15).

Fig. 11.15 16-channel millimeter-wave radio imager

The millimeter-wave rays are focused by an elliptical antenna designed to have another object focus at the 5-meter distance. Being equipped with a parabolic antenna, this radiometric system can visualize the objects placed at a larger distance.

Additionally to the radiometric regime, it works as a scatterometer, and a radiating horn antenna can illuminate the scene. In the near field, the phase can be registered, and a holographic regime is realized by this hardware. Although, the radiometric and scattering regimes could be realized in the stereoscopic variant by this hardware, the second necessary complex of this system was not manufactured at those times due to the tight budget. This manufactured hardware was created for experimentations in the security area to register the hidden weapons and dangerous stuff carried on the human bodies.

Detailed description of this imager is in [12],[13], and the results of some initial experimentations are in [14],[15]. Each channel of this radio imager has sensitivity around 0.05 K at the room temperature, and the integration time constant is 1.2 s in the radiometric regime. The frequency band of each channel is 3 GHz, and switching frequency of these Dicke-type radiometers is 9.2 KHz.

Before the measurements, the electronically controlled amplifiers should be tuned using the calibration noise generators or a plate of a known temperature, e.g. of liquid nitrogen. The electronics should be placed in a thermostat or be cooled by liquid nitrogen. In the last case, the calculated sensitivity is around 0.016 K.

The radiometer provides only 16 pixels of a 2-D image (1m x 0.5 m), and the developed software tool allows extending the number of points up to 64. The applied spline method smoothes the images, and the first results of experimentations are shown in Figs. 11.16-11.18.

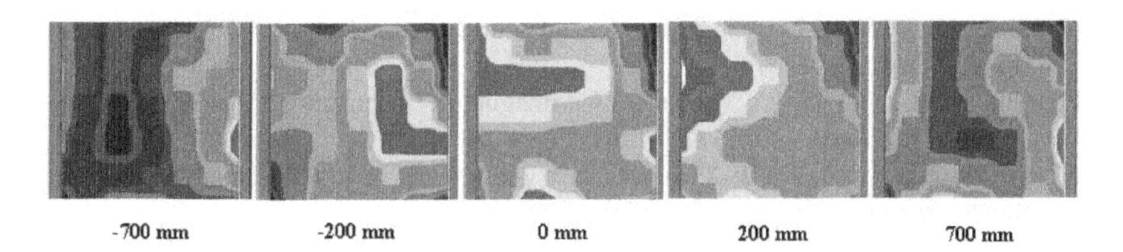

Fig. 11.16 Series of images of the chest of a human moving in front of the imager at the distance 3 meters. The brightest thermal spot is observed when the object is on the axis of the system. (Author's archive)

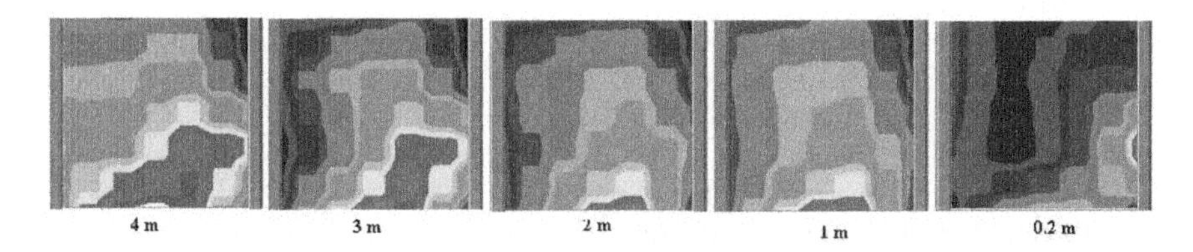

Fig. 11.17 Series of images of the chest of a human moving along the axis of the imager from 4 m to 0.2 m. The brightest image is for the 4-m distance due the proximity to the focus (5 m) of the elliptic antenna (Author's archive)

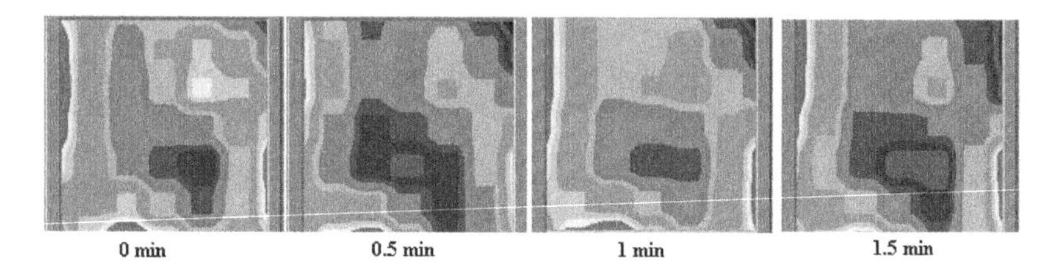

Fig. 11.18 Measured EM radiation level in the laboratory (Author's archive)

The first experimentations confirmed the ability of this system to work in all the three mentioned regimes, although further research was terminated due to financial circumstances at those times. Overall research performed in 1994-1998 allowed us to gain an interesting and intensive experience in the development of high-sensitive hardware and software on the de-noising of the radiometer signals and imaging of millimeter-wave radiating objects.

References

[1] Ulaby, F.T., Moore, R.K., Fung, A.K.: Microwave Remote Sensing. Active and Passive, pp. 1–3. Addison-Wesley Publishing Comp., Reading (1981)
[2] Sharkov, E.A.: Passive Microwave Remote Sensing of the Earth. Physical Foundations. Physical Foundations. Springer, Berlin (2003)
[3] Armand, N.A., Polyakov, V.M.: Radiowave Propagation and Remote Sensing of the Environment. CRC, Boca Raton (2004)
[4] Skou, N., Le Vine, D.: Microwave Radiometer Systems: Design and Analysis. Artech House, Norwood (2006)
[5] Godik, E.E., Gulyaev, Y.V.: Functional imaging of the human body. IEEE Eng. Medicine and Biology 10(4), 21–29 (1991)
[6] Foster, K.R., Cheever, E.A.: Microwave radiometry in biomedicine. A reappraisal. Bioelectromagnetics 13(6), 567–579 (1992)
[7] Klementsen, Ø., Birkelund, Y., Jacobsen, S.K., et al.: Design of medical radiometer front-end for improved performance. Progress in Electromagnetics Research B 27, 289–306 (2011)
[8] Heinz, E., May, T., Zieger, G., et al.: Passive submillimeter-wave stand-off video camera for security applications. J. Infrared Milli. Teraherz Waves 31, 1355–1369 (2010)
[9] Fedoseev, L.I., Bystrov, R.P., Krasnaynsky, A.F., et al.: Experimental study of radiothermal selectivity in millimeter-waves. Zhurnal Radioelektroniki (J. Radio Electronics) (12), 1–17 (2010) (in Russian)
[10] Chen, K., Zhu, Y., Guo, X., et al.: Design of 8-mm-band aperture synthetic radiometer and imaging experiment. J. Infrared Milli. Terahertz Waves 31, 724–734
[11] Peichl, M., Dill, S., Jirousek, M., Süß, H.: Microwave radiometry – imaging technologies and applications. In: Proc. WFMN 2007, pp. 75–83 (2007)
[12] Gvozdev, V.I., Kouzaev, G.A., Krivoruchko, V.I., Turygin, S.Y.: Multifunctional radio vision system (Imager). Russian Federation Patent, No. 2139522 dated, July 30 (1998)

[13] Gvozdev, V.I., Krivoruchko, V.I., Kouzaev, G.A., Turygin, S.Y.: A microwave imager. Measurement Techniques 43, 270–275 (2000)

[14] Kouzaev, G.A., Turygin, S.Y., Tchernyi, V.V., et al.: Millimeter-wave high-sensitive radiovision system for study of the human-body radiation. In: Proc. Int. SPIE Conf. EBIOS 2000, Amsterdam, Netherlands, Paper No 4158-50 (July 2000)

[15] Kouzaev, G.A., Kulevatov, M.A., Turygin, S.Y., et al.: A millimeter-wave high-sensitivity radio-vision system and a study of bio-objects' electromagnetic fields. Medical Physics (5), 70–71, in Russian (1998)

[16] Stephan, K.D.: Radiometry before World War II: Measuring infrared and millimeter-wave radiation 1800-1925. IEEE Antennas and Propag. 47(6), 28–37 (2005)

[17] Chen, Y.-F., Hover, D., Sendelbach, S., et al.: Microwave photon counter based on Josephson junction. Phys. Rev. Lett. 107(21), 217401 (2011)

[18] Dicke, R.H.: The measurement of thermal radiation at microwave frequencies. Rev. Sci. Instr. 3, 268–279 (1946)

[19] Jie, L., Zhesi, W., Yunnei, C., et al.: Research on the microwave radiometer signal waveform simulation. In: Proc. of 2011 4th Int. Conf. Intelligent Comput. Techn. Automation, pp. 679–682 (2011)

[20] Randa, J., Lahtinen, J., Camps, A., et al.: Recommended terminology for microwave radiometry. NIST Technical Note 1551 (2008)

[21] Hand, J.W., Van Leewen, G.M.J., Mizushina, S., et al.: Monitoring of deep brain temperature in infants using multi-frequency microwave radiometry and thermal modeling. Phys. Med. Biol. 46, 1885–1903 (2001)

[22] Grassberger, P., Procaccia, I.: Measuring the strangeness of strange attractors. Physica D 9(1-2), 189–208 (1983)

[23] Kouzaev, G.A.: The use of a data reconstruction algorithm to electromagnetic biosignals. In: Proc. Int. SPIE Conf. EBIOS 2000, pp. 4158–4149, Paper No 4158-49 (July 2000)

[24] Kouzaev, G.A., Bedenko, E.A.: Study of radiometric signals with chaotic dynamics methods. Medical Physics (5), 72–75 (1998) (in Russian)

[25] Box, G.E.P., Jenkins, G.M., Reinsel, G.C.: Time Series Analysis: Forecasting and Control. Prentice-Hall, Upper Saddle River (1994)

[26] Theiler, J.: Estimating fractal dimension. J. Opt. Soc. Am. A 7(6), 1055–1073 (1990)

[27] Potapov, A.: Fractals, scaling and fractional operators in radio techniques and electronics: contemporary state and developments. J. Radio Electronics (1), 1–99 (2010) (in Russian)

[28] Maasch, K.A.: Calculating climate attractor dimension from δ^{18}O records by the Grassberger-Procaccia algorithm. Climate Dynamics 4(1), 45–55 (1989)

[29] Fraedrich, K.: Estimating the dimensions of weather and climate attractors. J. Atmospher. Sci. 43(5), 419–432 (1986)

[30] Leok, M., B.T.: Estimating the attractor dimension of the equatorial weather system. Acta Phys. Polonica A 85(suppl. S-27) (1994)

[31] Tsonis, A., Elsner, J.B., Georgakakos, K.P.: Estimating the dimension of weather and climate attractors: Important issues about the procedure and interpretation. J. Atmospher. Sci. 50(15), 2549–2555 (1993)

[32] Ruelle, D., Takens, F.: On the nature of turbulence. Comm. Math. Phys. 20(3), 167–192 (1971)

[33] Malraison, B., Atten, P., Berge, P., Dubois, M.: Dimension of strange attractors: an experimental determination for the chaotic regime of two convective systems. J. Physique – Lett. 44(22), L-897–L-902 (1983)

[34] Torkamani, M.A., Asgari, J., Lucas, C.: Estimating strange attractor's dimension in very noisy data. In: 2nd Int. Conf. Information and Communication Technologies, ICTTA 2006, Damascus, Syria, pp. 1944–1947 (2006)
[35] Weber, R.O., Talkner, P., Stefanicki, G., Arvisais, L.: Search for finite dimensional attractors in atmospheric turbulence. Boundary-Layer Meteorology 73(1-2), 1–14 (1995)
[36] Lehnertz, K., Elger, C.: Can epileptic seizures be predicted? Evidence from nonlinear time series analysis of brain electrical activity. Phys. Rev. Lett. 80(22), 5019–5022 (1998)
[37] Kannathal, N., Choo, M.L., Acharya, U.R., Sadasivan, P.K.: Entropies for detection of epilepsy in EEG. Comput. Methods Prog. Biomed. 80(3), 187–194 (2005)
[38] Franca, L.F.P., Savi, M.A.: Estimating attractor dimension on the nonlinear pendulum time series. J. Braz. Soc. Mech. Sci. 23(4), 427–439 (2001)
[39] Baker, G.L., Gollub, J.P.: Chaotic Dynamics: an Introduction. Cambridge University Press, New York (1996)
[40] Mudelsee, M., Stattegger, K.: Plio-/Pleistocene climate modeling based on oxygen isotope time series from deep-sea sediment cores: The Grassberger-Procaccia algorithm and chaotic climate systems. Math. Geol. 26(7), 799–815 (1994)
[41] Takens, F.: Detecting strange attractors in turbulence. In: Rand, D., Young, B.S. (eds.). Lecture Notes in Mathematics, vol. 898, pp. 366–381. Springer, Berlin (1981)
[42] Fraser, A.M., Swinney, H.L.: Independent coordinates for strange attractors from mutual information. Phys. Rev. A 33(2), 1134–1140 (1986)
[43] McGuinness, M.J.: A computation of the limit capacity of the Lorenz attractor. Physica D 16(2), 265–275 (1985)
[44] Smith, L.A.: Intrinsic limits on dimension calculations. Phys. Lett. A 133(6), 283–288 (1988)
[45] Havstad, J.W., Ehlers, C.L.: Attractor dimension of nonstationary dynamical systems from small data sets. Phys. Rev. A 39(2), 845–853 (1989)
[46] Lorenz, E.N.: Deterministic nonperiodic flow. J. Atmospher. Sci. 20(2), 130–141 (1963)
[47] http://www.mathworks.com/support/tech-notes/1500/1510_files/ODE_non_adaptive/ode4.m
[48] Glazier, J.A., Libchaber, A.: Quasi-periodicity and dynamical systems: An experimentalist's view. IEEE Trans. Circuits Syst. 35(7), 790–809 (1988)
[49] Halsey, T.C., Jensen, M.H., Kadanoff, L.P., et al.: Fractal measures and their singularities: The characterization of strange sets. Phys. Rev. A 33(2), 1141–1151 (1986)
[50] Richter, H.: On a family of maps with multiple chaotic attractors. Chaos, Solitons Fractals 36(3), 559–571 (2008)
[51] Jensen, M.H., Kadanoff, L.P., Libchaber, A., et al.: Global universality at the onset of chaos: Results of a forced Rayleigh-Bénard experiment. Phys. Rev. Lett. 55(25), 2798–2801 (1985)
[52] Consolini, G., Marcucci, M.F., Candidi, M.: Multifractal structure of auroral electrojet index data. Phys. Rev. Lett. 76(21), 4082–4085 (1996)
[53] Luo, X., Small, M., Danca, M.-F., Chen, G.: On a dynamical system with multiple chaotic attractors. Int. J. Bifurcation Chaos 17(9), 3235–3251 (2007)
[54] Dong, L., Zhi-Gang, Z.: Multiple attractors and generalized synchronization in delayed Mackey-Glass systems. Chinese Phys. B 17(11), 4009–4013 (2008)
[55] Thode, H.C.: Testing for Normality. Marcel Dekker, New York (2002)

[56] Royston, J.P.: Approximating the Shapiro-Wilk W-test for non-normality. Statist. Comput. 2(3), 117–119 (1992)
[57] Royston, J.P.: Remark AS R94: A remark on algorithm AS 181: The W-test for normality. J. Royal Statist. Soc. C 44(4), 547–551 (1995)
[58] D'Agostino, R.B., Belanger, A., D'Agostino, R.B.: A suggestion for using powerful and informative tests of normality. Amer. Statist. 44(4), 316–320 (1990)
[59] `http://www.mathworks.com/matlabcentral/fileexchange/13964`
[60] `http://www.mathworks.com/matlabcentral/fileexchange/3954`
[61] Kapranov, S.V., Kouzaev, G.A.: Characterization of microwave radiometers and study of human body radiation by the means of the state space reconstruction algorithms Int. J. Signal Process. Image Process. Pattern Recognit. (submitted, 2012)
[62] Kouzaev, G.A.: A projective approach to the problems of processing of complex signals. Radioelektronika (Radioelectroniks). Izvestia Vysshikh Uchebnykh Zavedenyi (1), 53–57 (2001)
[63] Peichl, M., Dill, S., Jirouseck, M., et al.: Microwave radiometry - imaging technologies and applications. In: Proc. of WFMN 2007, Chemnitz, Germany, pp. 75–83 (2007)
[64] Watabe, K., Shimizu, K., Yoneyama, M., et al.: Millimeter-wave active imaging using neural networks for signal processing. IEEE Trans., Microw. Theory Techn. 51, 1512–1516 (2003)
[65] Nanzer, J.A., Rogers, R.L.: Analysis of the signal response of a scanning-beam millimeter-wave correlation radiometer. IEEE Trans., Microw. Theory Techn. 59, 2357–2368 (2011)
[66] Weinreb, S.: Monolithic integrated circuit imaging radiometers. In: 1991 IEEE MTT-S Dig., vol. L-8, pp. 405–408 (1991)
[67] Corbella, I., Tores, F., Camps, A., et al.: L-band aperture synthesis radiometry: hardware requirements and system performance. In: Proc. IGARSS 2000 Symp., vol. 7, pp. 2975–2977 (2000)
[68] Kim, W.-G., Moon, N.-W., Chang, Y.-S., et al.: System design of focal plane array based millimeter-wave imaging radiometer for concealed weapon detection. In: Proc. IGARSS 2011 Symp., pp. 2258–2261 (2011)
[69] Liao, S., Gopalsami, N., Elmer, T.W., et al.: Passive millimeter-wave dual-polarization imagers. IEEE Trans., Instr. Measur., 1–9 (2012)
[70] Lutchi, T., Matzler, C.: Stereoscopic passive millimeter-wave imaging and ranging. IEEE Trans., Microw. Theory Techn. 53, 2594–2599 (2005)
[71] Moon, N.-W., Singh, M.K., Kim, Y.-H.: Passive range measurement and discrepancy effects of distance for stereo scanning W-band radiometer. In: Proc. 11th Specialist Meeting Microwave Radiometry and Remote Sensing of the Environment, MicroRad 2010, pp. 217–220 (2010)
[72] Wang, B., Li, X., Qian, S.: Near range MNV synthetic aperture radiometer 3D passive imaging. In: Proc. 8th Int. Symp. Antennas, Propagation and EM Theory, pp. 233–236 (2008)
[73] Arakelyan, A., Grigorian, M., Hambaryan, A., et al.: Combined active and passive measurements of snow, bare and vegetated soils microwave reflective and emissive characteristics by Ka-band, combined scatterometer-radiometer system. In: Proc. IEEE Int. Symp. IGARSS 2010, pp. 4462–4465 (2010)

CPSIA information can be obtained at www.ICGtesting.com
Printed in the USA
LVOW05*1128161114

413959LV00003B/268/P

9 783642 303098